Interdisciplinary Applied Mathematics

T0189182

Springer
New York
Berlin
Heidelberg
Hong Kong
London
Milan
Paris
Tokyo

Interdisciplinary Applied Mathematics
Volume 10

Editors
J.E. Marsden **L. Sirovich** **S. Wiggins**

Fluid Dynamics and Nonlinear Physics
K.R. Sreenivasan, G. Ezra

Mathematical Biology
L. Glass, J.D. Murray

Mechanics and Materials
S.S. Antman, R.V. Kohn

Systems and Control
S.S. Sastry, P.S. Krishnaprasad

Problems in engineering, computational science, and the physical and biological sciences are using increasingly sophisticated mathematical techniques. Thus, the bridge between the mathematical sciences and other disciplines is heavily traveled. The correspondingly increased dialog between the disciplines has led to the establishment of the series: *Interdisciplinary Applied Mathematics*.

The purpose of this series is to meet the current and future needs for the interaction between various science and technology areas on the one hand and mathematics on the other. This is done, firstly, by encouraging the ways that mathematics may be applied in traditional areas, as well as point towards new and innovative areas of applications; secondly, by encouraging other scientific disciplines to engage in a dialog with mathematicians outlining their problems to both access new methods and to suggest innovative developments within mathematics itself.

The series will consist of monographs and high-level texts from researchers working on the interplay between mathematics and other fields of science and technology.

Shankar Sastry

Nonlinear Systems
Analysis, Stability, and Control

With 193 Illustrations

 Springer

Shankar Sastry
Department of Electrical Engineering
  and Computer Science
University of California, Berkeley
Berkeley, CA 94720-1770
USA

Editors
J.E. Marsden
Control and Dynamical Systems
Mail Code 107-81
California Institute of Technology
Pasadena, CA 91125
USA

L. Sirovich
Division of
  Applied Mathematics
Brown University
Providence, RI 02912
USA

S. Wiggins
Control and Dynamical Systems
Mail Code 107-81
California Institute of Technology
Pasadena, CA 91125
USA

Mathematics Subject Classification (1991): 93-01, 58F13, 34-01, 34Cxx, 34H05, 93c73, 93c95

Library of Congress Cataloging-in-Publication Data
Sastry, Shankar.
    Nonlinear systems : analysis, stability, and control / Shankar
Sastry.
        p.   cm. — (Interdisciplinary applied mathematics ; v. 10)
    Includes bibliographical references and index.
    ISBN 978-1-4419-3132-0
    1. Nonlinear systems.   2. System analysis.   I. Title.
II. Series.
QA402.S35157   1999
003′.75—dc21   99-11798

Printed on acid-free paper.

Production managed by Timothy Taylor; manufacturing supervised by Jacqui Ashri.

Printed in the United States of America.

9 8 7 6 5 4 3 2

Springer-Verlag New York Berlin Heidelberg

This book is gratefully dedicated to
Charles Desoer, Jerrold Marsden, and Roger Brockett
Visionaries of a new and beautiful world of nonlinear science

Preface

There has been a great deal of excitement in the last ten years over the emergence of new mathematical techniques for the analysis and control of nonlinear systems: Witness the emergence of a set of simplified tools for the analysis of bifurcations, chaos, and other complicated dynamical behavior and the development of a comprehensive theory of geometric nonlinear control. Coupled with this set of analytic advances has been the vast increase in computational power available for both the simulation and visualization of nonlinear systems as well as for the implementation in real time of sophisticated, real-time nonlinear control laws. Thus, technological advances have bolstered the impact of analytic advances and produced a tremendous variety of new problems and applications that are nonlinear in an essential way. Nonlinear control laws have been implemented for sophisticated flight control systems on board helicopters, and vertical take off and landing aircraft; adaptive, nonlinear control laws have been implemented for robot manipulators operating either singly, or in cooperation on a multi-fingered robot hand; adaptive control laws have been implemented for jet engines and automotive fuel injection systems, as well as for automated highway systems and air traffic management systems, to mention a few examples. Bifurcation theory has been used to explain and understand the onset of flutter in the dynamics of aircraft wing structures, the onset of oscillations in nonlinear circuits, surge and stall in aircraft engines, voltage collapse in a power transmission network. Chaos theory has been used to predict the onset of noise in Josephson junction circuits and thresholding phenomena in phase-locked loops. More recently, analog computation on nonlinear circuits reminiscent of some simple models of neural networks hold out the possibility of rethinking parallel computation, adaptation, and learning.

It should be clear from the preceding discussion that there is a tremendous breadth of applications. It is my feeling, however, that it is possible at the current time to lay out in a concise, mathematical framework the tools and methods of analysis that underly this diversity of applications. This, then, is the aim of this book: I present the most recent results in the analysis, stability, and control of nonlinear systems. The treatment is of necessity both mathematically rigorous and abstract, so as to cover several applications simultaneously; but applications are sketched in some detail in the exercises.

The material that is presented in this book is culled from different versions of a one-semester course of the same title as the book that I have taught once at MIT and several times at Berkeley from 1980 to 1997. The prerequisites for the first year graduate course are:

- An introduction to mathematical analysis at the undergraduate level.
- An introduction to the theory of linear systems at the graduate level.

I will assume these prerequisites for the book as well. The analysis prerequisite is easily met by Chapters 1–7 of Marsden's *Elementary Classical Analysis*, (W. H. Freeman, 1974) or similar books. The linear systems prerequisite is met by Callier and Desoer's *Linear Systems Theory*, (Springer Verlag, 1991) or Rugh's *Linear System Theory*, (Prentice Hall, 1993), Chen's *Linear System Theory and Design*, (Holt Reinhart and Winston, 1984); or Kailath's *Linear Systems*, (Prentice Hall, 1980) or the recent *Linear Systems* by Antsaklis and Michel, (McGraw Hill, 1998).

I have never succeeded in covering all of the material in this book in one semester (45 classroom hours), but here are some packages that I have covered, along with a description of the style of the course

- *Analysis, Stability and some Nonlinear Control*
  Chapters 1–7 and part of Chapter 9.
- *Analysis, Some Stability and Nonlinear Control*
  Chapters 1–3, 5–6 followed by Chapters 9, 10 with supplementary material from Chapter 8.
- *Mathematically Sophisticated Nonlinear Control Course*
  Chapters 1, 2, 4, 5–7, with supplementary material from Chapter 3, and Chapters 9–11 with supplementary material from Chapter 8.

Alternatively, it is possible to use all the material in this book for a two-semester course (90 classroom hours) on nonlinear systems as follows:

- (45 hours Semester 1) Chapters 1–7.
- (45 hours Semester 2) Chapters 8–12.

For schools on the quarter system, 80 classroom hours spread over two quarters can be used to cover roughly the same material, with selective omission of some topics from Chapters 3, 6, and 7 in the first quarter and the omission of some topics from Chapters 8, 11, and 12 in the second quarter. A shorter 60 classroom hour long two quarter sequence can also be devised to cover

1. (30 hours) Introductory course on Nonlinear Systems. Chapters 1, 2, 3 (Sections 3.1—3.5), Chapter 4 (Sections 4.1—4.6), and Chapter 5.
2. (30 hours) Intermediate course on Nonlinear Control. Chapter 3 (Section 3.9), Chapter 8, Chapter 9, 10, and parts of Chapter 11.

The structuring of courses at Berkeley favors the two semester structure, with the first course for second-semester graduate students (taught in the spring semester), and the second course called "Advanced Topics in Nonlinear Control" for second-year graduate students (taught in the fall). However, I wish to emphasize that we frequently see undergraduate students taking this course and enjoying it.

Access to a simulation package for simulating the dynamics of the nonlinear systems adds a great deal to the course, and at Berkeley I have made available Matlab, Simnon and Matrix-X at various times to the students as simulation toolkits to use to help stimulate the imagination and help in the process of "numerical experimentation." While I have usually had take home final examinations for the students, I think that it is useful to have "project-based" final examinations with numerical examples drawn from a set of particularly topical applications. A word about the problem sets in this book; they are often not procedural, and frequently need thought and sometimes further reference to the literature. I have found that this is a nice way to draw oneself into what is a very exciting, dynamic and rapidly evolving area of research. I have included these also because over the years, it has been a pleasant surprise to me to see students solve problem sets based on the archival literature with ease, when they are given adequate background. I have chosen applications from a wide variety of domains: mechatronic systems, classical mechanical systems, power systems, nonlinear circuits, neural networks, adaptive and learning systems, flight control of aircraft, robotics, and mathematical biology, to name some of the areas covered. I invite the reader to enjoy and relate to these applications and feel the same sense of scientific excitement that I have felt for the last twenty odd years at the marvels and mysteries of nonlinearity.

The author would be grateful for reports of typographic and other errors electronically through the WWW page for the book:

`robotics.eecs.berkeley.edu/~sastry/nl.book`

where an up-to-date errata list will be maintained along with possible additional exercises.

Shankar Sastry
Berkeley, California
March 1999

Acknowledgments

In any large undertaking there are a number of people on whose shoulders we stand. This is certainly the case for me in this book, and the particular shoulders on which I stand are those of my teachers and my students. I owe an immense debt of gratitude to Charles Desoer, Jerrold Marsden, and Roger Brockett for having given me the love for and curiosity about nonlinear systems. Pravin Varaiya, Sanjoy Mitter, and Petar Kokotović have all been my early (and continuing mentors) in this nonlinear endeavor as well. My students have provided me with some of the most exciting moments of discovery over the years that I have worked in nonlinear control and I wish to acknowledge, especially and gratefully, those that have worked with me on subject matter that is represented in this book in chronological order of completion of their academic careers at Berkeley: Brad Paden, Stephen Boyd, Marc Bodson, Li Chen Fu, Er Wei Bai, Andrew Packard, Zexiang Li, Ping Hsu, Saman Behtash, Arlene Cole, John Hauser, Arlene Cole, Richard Murray, Andrew Teel, Raja Kadiyala, A. K. Pradeep, Linda Bushnell, Augusto Sarti, Gregory Walsh, Dawn Tilbury, Datta Godbole, John Lygeros, Jeff Wendlandt, Lara Crawford, Claire Tomlin, and George Pappas. Indeed, these folks will find much in this book that is rather familiar to them since their research work (both with me and after they left Berkeley and set up their own research programs) is prominently featured in this book. In Chapters 8, 10, 11, and 12 I have explicitly pointed out the contributions of Claire Tomlin, Yi Ma, John Hauser, Richard Murray, Dawn Tilbury, George Pappas, and John Lygeros in writing parts of these chapters.

This book has been classroom tested in its different formative phases by John Hauser at the University of Southern California, Richard Murray and Jerrold Marsden at Caltech, Hsia Dong Chiang at Cornell, and Ken Mease at University of

California, Irvine, Claire Tomlin at Berkeley and Stanford, and Shahram Shahruz and George Pappas at Berkeley. I am grateful to them for their painstaking comments. I would also like to thank Claire Tomlin, Nanayaa Twum Danso, George Pappas, Yi Ma, John Koo, Claudio Pinello, Jin Kim, Jana Kosecka, and Joao Hespanha for their help with proofreading the manuscript. I thank Christine Colbert for her superb drafting of the figures and thank Achi Dosanjh of Springer-Verlag, for her friendly management of the writing, and her patience with getting the manuscript reviewed.

Colleagues who have worked with me and inspired me to learn about new areas and new directions abound. Especially fresh in my mind are the set of lectures on converse Lyapunov theorems given by M. Vidyasagar in Spring 1979 and the short courses on nonlinear control taught by Arthur Krener in Fall 1984 and by Alberto Isidori in the Fall 1989 at Berkeley that persuaded me of the richness of nonlinear control. I have fond memories of joint research projects in power systems and nonlinear circuits with with Aristotle Arapostathis, Andre Tits, Fathi Salam, Eyad Abed, Felix Wu, John Wyatt, Omar Hijab, Alan Willsky, and George Verghese. I owe a debt of gratitude to Dorotheé Normand Cyrot, Arthur Krener, Alberto Isidori, Jessy Grizzle, L. Robert Hunt, and Marica di Benedetto for sharing their passion in nonlinear control with me. Richard Montgomery and Jerry Marsden painstakingly taught me the amazing subtleties of nonholonomic mechanics. I thank Georges Giralt, Jean-Paul Laumond, Ole Sordalen, John Morten Godhavn, Andrea Balluchi, and Antonio Bicchi for their wonderful insights about non-holonomic motion planning. Robert Hermann, Clyde Martin, Hector Sussmann, Christopher Byrnes, and the early Warwick lecture notes of Peter Crouch played a big role in shaping my interests in algebraic and geometric aspects of nonlinear control theory. P. S. Krishnaprasad, John Baillieul, Mark Spong, N. Harris Mc Clamroch, Gerardo Lafferriere, T. J. Tarn, Dan Koditschek were co-conspirators into unlocking the mysteries of nonlinear problems in robotics. Stephen Morse, Brian Anderson, Karl Astrom, and Bob Narendra played a considerable role in my understanding of adaptive control.

The research presented here would not have been possible without the very consistent support, both technical and financial, of George Meyer of NASA Ames, who has had faith in the research operation at Berkeley and has painstakingly explained to me and the students here over the years the subtleties of nonlinear control and flight control. Jagdish Chandra, and then Linda Bushnell, at the Army Research Office have supported my work with both critical technical and financial inputs over the years, which I have most grateful for. Finally, Howard Moraff, at the National Science Foundation believed in non-holonomic motion planning when most people thought that non-holonomy was a mis-spelled word and supported our research into parking cars! The list of grants that supported our research and the writing of this book is NASA under grant NAG 2-243 (1983-1995), NAG 2-1039 (1995 onwards), ARO under grants DAAL-88-K0106 (1988-1991), DAAL-91-G0171 (1991-1994), DAAH04-95-1-0588 (1995-1998), and DAAH04-96-1-0341 (1996 onwards), NSF under grant IRI-9014490 (1990-1995).

Finally, on a personal note, I would like to thank my mother and late father for the courage of their convictions, selfless devotion, and commitment to excellence.

Shankar Sastry
Berkeley, California
March 1999

Contents

Standard Notation

The following notation is standard and is used throughout the text. Other non–standard notation is defined when introduced in the text and is referenced in the index. A word about the numbering scheme: Not all equations are numbered, but those that are frequently referenced are. Theorems, Claims, Propositions, Corollaries, Lemmas, Definitions, Examples are numbered consecutively in the order in which they appear and they are *all numbered*. Their text is presented in an *emphasized font*. If the theorems, claims, propositions, etc. are specifically noteworthy they are named in **bold font** before the statement. Exercises are at the end of each chapter and are all numbered consecutively, and if especially noteworthy are named like the theorems, claims, propositions, etc. Proofs in the text end with the symbol □ to demarcate the proof from the following text.

Sets

$a \in A$	a is an element of the set A
$A \subset B$	set A is contained in set B
$A \cup B$	union of set A with set B
$A \cap B$	Intersection of set A with set B
$\ni$	such that
$p \Rightarrow q$	p implies q
$p \Leftarrow q$	q implies p
$p \Leftrightarrow q$	p is equivalent to q
M°	interior of a set M

$\overline{M}$	closure of M
$]a, b[$	open subset of the real line
$[a, b]$	closed subset of the real line
$[a, b[$	subset of the real line closed at a, and open at b
$a \rightarrow b$	a *tends* to b
$a \downarrow b$	a *decreases* towards b
$a \uparrow b$	a *increases* towards b
$\oplus$	direct sum of *subspaces*

Algebra

$\mathbb{N}$	set of non-negative integers, namely, $(0, 1, 2, \ldots)$
$\mathbb{R}$	field of real numbers
$\mathbb{Z}$	ring of integers, namely, $(\ldots, -1, 0, 1, \ldots)$
j	square root of -1
$\mathbb{C}$	field of complex numbers
$\mathbb{R}_+ (\mathbb{R}_-)$	set of non-negative (non-positive) reals
$\mathbb{C}_+ (\mathbb{C}_-)$	set of complex numbers in the right (left) half plane, including the imaginary axis
$j\omega$ axis	set of purely imaginary complex numbers
$\mathbb{C}_-^\circ$	$\{s \in \mathbb{C} : \text{Re } s < 0\}$ = interior of $\mathbb{C}_-$
$\mathbb{C}_+^\circ$	$\{s \in \mathbb{C} : \text{Re } s > 0\}$ = interior of $\mathbb{C}_+$
A^n	set of n-tuples of elements belonging to the set A (e.g., $\mathbb{R}^n$, $\mathbb{R}[s]^n$)
$A^{m \times n}$	set of $m \times n$ arrays with entries in A.
$\sigma(A)$	set of eigenvalues (spectrum) of a square matrix A
$(x_k)_{k \in K}$	family of elements with K, an index set.
$F[x]$	ring of polynomials in one variable x with coefficients in a field F
$F(x)$	field of rational functions in one variable x with coefficients in a field F

Analysis

$f : A \mapsto B$	f maps the domain A into the codomain B
$f(A)$	range of $f := \{y \in B : y = f(x) \text{ for some } x \in A\}$
A°	interior of A
$\bar{A}$	closure of A
∂A	boundary of A

$C([t_0, t_1], \mathbb{R})$	vector space of continuous functions $[t_0, t_1] \mapsto \mathbb{R}$
$C([t_0, t_1])$	vector space of continuous functions $[t_0, t_1] \mapsto \mathbb{R}$
$C^k([t_0, t_1], \mathbb{R})$	vector space of continuous functions $[t_0, t_1] \mapsto \mathbb{R}$ with k continuous derivatives
$C^k([t_0, t_1], \mathbb{R}^n)$	vector space of continuous functions $[t_0, t_1] \mapsto \mathbb{R}^n$ whose first k derivatives are continuous
$|x|$	norm of an element x in a vector space
$\langle x, y \rangle$	inner-product of two vectors x, y in a Hilbert space
$\hat{f}, \hat{G}$	Laplace transform of scalar (or vector) function f or matrix function G both defined on $\mathbb{R}_+$
$\dot{f}, \dot{G}$	time derivative of scalar (or vector) function f or matrix function G both defined on $\mathbb{R}_+$
$Df(x)$	derivative of a function $f : \mathbb{R}^n \mapsto \mathbb{R}^m$ a matrix $\in \mathbb{R}^{m \times n}$
$D_i f(x_1, x_2, \ldots, x_p)$	Derivative of $f : \mathbb{R}^{n_1} \times \cdots \times \mathbb{R}^{n_p}$ with respect to the i-th argument
$D^2 f(x)$	second derivative of $f : \mathbb{R}^n \mapsto \mathbb{R}^m$ with respect to its argument, a bi-linear map from $\mathbb{R}^n \times \mathbb{R}^n \mapsto \mathbb{R}^m$
$D^k_{i_1,\ldots,i_k} f(x_1, x_2, \ldots, x_p)$	k-th partial derivative of $f(x_1, \ldots, x_p)$ with respect to $x_{i_1}, \ldots, x_{i_k}$ a k-linear map from $\mathbb{R}^{n_{i_1}} \times \cdots \times \mathbb{R}^{n_{i_k}} \mapsto \mathbb{R}^m$
$L_p[t_0, t_1]$	vector space of $\mathbb{R}$ valued functions with p-th power integrable over $[t_0, t_1]$
$L_p^k[t_0, t_1]$	vector space of $\mathbb{R}^k$ valued functions with p-th power integrable over $[t_0, t_1]$
$o(x)$	little "o" of x, that is a function $g(x)$, such that $\lim_{|x| \to 0} |g(x)|/|x| = 0$
$O(x)$	capital "O" of x, that is a function $h(x)$, such that $\lim_{|x| \to 0} |h(x)|$ is well-defined and $\neq 0$

1
Linear vs. Nonlinear

1.1 Nonlinear Models

Why do we need to have a nonlinear theory and why bother to study a qualitative nonlinear theory? After all, most models that are currently available are linear, and if a nonlinear model is to be used, computers are getting to be ever more powerful at simulating them. Do we really need a nonlinear theory? This is not a naive question, since linear models are so much more tractable than nonlinear ones and we can analyze quite sophisticated and high dimensional linear systems. Further, if one uses linear models with some possibly time-varying parameters, one may model real systems surprisingly well. Moreover, although nonlinear models may be conceptually more satisfying and elegant, they are of little use if one cannot learn anything from their behavior. Certainly, many practitioners in industry claim that they can do quite well with linear time varying models. Of course, an opposing argument is that we may use the ever increasing power of the computer to qualitatively understand the behavior of systems more completely and not have to approximate their behavior by linear systems.

However, the compelling reason that we *do* use nonlinear models is that the dynamics of linear systems *are not rich enough to describe many commonly observed phenomena.* Here are a few examples of such phenomena:

1. *Multiple equilibria or multiple operating points*
   Systems with many equilibria abound in practice; consider, for example:

   - Digital circuits for binary logic have at least two stable states,
   - Chemical reaction kinetics allow for multiple equilibria,

- Power flow equations modeling the flow of real and reactive power in a transmission network have multiple steady state operating points,
- A buckled beam has two stable buckled states,
- In population ecology, there are multiple equilibrium populations of competing species.

Now, consider the linear differential equation model

$$\dot{x} = Ax \tag{1.1}$$

with $x \in \mathbb{R}^n$ and $A \in \mathbb{R}^{n \times n}$, a constant matrix. The point $x = 0$ is an equilibrium point of the system. That is, if the initial state at time $t = 0$ of the differential equation of (1.1) is 0, i.e., $x(0) = 0$, then the state of the equation remains 0 for all t, i.e., $x(t) \equiv 0$. If A is a nonsingular matrix, then $x = 0$ is the only equilibrium point of the linear system. If A is singular, the set of equilibria for the system is the null space of the matrix A. This is an uncountably infinite set of equilibrium points. However, "generically," linear systems have only one equilibrium point, since if A were singular, an infinitesimal perturbation of the entries of A will "almost surely" cause it to become nonsingular. Consequently, linear systems of the form of (1.1) cannot robustly (here robustly is used in the sense of being qualitatively insensitive to modeling errors) allow for more than one equilibrium solution.

2. *Periodic variation of state variables or limit cycles.*
Again, instances of systems with periodic variations of a robust nature abound in practice:

- Differential equations for modeling the heartbeat and nerve impulse generation are of the so-called van der Pol kind, which we will study later. The cyclic nature of these phenomena is obtained from a single stable limit cycle. Other related examples model periodic muscular contractions in the oesophagus and intestines.
- Digital clock circuits or astable multivibrators exhibit cyclic variation between the logical 1 and 0 states and may also be modeled as a degenerate limit of the van der Pol equation mentioned above (this is investigated in detail in Chapter 7).
- Josephson junction circuits can be shown to have a limit cycle (of very high frequency) when the biasing current is greater than a certain critical value, and this oscillation marks the transition away from the superconducting to the conducting region (this is studied in detail in Chapter 2).

By way of contrast, consider once again the linear system of (1.1). If $A \in \mathbb{R}^{n \times n}$ has eigenvalues on the imaginary axis (referred to as the $j\omega$ axis) then the linear system admits of a continuum of periodic solutions. However, small perturbations of the entries of A will cause the eigenvalues to be displaced off the $j\omega$ axis and destroy the existence of periodic solutions. Thus, a small parameter variation in linear system models will destroy the continuum of periodic solutions, and instead produce either an unstable or stable equilibrium point. Thus, linear dynamical systems with periodic solutions are non robust models.

As in the case of multiple equilibrium points, there exist systems with multiple, periodic solutions. These can be finite in number in contrast to an infinite number for linear systems with $j\omega$ axis eigenvalues and consequently can not be described by a linear differential equation model.

3. *Bifurcations*

There are many examples of systems whose qualitative features, such as the number of equilibrium points, the number of limit cycles, and the stability of these features, changes with parametric variation in the model. For example:

a. A rod under axial loading has one unbuckled state as its equilibrium state till the loading reaches a critical value, at which point it acquires two stable buckled states and an unstable unbuckled state.

b. The compressor of a jet engine changes from steady state operation to a mode of operation where there is rotating stall when the angle of attack of the blades reaches a critical value.

c. The Josephson junction changes from the superconducting state, or zero resistance state (when there is a single stable equilibrium), to the conducting state (when there is a periodic orbit and an equilibrium point or just a periodic orbit) as the amount of current forcing goes through a critical value.

d. Hysteretic behavior in the current values of certain one port resistors, such as tunnel diodes, occurs when the voltage across them is raised or lowered.

e. As the angle of attack of an aircraft changes, a constant flight path angle trajectory becomes unstable and is replaced by a cyclic mode with constant roll rate.

Linear systems with parameters have behavior that is considerably less subtle. As parameters of the system change, the system can change from stable to unstable. The number of equilibria can go through infinity if any of the eigenvalues go through the origin, However, none of the changes described above can be captured by a parameterized linear model.

4. *Synchronization and Frequency Entrainment*

Oscillators when coupled weakly pull into frequency and phase synchronism, for example in

a. Heart muscle cells, muscular cells causing peristalsis in the intestines and the oesophagus,

b. Phase locked loops for tracking.

In addition, as parameters change the loss of synchronism or frequency entrainment is characterized by the onset of complicated behavior, for example, arrhythmias (for the heart muscle) or loop skipping (for the phase locked loop). There is no linear model for this phenomenon.

5. *Complex Dynamical Behavior*

Let us contemplate the dynamical behavior of the linear system of (1.1). It is extremely simple: the responses are sums of exponentials, with exponents given by the eigenvalues of the matrix A that either decay to zero or blow up as $t \to \infty$ when the eigenvalues of A are not on the $j\omega$ axis. In this case, exactly the same behavior is manifested for the system independent of initial conditions. When A has eigenvalues on the $j\omega$ axis, the solutions of equation (1.1) neither decay nor blow up for any initial condition. As we have discussed above, this model is nonrobust, since small perturbations will knock the eigenvalues off the $j\omega$ axis.

In contrast, the dynamics of many physical systems can be a complex and sensitive function of the initial conditions; for instance, the dynamics of population models, climatic models and turbulent fluid flow models. These systems evolve in ways that are quite a bit more subtle than diverging or contracting exponentials. Small changes in the initial conditions can make the trajectories vastly different over time. These are referred to as chaotic or complex dynamics. In the popular science literature, these are advanced as reasons for the famous "butterfly effect": The beating of the wings of a flock of butterflies in Central Park may cause variations in the global climate of such magnitude as to allow for typhoons in China. Chaotic or complex dynamics are examples of dynamics that cannot be generated by linear models.

1.2   Complexity in Nonlinear Dynamics

Before we begin a systematic study of nonlinear systems and their astoundingly rich behavior, we consider the dynamics of a very simple first-order difference equation, which reveals some amazing subtlety. We will, for the moment, consider a discrete system:

$$x_{t+1} = f(x_t). \tag{1.2}$$

This equation models the evolution of a population of blowflies numbering $x_t \in \mathbb{R}$ at time t in a box of fixed size with a fixed food supply. Admittedly, in reality, the population x is discrete, but the model with x continuous is surprisingly robust enough to overcome this drawback. We assume that the population is being measured at discrete times and x is large enough so that we may assume it to be a continuous variable. This model was the source of a great deal of excitement when it was studied by May and Oster [203] and validated by experiments with a blowfly population at Berkeley. This example will also serve as a vehicle to introduce an interesting graphical technique for understanding the dynamics of discrete nonlinear systems. The qualitative shape of the function f is depicted in Figure 1.1. For definiteness the reader may wish to consider the function

$$f(x) = hx(1 - x), \qquad x \in [0, 1]. \tag{1.3}$$

This function has a single maximum at $x = \frac{1}{2}$ of magnitude $h/4$. It follows that for $h \leq 4$, it maps the interval $[0, 1]$ into itself. With the f of equation (1.3), the model 1.2 is referred to as the *logistic map*.

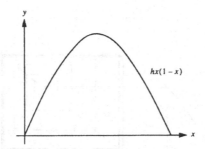

FIGURE 1.1. The one-hump function

The function f models the fact, that $x_{t+1} > x_t$ when the population is small, so that the population increases (abundance of food and living space). Also, $x_{t+1} < x_t$ when the population is large (competition for food and increased likelihood of disease). Thus $f(x) > x$ for x small and $f(x) < x$ for x large. Figure 1.2 shows how one solves the equation (1.3) graphically: One starts with the initial state x_0 on the horizontal axis and reads off $x_1 = f(x_0)$ on the vertical axis. The horizontal line intersecting the 45° line corresponds to the iterative process of replacing x_0 on the horizontal axis by x_1. The iteration now continues by progressively reading off x_{t+1} on the vertical axis and using the intersection of the horizontal line through x_{t+1} with the 45° line to reflect x_{t+1} onto the horizontal axis. The consequent appearance of the evolution of the population state variable is that of a *cobweb*. This term is, in fact, the terminology used by economists for simple one dimensional models of macroeconomic GNP growth and certain microeconomic phenomena [315].

We will study the effect of raising the height of the hump without increasing its support (i.e., the range of values (x) for which $f(x) \neq 0$) on the dynamics of the blow fly population (modeling the effect of an increase in the food supply). For definiteness, this corresponds to increasing the parameter h in the function of (1.3).

- The case $0 \leq h < 1$.
  A few moments reflection (verify this for yourself) will make it clear that for $0 \leq h < 1$ the graph of f never crosses the 45 degree line $x_{t+1} = x_t$, that is, the curve $y = f(x)$ lies below the line $y = x$. In this case x_t tends to zero monotonically in the time step t. This corresponds to a situation in which the population dies out asymptotically for lack of adequate food.
- The case $1 \leq h \leq 3$.
  We now increase the height of the hump, i.e., $h \geq 1$. Consider the scenario of Figure 1.3, showing the graph of f crossing the 45 degree line. Notice that if the population were to start at the intersection point it would stay there since at the intersection point, $x_{t+1} = f(x_t) = x_t$. In other words, the intersection point denoted by x_0^* is an equilibrium. Also (verify this graphically for yourself) all nonzero initial conditions, no matter how small they are, converge to x_0^* for $1 \leq h \leq 3$. Thus x_0^* is a stable equilibrium population. The zero population

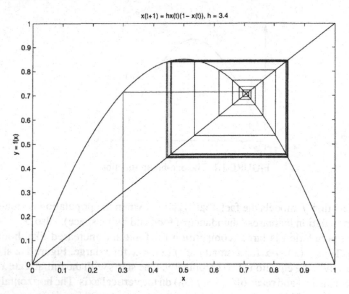

FIGURE 1.2. Graphical solution of a one dimensional system

equilibrium is unstable. The parameter value at which the zero population becomes unstable is $h_0 = 1$. It should be kept in mind that the stable population
value x_0^* is a function of h, given by

$$x_0^* = 1 - \frac{1}{h}.$$

Thus, at $h = 2$ the equilibrium point moves from the left side of the hump to
the right side.

• The case $3 < h \le 1 + \sqrt{6} = 3.449$.
Increasing the height of the hump even further, a little graphical experimentation
shows that when the slope of the function f near the intersection point becomes
sufficiently negative, more precisely, less than or equal to -1, a limit cycle of
period 2 shows up, as in the Figure 1.4. Additionally, Figure 1.4 shows the
cobweb at $h_1 = 3.25$, demonstrating the onset of the period 2 limit cycle.
The value h_1 is called a *period-doubling bifurcation point*. A period 2 limit
cycle is equivalent to a closed square in the cobweb in Figure 1.4 involving the
population alternating between two values x_1^* and x_2^*.
Further, the period 2 limit cycle is stable; that is, all nonzero initial populations
tend to an asymptotic pattern of alternating between the two values, x_1^* and x_2^*.
Another way of understanding these results is by examining the dynamics of
the two-step evolution, namely,

$$x_{t+2} = f(f(x_t)) =: f^2(x_t). \tag{1.4}$$

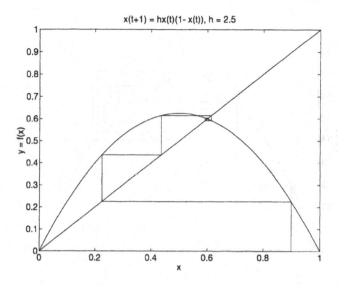

FIGURE 1.3. A single stable equilibrium

The form of the function $f^2(x) = f(x) \circ f(x)$, that is, f composed with f, (not to be confused with the square of f) is a two humped curve when f is steep enough, as shown in Figure 1.5. In fact, f^2 acquires its double hump character precisely when the value of the parameter $h = 3$.

Note that the 45 degree line intersects the two humped curve of Figure 1.5 in three points for $h > 3$. The middle intersection corresponds to the period 1 solution, which is now unstable in the sense that all populations not exactly equal to x_0^* tend away from it (as may be verified graphically) and towards the period 2 solution (see also Problem 1.4). Also, note that both the points x_1^*, x_2^* are equilibria for the system of equation (1.4). Thus, the interpretation of the dynamics of the system of (1.4) is one of strobing or sampling the dynamics of the system of (1.3) and more visually, the portrait of Figure 1.4 every 2 time steps.

• The case $3.449 < h \leq 3.570$.

Extrapolating from these observations, one may conjecture that as the height of the original hump is increased, the period 2 limit cycle becomes unstable and is replaced by a stable period 4 limit cycle involving four points. Associated with this response is the system corresponding to f^4, which would be 4 humped for this case. Also, the 45° line would intersect this curve in seven points corresponding respectively to one unstable period 1 point, two unstable period 2 points, and four stable period 4 points (forming a period 4 limit cycle). This intuition is indeed correct, and a period 4 limit cycle appears at $h_2 = 3.449$ but its formal verification needs systematic calculation and is a considerable undertaking. Thus, $h_2 = 3.449$ is the second period doubling bifurcation point.

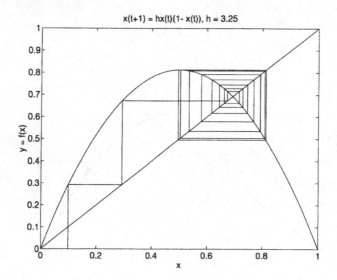

FIGURE 1.4. A period 2 limit cycle

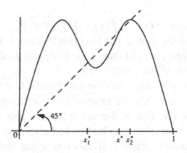

FIGURE 1.5. The graph of f^2

If this program is carried forward even further, points of period 2^k start appearing increasingly more frequently in h, for example eight period $8 = 2^3$ points appear at $h_3 = 3.544$, sixteen period $16 = 2^4$ points appear at $h_4 = 3.564$ and so on. In fact, Feigenbaum [96] has shown that,

$$\lim_{k \to \infty} \frac{h_k - h_{k-1}}{h_{k+1} - h_k} = 4.6692\ldots,$$

so that the sequence of period doubling bifurcations gets closer geometrically rapidly with periodic points of all periods 2^n for $n = 0, 1, 2, \ldots$ present at $h_\infty = 3.570$. The appearance of stable limit cycles of period 2^n, $n = 1, 2, \ldots$, and the corresponding destabilization of the limit cycles of periods 2^{n-1} are referred to as *the sequence of period-doubling bifurcations*. Actually the remarkable feature of Feigenbaum's theorem is that the ratio of the bifurcation

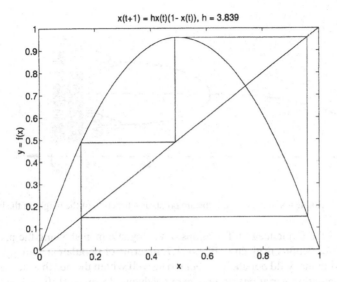

FIGURE 1.6. A period 3 limit cycle

values is exactly 4.6692..., not only for the specific logistic map that we have considered, but also for other one parameter maps that are "close" to it.

- The case $3.570 < h < 3.839$.
  In this range of the parameter h, the dynamics of the logistic map are quite complicated. For some initial conditions, the iterates are aperiodic and the trajectories appear to wander. In fact, the logistic map has been used as a random number generator in this regime (in fact in the next regime as well). There are also all the periodic points of period 2^n as well as some other periodic points.
- The case $3.839 \leq h \leq 4$.
  At $h = 3.839$, there is for the first time a periodic point of period 3, i.e., a fixed point of $f^3 = f \circ f \circ f$, (see Problem 1.5). The cobweb diagram showing the period 3 point is shown in Figure 1.6. In fact, at this value of h, there are periodic points of arbitrary period as well as some aperiodic points. The discovery of this value was the title of a landmark paper by Li and Yorke, "Period Three Implies Chaos", [182]. The period 3 orbit is the only orbit which is stable to small perturbations and is the only one that is easy to obtain from numerical experiments. As the value of h is increased there is a period doubling bifurcation of the period 3 orbit to a stable period 6 orbit. This, in turn, changes by period doubling to a stable period 12 orbit and so on and remarkably, the ratio of the parameter values for the period doubling bifurcations is remarkably once again the Feigenbaum number: 4.6692..... As it turns out, the result of Li and Yorke was a special case of an earlier result of Sharkovskii which we will describe shortly.

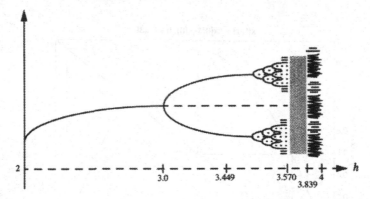

FIGURE 1.7. Stable and unstable equilibrium points as a function of the height of the hump

A plot of the sequence of bifurcations of the logistic map showing the period-doubling bifurcations is given in Figure 1.7. In terms of population ecology, it is not surprising that wild populations seem to lie well within the nonchaotic region; chaotic population variations do not seem conducive to survival. It is, however, possible to produce conditions for chaotic population fluctuation, as has been done by May and Oster in a laboratory using blowflies.

It is of obvious interest to understand which qualitative features of this incredibly delicate and complex behavior are preserved in one-hump maps more general than the logistic map. Surprisingly, most of the qualitative characteristics are preserved, and the sequence of bifurcations described above is referred to as the *period doubling route to chaos*. In an interesting generalization of the analysis of the logistic map by Sharkovskii, which actually preceded the work of Li and Yorke, Sharkovskii ordered the integers as $3 \rhd 5 \rhd 7 \rhd \cdots \rhd 2.3 \rhd 2.5 \rhd 2.7 \cdots \rhd 2^2.3 \rhd 2^2.5 \rhd \cdots \rhd \cdots \rhd 2^n \rhd 2^{n-1} \rhd \cdots \rhd 2^2 \rhd 2 \rhd 1$, that is, the odd integers except 1, followed by twice the odd integers except 1, followed by 2^2 times the odd integers except 1, followed by 2^3 times the odd integers followed except 1, and so on. The tail end of the ordering is made up of the decreasing powers of 2. He then showed that for an arbitrary system of the form of (1.2), if the discrete time system has a point of a certain period, it has points of all periods lower than that one in the Sharkovskii ordering. Thus in particular if the system has a point of period 3, it has points of arbitrary period. In the context of our example of (1.3), this happens at $h = 3.839$. Thus, at this value of h or higher, the population x_t can vary in a very complicated manner with great sensitivity to initial conditions.

1.2.1   Subtleties of Nonlinear Systems Analysis

In this book, we will be interested in studying more benign forms of nonlinear behavior, but it is worth keeping in mind that nonlinear systems are capable of very diverse and complex behavior. Nonlinear systems analysis differs from that of linear systems in two ways:

1. Generally speaking, one can obtain *closed form* solutions for linear systems. This is seldom the case for nonlinear systems. Consequently, there is a need for both qualitative insight and repeated simulation for quantitative verification. Qualitative understanding is important since, even if we had the most powerful computers at our disposal, exhaustive simulation will be prohibitively expensive. (The reader may wish to amuse herself with considering how much time would be needed to simulate an equation with about 10 states and a discretization grid of initial conditions of about 10 per unit along each axis for the initial conditions.)

2. The analysis involves mathematical tools that are more advanced in concept and involved in detail.

Consider for example, nonlinear systems whose dynamics can be described by first order vector differential equations of the form

$$\dot{x} = f(x, u, t), \quad x(0) = x_0. \tag{1.5}$$

Here $x \in \mathbb{R}^n$ is the state vector, x_0 is the state of the system at time 0 and $u \in \mathbb{R}^m$ is the control or forcing function. If equation (1.5) were affine in x and u for each t, one would expect that corresponding to each input $u(\cdot)$:

- (1.5) has *at least one* solution belonging to some reasonable class of functions (*existence of a solution*).
- (1.5) has *exactly one* solution in the same class as above (*uniqueness of the solution*).
- (1.5) has exactly one solution *for all time*, i.e., on $[0, \infty[$ (*extension of the solution up to $t = \infty$*).

Note that the latter requirements in this list successively subsume the former: They get more demanding. In the full generality of (1.5), none of the above statements are true for nonlinear systems. We will now give examples of systems that violate the three preceding requirements.

1. Lack of existence of solutions.

$$\dot{x} = -\text{sign}(x), \quad x(0) = 0. \tag{1.6}$$

Here $\text{sign}(x)$ is defined to be 1 if $x \geq 0$, and $\text{sign}(x) = -1$ if $x < 0$. Consequently, there can be no continuously differentiable function satisfying (1.6). Nevertheless, the system is an acceptable model of the dynamics of a thermostat about a set point temperature, modeled by $x = 0$ (one can imagine the furnace turned on full blast when the temperature drops below the set point and the air conditioner turned on full strength when the temperature rises above the set point). Indeed, you may have noticed that some thermostats tend to chatter about their fixed points. Have you thought about how you might quench this chattering?

2. Lack of uniqueness of solutions.
Consider the differential equation

$$\dot{x} = 3x^{2/3}, \quad x(0) = 0. \tag{1.7}$$

It may be verified that the family of functions described by

$$x_\alpha(t) = (t - \alpha)^3, \quad t \geq \alpha,$$

$$x_\alpha(t) = 0, \quad\quad\quad t < \alpha,$$

satisfies the differential equation and initial condition for arbitrary values of α.

3. Finite escape time.

Consider the system

$$\dot{x} = 1 + x^2, \quad x(0) = 0. \tag{1.8}$$

It has a solution $x(t) = \tan(t)$, so that there is no solution defined outside of the interval $[0, \pi/2[$.

We shall not resolve these issues immediately, but rather draw from them the lesson that the preceding requirements on the existence, uniqueness, and extension of solutions of (1.5), while obvious for linear systems, require more careful consideration in a nonlinear context.

1.2.2   Autonomous Systems and Equilibrium Points

Consider systems of the form of (1.5) with the input $u(t)$ a fixed function of time. Then, these systems take the form

$$\dot{x} = f(x, t). \tag{1.9}$$

Definition 1.1 Autonomous.   *The system (1.9) is said to be* autonomous *if $f(x, t)$ is not explicitly dependent on time t.*

Definition 1.2 Equilibrium Point at t_0. $x_0 \in \mathbb{R}^n$ *is said to be an* equilibrium point at time t_0 *iff*

$$f(x_0, t) = 0 \quad \forall \quad t \geq t_0.$$

Note that if (1.9) does, in fact, have unique solutions then if x_0 is an equilibrium point at t_0 and $x(t_0) = x_0$ then $x(t) \equiv x_0 \; \forall \; t \geq t_0$. Also if $f(x_0, t) = 0$ for all t, then x_0 is referred to simply as an equilibrium point. Below is a simple example to illustrate the definitions just introduced.

Example 1.3 Pendulum equation.   *The following equation models the dynamics of a pendulum with no forcing:*

$$\dot{x}_1 = x_2,$$

$$\dot{x}_2 = -k_1 x_2 - k_2 \sin(x_1).$$

In these equations k_1 is proportional to frictional damping and k_2 the length of the pendulum. The system is autonomous and has equilibria at

$$x_2 = 0, \quad x_1 = n\pi, \quad n = 0, \pm 1, \pm 2, \dots.$$

Note that the system has multiple equilibria.

If the system (1.9) is autonomous, then finding the equilibrium points corresponds to solving a nonlinear algebraic equation

$$f(x) = 0. \tag{1.10}$$

(1.10) may have no solution, several solutions, or a continuum of solutions. In the linear case the equation

$$Ax = 0.$$

has a unique solution $x = 0$ iff the matrix A is nonsingular. If A is singular, then it has a continuum of solutions, namely the null space of A.

Definition 1.4 Isolated Equilibria. *An equilibrium point x_0 of an autonomous system (1.9) is isolated if there exists some $\delta > 0$ such that there is no other equilibrium point in the ball $B(x_0, \delta) = \{x : |x - x_0| < \delta\}$.*

To give the reader an early feel for the methods of nonlinear analysis, we show how sufficient conditions for the existence of isolated equilibria in a system of the form (1.9) can be given. The proof will use concepts of norms ($|\cdot|$ on $\mathbb{R}^n$), and inequalities associated with norms: These are developed more completely in Chapter 3. Assume that x_0 be an equilibrium point of (1.9) for all time, and assume that $f(x, t)$ is a C^1 function (i.e., a function that is continuously differentiable in x), and define its linearization at t_0 as a matrix in $\mathbb{R}^{n \times n}$:

$$A(t_0) := \frac{\partial f}{\partial x}(x_0, t_0).$$

Here $A_{ij}(t_0) = \frac{\partial f_i}{\partial x_j}(x_0, t_0)$.

Proposition 1.5 Sufficient Condition for Isolated Equilibria. *Consider a C^1 system (1.9) with equilibrium point x_0 for all t and its linearization $A(t_0)$. Then, if $A(t_0)$ is nonsingular, x_0 is an isolated equilibrium.*

Proof: Since $A(t_0)$ is nonsingular, there exists $c > 0$ such that

$$|A(t_0)x| \geq c|x|$$

Further, since $f(x, t_0)$ is C^1, we may write down its Taylor series about $x = x_0$ as

$$f(x, t_0) = f(x_0, t_0) + A(t_0)(x - x_0) + r(x, t_0), \tag{1.11}$$

where $f(x_0, t_0)$ is zero, since x_0 is an equilibrium, the linear term is $A(t_0)(x - x_0)$, and the remainder $r(x, t_0)$ is the sum of the quadratic, cubic and higher order terms, i.e., the *tayl*[1] of the Taylor series. $r(x, t_0)$ is of order $|x - x_0|^2$, i.e.,

$$\lim_{|x-x_0| \to 0} \frac{|r(x, t_0)|}{|x - x_0|} = 0. \tag{1.12}$$

---

[1]This misspelling of "tail" is standard in the literature for comic intent.

In view of (1.12), it follows that there exists $\delta > 0$ such that

$$|r(x, t_0)| \leq \frac{c}{2}|x - x_0| \quad \forall \ |x - x_0| < \delta.$$

Using this estimate in (1.11) along with the bound on $|A(t_0)x|$ yields

$$|f(x, t_0)| \geq |A(t_0)(x - x_0)| - |r(x, t_0)|$$
$$\geq c/2|x - x_0| \qquad \qquad \forall \ |x - x_0| < \delta$$
$$> 0 \qquad \qquad \forall \ x \neq x_0 \ \in B(x_0, \delta).$$

This completes the proof. $\qquad\qquad\qquad\qquad\qquad\qquad\qquad\qquad\qquad\qquad\qquad$ □

Remarks: The conditions of the preceding proposition are sufficient, but not necessary, for the existence of isolated equilibria: For example, the system

$$\dot{x} = x^3, \quad x \in \mathbb{R},$$

has linearization 0 at the equilibrium 0, which is nonetheless isolated. However, the conditions of the proposition are tight in the sense that there are examples of nonlinear systems which do not have isolated equilibria when they have a singular linearization. In the following model for the spread of disease in a population, x_1 is the number of infected people, x_2 the number not infected, a the death rate, and b the infection rate:

$$\dot{x}_1 = -ax_1 + bx_1x_2,$$
$$\dot{x}_2 = -bx_1x_2.$$

The point $(0, x_2)$ is a continuum of equilibria for this system and the second column of the corresponding linearization is zero.

1.3   Some Classical Examples

There is a large number of classical and simple examples of nonlinear dynamical systems, that have been studied in detail. Before diving into the formal theory required to study their dynamics, we will describe some of them briefly:

1.3.1   The Tunnel Diode Circuit

A simple circuit example is useful for illustrating an example of a circuit with multiple equilibria. Consider the tunnel diode circuit of Figure 1.8. The equations of the circuit obtained from using the two relations $i_C = -i_R + i_L$, $v_C = v_R$ along with Faraday's law applied to the inductor and Coulomb's law applied to the capacitor are

$$C\dot{v}_C = -i_R(v_C) + i_L,$$
$$L\dot{i}_L = -v_C - Ri_L + E. \qquad\qquad\qquad (1.13)$$

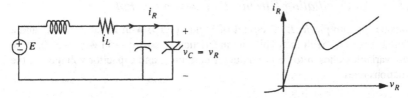

FIGURE 1.8. A tunnel diode circuit with the tunnel diode characteristic

Here C, L, and R stand for the linear capacitance, inductance and resistance values respectively and E stands for the battery voltage. The only nonlinear element in the circuit is the tunnel diode with characteristic $i_R(v_R)$. Since $v_R = v_C$, equation (1.13) uses as characteristic $i_R(v_C)$. The equilibria of the circuit are the values at which the right-hand side vanishes; namely, when $i_R = i_L$ and $v_C = E - Ri_L$. Combining these two equations gives $v_C = E - Ri_R(v_C)$. This equation may be solved graphically as shown in Figure 1.9. The resistive characteristic of a tunnel diode has a humped part followed by an increasing segment, as shown in the figure. Thus, as seen from the figure, there are either three, two or even one solutions of the equation, depending on the value of E. The situation corresponding to two solutions is a special one and is "unstable" in the sense that small perturbations of E or the nonlinear characteristic will cause the formation of either one solution or three solutions.

Consider the thought experiment of increasing the voltage E and letting the circuit equilibrate. For small values of E, the equilibrium current follows the left branch of the $i - v$ characteristic of the tunnel diode till $E = E_3$ at which point, there is no longer an equilibrium point on the left leg of the characteristic, but there is one on the right branch of the characteristic. Now, as the voltage is reduced, the equilibrium will stay on the right branch of the characteristic till $E = E_2$, at which point it will jump back to the right hand branch. The hysteresis in the equilibrium values in this *quasi-static* experiment is characteristic of nonlinear systems.

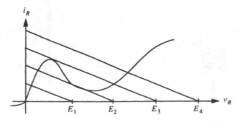

FIGURE 1.9. Equilibrium points of a tunnel diode circuit shown by the method of load lines

1.3.2 An Oscillating Circuit: Due to van der Pol

Consider the simple R, L, C circuit of Figure (1.10) with linear L and C and a nonlinear resistor with the cubic characteristic $i = \alpha v(v^2 - 1)$ with $\alpha > 0$. With state variables chosen to be the inductor current i_L and capacitor voltage v_C, the equations read

$$C\dot{v}_C = -i_L - i_R(v_C),$$
$$L\dot{i}_L = v_C. \tag{1.14}$$

A simulation as shown in Figure 1.11 reveals that the equilibrium point at the origin is unstable and is surrounded by a limit cycle. All nonzero initial conditions appear to converge to this limit cycle. It is of interest to see the nature of the changes in the phase portrait as C is gradually decreased. For extremely small values of the capacitance, the shape of the oscillation takes the form of Figure (1.12) consisting of two extremely fast pieces and two slow pieces. This is the prototype for a so-called *astable multivibrator circuit*, with the two fast pieces of the trajectory representing transitions between two logic states. The circuit was first analyzed by an electrical engineer, van der Pol, who purportedly thought that the circuit was capable of oscillation, because the nonlinear resistor is "active" (that is, its vi product, with both v and i pointing into the resistor, is non-positive) for small v and i and "passive' (i.e., vi is non-negative) for large v and i.

One way to understand the fast segments of the van der Pol oscillator limit cycle for small values of capacitance is to note that when $C = 0$ the equations (1.14) reduce to a constrained differential equation

$$0 = i_L + i_R(v_C),$$
$$L\dot{i}_L = v_C. \tag{1.15}$$

Thus the equation spends most of its time on segments where $i_L = i_R(v_C)$, except when it jumps (i.e., makes an instantaneous transition) from one leg of the characteristic to the other. In Chapter 6 we will see why this system of equations does not have solutions that are continuous functions of time, see also [337].

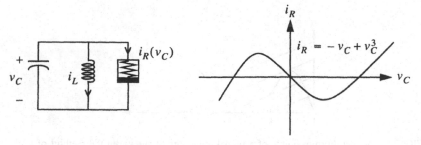

FIGURE 1.10. The van der Pol oscillator circuit

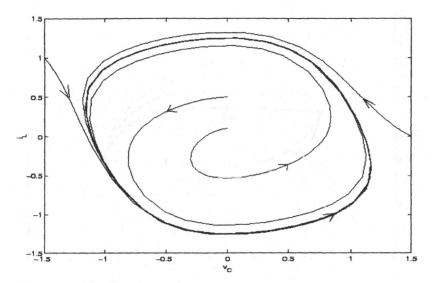

FIGURE 1.11. The phase portrait of the van der Pol oscillator

1.3.3 The Pendulum: Due to Newton

Consider the pendulum of Figure 1.13 swinging under gravity. Assume that the length of the pendulum is l, that it has a mass m concentrated at the tip and that the frictional damping is assumed to be viscous with damping coefficient d. From elementary mechanics, the equation of motion of the pendulum is given by

$$ml^2\ddot{\theta} + d\dot{\theta} + mgl\sin(\theta) = 0. \tag{1.16}$$

In (1.16) θ is the angle made by the pendulum with the vertical, and g is the gravitational acceleration. This equation has two state variables, θ and $\dot{\theta}$. It is possible to depict some of the trajectories of this dynamical system on a *phase plane*, namely, the plane consisting of the state variables, with time as an implicit parameter. Classical books in nonlinear systems spent a great deal of time in describing graphical techniques for generating phase plane portraits by hand, but with the advent of simulation packages an efficient way of generating a phase portrait is by interactively integrating the solutions of the given differential equations both forwards and backwards in time from different initial conditions. Some particularly nice packages that automatically generate such phase plane portraits are DSTOOL and KAOS [123]. Such a phase plane for the case that the damping d is zero is shown in Figure 1.14.

It is instructive to understand features of this phase portrait. First consider the equilibrium points: These are the points at which the trajectory is stationary for all time, as defined in the previous section. The equilibrium points of this system are at $\dot{\theta} = 0$, $\theta = \pm n\pi$. They are all isolated and appear to be infinitely many in number. Another striking characteristic is that the phase portrait is periodic in

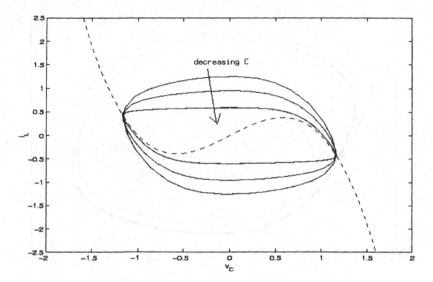

FIGURE 1.12. Changes in the limit cycle as the capacitance decreases

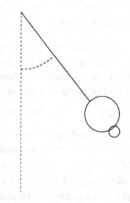

FIGURE 1.13. A pendulum swinging under gravity

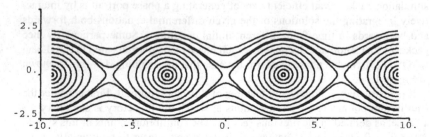

FIGURE 1.14. Phase portrait of an undamped pendulum

the θ direction with period 2π. This again is not surprising, given that the right hand side of the differential equation is periodic in the variable θ with period 2π. Further, the equilibrium points are surrounded by a continuum of trajectories that are closed orbits corresponding to periodic orbits in the state space. A little numerical experimentation shows that the closed orbits are of a progressively lower frequency as one progresses away from the equilibrium point. One particularly curious trajectory is the trajectory joining the two saddles, which are 2π radians apart. This trajectory corresponds to one where a slight displacement from the equilibrium position ($\theta = -\pi$, $\dot{\theta} = 0$) results in the pendulum building up kinetic energy as it swings through $\theta = 0$ and just back up to another equilibrium point at $\theta = \pi$ trading the kinetic energy for potential energy. Actually, since the equations of the pendulum are periodic in the θ variable with period 2π it is instructive to redraw the phase portrait of (1.14) on a state space that is cylindrical, i.e., the θ variable evolves on the interval $[0, 2\pi]$ with the end points identified as shown in Figure 1.15. Note that, on this state space, the points $\theta = -\pi$ and $\theta = \pi$ are identical. Also, as mentioned above, each of the equilibrium points of the pendulum is surrounded by a set of closed orbits of increasing amplitude and decreasing frequency. The trajectory joining the saddle at $(\pi, 0)$ to the saddle at $(-\pi, 0)$ which lies in the upper half plane (that is, with $\dot{\theta} \geq 0$), and its mirror image about the θ axis, taken together may be thought of as constituting an infinite period closed orbit. This simple example manifests many of the features of a nonlinear system that we mentioned at the outset: multiple equilibrium points, periodic orbits (in fact, a continuum of periodic orbits), and an interesting dynamic feature corresponding to the union of the upper and lower saddle connection trajectories, which bears some resemblance to an infinite period closed orbit.

This equation shows up in several other contexts as well. In Chapter 2 we study the version of this system with non-zero damping in the context of the dynamics of a Josephson junction circuit and in Chapter 5 we study it in the context of the so called *swing equations* modeling the dynamics of an electrical generator coupled to a power transmission network. The continuum limit of this equation also appears in many problems of classical and quantum physics, and this partial differential equation is referred to as the sine-Gordon equation.

1.3.4   The Buckling Beam: Due to Euler

Consider the following qualitative description of the effect of axial loading on a thin metal bar balanced on end. When the axial load is small, the bar is slightly compressed but unbuckled. In this unbuckled configuration, when the bar is pushed to one side and released it will oscillate back and forth. Under a heavier axial load, it will buckle. If the beam is symmetric about the axis of loading, then there are two symmetric buckled states. When pushed away from one of the buckled states it will oscillate about that state. To obtain this qualitative feature a good mathematical model (see Stoker [286] for an exposition of Euler's model) is

$$m\ddot{x} + d\dot{x} - \mu x + \lambda x + x^3 = 0. \qquad (1.17)$$

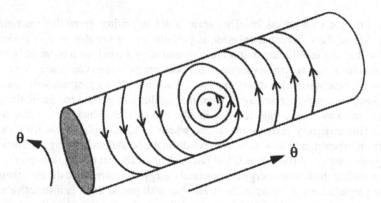

FIGURE 1.15. Phase portrait of the pendulum on a cylinder

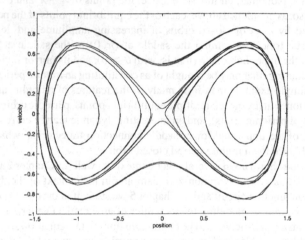

FIGURE 1.16. Phase portrait of a buckled beam

Here, x is the variable representing the one-dimensional deflection of the beam normal to the axial direction, μ stands for the applied axial force, $\lambda x + x^3$ models the restoring spring force in the beam, and d is the damping. The state variables are x and $\dot{x}$. Further, it is interesting to note that the system has one equilibrium point $x = \dot{x} = 0$ when $\lambda \le \mu$ and three equilibrium points $x = 0, \dot{x} = 0$ and $x = \pm\sqrt{\lambda - \mu}, \dot{x} = 0$ when $\lambda > \mu$. A phase portrait for the undamped case $(d = 0)$ when $\mu - \lambda = 1$ showing the three equilibria is given in Figure 1.16. Note the existence of the three equilibrium points of this system at $\dot{x} = 0, x = 0, \pm1$, and the orbits connecting the equilibrium point $\dot{x} = 0, x = 0$ to itself, and a continuum of closed orbits between the equilibrium point $\dot{x} = 0, x = -1$ and the trajectory connecting the saddle at $\dot{x} = 0, x = 0$ to itself as well as between the equilibrium point $\dot{x} = 0, x = 1$ and the other trajectory connecting the saddle to itself. Each of

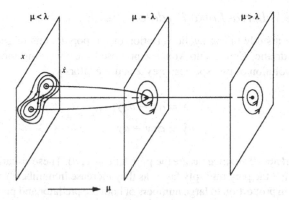

FIGURE 1.17. A bifurcation diagram for the Euler beam buckling showing the transition from the unbuckled state to the buckled state

the saddle trajectories bears resemblance to the preceding example, though each of them connects the saddle to itself, rather than a different saddle. Again the continuum of closed orbits contained inside each of the saddle connections is of progressively decreasing frequency away from the equilibrium point towards the "infinite" period saddle connections. A bifurcation diagram showing the transition from the unbuckled state to the buckled state with μ as the bifurcation parameter is shown in Figure 1.17. There are several qualitative features of the phase portrait of the buckling beam that are like those of the undamped pendulum. These features are common to Hamiltonian systems, that is, systems of the form

$$\dot{x}_1 = \frac{\partial H(x_1, x_2)}{\partial x_2},$$
$$\dot{x}_2 = -\frac{\partial H(x_1, x_2)}{\partial x_1}.$$
(1.18)

for some Hamiltonian function $H(x_1, x_2)$. In the case of the undamped beam ($d = 0$), with $x = x_1, \dot{x} = x_2$, we have that

$$H(x_1, x_2) = \frac{1}{2}x_2^2 + \frac{1}{m}\left((\lambda - \mu)\frac{x_1^2}{2} + \frac{x_1^4}{4}\right).$$

The phase portraits of Hamiltonian systems are characterized by equilibrium points surrounded by a continuum of closed orbits (of progressively longer and longer time period), and if there are multiple equilibrium points, trajectories connecting equilibrium points to themselves. The equation of the buckling beam is referred to as the undamped Duffing equation. It is instructive to study what happens to the phase portraits of Figures 1.14 and 1.16 when some damping is present (the system then ceases to be Hamiltonian), i.e., $d > 0$ in equations (1.16) and (1.17) (cf. Problem (1.9)). Problem 1.10 explores a scenario involving a pendulum with magnets, which is related to the buckled beam.

1.3.5   The Volterra–Lotka Predator–Prey Equations

Motivated by a study of the cyclic variation of the populations of certain small fish in the Adriatic, Count Vito Volterra proposed the following model for the population evolution of two species: prey x and predators y.

$$\dot{x} = ax - bxy,$$
$$\dot{y} = cxy - dy. \tag{1.19}$$

Here a, b, c, d are all positive (as are the populations x, y). These equations follow from noting that the prey multiply faster as they increase in number (the term ax) and decrease in proportion to large numbers of both the predator and prey (the term bxy). The predator population is assumed to increase at a rate proportional to the number of predator–prey encounters modeled as $c(x-d/c)y$. Here c is the constant of proportionality, and the number of predator prey encounters should be xy. However, to model the fact that a steady state prey population is required to sustain growth, this term is modified using an offset in x as $(x-d/c)$. Admittedly, the model is oversimplified but is nevertheless ecologically meaningful. A more meaningful model is given in the exercises as a refinement of this one. A representative phase portrait of this system is given in Figure 1.18. Other scenarios could occur as well: for some values of the parameters, the Volterra–Lotka equations predict convergence of the solutions to a single stable equilibrium population. These are explored in the exercises (Problems 1.11–1.13). In particular, Problem 1.13 explores how robust the model is to uncertainties in modeling.

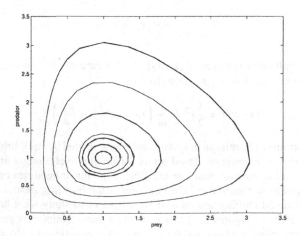

FIGURE 1.18. Phase portrait of the Volterra–Lotka equations showing cyclic variation of the populations

1.4   Other Classics: Musical Instruments

The theory of dynamical systems owes a great deal to the study of musical instruments, begun by Chladni, whose work attracted the attention of Napoleon. It was continued and considerably embellished by Rayleigh, who in a book entitled "The Theory of Sound" [245], separated musical instruments into two classes:

- *Percussion instruments* such as drums, guitars, and pianos which he modeled by damped oscillations.
- *Sustained instruments* such as bowed strings, and winds which he modeled by "self sustained oscillations", or closed orbits.

While the models of the first class are interesting, in terms of dynamics they correspond to transient behavior of the system returning to an equilibrium point. Models of the second class are more interesting and we will discuss two of them here.

1.4.1   Blowing of a Clarinet Reed: Due to Rayleigh

Rayleigh modeled the clarinet reed as a linear oscillator, a second order system of the form

$$\ddot{x} + kx = 0.$$

The effect of the clarinetist is modeled by introducing a term on the right hand side of the form $\alpha\dot{x} - \beta(\dot{x})^3$, with $\alpha, \beta > 0$, indicating negative damping for small $\dot{x}$ and positive damping for high $\dot{x}$. This yields a composite model of the form

$$\ddot{x} - \alpha\dot{x} + \beta(\dot{x})^3 + kx = 0. \tag{1.20}$$

This is the dynamical model for the sustained oscillation of the blown clarinet reed. The phase portraits are shown in Figure 1.19. There are three different oscillations in the figure. The one in the middle corresponds to $k = 1, \alpha = 1, \beta = 1$. The equilibrium point at the origin is unstable, and so trajectories converge to the inner limit cycle. Let us consider the relationship between the parameters of the model and the sound of the clarinet. A stiffer reed is modeled by a stiffer spring with larger k. The innermost limit cycle corresponds to the situation where $k = 3$. The limit cycle has a somewhat different shape, and the tone (timbre) of the clarinet is richer. Blowing harder is modeled as a somewhat broader friction characteristic, by setting $\beta = 0.5$ and restoring $k = 1$. Then the limit cycle is the outermost. It is somewhat larger, and the sound of the clarinet is louder.

1.4.2   Bowing of a Violin String: Due to Rayleigh

A model of the bowing of a violin string proposed by Rayleigh has a mechanical analogue consisting of a spring–mass system resting on a conveyor belt moving at a constant velocity b as shown in the figure (1.20). The model of the unbowed string is simply the spring mass system and the bowing is modeled in this analogue

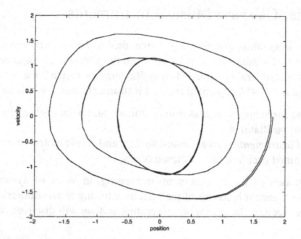

FIGURE 1.19. Phase portrait of the clarinet model, with different spring constants and different blowing characteristics

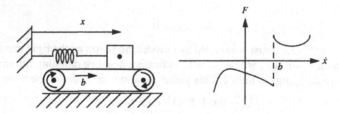

FIGURE 1.20. The mechanical analogue of the bowing of a violin string

by having a friction characteristic between the body and the conveyor belt, that is sticky friction or stiction. The friction force between the conveyor belt has the qualitative form shown in Figure 1.20. The equations of the system are given by;

$$M\ddot{x} + kx + f(\dot{x} - b) = 0. \tag{1.21}$$

The equilibrium of the system is given by $\dot{x} = 0, x = -f(-b)/k$. It is unstable in the sense that initial conditions close to it diverge, as shown in the phase portrait of Figure 1.21. The phase portrait however has an interesting limit cycle to which trajectories starting from other initial conditions appear to be attracted. The limit cycle shows four distinct regimes, with the following physical interpretation:

1. *Sticking.* The block is stuck to the conveyor, and the friction force matches the spring force. The velocity of the mass is equal to the velocity of the conveyor belt. Thus, the block sticks to the conveyor belt and moves forward.
2. *Beginning to slip.* The spring has been extended sufficiently far that the force on the block is strong enough to cause the block to start slipping. That is, the spring

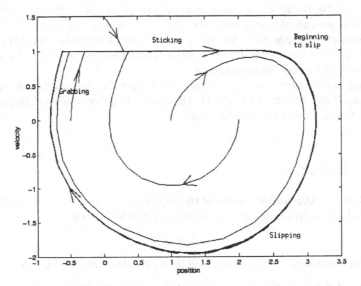

FIGURE 1.21. Phase portraits of a violin string starting from different initial conditions

force exceeds the frictional force. Hence the acceleration becomes negative and the velocity starts to decrease.

3. *Slipping*. The sudden drop of friction caused by the onset of slipping yields the tug of war to the spring, causing the block to slip faster—the acceleration becomes negative, the velocity becomes negative, and the block starts moving to the left.

4. *Grabbing*. When the leftward motion of the block has decreased the spring force to a value at which it is less than the slipping friction, the motion to the left slows, and the velocity starts going positive. When the velocity reaches a critical value we return to the sticking phase.

Convince yourself by simulation that initial conditions starting from far away also converge to this limit cycle. It is once again of interest to see that a stiffer string may be modeled by a stronger spring with a consequent greater tone or richness of the note. A broader friction characteristic of the stiction or higher velocity of the belt will correspond to a louder note.

1.5 Summary

In this chapter, we have seen the richness of dynamic behavior predicted by non-linear differential equations and difference equations. To warm up for the sort of detailed mathematical analysis that will be necessary in the chapters to come, we have given, by way of example, sufficient conditions for the existence of isolated equilibria. The origins of the endeavor to build simple nonlinear models of physical phenomena are old. Here we have tried to give a modern justification for some important classical equations. We encourage the reader to begin his own voyage of discovery in the literature and to use a simulation package to play with the models, especially in the two-dimensional case.

Some excellent references are Abraham and Shaw [3], Hale and Koçak [126], Wiggins [329], Hirsch and Smale [141], Arnold [11], Nemytskii and Stepanov [229], Guckenheimer and Holmes [122].

1.6   Exercises

Problem 1.1   Alternative version of the logistic map.   Find the transformation required to convert the logistic map of (1.2), (1.3) into the form

$$y_{k+1} = \mu - y_k^2.$$

Find the transformation from x to y and the relationship between h and μ.

Problem 1.2   Computing the square root of 2.   Your friend tells you that a good algorithm to compute the square roots of 2, is to iterate on the map

$$x_{k+1} = \frac{x_k}{2} + \frac{1}{x_k}.$$

Do you believe her? Try the algorithm out, with some random guesses of initial conditions. Do you get both the square roots of 2?

Problem 1.3.   It is instructive to derive conditions for the existence of limit cycles for linear systems of the form (1.1). Indeed, let $x_0 \in \mathbb{R}^n$ be a point lying on such a limit cycle and further assume that the period of the limit cycle is T. Then derive the equation that x_0 should satisfy and derive necessary and sufficient conditions for the existence of such a limit cycle in terms of the eigenvalues of A. Further, show that if the conditions are satisfied, there is a continuum of x_0 satisfying the equation you have derived. Characterize this set (subspace).

Problem 1.4.   Consider the one hump function of (1.3). Derive the dynamical system for the two step evolution of this function. Now determine the minimum value of the parameter h for which the period 2 limit cycle exists by looking for equilibrium points of f^2. Note that equilibrium points of f will also be equilibrium points of f^2. Thus you are looking for the parameter value h at which the equation

$$f^2(x) = x$$

has 4 solutions. Determine the values of the equilibria explicitly. The critical value of h at which the period 2 limit cycle appears is the value of h at which the map $f^2(x)$ becomes *two humped*. Explain this and use this as an alternate method of doing this problem.

Problem 1.5 Period 3 points. Consider the one hump equation of (1.3). Find the value of h for which it has a period three point. That is, find the value of h for which there exist $a < b < c$, such that $f(a) = b$, $f(b) = c$, $f(c) = a$. (Hint: try $h = 3.839$).

Problem 1.6. Read more about the details of the dynamics of the one-hump map, you may wish to study either the original paper of Li and Yorke [182] or read about it in the book of Devaney [84].

Problem 1.7 One Hump Map Game. Consider the discrete time system of the one hump map, namely

$$x_{k+1} = hx_k(1 - x_k).$$

For $0, h < 1$ find all the fixed points of the equation, that is, for which $x_{k+1} = x_k$, and determine which are stable and which are unstable (if any!).

Let $1 < h < 3$. Find all the equilibrium points. Show that

1. If $x_0 < 0$, then $x_k \to -\infty$ as $k \to \infty$.
2. If $x_0 > 1$, then $x_k \to -\infty$ as $k \to \infty$.
3. For some interval $I \subset]0, 1[$, if $x_0 \in I$, then $x_k \to x_h := \frac{h-1}{h}$.

Determine the open interval of attraction of x_h. For a game, program the above recursion on your calculator or computer for $h = 4$ and note that initial conditions starting in $]0, 1[$ stay in $]0, 1[$. Watch the sequence x_k wander all over the interval except for some special initial conditions x_0 corresponding to periodic solutions.

Problem 1.8. Use your favorite simulation environment to simulate the pendulum equation, the buckling beam equation, the Volterra–Lotka predator prey equations, the Rayleigh equation and the van der Pol equation with several of the parameters of the equations being varied. Write a small macro to run the simulation for a set of initial conditions and plot the resulting phase diagram. Then, think about how you may look for interesting features in the phase portraits. How do you know where to look for periodic solutions? How do you get at unstable equilibrium points? Remember that you can integrate equations for $t \le 0$ by changing the sign of the right hand side. What does this do to the stability of equilibrium points?

Problem 1.9. Consider the equations of the pendulum and the buckling beam namely (1.16) and (1.17) with positive non zero damping d. Use a numerical simulation package to draw phase portraits for these two cases. What happens to the continuum of limit cycles and the orbits connecting the saddles? The phase portrait of Figure 1.22 is that of the damped buckling beam equation with $m = 1$, $\mu = 1.2$, $\lambda = 0.2$, $d = 0.3$. Use it to stimulate your imagination!

Problem 1.10. Consider the physical scenario corresponding to the pendulum of Figure 1.14 with damping $d > 0$ as in the previous problem and two magnets of unequal strength as shown in Figure 1.23 near the bottom of the arc of the swing of the pendulum. Conjecture (it is probably not very easy to derive the equations of motion) what the phase portrait of this system might look like. In particular,

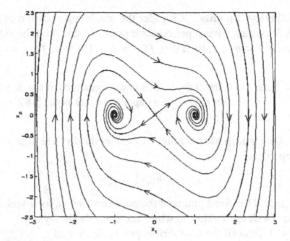

FIGURE 1.22. Segments of the phase portrait of the damped buckling beam equation with $d = 0.3$. The reader is invited to fill in the location of the stable manifold of the saddle at the origin of the picture.

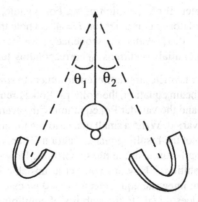

FIGURE 1.23. Showing a pendulum with two magnets of unequal strength

how many equilibria do you expect (for $-\pi \leq \theta < \pi$)? How many of them are stable and how many unstable? What can you say about what volume of initial conditions gets attracted to each equilibrium point? How does this phase portrait relate to that of the Euler buckled beam?

Problem 1.11. Consider the Volterra–Lotka equations of (1.19). Experiment with values of $a, b, c, d > 0$. Do you always get cyclic variations in the population or is it possible to have trajectories converging to a non-zero predator-prey steady state population. How could you modify the model to have trajectories converging to a steady state non-zero population of predator and prey.

Problem 1.12 Predator–Prey with limited growth.   The Volterra–Lotka model of (1.19) assumes unlimited growth of the prey species if the predators are absent. Frequently, "social factors" (overcrowding) will make this impossible. A model showing the saturation of the prey population in the absence of predators has an extra term $-\lambda x^2$ as shown below. A similar consideration for the predator population gives the term $-\mu y^2$, and the composite equations are

$$\dot{x} = (a - by - \lambda x)x,$$
$$\dot{y} = (cx - d - \mu y)y.$$
(1.22)

Study the equilibria of this system and try to simulate it for various values of the parameters $a, b, c, d, \lambda, \mu > 0$.

Problem 1.13 Volterra–Lotka generalized.   A more meaningful Volterra–Lotka model is obtained by considering the equations

$$\dot{x} = xM(x, y),$$
$$\dot{y} = yN(x, y).$$
(1.23)

Think about what minimum requirements M, N should have in order for the equations to model predator–prey characteristics (with the bounded growth phenomenon discussed above). From drawing the curves $\{(x, y) : M(x, y) = 0\}$ and $\{(x, y) : N(x, y) = 0\}$ qualitatively determine the nature (stability characteristics) of the equilibrium points. Further, in the positive quadrant draw the qualitative shape of the direction of population evolution in the four regions obtained by drawing the curves above. Give conditions for convergence of all prey populations to zero, all predator populations to zero, and for the nonexistence of periodic solutions. Simulate the systems for some guesses of functions M, N.

Problem 1.14 Competing Populations.   Let us say that x and y model the populations of two competing species. We will consider a somewhat different kind of population dynamics for them. Let a, b, c, d be positive where a and d represent the unimpeded population growth, and b, c represent the competition: Species 1 eats species 2 and vice versa. Thus, we have

$$\dot{x} = ax - bxy,$$
$$\dot{y} = dy - cxy.$$
(1.24)

Find all the equilibrium points and sketch a plausible phase portrait. Try to generalize this example as in the previous one to the case when the right hand sides are respectively $xM(x, y)$ and $yN(x, y)$, with appropriate hypotheses on M, N.

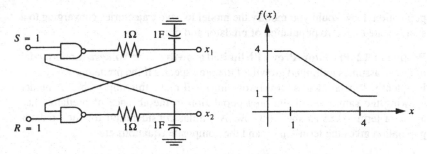

FIGURE 1.24. An SR latch

Problem 1.15 The SR Latch.   The figure (1.24) shows a dynamic model of a standard SR (Set-Reset) latch. Assume that the NAND gates are identical, and that S,R are both high (represented in the figure as the binary 1), so that the NAND gates function as inverters. Let the state variables x_1, x_2 be the voltages on the capacitors and draw an approximate phase portrait showing the (three!) equilibria of the system. Two of them are the logic one and logic zero output of the latch. What can you say about the third?

2
Planar Dynamical Systems

2.1 Introduction

In the previous chapter, we saw several classical examples of planar (or 2 dimensional) nonlinear dynamical systems. We also saw that nonlinear dynamical systems can show interesting and subtle behavior and that it is important to be careful when talking about solutions of nonlinear differential equations. Before dedicating ourselves to the task of building up in detail the requisite mathematical machinery we will first study in semi-rigorous but detailed fashion the dynamics of planar dynamical systems; that is, systems with two state variables. We say semi-rigorous because we have not yet given precise mathematical conditions under which a system of differential equations has a unique solution and have not yet built up some necessary mathematical prerequisites. These mathematical prerequisites are deferred to Chapter 3.

The study of planar dynamical systems is interesting, not only for its relative simplicity but because planar dynamical systems, continuous and discrete, contain in microcosm a great deal of the variety and subtlety of nonlinear dynamics. The intuition gained in this study will form the heart of the general theory of dynamical systems which we will develop in Chapter 7.

2.2 Linearization About Equilibria of Second-Order Nonlinear Systems

The study of nonlinear systems in the plane begins with a study of their equilibria and the flows (namely, the trajectories of the equation) in a neighborhood of the equilibrium point. One can say a great deal about the qualitative behavior

of nonlinear systems in the vicinity of an equilibrium point from a study of their
linearization at the equilibrium point. Consequently, we will begin with a study of
the phase portraits of linear systems in the plane.

2.2.1   Linear Systems in the Plane

A good starting point for analyzing solutions of linear systems in the plane of the
form

$$\dot{x} = Ax, \quad x \in \mathbb{R}^2, \tag{2.1}$$

is the study the eigenvalues of A. Recall that in order for the system (2.1) to have
an isolated equilibrium point at the origin, it should have no eigenvalues at the
origin. Now, consider the similarity transformation $T \in \mathbb{R}^{2 \times 2}$,

$$T^{-1}AT = J,$$

that determines the real Jordan form[1] of the matrix A. Applying the coordinate
transformation $z = T^{-1}x$, the system is transformed into the form

$$\dot{z} = T^{-1}ATz = Jz.$$

The following three cases, having qualitatively different properties, now arise.

- *Real Eigenvalues*

$$J = \begin{bmatrix} \lambda_1 & 0 \\ 0 & \lambda_2 \end{bmatrix} \tag{2.2}$$

with $\lambda_1, \lambda_2 \in \mathbb{R}$ results in the following formula for the relationship between
z_1, z_2 obtained by eliminating time from the formulas

$$z_1(t) = z_1(0)e^{\lambda_1 t}, \quad z_2(t) = z_2(0)e^{\lambda_2 t}$$

for the system dynamics:

$$\frac{z_1}{z_{10}} = \left(\frac{z_2}{z_{20}} \right)^{\frac{\lambda_1}{\lambda_2}}. \tag{2.3}$$

Figure 2.1 shows the phase portraits for the two cases that $\lambda_1, \lambda_2 < 0$ (called
a *stable node*) and $\lambda_1 < 0 < \lambda_2$ (called a *saddle*). It is interesting to observe
that equation (2.3) holds for both these cases, with the only difference being
that the exponent is positive for the stable node (resulting in parabolic curves)
and negative for the saddle (resulting in hyperbolic curves) as shown in Figure
2.1. The arrows denoting the progress of time are determined by the detailed

---

[1]A real Jordan form for a real-valued matrix is one for which the matrix T is constrained
to be real. Consequently, real-valued matrices with complex eigenvalues cannot be diagonalized, but rather only block diagonalized by using the real and imaginary parts of the
eigenvectors associated with a complex eigenvalue as columns of T. For details on real
Jordan forms, see Chapter 6 of [141].

Node   Saddle

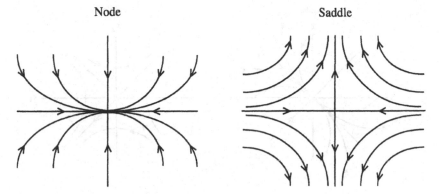

FIGURE 2.1. Phase portraits for a node and saddle in the z coordinates

formulas for z_1, z_2. The stable node has particularly simple dynamics. All the trajectories decay to the origin. An *unstable node* corresponding to the case that $\lambda_1, \lambda_2 > 0$ has equally simple dynamics. All trajectories take off from the neighborhood of the origin exponentially (alternatively, one could say that trajectories tend to the origin in reverse time (or, as $t \to -\infty$). Thus, the phase portrait is the same as that of the stable node of Figure 2.1 but with the sense of time reversed. The z_1, z_2 axes are the eigenspaces associated with the two eigenvalues. They are invariant under the flow of (2.1), in the sense that trajectories starting on the z_1 or z_2 axes stay on the z_1, or z_2 axes, respectively. The term saddle is particularly evocative in the description of the dynamics at an equilibrium point which has one positive (unstable) and one negative (stable) eigenvalue. The only initial conditions that are attracted to the origin asymptotically are those on the z_1 axis (the bow of the saddle). All other initial conditions diverge asymptotically along the hyperbolas described by (2.3). Of course, when $\lambda_1 > 0 > \lambda_2$, the equilibrium point is still a saddle. (Think about the use of the terminology *saddle* and justify it in your mind by rolling small marbles from different points on a horse's saddle.)

In the original x coordinates, the portraits of Figure 2.1 are only slightly modified since the relationship between x and z is linear, as shown in Figure 2.2.

- *Single repeated eigenvalue with only one eigenvector*
In this case the Jordan form J is not diagonal but has the form

$$J = \begin{bmatrix} \lambda & 1 \\ 0 & \lambda \end{bmatrix}. \tag{2.4}$$

This, in turn, results in a solution of the form:

$$z_1(t) = z_{10}e^{\lambda t} + z_{20}te^{\lambda t},$$
$$z_2(t) = z_{20}e^{\lambda t}.$$

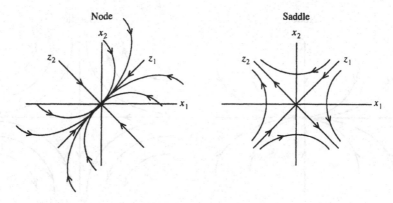

FIGURE 2.2. Phase portraits for a node and saddle in the x coordinates

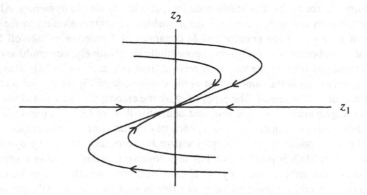

FIGURE 2.3. An improper stable node

Observe that by eliminating t the form of the trajectories in the phase plane is as shown in Figure 2.3 and is given by

$$z_1 = \frac{z_{10}}{z_{20}} z_2 + \frac{1}{\lambda} z_2 \ln \left(\frac{z_2}{z_{20}} \right).$$

Thus, the trajectories converge to the origin as in Figure 2.1, but they do so in a more complex fashion. This kind of an equilibrium is referred to as an *improper stable node (or simply improper node)*. For $\lambda > 0$, the improper node is referred to as an *improper unstable node*. In this case the z_1 axis is the eigenspace associated with the eigenvalue λ and is invariant. However, the z_2 axis is not invariant, as evidenced by the fact that trajectories cross the z_2 axis in the phase portrait of Figure 2.3.

- *Complex pair of eigenvalues*
Although the eigenvalues are distinct, unless the transformation matrix T is complex, the A matrix cannot be diagonalized; however, its real Jordan form is

$$J = \begin{bmatrix} \alpha & \beta \\ -\beta & \alpha \end{bmatrix}. \tag{2.5}$$

To obtain solutions for this case, it is useful to perform a polar change of coordinates of the form

$$r = (z_1^2 + z_2^2)^{1/2}, \quad \phi = \tan^{-1}\left(\frac{z_2}{z_1}\right).$$

In these coordinates, the equations are

$$\dot{r} = \alpha r,$$
$$\dot{\phi} = -\beta.$$

The phase portraits are now easy to visualize: The angular variable ϕ increments at a fixed rate (the trajectory spins around in a clockwise direction if β is positive and counterclockwise otherwise). Also, the trajectory spirals inward toward the origin if $\alpha < 0$, in which case the equilibrium is referred to as a *stable focus*. When $\alpha > 0$, trajectories spiral away from the origin, and the equilibrium is called an *unstable focus*. If $\alpha = 0$, the origin is surrounded by an infinite number of circular closed orbits and the equilibrium referred to as a *center*. Figure 2.4 shows a stable focus and a center. When the phase portrait is drawn in the original x coordinates, it is rotated as in the instance of the node and saddle.

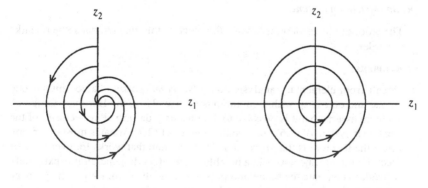

FIGURE 2.4. A stable focus and center

2.2.2   Phase Portraits near Hyperbolic Equilibria

The first feature of second order nonlinear systems that we study is equilibrium points. Consider the following autonomous second order system:

$$\dot{x}_1 = f_1(x_1, x_2),$$
$$\dot{x}_2 = f_2(x_1, x_2). \tag{2.6}$$

Let $x_0 \in \mathbb{R}^2$ be an isolated equilibrium point of (2.6). As we discussed in Chapter 1, this means that $f_1(x_0) = f_2(x_0) = 0$ and there exists a $\delta > 0$ such that there are no other equilibrium points in the ball $B(x_0, \delta)$ centered at x_0 of radius δ. Further, let

$$A(x_0) = \frac{\partial f}{\partial x}(x_0) = \begin{bmatrix} \dfrac{\partial f_1}{\partial x_1} & \dfrac{\partial f_1}{\partial x_2} \\ \dfrac{\partial f_2}{\partial x_1} & \dfrac{\partial f_2}{\partial x_2} \end{bmatrix} (x_0) \in \mathbb{R}^{2 \times 2}$$

be the Jacobian matrix of f at x_0 (linearization of the vector field $f(x)$ about x_0). A basic theorem of Hartman and Grobman establishes that the phase portrait of the system (2.6) resembles that of the linearized system described by

$$\dot{z} = A(x_0)z \tag{2.7}$$

if none of the eigenvalues of $A(x_0)$ are on the $j\omega$ axis. Such equilibrium points are referred to as *hyperbolic equilibria*. A precise statement of the theorem follows:

Theorem 2.1 Hartman–Grobman Theorem.   *If the linearization of the system (2.6), namely $A(x_0)$, has no zero or purely imaginary eigenvalues then there exists a homeomorphism (that is, a continuous map with a continuous inverse) from a neighborhood U of x_0 into $\mathbb{R}^2$,*

$$h : U \mapsto \mathbb{R}^2,$$

taking trajectories of the system (2.6) and mapping them onto those of (2.7). In particular, $h(x_0) = 0$. Further, the homeomorphism can be chosen to preserve the parameterization by time.

The proof of this theorem is beyond the scope of this chapter, but some remarks are in order.

Remarks:

1. Recall from elementary analysis that a *homeomorphism* is a continuous one to one and onto map with a continuous inverse. Thus, the Hartman–Grobman theorem asserts that it is possible to continuously deform all trajectories of the nonlinear system of (2.6) onto the trajectories of (2.7). What is most surprising about the theorem is that there exists a single map that works for every one of a continuum of trajectories in a neighborhood of x_0. In general, it is extremely difficult to compute the homeomorphism h of the theorem. Hence, the primary interest of this theorem is conceptual; it gives a sense of the dynamical behavior

of the system near x_0. In modern dynamical systems parlance, this is referred to as the *qualitative dynamics* about x_0. The theorem is illustrated in Figure 2.5 for the case that the linearized system has a saddle equilibrium point.

2. To state the Hartman–Grobman theorem a little more precisely, we will need the following definition:

Definition 2.2 Flow. *The state of the system (2.6) at time t starting from x at time 0 is called the* flow *and is denoted by $\phi_t(x)$.*

Using this definition, the Hartman–Grobman theorem can be expressed as follows: If $x \in U \subset \mathbb{R}^2$ and $\phi_t(x) \in U$, then

$$h(\phi_t(x)) = e^{A(x_0)t} h(x). \tag{2.8}$$

The right hand side is the flow of the linearized system (2.1), starting from the initial condition $h(x)$ at time 0. The preceding equation states that, so long as the trajectory of (2.6) stays inside $B(x_0, \delta)$ and the homeomorphism h is known, it is not necessary to integrate the nonlinear equation, but it is possible simply to use the linear flow $e^{A(x_0)t}$ as given in (2.8) above. Equation (2.8) may be rewritten as

$$\phi_t(x) = h^{-1}(e^{A(x_0)t} h(x)).$$

3. The Hartman–Grobman theorem says that the qualitative properties of nonlinear systems in the vicinity of isolated equilibria are determined by their linearization if the linearization has no eigenvalues on the $j\omega$ axis.

4. When the linearization has an eigenvalue at the origin, then the linearization predicts the existence of a continuum of equilibria. However, the original nonlinear system may or may not have a continuum of equilibria. Also, the qualitative behavior may vary greatly depending on the higher order nonlinear terms. The following example shows what can happen when there is an eigenvalue at the origin:

Example 2.3. *The system*

$$\dot{x} = -x^3$$

is a scalar system with a single equilibrium point at 0. The linearization of this system is zero about $x = 0$, but all nonzero initial conditions tend to it as $t \to \infty$. However

$$\dot{x} = x^2$$

is also a scalar system with a zero linearization and also a single equilibrium point at 0. However, only negative initial conditions converge to the origin as $t \to \infty$.

Thus when the linearization has an eigenvalue at the origin, the linearization is *inconclusive* in determining the behavior of the nonlinear system.

5. When the linearization has nonzero eigenvalues on the $j\omega$ axis, the linearization predicts the flow in the neighborhood of the equilibrium point to resemble

the flow around a focus. However, the original nonlinear system may have trajectories that either spiral towards the origin or away from it depending on the higher order nonlinear terms. Thus, the linearization is *inconclusive* in determining the qualitative behavior close to the equilibrium point. The following example illustrates this point:

Example 2.4. *Consider the nonlinear system*

$$\dot{x}_1 = x_2,$$
$$\dot{x}_2 = -x_1 - \epsilon x_1^2 x_2.$$

The linearized system has eigenvalues at $\pm j1$ but for $\epsilon > 0$ the system has all trajectories spiraling in to the origin (much like a sink), and for $\epsilon < 0$ the trajectories spiral out (like a source). Think about how one might prove these statements! (Hint: Determine the rate of change of $x_1^2 + x_2^2$ along trajectories of the equation.)

It is important to try to understand what the Hartman–Grobman theorem means. If the linearization of an equilibrium point is a saddle, the conclusions of the theorem are as shown in Figure 2.5. As before, note that the same homeomorphism h maps all the trajectories of the nonlinear system onto those of the linearized system. Thus the homeomorphism h has the effect of straightening out the trajectories of the nonlinear system.

Another question that comes to mind immediately in the context of the Hartman Grobman theorem is: What do the eigenvectors of the linearization correspond to? For the linearized system (2.1) they correspond to the eigenspaces which are invariant under the flow of the differential equation (2.1), as was discussed in the previous section. The nonlinear system (2.6) also has "nonlinear eigenspaces"—invariant manifolds. Invariant manifolds are the nonlinear equivalents of subspaces. [2] It is quite striking that these manifolds are, in fact, locally tangent to the eigenspaces of the linearized system at x_0. To make this more precise, consider the following

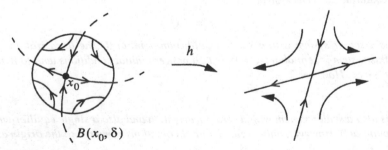

FIGURE 2.5. Illustrating the Hartman Grobman theorem

---

[2]A more precise definition of a manifold will be given in Chapter 3; for now, think of manifolds as being bent or curved pieces of subspaces.

definitions:

$$W^s(x_0) := \{x \in U : \phi_t(x) \to x_0 \text{ as } t \to \infty\},$$

$$W^u(x_0) := \{x \in U : \phi_t(x) \to x_0 \text{ as } t \to -\infty\}.$$

Here $\phi_t(x)$ is the flow of (2.6) starting from $x(0) = x$, i.e., the location of the state variable $x(t)$ at time t starting from $x(0) = x$ at time 0, and U is an open set containing x_0. The set $W^s(x_0)$ is referred to as the *local inset* of x_0, since it is the set of initial conditions in U that converge to x_0 as $t \to \infty$, while the set $W^u(x_0)$ is referred to as the *local outset* of x_0, since it is the set of points in U which is anti-stable, i.e., the set of initial conditions which converge to x_0 as $t \to -\infty$. A theorem called the stable–unstable manifold theorem states that, when the equilibrium point is hyperbolic (that is, the eigenvalues of the linearization are off the $j\omega$ axis), the inset and outset are the nonlinear versions of the eigenspaces and are called, respectively, the *stable invariant manifold* and *unstable invariant manifold*. In Chapter 7 we will see that if the equilibrium is not hyperbolic, the inset and outset need not be manifolds (this distinction is somewhat more technical than we have machinery to handle here, we will develop the concept of manifolds in Chapter 3). Figure 2.6 shows the relationship between the subspaces associated with the eigenvalues of the linearized system and $W^s(x_0)$, $W^u(x_0)$ close to x_0. The eigenspaces of the linearized system are tangent to $W^s(x_0)$, $W^u(x_0)$ at x_0.

The tangency of the sets W^s, W^u to the eigenspaces of the linearization needs proof that is beyond the level of mathematical sophistication of this chapter. We will return to this in Chapter 7. For now, the preceding discussion is useful in determining the behavior around hyperbolic equilibrium points of planar systems. As the following example shows, qualitative characteristics of W^s, W^u are also useful in determining some plausible global phase portraits:

Example 2.5 Unforced Duffing Equation. *Consider the second order system*

$$\ddot{x} + \delta\dot{x} - x + x^3 = 0.$$

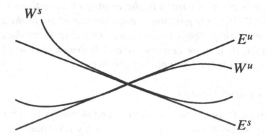

FIGURE 2.6. Showing the relationship between the eigenspaces and stable, unstable manifolds

written in phase plane form as:

$$\dot{x}_1 = x_2,$$
$$\dot{x}_2 = x_1 - x_1^3 - \delta x_2. \qquad (2.9)$$

The equilibrium points are $(0, 0)$, $(\pm 1, 0)$. *The Jacobian of the linearization is*

$$\begin{bmatrix} 0 & 1 \\ 1 - 3x_1^2 & -\delta \end{bmatrix}.$$

A small calculation shows that the eigenvalues of the linearization at the equilibria $(\pm 1, 0)$ *are*

$$\frac{-\delta \pm \sqrt{\delta^2 - 8}}{2}.$$

Thus, for $\delta > 0$ *they are both stable. The eigenvalues of the linearization at* $(0, 0)$ *are*

$$\frac{-\delta \pm \sqrt{\delta^2 + 4}}{2}.$$

For $\delta > 0$, *one of these is positive and the other negative corresponding to a saddle equilibrium. The corresponding eigenvectors are given by*

$$\left(1, \frac{-\delta \pm \sqrt{\delta^2 + 4}}{2}\right)^T.$$

At $\delta = 0$, *the system is Hamiltonian, i.e.,*

$$\dot{x}_1 = \frac{\partial H}{\partial x_2},$$
$$\dot{x}_2 = -\frac{\partial H}{\partial x_1}.$$

and we can plot the phase portrait as the level sets of the Hamiltonian

$$H(x_1, x_2) = \frac{1}{2}x_2^2 - \frac{1}{2}x_1^2 + \frac{1}{4}x_1^4.$$

Thus a few plausible phase portraits in the vicinity of these three equilibria are shown in Figure 2.7. The only part that we are certain about in these portraits is the behavior near the (hyperbolic) equilibria. The existence of the closed orbits in some of the portraits is more conjectural and is drawn only to illustrate their possible existence. This is the topic of the next section.

An Aside: Poincaré Linearization

Poincaré asked himself a question similar to that answered by Hartman–Grobman: If the nonlinear function f were a polynomial (analytic), under what conditions could one find a polynomial (analytic) change of coordinates such that the system had the form of its linearization in the new coordinates? This kind of linearization

(i)

(ii)

(iii)

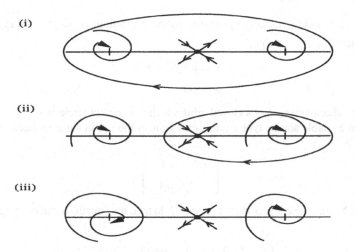

FIGURE 2.7. Plausible phase portraits for the Duffing equation from the linearization

(certainly more demanding than the homeomorphism of Hartman–Grobman) is known as the Poincaré linearization. A sufficient condition for its existence, called the Poincaré–Sternberg theorem is that the eigenvalues of the linearization be independent over the ring of integers, i.e., that there is no integer combination of the eigenvalues which sums to zero (the so called *non-resonance* condition). See the exercises in Chapter 9 (Problem 9.5) for details on how to get a feel for this result, or refer to [12].

2.3   Closed Orbits of Planar Dynamical Systems

Definition 2.6 Closed Orbit.   *A closed orbit of a dynamical system is the trace of the trajectory of a non-trivial (i.e., not a point) periodic solution. Thus $\gamma \subset \mathbb{R}^2$ is a closed orbit if γ is not an equilibrium point and there exists a time $T < \infty$ such that for each $x \in \gamma$, $\phi_{nT}(x) = x$ for all $n \in \mathbb{Z}$.*

By a closed orbit is meant a trajectory in the phase space that is periodic with a finite period. Thus, a collection of trajectories connecting a succession of saddle points is not referred to as a closed orbit; consider for example the trajectories connecting the saddles in Figure 1.14. These trajectories are, roughly speaking, infinite period closed orbits (reason about this before you read on, by contemplating what happens on them). In the sequel, we will refer to such trajectories as saddle connections, and not as closed orbits. Consider a simple example of a system with a closed orbit:

$$\dot{x}_1 = x_2 + \alpha x_1(\beta^2 - x_1^2 - x_2^2),$$
$$\dot{x}_2 = -x_1 + \alpha x_2(\beta^2 - x_1^2 - x_2^2). \tag{2.10}$$

The vector field is radially symmetric so that by defining $r = \sqrt{(x_1^2 + x_2^2)}$, and $\phi = \tan^{-1}(x_2/x_1)$, we get

$$\dot{r} = \alpha r (\beta^2 - r^2),$$
$$\dot{\phi} = -1.$$

The resultant system has a closed orbit which is a perfect circle at $r = \beta$. It also has an unstable source type equilibrium point at the origin since its linearization there is

$$\begin{bmatrix} \alpha\beta^2 & 1 \\ -1 & \alpha\beta^2 \end{bmatrix},$$

which has unstable eigenvalues at $\alpha\beta^2 \pm j$. In fact, the equations can be integrated to give

$$r(t) = \beta\left(1 + \left(\frac{\beta}{r(0)} - 1\right)e^{-2\beta^2\alpha t}\right)^{-1/2} \quad \text{for} \quad r(0) \neq 0$$

and $\phi(t) = \phi(0) - t$. Thus all nonzero initial conditions converge to $r = \beta$.

This example is somewhat contrived, since in general it is not easy to find closed orbits for nonlinear systems, even by detailed simulation, as you may have noticed in working through the simulation exercises after Chapter 1. Thus, it is of great interest to give conditions under which planar dynamical systems either have or do not have closed orbits, without doing simulation. The following celebrated theorem gives conditions for the *nonexistence of closed orbits*.

Theorem 2.7 Bendixson's theorem—absence of closed orbits. *Let D be a simply connected region[3] in $\mathbb{R}^2$, such that $div(f) := \frac{\partial f_1}{\partial x_1} + \frac{\partial f_2}{\partial x_2}$ is not identically zero in any sub-region of D and does not change sign in D. Then D contains no closed orbits of the system (2.6).*

Proof: Assume, for the sake of contradiction, that J is a closed orbit of (2.6) as shown in the Figure 2.8. $f(x)$ then is tangent to J. If $n(x)$ denotes the outward directed unit normal as shown, then it follows that $f(x) \cdot n(x) = 0$. But, the divergence theorem of Gauss can be used to relate the integral of $f(x) \cdot n(x)$ on J to the div(f) in the region bounded by J to write

$$\int_J f(x) \cdot n(x)\, dl = 0 = \iint_S \text{div } f\, dx\, dy.$$

But div f does not change sign and is not identically zero in S by assumption. This establishes the contradiction.[4]   □

---

[3]A simply connected region in the plane is one that can be (smoothly) contracted to a point. Thus, a simply connected region cannot have more than one "blob", and that blob should not have any holes in its interior.

[4]We have been a little cavalier with the smooth dependence of $f(x)$ on x needed to use Gauss's theorem. We will be more careful about stating such assumptions more precisely

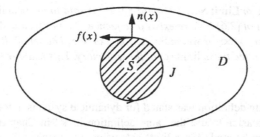

FIGURE 2.8. The proof of Bendixson's theorem

Example 2.8. *Consider the system*

$$\dot{x}_1 = x_2 + x_1 x_2^2,$$
$$\dot{x}_2 = -x_1 + x_1^2 x_2.$$

The linearization has eigenvalues at $\pm j$, but the nonlinear system has

$$\text{div } f = x_1^2 + x_2^2 > 0 \quad \text{when } x \neq 0$$

so that it is not identically zero and does not change sign. Thus this system has no closed orbits in all of $\mathbb{R}^2$.

Example 2.9 Duffing Equation. *For the Duffing equation of (2.9) we have that*

$$\text{div}(f) = -\delta.$$

Thus, for $\delta > 0$ the Duffing equation has no closed orbits in all of $\mathbb{R}^2$. This rules out all the phase portraits of Figure 2.7 which have closed orbits.

Example 2.10. *It is important that the region D of the theorem be simply connected. For example, consider the system of (2.10). It has*

$$\text{div}(f) = 2\alpha\beta^2 - 4\alpha(x_1^2 + x_2^2)$$

which is less than 0 in the not simply connected domain $D = \{(x_1, x_2) : \frac{2}{3}\beta^2 \leq x_1^2 + x_2^2 \leq 2\beta^2\}$. However, we know that the system has a closed orbit of radius β which is contained in D.

In order to be able to state the next result, concerning the existence of closed orbits in the plane we need the following definition:

---

after the next chapter. It is also important to notice that in the hypotheses of Bendixson's theorem div f may be allowed to vanish on sets of measure 0 in D, since it will not change the surface integral of div f: this indeed is the import of the assumption on div f not vanishing on any sub-region of D.

Definition 2.11 ω Limit Set. *A point $z \in \mathbb{R}^2$ is said to be an ω limit point of a trajectory $\phi_t(x)$ of (2.6) if there exists a sequence of times t_n, $n = 1, \ldots, \infty$ such that $t_n \uparrow \infty$ as $n \to \infty$ for which $\lim_{n \to \infty} \phi_{t_n}(x) = z$. The set of all limit points of a trajectory is called the ω limit set of the trajectory. This is abbreviated as $\omega(x)$.*

Remarks:

1. The preceding definition was stated for dynamical systems in the plane but we will have occasion to use the same definition in $\mathbb{R}^n$ in Chapter 7. A similar definition can be made for α **limit sets** using a sequence of times tending to $-\infty$. The terms α and ω, the first and last letters of the Greek alphabet, are use to connote the limits at $-\infty$, and $+\infty$, respectively. While we have defined ω and α limit sets of trajectories, the notation $\omega(x)$ or $\alpha(x)$ will be used to mean the ω and α limit sets of trajectories going through x at $t = 0$. We will also refer to $\omega(x)$ and $\alpha(x)$ as ω (respectively, α) limit sets of the point x.

2. If x_0 is an equilibrium point, then $\omega(x_0) = x_0$ and $\alpha(x_0) = x_0$. Also, if $x_0 \in \gamma$ is a point on a closed orbit, then $\omega(x_0) = \gamma$ and $\alpha(x_0) = \gamma$.

3. The ω and α limit sets can sometimes be quite complicated and non-intuitive, as this next example shows:

Example 2.12 Dynamics on a torus. *Consider a manifold called a torus, which looks like a hollow doughnut. It can be constructed by taking a unit square and gluing the top edge with the bottom edge, and the left edge to the right one as shown in Figure 2.9.*
It is easier to draw the trajectories on a plane than on the torus per se. Note that trajectories exiting the top edge of the square resume on the bottom and those on the right edge resume on the left edge. Now, consider the constant vector field:

$$\dot{x}_1 = p$$
$$\dot{x}_2 = q$$

where p, q are both integers. Verify that, if the ratio p/q is rational, the ω limit set of a trajectory is the trajectory itself. If, however, the ratio is irrational, the

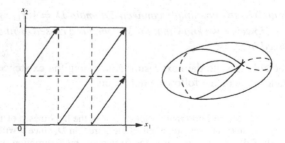

FIGURE 2.9. A constant rational flow on a torus

ω limit set is all of the torus. (Be sure that you can convince yourself of this fact. Can you determine the α limit sets in these two instances?)

4. If the α and ω limit set of a point are the same, they are referred to simply as the limit set.

Using these definitions, a distinction is sometimes made between a *closed orbit* and a *limit cycle* as follows.

Definition 2.13 Limit Cycle. *A limit cycle is a closed orbit γ for which there exists at least one x not in γ such that either γ is the α limit set of x, or γ is the ω limit set of x. Alternatively, this means that*

$$\lim_{t \to \infty} \phi_t(x) \to \gamma$$

or

$$\lim_{t \to -\infty} \phi_t(x) \to \gamma.$$

The convergence is assumed to mean convergence to the set represented by γ.

Thus, every limit cycle is a closed orbit. The converse, however, is not true: for instance, linear systems of the form of equation (2.1) have closed orbits, when A has eigenvalues on the $j\omega$ axis, but not limit cycles. In fact, the critical difference between linear and nonlinear systems that was pointed out in Chapter 1 was that nonlinear systems have limit cycles, but linear systems do not. However, unfortunately, neither Bendixson's theorem nor the theorem to follow enable us to establish the existence or non-existence of a limit cycle rather than a closed orbit.

Definition 2.14 Positively (Negatively) Invariant Regions. *A region $M \subset \mathbb{R}^2$ is said to be positively (negatively) invariant for the flow $\phi_t(x)$ if for each $x \in M$, $\phi_t(x) \in M$ for all $t \geq 0$ ($t \leq 0$).*

Informally, a region is positively (negatively) invariant if trajectories beginning in it are confined to it for all future (past) times. Often, we will use the term invariant to mean positively invariant, by abuse of terminology. The next theorem, giving conditions for the existence of closed orbits for equations in the plane, is a classic. Our treatment follows the style of [329].

Theorem 2.15 Poincaré–Bendixson Theorem. *Consider the continuous time planar dynamical system of (2.6) and let M be a compact, positively invariant set for the flow $\phi_t(x)$. Let $p \in M$. Then, if M contains no equilibrium points, $\omega(p)$ is a closed orbit of (2.6).*

Proof: The proof uses some technical facts about ω limit sets which we first state. We prove all the facts except for Fact 1, since it is a standard exercise in analysis (our development follows pages 46–50 of [329] to which we refer the reader for additional details).

Fact 1: Let M be a compact, invariant set and $p \in M$. Then $\omega(p)$, the ω limit set of p satisfies the following properties:

1. $\omega(p) \neq \varnothing$, that is, the ω limit set of a point is not empty.
2. $\omega(p)$ is closed.
3. $\omega(p)$ in an invariant set.
4. $\omega(p)$ is connected.

A further definition which is used in the construction is a *transverse section*. A transverse section Σ is a continuous, connected arc such that the dot product of the unit normal to Σ and the vector field is not zero and does not change sign on Σ. That is, the vector field has no equilibrium points on Σ and is never tangent to Σ.

Fact 2: Let Σ be a transverse section in M. The flow of any point $p \in M, \phi_t(p)$ intersects Σ in a monotone sequence for $t \geq 0$. Thus, if p_i is the i-th intersection of $\phi_t(p)$ with Σ, then $p_i \in [p_{i-1}, p_{i+1}]$.

Proof of Fact 2: Consider the piece of the orbit $\phi_t(p)$ from p_{i-1} to p_i along with the segment $[p_{i-1}, p_i] \subset \Sigma$ (see Figure 2.10). Of course, if the orbit $\phi_t(p)$ intersects Σ only once, then we are done. If it intersects more than once, consider the piece of $\phi_t(p)$ from p_{i-1} to p_i. This piece along with the segment of Σ from p_{i-1} to p_i form the boundary of a positively invariant region D as shown in Figure 2.10. Hence, $\phi_t(p_i) \subset D$, and it follows from the invariance of D that p_{i+1} (if it exists) $\subset D$. Thus, $p_i \in [p_{i-1}, p_{i+1}]$.

Fact 3: The ω limit set of a point p intersects Σ in at most one point.

Proof of Fact 3: The proof is by contradiction. Suppose that $\omega(p)$ intersects Σ in two points, q_1, q_2. Then, by the definition of ω limit sets we can find two sequences of points along $\phi_t(p)$, say p_n and $\bar{p}_n$, such that as $n \to \infty$, $p_n \in \Sigma \to q_1$ and $\bar{p}_n \in \Sigma \to q_2$. However, this would contradict the monotonicity of intersections of $\phi_t(p)$ with Σ.

Fact 4: In the plane, if the ω limit set of a point does not contain equilibrium points, then it is a closed orbit.

Proof of Fact 4: The proof of Fact 4 proceeds by choosing a point $q \in \omega(p)$ and showing that the orbit of q is the same as $\omega(p)$. Let $x \in \omega(q)$; then x is not a fixed point since $\omega(q) \subset \omega(p)$ (this actually uses the fact that $\omega(p)$ is closed).

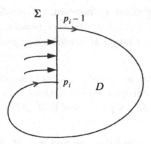

FIGURE 2.10. Construction for the proof of Fact 2 in the Poincare–Bendixson Theorem

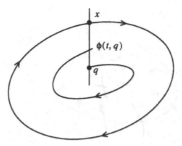

FIGURE 2.11. Construction for the proof of the Poincare–Bendixson theorem

Construct the line section S transverse (i.e., not tangential) to the vector field at x, as shown in Figure 2.11. Now, there exists a sequence of $t_n \uparrow \infty$ such that $\phi(t_n, q) \in S \to x$ as $n \to \infty$. By the Fact 3 above, this sequence is a monotone sequence tending to x. But, since $\phi(t_n, q) \in \omega(p)$, we must have by Fact 3 above that $\phi(t_n, q) \equiv x$. Thus, the orbit of q must be a closed orbit. It remains to be shown that the orbits of q and $\omega(p)$ are the same. But, this follows from an argument identical to the one given above by choosing a local section transverse to the trajectory at q and noting from Fact 3 above that $\omega(p)$ can intersect Σ only at q. Since $\omega(p)$ is an invariant set, contains no fixed points and is connected, it follows that the orbit of q is $\omega(p)$.[5]

To finish the proof of the theorem we simply put the above facts together. First, since M is invariant for the flow, $p \in M \Rightarrow \omega(p) \in M$. Further, since M has no equilibrium points, by Fact 4 it follows that $\omega(p)$ is a closed orbit.

Remarks and Applications:

1. In practice, it is difficult to verify that a region M is invariant. However, if a set M is closed and trajectories entering it do not leave it, then M is positively invariant, and the ω limit set of all trajectories that enter M is contained in M. In turn, it is easy to verify this graphically by observing whether f points inwards into M on the boundaries of M as shown in Figure 2.12. Similar conclusions can be reached about negatively invariant sets when the trajectories point out of the set.

2. From the preceding discussion, we can state that every compact,[6] nonempty positively invariant set K contains either an equilibrium point or a closed orbit. At this point, we may be tempted to think that the only ω limit sets of points inside K are closed orbits or equilibrium points. This is false, and in the instance that K does contain equilibrium points, the ω limit sets of points in K may be either an equilibrium point, a closed orbit or a *union of trajectories connecting*

---

[5]The arguments involving crossing a local section transverse to a flow are specific to planar systems and are also used in Remark 3. (What about the plane is so specific? Is it the fact that a closed curve in the plane divides the plane into an inside and an outside?)

[6]Compact sets in $\mathbb{R}^n$ are closed and bounded sets.

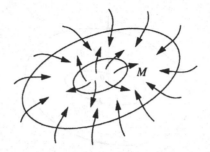

FIGURE 2.12. Checking the invariance of a set M

FIGURE 2.13. A union of saddle connections as an ω limit set

equilibrium points, i.e., a union of finitely many saddle connections (see Figure 2.13 for an example of this). As we will see in Section 7.5 a theorem of Andronov establishes that these are the only possible ω limit sets for dynamical systems in the plane.

3. Let γ be a closed orbit of the system enclosing an open set U. Then U contains either an equilibrium point or a closed orbit.

Proof: Define the set $D = U \cup \gamma$. Then D is a compact invariant set. Assume, for the sake of contradiction, that there are no equilibrium points or closed orbits inside D. Since the vector field on the boundary of D is tangential to γ, it follows that the ω limit set of each point inside U is in D. By the Poincaré–Bendixson theorem, then, the ω limit set is γ. By applying the same reasoning to the system

$$\dot{x} = -f(x)$$

we may conclude that for each $x_0 \in U$, there exists a sequence $t_n \uparrow \infty$, $s_n \downarrow -\infty$ such that $x(t_n) \to z \in \gamma$ and $x(s_n) \to z \in \gamma$ as $n \to \infty$ starting from the same initial condition x_0. To see that this results in a contradiction draw a line section S transverse to the solution starting from x_0 (i.e., not tangent to the trajectory). We now claim that the sequences $x(t_n)$, $x(s_n)$ are monotone on the section S, i.e., if $x(s_1)$ is above x_0, then so are $x(s_2), x(s_3), \ldots$. Also in this instance $x_{t_1}, x_{t_2}, \ldots$ lie below x_0 on S. (This is intuitively clear from studying

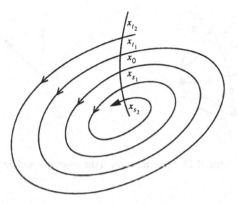

FIGURE 2.14. The monotonicity of the trajectory on a section S

Figure 2.14.) Thus, the region U should contain either an equilibrium point or a closed orbit.

4. Actually, in the previous application we can say more: not only does the region U contain either a closed orbit or an equilibrium point, but it certainly includes an equilibrium point. Thus, every closed orbit in the plane encloses an equilibrium point. The proof of this result uses a very interesting technique called index theory, which is the topic of the next section.

2.4   Counting Equilibria: Index Theory

Definition 2.16 Index. *Let J be a simple,[7] closed, positively oriented contour in $\mathbb{R}^2$ enclosing a region D. Then the index of D with respect to f is defined on J for $f \neq 0$ on J as*

$$I_f(D) = \frac{1}{2\pi} \oint_J d\theta_f(x_1, x_2), \tag{2.11}$$

where

$$\theta_f(x_1, x_2) := \tan^{-1} \frac{f_2}{f_1}.$$

More explicitly, we have

$$I_f(D) = \frac{1}{2\pi} \oint_J \frac{f_1 \, df_2 - f_2 \, df_1}{f_1^2 + f_2^2}. \tag{2.12}$$

---

[7]A simple curve means one which can be contracted to a point. This definition of a simple curve is reminiscent of the definition of a simply connected region in the statement of the Bendixson theorem, Theorem 2.7.

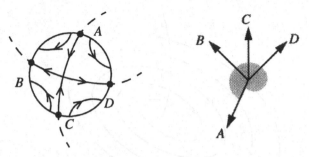

FIGURE 2.15. Determination of the index of a saddle

Remarks:

1. θ_f is the angle made by f with the x_1 axis, and $I_f(D)$ is the net change in the direction of f as one traverses J divided by 2π. It is easy to verify that $I_f(D)$ is an integer.

2. If x_0 is an equilibrium point inside D and D encloses no other equilibrium points then $I_f(D)$ is called the *index of the equilibrium point* x_0 and is denoted $I_f(x_0)$.

3. Prove that I_f of a center, focus, and node is 1 and that of a saddle is -1. The best way to verify this is to keep track of the rotation of the vector field f as one traverses the curve J on the boundary of D as shown in Figure 2.15. That is, if the linearization of the vector field f at an equilibrium point has no eigenvalues at the origin, then one can write its index by inspection of the eigenvalues and it is ± 1. The exercises explore situations in which equilibrium points have index other than ± 1. Here is a simple example of an equilibrium point of index 2:

$$\dot{x}_1 = x_1^2 - x_2^2,$$

$$\dot{x}_2 = 2x_1 x_2.$$

A phase portrait of this system is shown in Figure 2.16.

4. If D contains finitely many equilibrium points $x_i, i = 1, \ldots, p$, then

$$I_f(D) = \sum_{i=1}^{p} I_f(x_i).$$

The utility of index theory is that it enables one to predict the existence of equilibrium points in a region D without doing detailed calculations. In particular, one never does the calculation of the line integral of formula (2.12) for computing the index of a region. Rather, one does the sort of graphical calculation of Figure 2.15. For instance, if the vector field f points inwards everywhere on the boundary of a region D, then one can verify graphically that the index of D is 1. This could result from either of the situations shown in Figure 2.17. *It is a basic theorem that if a region D has non-zero index, then it has at least one equilibrium point.* If J is a closed orbit enclosing a region D one can verify graphically that the index of

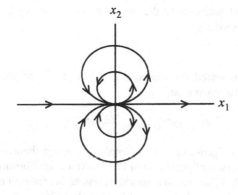

FIGURE 2.16. An equilibrium point of index 2

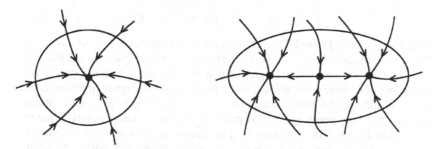

FIGURE 2.17. The interiors of two regions of index 1

D is 1 (whether the orbit is oriented in a counterclockwise sense or a clockwise sense). Thus, there is an equilibrium point inside region D.

There is a great deal more that can be said about index theory: for example, the index is a continuous function of the region D and the function f (homotopy invariance), as long as there are no points where f vanishes on the boundary of D. This has been used very skillfully in many applications, to deform a complicated f into a simple one whose index is easy to compute and thus determine the index of a region. There is also a generalization of index theory to $\mathbb{R}^n$ called the Poincaré–Hopf index. A related notion is the concept of *degree* for maps, which is discussed in Chapter 3.

2.5   Bifurcations

The term *bifurcation* was originally used by Poincaré to describe branching of equilibrium solutions in a family of differential equations

$$\dot{x} = f_\mu(x) \quad x \in \mathbb{R}^n, \quad \mu \in \mathbb{R}^k. \tag{2.13}$$

Here, we will restrict ourselves to the rudiments of the theory for $n = 2$. The equilibrium solutions satisfy

$$f_\mu(x) = 0. \tag{2.14}$$

As the parameter μ is varied, the solutions $x^*(\mu)$ of (2.14) are smooth functions of μ, so long as the Jacobian matrix

$$D_x f_\mu(x^*(\mu)) := \frac{\partial f_\mu}{\partial x}(x^*(\mu))$$

does not have a zero eigenvalue (by the implicit function theorem). The graph of each $x^*(\mu)$ is a *branch* of equilibria of (2.13). At an equilibrium point $(x_0 := x^*(\mu_0), \mu_0)$ where $D_x f_\mu$ has a zero eigenvalue, several branches of equilibria may come together, and one says that (x_0, μ_0) is a *point of bifurcation*. Consider, for example,

$$f_\mu(x) = \mu x - x^3 \quad x \in \mathbb{R}, \quad \mu \in \mathbb{R}. \tag{2.15}$$

Here $D_x f_\mu = \mu - 3x^2$ and the only bifurcation point is $(0, 0)$. It is easy to verify that the equilibrium point $x = 0$ is stable for $\mu \le 0$ and that it is unstable for $\mu > 0$. Also, the new bifurcating equilibria at $\pm\sqrt{\mu}$ for $\mu > 0$ are stable. We may show all of these solution branches in the bifurcation diagram of Figure 2.18. Note that although the bifurcation diagram is intended only to be a plot of stable and unstable equilibrium points, it is possible to show the entire family of phase portraits in Figure 2.18. This kind of a bifurcation is referred to as a *pitchfork bifurcation* (for obvious reasons). Bifurcations of equilibria only happen at points where $D_x f_\mu$ has zero eigenvalues. However, sometimes solution branches do not bifurcate at these points. Consider, for example:

$$\dot{x} = \mu x - x^2. \tag{2.16}$$

At $(0,0)$ this system has $D_x f_\mu = 0$. However, its solution branches do not change in number and only exchange stability as shown in Figure 2.19. This phenomenon is referred to simply as *transcritical* or *exchange of stability*. In some sense, the canonical building block for bifurcations is the *saddle-node* bifurcation described

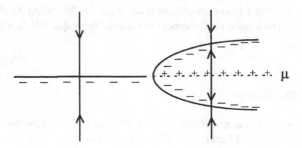

FIGURE 2.18. The pitchfork bifurcation

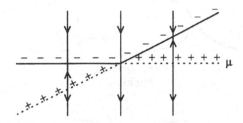

FIGURE 2.19. Transcritical or Exchange of Stability

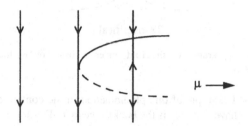

FIGURE 2.20. A fold bifurcation

by

$$\dot{x} = \mu - x^2. \tag{2.17}$$

For $\mu > 0$ this equation has two distinct equilibrium points, one stable and the other unstable. These two equilibria fuse together and disappear for $\mu < 0$, as shown in Figure 2.20. The pitchfork of Figure 2.18 appears to be different, but it is not hard to visualize that if the system of equation (2.15) is perturbed on the right-hand side with additional terms, then the pitchfork changes into one unbifurcated branch and one fold as shown in Figure 2.21. Further, though the transcritical appears not to resemble the fold, it is not difficult to visualize that under perturbation the transcritical changes into one of the two portraits shown in Figure 2.21. Note that under perturbation we either get two folds or else a fold and one unbifurcated branch.

The preceding discussion has focused exclusively on scalar examples of bifurcations. While we will have to defer a detailed discussion of a study of the forms of bifurcations to Chapter 7, a few brief observations about bifurcations of planar dynamical systems can be made here. First, bifurcations like the fold, pitchfork, and transcritical may be embedded in planar dynamical systems. When the eigenvalues of the linearization of an equilibrium point cross the $j\omega$ axis at the origin as a function of the *bifurcation parameter* μ then solution branching may occur. Folds are characterized by the appearance or disappearance of a pair of equilibria, one a saddle and the other a (stable or unstable) node. They are referred to as *saddle node bifurcations*. Pitchforks are characterized by the appearance of either two nodes and a saddle from a node or two saddles and a node from a saddle. (It is useful to notice the conservation of index as one passes through a point of

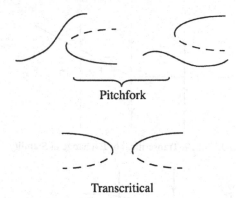

Pitchfork

Transcritical

FIGURE 2.21. The appearance of the fold under perturbation of the pitchfork and transcritical bifurcations

bifurcation). A good example of this phenomenon in the context of some of the examples which we have just seen is the buckled beam or Duffing equation of (2.9) with the parameter μ:

$$\dot{x}_1 = x_2,$$
$$\dot{x}_2 = \mu x_1 - x_1^3 - \delta x_2.$$

This equation has one equilibrium for $\mu \leq 0$, namely $(x_1 = 0, x_2 = 0)$, and three equilibria for $\mu > 0$, namely $(x_1 = 0, x_2 = 0)$, $(x_1 = \pm\sqrt{\mu}, x_2 = 0)$. This is a pitchfork bifurcation. Indeed, it is not difficult to see how equation (2.15) is embedded in it. By checking the linearization at the equilibria, one observes that the stable node for $\mu < 0$ bifurcates into a saddle and two stable nodes for $\mu > 0$.

One other new phenomenon that does not occur in scalar dynamical systems appears when the eigenvalues of the linearization of an equilibrium point cross from $\mathbb{C}_-^\circ$ to $\mathbb{C}_+^\circ$ and not through the origin. A canonical example of such a system is furnished by

$$\dot{x}_1 = -x_2 + x_1(\mu - x_1^2 - x_2^2),$$
$$\dot{x}_2 = x_1 + x_2(\mu - x_1^2 - x_2^2).$$
(2.18)

In polar coordinates the equation reads

$$\dot{r} = r(\mu - r^2),$$
$$\dot{\theta} = 1.$$
(2.19)

Note that the eigenvalues of the linearization of (2.18 around the equilibrium at the origin are respectively $\mu \pm j$. Thus, as μ goes through zero, the equilibrium point at the origin changes from being a stable node to being an unstable node. A simple polar change of coordinates also establishes that the equilibrium point is surrounded by a stable circular closed orbit of radius $\sqrt{\mu}$ for $\mu > 0$. This is referred to as a *supercritical Hopf bifurcation*.[8] The *subcritical* Hopf bifurcation refers to the fusion of an unstable limit cycle with a stable node to yield an unstable node. We will study these in far greater detail in Chapter 7.

In the next section we will see a number of bifurcations and how they arise in the context of a simple circuit model; however, we will end this section with an interesting example of the occurrence of a saddle node bifurcation of closed orbits.

Example 2.17 Saddle Node Bifurcation of Periodic Orbits. *Consider the planar dynamical system*

$$\dot{x}_1 = -x_1 \sin \mu - x_2 \cos \mu + (1 - x_1^2 - x_2^2)(x_1 \cos \mu - x_2 \sin \mu),$$
$$\dot{x}_2 = x_1 \cos \mu - x_2 \sin \mu + (1 - x_1^2 - x_2^2)(x_1 \sin \mu + x_2 \cos \mu). \tag{2.20}$$

where $\mu \in \mathbb{R}$ is the (small) bifurcation parameter. Converting this equation into polar coordinates yields:

$$\dot{r} = r\left[(1 - r^2)^2 \cos \mu - \sin \mu\right],$$
$$\dot{\theta} = (1 - r^2)^2 \sin \mu + \cos \mu. \tag{2.21}$$

*For $\mu < 0$, the system has one unstable equilibrium point. At $\mu = 0$, somewhat amazingly a periodic orbit of radius 1 appears. For $\mu > 0$, this breaks up into two periodic orbits, a stable one of radius $\sqrt{1 - \sqrt{\tan \mu}} < 1$ and an unstable one of radius $\sqrt{1 + \sqrt{\tan \mu}} > 1$. This bifurcation is one in which a stable and unstable orbit fuse together at $\mu = 0$. It is sometimes referred to as the **blue sky catastrophe**, since the orbits for small $\mu > 0$ appear suddenly. This example also shows a **subcritical Hopf bifurcation** at $\mu = \pi/4$ in which the stable closed orbit merges into the unstable equilibrium point, yielding a stable equilibrium point at the origin for $\mu > \pi/4$. At $\mu = \pi/2$, the unstable closed orbit disappears, since it has infinite radius.*

2.6 Bifurcation Study of Josephson Junction Equations

This section is an application of the methods of this chapter along with some numerical experimentation to understand the *v-i* characteristic of a Josephson junction. The equations of the Josephson junction are the same as those modeling either a damped pendulum or a single electrical generator coupled to an electrical distribution network. Thus, in effect we are also studying bifurcations of these systems.

While a complete model for the Josephson junction is quite involved, a satisfactory model of the junction may be obtained by modeling it as an RLC circuit with a linear resistor (representing the leakage across the junction), a linear capacitance (modeling the junction itself), and a nonlinear inductance modeling the tunneling current through the junction. In the circuit analogue, if ϕ represents the flux

---

[8]More properly the supercritical Poincare–Andronov–Hopf bifurcation.

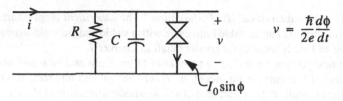

$$v = \frac{\hbar}{2e}\frac{d\phi}{dt}$$

$$-I_0 \sin\phi$$

FIGURE 2.22. A circuit diagram of the Josephson junction

through the inductor the "tunneling current" is given by $I_0 \sin(\phi)$, where I_0 is the maximum tunneling current. (in the quantum mechanical model of the Josephson junction the quantity ϕ has an interpretation in terms of the difference of phase of the wave function across the junction). Josephson's basic result was that the voltage across the junction is given by

$$v = \frac{\hbar}{2e}\dot{\phi},$$

where $\hbar$ is Planck's constant (divided by 2π) and e is the charge of an individual electron [310]. A circuit diagram of the junction is as shown in Figure 2.22. The circuit equations are given by

$$i = I_0 \sin(\phi) + \frac{\hbar}{2eR}\dot{\phi} + \frac{\hbar C}{2e}\ddot{\phi}. \tag{2.22}$$

Here i is the forcing current which we will assume to be constant. Following the convention in the Josephson junction literature [310], define

$$\alpha = \frac{1}{RC},$$

$$\beta = \frac{2eI_0}{\hbar C},$$

$$\gamma = \frac{2ei}{\hbar C}.$$

to get the following equation

$$\ddot{\phi} + \alpha\dot{\phi} + \beta \sin(\phi) = \gamma. \tag{2.23}$$

To express these equations in the form of a planar dynamical system, define $x_1 = \phi$ and $x_2 = \dot{\phi}$:

$$\dot{x}_1 = x_2,$$

$$\dot{x}_2 = -\alpha x_2 - \beta \sin(x_1) + \gamma.$$

We will draw phase portraits for a fixed junction, i.e., α, β are fixed as the parameter γ is increased from zero in the positive direction and keep track of the average voltage at the junction terminals, that is, the average value of $\frac{\hbar}{2e}\dot{\phi}$ at the terminals. The same experiment can also be repeated for negative values of γ and yields similar conclusions. Thus, we will derive the i-v characteristics of the junction intuitively.

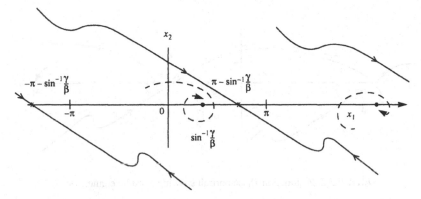

FIGURE 2.23. Junction phase portrait for small forcing

1. *Small forcing.* When γ is small, the phase portrait has the form shown in Figure 2.23. Note the presence of the two kinds of equilibria with $x_2 = 0$ and $x_1 = \sin^{-1}(\frac{\gamma}{\beta})$ and $x_1 = \pi - \sin^{-1}(\frac{\gamma}{\beta})$. The first is a stable focus and the second a saddle (verify!). These are also repeated every 2π because the right hand side of (2.6) is periodic with period 2π in x. The phase portrait shows the appearance of a "washboard" like stable manifold for the saddle; of course only the local behavior of the stable manifold of the saddle is obtained from the eigenvector of the linearization: the exact nature of the stable manifold is determined only by painstaking simulation. Also almost all initial conditions will converge to one of the stable foci, and the resulting steady state voltage (x_2) across the junction will be zero. *This is the superconducting region of the junction, since a nonzero applied current γ produces zero average voltage across the junction (proportional to $\dot{\phi}$).* This qualitative feature persists for $\gamma < \beta$ small, though as γ increases, the stable manifold of the saddle equilibrium at $\pi - \sin^{-1}(\frac{\gamma}{\beta})$ dips closer to the stable manifold of the saddle equilibrium at $-\pi - \sin^{-1}(\frac{\gamma}{\beta})$ in the upper half plane.

2. *Saddle connection.* As γ is increased, the upper part of the "washboard" shaped stable manifold of the saddle dips lower and lower to the preceding saddle till it finally touches it at a critical value of $\gamma = \gamma_c$. The critical value γ_c is an involved function of the parameters α, β of the junction, but $\gamma_c(\alpha, \beta) < \beta$, so that the two kinds of equilibria of the previous region persist. At this point we have the phase portrait of Figure 2.24. Note the cap shaped regions of attraction of the stable foci and the saddle connection. The caps are in the upper half plane for positive forcing $\gamma > 0$, which we have considered first. For negative forcing, $\gamma < 0$, the caps would have been in the lower half plane. Finally, note that if this phase portrait had been drawn on the cylinder, the saddle would have been connected to itself. Trajectories starting inside the cap reach the stable equilibrium point. Trajectories starting above the saddle connection converge to the saddle connection; that is, their ω limit set is the set of trajectories joining the (infinitely many) saddle equilibrium points along with the saddle equilibria. If

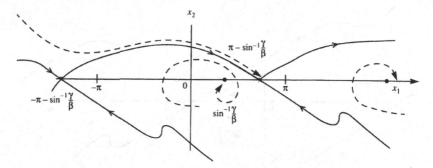

FIGURE 2.24. Junction phase portrait showing a saddle connection

the phase portrait were drawn on the cylinder, then the ω limit set would be a single saddle and a trajectory connecting the saddle to itself. This trajectory represents the confluence of one leg of the stable manifold of the saddle with a leg of the unstable manifold of the saddle—the *saddle connection*. This saddle connection trajectory takes infinite time to complete. Thus it resembles a closed orbit, but it has infinite period and so is not technically one. However, the average value of x_2 on this ω limit set is zero. Thus, for all initial conditions, the steady state value of the voltage across the junction is 0.

3. *Coexistent limit cycle and equilibria.* As the magnitude of the forcing increases beyond γ_c, the saddle connection bifurcates into the trajectory LI. This trajectory LI is a limit cycle for the trajectory on a cylindrical state space since it is periodic in the x_1 variable as shown in Figure 2.25. Note the cap shaped domains of attraction for the stable foci. Initial conditions in the cap regions tend to the focus where the steady state value of the voltage is zero. Other initial conditions tend to the limit cycle, which has nonzero average value, since it is in a region where x_2 is positive.

4. *Saddle node bifurcation.* As γ increases up to β, the saddle and the focus move closer together and finally coincide at $\gamma = \beta$. At this point, the saddle and node

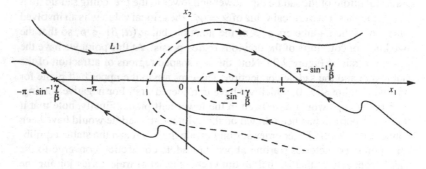

FIGURE 2.25. Phase portrait showing coexistent limit cycle and equilibria

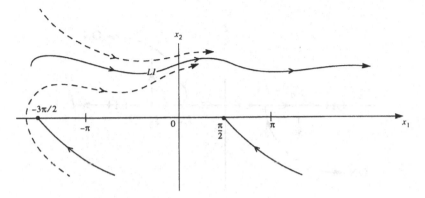

FIGURE 2.26. Junction portrait showing the saddle node bifurcation

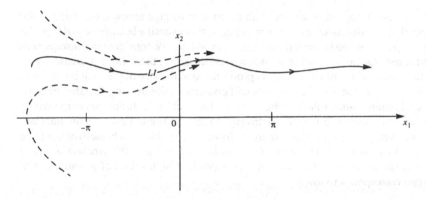

FIGURE 2.27. Junction portrait showing only a limit cycle

fuse together and pinch together the cap shaped domain of attraction to a line, as shown in Figure 2.26. Almost all initial conditions now converge to the limit cycle with its nonzero average value of x_2 and attendant nonzero voltage across the junction.

5. *Limit cycle alone.* For values of $\gamma > \beta$ the system of equation (2.22) has only a limit cycle and no equilibria, as shown in Figure 2.27. All initial conditions are attracted to the limit cycle with non-zero average value of the voltage. As γ increases the average value increases. This is the resistive region of the Josephson junction.

Using this succession of phase portraits, one can estimate the form of the i-v characteristic of the junction. It is as shown in Figure 2.28. In an experiment in which the current is gradually or quasi-statically increased, i.e., the current is increased and the system allowed to settle down to its equilibrium value from zero the steady state voltage stays zero for low values of current forcing i. Since i is a scaled version of γ we will conduct the discussion in terms of γ. After γ exceeds

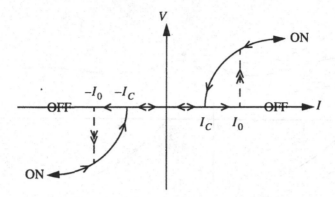

FIGURE 2.28. The i-v characteristic of the Josephson junction

γ_c, a new steady state solution with a non zero voltage appears, but small initial conditions still converge to the zero voltage solution until γ becomes equal to β. For $\gamma > \beta$ the voltage is nonzero and increases with γ. We consider the consequences of quasi-statically reducing γ. When γ is decreased, the voltage decreases since the average value of the voltage is proportional to the size of the limit cycle. Even when $\gamma < \beta$, the initial conditions will converge to the limit cycle rather than the equilibrium points. Finally, when $\gamma < \gamma_c$, the limit cycle disappears in the saddle connection and all trajectories converge to the equilibrium points with attendant zero average voltage. Thus, there is hysteresis in the i-v characteristic of the Josephson junction. This hysteresis can be used to make the junction a binary logic storage device: The ON state corresponds to high values of γ, and the OFF state corresponds to low γ.

2.7   The Degenerate van der Pol Equation

Consider a planar differential equation corresponding to the dynamics of the non-linear RC circuit in Figure 2.29 with a one farad linear capacitor and a cubic nonlinear resistor with v-i characteristic given by $v = i - i^3$. The equations of this system are given by

$$\dot{x}_1 = x_2,$$
$$0 = x_1 - x_2 + x_2^3. \qquad (2.24)$$

The model of equation (2.24) is not a differential equation in the plane, it is a combination of a differential equation and an algebraic equation. The algebraic equation constrains the x_1, x_2 values to lie on the curve shown in Figure 2.30. In the region where $x_2 \geq 0$, the variable x_1 is constrained to increase as shown in the figure. From an examination of the figure, it follows that there are difficulties with drawing the phase portrait at $x_2 = \pm\frac{1}{\sqrt{3}}$. At these points, it does not appear to be

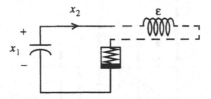

FIGURE 2.29. Nonlinear RC circuit model for the degenerate van der Pol oscillator

possible to continue to integrate the differential equation $\dot{x}_1 = x_2$ and still maintain consistency with the constraint. As a consequence these points are referred to as *impasse points*. To be able to resolve this situation as well as to model the behavior of the system starting from initial conditions not compatible with the constraint, we *regularize* the system description by admitting certain parasitic elements neglected in the initial modeling. To be specific, we consider the effects of the lead inductance of the wires of the RC circuit, shown dotted in Figure 2.29 as an inductor of magnitude ϵ. The dynamics of the resulting RLC circuit are:

$$\dot{x}_1 = x_2,$$
$$\epsilon \dot{x}_2 = x_1 - x_2 + x_2^3. \tag{2.25}$$

Here $\epsilon > 0$ is a small parameter modeling the small parasitic inductance. This system is well defined everywhere in the plane. Qualitatively, it is important to note that, when ϵ is small, then outside a small region where $x_2 - x_2^3 - x_1$ is of the order of ϵ, $\dot{x}_2$ is much larger than $\dot{x}_1$, making the trajectories almost vertical. The phase portrait of this system for $\epsilon = 0.1$ is shown in Figure 2.31 with an unstable equilibrium point at the origin surrounded by a limit cycle. The limit cycle is exceptional in that it has two slow segments close to the curve $x_1 = x_2 - x_2^3$ and two fast almost vertical segments close to $x_1 = \pm \frac{1}{\sqrt{3}}$. To return to the original system of (2.24), we define its trajectories to be the limit trajectories of (2.25) as $\epsilon \downarrow 0$. These trajectories are as shown in Figure 2.31. Several points about this portrait are worth noting:

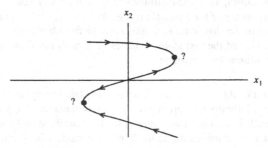

FIGURE 2.30. The dynamics of the RC circuit with impasse points

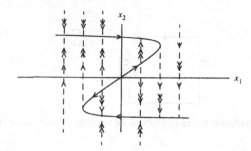

FIGURE 2.31. Showing the trajectories of the RC circuit obtained from the limit trajectories of the augmented RLC circuit

1. The trajectories continue through impasse points by having a jump or a discontinuous variation in the x_2 variable between the upper and lower leg of the curve $x_1 = x_2 - x_2^3$. The resultant trajectory is a "limit cycle" involving two jump segments. This sort of a limit cycle is referred to as a *relaxation oscillation*, and the impasse points are referred to as *turning points* for the oscillation.
2. When the initial conditions are not on the curve $x_1 = x_2 - x_2^3$ the trajectory instantaneously transports the initial condition onto the curve through a zero time transition in the x_2 variable. All initial conditions except those starting at the origin converge to the relaxation oscillation.
3. The limit trajectories of the regularized system seem to render the central section of the curve $x_1 = x_2 - x_2^3$ unstable in the sense that initial conditions not starting right on the curve tend to drift away from it, as shown in Figure 2.31.

The preceding example is very important in applications. The equation models how **hysteretic behavior** arises in nonlinear dynamical systems:

1. The *sharp transitions of a hysteresis* are modeled as arising from taking the limit as the dynamics of a subset of the state variables (referred to as parasitic or fast state variables, x_2 in the example) become infinitely fast.
2. The *lack of symmetry of the transitions*, i.e., *hysteresis*, for increasing and decreasing values of the slow state variable (x_1) arise from bifurcations associated with the dynamics of the fast state variables (x_2), with the slow state variable treated as the bifurcation parameter.

A simple variant of this model may be used to explain the dynamics of a clock circuit (astable multivibrator), where the oscillation is considered to jump between the 0 and 1 states. It is also used as a model for the heartbeat and nerve impulse [337]. Generalizations of this example to $\mathbb{R}^n$ provide models for jump behavior in many classes of circuits and systems (see, for example, [263]).

2.8   Planar Discrete-Time Systems

By a planar discrete time system we mean a system of the form

$$x_{n+1} = G(x_n). \tag{2.26}$$

Here $G : \mathbb{R}^2 \to \mathbb{R}^2$ is a smooth map, and $x_n \in \mathbb{R}^2$, $n \in \mathbb{Z}$. The discrete time system has a flow $\Phi_n(x) = G \circ \cdots \circ G(x)$ (the composition being taken n times). This is frequently referred to as $G^n(x)$ and the notation $\phi_n(x)$ dropped.[9] The flow can be defined for $n \leq 0$ if the map G is invertible. In this section, we will assume this to be the case.

2.8.1   Fixed Points and the Hartman–Grobman Theorem

We began the study of planar dynamical systems with the study of linear differential equations in the plane. The qualitative behavior of invertible, linear maps closely parallels that of continuous systems: with nodes, saddles, foci, and centers as we explore in Problem 2.19. Of course, trajectories in this context are really sequences of points rather than a continuous curve. Further, if either of the eigenvalues is negative, the trajectory shows *reflection* corresponding to the change in sign of the components of the trajectory.

An *equilibrium point* $\bar{x}$ of (2.26) satisfies

$$\bar{x} = G(\bar{x}).$$

Thus an equilibrium point of (2.26) is a fixed point of the map G. It is not surprising, therefore, that the study of planar dynamical systems is closely associated with the study of maps (especially invertible ones). Just as in the continuous time case the linearization of the dynamical system at an equilibrium point provides clues to the qualitative behavior in a vicinity of the point. Thus if $z = x - \bar{x}$, then the linearization of (2.26) is the system

$$z_{n+1} = DG(\bar{x})z_n. \tag{2.27}$$

The behavior of the linear system (2.27) is determined by the eigenvalues of $DG(\bar{x})$. In particular, if the eigenvalues have modulus less than 1, the equilibrium point at the origin is stable, in the sense that all initial conditions are attracted to the origin as $n \to \infty$. If the eigenvalues have modulus greater than 1, then all non-zero initial conditions are repelled from the origin. The behavior of initial conditions belonging to the eigenspaces of the different eigenvalues is very similar to the cases described for the continuous time case. (See Problem 2.16 for the case that the linear map is *hyperbolic*; that is its eigenvalues are off the unit disc and Problem 2.17 for the case that the linear map has eigenvalues on the boundary of the unit disc.) The chief difference between the dynamics of differential equations and the dynamics of maps is that the trajectories of maps are sequences of points, rather than curves.

---

[9] G^n, the n-th iterate of the map G, should not be confused with the n-th power of G.

However, one can mimic the discussion for continuous time linear systems to get the stable and unstable invariant subspaces (corresponding to eigenvalues in the interior or exterior of the unit disc). A version of the Hartman–Grobman theorem also applies for nonlinear systems with hyperbolic equilibria.

Theorem 2.18 Hartman–Grobman for Maps. *Consider the discrete time planar dynamical system of equation (2.26). Assume that $\bar{x}$ is a fixed point of the system, and let its linearization be given by (2.27). If the linearization is hyperbolic, that is to say that it has no eigenvalues on the boundary of the unit disk, then there exists a homeomorphism*

$$h : B(\bar{x}, \delta) \to \mathbb{R}^2$$

taking trajectories of (2.26) onto those of (2.27). Further, the homeomorphism can be chosen to preserve the parametrization by time.

Exactly the same comments as in the case of the differential equation apply. Thus, if the linearization is $A := DG(\bar{x})$, the diffeomorphism maps $\bar{x}$ into the origin and $G^n(x) \in B(\bar{x}, \delta)$, it follows that

$$h(G^n(x)) = A^n h(x).$$

Also, when the linearization is not hyperbolic, the linearization does not accurately predict the behavior of the nonlinear system.

As in the continuous time case the eigenvectors of the linearization correspond to tangents of the "nonlinear eigenspaces." Mimicking the definitions of the continuous time case, we define:

$$W^s(\bar{x}) := \{x \in U : G^n(x) \to \bar{x} \text{ as } n \to \infty\},$$
$$W^u(\bar{x}) := \{x \in U : G^n(x) \to \bar{x} \text{ as } n \to -\infty\}.$$

Here U is a neighborhood of $\bar{x}$. The *local inset W^s and the local outset W^u* are respectively the *local stable, invariant manifold* and the *local unstable, invariant manifold* and are tangent to the eigenspaces of the linearization $A = DG(\bar{x})$, provided none of the eigenvalues lie on the unit disk, i.e., that the equilibrium point is *hyperbolic*.

Example 2.19 Delayed Logistic Map. *Consider the logistic map (one-hump map) of Chapter 1 with a one step delay as an example of a planar dynamical system:*

$$x_{n+1} = y_n,$$
$$y_{n+1} = \lambda y_n(1 - x_n). \tag{2.28}$$

with $\lambda > 0$ corresponding to the delayed logistic equation; compare this with (1.3):

$$y_{n+1} = \lambda y_n(1 - y_{n-1}).$$

This map has two fixed points, one at the origin and the other at $(1 - 1/\lambda, 1 - 1/\lambda)$. *The linearization of the map at the origin is*

$$\begin{pmatrix} 0 & 1 \\ 0 & \lambda \end{pmatrix}$$

with eigenvalues 0 *and* λ. *Thus, the origin is asymptotically stable if* $\lambda < 1$ *and unstable if* $\lambda > 1$. *The linearization at the other equilibrium is*

$$\begin{pmatrix} 0 & 1 \\ 1 - \lambda & 1 \end{pmatrix}$$

with eigenvalues $\frac{1}{2}(1 \pm \sqrt{5 - 4\lambda})$. *This equilibrium is a saddle for* $0 < \lambda < 1$ *and is a stable node for* $1 < \lambda < 2$. *The phase portrait of this system at* λ *close to 2 is particularly pretty: Generate it for yourself!*

2.8.2 Period N Points of Maps

The discrete time counterpart of a periodic orbit is a period N point.

Definition 2.20 Period N Points. *A set of distinct points* $x_1, x_2, \ldots, x_N \in \mathbb{R}^2$ constitutes a *set of period* N points *for the system of (2.26) if*

$$x_2 = G(x_1), \quad x_3 = G(x_2), \quad \ldots, x_N = G(x_{N-1}), \quad x_1 = G(x_N). \quad (2.29)$$

Note that each point x_i of a period N orbit is a fixed point of the map $G^N(x)$. Thus, if one considers the new map $F(x) = G^N(x)$, its fixed points are points of period N or submultiples thereof (thus, if $N = 6$, the fixed points of F are points of period 1, 2, 3, and 6 of G). The big simplification that one encounters in discrete time systems is that the stability of period N points is determined by linearizing F about x_i. It should be of immediate concern that the stability of the period N points may be different. Indeed even the linearization of F is different at the points $x_i, i = 1, \ldots, N$. In Problem 2.20 you will show that though the Jacobians

$$DF(x_i) = DG^N(x_i) \quad i = 1, \ldots, N,$$

are different, they have the same eigenvalues. In fact, more subtly one can establish that the stability of the x_i is the same even when they are not hyperbolic! Since period N points are discrete orbits of the system (2.26) there is no hope of finding discrete counterparts to the Bendixson and Poincaré–Bendixson theorems.

Also, it is worth noting that it is not automatic that even linear maps with eigenvalues on the boundary of the unit disk have period N points. Consider the linear map

$$z_{n+1} = \begin{pmatrix} \cos\theta & \sin\theta \\ -\sin\theta & \cos\theta \end{pmatrix} z_n \quad (2.30)$$

with eigenvalues $\cos\theta \pm j\sin\theta$ on the unit disk. Verify that the system has period N points if and only if

$$\frac{\theta}{2\pi} = \frac{p}{q}, \quad p, q \in \mathbb{Z},$$

is a *rational* number. In this case, all initial conditions are points of period q (provided p, q are relatively prime). Further every set of period q points lies on a circle of constant radius. When this number is not rational, points wind around on the edge of each disk, densely covering it!

2.8.3   Bifurcations of Maps

If the fixed point of a nonlinear map is not hyperbolic, that is, at least one of its eigenvalues has modulus 1, then the stability type of the fixed point cannot be determined from the linearization alone. Further, if the map depends on parameters, then we expect the fixed point to undergo a bifurcation as the parameters are varied.

Bifurcations with real eigenvalues: Saddle Nodes, Transcritical and Pitchforks

If one eigenvalue of the linearization at a fixed point of a planar map is $+1$ and the other one is different from ± 1, then the fixed point typically undergoes a **saddle node** bifurcation (or a pitchfork or transcritical, depending on symmetries). As in the case of the continuous time system, the prototypical behavior is best understood on some scalar examples:

1. **Saddle Node Bifurcation** Consider the scalar map $x \mapsto f(x, \mu) = x + \mu \pm x^2$. It is easy to verify that at $\mu = 0$, the fixed point $x = 0$ is non-hyperbolic with linearization equal to 1. Depending on the sign of the quadratic term, we have two fixed points for $\mu > 0$ or two for $\mu < 0$, one stable and the other unstable (the reader is invited to verify these facts).
2. **Transcritical Bifurcation.** Consider the scalar map $x \rightarrow f(x, \mu) = x + \mu x \pm x^2$. The fixed points are at $x = 0$ and $x = \pm \mu$. At $x = 0, \mu = 0$, the fixed points trade or exchange stability type.
3. **Pitchfork Bifurcation.** The map $x \rightarrow f(x, \mu) = x + \mu x - x^3$ has fixed points at $x = 0$ and for $x = \sqrt{\mu}$ for $\mu > 0$. Similarly the map $x \rightarrow f(x, \mu) = x + \mu x + x^3$ has fixed points at $x = 0$, and at $x = \sqrt{-\mu}$ for $\mu < 0$.

Bifurcation with Eigenvalue -1: Period–Doubling Bifurcation

If one eigenvalue of the linearization at a fixed point of a planar map is -1 and the other one is different from ± 1, then the fixed point typically undergoes a period–doubling bifurcation. The easiest example of this was discussed in Chapter 1 for the one hump map

$$x \rightarrow f(x, \mu) = \mu x(1 - x).$$

It is a good exercise for the reader to revisit this example and show that the fixed point at $x = 1 - \frac{1}{\mu}$ undergoes a pitchfork bifurcation at $\mu = 3$, with two stable period 2 point and one unstable fixed point emerging from the stable fixed point. The period doubling continues at $\mu = 3.449, 3.570, \ldots$. Period–doubling is best studied by considering the fixed points of $f(x, \mu)$, $f^2(x, \mu)$, ... The calculations of these fixed points can be quite involved, and so we consider, as a (simpler) prototypical example the map $x \to f(x, \mu) = -x - \mu x + x^3$. Easy calculations show that for this system:

1. The fixed point $x = 0$ is unstable for $\mu \le -2$, $\mu > 0$ and stable for $-2 < \mu \le 0$.
2. $x = \pm\sqrt{2 + \mu}$ are two fixed points defined for $\mu \ge -2$ which are unstable.

The map has a pitchfork bifurcation at $\mu = -2$ when two unstable fixed points, and a stable fixed point emerge from an unstable fixed point. However at $\mu = 0$, the stable fixed point becomes unstable so that for $\mu \ge 0$ the map has exactly three fixed points, all of which are unstable. To explore the map further, consider the second iterate of f, namely the map

$$x \to f^2(x, \mu) = x + \mu(2 + \mu)x - 2x^3 + O_4(x, \mu),$$

where $O_4(x, \mu)$ stands for terms of order 4 and higher in x, μ. Note now that if one neglects the fourth order terms, the second iterate map has an additional pitch fork bifurcation ("additional" meaning in addition to the one at $\mu = -2$) at $\mu = 0$. Thus, at $\mu > 0$, two stable fixed points and one unstable fixed point of $f^2(x, \mu)$ are created from the stable fixed point at the origin for $-2 < \mu < 0$. These new fixed points of f^2 which are at approximately $\pm\sqrt{\mu + 1/2\mu^2}$, are not fixed points of $f(x, \mu)$ and hence are period 2 points. Thus, the complete bifurcation diagram of this system is as shown in Figure 2.32.

Bifurcation with Complex Eigenvalues of modulus 1: Naimark–Sacker Bifurcation

Roughly speaking, this is the equivalent of the Poincaré–Andronov–Hopf bifurcation for maps: when a pair of complex conjugate eigenvalues of the linearization cross the unit disk, then an invariant circle appears or disappears in the phase portrait. The prototypical example is obtained in the plane by defining $z = x_1 + jx_2 \in$

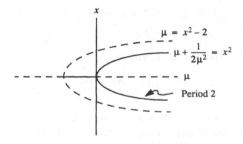

FIGURE 2.32. The "prototypical" period doubling bifurcation

$\mathbb{C}$, and considering the map from $\mathbb{C} \to \mathbb{C}$ given by

$$z \to \lambda(\mu)z + c(\mu)z^2 z^* + O_4(z, \mu).$$

In polar coordinates, $z = re^{j\theta}$, the map becomes

$$r \mapsto r + (d\mu + ar^2)r + O_3(r, \mu),$$
$$\theta \mapsto \theta + \phi_0 + \phi_1\mu + br^2 + O_3(r, \mu). \tag{2.31}$$

A normal form is obtained by truncating the O_3 terms above. For the system of (2.31) note that $r = 0$ is a fixed point which is stable for $d\mu < 0$, unstable for $d\mu > 0$. When $\mu = 0$, the fixed point is unstable for $a > 0$ and stable for $a < 0$. Further, for

$$-\frac{\mu d}{a} > 0,$$

the circle given by

$$\left\{ (r, \theta) \in \mathbb{R}^+ \times [0, 2\pi] : r = \sqrt{-\frac{\mu d}{a}} \right\}$$

is invariant. Further the following four cases arise:

1. $d > 0, a > 0$. The origin is an unstable fixed point for $\mu > 0$ and a stable fixed point surrounded by an unstable invariant circle for $\mu < 0$.
2. $d > 0, a < 0$. The origin is an unstable fixed point surrounded by a stable invariant circle for $\mu > 0$ and a stable fixed point for $\mu < 0$.
3. $d < 0, a > 0$. The origin is an unstable fixed point for $\mu < 0$ and a stable fixed point surrounded by an unstable invariant circle for $\mu > 0$.
4. $d < 0, a < 0$. The origin is an unstable fixed point surrounded by a stable invariant circle for $\mu < 0$ and a stable fixed point for $\mu > 0$.

This bifurcation is referred to as the Naimark–Sacker bifurcation, a discrete counterpart of the Poincaré–Andronov–Hopf bifurcation.

Example 2.21 Saddle Node in the Henon map. *A celebrated planar map is the Henon map, given by:*

$$x_{n+1} = 1 + y_n - ax_n^2,$$
$$y_{n+1} = bx_n. \tag{2.32}$$

The fixed points of the Henon map satisfy

$$\bar{x} = \frac{-(1-b) \pm \sqrt{(1-b)^2 + 4a}}{2a}, \quad \bar{y} = b\bar{x}.$$

Thus, it follows that when $a < -(1-b)^2/4$, there are no fixed points. There are two fixed points, one stable and the other unstable, when $a > (1-b)^2/4$, and there is a saddle node at $a = (1-b)^2/4$.

2.9 Summary

In this chapter, we described some important techniques for the analysis of planar continuous time and discrete time dynamical systems. Some of the techniques and theorems generalize readily to higher dimensions, for example, the Hartman Grobman theorem, bifurcations (see Chapter 7), and index theory (see degree theory in Chapter 3). However, some of the main results of this chapter such as the Bendixson theorem and the Poincaré–Bendixson theorem are unique to two dimensional continuous time systems, since they rely on the Jordan curve theorem. For further reading on the material in this chapter, see Wiggins [329], Hale and Koçak [126], Abraham and Shaw [3], Guckenheimer and Holmes [122], and Hirsch and Smale [141].

2.10 Exercises

Problem 2.1. Consider scalar differential equations of the form

$$\dot{x} = kx^n.$$

For all integral values of $n > 0$ they have an isolated equilibrium at the origin. Also, except for $n = 1$ their linearization is inconclusive. What is the qualitative behavior of trajectories for different $n \in Z$ and $k \in \mathbb{R}$.

Use this to conjecture the behavior of trajectories of planar dynamical systems when the linearization has a *single, simple zero* eigenvalue. More precisely think through what would happen in the cases that the other eigenvalue is positive, negative.

Problem 2.2 Weak attractors and repellors. Consider the system of the previous problem:

$$\dot{x} = kx^n$$

for $n > 1$. The equation can be integrated explicitly. Use this to determine the rate of convergence and divergence from the origin of these systems. Reconcile this result with the title of this problem.

Problem 2.3 Slight generalization of Bendixson's theorem due to Dulac. Let $q : \mathbb{R}^2 \mapsto \mathbb{R}$ be a given smooth function. For a planar dynamical system show that if div (qf) does not change sign or vanish on a region D, there are no orbits in that region. Apply Dulac's theorem to the system

$$\dot{x}_1 = -x_2 + x_1^2 - x_1 x_2,$$
$$\dot{x}_2 = x_1 + x_1 x_2,$$

with a choice of the Dulac function

$$q(x_1, x_2) = (1 + x_2)^{-3}(-1 - x_1)^{-1}.$$

Actually, div (qf) does change sign at $x_2 = -1$ and $x_1 = 1$, but find appropriate invariant sets to rule out periodic orbits which cross from one region where div (qf) has one sign into another region where it has a different sign.

Problem 2.4. Consider the system of equation (3) with its equilibrium point of index 2. How would you construct equilibrium points of index $+n$ and $-n$?

Problem 2.5. Investigate the bifurcations of the equilibria for the scalar differential equation:

$$\dot{x} = f_\mu(x)$$

for the following $f_\mu(x)$:

$$f_\mu(x) = \mu - x^2,$$
$$f_\mu(x) = \mu^2 x - x^3,$$
$$f_\mu(x) = \mu^2 x + x^3,$$
$$f_\mu(x) = \mu^2 \alpha x + 2\mu x^3 - x^5.$$

For the last example, use different values of α as well as of μ.

Problem 2.6 Modification of Duffing's equation [329].   Consider the modified Duffing equation

$$\dot{x}_1 = x_2,$$
$$\dot{x}_2 = x_1 - x_1^3 - \delta x_2 + x_1^2 x_2.$$

Find its equilibria. Linearize about the equilibria. Apply Bendixson's theorem to rule out regions of closed orbits. Use all this information to conjecture plausible phase portraits of the equation.

Problem 2.7 Examples from Wiggins [329].   Use all the techniques you have seen in this chapter to conjecture phase portraits for the following systems;

1.

$$\dot{x}_1 = -x_1 + x_1^3,$$
$$\dot{x}_2 = x_1 + x_2.$$

2.

$$\dot{x}_1 = x_2,$$
$$\dot{x}_2 = -\sin x_1 - \delta x_2 + \mu.$$

3.

$$\dot{x}_1 = -\delta x_1 + \mu x_2 + x_1 x_2,$$
$$\dot{x}_2 = -\mu x_1 - \delta x_2 + x_1^2/2 - x_2^2/2.$$

Assume that $\delta, \mu > 0$.

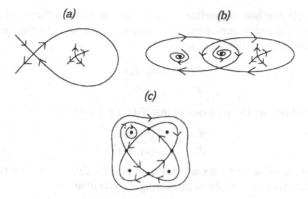

FIGURE 2.33. Phase portraits of planar systems

Problem 2.8 Plausible phase portraits.   Use all the techniques that you have learned in this chapter to make the phase portraits of Figure 2.33 plausible. You are allowed to only change the directions of some of the trajectories, or add trajectories including orbits.

Problem 2.9 First integrals.   One way of drawing phase portraits for planar dynamical systems is to find a first integral of the system. A first integral of (2.6) is a function $H : \mathbb{R}^2 \to \mathbb{R}$, conserved along the flow, that is,

$$\dot{H}(x_1, x_2) = \frac{\partial H}{\partial x_1} f_1(x_1, x_2) + \frac{\partial H}{\partial x_2} f_2(x_1, x_2) = 0.$$

Hence H is constant along the trajectories of the differential equation. How do first integrals help draw trajectories of the system? Find first integrals of the following systems:

1. Volterra–Lotka model:

$$\dot{x}_1 = ax_1 - bx_1x_2,$$
$$\dot{x}_2 = cx_1x_2 - dx_2.$$

2. Beam buckling model with no damping:

$$\dot{x}_1 = x_2,$$
$$\dot{x}_2 = \mu x_1 - x_1^3.$$

3. Pendulum model with forcing but no damping:

$$\dot{x}_1 = x_2,$$
$$\dot{x}_2 = a - b \sin x_1.$$

Problem 2.10 Method of Isoclines. A simple but useful technique for the approximation of solution curves in the phase plane is provided by the method of isoclines. Given the nonlinear system

$$\dot{x}_1 = f_1(x_1, x_2), \tag{2.33}$$

$$\dot{x}_2 = f_2(x_1, x_2), \tag{2.34}$$

we can write (assume for the moment that $f_2(x_1, x_2) \neq 0$)

$$\frac{dx_1}{dx_2} = \frac{f_1(x_1, x_2)}{f_2(x_1, x_2)}. \tag{2.35}$$

We seek curves $x_2 = h(x_1)$ on which the slope $dx_2/dx_1 = c$ is constant. Such curves, called *isoclines*, are given by solving the equation

$$f_2(x_1, x_2) = cf_1(x_1, x_2). \tag{2.36}$$

Now, consider the following system:

$$\dot{x}_1 = x_1^2 - x_1 x_2,$$

$$\dot{x}_2 = -x_2 + x_1^2.$$

- Find the equilibria of this system and show that the linearization of the system around the equilibria is inconclusive in terms of determining the stability type.
- Show that the x_2-axis is invariant and that the slope dx_2/dx_1 is infinite (vertical) on this line; find other lines in the plane on which the slope dx_2/dx_1 is infinite.
- Now seek isoclines on which $dx_2/dx_1 = c$ for finite c (try $c = 0, .5, 1, 2$).
- Sketch these curves, and the associated slopes dx_2/dx_1 on top of these curves, in the (x_1, x_2) plane.
- Conjecture the phase portrait from this information, and verify by simulation.

This problem was formulated by C. Tomlin, see also [122; 317].

Problem 2.11. Try using the Hartman–Grobman theorem and some numerical experimentation to determine the insets of the three equilibria of the Duffing equation, with damping:

$$\dot{x}_1 = x_2,$$

$$\dot{x}_2 = -kx_2 + x_1 - x_1^3.$$

Problem 2.12 Application of the Poincaré–Bendixson theorem [126]. Consider the system

$$\dot{x}_1 = x_2,$$

$$\dot{x}_2 = -x_1 + x_2(1 - x_1^2 - 2x_2^2).$$

Use the Poincaré–Bendixson theorem to prove that the annulus

$$\left\{ (x_1, x_2) : \frac{1}{2} \leq x_1^2 + x_2^2 \leq 1 \right\}$$

contains a closed orbit. (Hint: consider the rate of change of $v(x_1, x_2) = x_1^2 + x_2^2$ on the boundary of the annulus).

Problem 2.13. Use the techniques of this chapter to draw the phase portrait of the system

$$\dot{x}_1 = (2 - x_1 - 2x_2)x_1,$$
$$\dot{x}_2 = (2 - 2x_1 - x_2)x_2.$$

in the positive quadrant. Find lines in the x_1-x_2 plane which are invariant, locate equilibria, and go forward from there.

Problem 2.14 Recurrent points. A point $x^* \in \mathbb{R}^2$ is said to be *recurrent* for the flow $\phi_t(x)$ if x belongs either to the α limit set or to the ω limit set of the trajectory $\phi_t(x)$, for some x. Show that for systems in $\mathbb{R}^2$ all recurrent points are either equilibrium points or belong to a closed orbit.

Problem 2.15 Minimal sets. A set $M \subset \mathbb{R}^2$ is said to be *minimal* for the flow $\phi_t(x)$ if it is a non-empty, closed, invariant set with the property that no non-empty closed subset of it is invariant. Show that if a minimal set is unbounded, then it consists of a single trajectory that has empty α and ω limit sets.

Problem 2.16 Hyperbolic linear maps. Consider the following hyperbolic linear maps on $\mathbb{R}^2$:

1.

$$G = \begin{pmatrix} \lambda & 0 \\ 0 & \mu \end{pmatrix}.$$

Determine some sample trajectories (they will all be discrete sequences) for $|\lambda| < 1, |\mu| > 1$. Distinguish between the cases when $\lambda, \mu > 0$, and $\lambda, \mu < 0$. Repeat the problem for $|\lambda|, |\mu| < 1$, and assorted choices for the signs of λ, μ.

2.

$$G = \begin{pmatrix} \lambda & -\omega \\ \omega & \lambda \end{pmatrix}.$$

Distinguish between the cases $\lambda^2 + \omega^2 < 1$ and $\lambda^2 + \omega^2 > 1$, and $\lambda + j\omega = e^{j\alpha}$ for α rational or irrational.

Problem 2.17 Nonhyperbolic linear maps. Consider the following non-hyperbolic linear maps on $\mathbb{R}^2$:

1.

$$G = \begin{pmatrix} 1 & 0 \\ 0 & \lambda \end{pmatrix}$$

with $|\lambda| < 1$. Draw some sample trajectories.

2.

$$G = \begin{pmatrix} 1 & \lambda \\ 0 & 1 \end{pmatrix}$$

with $\lambda > 0$. Again, draw some sample trajectories.

What happens to these linear maps under small linear perturbations?

Problem 2.18. Give examples of nonlinear maps from $\mathbb{R}^2 \to \mathbb{R}^2$ where the linearization does not accurately predict the behavior of the nonlinear system.

Problem 2.19. Identify nodes, saddles, and foci for linear discrete time dynamical systems

$$x_{n+1} = Ax_n$$

Do this by first considering the case that A is convertible by real transformations to diagonal form,

$$\begin{pmatrix} \lambda_1 & 0 \\ 0 & \lambda_2 \end{pmatrix}$$

with λ_1, λ_2 both real, and then consider the case of a Jordan form with one real eigenvalue. Finally, consider the case that A has two complex eigenvalues $\alpha \pm \beta$, so that it can be converted into its real Jordan form

$$\begin{pmatrix} \alpha & \beta \\ -\beta & \alpha \end{pmatrix}.$$

Pay special attention to the case where one or both of the eigenvalues is negative or $\alpha < 0$.

Problem 2.20. Consider a set of period N points of a planar dynamical system, given by $x_1, x_2, \ldots, x_N$. Prove using the the chain rule from calculus that the Jacobian of G^N at each of the points x_i, $i = 1, \ldots, N$, has the same eigenvalues. You may find it instructive to begin with the case $N = 2$.

Problem 2.21. Study the bifurcations of the following planar maps as a function of the parameter λ near $\lambda = 0$:

1.

$$\begin{pmatrix} x \\ y \end{pmatrix} \mapsto \begin{pmatrix} \lambda + x + \lambda y + x^2 \\ 0.5y + \lambda x + x^2 \end{pmatrix}.$$

2.

$$\begin{pmatrix} x \\ y \end{pmatrix} \mapsto \begin{pmatrix} \lambda - x + xy + y^2 \\ 2x - xy - y^2 \end{pmatrix}.$$

3.

$$\begin{pmatrix} x \\ y \end{pmatrix} \mapsto \begin{pmatrix} \lambda + x + x^2 - y^2 \\ \lambda + x^2 + y^2 \end{pmatrix}.$$

3

Mathematical Background

In this chapter we will review very briefly some definitions from algebra and analysis that we will periodically use in the book. We will spend a little more time on existence, and uniqueness theorems for differential equations, fixed point theorems and some introductory concepts from degree theory. We end the chapter with an introduction to differential topology, especially the basics of the definitions and properties of manifolds. More details about differential geometry may be found in Chapter 8.

3.1 Groups and Fields

Definition 3.1 Group. *A group G is a set with a binary operation $(\cdot) : G \times G \mapsto G$, such that the following properties are satisfied:*

1. associativity: $(a \cdot b) \cdot c = a \cdot (b \cdot c)$ for all $a, b, c \in G$.
2. $\exists$ an identity $e \ni a \cdot e = e \cdot a = a$ for all $a \in G$.
3. For all $a \in G$, there exists an inverse $a^{-1} \ni a \cdot a^{-1} = a^{-1} \cdot a = e$.

A group G is called *abelian* if $a \cdot b = b \cdot a, \forall a, b$ in G.

Definition 3.2 Homomorphism. *A homomorphism between groups, $\phi : G \mapsto H$, is a map which preserves the group operation:*

$$\phi(a \cdot b) = \phi(a) \cdot \phi(b).$$

Definition 3.3 Isomorphism. *An isomorphism is a homomorphism that is bijective.*

Definition 3.4 Field. *A field K is a set with two binary operations: addition $(+)$, and multiplication $(\cdot)$, such that:*

1. *K is an abelian group under $(+)$, with identity 0.*
2. *$K - \{0\}$ is an abelian group under $(\cdot)$, with identity 1. If $k - \{0\}$ is a noncommutative group under $(\cdot)$, we refer to the field as a skew field.*
3. *$(\cdot)$ distributes over $(+) \ni a \cdot (b + c) = a \cdot b + a \cdot c$.*

Some examples of fields:

1. $\mathbb{R}$ is a field with addition and multiplication defined in the usual way.
2. $\mathbb{R}^2$, with addition defined in the usual way and with multiplication defined as

$$(x_1, x_2) \cdot (y_1, y_2) = (x_1 y_1, x_2 y_2)$$

for $\{x_1, x_2, y_1, y_2\}$ in $\mathbb{R}$, is not a field. Why not? If it were, we would have

$$(1, 0) \cdot (0, 1) = (0, 0)$$
$$(1, 0)^{-1} \cdot (1, 0) \cdot (0, 1) = (1, 0)^{-1} \cdot (0, 0)$$
$$(0, 1) = (0, 0).$$

This is clearly a contradiction. $\mathbb{R}^2$ can be made into a field if we define $(\cdot)$ as $(x_1, x_2) \cdot (y_1, y_2) = (x_1 y_1 - x_2 y_2, x_1 y_2 + x_2 y_1)$. We denote this field as $\mathbb{C}$, the set of complex numbers, with the understanding that $(x_1, x_2) = x_1 + \iota x_2$.

3. The *quaternions* $\mathbb{H}$ comprise the set of 4-tuples $(x_1, x_2, x_3, x_4) = (x_1 + \iota x_2 + J x_3 + k x_4)$ with addition defined in the usual way, and multiplication defined according to the following table:

$(\cdot)$	1	ι	J	k
1	1	ι	J	k
ι	ι	-1	k	$-J$
J	J	$-k$	-1	ι
k	k	J	$-\iota$	-1

Note that $K - \{0\}$ in the definition of the field is only a group, which is not abelian under multiplication. The skew field of quaternions is denoted $\mathbb{H}$, for *Hamiltonian field*.

In the rest of this section, we will be defining similar constructions for each of the fields $\mathbb{R}$, $\mathbb{C}$, and $\mathbb{H}$. For ease of notation, we denote the set as $\mathbb{K} \in \{\mathbb{R}, \mathbb{C}, \mathbb{H}\}$. We write $\mathbb{K}^n$ as the set of all n-tuples whose elements are in $\mathbb{K}$.

3.2 Vector Spaces, Algebras, Norms, and Induced Norms

Definition 3.5 Vector Space. *A vector space V (over a field $\mathbb{K}$) is a space with two operations, an addition and a scalar multiplication, defined on it. It is a* *commutative group under the addition operation* $(+)$ *and the scalar multiplication operation defined on it from* $(V, \mathbb{K}) \mapsto V$ *with scalar multiplication distributing over the addition.*

Examples of vector spaces are $\mathbb{R}^n$, n tuples of reals over the field $\mathbb{R}$; $\mathbb{C}^n$, n tuples of complex numbers over the complex field $\mathbb{C}$, the set of real valued functions on an interval $[a, b]$ over the field $\mathbb{R}$, the set of maps from a set $M \subset \mathbb{R}^m \mapsto \mathbb{R}^n$ (over $\mathbb{R}$). More generally $\mathbb{K}^n$ is a vector space over the field $\mathbb{K}$ for each of the fields defined in the previous section. If we define $\psi : \mathbb{K}^n \mapsto \mathbb{K}^n$ as a linear map, then ψ has matrix representation $M_n(\mathbb{K}) \in \mathbb{K}^{n \times n}$. We note that $\mathbb{K}^n$ and $M_n(\mathbb{K})$ are both vector spaces over $\mathbb{K}$.

Definition 3.6 Algebra. *An algebra is a vector space with a multiplication operation that distributes over addition.*

$M_n(\mathbb{K})$ is an algebra with multiplication defined as the usual multiplication of matrices: For $A, B, C \in M_n(\mathbb{K})$,

$$A(B + C) = AB + AC;$$
$$(B + C)A = BA + CA.$$

Definition 3.7 Unit. *If A is an algebra, $x \in A$ is a unit if there exists $y \in A$ such that $xy = yx = 1$.*

If A is an algebra with an associative multiplication operation, and $U \in A$ is the set of units in A, then U is a group with respect to this multiplication operation. More details on algebras are given in Chapter 12.

In order to define distances in a vector space, one introduces the concept of a normed vector space:

Definition 3.8 Normed Vector Space. *A vector space V over the field of reals is said to be a* normed vector space *if it can be endowed with a norm, denoted by $|\cdot|$, a function from $V \mapsto \mathbb{R}_+$ satisfying:*

1. $|x| \geq 0 \;\; \forall x \in X$ *and* $|x| = 0 \Leftrightarrow x = 0$.
2. $|\alpha x| = |\alpha||x| \;\; \forall \alpha \in \mathbb{R}$.
3. $|x + y| \leq |x| + |y|$.

Here are some examples of norms on $\mathbb{R}^n$, a vector space over the reals ($\mathbb{R}$).

1. $|x|_\infty = \max(|x_i|, i = 1, \ldots, n)$.
2. $|x|_1 = \sum_{i=1}^n |x_i|$.
3. $|x|_p = (\sum_{i=1}^n |x_i|^p)^{1/p}$ for $1 \leq p < \infty$. This norm with $p = 2$ is referred to as the Euclidean norm.

Definition 3.9 Independence and Basis Sets. *A set of vectors (elements) $v_1, \ldots, v_m \in V$, a vector space, is said to be* independent *if there exists no set of scalars α_i, not all zero in the associated field F such that*

$$\sum_{i=1}^n \alpha_i v_i = 0.$$

A set of vectors $B \subset V$ is said to be a basis[1] *if every element $v \in V$ can be uniquely written as*

$$v = \sum_{v_b \in B} \alpha_b v_b.$$

It follows from the requirement that v be represented uniquely that a basis set is an independent set of vectors. A vector space is said to be *finite dimensional* if it has a finite basis set. It is a basic fact that all basis sets of a vector space have the same dimension. Thus the *dimension* of the vector space is the cardinality of a (any) basis set. It may be shown that all finite dimensional real vector spaces are isomorphic to $\mathbb{R}^n$. (Isomorphic to $\mathbb{R}^n$ means that one can find a linear map from $\mathbb{R}^n$ to the space which is one to one and onto). Thus, $\mathbb{R}^n$ will be our prototype finite dimensional vector space and when we state a property of $\mathbb{R}^n$, we will mean that it holds for all finite dimensional vector spaces. It will frequently be of interest to study infinite dimensional vector spaces, i.e., those with no finite basis set. Consider the space of all continuous functions from $[0, 1]$ to $] - \infty, \infty[$. We will denote elements in this space by $f(\cdot)$ to emphasize the fact that they are functions. The space is a real vector space denoted by $C[0, 1]$ and may be endowed with a norm,

$$|f(\cdot)| := \sup|f(t)| \quad t \in [0, 1],$$

for functions $f(\cdot) \in C[0, 1]$. Other examples of norms on $C[0, 1]$ follow by analogy to the norms in $\mathbb{R}^n$:

$$|f(\cdot)|_p := \left(\int_0^1 |f(t)|^p dt \right)^{1/p} \quad \text{for} \quad p = 1, 2, \ldots.$$

Similarly, the space of continuous maps from $[0, 1]$ to $\mathbb{R}^n$, denoted $C^n[0, 1]$ may be endowed with norms by mixing norms on $\mathbb{R}^n$ and the function space norms just presented. For example, we may have:

1. $|f(\cdot)|_\infty = \sup(|f(t)|_2 \quad t \in [0, 1])$.
2. $|f(\cdot)|_2 = (\int_0^1 |f(t)|_\infty^2 dt)^{1/2}$.

The *open unit ball* in a given norm, say $|.|_\alpha$ is defined by

$$B(0, 1) := \{x \in X \text{ such that } |x|_\alpha < 1\}$$

Figure 3.1 shows unit balls in $\mathbb{R}^n$ under various norms.

While the balls shown in Figure 3.1 may appear different, from a conceptual and analytical point of view they are not terribly different, in the sense of the following proposition, which we state without proof (see [246] for a proof).

---

[1]If the basis set is not countable, then the summation in the definition needs to be specified to be a countable sum. In all of the examples that we will consider the basis sets are at most countable.

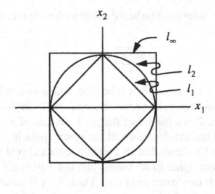

FIGURE 3.1. Unit balls in various norms

Proposition 3.10 Equivalence of Norms. *All norms are equivalent on $\mathbb{R}^n$, i.e., if $|\cdot|_\alpha$ and $|\cdot|_\beta$ are two norms on $\mathbb{R}^n$, then $\exists\ k_1, k_2$ such that for all $x \in \mathbb{R}^n$,*

$$k_1|x|_\alpha \leq |x|_\beta \leq k_2|x|_\alpha,$$

and $\exists\ k_3, k_4$ such that

$$k_3|x|_\beta \leq |x|_\alpha \leq k_4|x|_\beta.$$

The preceding proposition allows us to be somewhat careless in $\mathbb{R}^n$ about the choice of norms, especially in checking continuity, convergence of sequences and other such qualitative properties of functions and sequences. To illustrate this point consider the following definition:

Definition 3.11 Convergence of a Sequence. *A sequence $(x_n,\ n = 1, 2, 3, \ldots)$ in a normed linear space is said to* converge *to an element $x_0 \in X$ if for every $\epsilon > 0$ there exists $N(\epsilon)$ such that*

$$|x_n - x_0| < \epsilon \ \text{ when } \ n \geq N(\epsilon).$$

It follows that if a sequence converges in one norm, it converges in an equivalent norm, as well. However, in many instances we do not know whether a sequence converges, and if it does we do not know its limit. To be able to determine the convergence of sequences without knowing their limits, we have the following definition:

Definition 3.12 Cauchy Sequences. *A sequence $(x_n, n = 1, 2, 3, \ldots)$ is said to be* Cauchy *if given $\epsilon > 0$, there exists N such that*

$$|x_n - x_m| < \epsilon \ \forall \ n, m \geq N.$$

A sequence is a Cauchy sequence if the elements of the sequence get closer and closer towards the tail of the sequence. It is useful to note that every convergent sequence is a Cauchy sequence. However, the converse is not necessarily true. A normed vector space is said to be a *complete* or *Banach* space in a specified norm if all Cauchy sequences converge to limit points in the space. The virtue of the definition of Cauchy sequences is that it enables us to check the convergence of sequences without knowing a priori their limits. Note that sequences that are Cauchy under one norm need not be Cauchy under another norm, unless the two norms are equivalent in the sense of the proposition above.

The space of continuous functions $C^n[0, 1]$ is not a Banach space under any of the norms defined above, namely

$$|f(\cdot)|_p := \left(\int_0^1 |f(t)|^p dt \right)^{\frac{1}{p}}.$$

The completion of this space under this norm is referred to as $L_p^n[0, 1]$ for $p = 1, 2, \ldots$. We are sloppy about what norm is used at a fixed value of t, namely on $f(t)$, since all norms are equivalent on $\mathbb{R}^n$.

Definition 3.13 Inner Product Spaces.   *A real vector space X is said to be an* inner product space *if $\exists$ a function $\langle \cdot, \cdot \rangle : X \times X \mapsto \mathbb{R}$ with the following properties:*

1. $\langle x, y \rangle = \langle y, x \rangle$ *for all $x, y \in X$.*
2. $\langle x, y + z \rangle = \langle x, y \rangle + \langle x, z \rangle$ *for all $x, y, z \in X$.*
3. $\langle x, \alpha y \rangle = \alpha \langle x, y \rangle$ *for all $x, y \in X$ and $\alpha \in \mathbb{R}$.*
4. $\langle x, x \rangle = |x|^2$.

Note that the last property shows that every inner product induces a norm on the vector space. The following important inequality (sometimes called the Cauchy–Bunyakovsky–Schwartz inequality) relating the inner product of two vectors to their norms may be derived from the properties in the foregoing definition:

$$|\langle x, y \rangle| \leq |x||y|. \tag{3.1}$$

A complete inner product space is called a *Hilbert* space. Some examples of inner product spaces are:

1. On $\mathbb{R}^n$ we may define the inner product of two vectors x, y as

$$\langle x, y \rangle = \sum_{i=1}^n x_i y_i.$$

2. On $C^n[0, 1]$ one may define the inner product of two elements $f(\cdot), g(\cdot)$ as

$$\langle f(\cdot), g(\cdot) \rangle = \int_0^1 \langle f(t), g(t) \rangle \, dt,$$

where the pointwise inner product is as in the previous example. It may be shown that $C^n[0, 1]$ is not a Hilbert space. Its completion, $L_2^n[0, 1]$, the space of square integrable functions is a Hilbert space.

The space of linear maps from a vector space X to a vector space Y is a vector space in its own right. If both X, Y are real, normed vector spaces with norms

$|\cdot|_X, |\cdot|_Y$, then these two norms induce a norm on the vector space of linear maps from $X \mapsto Y$, denoted by $L(X, Y)$.

Definition 3.14 Induced Norms. *Let X, Y be normed linear spaces with norms $|.|_X, |.|_Y$ respectively. Then the space of linear maps from $X \mapsto Y$, $L(X, Y)$ has a norm induced by the norms on X, Y as follows. Let $A \in L(X, Y)$. Then*

$$|A|_i = \sup \left(\frac{|Ax|_Y}{|x|_X} : |x| \neq 0, x \in X \right)$$

$$= \sup(|Ax|_Y : |x|_X = 1).$$

Since $m \times n$ matrices are representations of linear operators from $\mathbb{R}^n$ to $\mathbb{R}^m$, norms on $\mathbb{R}^n$ and $\mathbb{R}^m$ induce matrix norms on $\mathbb{R}^{m \times n}$. Consider the following examples of matrix norms on $\mathbb{R}^{m \times n}$. For the purpose of these examples we will use the same norms on $\mathbb{R}^m$ and $\mathbb{R}^n$.

$$
\begin{array}{ll}
\text{Norm on } \mathbb{R}^n, \mathbb{R}^m & \text{Induced norm on } \mathbb{R}^{m \times n} \\[2mm]
|x|_\infty = \max_i |x_i| & |A|_{i\infty} = \max_i \sum_j |a_{ij}| \\[2mm]
|x|_1 = \sum_i |x_i| & |A|_{i1} = \max_j \sum_i |a_{ij}| \\[2mm]
|x|_2 = (\sum_i |x_i|^2)^{1/2} & |A|_{i2} = (\lambda_{\max}(A^T A))^{1/2}
\end{array}
\tag{3.2}
$$

An added feature of induced norms is that they are submultiplicative. More precisely:

Proposition 3.15 Submultiplicative Induced Norms. *Let X be a normed vector space and A, B be linear operators from $X \mapsto X$. Then AB, the composition of the two linear operators, is also a linear operator from X to X and its induced norm satisfies*

$$|AB|_i \leq |A|_i |B|_i.
\tag{3.3}$$

Proof: See Problem 3.5.

3.3   Contraction Mapping Theorems

There is a very large class of theorems which is useful for determining the existence of fixed points of mappings from a Banach space into itself. These theorems are, in some ways, workhorse theorems of nonlinear analysis and are used repeatedly in a variety of different ways. There are several fixed point theorems, and they bear the names of their inventors: the Brouwer, Leray–Schauder and Kakutani fixed point theorem are some examples. The interested reader should pursue these theorems (an excellent reference is [278]), for they are indeed invaluable in applications. In this section we will establish a rather simple fixed point theorem and illustrate its use. For preciseness, let us define the fixed point of a map.

Definition 3.16 Fixed Point.   *A map T from a vector space X to X is said to have a* fixed point *at x^* if*

$$T(x^*) = x^*.$$

By abuse of notation in what follows we will refer to $T(x^*)$ as Tx^*. The next theorem also known as the Banach fixed point theorem, gives conditions for the existence of a fixed point of an operator T.

Theorem 3.17 Global Contraction Mapping Theorem.   *Let $(X, |\cdot|)$ be a Banach space and $T : X \mapsto X$ be a mapping for which $\exists\ \rho < 1$ such that*

$$|Tx - Ty| \le \rho|x - y|\ \forall x, y \in X. \tag{3.4}$$

Then $\exists$ a unique $x^* \in X$ such that

$$Tx^* = x^*.$$

Proof: Let $x_0 \in X$ be arbitrary. Define the sequence $x_{n+1} = Tx_n$ for $n = 0, 1, 2, \ldots$. Now using equation (3.4) repeatedly, we see that

$$|x_{n+1} - x_n| \le \rho|x_n - x_{n-1}| \le \cdots \le \rho^n|x_1 - x_0|. \tag{3.5}$$

Further, if $m = n + r$, then using the triangle inequality on norms we have

$$|x_m - x_n| \le \sum_{i=0}^{r-1} |x_{n+i+1} - x_{n+i}|$$

$$\le \sum_{i=0}^{r-1} \rho^{n+i}|x_1 - x_0|$$

$$\le \frac{\rho^n}{1-\rho}|x_1 - x_0|.$$

Since $\rho < 1$, given $\epsilon > 0$, one sees by choosing N large enough that

$$|x_m - x_n| < \epsilon\ \forall\ m > n \ge N;$$

hence the sequence $(x_n)_1^\infty$ is Cauchy. Since X is a Banach space the sequence converges to a point in X. Call its limit x^*. We claim that x^* is the desired fixed point of T. Indeed, equation (3.4) guarantees the continuity of T so that

$$Tx^* = \lim_{n\to\infty} Tx_n = \lim_{n\to\infty} x_{n+1} = x^*.$$

As for the uniqueness of x^*, we assume for the sake of contradiction that x^{**} is another fixed point. Then, we have

$$|x^* - x^{**}| = |Tx^* - Tx^{**}| \le \rho|x^* - x^{**}|.$$

Since $\rho < 1$, this establishes that $x^* = x^{**}$. $\square$

Remark: It is important to note that it is not possible to weaken the hypothesis of the theorem to

$$|Tx - Ty| < |x - y|\ \forall x, y \in X. \tag{3.6}$$

Indeed, consider the function from $\mathbb{R} \mapsto \mathbb{R}$ given by $T(x) = x + \pi/2 - \tan^{-1} x$. By noting that its derivative is given by $\frac{x^2}{1+x^2}$, which is less than 1 for all x, it may be verified that it satisfies the condition of (3.6). However, there is no finite value of x at which $T(x) = x$. The hypothesis on T of the preceding theorem is referred to as a *global contraction* condition. The condition is called global, since it holds uniformly in the whole space. Under suitable conditions, this hypothesis may be weakened to give the following local version of this theorem:

Theorem 3.18 Local Contraction Mapping Theorem. *Consider a subset M of a Banach space $(X, |.|)$. Let $T : X \mapsto X$ be such that for some $\rho < 1$,*

$$|Tx - Ty| \le \rho|x - y| \quad \forall x, y \in M. \tag{3.7}$$

Then, if there exists $x_0 \in X$ such that the set

$$B = \left\{ x \in X : |x - x_0| \le \frac{|Tx_0 - x_0|}{1 - \rho} \right\} \subset M,$$

Then T has exactly one fixed point in M.

Proof: The proof is a replica of the proof of the preceding theorem with the x_0 of the hypothesis being the starting point of the sequence defined as before by $x_{n+1} = Tx_n$. The estimates of the previous theorem guarantee that

$$|x_n - x_0| \le \frac{|Tx_0 - x_0|}{1 - \rho}.$$

Thus the sequence is confined to the set B. Since the sequence is Cauchy and the set B is closed the limit x^* belongs to B. The proof of the uniqueness of the fixed point in M mimics the corresponding part of the preceding proof.   □

Remark: In applications another useful version of a local contraction mapping theorem is as follows: Let S be a closed subset of X and T be a contraction map on S, then T has a unique fixed point in S. The proof is exactly like the proofs of the previous theorems. The fact that S is closed guarantees that the fixed point belongs to S.

3.3.1   Incremental Small Gain Theorem

Consider the feedback loop of Figure 3.2. Assume that the plant and compensator N_1, N_2 are maps from $L_p^{n_i}[0, \infty[\mapsto L_p^{n_o}[0, \infty[$ and from $L_p^{n_o}[0, \infty[\mapsto L_p^{n_i}[0, \infty[$, respectively.

Proposition 3.19 Incremental Small Gain Theorem. *Consider the feedback system of Figure 3.2 with input $u_1 \in L_p^{n_i}$, $u_2 \in L_p^{n_o}$. Assume that the maps N_1, N_2 satisfy*

$$|N_1(x_1) - N_1(x_2)| \le \gamma_1|x_1 - x_2| \quad \forall x_1, x_2 \in L_p^{n_i},$$
$$|N_2(x_3) - N_2(x_4)| \le \gamma_2|x_3 - x_4| \quad \forall x_3, x_4 \in L_p^{n_o}. \tag{3.8}$$

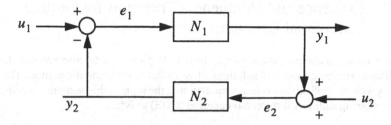

FIGURE 3.2. Showing a feedback loop

Then, if

$$\gamma_1\gamma_2 < 1, \tag{3.9}$$

the system has unique solutions y_1, y_2 for arbitrary inputs u_1, u_2.

Proof: The feedback loop equations are:

$$e_1 = u_1 - N_2(e_2),$$
$$e_2 = u_2 + N_1(e_1),$$
$$y_1 = N_1(e_1),$$
$$y_2 = N_2(e_2).$$

Now fix u_1, u_2 and note that we have unique solutions y_1, y_2 if and only if there are unique solutions e_1, e_2. In turn the equations for e_1, e_2 can be combined to yield

$$e_1 = u_1 - N_2(u_2 + N_1(e_1)). \tag{3.10}$$

The solution of the equation can now be equated to the existence of a fixed point of the map:

$$T(e_1) = u_1 - N_2(u_2 + N_1(e_1)).$$

Now note that

$$|Tx_1 - Tx_2| = |N_2(u_2 + N_1(x_1)) - N_2(u_2 + N_1(x_2))|$$
$$\leq \gamma_2|N_1(x_1) - N_1(x_2)|$$
$$\leq \gamma_2\gamma_1|x_1 - x_2|.$$

Hence, by equation (3.9) the map T is a global contraction map on $L_p^{n_i}[0, \infty[$ and so has a unique fixed point. Consequently there exist unique e_1, e_2 solving the feedback loop. The existence of unique y_1, y_2 now follows.   □

3.4   Existence and Uniqueness Theorems for Ordinary Differential Equations

The contraction mapping theorem may be used to give a constructive proof of the existence and uniqueness of solutions of an ordinary differential equation. This along with its application to circuit simulation is the topic of this section. Consider the following ordinary differential equation (ODE) in $\mathbb{R}^n$:

$$\dot{x} = f(x, t) \quad t \geq 0 \quad x(0) = x_0. \tag{3.11}$$

By a solution of equation (3.11) we mean a continuously differentiable function of time $x(t)$ satisfying

$$x(t) = x_0 + \int_0^t f(x(\tau), \tau) d\tau. \tag{3.12}$$

Such a solution to (3.11) is called a solution in the *sense of Caratheodory*,

Theorem 3.20 Local Existence and Uniqueness.   *Consider the system of (3.11). Assume that $f(x, t)$ is continuous in t and x, and that there exist T, r, k, h such that for all $t \in [0, T]$ we have*

$$|f(x, t) - f(y, t)| \leq k|x - y| \quad \forall x, y \in B(x_0, r),$$
$$|f(x_0, t)| \leq h, \tag{3.13}$$

with $B(x_0, r) = \{x \in \mathbb{R}^n : |x - x_0| \leq r\}$ is a ball of radius r centered at x_0. Then equation (3.11) has exactly one solution of the form of (3.12) on $[0, \delta]$ for δ sufficiently small.

The proof of this theorem needs a lemma called the Bellman–Gronwall lemma which is an important result in its own right.

Proposition 3.21 Bellman–Gronwall Lemma.   *Let $z(\cdot), a(\cdot), u(\cdot) : \mathbb{R}_+ \mapsto \mathbb{R}$ be given positive functions defined on $[0, \infty[$ and let $T > 0$. Then, if for all $t \leq T$,*

$$z(t) \leq u(t) + \int_0^t a(\tau)z(\tau)d\tau, \tag{3.14}$$

we have that for $t \in [0, T]$,

$$z(t) \leq u(t) + \int_0^t a(\tau)u(\tau)e^{\int_\tau^t a(\sigma)d\sigma} d\tau. \tag{3.15}$$

Proof of Bellman–Gronwall lemma: Define

$$r(t) = \int_0^t a(\tau)z(\tau)d\tau;$$

so that by (3.14) we have

$$\dot{r}(t) = a(t)z(t) \leq a(t)r(t) + a(t)u(t).$$

Hence, for some positive function $s(t)$, we have

$$\dot{r}(t) - a(t)r(t) - a(t)u(t) + s(t) = 0.$$

Integrating this scalar differential equation with initial condition $r(0) = 0$, we get

$$r(t) = \int_0^t e^{\int_\tau^t a(\sigma)d\sigma}(a(\tau)u(\tau) - s(\tau))d\tau.$$

since the exponential is positive and $s(.) \geq 0$, the inequality (3.15) follows directly.
$\square$

Proof of Existence and Uniqueness theorem: Let $x_0(\cdot)$ be a function on $C^n[0, \delta]$, defined by abuse of notation as

$$x_0(t) \equiv x_0 \quad \text{for } t \in [0, \delta].$$

Let S denote the closed ball of radius r about the center $x_0(.)$ in $C^n[0, \delta]$ defined by

$$S = \{x(\cdot) \in C^n[0, \delta] : |x(\cdot) - x_0(\cdot)| \leq r\}.$$

Define $P : C^n[0, \delta] \mapsto C^n[0, \delta]$ by

$$(Px)(t) := x_0 + \int_0^t f(x(\tau), \tau)d\tau.$$

We are going to apply the contraction mapping theorem on a a closed subset of the Banach space $C^n[0, \delta]$. The reason that the map is well defined on $C^n[0, \delta]$ is that we are assured that any solution in the sense of Caratheodory (3.12) has to be a continuous function.

Note that $x(t)$ is a fixed point of the map P, if it is a solution of (3.11) in the sense of (3.12). We now choose δ to be small enough to make P a contraction on S. Let $x(\cdot), y(\cdot) \in S$. Then, using the hypotheses of equation (3.13) we get

$$Px(t) - Py(t) = \int_0^t [f(x(\tau), \tau) - f(y(\tau).\tau)]d\tau,$$

$$|Px(t) - Py(t)| \leq \int_0^t k|x(\tau) - y(\tau)|d\tau, \tag{3.16}$$

$$|Px(\cdot) - Py(\cdot)| \leq k\delta|x(\cdot) - y(\cdot)| \quad \text{for } t \leq \delta.$$

Choose δ such that $k\delta =: \rho < 1$. Now we show that P maps S onto itself. Indeed, let $x(\cdot) \in S$. Then,

$$Px(t) - x_0 = \int_0^t f(x(\tau).\tau)d\tau$$

$$= \int_0^t [f(x(\tau), \tau) - f(x_0, \tau) + f(x_0, \tau)]d\tau,$$

$$|Px(t) - x_0| \leq \int_0^t k|x(\tau) - x_0|d\tau + \int_0^t |f(x_0, \tau)|d\tau$$

$$\leq kr\delta + h\delta,$$

using hypotheses (3.13). For $\delta \leq \frac{r}{kr+h}$, P maps S into itself. Thus we choose $\delta \leq \min(\frac{r}{kr+h}, \frac{\rho}{k})$ to guarantee that P is a contraction on S. Hence, (3.11) has exactly one solution in S. But does it have only one solution on $C^n[0, \delta]$? Indeed, if $x(\cdot)$ is a solution of (3.12), we have

$$x(t) - x_0 = \int_0^t f(x(\tau), \tau)d\tau,$$

$$|x(t) - x_0| \leq |\int_0^t [f(x(\tau), \tau) - f(x_0, \tau) + f(x_0, \tau)]d\tau$$

$$\leq \int_0^t |f(x(\tau), \tau) - f(x_0, \tau)| + |f(x_0, \tau)|d\tau$$

$$\leq \int_0^t (k|x(\tau) - x_0| + h)d\tau$$

$$\leq h\delta + k \int_0^t |x(\tau) - x_0|d\tau.$$

By the Bellman Gronwall lemma (with $a(t) \equiv k$, $z(t) = |x(t) - x_0|$ and $u(t) = h\delta$ in the notation of the lemma), we have

$$|x(t) - x_0| \leq h\delta e^{kt} \leq h\delta e^{k\delta}. \tag{3.17}$$

Thus if δ were chosen such that $\delta \leq \min(\frac{r}{kr+h}, \frac{\rho}{k}, T)$ and $h\delta e^{k\delta} \leq r$, then any solution to (3.11) has to lie in S, so that the unique fixed point of P in S is indeed the unique solution in $C^n[0, \delta]$ (actually also $L_\infty^n[0, \delta]$). □

Theorem 3.22 Global Existence and Uniqueness Theorem. *Consider the system of (3.11) and assume that $f(x, t)$ is piecewise continuous with respect to t and for each $T \in [0, \infty[$ there exist finite constants k_T, h_T such that for all $t \in [0, T]$,*

$$|f(x, t) - f(y, t)| \leq k_T|x - y| \quad \forall x, y \in \mathbb{R}^n,$$
$$|f(x_0, t)| \leq h_T. \tag{3.18}$$

Then equation (3.11) has exactly one solution on $[0, T]$ for all $T < \infty$.

Proof: Let T be specified. Then by the preceding theorem, the system of (3.11) has a solution on some interval $[0, \frac{\rho}{k_T}]$ for some $\rho < 1$. if $T \leq \frac{\rho}{k_T}$, we are done. If not, set $\delta := \frac{\rho}{k_T}$ and define $y_1(\cdot)$ to be the solution on $[0, \frac{\rho}{k_T}]$ and consider the system

$$\dot{x} = f(x, t + \delta) \quad x(0) = y_1\left(\frac{\rho}{k_T}\right). \tag{3.19}$$

The right hand side of (3.19) satisfies the conditions of (3.18) above, so we have a unique solution on the interval $[0, \frac{\rho}{k_T}]$. Call this $y_2(.)$. Then the solution $x(t) = y_1(t)$ for $t \in [0, \delta]$ and $x(t) = y_2(t - \delta)$ for $t \in [\delta, 2\delta]$ is the unique solution of (3.11) on $[0, 2\delta]$. Continue this process till $m\delta > T$. □

Theorem 3.23 Continuous Dependence on Initial Conditions. *Consider the system of (3.11) and let $f(x, t)$ satisfy the hypotheses (3.18) of the preceding theo-*

rem. Let $x(\cdot)$, $y(\cdot)$ be two solutions of this system starting from x_0, y_0 respectively. Then, for given $\epsilon > 0$ there exists $\delta(\epsilon, T)$ such that

$$|x_0 - y_0| \leq \delta \Rightarrow |x(\cdot) - y(\cdot)| \leq \epsilon. \tag{3.20}$$

Proof: Since $x(t)$, $y(t)$ are both solutions of (3.11) we have

$$|x(t) - y(t)| \leq |x_0 - y_0| + \int_0^t |f(x(\tau), \tau) - f(y(\tau), \tau)| d\tau$$

$$\leq |x_0 - y_0| + k_T \int_0^t |x(\tau) - y(\tau)| d\tau$$

By the Bellman Gronwall lemma we have that for all $t \in [0, T]$

$$|x(t) - y(t)| \leq |x_0 - y_0| e^{k_T t}.$$

As a consequence, given $\epsilon > 0$ one may choose $\delta = \epsilon / e^{k_T T}$ to prove the theorem.
□

The most important hypothesis in the proof of the existence-uniqueness theorems given above is the assumption that $f(x, t)$ satisfies the first condition of (3.13) referred to as *Lipschitz continuity*.

Definition 3.24 Lipschitz Continuity.   *The function f is said to be* locally Lipschitz continuous *in x if for some $h > 0$ there exists $l \geq 0$ such that*

$$|f(x_1, t) - f(x_2, t)| \leq l|x_1 - x_2| \tag{3.21}$$

for all $x_1, x_2 \in B_h$, $t \geq 0$. The constant l is called the Lipschitz *constant. A definition for* globally Lipschitz *continuous functions follows by requiring equation (3.21) to hold for $x_1, x_2 \in \mathbb{R}^n$. The definition of* semi-globally Lipschitz *continuous functions holds as well by requiring that equation (3.21) hold in B_h for arbitrary h but with l possibly a function of h. The Lipschitz property is by default assumed to be uniform in t.*

If f is Lipschitz continuous in x, it is continuous in x. On the other hand, if f has bounded partial derivatives in x, then it is Lipschitz. Formally, if

$$D_1 f(x, t) := \left[\frac{\partial f_i}{\partial x_j} \right]$$

denotes the partial derivative matrix of f with respect to x (the subscript 1 stands for the first argument of $f(x, t)$), then $|D_1 f(x, t)| \leq l$ implies that f is Lipschitz continuous with Lipschitz constant l (again locally, globally or semi-globally depending on the region in x that the bound on $|D_2 f(x, t)|$ is valid). The reader may also now want to revisit the contraction mapping theorem (Theorem 3.17) to note that the hypothesis of the theorem requires that $T : X \mapsto X$ is Lipschitz continuous with Lipschitz constant less than 1! In light of the preceding discussion if $X = \mathbb{R}^n$, a sufficient conditions for T to be a contraction map would be that the norm of the matrix of partial derivatives of T with respect to its arguments is less than 1.

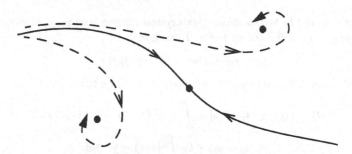

FIGURE 3.3. Showing lack of continuous dependence on non compact time intervals

3.4.1   Dependence on Initial Conditions on Infinite Time Intervals

It is *very* important to note that the preceding theorems, which state that the solutions of the system of (3.11) depend continuously on their initial conditions, holds only for compact intervals of time. To show that it is too much to hope for on infinite intervals of time consider the behavior of trajectories starting on either side of the stable manifold of a saddle, as shown in Figure (3.3). Three trajectories which are arbitrarily close at the initial time tend to three different equilibria as time $T \to \infty$. To estimate the convergence and divergence of two trajectories over infinite time intervals, one defines the **Lyapunov exponent** λ_+ associated with the divergence of two trajectories $x(\cdot)$, $y(\cdot)$ as

$$\lambda_+ = \limsup_{t \to \infty} \left(\frac{1}{t} \log \left(\frac{|x(t) - y(t)|}{|x_0 - y_0|} \right) \right), \tag{3.22}$$

From the preceding theorem it follows that if $k_T \le k$ for all T, that

$$-k \le \lambda_+ \le k.$$

In general, the estimates of (3.20) are quite crude. Usually, two trajectories will not diverge at the exponential rate predicted by this estimate, especially if both trajectories are in the basin of the same attracting set. If trajectories in the basin of the same attractor do in fact diverge at an exponential rate, in the sense that λ_+ as defined above is close to the upper bound k, then the system shows extremely sensitive dependence on initial conditions. This is taken to be indicative of chaotic behavior. In fact some definitions of chaos are precisely based on the closeness of the Lyapunov exponent to the Lipschitz constant in the basin of attraction of the same attracting set.

3.4.2   Circuit Simulation by Waveform Relaxation

The process of proving Theorem 3.22 is constructive and is referred to as the Picard–Lindelöf construction. While it seems to be a somewhat simple minded iteration to calculate the solution of the differential equation it turns out to be extremely useful as a technique for simulating the dynamics of digital VLSI circuits. In this literature the procedure is referred to as waveform relaxation. The details of the application to circuits is involved; we only sketch the facts most relevant to illustrate the theory.

The dynamics of a large class of nonlinear circuits may be written in the so-called tableau form and read as

$$\dot{x} = \psi(\dot{x}, x, u, t). \tag{3.23}$$

In equation (3.23) the state variable x is a list of all the state variables in the circuit (usually, the inductor fluxes and capacitor charges) and the equations are Faraday's law and Coulomb's law with the resistive constitutive relationships built in to the right hand side. The input u represents the independent voltage and current sources to the circuit. It is instructive to note that these equations are not explicit differential equations like (3.11), but are implicitly defined. That is, $\dot{x}$ is not explicitly specified as a function of $x(t)$, $u(t)$, t. In general, implicitly defined differential equations are an involved story, in and of themselves, since there is the possibility that there are either many or none or even uncountably many solutions to $\dot{x}$ in (3.23) and it is not clear which solution is to be chosen when there are many. Here, we use the contraction mapping theorem to conclude the following proposition:

Proposition 3.25 Solution of Implicit ODEs. *Consider the system of equation (3.23). Further assume that*

$$|\psi(y, x, u, t) - \psi(z, x, u, t)| \leq k_1 |y - z| \ \forall x \in \mathbb{R}^n, \ u \in \mathbb{R}^m, \ t \in [0, T].$$

Then if $k_1 < 1$ the implicit system (3.23) can be transformed to an explicit system of the usual form, (3.11).

Proof: See Exercises.

Proposition (3.25) may be used to convert an implicit differential equation into an explicit one, namely one of the form of (3.11). Further, the uniqueness part of the contraction mapping theorem guarantees that there is only one solution for $\dot{x}$ in (3.23).

The waveform relaxation algorithm to solve the system of (3.23) starting from an initial condition x_0 is based on the following iteration which is in the spirit of the Picard–Lindelöf iteration of the preceding section

$$x^k(t) = x_0 + \int_0^t \psi(\dot{x}^{k-1}(\tau), x^{k-1}(\tau), u(\tau), \tau)d\tau. \tag{3.24}$$

The initial function $x^0(t)$ is arbitrary except that $x^0(0) = x_0$. We will assume that the conditions of Proposition 3.25 hold so that the equation (3.23) may be written as the normal form equation of the form of (3.11), namely

$$\dot{x} = f(x, u, t) \tag{3.25}$$

In this instance the iteration of (3.24) reads:

$$x^k(t) = x_0 + \int_0^t f(x^{k-1}(\tau), u(\tau), \tau)d\tau. \tag{3.26}$$

In the language of the proof of the existence–uniqueness theorem of the previous section the iteration is

$$x^k(\cdot) = P(x^{k-1}(\cdot)).$$

Since the proof of that theorem involved proving that the map P was a contraction map, the sequence constructed by the Picard–Lindelöf construction is exactly that of the proof of the contraction mapping theorem. Hence, we are guaranteed convergence of the iterates. The only extra feature of waveform relaxation beyond what was elucidated in the previous section is the convergence of the iterates on a given interval $[0, T]$. It was shown in the previous section that P was a contraction map on an interval of length δ for δ sufficiently small. However, this situation may be improved to get convergence on the entire interval $[0, T]$ by defining the new norm

$$|x(\cdot)|_w = \sup_{t\in[0,T]}|x(t)e^{-wt}|. \tag{3.27}$$

Norms of the form of (3.27) are called (exponentially) weighted norms. It is instructive to verify that $C^n[0, T]$ is a Banach space under this norm. Now, if the hypotheses of the global existence and uniqueness of the theorem of the previous section held for the normal form system, then equation (3.23) may be transformed into (3.25) under the Lipschitz continuity hypothesis on f uniformly in u, t of Proposition 3.25 then the iteration map P of that theorem (implemented by the iteration of (3.26)) may be shown to be a contraction under the norm of (3.27). Indeed using the estimate of (3.16),

$$|Px(t) - Py(t)| \le \int_0^t k|x(\tau) - y(\tau)|d\tau, \tag{3.28}$$

we have that

$$|Px(t) - Py(t)| \le \int_0^t ke^{w\tau}|x(\cdot) - y(\cdot)|_w d\tau. \tag{3.29}$$

Using this equation we see that

$$e^{-wt}|Px(t) - Py(t)| \le e^{-wt}\int_0^t ke^{w\tau}|x(\cdot) - y(\cdot)|_w d\tau,$$

$$|Px(\cdot) - Py(\cdot)|_w \le \frac{k}{w}(1 - e^{-wT})|x(\cdot) - y(\cdot)|_w, \tag{3.30}$$

$$|Px(\cdot) - Py(\cdot)|_w \le \frac{k}{w}|x(\cdot) - y(\cdot)|_w.$$

Thus, it is possible to choose w large enough to make P a contraction. The convergence of the sequence of Picard–Lindelöf iterates is in the $|.|_w$ norm; hence time instants earlier in the interval are weighted more heavily than later ones. Now, we have that the iteration of (3.26) converges in the $|.|_w$ norm on all intervals of the form $[0, T]$ for w large enough. But since

$$|x(\cdot)|_w \le |x(\cdot)| \le e^{wT}|x(\cdot)|_w$$

where $|x(\cdot)|$ is the sup or L_∞ norm on $C^n[0, T]$, it follows that the sequence of Picard–Lindelöf iterates converges in the L_∞ norm as well!

In practice, the way that waveform relaxation is used in the simulation (timing analysis) of digital VLSI circuits is as follows: A logic level simulator provides a digital waveform for the response of different state variables in the circuit. The output of the logic level simulator is used as the initial guess for the waveform relaxation algorithm to provide the final analog waveform for the circuit simulation. Of course, in a practical context the large dimensional state variable (i.e., $x \in \mathbb{R}^n$ with n large for VLSI circuits) is partitioned according to subsystems of the original system and each subsystem is iterated to converge, thus the simulation is rippled through from stage to stage to speed up the computation. For this and other interesting details on how the circuit equations can be made to satisfy the conditions of the contraction mapping theorem see [177] and Problem 3.7.

3.5   Differential Equations with Discontinuities

While the existence uniqueness theory for ordinary differential equations is powerful and comprehensive, there are several examples in practice of differential equations with discontinuous right hand sides. They arise frequently in a control context. Indeed, the simplest kind of controllers are on–off controllers. On–off controllers have also been studied in the context of what is known as bang-bang control[2]. The basic mathematical problem in studying the dynamics of systems with switched control laws is that they represent differential equations with discontinuous right hand sides. To illustrate, consider the example of a discontinuous differential equation in $\mathbb{R}^n$. Define, using a function $s : \mathbb{R}^n \mapsto \mathbb{R}$, $S_0 = \{x \in \mathbb{R}^n : s(x) = 0\}$ to be a surface (informally, an $n - 1$ dimensional surface) called the *switching boundary*. Now define the differential equation by

$$\dot{x} = f_+(x) \quad \text{for } \{x : s(x) > 0\} =: S_+$$
$$\dot{x} = f_-(x) \quad \text{for } \{x : s(x) < 0\} =: S_- \tag{3.31}$$

where f_+, f_- are smooth functions from $\mathbb{R}^n \mapsto \mathbb{R}^n$. In general, f_+, f_- do not match on S_0 so that the dynamics are discontinuous at S_0 (more precisely, there is a step discontinuity in the differential equation at S_0)[3]. In Figure 3.4 we show some of the possible phase portraits associated with the discontinuity. In the figure on the upper left hand side, the trajectories both point towards the discontinuity surface S_0. Intuitively, one would expect that imperfections in the switching should cause the state trajectory to "chatter" or zig-zag across the discontinuity surface, as suggested by the jagged line in Figure. In the case of the figure on the top right,

---

[2]This frequently arises from optimal control problems, usually minimum time problems with limits on how large the controls can be. For details, see [50].

[3]Recall that a step discontinuity in a function is one where the function has both a right limit and a left limit and the two are not equal.

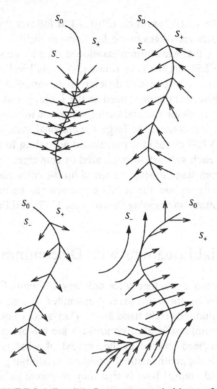

FIGURE 3.4. Possible flows near a switching surface

the trajectories of f_+ point toward S_0 and those of f_- away from it. There appears to be no problem with continuing the solution trajectories in this instance through S_0. In the bottom left figure the trajectories of f_+, f_- both point away from S_0. It would appear that the initial conditions on S_0 would follow either one of the trajectories of f_+ or f_- and which one specifically appears to be ambiguous. In the last figure on the bottom right we have a combination of the circumstances represented in the three other figures. The standard technique in the differential equations literature for dealing with this and other breakdowns of assumptions needed to guarantee the existence and uniqueness of solutions is to *regularize* the system. This means adding a small perturbation to the given system so as to make the system a well defined differential equation (satisfying the standard existence and uniqueness conditions) and then studying the behavior of the well-defined systems in the limit that the perturbation goes to zero. One common regularization for the case of step discontinuities in the differential equation is to assume that (3.31) is the limit as $\Delta \downarrow 0^4$ of the hysteretic switching mechanism shown in Figure 3.5.

---

[4] $\downarrow$ means that Δ goes down to zero, through positive values

The variable y represents the switching variable: when $y = +1$, the dynamics are described by f_+, and when $y = -1$ they are described by f_-. Applying this regularization yields the phase portraits shown in Figure 3.6 for the scenario of Figure 3.4(a) for successively smaller values of Δ. The frequency of crossing S_0 (chattering) increases as $\Delta \downarrow 0$. Also, it appears that in the limit $\Delta = 0$, that the trajectory is confined to the surface S_0. Other forms of regularization for (3.31) representing various imperfections in the switching mechanisms such as, for instance, time delays associated with the switching or neglected "fast" dynamics associated with the switching mechanism may also be used. Consistent with the foregoing intuition, Filippov [102] proposed a solution concept for differential equations with discontinuous right hand sides. His definition was for general discontinuous differential equations, of which (3.31) is a special case (only step discontinuous), of the form

$$\dot{x} = f(x) \tag{3.32}$$

with $f(x)$ a general discontinuous function:

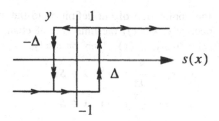

FIGURE 3.5. Showing the hysteretic switching mechanism

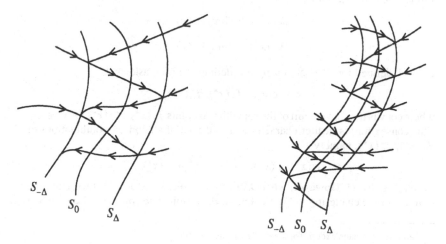

FIGURE 3.6. Effect of regularization of switching

Definition 3.26 Solution in the Sense of Filippov. *An absolutely continuous function $x(t) : [0, T] \mapsto \mathbb{R}^n$ is said to be a solution of (3.32) in the sense of Filippov if for almost all[5] $t \in [0, T]$,*

$$\dot{x} \in \cap_{\delta>0} \cap_N \text{ conv } f(B(x(t), \delta) - N), \qquad (3.33)$$

where $B(x(t), \delta)$ is a ball of radius δ centered at $x(t)$ and the intersection is taken over all sets N of zero measure. Here, conv refers to the convex hull, that is, the smallest convex set containing $f(B(x(t), \delta) - N)$.

Remarks:

1. The definition of (3.33) is quite general—it includes more general classes of discontinuous differential equations than those with a piecewise continuous right hand side, which are the systems of greatest current interest.
2. The reason for the sets N of measure 0 in the definition is to be able to exclude sets on which $f(x)$ is not defined, such as S_0.
3. It is of interest to note that the definition requires only that $\dot{x}$ belong to a set. Thus, equation (3.33) is called a *set valued differential equation* or *differential inclusion*.

We will now study the application of our definition to the system of (3.31). Denote by $\lambda_+(x)$ (respectively, $\lambda_-(x)$) the time rate of change of $s(x)$ along trajectories of $f_+(x)$ (respectively, $f_-(x)$). More precisely,

$$\lambda_+(x) = \frac{\partial s}{\partial x} f_+(x) \quad x \in S_+,$$
$$\lambda_-(x) = \frac{\partial s}{\partial x} f_-(x) \quad x \in S_-. \qquad (3.34)$$

Since $f_+(x)$, $f_-(x)$ are all smooth functions of x assumed to have left and right limits, respectively, both

$$\lambda_+(x^*) = \lim_{x \to x^*} \lambda_+(x)$$
$$\lambda_-(x^*) = \lim_{x \to x^*} \lambda_-(x)$$

are well defined for $x^* \in S_0$. Filippov's definition (3.33) asks that

$$\dot{x} \in \text{ conv } \{f_+(x^*), f_-(x^*)\},$$

where conv is the convex hull of the set with two points $f_+(x^*)$, $f_-(x^*)$ for $x^* \in S_0$. This convex hull is further characterized as the set of all convex combinations of $f_+(x^*)$, $f_-(x^*)$, namely

$$\alpha(x^*) f_+(x^*) + (1 - \alpha(x^*)) f_-(x^*)$$

for $\alpha(x) \in [0, 1]$. However, when $\lambda_+(x^*) < 0$ and $\lambda_-(x^*) > 0$, then from the intuition of the chattering in Figure 3.6, the definition yields more (cf. Lemma 3 of

---

[5]almost all means except for a set of t of measure 0.

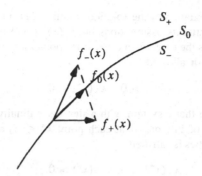

FIGURE 3.7. Construction of the f_0 in the convex hull of f_+, f_-

[102]): It yields that $\dot{x}$ is that unique $f_0(x)$ in the convex hull of the set containing $f_+(x)$, $f_-(x)$ chosen to make the trajectory lie on S_0, that is,

$$\frac{\partial s}{\partial x}(x^*)f_0(x^*) = 0.$$

It is easy to verify that the specific convex combination of $f_+(x^*)$, $f_-(x^*)$ required to achieve this is

$$f_0(x^*) = \frac{\lambda_-(x^*)}{\lambda_-(x^*) - \lambda_+(x^*)} f_+(x^*) + \frac{-\lambda_+(x^*)}{\lambda_-(x^*) - \lambda_+(x^*)} f_-(x^*). \qquad (3.35)$$

This construction is shown in Figure 3.7. The construction is consistent with the intuition that in the limit as the regularization Δ of (3.31) goes to 0, the chattering becomes infinitely rapid and of infinitesimal amplitude. Thus, f_0 is the *averaging* of the chattering. Further, the trajectory of the system slides along the surface of S_0 once it hits S_0. This is referred to as the **sliding mode**. The combination of the conditions $\lambda_+(x^*) < 0$, $\lambda_-(x^*) > 0$ is realized by

$$\frac{d}{dt}s^2(x) < 0 \quad \text{for } x \in B(x^*, \delta) - S_0, \qquad (3.36)$$

where δ is a small positive number. The understanding is that $(d/dt)s^2(x)$ is evaluated along trajectories of $f_+(x)$ in S_+ and along those of $f_-(x)$ in S_-. Equation (3.36) is referred to as the *local sliding condition* and guarantees that trajectories are attracted to the sliding surface. The condition is said to be a *global sliding condition* if the ball $B(x^*, \delta)$ can be chosen to be arbitrarily large.

When, $\lambda_+(x^*) > 0$ and $\lambda_-(x^*) > 0$ then, consistent with the intuition of Figure 3.4, it may be shown (cf. Lemma 9 of Filippov [102]) that the trajectory defined in the sense of Filippov has only x^* in common with S_0 and goes from S_- to S_+ through x^*. When $\lambda_+(x^*) < 0$, $\lambda_-(x^*) < 0$, one may similarly show that the trajectory goes from S_+ to S_- through x^*. Just as in the case of the solution concept in the sense of Caratheodory, one can state and prove a theorem about the existence and uniqueness for solutions in the sense of Filippov. The theorem

on existence and continuation of the solutions requires (Theorems 4, 5 of [102])
only the Lipschitz continuity assumptions on $f_+(x)$, $f_-(x)$ away from S_0. For
uniqueness of solutions the case when trajectories point away from S_0 along both
f_+ and f_- needs to be disallowed:

$$\lambda_+(x^*) > 0, \quad \lambda_-(x^*) < 0.$$

Thus, it may be shown that a system with a step discontinuity at S_0 has unique
solutions in the sense of Filippov if at each point $x^* \in S_0$ at least one of the
following two inequalities is satisfied:

$$\lambda_+(x^*) < 0, \quad \lambda_-(x^*) > 0.$$

Furthermore, the solution depends continuously on the initial conditions. The one
case not covered by the foregoing discussion is when both λ_+, λ_- are equal to 0.
This is the case when both f_+, f_- are tangential to the surface S_0 at at point x.
In this case one can have more complicated behavior, referred to as higher-order
sliding, see for example, [198].

The preceding development was for the time-invariant case, that is, s, f_+, f_- are
not explicitly functions of time. When s, f_+, f_- are functions of x, t, the preceding
development generalizes as follows: define the sliding surface M_0 in (x, t) space
as

$$M_0 := \{(x, t) : s(x, t) = 0\} \subset \mathbb{R}^{n+1}$$

and the functions

$$\lambda_+(x, t) = \frac{\partial s(x, t)}{\partial t} + \frac{\partial s(x, t)}{\partial x} f_+(x, t),$$

$$\lambda_-(x, t) = \frac{\partial s(x, t)}{\partial t} + \frac{\partial s(x, t)}{\partial x} f_-(x, t).$$

In the instance that $\lambda_+(x^*, t^*) < 0$ and $\lambda_-(x^*, t^*) > 0$ a formula completely
analogous to (3.35) may be obtained. One easy way of proving this from the
preceding development is to convert the time variant case to the time invariant case
by augmenting the state space with the t variable and augmenting the dynamics
with $\dot{t} = 1$. This is a very useful technique known as *suspension*. We will have
occasion to use it again. As before, the (x, t) trajectory slides along the manifold
M_0 once it reaches M_0. The sliding condition (3.36) is modified to

$$\frac{d}{dt} s^2(x, t) < 0 \quad \text{for } (x, t) \in B((x^*, t^*), \delta) - M_0$$

with the understanding that $(d/dt)s^2(x, t)$ is evaluated along trajectories of
$f_+(x, t)$ in $M_+ = \{(x, t) : s(x, t) > 0\}$ and along those of $f_-(x, t)$ in
$M_- = \{(x, t) : s(x, t) < 0\}$. As before, the uniqueness theorem is also valid
provided that at least one of the two inequalities

$$\lambda_-(x^*, t^*) > 0 \quad \lambda_+(x^*, t^*) < 0$$

is satisfied for each $(x^*, t^*) \in M_0$. Some authors prefer to deal not with M_0 but with a time varying sliding surface

$$S(t) := \{x : s(x, t) = 0\}.$$

Of course, this does not change any of the preceding formulation.

3.6   Carleman Linearization

We begin this section with some notation: Given matrices $A \in \mathbb{R}^{n_1 \times m_1}$, $B \in \mathbb{R}^{n_2 \times m_2}$, we define their *Kronecker product*, denoted by $A \otimes B \in \mathbb{R}^{n_1 n_2 \times m_1 m_2}$ defined by

$$A \otimes B := \begin{bmatrix} a_{11} B & \cdots & a_{1m_1} B \\ \vdots & \vdots & \vdots \\ a_{n_1 1} B & \cdots & a_{n_1 m_1} B \end{bmatrix}.$$

The following properties of the Kronecker product are left as Problem 3.13.

Proposition 3.27 Properties of the Kronecker Product.

1. $(A + B) \otimes (C + D) = (A \otimes C) + (A \otimes D) + (B \otimes C) + (B \otimes D)$.
2. $(AB) \otimes (CD) = (A \otimes C)(B \otimes D)$.
3. $A \otimes B = 0 \Leftrightarrow A = 0$ *or* $B = 0$.
4. *If A, B are square and invertible, so is $A \otimes B$, and*

$$(A \otimes B)^{-1} = A^{-1} \otimes B^{-1}.$$

5. *If rank A is r_1 and rank B is r_2, the rank of $A \otimes B = r_1 r_2$.*

The Kronecker product may be used to write the *Taylor series* of an analytic function $f : \mathbb{R}^n \mapsto \mathbb{R}^m$ as follows:

$$f(x) = F_0 + F_1 x + F_2 x \otimes x + F_3 x \otimes x \otimes x + \cdots, \qquad (3.37)$$

where $x \in \mathbb{R}^{n(\times 1)}$ and $F_i \in \mathbb{R}^{m \times n^i}$. By convention, $x \otimes \cdots \otimes x$ (repeated i times) is abbreviated $x^{(i)}$. The Taylor series of an analytic function is *convergent* in the domain of convergence $U \subset \mathbb{R}^n$. Thus, given $U, \epsilon > 0$, there exists $N(U, \epsilon)$ such that

$$\forall x \in U \quad \left| f(x) - \sum_{i=1}^{N} F_i x^{(i)} \right| < \epsilon.$$

Some notational savings may be gained by noting that some of the terms in $x^{(i)}$ are repeated. For example if $x \in \mathbb{R}^3$ and $i = 2$, we have that

$$x^{(2)} = (x_1^2 \; x_1 x_2 \; x_1 x_3 \; x_2 x_1 \; x_2^2 \; x_2 x_3 \; x_3 x_1 \; x_3 x_2 \; x_3^2)^T.$$

A more efficient notation uses a lexicographic ordering to eliminate the repeated entries, with the notation $x^{[2]}$:

$$x^{[2]} := (x_1^2 \; x_1 x_2 \; x_1 x_3 \; x_2^2 \; x_2 x_3 \; x_3^2)^T,$$

and the Taylor series may then be written as

$$f(x) = \sum_{i=0}^{\infty} \bar{F}_i x^{[i]},$$

where $\bar{F}_i$ has m rows and

$$\binom{n+i-1}{i}$$

columns.

Now, consider the linear differential equation in $\mathbb{R}^n$

$$\dot{x} = Ax, \quad x(0) = x_0. \tag{3.38}$$

It is easy to see that the differential equation satisfied by $x^{(2)} \in \mathbb{R}^{n^2}$ is

$$\frac{dx^{(2)}}{dt} = [A \otimes I + I \otimes A] x^{(2)}, \quad x^{(2)}(0) = x_0 \otimes x_0. \tag{3.39}$$

Thus, $x^{(2)}$ satisfies a linear differential equation as well. More generally, it may be shown (see Problem 3.14) that $x^{(i)}$ and $x^{[i]}$ satisfy linear differential equations as well.

We will now use Taylor series expansions for solutions of general analytic differential equations of the form

$$\dot{x} = \sum_{i=1}^{\infty} A_i x^{(i)}, \quad x(0) = x_0. \tag{3.40}$$

It follows that the differential equation satisfied by $x^{(2)}$ is

$$\frac{dx^{(2)}}{dt} = \sum_{i=1}^{\infty} [A_i \otimes I + I \otimes A_i] x^{(i+1)} \quad x^{(2)}(0) = x_0 \otimes x_0$$

Iterating on this procedure yields

$$\frac{dx^{(k)}}{dt} = \sum_{i=1}^{\infty} A_{k,i} x^{i+k-1}, \quad x^{(k)}(0) = x_0 \otimes \cdots \otimes x_0.$$

with the understanding that $A_{1,i} = A_i$ and for $k > 1$

$$A_{k,i} = A_i \otimes I \cdots \otimes I + I \otimes A_i \cdots \otimes I + \cdots + I \otimes \cdots \otimes A_i,$$

that is, $A_{k,i}$ has k terms each of which has $k-1$ Kronecker products.

Now define for given N the vector $x^{\oplus} \in \mathbb{R}^{n+n^2+\cdots+n^N}$ by

$$x^{\oplus} = \begin{bmatrix} x \\ x^{(2)} \\ \vdots \\ x^{(N)} \end{bmatrix} .$$

The differential equation satisfied by $x^{\otimes}$ is

$$\frac{dx^{\oplus}}{dt} = \begin{bmatrix} A_{11} & A_{12} & \cdots & A_{1N} \\ 0 & A_{21} & \cdots & A_{2,N-1} \\ 0 & 0 & \cdots & A_{3,N-2} \\ \vdots & \vdots & \vdots & \vdots \\ 0 & 0 & \cdots & A_{N1} \end{bmatrix} x^{\oplus} + \text{h.o.t.} \tag{3.41}$$

In the preceding expression h.o.t. stands for higher order terms (i.e., terms involving polynomials of degree greater than N). Formally, we can continue the process by defining $x^{\oplus}$ to be an infinitely long vector, and then the equation (3.41) will not have any higher order terms, and the resulting infinite dimensional system ($\sum_{i=1}^{\infty} n^i$) is *linear*. This is referred to as the process of *Carleman linearization*. Sometimes, the truncation of the terms of (3.41) to terms of order N (i.e., dropping the higher order terms) is also referred as the (approximate) Carleman linearization. For compact intervals of time, bounds for the distance between the solutions of the approximate Carleman linearization and the original differential equation (3.40) may be derived (see Problem 3.15).

3.7   Degree Theory

Degree theory is a generalization of the index theory which we encountered in the previous chapter in the context of planar dynamical systems. Consider a smooth map $f : \overline{D} \mapsto \mathbb{R}^n$ with D an open, bounded set in $\mathbb{R}^n$ and $\overline{D}$ its closure. For $p \in \mathbb{R}^n$, we have the following definition:

Definition 3.28 Degree.   *The* degree *of f with respect to D at the point p is defined by*

$$d(f, p, D) = \sum_{x \in f^{-1}(p) \cap D} \text{sgn} \det Df(x). \tag{3.42}$$

It is required that $f^{-1}(p)$ be a finite set and that $Df(x)$ be nonsingular for $x \in f^{-1}(p)$ and that $f^{-1}(p) \cap \partial D = \varnothing$, that is, that there are no solutions on the boundary of D.[6]

If p is such that $Df(x)$ is singular for $x \in f^{-1}(p)$, then perturb p to p_ϵ. Now, if $f^{-1}(p_\epsilon)$ is a finite set define

$$d(f, p, D) := \lim_{p_\epsilon \to p} d(f, p_\epsilon, D).$$

The degree of f with respect to D at the point p is a mod 2 count of the number of solutions of the equation $f(x) = p$. We say a mod 2 count, since each solution is given a sign, either $+1$ or -1 depending on the sign of the determinant of $Df(x)$.

There is a "volume integral" formula for the degree of a map, namely

$$d(f, p, D) = \int_D \psi_\epsilon(|f(x) - p|) \det Df(x)\, dx, \qquad (3.43)$$

where $\epsilon > 0$ is small and $\psi_\epsilon : \mathbb{R}_+ \mapsto \mathbb{R}_+$ is a continuous "bump" function, satisfying:

1. $\psi_\epsilon(s) = 0$ for $s \geq \epsilon$.
2.

$$\int_{\mathbb{R}^n} \psi_\epsilon(|x|)\, dx = 1.$$

Proof of volume integral formula: Assume that $f^{-1}(p) = \{x_1, x_2, \ldots, x_k\}$ is the finite set of solutions to $f(x) = p$. There exist ϵ, and $\epsilon_i, i = 1, \ldots, k > 0$ such that f is a homeomorphism from each ball $B(x_i, \epsilon_i) \mapsto B(p, \epsilon)$. (This result, somewhat generalized, is referred to as the stack of records theorem in the Exercises.) We may rewrite (3.43) as

$$\int_D \psi_\epsilon(|f(x) - p|) \det Df(x)\, dx = \sum_{i=1}^{k} \int_{B(x_i, \epsilon_i)} \psi_\epsilon(|f(x) - p|) \det Df(x)\, dx.$$

Defining the change of variables $z(x) = f(x) - p$ with Jacobian $Dz = Df(x)$ and writing $\det Df(x) = |\det Df(x)| \operatorname{sgn}(\det Df(x))$, we obtain from the standard change of variables formula for volume integral

$$d(f, p, D) = \sum_{i=1}^{k} \int_{B(x_i, \epsilon_i)} \psi_\epsilon(|z(x)|)|\det Df(x)| \operatorname{sgn} \det Df(x)\, dx,$$

$$= \sum_{i=1}^{k} \left[\int_{B(0, \epsilon)} \psi_\epsilon(|x|) dx \right] \operatorname{sgn} \det Df(x_i),$$

$$= \sum_{i=1}^{k} \operatorname{sgn} \det Df(x_i).$$

---

[6]In the terminology of the next section, p is said to be a *regular value* of the function f.

In the first step we used the change of variable formula and the fact that for ϵ_i sufficiently small the sign of the determinant of $Df(x)$ is constant and in the second step we used the fact that

$$\int_{B(0,\epsilon)} \psi_\epsilon(|x|)\, dx = \int_{\mathbb{R}^n} \psi_\epsilon(|x|)\, dx = 1.$$

We may group the properties of degree into two categories:

1. Topological Properties of Degree

1. *Boundary Dependence.*
   $d(f, p, D)$ is uniquely determined by f on ∂D, the boundary of the open set D.
2. *Homotopy Invariance.*
   Suppose that $H(x, t) = p$ has no solution $x \in \partial D$ for any $t \in [0, 1]$; then $d(H(x, t), p, D)$ is a constant for $t \in [0, 1]$.
3. *Continuity.*
   $d(f, p, D)$ is a continuous function of f, that is,

   $$|f(x) - g(x)| \leq \epsilon \ \forall x \in D \Rightarrow d(f, p, D) = d(g, p, D)$$

   for ϵ small enough. $d(f, p, D)$ is a continuous function of p also.
4. *Poincaré–Bohl.*
   If $f(x) - p$ and $g(x) - p$ never point in opposite directions for $x \in \partial D$, then $d(f, p, D) = d(g, p, D)$, provided that each is defined.

Proof:

- (3) follows from the volume definition of degree, since all the quantities in the integrand of (3.43) are continuous functions of the function f and the point p. Further, ϵ is to be chosen small enough so that no solutions leave through the boundary ∂D.
- (3) $\Rightarrow$ (2). Define $h(t) = d(H(x, t), p, D)$. Then, so long as there are no solutions to $H(x, t) = p$ on ∂D, $h(t)$ is a continuous function of t. Since it is an integer valued function, it is constant.
- (2) $\Rightarrow$ (1) Let $g = f$ on ∂D. Then consider

$$H(x, t) = tf(x) + (1 - t)g(x).$$

Since neither $f(x)$ or $g(x)$ equals p and $H(x, t) \equiv f(x) \equiv g(x)$ on ∂D, we have $H(x, t) \neq p$ for $x \in \partial D$ and we have

$$d(H(x, 0), p, D) = d(H(x, 1), p, D),$$

$$d(f, p, D) = d(g, p, D).$$

- (2) $\Rightarrow$ (4) Define $H(x, t) = t(f(x) - p) + (1 - t)(g(x) - p)$. Since $f(x) - p \neq -\alpha(g(x) - p) \ \forall x \in \partial D$ and any $\alpha \in \mathbb{R}_+$, $H(x, t) \neq 0 \ \forall x \in \partial D, t \in [0, 1]$.

Hence, we have

$$d(H(x, 0), p, D) = d(H(x, 1), p, D),$$
$$d(f, p, D) = d(g, p, D).$$

2. Arithmetic Properties of Degree

1. If $d(f, p, D) \neq 0$, then $f(x) = p$ has solutions in D.
2. If $\{D_i, i = 1, \ldots, k\}$ is a finite collection of open disjoint subsets of D and $f(x) \neq p$ for $x \in \overline{D} - \cup_i D_i$, then

$$d(f, p, D) = \sum_{i=1}^{k} d(f, p, D_i) \tag{3.44}$$

3. If $f(x) \neq p$ in $\overline{D}$ then $d(f, p, D) = 0$.
4. The map f is said to *omit the direction* $c \in \mathbb{R}^n$ (c will be chosen to be normalized to unit length) if

$$\frac{f(x)}{|f(x)|} \neq c \quad \text{for all } x \in \partial D.$$

If f omits any direction c then $d(f, 0, D) = 0$.
5. Let D be symmetric with respect to the origin, that is, $x \in D \Rightarrow -x \in D$, and $f(x) = -f(-x)$ on ∂D with $f \neq 0$. Then $d(f, 0, D)$ is odd.
6. Let D be symmetric with respect to the origin and let $f(x)$ and $f(-x)$ not point in the same direction for all $x \in \partial D$, then $d(f, 0, D)$ is an odd integer.
7. If $0 \in D$ and $d(f, 0, D) \neq \pm 1$, then

   a. There exists at least one $x \in \partial D$ such that $f(x)$ and x point in the same direction.
   b. There exists at least one $x^1 \in \partial D$ such that $f(x^1)$ and x^1 point in opposite directions.

Proof:

1. This simply follows from the definition of degree.
2. This simply follows from the definition of degree.
3. Also a direct consequence of the definition of degree.
4. Define the homotopy $H(x, t) = t f(x) + (1 - t)c$ between f and the constant map $-c$ on $\overline{D}$. Since f omits the direction c, it follows that $H(x, t) \neq 0$ on ∂D for all $t \in [0, 1]$. Thus $d(f, 0, D) = d(\tilde{f}, 0, D)$ where $\tilde{f}(x) \equiv -c$. Since $c \neq 0$, it follows that $d(f, 0, D) = d(\tilde{f}, 0, D) = 0$.
5. Since f is odd for every solution x_i of $f(x) = 0$, there is a solution $-x_i$. Also, $f(0) = 0$. Further $\det Df(x_i) = \det Df(-x_i)$. Consequently, if all the

$\det Df(x_i) \neq 0$ it follows that

$$d(f, 0, D) = \sum_{i=1}^{k} [\text{sgn det} Df(x_i) + \text{sgn det} Df(-x_i)] + \text{sgn det} Df(0)$$

$$= \text{even integer} \pm 1$$

$$= \text{odd integer}.$$

If some of the $\det Df(x_i) = 0$ perturb 0 as in the definition of degree, the resulting perturbed solutions may no longer be negatives of each other and the signs of their determinants may not be identical but nonetheless the qualitative conclusion of the preceding equation will hold.

6. Define the homotopy $H(x, t) = t(f(x) - f(-x)) + (1-t)f(x)$. Since $f(x) \neq \alpha f(-x) \; \forall \alpha \in \mathbb{R}_+ \; \forall x \in \partial D$, it follows that $H(x, t) \neq 0 \forall x \in \partial D$. Hence $d(H(x, 0), 0, D) = d(H(x, 1), 0, D)$. As a consequence $d(f(x), 0, D) = d(f(x) - f(-x), 0, D)$. But since $g(x) = f(x) - f(-x)$ is an odd function it follows from item 5 above that $d(g, 0, D)$ is an odd integer. Hence $d(f, 0, D)$ is also odd.

7. If either of the conclusions is violated, there is a homotopy between the map f and the identity map on D, which has degree 1.

The following theorem, called the Brouwer fixed point theorem is an easy consequence of these definitions and is an extremely important fixed point theorem (especially in microeconomics, see [80]).

Theorem 3.29 Brouwer Fixed Point.   *Let* $f : \overline{B}(0, 1) \mapsto \overline{B}(0, 1)$ *be a continuously differentiable map. Then f has a fixed point in $\overline{B}$.*

Proof: For the sake of contradicting the conclusion assume that $f(x) - x$ has no zeros for $x \in \partial B$. Further, since $f(x) \neq x$ for $x \in B$ we have $d(x - f(x), 0, B) = 0$. Since $f(x) - x$ has no zeros for $|x| = 1$, $|tf(x)| < 1 \; \forall x \in \partial B$ and $t \in [0, 1[$. Thus $x - tf(x)$ has no zeros on ∂B for $t \in [0, 1]$. Hence,

$$d(x - f(x), 0, B) = d(x - tf(x), 0, B) = d(x, 0, B) = 1$$

This establishes the contradiction.   □

3.8   Degree Theory and Solutions of Resistive Networks

One of the nicest applications of degree theory is to study the existence or non-existence of solutions to resistive networks. We give a brief description of the application. Consider a resistive nonlinear network with some linear components, nonlinear components and dc sources as shown in Figure 3.8. The nonlinear elements are assumed to be extracted outside the linear resistive box as are the constant independent source elements. The independent variables of the nonlinear resistors (either voltages or currents) are labeled $z_i \in \mathbb{R}$, $i = 1, \dots, k$. It is assumed that the characteristics of the linear part of the network along with the independent sources

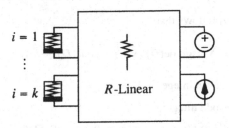

FIGURE 3.8. A Resistive Nonlinear Network

can be expressed in terms of the independent variables of the nonlinear resistors, so that the resistive constitutive relations may be written as

$$f(z) := g(z) + Gz + s = 0. \tag{3.45}$$

Here $g : \mathbb{R}^k \mapsto \mathbb{R}^k$ is a "diagonal nonlinearity" of resistor constitutive relations, that is,

$$g = \begin{bmatrix} g_1(z_1) \\ \vdots \\ g_k(z_k) \end{bmatrix}.$$

We will assume that $G \in \mathbb{R}^{k \times k}$, the constitutive relation matrix of the linear resistive network, is positive definite; that is, the linear resistors are passive. The vector $s \in \mathbb{R}^k$ models the effect of the sources on the network.

Proposition 3.30 Eventually Passive Resistors.   *Consider a nonlinear circuit with constitutive relation given by (3.45). Assume that*

- *$g(\cdot)$ is eventually passive, that is to say $z^T g(z) \geq 0$ for $|z| \geq R$,*
- *The constitutive matrix G is positive definite.*

Then, equation (3.45) has at least one solution for all s.

Proof: The proof consists in establishing a homotopy to the identity map. Define the homotopy map

$$H(z, t) = tf(z) + (1 - t)z.$$

Now, we have that

$$z^T H(z, t) = tz^T f(z) + (1 - t)|z|^2 = tz^T g(z) + tz^T Gz - tz^T s + (1 - t)|z|^2.$$

Since $g(\cdot)$ is eventually passive $z^T g(z) \geq 0$ for $|z| \geq R$. Also, since G is positive definite, $z^T Gz \geq \gamma |z|^2$ and thus for $|z| \geq R_1$ we have that $z^T Gz - z^T s > 0$. For $|z| = \max (R, R_1)$ we have that $H(z, t) \neq 0$ and

$$z^T H(z, t) > 0 \quad \forall |z| = \max(R, R_1) \quad t \in [0, 1]$$

Since the set $\{z : |z| = \max(R, R_1)\}$ is the boundary of the open ball $\{z : |z| < \max(R, R_1)\}$, it follows from arithmetic property 7 above that

$$d(f, 0, B(0, \bar{R})) = 1$$

with $\bar{R} = \max(R, R_1)$. Thus, the equation $f(z) = 0$ has at least one solution inside $B(0, \bar{R})$. $\square$

Proposition 3.31 Saturating Resistors. *Consider a nonlinear circuit with constitutive relation given by (3.45). Assume that*

- *$g(\cdot)$ eventually saturates, that is, $|g(z)| \leq b$ for $z \in \mathbb{R}^k$*
- *The hybrid matrix H is nonsingular.*

Then, equation (3.45) has at least one solution for all s.

Proof: Consider premultiplying (3.45) by $z^T H^T$ to get

$$z^T H^T f(z) = z^T H^T g(z) + z^T H^T H z - z^T H^T s.$$

Since H is nonsingular, we have that $z^T H^T H z \geq \gamma |z|^2$. Since $g(\cdot)$ is bounded there exists R such that

$$z^T H^T f(z) > 0 \quad \forall |z| = R.$$

On the boundary of the sphere of radius R, $f(z)$ and $f(-z)$ are both nonzero and do not point in the same direction. For, if there exists $k > 0$ such that $f(z) = kf(-z)$, then we would have $-z^T H^T f(z) = -kz^T H^T f(-z)$. This contradicts the previous equation and establishes the conclusion that $f(z)$, $f(-z)$ do not point in the same direction. By Arithmetic Property 6 above, it follows that $d(f, 0, B(0, R))$ is an odd integer. $\square$

3.9   Basics of Differential Topology

Roughly speaking, manifolds are "locally" like vector spaces but are globally curved surfaces. Examples abound around us: the surface of the earth is "locally flat" but globally curved, the surface of a doughnut is "locally flat" but globally not a vector space. The question naturally arises as to how to make this intuition into a precise set of definitions. In particular, what sense can one make of a "locally flat" space? One sense that one could make is that a "locally flat" space can be "gently" flattened into a vector space (see Figure 3.9). By this sense, a cone does not classify as being able to be gently flattened into a plane since no neighborhood of the vertex looks like the plane. In the following sections we will translate this intuitive discussion into a set of formal definitions.

3.9.1   Smooth Manifolds and Smooth Maps

Let $U \subset \mathbb{R}^k$ and $V \subset \mathbb{R}^l$ be open sets. A mapping $f : U \mapsto V$ is called *smooth*, if all of the partial derivatives $\partial^n f / \partial x_{i_1}, \ldots, \partial x_{i_n}$ exist and are continuous. More

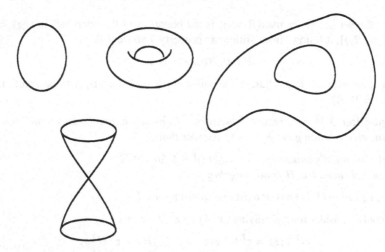

FIGURE 3.9. Surfaces in 3 dimensional space

generally, if $X \subset \mathbb{R}^k$ and $Y \subset \mathbb{R}^l$ are arbitrary subsets of Euclidean spaces (not necessarily open), then $f : X \mapsto Y$ is called smooth if there exists an open set $U \subset \mathbb{R}^k$ containing X and a smooth mapping $F : U \mapsto \mathbb{R}^l$ that coincides with f in $U \cap X$. If $f : X \mapsto Y$ and $g : Y \mapsto Z$ are smooth, then so is $g \circ f : X \mapsto Z$.

Definition 3.32 Diffeomorphism[7]. *A map $f : X \mapsto Y$ is said to be a diffeomorphism if f is a homeomorphism (i.e., a one-to-one continuous map with continuous inverse) and if both f and f^{-1} are smooth.*

The sets X and Y are said to be diffeomorphic if there exists a(ny) diffeomorphism between them. Some examples of sets diffeomorphic to and not diffeomorphic to the closed interval, circle, and sphere are shown in Figure 3.10.

Definition 3.33 Smooth Manifold of Dimension m. *A subset $M \subset \mathbb{R}^k$ is called a smooth manifold of dimension m if for each $x \in M$ there is a neighborhood $W \cap M$ ($W \subset \mathbb{R}^k$), that is diffeomorphic to an open subset $U \subset \mathbb{R}^m$.*

A diffeomorphism ψ from $W \cap M$ into U is called a *system of coordinates* on $W \cap M$ and its inverse ψ^{-1} is called a *parameterization*. The map itself is referred to as a *coordinate map*. These definitions are illustrated in Figure 3.11.

Examples:

1. The unit circle $S^1 \subset \mathbb{R}^2$ defined by $\{(\cos\theta, \sin\theta), \theta \in [0, 2\pi]\}$.
2. The unit sphere $S^2 \subset \mathbb{R}^3$ defined by $\{(x_1, x_2, x_3) : x_1^2 + x_2^2 + x_3^2 = 1\}$ is a smooth manifold of dimension 2. Indeed, the diffeomorphism

$$(x_1, x_2) \mapsto \left(x_1, x_2, \sqrt{(1 - x_1^2 - x_2^2)}\right)$$

---

[7]This is the gentle flattening discussed previously.

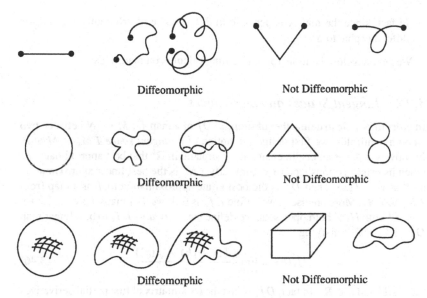

FIGURE 3.10. Surfaces diffeomorphic to and not diffeomorphic to the interval, circle, and sphere

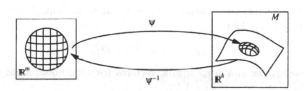

FIGURE 3.11. Illustrating the definition of a manifold

for $x_1^2 + x_2^2 < 1$ parameterizes the region $S^2 \cap \{x_3 > 0\}$. By interchanging the roles of x_1, x_2, x_3 and the sign of the radical we can cover all of S^2. More generally, the *n sphere* $S^n \subset \mathbb{R}^{n+1}$ is the set of $x \in \mathbb{R}^{n+1}$ with $\sum_{i=1}^{n+1} x_i^2 = 1$.

3. The 2 torus $T^2 \in \mathbb{R}^3$, which may be defined either analytically or as $S^1 \times S^1$. Its generalization is the n-torus $T^n \subset \mathbb{R}^{n+1}$ defined by $S^1 \times \cdots (n\text{ times}) \times S^1$.

4. A cone in $\mathbb{R}^3$ defined by $\{(x_1, x_2, x_3) : x_3^2 = x_1^2 + x_2^2\}$ is *not a manifold of dimension 2*, since there is no diffeomorphism which will map a neighborhood of the vertex of the cone onto an open subset of $\mathbb{R}^2$.

5. The space of orthogonal matrices in $\mathbb{R}^{2 \times 2}$ with determinant 1, is a manifold of dimension 1, since every 2×2 unitary matrix of determinant 1 can be written as:

$$\begin{bmatrix} \cos\theta & -\sin\theta \\ \sin\theta & \cos\theta \end{bmatrix}.$$

In fact, since the matrix is periodic in θ, the space of orthogonal matrices is diffeomorphic to S^1.

We next develop the tools for performing calculus on manifolds:

3.9.2   Tangent Spaces and Derivatives

In order to be able to define the differential Df of a map $f : M \mapsto N$ between two smooth manifolds, we first define the notion of a *tangent space* TM_x to M at x. Intuitively TM_x is an m-dimensional hyperplane in $\mathbb{R}^k$ that best approximates M when its origin is translated to x. Similarly TN_y is the best linear approximation to N at $y = f(x)$. Then Df is the best affine approximation to f as a map from TM_x to TN_y. More precisely, we define Df_x as follows: For maps $f : U \subset \mathbb{R}^k \mapsto V \subset \mathbb{R}^l$ with U, V both open sets, we define the *derivative* Df_x to be a linear map $Df : \mathbb{R}^k \mapsto \mathbb{R}^l$ satisfying

$$Df_x(h) = \lim_{t \to 0} \frac{(f(x+th) - f(x))}{t} \tag{3.46}$$

for $x \in U$ and $h \in \mathbb{R}^k$. In fact, Df_x is just the $l \times k$ matrix of first partial derivatives $\frac{\partial f_i}{\partial x_j}$ evaluated at x. It is easy to check by direct calculation that the derivative satisfies three fundamental properties:

1. *Chain rule.* If $f : U \mapsto V$ and $g : V \mapsto W$ are smooth maps between open sets and $y = f(x)$, then

$$D(g \circ f)_x = Dg_y \circ Df_x. \tag{3.47}$$

In other words, we can draw similar pictures for the smooth maps and their derivatives as shown in Figure 3.12.

2. *Linearity.* If $f_1, f_2 : U \mapsto V$, then

$$D(f_1 + f_2)_x = Df_{1x} + Df_{2x}.$$

3. *Inclusion.* If $U \subset U_1 \subset \mathbb{R}^k$ are open sets and i is the inclusion map of U into U_1 (that is $i : U \mapsto U_1$ with $i(x) = x$ for all $x \in U$), then Di_x is the identity map of $\mathbb{R}^k$.

4. If L is a linear map from $U \mapsto V$, then $DL_x = L$.

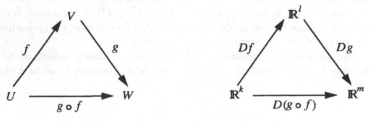

FIGURE 3.12. Commutative diagram of smooth maps and their derivatives

A simple application of these properties is the following proposition:

Proposition 3.34. *If f is a diffeomorphism between open sets $U \subset \mathbb{R}^k$ and $V \subset \mathbb{R}^l$, then $k = l$ and the linear map $Df_x : \mathbb{R}^k \mapsto \mathbb{R}^l$ is nonsingular.*

Proof: The composition of $f \circ f^{-1}$ is the identity map of $\mathbb{R}^k$. Consequently, $Df_x \circ Df_y^{-1}$ is the identity map on $\mathbb{R}^k$ and we must have that $k = l$ and that Df_x is nonsingular. $\square$

A partial converse to this proposition is the inverse function theorem, stated here without proof (see [200]):

Theorem 3.35 Inverse Function Theorem.   *Consider a map $f : U \mapsto V$ between open sets in $\mathbb{R}^k$. If the derivative Df_x is nonsingular then f maps a sufficiently small neighborhood U_1 of x diffeomorphically onto an open set $f(U_1)$.*

Remark: f need not be a global diffeomorphism on noncompact sets U even if Df_x is bounded away from non-singularity. For example, the map $f : (x_1, x_2) \mapsto (e^{x_1} \cos x_2, e^{x_1} \sin x_2)$ has a non-singular Jacobian at every point (verify this) but is not a global diffeomorphism, since it is periodic in the x_2 variable. However, a theorem due to Palais ([240], for the proof refer to Wu and Desoer [333]), states that if f is a *proper* map; that is, the inverse image of compact sets is compact, then Df_x bounded away from singularity on $\mathbb{R}^k$ guarantees that f is a global diffeomorphism. We may now define the *tangent space* TM_x to M at x.

Definition 3.36 Tangent Space TM_x.   *Choose a parameterization $\psi : U \subset \mathbb{R}^m \mapsto M \subset \mathbb{R}^k$ of a neighborhood $\psi(U)$ of x in M. Now $D\psi_u : \mathbb{R}^m \mapsto \mathbb{R}^k$ is well defined. The tangent space to M at $x = \psi(u)$ is defined to be the image of $D\psi_u(\mathbb{R}^m)$.*

The first order of business is to verify that TM_x as defined above is indeed well defined, i.e., it does not depend on the specific parameterization map ψ chosen. Indeed let $\phi : V \mapsto M$ be another parameterization of M with $x = \phi(v)$. Now the map $\phi^{-1} \circ \psi : U_1 \mapsto V_1$ defined from some neighborhood of $u \in U$ to a neighborhood of $v \in V$ is a diffeomorphism. Thus, we have the commutative diagrams of ϕ, ψ and their derivatives as shown in Figure 3.13. Hence by Proposition 3.34 the linear map $D(\phi_v)^{-1} \circ D\psi_u$ is nonsingular. Consequently, the image

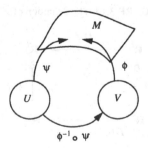

FIGURE 3.13. Commutative diagram of parameterizations

of $D\psi_u$ is equal to the image of $D\phi_v$, and the tangent space to M at x is well defined.

The second order of business is to verify that TM_x is an m dimensional vector space. Now ψ^{-1} is a smooth map from $\psi(U)$ onto U. Hence, it is possible to find a map F defined on an open set $W \in \mathbb{R}^k$ containing x such that F coincides with ψ^{-1} on $W \cap \psi(U)$. Then the map $F \circ \psi : \psi^{-1}(W \cap \psi(U)) \mapsto \mathbb{R}^m$ is an inclusion. Consequently, $DF_x \circ D\psi_u$ is the identity so that the map $D\psi_u$ has rank m, whence its image has dimension m.

We are now ready to define the derivative of a smooth map $f : M \subset \mathbb{R}^k \mapsto N \subset \mathbb{R}^l$ with M, N smooth manifolds of dimension m, n respectively. If $y = f(x)$, then the derivative Df_x is a linear map from TM_x to TN_y defined as follows: since f is smooth there exists a map F defined on a n open set $W \in \mathbb{R}^k$ containing x to $\mathbb{R}^l$ which coincides with f on $W \cap M$. *Now define $Df_x(v)$ to be equal to $DF_x(v)$ for all $v \in TM_x$.* To justify this definition, we must prove that $DF_x(v)$ belongs to TN_y and that it is the same no matter which F we choose. Indeed, choose parameterizations

$$\phi : U \subset \mathbb{R}^m \mapsto M \subset \mathbb{R}^k$$
$$\psi : V \subset \mathbb{R}^n \mapsto N \subset \mathbb{R}^l$$

for neighborhoods $\phi(U)$ of x and $\psi(V)$ of y. By replacing U by a smaller set, if necessary, we may assume that $\phi(U) \subset W$ and that $f : \phi(U) \mapsto \psi(V)$. it follows that the relationships may be represented on a diagram called a commutative diagram as shown in Figure 3.14. Taking derivatives, we see that we have a

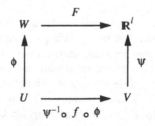

FIGURE 3.14. The definition of F

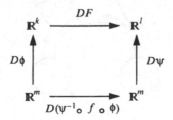

FIGURE 3.15. Commutative Diagram of Derivatives

commutative diagram of linear maps as shown in Figure 3.15. If $u = \phi^{-1}(x)$ and $v = \psi^{-1}(y)$ then it follows that DF_x maps $TM_x = $ Image $D\phi_u$ onto $TN_y = $ Image $(D\psi_v)$. The diagram 3.15 also establishes that the resulting map Df_x does not depend on F by following the lower arrow in the diagram, which shows that

$$Df_x = D\psi_v \circ D(\psi^{-1} \circ f \circ \phi)_u \circ (D, \phi_u)^{-1},$$

where $(D\phi_u)^{-1}$ is the inverse of $D\phi_u$ restricted to the image of $D\phi_u$. It still remains to be shown that the definition of Df_x is independent of the coordinate charts ϕ, ψ; the details of this are left to the reader.

As before, the derivative map satisfies two fundamental properties:

1. *Chain Rule* If $f : M \mapsto N$ and $g : N \mapsto P$ are smooth with $y = f(x)$, then

$$D(g \circ f)_x = Dg_y \circ Df_x. \tag{3.48}$$

2. If $M \subset N$ and i is the inclusion map then $TM_x \subset TN_x$ with inclusion map Di_x (see Figure 3.16).

These properties are easy (though a little laborious) to verify, and in turn lead to the following proposition:

Proposition 3.37 Derivatives of Diffeomorphisms are Isomorphisms. *If $f : M \mapsto N$ is a diffeomorphism then $Df_x : TM_x \mapsto TN_y$ is an isomorphism of vector spaces. In particular the two manifolds M, N have the same dimension.*

Proof: See Problem 3.18.

3.9.3 Regular Values

Let $f : M \mapsto N$ be a smooth map between manifolds of the same dimension. A point $x \in M$ is said to be a *regular point* of f if the derivative Df_x is nonsingular. If Df_x is singular, x is called a *critical point*. If x is a regular point, it follows from the inverse function theorem that f maps a neighborhood of x diffeomorphically onto a neighborhood of $y = f(x)$.

Definition 3.38 Regular Values. *Given a map f from $M \mapsto N$, $y \in N$ is said to be a* regular value *if every point in the set $f^{-1}(y)$ is a regular point.*

A point $y \in N$ is said to be a *critical value* if it is the image of at least one critical point. Thus each $y \in N$ is either a critical value or a regular value according as $f^{-1}(y)$ does or does not contain a critical point. If M is compact and y is a

FIGURE 3.16. The tangent space of a submanifold

regular value then the set $f^{-1}(y)$ is finite. Further, if y is a regular value, then the number of points in the set $f^{-1}(y)$ is locally constant in a neighborhood of y. Proofs of these facts are left to the reader as exercises. A basic theorem known as *Sard's theorem* establishes that the set of critical values is nowhere dense and consequently a set of measure 0 in the co-domain. The impact of this theorem is that a point in the co-domain of a map $f : M \mapsto N$ is generically a regular value, and consequently it's inverse image contains only finitely many points, i.e., if a point y in the co-domain of the map is a critical value, then there are (several) regular values in arbitrarily small neighborhoods of it. It is important to keep in mind while applying this theorem that the theorem does not assert that the set of critical values is nowhere dense in the *image* of f, rather than the co-domain of f. Indeed, if the image of f is a submanifold of N of dimension less than n, then every point in the range of f is a critical value without contradicting Sard's theorem.

We will now define regular values for maps between manifolds of arbitrary dimensions: we will be mainly interested in the case that $m \geq n$. The set of critical points $C \subset M$ is defined to be the set of points x such that

$$Df_x : TM_x \mapsto TN_y$$

has rank less than n, i.e., it is not onto. The image of the set of critical points is the set of critical values of f. An easy extension of Sard's theorem yields that the set of regular values of f is everywhere dense in N. The following theorem is extremely useful as a way to construct manifolds :

Theorem 3.39 Inverse Images of Regular Values. *If $f : M \mapsto N$ is a smooth map between manifolds of dimension $m \geq n$ and if $y \in N$ is a regular value, then the set $f^{-1}(y) \subset M$ is a smooth manifold of dimension $m - n$.*

Proof: Let $x \in f^{-1}(y)$. Since y is a regular value, the derivative Df_x maps TM_x onto TN_y. Consequently, Df_x has a $m - n$ dimensional null space $\mathcal{N} \subset TM_x$. If $M \subset \mathbb{R}^k$, choose a linear map $L : \mathbb{R}^k \mapsto \mathbb{R}^{m-n}$ which is one to one on the subspace $\mathcal{N}$. Now define the map

$$F : M \mapsto N \times \mathbb{R}^{m-n}$$

by $F(\xi) = (f(\xi), L\xi)$. Its derivative DF_x is equal to (Df_x, L). Thus, DF_x is nonsingular; as a consequence F maps some neighborhood U of x onto a neighborhood V of (y, Lx). Also note that $F \circ f^{-1}(y) = \{y\} \times \mathbb{R}^{m-n}$. In fact, F maps $f^{-1}(y) \cap U$ diffeomorphically onto $(\{y\} \times \mathbb{R}^{m-n}) \cap V$. Thus we have established that $f^{-1}(y)$ is a smooth manifold of dimension $m - n$.

The preceding theorem enables us to construct numerous examples of manifolds, for example:

1. The unit sphere S^{n-1} in $\mathbb{R}^n$ defined by

$$S^{n-1} := \{x \in \mathbb{R}^n : x_1^2 + \cdots + x_n^2 = 1\}$$

may be seen to be a manifold of dimension $n - 1$ by verifying that 1 is a regular value of the map $f = x_1^2 + \cdots + x_n^2 : \mathbb{R}^n \mapsto \mathbb{R}$.

2. The set of all matrices in $\mathbb{R}^{n \times n}$ of determinant 1, denoted by $SL(n)$ is a manifold of dimension $n^2 - 1$ as may be shown by verifying that 1 is a regular value of the determinant function det: $\mathbb{R}^{n \times n} \mapsto \mathbb{R}$.

3. The set of all orthogonal matrices in $\mathbb{R}^{3 \times 3}$ of determinant 1, denoted by $SO(3)$, is a manifold of dimension 3, as may be verified by noting that matrices $A \in SO(3)$ are characterized by

$$\sum_{j=1}^{3} a_{ij}^2 = 1 \quad \text{for } i = 1, 2, 3,$$

$$\sum_{j=1}^{3} a_{1j} a_{2j} = 0,$$

$$\sum_{j=1}^{3} a_{1j} a_{3j} = 0,$$

$$\sum_{j=1}^{3} a_{2j} a_{3j} = 0.$$

This manifold is particularly important since it models the space of all orientations of a rigid body in space. Its generalization to the space of $n \times n$ matrices, denoted by $SO(n)$, is a manifold of dimension $n(n-1)/2$.

3.9.4 Manifolds with Boundary

Manifolds that we have defined above do not have boundaries in the sense that the neighborhood of every point is diffeomorphic to $\mathbb{R}^n$. In particular, this rules out objects such as solid spheres or solid cylinders whose interior has open neighborhoods diffeomorphic to open sets of $\mathbb{R}^n$ but whose surface (or boundary) is not diffeomorphic to $\mathbb{R}^n$. Roughly speaking, manifolds with boundary are manifolds that are locally like closed half spaces of $\mathbb{R}^n$ (rather than open subsets of $\mathbb{R}^n$). More precisely, define

$$H^n = \{(x_1, \ldots, x_n) : x_n \geq 0\}$$

and its boundary ∂H^n is defined to be the hyperplane $R^{n-1} \times \{0\} \subset \mathbb{R}^n$.

Definition 3.40 Manifolds with Boundary. *A subset $X \subset \mathbb{R}^k$ is called a smooth m-dimensional manifold with boundary if each $x \in X$ has a neighborhood $U \cap X$ diffeomorphic to an open subset $V \cap H^n$ of H^n. The boundary ∂X is the set of all points in X which correspond to ∂H^n under such a diffeomorphism.*

It is easy to verify that the *boundary* ∂X is a well defined smooth manifold of dimension $n - 1$ and the *interior* $X - \partial X$ is a smooth manifold of dimension n. The tangent space TX_x is defined to be a full n dimensional tangent space, even if x is a boundary point. A canonical way of generating manifolds with boundary is as follows:

Proposition 3.41. *Let M be a manifold without boundary and let* $g : M \mapsto \mathbb{R}$ *have 0 as a regular value. Then*

$$X = \{x \in M : g(x) \geq 0\}$$

is a smooth manifold, with boundary equal to $g^{-1}(0)$.

Proof: The proof follows the lines of Theorem 3.39.

Example 3.42.

1. *The* unit disk D^n *consisting of all* $x \in \mathbb{R}^n$ *with*

$$1 - \sum_{i=1}^{n} x_i^2 \geq 0$$

is a smooth manifold with boundary given by S^{n-1}.
2. *There are two standard ways of making manifolds with boundary out of a rectangular strip of paper by gluing opposite edges (see Figure 3.17). Simple gluing produces a* cylinder, *whereas gluing after one twist produces the closed* Möbius band. *Verify that the boundary of the cylinder is two copies of* S^1 *and that of the Möbius band is one copy of* S^1. *(Exercise for the reader: Think through what would happen if we had n twists before gluing).*

Theorem 3.43 Inverse Image of Regular Values. *Consider a smooth map* $f : X \mapsto N$ *from an m manifold with boundary, X to an n manifold N with* $m > n$. *If* $y \in N$ *is a regular value both for f and the restriction* $f|\partial X$, *then* $f^{-1}(y) \subset X$ *is a smooth* $(m - n)$ *manifold with boundary. Furthermore, the boundary* $\partial(f^{-1}(y))$ *is precisely equal to the intersection of* $f^{-1}(y)$ *with* ∂X.

Proof: Since the property to be proved is local, it suffices to consider the instance that f actually maps only the half space H^m to $\mathbb{R}^n$ with a regular value y. If $\bar{x} \in f^{-1}(y)$ is an interior point of H^m then as in Theorem 3.39 it follows that $f^{-1}(y)$ is a smooth manifold in the neighborhood of $\bar{x}$. Now, suppose that $\bar{x}$ is a boundary point. Choose a smooth map $g : U \mapsto \mathbb{R}^n$ that is defined throughout a neighborhood of $\bar{x} \in \mathbb{R}^m$ and coincides with f on $U \cap H^m$. Replacing U by a

FIGURE 3.17. Construction of a cylinder and a Möbius strip

smaller neighborhood, if necessary, we may assume that g has no critical points. Thus, $g^{-1}(y)$ is a smooth manifold of dimension $m - n$. Let $\pi : g^{-1}(y) \mapsto \mathbb{R}$ denote the coordinate projection,

$$\pi(x_1, \ldots, x_n) = x_n.$$

We claim that π has 0 as a regular value. Indeed, the tangent space of $g^{-1}(y)$ at a point $x \in \pi^{-1}(0)$ is equal to the null space of

$$dg_x = df_x : \mathbb{R}^m \mapsto \mathbb{R}^n$$

but the hypothesis that $f|\partial H^m$ is regular at x guarantees that this null space cannot be completely contained in $R^{n-1} \times 0$.

Hence, the set $g^{-1}(y) \cap H^m = f^{-1}(y) \cap U$, consisting of all $x \in g^{-1}(y)$ with $\pi(x) \geq 0$ is a smooth manifold by the previous proposition with boundary equal to $\pi^{-1}(0)$. $\qquad \Box$

We may now prove once again, by a different method, the Brouwer fixed point theorem, that we proved using techniques of degree theory. The proof technique is of independent interest.

Lemma 3.44. *Let X be a manifold with boundary. There is no smooth map f : $X \mapsto \partial X$ that leaves ∂X pointwise fixed.*

Proof: Suppose for the sake of contradiction that $f : X \mapsto \partial X$ is a map that leaves ∂X fixed. Let $y \in \partial X$ be a regular value for f. Since y is certainly a regular value for the identity map $f|\partial X$ also, it follows that $f^{-1}(y)$ is a smooth 1 dimensional manifold, with the boundary consisting of the single point

$$f^{-1}(y) \cap \partial X = \{y\}.$$

However, $f^{-1}(y)$ is also compact and the only compact one dimensional manifolds are disjoint unions of circles and segments. Hence, $\partial f^{-1}(y)$ must have an even number of points. This establishes the contradiction and proves that there is no map from X to ∂X that leaves the boundary pointwise fixed. $\qquad \Box$

We define the unit disk D^n in $\mathbb{R}^n$ as as

$$D^n = \{x \in \mathbb{R}^n | x_1^2 + \cdots + x_n^2 \leq 1\},$$

a smooth n–dimensional manifold with boundary given by the unit sphere

$$S^{n-1} = \{x \in \mathbb{R}^n | x_1^2 + \cdots + x_n^2 = 1\}.$$

Thus, by the preceding proposition it is not possible to extend the identity map on S^{n-1} to a smooth map $D^n \mapsto S^{n-1}$.

Lemma 3.45 Brouwer's Theorem for Smooth Maps. *Any smooth map g : $D^n \mapsto D^n$ has a fixed point $x \in D^n$ with $g(x) = x$.*

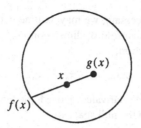

FIGURE 3.18. Construction for the proof of Brouwer's theorem for smooth maps

Proof: Suppose that g has no fixed point. For $x \in D^n$, let $f(x) \in S^{n-1}$ be the point nearer x than $g(x)$ on the line through x and $g(x)$ (see Figure 3.18). Then f is a smooth map from $D^n \mapsto S^{n-1}$ leaving S^{n-1} invariant. This is impossible by the preceding discussion ($f(x)$ can be written out explicitly as

$$f(x) = x + tu$$

where

$$u = \frac{x - g(x)}{|x - g(x)|}, \quad t = -x^T u + \sqrt{1 - |x|^2 + (x^T u)^2}$$

with $t > 0$). □

Theorem 3.46 Brouwer's theorem for continuous maps. *Any continuous map* $G : D^n \mapsto D^n$ *has a fixed point.*

Proof: This theorem is reduced to the preceding lemma by approximating G by a smooth mapping. Given $\epsilon > 0$, according to the Weierstrass approximation theorem [91], there is a polynomial function $P : \mathbb{R}^n \mapsto \mathbb{R}^n$ with $|P(x) - G(x)| < \epsilon$ for $x \in D^n$. To cut down the range of P to D^n, we define

$$P_1 := \frac{P(x)}{1 + \epsilon}.$$

Then $P_1 : D^n \mapsto D^n$ and $|P_1(x) - G(x)| < 2\epsilon$ for $x \in D^n$. Suppose that $G(x) \neq x$ for $x \in D^n$, then the function $|G(x) - x|$, which is continuous on D^n must take on a maximum value on D^n, say M. Choose $\epsilon < M/2$. Then $P_1(x) \neq x$ for $x \in D^n$. But $P_1(x)$ is a smooth map, since it is a polynomial. Thus a non-existence of a fixed point for P_1 contradicts the preceding lemma. □

3.10   Summary

In this chapter we have presented a set of mathematical tools for the study of nonlinear systems. The methods have been presented in abbreviated form and here we include some additional references. These methods included a review of linear algebra (for more on linear algebra see for example [287], [153], [233], [275]. We presented an introduction to the existence and uniqueness theorems for differential equations (for more on this subject see [125], [68]), including a study of how to "regularize", i.e., to define solutions in a relaxed sense for differential equations which do not meet the existence uniqueness conditions in the sense of Caratheodory [102]. We did not have a detailed discussion on numerical methods of integrating differential equations, for a good review of numerical solution techniques see [65]. However, we discussed a method of solution of differential equations called waveform relaxation [177], which is useful in circuit theoretic applications. We discussed the generalization of degree theory to $\mathbb{R}^n$ referred to as index theory and gave an introduction to differential topology. Our treatment of degree theory followed that of [278]. In particular, the reader may wish to study fixed point theorems in general Banach spaces, which are invaluable in applications. The introduction to differential topology is standard, and followed the style of [211]. There are several other additional good introductions to differential topology, such as [221], [34], [282]. There are some very interesting generalizations of Brouwer's fixed point theorem, such as the Kakutani fixed point theorem, which is of great importance in mathematical economics, see for example [80]. Also, the infinite dimensional counterpart of the Brouwer fixed point theorem, called the Rothe theorem is useful for proving certain kinds of nonlinear controllability results [260].

3.11 Exercises

Problem 3.1. Show that the norms $|.|_p$, $|.|_\infty$ are equivalent on $\mathbb{R}^n$.

Problem 3.2 Measure of a matrix [83].   Given a matrix $A \in \mathbb{C}^{n \times n}$, the *measure of matrix* A is defined as

$$\mu(A) := \lim_{s \downarrow 0} \frac{|I + sA| - 1}{s}. \tag{3.49}$$

By checking that for a given induced norm $| \cdot |$,

$$-|A| \leq \lim_{s \downarrow 0} \frac{|I + sA| - 1}{s} \leq |A| \quad \forall s,$$

show that the measure is well defined. Also, note that the measure of a matrix is the (one-sided) directional derivative of the $| \cdot |$ function along the direction A at I. Now prove that

1. For all $c \in \mathbb{R} \geq 0$, $\mu(cA) = c\mu(A)$.
2. $mu : \mathbb{C}^{n \times n} \mapsto \mathbb{R}$ is a convex function, that is, for all $A, B \in \mathbb{C}^{n \times n}$, $\lambda \in [0, 1]$

$$\mu(\lambda A + (1 - \lambda)B) \leq \lambda\mu(A) + (1 - \lambda)\mu(B).$$

3. $-\mu(-A) \leq \text{Re}\ (\lambda_i(A)) \leq \mu(A)$.

Problem 3.3 Matrix measures for l_1, l_2, l_∞ induced norms.   Prove that for the matrix norms induced by the l_1, l_2, l_∞ norms, we have

$$\mu_1(A) = \max_j \left[\mathrm{Re}\, (a_{jj} + \sum_{i, i \neq j} |a_{ij}| \right],$$

$$\mu_2(A) = \max_i \left[\lambda_i (A + A^*)/2 \right],$$

$$\mu_\infty(A) = \max_i \left[\mathrm{Re}\, (a_{ii} + \sum_{j, j \neq i} |a_{ij}| \right].$$

Problem 3.4. Use the definition of the matrix measure to prove that the solution of a linear time varying differential equation

$$\dot{x} = A(t)x$$

satisfies

$$|x(t_0)| \exp \left\{ - \int_{t_0}^t \mu(-A(-\tau)) d\tau \right\} \leq |x(t)| \leq |x(t_0)| \exp \left\{ \int_{t_0}^t \mu(A(\tau)) d\tau \right\}.$$

Problem 3.5. Give a proof of Proposition 3.15.

Problem 3.6. Give a proof of Proposition 3.25.

Problem 3.7 Waveform relaxation for implicit systems.   Work out the details required to verify that the Picard–Lindelöf iteration on (3.24) actually converges when ψ is a contraction map in its first argument. Assume that ψ is Lipschitz continuous in both its arguments and use a norm for functions on $[0, T]$ given by

$$|x(\cdot)|_w = \sup_{t \in [0,T]} \left(e^{-wt}(k_1 |x(t)| + |\dot{x}(t)|) \right).$$

Now, choose k_1, w so as to make the iteration map a contraction map.

Problem 3.8. Proposition 3.25 gave conditions under which implicit differential equations of the form of (3.23) could be reduced to normal form equations. Here we explore what could happen if the implicit equation cannot be made explicit. Consider:

$$\dot{x}_1 = x_2,$$
$$0 = x_2 - x_2^3 - x_1. \tag{3.50}$$

Draw a phase portrait of this system as follows: represent the algebraic equation as a curve in the x_1, x_2 plane. From the first equation trace the behavior of the state variable on the curve. Conjecture as to what the differential equation does at the state values $(\pm \frac{2}{3\sqrt{3}}, \pm \frac{1}{\sqrt{3}})$. The preceding differential equation could be thought of as a long hand version of

$$\dot{x}_1 - \dot{x}_1^3 - x_1 = 0.$$

It is a physically important system modeling, for instance, differential equations for the heartbeat and nerve impulse. Draw an RC circuit with a linear capacitor and a nonlinear resistor whose equation is (3.50).

Problem 3.9. Consider the effect of hysteretic switching on other phase portraits in Figure 3.4. Draw the resultant phase portraits for $\Delta \downarrow 0$.

Problem 3.10 Sliding mode control of uncertain linear systems. Consider the linear (time varying) control system

$$
\dot{x} = \begin{bmatrix} 0 & 1 & \cdots & 0 \\ 0 & 0 & \ddots & 0 \\ \vdots & \vdots & & 1 \\ -a_1(t) & -a_2(t) & \cdots & -a_n(t) \end{bmatrix} x + \begin{bmatrix} 0 \\ 0 \\ \vdots \\ 1 \end{bmatrix} u \qquad (3.51)
$$

The $a_i(t)$ are not known exactly, we know only that $\alpha_i \le a_i(t) \le \beta_i$. Also we have a bound v on the magnitude of $y_d^{(n)}$. The control objective is to get the output of the system $y(t) = x_1(t)$ to track a specified n times differentiable function $y_d(t)$. We define a sliding surface $S(t) \subset \mathbb{R}^n$ by

$$
s(x, t) = (x_n - y_d^{(n-1)}(t)) + c_1(x_{n-1} - y_d^{(n-2)}(t)) + \cdots + c_n(x_1 - y_d(t)) \quad (3.52)
$$

We will choose a control law of the form

$$
u = [\gamma_1(x), \gamma_2(x), \ldots, \gamma_n(x)] x + \sum_{i=1}^{n-1} k_i(x, t)(x_{i+1} - y_d^{(i)}) - k_n \operatorname{sgn}(s(x, t)),
$$

where sgn stands for the sign of $s(x, t)$, defined to be $+1$ when $s(x, t) > 0$ and -1 when $s(x, t) < 0$. Choose the constants $\gamma_i(x), k_i(x, t)$ so as to make $S(t)$ a sliding surface. Thus, choose u so that the system of (3.51) satisfies the global sliding condition, namely,

$$
\frac{d}{dt} s^2(x, t) \le \epsilon |s(x, t)|
$$

for some ϵ. This will guarantee that we reach the sliding surface in finite time. Explain why. Now show that once on the sliding surface, we can make the tracking error $e(t) = x_1(t) - y_d(t)$ go to zero asymptotically by a clever choice of the c_i.

Simulate this example for several choices of $a_i(t)$ and the bounds α_i and β_i. What do you notice if the bounds are far apart, that is, $\beta_i - \alpha_i$ is large? Can you comment on this.

Problem 3.11 Peano–Baker Series. Consider a linear time varying dynamical system

$$
\dot{x} = A(t)x(t), \quad x(t_0) = x_0.
$$

Verify that the solution of this differential equation is of the form $x(t) = \Phi(t, t_0)x_0$ where $\Phi(t, t_0) \in \mathbb{R}^{n \times n}$, called the fundamental matrix is the solution of the matrix differential equation

$$\dot{X} = A(t)X, \quad X(0) = I.$$

Now prove that if $A(\cdot)$ is bounded, we have the following convergent series expansion for $\Phi(t, \tau)$, called the Peano–Baker series:

$$\Phi(t, \tau) = I + \int_\tau^t A(\sigma_1)d\sigma_1 + \int_\tau^t A(\sigma_1) \int_\tau^{\sigma_1} A(\sigma_2)d\sigma_2 d\sigma_1$$
$$+ \cdots + \int_\tau^t A(\sigma_1) \int_\tau^{\sigma_1} A(\sigma_2) \cdots \int_\tau^{\sigma_{k-1}} A(\sigma_k)d\sigma_k \cdots d\sigma_1 + \cdots.$$

Problem 3.12 Numerical solution techniques: 4th-order Runge–Kutta. The Taylor series expansion for the solution of the differential equation

$$\dot{x} = f(x, t) \quad x(0) = x_0$$

at time $t_n = nh$ with $x_n := x(nh)$ is given by

$$x_{n+1} = x_n + h\dot{x}_n + \frac{h^2}{2!}\ddot{x}_n + \cdots$$

with

$$\dot{x}_n = f(x_n, t_n), \quad \ddot{x}_n = D_1 f(x_n, t_n) f(x_n, t_n) + D_2 f(x_n, t_n) + \cdots,$$

where D_1, D_2 stand for the derivatives with respect to the first and second argument respectively. Nth-order *Runge–Kutta* methods seek to approximate the first N terms of the Taylor series using not the derivatives of f (which may be hard to compute, especially if f is not given in functional form), but the values of f at some intermediate values of x, t between (x_n, t_n) and (x_{n+1}, t_{n+1}). Prove that the scheme described by

$$x_{n+1} = x_n + \frac{h}{6}(g_1 + 2g_2 + 2g_3 + g_4), \tag{3.53}$$

where

$$g_1 = f(x_n, t_n),$$
$$g_2 = f\left(x_n + \frac{hg_1}{2}, t_n + \frac{h}{2}\right),$$
$$g_3 = f\left(x_n + \frac{hg_2}{2}, t_n + \frac{h}{2}\right),$$
$$g_4 = f(x_n + hg_3, t_n + h),$$

matches the first four coefficients of the Taylor series for x_{n+1}.

Problem 3.13. Prove Proposition 3.27.

Problem 3.14. Prove that if $x \in \mathbb{R}^n$ satisfies a linear differential equation of the form of equation (3.38), then $x^{(i)}$ also satisfies a linear differential equation. Prove the same conclusion for $x^{[i]}$ as well.

Problem 3.15 Carleman Linearization solutions. Consider the approximate Carleman linearization of (3.40) given by

$$\frac{dx^{\oplus}}{dt} = \begin{bmatrix} A_{11} & A_{12} & \cdots & A_{1N} \\ 0 & A_{22} & \cdots & A_{2N} \\ 0 & 0 & \cdots & A_{3N} \\ \vdots & \vdots & \vdots & \vdots \\ 0 & 0 & \cdots & A_{N1} \end{bmatrix} x^{\oplus} \tag{3.54}$$

with the appropriate initial conditions on $x^{\oplus}$. Show, given $T > 0$, that if the initial conditions are small enough that for $t \in [0, T]$ the difference between the first entry of $x^{\oplus}(t)$ solving (3.54) and the exact solution $x(t)$ of (3.40) can be bounded by a growing exponential.

Problem 3.16 Carleman (bi-)linearization of control systems. Repeat the steps of the Carleman linearization for the *control system* with single input $u \in \mathbb{R}$ given by

$$\dot{x} = \sum_{i=1}^{\infty} A_i x^{(i)} + \sum_{i=0}^{\infty} B_i x^{(i)} u \tag{3.55}$$

to get:

$$\frac{dx^{\oplus}}{dt} = \begin{bmatrix} A_{11} & A_{12} & \cdots & A_{1N} \\ 0 & A_{22} & \cdots & A_{2N} \\ 0 & 0 & \cdots & A_{3N} \\ \vdots & \vdots & \vdots & \vdots \\ 0 & 0 & \cdots & A_{N1} \end{bmatrix} x^{\oplus}$$

$$+ \begin{bmatrix} B_{11} & B_{12} & \cdots & B_{1N} \\ 0 & B_{22} & \cdots & B_{2N} \\ 0 & 0 & \cdots & B_{3N} \\ \vdots & \vdots & \vdots & \vdots \\ 0 & 0 & \cdots & B_{N1} \end{bmatrix} x^{\oplus} u + \begin{bmatrix} B_{10} \\ 0 \\ 0 \\ \vdots \\ 0 \end{bmatrix} u. \tag{3.56}$$

This is referred to as the *Carleman bilinearization* of the control system, since it is not linear in x but rather bi-linear in x, u. Give formulas for the matrices B_{ij} in (3.56).

Problem 3.17. Prove that the definition of the derivative Df_x given in this chapter does not depend on the choice of coordinate charts in its definition.

Problem 3.18. Prove Proposition 3.37.

Problem 3.19. Prove that if $f : M \mapsto N$ is a smooth map and y a regular value of the map and M a compact manifold, then the set $f^{-1}(y)$ is finite.

Problem 3.20. Prove that if $f : M \mapsto N$ is a smooth map and y a regular value of the map and M a compact manifold, then the set $f^{-1}(y)$ is locally constant, i.e., constant in a neighborhood of y.

Problem 3.21 Stack of records theorem.   Prove that if $f : M \mapsto N$ is a smooth map and y a regular value of the map and M a compact manifold and the set $f^{-1}(y) = \{x_1, \ldots, x_k\}$, then there exist neighborhoods $U_1, \ldots, U_k$ of $x_1, \ldots, x_k$ respectively and a neighborhood V of y such that f is a diffeomorphism of U_i into V.

Problem 3.22. Verify that $SL(n)$ is a manifold of dimension $n^2 - 1$. In the process, determine the critical values of the determinant function of a matrix.

Problem 3.23. Prove that $SO(n)$ is a manifold of dimension $n(n - 1)/2$.

Problem 3.24. Show that the subset of $\mathbb{R}^{n \times m}$ matrices of rank r is a manifold of dimension $mr + nr - r^2$.

Problem 3.25 Global inverse function theorem [82].   A theorem of Palais states that a map $f : \mathbb{R}^n \mapsto \mathbb{R}^n$ is globally surjective if it is locally surjective, that is $Df(x)$ is surjective as a linear map from $T_x \mathbb{R}^n \mapsto T_x \mathbb{R}^n$, and f is *proper*, that is, $f^{-1}(K)$ is compact when $K \subset \mathbb{R}^n$ is compact.

Now, show that if there exists a function $m(\cdot) : \mathbb{R}_+ \mapsto \mathbb{R}_+$ such that for some matrix measure μ (see Problem 3.2)

$$\mu(Df(x)) \le -m(|x|) \quad \forall x,$$

$$\int_0^\infty m(\alpha) \, d\alpha = \infty.$$

then f is globally surjective.

Problem 3.26 Lie Group $SO(3)$.   Give a geometric description of the tangent space to $SO(3)$ the space of unitary matrices in $\mathbb{R}^{3 \times 3}$, that is matrices $U \in \mathbb{R}^{3 \times 3}$ satisfying $U^T U = I$ and determinant $(U) = +1$. Try to relate the tangent space at an arbitrary $U \in SO(3)$ to the tangent space at the identity. More generally, $SO(3)$ is an example of a manifold which is also a group. Such manifolds are called **Lie groups** and their tangent space at the identity (I) is called the associated **Lie algebra**. This is discussed in greater detail in Chapter 8.

Problem 3.27 More matrix Lie groups.   In this problem we will discuss some interesting matrix Lie groups:

1. Consider the set of all nonsingular $n \times n$ matrices. Denote it $GL(n)$ (for general linear group). Prove that $GL(n)$ is a Lie group. What is its dimension?
2. Consider the set of all $n \times n$ unitary matrices, i.e., matrices such that $A^T = A^{-1}$ with determinant $+1$. Denote it by $SO(n)$ (for special orthogonal group). Prove

that $SO(n)$ is a Lie group. What is its dimension? What can you say about unitary matrices of determinant -1?

3. Consider matrices in $\mathbb{R}^{(n+1) \times (n+1)}$ of the form

$$\begin{bmatrix} A & b \\ 0 & 1 \end{bmatrix}$$

with $A \in SO(n)$, $b \in \mathbb{R}^n$. Denote it by $SE(n)$ (for special Euclidean group). Prove that $SE(n)$ is a Lie group; in particular, be sure to write a formula for the inverse of elements in $SE(n)$. What is the dimension of $SE(n)$?

Problem 3.28. Prove Proposition 8.3.

Problem 3.29. Let U, ψ, and V, ϕ be two different coordinate charts for neighborhoods of a point $p \in M$, a smooth n-dimensional manifold with local coordinates x, z. Denote the coordinate transformation $z = \phi \circ \psi^{-1}$ by $S(x)$. If a given $X_p \in T_p M$ may be represented as $\sum_{i=1}^{n} \alpha_i \frac{\partial}{\partial x_i}|_p$ and as $\sum_{i=1}^{n} \beta_i \frac{\partial}{\partial z_i}|_p$, then prove that the coefficients α_i and β_i are related by

$$\beta_i = \sum_{j=1}^{n} \frac{\partial S_i}{\partial x_j} \alpha_j.$$

Problem 3.30. Find a geometric description of the tangent space of a circle S^1, a torus $S^1 \times S^1$ and a cylinder $S^1 \times \mathbb{R}$. More generally, what can you say about the tangent space of the product of two manifolds $M_1 \times M_2$.

Problem 3.31 Grassmann and Stiefel Manifolds. Define the *Grassmann manifold* $G(m, n)$ to be the space of all m dimensional subspaces of $\mathbb{R}^n$. Verify that an element of $G(m, n)$ may be represented by an equivalence class of matrices $\mathbb{R}^{n \times m}$ of rank m with the equivalence class defined by

$$A \sim B \Leftrightarrow \exists T \in \mathbb{R}^{m \times m} \quad \text{non-singular, such that} \quad A = BT$$

In the equivalence class above T refers to change of basis of the subspace. Prove that in the topology inherited from $\mathbb{R}^n$, $G(m, n)$ is a manifold. What is its dimension? Can you identify $G(1, 2)$, $G(n, n + 1)$?

Define a *Stieffel manifold* $S(m, n)$ to be a Grassmann manifold with orthogonal bases. Thus, elements of $S(m, n)$ are an equivalence class of matrices $U \in \mathbb{R}^{n \times m}$ satisfying $U^T U = I \in \mathbb{R}^{m \times m}$ with the equivalence class defined by

$$U \sim V \Leftrightarrow \exists W \in SO(m) \quad \text{such that} \quad U + VW$$

What is the dimension of $S(m, n)$? Can you identify $S(n - 1, n)$, $S(n, n)$?

Problem 3.32 Morse lemma. Let $f : \mathbb{R}^n \mapsto \mathbb{R}$ and $f(0) = 0$. Prove that in a neighborhood of $x = 0$ the function f may be written as

$$f(x) = \sum_{i=1}^{n} x_i g_i(x), \tag{3.57}$$

where $g_i(0) = \frac{\partial f}{\partial x_i}|_0$. Now, let x_0 be a non-degenerate critical point of f at x_0 (that is the Hessian $D^2 f(x_0)$ is nonsingular). Prove using the preceding fact that in a neighborhood U of x_0 there is a diffeomorphism from $x \in \mathbb{R}^n \mapsto y \in \mathbb{R}^n$ such that

$$f(y) = f(x_0) - y_1^2 - \cdots - y_p^2 + y_{p+1}^2 + \cdots + y_n^2. \tag{3.58}$$

This lemma, called the Morse lemma, says that non degenerate critical points are locally quadratic. The signature of the Hessian is contained in the p minuses and $n - p$ pluses above. Think about what may happen if the critical point is degenerate. Consider on $\mathbb{R}^2$ the functions x_1^2 (pig trough), $x_1^3 - 3x_1x_2$ (monkey saddle), $x_1^2 x_2^2$ (crossed pig trough).

Problem 3.33 Degenerate fixed points of scalar functions.   Prove that if $f : \mathbb{R} \mapsto \mathbb{R}$ is such that

$$f(0) = Df(0) = \cdots = D^{k-1} f(0) = 0 \quad D^k f(0) \neq 0,$$

then there exists a local change of coordinates $x \in \mathbb{R} \mapsto y \in \mathbb{R}$ such that f takes the form

$$y^k, \quad k \text{ odd},$$

$$\pm y^k, \quad k \text{ even}.$$

The sign in the second equation above is that of $D^k f(0)$.

Problem 3.34 Splitting lemma.   The preceding problem 3.33 happily told that us that degenerate functions of scalar variables could be transformed to polynomials of their first non vanishing order. For multi-variable functions, the most that one can do is the following construction, known as the *splitting lemma*. Let $f : \mathbb{R}^n \mapsto \mathbb{R}$ and x_0 be a critical point of f. Let the Hessian of f have rank r. Use a modification of the proof of the Morse lemma to first transform the coordinates into one where the Hessian is diagonal $(\pm 1, \ldots, \pm 1, 0, \ldots 0)$ and then establish that there is a diffeomorphism from $\mathbb{R}^n \mapsto \mathbb{R}^n$ such that

$$f(y) = \pm y_1^2 \cdots \pm y_r^2 + \hat{f}(y_{r+1}, \ldots, y_n). \tag{3.59}$$

where the derivative and Hessian of $\hat{f}$ vanish at $y = 0$.

4
Input–Output Analysis

In this chapter we introduce the reader to various methods for the input-output analysis of nonlinear systems. The methods are divided into three categories:

1. *Optimal Linear Approximants for Nonlinear Systems.* This is a formalization of a technique called the describing function technique, which is popular for a quick analysis of the possibility of oscillation in a feedback loop with some nonlinearities in the loop.
2. *Input-output Stability.* This is an extrinsic view to the stability of nonlinear systems answering the question of when a bounded input produces a bounded output. This is to be compared with the intrinsic or state space or Lyapunov approach to stability in the next two chapters.
3. *Volterra Expansions for Nonlinear Systems.* This is an attempt to derive a rigorous "frequency domain" representation of the input output behavior of certain classes of nonlinear systems.

In some sense, this chapter represents the classical approach to input-output non-linear systems, with many fundamental contributions by Sandberg, Zames, Popov, Desoer, and others. An excellent classical reference to this is the book of Desoer and Vidyasagar [83]. The current modernized treatment of the topics, while abbreviated, reveals not only the delicacy of the results but also a wealth of interesting analysis techniques.

4.1   Optimal Linear Approximants to Nonlinear Systems

In this section we will be interested in trying to approximate nonlinear systems by linear ones, with the proviso that the "optimal" approximating linear system varies as a function of the input. We start with single-input single-output nonlinear systems. More precisely, we view a nonlinear system, in an input output sense, as a map N from $C[0, \infty[$, the space of continuous functions on $[0, \infty[$, to $C([0, \infty[)$. Thus, given an input $u \in C([0, \infty[)$, we will assume that the output of the nonlinear system N is also a continuous function, denoted by $y_N(\cdot)$, defined on $[0, \infty[$:

$$y_N = N(u) \in C([0, \infty[)$$

We will now optimally approximate the nonlinear system for a given reference input $u_0 \in C([0, \infty[)$ by the output of a linear system. The class of linear systems, denoted by W, which we will consider for optimal approximations are represented in convolution form as integral operators. Thus, for an input $u_0 \in C([0, \infty[)$, the output of the linear system W is given by

$$y_L(t) = (W(u_0))(t) := \int_{-\infty}^{\infty} w(t)u_0(t - \tau)d\tau. \tag{4.1}$$

with the understanding that $u(t) \equiv 0$ for $t \leq 0$. The convolution kernel is chosen to minimize the *mean squared error* defined by

$$e(w) = \lim_{T \to \infty} \frac{1}{T} \int_0^T [Wu_0(\tau) - Nu_0(\tau)]^2 d\tau \tag{4.2}$$

The following assumptions will be needed to solve the optimization problem of (4.2):

1. *Bounded-Input Bounded-Output (b.i.b.o.) Stability.* For given b, there exists $m_0(b)$ such that $|u(t)| < b \Rightarrow |Nu(t)| < m_0(b)$ for all $t \in [0, \infty[$. Thus, a bounded input to the nonlinear system is assumed to produce a bounded output.
2. *Causal, Stable Approximators.* The class of approximating linear systems is assumed *causal* and *bounded input bounded output stable*, i.e., $w(t) \equiv 0 \forall t < 0$, and

$$\int_0^{\infty} |w(\tau)| d\tau < \infty. \tag{4.3}$$

Equation (4.3) guarantees that a bounded input $u(\cdot)$ to the linear system W produces a bounded output.
3. *Stationarity of Input.* The input $u_0(\cdot)$ is stationary, i.e.,

$$\lim_{T \to \infty} \frac{1}{T} \int_s^{s+T} |u_0(t)|^2 dt \tag{4.4}$$

exists uniformly in s. The terminology of a *stationary* deterministic signal is due to Wiener in his theory of generalized harmonic analysis [327]:

Definition 4.1 Stationarity and Autocovariance. *A signal $u : \mathbb{R}_+ \mapsto \mathbb{R}$ is said to be* stationary *if the following limit exists uniformly in s:*

$$R_u(t) := \lim_{T \to \infty} \frac{1}{T} \int_s^{s+T} u(\tau)u(t + \tau)\,d\tau, \tag{4.5}$$

in which instance, the limit $R_u(t)$ (independent of s) is called the autocovariance *of u.*

The concept of autocovariance is well known in the theory of stochastic systems, where it is often referred to as *autocorrelation*. There is a strong analogy between (4.5) and $R_u^{stoch}(t)$ defined to be the expected value $E[u(\tau)u(t + \tau)]$ when u is a wide-sense stationary *ergodic* process.[1] But, we emphasize that the definition (4.5) is completely deterministic. The definition may also be replaced by

$$R_u(t) := \lim_{T \to \infty} \frac{1}{T} \int_s^{s+T} u(\tau - t)u(\tau)\,d\tau.$$

Thus, it follows that

$$R_u(t) = R_u(-t).$$

We need the convergence in the definition of equation (4.5) to be uniform in s in order to guarantee that the length of the window over which the integrand is averaged does not grow with s. Related to the definition of autocovariance is the following:

Definition 4.2 Cross Covariance. *Let $u(\cdot), y(\cdot) : \mathbb{R} \mapsto \mathbb{R}$ be stationary. The* cross covariance *between u and y is defined to be the following limit, uniformly in s:*

$$R_{yu}(t) := \lim_{T \to \infty} \frac{1}{T} \int_s^{s+T} y(\tau)u(t + \tau)\,d\tau. \tag{4.6}$$

As before, we may rewrite this as

$$R_{yu}(t) := \lim_{T \to \infty} \frac{1}{T} \int_s^{s+T} y(\tau - t)u(\tau)\,d\tau.$$

The autocovariance is a positive definite function. That is, for all $t_1, \ldots, t_k \in \mathbb{R}$, and $c_1, \ldots, c_k \in \mathbb{R}$ we have that

$$\sum_{i,j=1}^k c_i c_j R_u(t_i - t_j) \geq 0$$

(this can be proved starting from the definition (see Problem 4.2)). Thus, the Fourier transform of the autocovariance function $R_u(t)$ is a positive function $S_u(\omega)$,

---

[1] Indeed, for a wide-sense stationary stochastic process if $u(t, \omega)$ is a single sample function of a stochastic process (here, ω denotes a sample point of the underlying probability space), the autocovariance $R_u(t, \omega)$ exists and is equal to $R_u^{stoch}(t)$ for almost all ω or sample functions.

referred to as the *spectral measure* or *power spectral density* of the signal u:

$$S_u(\omega) = \int_{-\infty}^{\infty} e^{-j\omega\tau} R_u(\tau)\, d\tau, \tag{4.7}$$

and the corresponding inverse transform is

$$R_u(t) = \frac{1}{2\pi} \int_{-\infty}^{\infty} e^{j\omega t} S_u(\omega)\, d\omega, \tag{4.8}$$

Some properties of spectral measure are explored in the exercises. Here, we will prove the following proposition relating the autocovariance of the input and output signals of a stable, linear, time-invariant system:

Proposition 4.3 Linear Filter Lemma. *Let $y = H(u)$ where H is a causal, b.i.b.o. stable linear system, i.e., its convolution kernel $h(t)$ is identically 0 for $t \leq 0$, and $h(\cdot) \in L_1$. Then if u is stationary with autocovariance $R_u(t)$, then so is $y(t)$. Further, the cross correlation between u and y is given by*

$$R_{yu}(t) = \int_{-\infty}^{\infty} h(\tau_1) R_u(t + \tau_1)\, d\tau_1, \tag{4.9}$$

and the cross spectral measure is

$$S_{yu}(\omega) = \hat{h}^*(j\omega) S_u(\omega).$$

Proof: First, we establish that $y(\cdot)$ is stationary. By definition,

$$y(t) = \int_{-\infty}^{\infty} h(s) u(t - s)\, ds.$$

Thus, we have that $R_y(t)$ is given by

$$\lim_{T\to\infty} \frac{1}{T} \left(\int_s^{s+T} \left[\int_{-\infty}^{\infty} \int_{-\infty}^{\infty} h(\tau_1) u(\tau - \tau_1) h(\tau_2) u(\tau + t - \tau_2)\, d\tau_1\, d\tau_2 \right] d\tau \right).$$

Interchanging the order of integration between τ and τ_2 yields the following formula for $R_y(t)$:

$$\lim_{T\to\infty} \frac{1}{T} \left(\int_{-\infty}^{\infty} \int_{-\infty}^{\infty} \int_s^{s+T} h(\tau_1) h(\tau_2) u(\tau - \tau_1) u(t + \tau - \tau_2)\, d\tau\, d\tau_1\, d\tau_2 \right).$$

For all T, the integral above exists absolutely, since $h(\cdot) \in L_1$; hence, we can interchange the order of integration to obtain

$$R_y(t) = \int_{-\infty}^{\infty} \int_{-\infty}^{\infty} h(\tau_1) h(\tau_2) R_u(t + \tau_1 - \tau_2)\, d\tau_1\, d\tau_2.$$

establishing the stationarity of $y(\cdot)$. To show that there exists a cross correlation between y and u, one proceeds similarly:

$$\frac{1}{T} \int_s^{s+T} y(\tau) u(t + \tau)\, d\tau = \frac{1}{T} \int_s^{s+T} \left[\int_{-\infty}^{\infty} h(\tau_1) u(\tau - \tau_1)\, d\tau_1 \right] u(t + \tau)\, d\tau.$$

As before for all T, the integral above exists absolutely, since $h(\cdot) \in L_1$; hence, we can interchange the order of integration to obtain

$$\int_{-\infty}^{\infty} h(\tau_1) \, d\tau_1 \left[\frac{1}{T} \int_{s}^{s+T} u(t+\tau)u(\tau - \tau_1) \, d\tau \right].$$

The expression in the brackets converges uniformly to $R_u(t + \tau_1)$, so that we have

$$R_{yu}(t) = \int_{-\infty}^{\infty} h(\tau_1) R_u(t + \tau_1) \, d\tau_1. \qquad \square$$

Using these foregoing definitions and propositions, we have the following result:

Theorem 4.4 Optimal Causal, Stable Linear Approximation. *There exists an optimal causal, b.i.b.o. stable linear system approximant solving the optimization problem of (4.2), with convolution kernel $w^*(\cdot)$. Then w^* satisfies*

$$\phi_{w^*}(\tau) \equiv \phi_N(\tau), \qquad (4.10)$$

where $\phi_{w^*}(\tau)$ is short hand for the cross correlation between $u_0(\cdot)$ and $W^*(u_0(\cdot))$ and $\phi_N(\tau)$ is the cross correlation between $u_0(\cdot)$ and $N u_0(\cdot)$ defined exactly as in the previous equation with $W^* u_0$ replaced by $N u_0$.

Proof: Let w^* be the convolution kernel of the optimal linear system. We know that an optimal linear causal b.i.b.o. stable approximant exists, since the criterion of (4.2) is quadratic in the convolution kernel $w(\cdot)$. Now, for any w we should have $e(w^*) \le e(w)$. Thus, we have that $e(w) - e(w^*)$ is given by

$$e(w) - e(w^*) = \lim_{T \to \infty} \frac{1}{T} \left[\int_0^T (N u_0(t) - W u_0(t))^2 - (N u_0(t) - W^* u_0(t))^2 \, dt \right],$$

Since w^* is the optimal approximant, we have that

$$0 \le \lim_{T \to \infty} \frac{1}{T} \left[\int_0^T (D u_0)(t)^2 - 2 D u_0(t)(N u_0(t) - W^* u_0(t)) \, dt \right], \qquad (4.11)$$

where $D u_0(t) = W u_0(t) - W^* u_0(t)$ is the linear difference operator between W and W^*. For the right hand side of (4.11) to be nonnegative for all $D u_0(t)$ we must have

$$\lim_{T \to \infty} \frac{1}{T} \int_0^T D u_0(t)(N u_0(t) - W^* u_0(t)) \, dt = 0. \qquad (4.12)$$

Now note that the convolution expression for $W u_0$ may be written as

$$\int_{-\infty}^{\infty} w(t-\tau)u_0(\tau) \, d\tau = \int_{-\infty}^{t} w(t-\tau)u_0(\tau) \, d\tau = \int_0^t w(\tau)u_0(t-\tau) \, d\tau \qquad (4.13)$$

using the fact that $w(t) \equiv 0$, $u_0(t) \equiv 0$ for $t \le 0$. Using (4.13) in (4.12) yields

$$\lim_{T \to \infty} \frac{1}{T} \int_0^T \left\{ \int_0^t (w(\tau) - w^*(\tau))u_0(t-\tau) \, d\tau \right\} (W^* u_0 - N u_0) \, dt \qquad (4.14)$$

To interchange the orders of integration it is convenient to define

$$g_T(\tau) = \frac{1}{T} \int_\tau^T u_0(t - \tau)(W^* u_0 - N u_0)dt.$$

Then equation (4.14) is

$$\lim_{T \to \infty} \int_0^T (w(\tau) - w^*(\tau))g_T(\tau)d\tau = 0. \tag{4.15}$$

Now, since both the linear systems W and W^* are b.i.b.o. stable, we have that

$$\int_0^T |w(\tau) - w^*(\tau)|d\tau < \infty.$$

Further, the assumption on N guarantees that if $u_0(t) < b$ for all t and that $u_0(\cdot)$ is stationary, then $|g_T(t)| < K$ for all t. Then, we may use Fatou's lemma to interchange the order of integration to yield

$$\int_0^\infty (w(\tau) - w^*(\tau)) \lim_{T \to \infty} g_T(\tau)d\tau = 0 \tag{4.16}$$

Since this equation should hold for arbitrary $w^* - w$, we have

$$\lim_{T \to \infty} \frac{1}{T} \int_0^T u_0(t - \tau)(W^* u_0 - N u_0)dt. \tag{4.17}$$

Equation (4.17) is a restatement of the optimality condition of (4.10). It is instructive to note that equation (4.17) is the usual outcome of a linear least squares optimization, namely, the approximation error $N u_0 - W^* u_0$ is orthogonal to the input u_0 (the inner product is given by the cross covariance, that is, the average integral as $T \to \infty$). □

4.1.1   Optimal Linear Approximations for Memoryless, Time-Invariant Nonlinearities

It is useful to apply the theorem above to obtain optimal linear approximations to some memoryless, time invariant nonlinear operators N, i.e., $N(u_0)(t) = n(u_0(t))$ for some function $n : \mathbb{R} \mapsto \mathbb{R}$.

- *Constant Reference Input*
  In this case we have that $u_0(t) \equiv k$ and

$$\phi_N(\tau) = \lim_{T \to \infty} \frac{1}{T} \int_0^T kn(k)d\tau = kn(k).$$

Thus W^* is *any causal, b.i.b.o. stable linear system* that has dc gain of $\frac{n(k)}{k}$ since $\phi_{W^*}(\tau) \equiv kn(k)$. A particularly easy choice is a linear operator with convolution kernel $\frac{n(k)}{k}\delta(t)$, though the kernel is not an integrable function as required in the preceding derivation. However, here, as in other places in the engineering literature, we will use these generalized functions (the reader

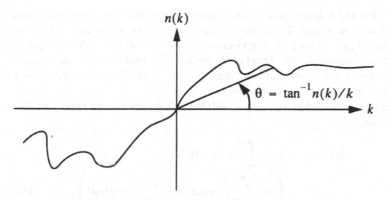

FIGURE 4.1. Quasi-linearization of a nonlinear operator

may justify their usage as limits of integrable functions). This approximation
is called the quasi-linearization and is shown in Figure 4.1.

- *Sinusoidal Reference Input*

Let $u_0(t) = a \sin(\omega t)$. Now, since N is a memoryless time-invariant operator
it is easy to verify that $(Nu_0)(t)$ is also periodic with the same period, and
consequently has a Fourier series (for $t \geq 0$) given by

$$(Nu_0)(t) = n(u_0(t)) = c_0 + \sum_{i=1}^{\infty} c_i \cos(i\omega t) + \sum_{i=1}^{\infty} d_i \sin(i\omega t). \quad (4.18)$$

We begin with the case that $n(\cdot)$ is an odd nonlinearity, that is, $n(x) = -n(-x)$.
In this instance, we have $c_i = 0$ for all i. Now, using the formula for the cross
correlation and the orthogonality of sinusoids at integrally related frequencies,
we get

$$\phi_N(\tau) = \lim_{T \to \infty} \frac{1}{T} \int_0^{\infty} a \sin \omega(t - \tau) \left(\sum_{i=1}^{\infty} d_i \sin(i\omega t) \right)$$
$$= \frac{d_1 a}{2} \cos(\omega \tau). \quad (4.19)$$

For an input $u_0(t) = a \sin(\omega t)$, the optimal linear system W^* will produce an
output of the form

$$W^* u_0(t) = \gamma_1(\omega) a \sin(\omega t) + \gamma_2(\omega) a \cos(\omega t),$$

where $\gamma_1(\omega) + j\gamma_2(\omega) = W^*(j\omega)$, with $W^*(j\omega)$ being the Fourier transform
of the convolution kernel $w^*(\cdot)$. Thus, the cross covariance $\phi_{w^*}(\tau)$ is

$$\phi_{w^*}(\tau) = \frac{\gamma_1(\omega) a^2}{2}.$$

By inspection it follows that the optimal linear system should have $W^*(j\omega) = \frac{d_1}{a} + j\gamma_2(\omega)$ with γ_2 being arbitrary. One may set $\gamma_2 = 0$ and take as choice of
convolution kernel $w^*(t) = \frac{d_1}{a} \delta(t)$. It is worthy of note, that *the optimal causal*

b.i.b.o. stable linear approximant depends on both the amplitude and frequency of the input signal. The optimal linear system is one whose equivalent gain at frequency ω is the first component of the Fourier series divided by a, that is, the first harmonic component of the output of the nonlinear system is balanced by the output of the optimal linear approximation. This is referred to as the **principle of harmonic balance**.

In the instance that $n(\cdot)$ is not odd, the calculation for the cross correlation yields

$$\phi_N(\tau) = \lim_{T \to \infty} \frac{1}{T} \int_0^\infty a \sin \omega(t - \tau)$$

$$\times \left(c_0 + \sum_{i=1}^\infty c_i \cos(i\omega t) + \sum_{i=1}^\infty d_i \sin(i\omega t) \right) \qquad (4.20)$$

$$= \frac{d_1 a}{2} \cos(\omega\tau) + \frac{c_1 a}{2} \sin(\omega\tau);$$

so that the optimal linear system needs to have Fourier transform

$$W^*(j\omega) = \frac{d_1}{a} + j\frac{c_1}{a}. \qquad (4.21)$$

Thus the optimal linear system maintains *harmonic balance to the first harmonic of the output* of the nonlinear system. If the approximating linear system is a memoryless linear system, it is characterized by its gain alone, referred to as *equivalent gain*. The equivalent gain or **describing function** is the complex number

$$\eta(a, \omega) = \frac{d_1(\omega, a)}{a} + j\frac{c_1(\omega, a)}{a}$$

Thus, the describing function is a complex number which is a function of both the frequency ω and amplitude a of the input necessary to match the first harmonic of the output of the nonlinear system.

Properties of Describing Functions

Proposition 4.5. *If the nonlinearity $n(\cdot)$ is memoryless and time-invariant, then $\eta(a, \omega)$ is independent of ω.*

Proof: See Exercises.

As a consequence of this proposition we can write the following formula for the Fourier series coefficients explicitly, showing the independence on ω:

$$c_1(a) = \frac{1}{\pi} \int_0^{2\pi} n(a \sin \psi) \sin \psi \, d\psi,$$

$$(4.22)$$

$$d_1(a) = \frac{1}{\pi} \int_0^{2\pi} n(a \sin \psi) \cos \psi \, d\psi.$$

Proposition 4.6. *Suppose that the function $n(\cdot)$ is odd and is sector bounded, i.e., it satisfies*

$$k_1\sigma^2 \leq \sigma n(\sigma) \leq k_2\sigma^2 \ \forall \sigma \in \mathbb{R}. \tag{4.23}$$

Then the describing function $\eta(a)$ satisfies

$$k_1 \leq \eta(a) \leq k_2. \tag{4.24}$$

Proof: See Exercises.

4.1.2 Optimal Linear Approximations for Dynamic Nonlinearities: Oscillations in Feedback Loops

We have studied how harmonic balance can be used to obtain the describing function gain of simple nonlinear systems – memoryless nonlinearities, hysteresis, dead zones, backlash, etc. The same idea may be extended to dynamic nonlinearities. Consider, for example,

$$\ddot{y} + 3y^2\dot{y} + y = u \tag{4.25}$$

with forcing $u = A\sin(\omega t)$. If the nonlinear system produces a periodic output (this is a very nontrivial assumption, since several rather simple nonlinear systems behave chaotically under periodic forcing), then one may write the solution $y(t)$ in the form

$$y(t) = \sum_{k=1}^{\infty} Y_k \sin k(\omega t + \theta_k).$$

Using this solution in (4.25) simplifying and equating first harmonic terms yields

$$(1 - \omega^2)Y_1 \sin(\omega t + \theta_1) + \frac{3}{4}\omega Y_1^3 \cos(\omega t + \theta_1) = A\sin(\omega t). \tag{4.26}$$

Simplifying this equation yields

$$(1 - \omega^2)Y_1^2 + \left(\frac{3}{4}\omega Y_1^3\right)^2 = A^2$$

$$\theta_1 = -\tan^{-1}\frac{3\omega Y_1^2}{4(1 - \omega^2)}. \tag{4.27}$$

Thus, if one were to find the optimal linear, causal, b.i.b.o. stable approximant system of the nonlinear system (4.25), it would be have Fourier transform at frequency ω given by what has been referred to as the *describing function gain*

$$\frac{Y_1 e^{j\theta_1}}{A}.$$

Definition 4.7 Describing Function Gain. *Consider a bounded-input bounded-output stable nonlinear system with sinusoidal input $u_0(t) = a\sin \omega t$. Under the* *assumption that the output of the nonlinear system is periodic with Fourier series*

$$y(t) = \sum_0^\infty Y_k \sin(k(\omega t + \theta_k)), \qquad (4.28)$$

the describing function gain *is the complex number representing the equivalent gain of the optimal approximating causal, b.i.b.o. stable linear system given by*

$$\frac{Y_1 e^{j\theta_1}}{A}. \qquad (4.29)$$

The key point to the definition is the major assumption that the output of the system is *periodic with the same period as the input*. Nonlinear systems that have sub-harmonic components thus do not qualify for this definition. The describing function defined above should more properly be referred to as the *sinusoidal input describing function*. Examples of other describing functions may be studied in Gelb and van der Velde [109]. Note that the definition of the describing function has built into it the notion of harmonic balance.

One of the major uses of the describing function is in predicting (nearly sinusoidal) oscillations in nonlinear control systems. Consider the class of unity feedback systems consisting of a nonlinear system and a linear system, as shown in Figure 4.2. The two systems N and G are described respectively by

$$u(t) = Ne(t),$$

$$y(t) = Gu(t) = \int_0^t g(t - \tau)u(\tau)d\tau.$$

Combining these equations, we get

$$e(t) = -y(t) = -\int_0^t g(t - \tau)(N(e))(\tau)d\tau. \qquad (4.30)$$

The problem to be addressed is whether (4.30) has a fixed point $e(\cdot)$. The first approach is to try

$$e(t) = a \sin(\omega t).$$

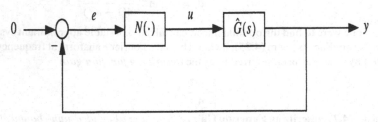

FIGURE 4.2. A unity feedback system with linear and nonlinear systems

If the describing function of N is $\eta(a, \omega) = \gamma_1 + j\gamma_2$, then $(Ne)(\cdot)$ has a fundamental harmonic of

$$\gamma_1 a \sin(\omega t) + \gamma_2 a \cos(\omega t).$$

Now, the fundamental harmonic of $y(\cdot)$ is given by

$$ua \sin(\omega t) + va \cos(\omega t)$$

if $u(a, \omega) + jv(a, \omega) = \hat{g}(j\omega)\eta(a, \omega)$. Now equating first harmonics on the left-hand side and right-hand side of (4.30) yields

$$1 + \hat{g}(j\omega)\eta(u, \omega) = 0. \tag{4.31}$$

Equation (4.31) has two unknowns, a and ω, and two equations. Separating it into its real and imaginary parts yields the two equations

$$1 + u(a, \omega) = 0, \quad v(a, \omega) = 0.$$

The describing function method claims that the solutions (a, ω) approximately represent the frequency and amplitude of the oscillation. Nonexistence of solutions is taken to mean non-oscillation. The test of (4.31) can be mechanized graphically, especially if the nonlinearity N is memoryless. Then, $\eta(a, \omega)$ is real, non-negative, and independent of ω, and we have

$$1 + \hat{g}(j\omega)\eta(a) = 0. \tag{4.32}$$

Graphically, this equation is solved on the Nyquist diagram of $\hat{g}(j\omega)$ as shown in Figure 4.3. The points of intersection of the Nyquist locus with the locus of $-1/\eta(a)$ yield the frequency of oscillation on the Nyquist locus and the amplitude of oscillation on the describing function locus. Even further simplification occurs when the nonlinearity is odd, in which case the describing function $\eta(a)$ is real. Then the imaginary part of the equation (4.31) namely $\hat{g}_i(j\omega) = 0$ ($\hat{g}_i$ stands for the imaginary part and $\hat{g}_r$ the real part of the transfer function $\hat{g}$), gives the frequencies $\omega_1, \ldots, \omega_m$, and $\eta(a) = -\frac{1}{\hat{g}_r(j\omega_i)}$ yields the amplitudes $a_1, \ldots, a_m$.

It is important to notice that this method may have a problem right off if the nonlinearity N generates a constant term in addition to the first harmonic, as it indeed will if the nonlinearity is not odd. In this instance the assumption of the

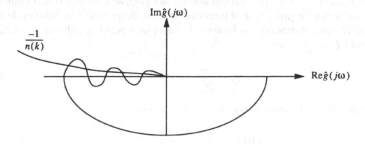

FIGURE 4.3. Graphical description of describing function method

form of the limit cycle needs to be modified to be of the form

$$e(t) = b + a \sin(\omega t).$$

The describing function is now computed to be the constant term plus the first harmonic of the output of the nonlinear system $Ne(\cdot)$:

$$\beta_0 + \beta_1 a \sin(\omega t) + \beta_2 \cos(\omega t).$$

Note that in general, β_1, β_2 will be different from γ_1, γ_2. For example, if N is a memoryless nonlinearity then we have

$$\beta_1(a, b) = \frac{1}{\pi} \int_0^{2\pi} n(b + a \sin \psi) \sin \psi \, d\psi,$$

$$\beta_2(a, b) = \frac{1}{\pi} \int_0^{2\pi} n(b + a \sin \psi) \cos \psi \, d\psi.$$

Note, in particular, that β_1, β_2 are both functions of a, b. Now, the constant and first harmonic terms of $z(\cdot)$ are given by

$$u_0(a, b) + u_1(a, b, \omega)a \sin(\omega t) + u_2(a, b, \omega)a \cos(\omega t)$$

where $u_0 = \beta_0 \hat{g}(0)$ and $u_1(a, b, \omega) + ju_2(a, b, \omega) = \hat{g}(j\omega)(\beta_1(a, b, \omega) + \beta_2(a, b, \omega))$. Equating coefficients with those of $-e$ yields

$$u_0(a, b) + b = 0,$$
$$1 + u_1(a, b, \omega) = 0, \qquad\qquad (4.33)$$
$$u_2(a, b, \omega) = 0.$$

Equation (4.33) has three unknowns, a, b, and ω and three equations. There is, however, no easy graphical way to mechanize solutions of the system (4.33).

4.1.3   Justification of the Describing Function

While the describing function technique is remarkably easy to use and has a certain "engineering" appeal to practitioners, it is difficult to know when it gives the correct answers and when it does not. The first analytic justification of the describing function method was given by Bergen and Franks [22], and we describe it briefly:

First, we refine the procedure of incorporating the dc term into the limit cycle to other higher-order harmonics as follows: Let $e(t)$ be a periodic solution of (4.22) with period $T = \frac{2\pi}{\omega}$. Then

$$e(t) = \sum_{k=-\infty}^{\infty} e_k e^{j\omega k t}. \qquad\qquad (4.34)$$

Assume that $Ne(t)$ is also periodic with Fourier series

$$(Ne)(t) = \sum_{k=-\infty}^{\infty} d_k(e, \omega) e^{j\omega k t}, \qquad\qquad (4.35)$$

where $e = (e_1, e_2, \ldots)$ is the vector of Fourier series coefficients of $e(t)$. From (4.34) and (4.35) it follows that

$$(GN)e(t) = \sum_{k=-\infty}^{\infty} d_k(e, \omega)\hat{g}(j\omega k)e^{j\omega k t}, \qquad (4.36)$$

to yield the set of equations

$$e_k = -d_k(e, \omega)\hat{g}(j\omega k), \quad k = 1, 2, \ldots. \qquad (4.37)$$

Equations (4.37) represent infinitely many equations in the unknowns e_k, $k = 0, 1, 2, \ldots$. Of course, (4.33) is the set of equations truncated at $k = 1$ in the sense that $e_2 = e_3 = \cdots = 0$. Intuitively, then, if equation (4.33) has a solution, and the equations (4.37) have a solution with $e_2, e_3, \ldots$ small then we would expect that the e_1 solving equation (4.33) would be a good approximation to the one obtained from (4.37). Also, in this instance if (4.33) does not have a solution, then we would also expect equations (4.37) to not have a solution.

Proceeding more formally, consider the space of continuous functions with period T. Since the functions repeat after T seconds, we may consider continuous functions on the interval $[0, T]$, with the understanding that they repeat after T. Let P_m be the projection operator from $C[0, T] \mapsto C[0, T]$ projecting such periodic functions onto their first m harmonics and let $\widetilde{P}_m = I - P_m$ be its complementary operator (projection onto the remaining harmonics). Then equation (4.37) may be rewritten as

$$\begin{aligned} P_m e &= -P_m(GN(P_m e + \widetilde{P}_m e), \\ \widetilde{P}_m e &= -\widetilde{P}_m(GN(P_m e + \widetilde{P}_m e)). \end{aligned} \qquad (4.38)$$

If $\widetilde{P}_m e$ is small, then (4.38) is approximated by setting $\widetilde{P}_m e = 0$ to get

$$P_m e = -P_m(GN(P_m e)). \qquad (4.39)$$

In turn, (4.39) may be written as

$$e_k = -d_k((e_1, e_2, \ldots, e_m, 0, 0, \ldots)\hat{g}(j\omega k) \qquad (4.40)$$

for $k = 1, \ldots, m$. Equation (4.40) is a set of m (complex) equations in the m (complex) unknowns $e_1, e_2, \ldots, e_m$. The remaining components $e_{m+1}, e_{m+2}, \ldots$ are set to zero.

Under suitable conditions, this approximation of setting $\widetilde{P}_m e = 0$ may be justified. (Roughly, that e has low enough high-frequency content; in turn, this is implied by G being sufficiently low-pass.) The existence of a solution of (4.39) can be shown to imply existence of solutions to (4.38). This technique of reducing the number of equations to be solved from infinitely many in (4.36) to finitely many in (4.40) is referred to as to the *Lyapunov–Schmidt method*.

We may use degree theory from Chapter 3 to study the existence of solutions to (4.38) as follows:

$$|Ne| \leq \gamma|e| \quad \text{and} \quad |\widetilde{P}_m G| \leq \epsilon.$$

Now, the second equation of (4.38) is Lipschitz continuous in $\tilde{P}_m e$ with Lipschitz constant $\epsilon \gamma < 1$, has a solution with

$$|\tilde{P}_m e| \leq \frac{\epsilon \gamma}{1 - \epsilon \gamma} |P_m e|. \tag{4.41}$$

Using this solution in the first equation of (4.38), we get

$$P_m e = -P_m(GN(P_m e + \tilde{P}_m e)) \tag{4.42}$$
$$= -V(P_m e).$$

This function V is to be compared with the function U of (4.39):

$$P_m e = -P_m(GN(P_m e)) \tag{4.43}$$
$$= -U(P_m e).$$

The idea is to show that $P_m e + U(P_m e)$ and $P_m e + V(P_m e)$ have the same degree with respect to zero in a bounded region in the space of $P_m e$, namely $\mathbb{R}^{2m+1}$. The details are as follows: We will assume that we represent $P_m e$ by the set of m complex numbers $e_1, \ldots, e_m$ and one real number e_0, since

$$P_m e = \sum_{k=-m}^{m} e_k e^{j\omega k t}.$$

We will refer to the space of e_k as $\mathbb{R}^{2m+1}$, though it consists of m complex numbers and one real number. We will assume that the "describing function" equation (4.43) has a single solution in a set $B \subset \mathbb{R}^{2m+1}$ for some $\omega = \frac{2\pi}{T} \in [\omega_l, \omega_h]$. The assumption that $|\tilde{P}_m G| \leq \epsilon$ can then be re-expressed as

$$\hat{g}(j\omega k) \leq \epsilon \text{ for } k > m \quad \omega \in [\omega_l, \omega_h].$$

To establish that the degree of $P_m e + U(P_M e)$ is the same as that of $P_m e + V(P_m e)$, we construct the homotopy

$$P_m e + t U(P_m e) + (1 - t)V(P_m e)$$

To establish that it is non-zero on the boundary of B, note that

$$P_m e + t U(P_m e) + (1 - t)V(P_m e)$$
$$= (P_m e + U(P_m e)) + (1 - t)(U(P_m e) - V(P_m e)). \tag{4.44}$$

In turn, if $\sup \hat{g}(j\omega k) \leq \beta$ for $k = 0, 1, \ldots, m$ and $\omega \in [\omega_l, \omega_h]$ then

$$|V(P_m e) - U(P_m e)| \leq \beta \gamma |\tilde{P}_m e| \leq \frac{\epsilon \beta \gamma^2}{1 - \epsilon \gamma} |P_m e|.$$

Using this in (4.44) yields

$$|P_m e + t U(P_m e) + (1 - t)V(P_m e)| \geq |P_m e + U(P_m e)| - \frac{\epsilon \beta \gamma^2}{1 - \epsilon \gamma} |P_m e|.$$

If ϵ were small enough that on the boundary of the set B the right hand side of this equation is greater than 0, then, by the homotopy invariance of degree, we have that

$$d(P_m e + V(P_m e), 0, B) = d(P_m e + U(P_m e), 0, B).$$

Since the right-hand side is not equal to zero (there is exactly one solution in the set B), there is at least one solution to the equation (4.42) in the set B.

We have been somewhat careless in not stating the results very precisely in the form of a theorem. This is partly because checking the condition involves verifying the bounds over a range of frequencies $[\omega_u, \omega_h]$ where one expects the solutions to lie. Roughly speaking, the *low pass condition* on G demands that the neglected harmonics be small enough. The best way in which this result can be applied is simply by checking whether the system is low pass enough to merit doing only a first order harmonic balance. The following example from [69] shows when the describing function method works and when it fails.

Example 4.8 Van der Pol's Equation. *Consider a form of the van der Pol equation described by*

$$\ddot{y} + \mu(3y^2 - 1)\dot{y} + y = 0 \qquad (4.45)$$

and let us model it in the form of a nonlinear feedback system as in Figure 4.2 with

$$\hat{g}(s) = \frac{\mu s}{s^2 - \mu s + 1} \qquad (4.46)$$

with N being represented by a memoryless nonlinearity of the form $n(x) = x^3$. Since the nonlinearity is odd, the first harmonic is the only one that needs to be accounted for, and $\eta(a) = 3a^2/4$. The Nyquist locus and the describing function locus of $-1/\eta(a)$ are shown in Figure 4.4. The Nyquist locus does not change in shape with μ though the scaling with frequency does indeed change. However the intersection with the imaginary axis is always at $\omega = 1$ so that the frequency of the oscillation and the amplitude a predicted by the intersection is $2/\sqrt{3}$ (also independent of μ). Let us compare these conclusions with an analysis of what happens in the limit $\mu \to \infty$. Indeed, in the state space we may write the equations in the scaled time $\tau = \frac{t}{\mu}$ as

$$\frac{dx_1}{d\tau} = x_2,$$

$$\frac{1}{\mu^2} \frac{dx_2}{d\tau} = -x_1 + x_2 - x_2^3. \qquad (4.47)$$

The formal limit of this system is the degenerate van der Pol system (see Problem 3.8).

$$\frac{dx_1}{d\tau} = x_2,$$

$$0 = -x_1 + x_2 - x_2^3, \qquad (4.48)$$

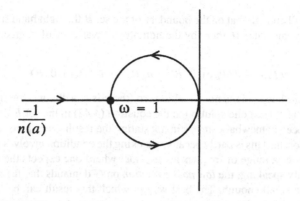

FIGURE 4.4. The describing function of the van der Pol oscillator

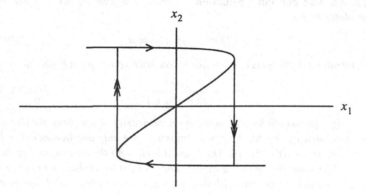

FIGURE 4.5. The relaxation oscillation of the van der Pol oscillator

which has an oscillation in the x_1, x_2 state space (y, $\dot{y}$ space) of the form shown
in Figure 4.5. The amplitude of this oscillation is not close to $2/\sqrt{3}$. Of course,
the describing function is not expected to give the correct answer as μ gets to be
large since the transfer function $\hat{g}(j\omega)$ is close to 1 over a very large frequency
band as μ gets to be large as may be verified by writing $g(j\omega)$ as

$$\hat{g}(j\omega) = \left\{ -1 + \frac{j}{\mu}\left(\omega - \frac{1}{\omega}\right) \right\}^{-1}.$$

4.2   Input–Output Stability

Up to this point, a great deal of the discussion has been based on a state space description of a nonlinear system of the form

$$\dot{x} = f(t, x, u),$$
$$y = h(t, x, u)$$

or

$$x_{k+1} = f(k, x_k, u_k),$$
$$y_k = h(k, x_k, u_k).$$

One can also think of this equation from the input–output point of view. Thus, for example, given an initial condition $x(0) = x_0$ and an input $u(\cdot)$ defined on the interval $[0, \infty[$, say piecewise continuous, and with suitable conditions on $f(\cdot, \cdot, \cdot)$ to make the differential equation have a unique solution on $[0, \infty[$ with no finite escape time, it follows that there is a map N_{x_0} from the input $u(\cdot)$ to the output $y(\cdot)$. It is important to remember that the map depends on the initial state x_0 of the system. Of course, if the vector field f and functions h are affine in x, u then the response of the system can be broken up into a part depending on the initial state and a part depending on the input. More abstractly, one can just define a nonlinear system as a map (possibly dependent on the initial state) from a suitably defined input space to a suitable output space. The input and output spaces are suitably defined vector spaces. Thus, the first topic in formalizing and defining the notion of a nonlinear system as a nonlinear operator is the choice of input and output spaces. We will deal with the continuous time and discrete time cases together. To do so, recall that a function $g(\cdot) : [0, \infty[\mapsto \mathbb{R}$ (respectively, $g(\cdot) : Z_+ \mapsto \mathbb{R}$) is said to belong to $L_p[0, \infty[$ (respectively, $l_p(Z_+)$) if it is measurable and in addition

$$\int_0^\infty |g(t)|^p dt < \infty, \quad \text{resp.} \quad \sum_{n=0}^\infty |g(n)|^p < \infty.$$

Also, the set of all bounded[2] functions is referred to as $L_\infty[0, \infty[$ (respectively, $l_\infty(Z_+)$). The L_p (l_p) norm of a function $g \in L_p[0, \infty[$ (respectively, $l_p(Z_+)$) is defined to be

$$|g(\cdot)|_p := \left(\int_0^\infty |g(t)|^p dt \right)^{\frac{1}{p}} \quad \text{resp.} \quad |g(\cdot)|_p := \left(\sum_{n=0}^\infty |g(n)|^p \right)^{\frac{1}{p}},$$

and the L_∞ (resp. l_∞) norm is defined to be

$$|g(\cdot)|_\infty := \sup_t |g(t)| \quad \text{resp.} \quad \sup_n |g(n)|.$$

---

[2]Actually, *essentially bounded*; essentially bounded means that the function is bounded except on a set of zero measure.

Unlike norms of finite dimensional spaces, norms of infinite dimensional spaces are not equivalent (cf. Chapter 3), and thus they induce different norms on the space of functions. Here are several examples of functions in these spaces (the examples are given in continuous time, but the reader should fill in the examples in discrete time):

Example 4.9.

- $f \in L_1$ but $f \notin L_2, L_\infty : t \to \frac{1}{\sqrt{t}(1+t)}$.
- $f \in L_2$ but $f \notin L_1, L_\infty : t \to \frac{(1+t)^{1/4}}{t^{1/4}(1+t)}$.
- $f \in L_\infty$ but $\notin L_1, L_2 : t \to \sin t$.

For $p \in [1, \infty]$, L_p is a *Banach space*, or a *complete normed vector space*, this result is sometimes refereed to as the Fischer Riesz theorem. Some equalities about L_p spaces are useful: we quote a few here (for more details, see Brezis [42]): The first is easy to verify

Proposition 4.10 Hölder, (Schwarz), and Minkowski Inequalities. *Given $p \in [1, \infty]$, define it's conjugate $q \in [1, \infty[$ to satisfy*

$$\frac{1}{p} + \frac{1}{q} = 1.$$

Then, given functions $f(.), g(.)$ such that $f \in L_p, g \in L_q$ (l_p, l_q, respectively) the following inequality, called the Hölder inequality holds:

$$\int_0^\infty |f(t)g(t)|dt \le |f|_p |g|_q \tag{4.49}$$

For $p = q = 2$, the inequality (4.49) is known as the Schwarz inequality. Further, if $f, g \in L_p$ (respectively, $f, g \in l_p$) then $f + g \in L_p(l_p)$, and the following inequality, called the Minkowski inequality holds:

$$|f + g|_p \le |f|_p + |g|_p \tag{4.50}$$

for all $p \in [1, \infty]$. In particular, the Minkowski inequality establishes that $L_p[0, \infty[$ is a vector space.

Proof: We start with the proof of the Hölder inequality. The inequality (4.49) is trivial when either $|f(\cdot)|_p = 0$, $|g(\cdot)|_p = 0$, hence we will define

$$A = \left\{ \int_0^\infty |f(t)|^p dt \right\}^{\frac{1}{p}}, \quad B = \left\{ \int_0^\infty |g(t)|^q dt \right\}^{\frac{1}{q}},$$

and the functions $F \in L_p[0, \infty[, g \in L_q[0, \infty[$ by

$$F(\cdot) = \frac{f(\cdot)}{A}, \quad G(\cdot) = \frac{g(\cdot)}{B}.$$

Now choose $u(t), v(t) \in \mathbb{R}$ such that

$$|F(t)| = e^{\frac{u(t)}{p}}, \quad |G(t)| = e^{\frac{v(t)}{q}}.$$

Since $1/p + 1/q = 1$, from the convexity of the exponential function we have that

$$e^{u(t)/p+v(t)/q} \leq p^{-1}e^{u(t)} + q^{-1}e^{v(t)}.$$

It follows that for all $t \in [0, \infty[$,

$$|F(t)G(t)| \leq p^{-1}F(t)^p + q^{-1}G(t)^q.$$

Integrating both sides of this equation from 0 to ∞ yields

$$\int_0^\infty |F(t)G(t)|dt \leq p^{-1} + q^{-1} = 1.$$

The Hölder inequality (4.49) now follows readily.

To prove (4.50), write for $f, g \in L_p([0, \infty[)$,

$$|f + g|^p \leq |f| \cdot |f + g|^{p-1} + |g| \cdot |f + g|^{p-1}. \qquad (4.51)$$

Now, since $(f + g)^{p-1} \in L_q([0, \infty[)$ (the reader should verify this), we may use the Hölder inequality on each of the terms on the right hand side of the preceding inequality to obtain

$$\int_0^\infty |f||f + g|^{p-1}dt = \left[\int_0^\infty |f|^p dt\right]^{1/p} \left[\int_0^\infty |f + g|^{(p-1)q}dt\right]^{1/q},$$

$$\int_0^\infty |g||f + g|^{p-1}dt = \left[\int_0^\infty |g|^p dt\right]^{1/p} \left[\int_0^\infty |f + g|^{(p-1)q}dt\right]^{1/q}.$$

$$(4.52)$$

Using (4.52) in (4.51) yields

$$\int_0^\infty |f + g|^p dt$$

$$\leq \left[\int_0^\infty |f + g|^p dt\right]^{1/q} \left[\left(\int_0^\infty |f(t)|^p dt\right)^{1/p} + \left(\int_0^\infty |g(t)|^p dt\right)^{1/p}\right].$$

$$(4.53)$$

Now, the Minkowski inequality (4.50) follows by dividing both sides of (4.53) by the first term on the right hand side of (4.53). $\qquad \square$

The function spaces L_p contain many bizarre functions. For instance, it is not true that every L_2 or L_1 function has to tend to 0 as $t \to \infty$, or that every L_2 or L_1 function is even bounded. However, if a function $f \in L_1$ or L_2 and its derivative $\dot{f}$ are bounded, it follows that $f(t) \to 0$ as $t \to \infty$. This fact is referred to as Barbalat's lemma and is useful in Lyapunov analysis: It is an exercise (Problem 5.9 in Chapter 5). Another interesting fact is that if $f \in L_1 \cap L_\infty$, then $f \in L_p$ for $p \in [1, \infty]$ (see 4.10).

We may also define **truncations** of functions as follows: given a function $f(\cdot) \in L_p[0, \infty[$ (respectively, $l_p(Z_+)$), the truncation up to time T, denoted $f_T(\cdot)$, is defined to be

$$f_T(t) = \begin{bmatrix} f(t), & 0 \le t \le T, \\ 0, & T < t. \end{bmatrix}$$

Definition 4.11 Extended L_p (l_p) spaces. *A function $f(\cdot)$ is said to belong to the extended L_p (l_p) space denoted $L_{pe}[0, \infty[$ (respectively, $l_{pe}(Z_+)$), if for every truncation $f_T(\cdot)$ we have $f_T(\cdot) \in L_p(l_p)$.*

Extended L_p (l_p) spaces capture in their definition functions which are potentially unbounded in an L_p (l_p) sense, but which take infinitely long to achieve an unbounded norm.

Example 4.12 Examples of functions in L_{pe}, l_{pe}.

1. *The functions t, e^t, $\log t$ are all in $L_{\infty e}$ but not in L_∞.*
2. *The functions $\sin t$, t^n, e^t are all in L_{pe} but not in L_p.*
3. *The sequences n, $\sin n$, 2^n are all in l_{pe} but not in l_p.*

The extension of the preceding definitions to vector valued time functions proceeds in the usual manner with some finite dimensional vector space norm replacing the absolute value. The notation that we use for a signal $f(\cdot) : [0, \infty[\mapsto \mathbb{R}^n$ is L_{pe}^n (an analogous definition holds for l_{pe}^n).

We will view nonlinear operators, abstractly, as maps N_{x_0} from $L_{pe}^{n_i} \mapsto L_{pe}^{n_o}$ or in the discrete time case as maps from $l_{pe}^{n_i} \mapsto l_{pe}^{n_o}$. Note that by this definition, we allow for the possibility of having *unstable* operators of a certain kind in the domain of discourse, namely, those which produce $L_{pe}^{n_o}$ output functions with infinite $L_p^{n_o}$ norm in response to $L_p^{n_i}$ inputs. For notational convenience, we will drop the dependence on initial condition in the subsequent development, but the reader would be well advised to keep in mind that every nonlinear system is represented by a *family of nonlinear operators* (indexed by the initial condition). When nonlinear operators are specified as arising from differential (or difference) equation descriptions, it follows readily that the output at time t given x_0 at time 0 depends only on $u(t)$ causally, i.e., defined on $[0, t]$. When we specify a nonlinear map more abstractly, however, it is important to insist that it be causal. In the development that follows we will state definitions for the continuous time case and refer to the analogous results for the discrete time case onl when they are different. Formally, we define causal nonlinear operators as follows:

Definition 4.13 Causal Nonlinear Operators. *A mapping $N : L_{pe}^n \mapsto L_{pe}^m$ is said to be* causal *if for every input $u(.) \in L_{pe}^n$ and every $T > 0$ it follows that*

$$(A(u))_T = (A(u_T))_T.$$

In other words the truncation of the output to $u(.)$ up to time T is the same as the output to $u_T(.)$ up to time T. The definition is depicted in Figure 4.6.

It is important to remember that in general, $A(u_1 + u_2) \ne A(u_1) + A(u_2)$. We are now ready to define L_p stability.

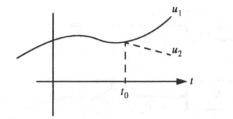
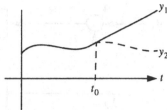

FIGURE 4.6. Visualizing the definition of causality

Definition 4.14 Finite Gain L_p Stability.   *A causal non linear operator A :*
$L_{pe}^n \mapsto L_{pe}^m$ *is said to be* finite gain L_p *stable if for given input $u \in L_p^n$, the output*
$A(u) \in L_p^m$. *Further, there exist $k, \beta > 0$ such that for given $u \in L_{pe}^n$ the output*
$A(u) \in L_{pe}^m$ *satisfies*

$$|A(u)|_T \leq k|u|_T + \beta \quad \forall T > 0. \tag{4.54}$$

It is important to note the two separate requirements of the definition. The first
insists that *bounded inputs*, i.e., those inputs $u \in L_p^n$ produce *bounded outputs*,
i.e., in L_p^m. The second requires that there is an affine bound involving two positive
numbers k, β involving possibly unbounded inputs and outputs, i.e., those residing
in L_{pe}^n, L_{pe}^m, respectively. Of course, the two numbers k, β are required to be independent of T so that the bound of equation (4.54) also holds for inputs in L_p^n. We
refer to the definition as *finite gain* stability in order to emphasize that it is stronger
than merely *bounded input bounded output* stability which requires merely that
bounded inputs produce bounded outputs. The smallest number k satisfying equation (4.54) qualifies for a definition of the *gain* of the operator A. The simplest
theorem that can be proven about the stability of two interconnected systems of
the form of Figure 4.7 is as follows:

Theorem 4.15 Small Gain Theorem.   *Consider the interconnected feedback
system of Figure 4.7 with inputs u_1, u_2 and outputs y_1, y_2. Assume that the systems
$N_1 : L_{pe}^{n_i} \mapsto L_{pe}^{n_o}, N_2 : L_{pe}^{n_o} \mapsto L_{pe}^{n_i}$ are both causal and finite gain stable, that is,
there exist $k_1, k_2, \beta_1, \beta_2$ such that*

$$|N_1(e_1)|_T \leq k_1|e_1|_T + \beta_1 \quad \forall e_1 \in L_{pe}^{n_i},$$

$$|N_2(e_2)|_T \leq k_2|e_2|_T + \beta_2 \quad \forall e_2 \in L_{pe}^{n_o}.$$

*Further, assume that the loop is well posed in the sense that for given $u_1 \in
L_{pe}^{n_i}, u_2 \in L_{pe}^{n_o}$ there are unique $e_1, y_2 \in L_{pe}^{n_i}; e_2, y_1 \in L_{pe}^{n_o}$ satisfying the loop
equations namely*

$$e_1 = u_1 - N_2(e_2), \quad y_1 = N_1(e_1),$$
$$e_2 = u_2 + N_1(e_1), \quad y_2 = N_2(e_2). \tag{4.55}$$

Then, the closed loop system is also finite gain stable from u_1, u_2 to y_1, y_2 if

$$k_1 k_2 < 1.$$

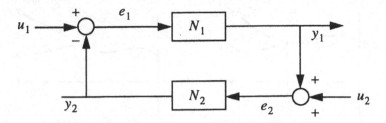

FIGURE 4.7. A feedback control system

Proof: Consider inputs $u_1 \in L_{pe}^{n_i}$, $u_2 \in L_{pe}^{n_o}$. From the well posedness assumption it follows that there are unique solutions e_1, e_2, y_1, y_2 to the feedback loop. Further from the feedback loop equations (4.55) it follows that

$$|e_1|_T \le |u_1|_T + k_2|e_2|_T + \beta_2.$$

Using the other estimate

$$|e_2|_T \le |u_2|_T + k_1|e_1|_T + \beta_1.$$

in this equation yields

$$|e_1|_T \le k_1 k_2 |e_1|_T + |u_1|_T + k_2|u_2|_T + k_2\beta_1 + \beta_2.$$

Analogously, one may obtain also,

$$|e_2|_T \le k_1 k_2 |e_2|_T + |u_2|_T + k_1|u_1|_T + k_1\beta_2 + \beta_1.$$

The hypothesis of the theorem guarantees that $(1 - k_1 k_2) > 0$ so that one obtains the following estimates:

$$|y_1|_T \le \frac{k_1|u_1|_T + k_1 k_2|u_2|_T + \beta_1 + k_1\beta_2}{1 - k_1 k_2},$$

$$|y_2|_T \le \frac{k_2|u_2|_T + k_1 k_2|u_1|_T + \beta_2 + k_2\beta_1}{1 - k_1 k_2}.$$

From these estimates it follows that bounded inputs, i.e., $u_1 \in L_p^{n_i}$, $u_2 \in L_p^{n_o}$ produce bounded outputs. Also, this establishes that the closed loop system is finite gain stable.   □

Remarks:

1. The preceding theorem is called a *small gain theorem* because it gives conditions under which the closed loop system is finite gain stable when the loop gain $k_1 k_2$ is less than 1 (small).
2. The small gain theorem does not guarantee the existence or uniqueness of solutions to the closed loop system, i.e., the existence and uniqueness of e_1, e_2, y_1, y_2 given u_1, u_2 in their respective L_{pe} spaces. This is one of the hypotheses of the theorem.

In order to state results concerning existence and uniqueness of solutions as well as stability, one needs the following definition:

Definition 4.16 Incremental Finite Gain Stability. *A nonlinear operator* N :
$L_{pe}^{n_i} \mapsto L_{pe}^{n_o}$ *is said to be* incrementally finite gain stable *if*

1. $N(0) \in L_p^{n_o}$, *where 0 is the identically zero input.*
2. *For all* $T > 0$ *and* $u_1, u_2 \in L_{pe}^{n_i}$ *there exists some* $k > 0$ *such that*

$$|N(u_1) - N(u_2)|_T \leq k|u_1 - u_2|_T$$

It is easy to see that an incrementally finite gain operator is also a finite gain operator, since

$$|N(u)|_T \leq k|u|_T + |N(0)|_T.$$

By hypothesis $N(0)$ is in $L_p^{n_o}$, so that we may define $\beta := |N(0)| \geq |N(0)|_T$. Also, it should be clear from the definition that incremental finite gain is a Lipschitz continuity condition on the operator N.

Theorem 4.17 Incremental Small Gain Theorem.   *Consider the feedback loop of Figure 4.7. Let the operators* $N_1 : L_{pe}^{n_i} \mapsto L_{pe}^{n_o}, N_2 : L_{pe}^{n_o} \mapsto L_{pe}^{n_i}$ *be incrementally finite gain stable with gains* k_1, k_2 *respectively. Then, if*

$$k_1 k_2 < 1,$$

it follows that for given $u_1 \in L_{pe}^{n_i}, u_2 \in L_{pe}^{n_o}$ *there exist unique* e_1, e_2, y_1, y_2 *solving the feedback loop equations (4.55). Further, the feedback system is incremental finite gain stable from* u_1, u_2 *to* y_1, y_2.

Proof: We will drop the subscript T for the truncated time functions in this proof for notational convenience. It is important, however, to keep in mind that the estimates do in fact hold for the appropriate L_{pe} spaces. The feedback loop equations are

$$e_1 = u_1 - N_2(e_2), \quad y_1 = N_1(e_1),$$
$$e_2 = u_2 + N_1(e_1), \quad y_2 = N_2(e_2).$$

Now, fix u_1, u_2 and note that we have unique solutions y_1, y_2, if and only if, there are unique solutions y_1, y_2. Combining the two equations above yields

$$e_1 = u_1 - N_2(u_2 + N_1(e_1)). \tag{4.56}$$

This equation has a unique solution if and only if the map $T : L_{pe}^{n_i} \mapsto L_{pe}^{n_i}$ has a unique fixed point. To establish this we use the contraction mapping theorem (Theorem 3.17). Note that

$$|T(e_1) - T(\bar{e}_1)| \leq |N_2(u_2 + N_1(e_1)) - N_2(u_2 + N_1(\bar{e}_1)|$$
$$\leq k_2|N_1(e_1) - N_1(\bar{e}_1)|$$
$$\leq k_2 k_1|e_1 - \bar{e}_1|.$$

By hypothesis $k_1 k_2 < 1$ establishing that the map T is a contraction map so that there are unique solutions e_1, e_2 given u_1, u_2. As for the incremental finite gain of the composite system, let u_1, u_2, and $\tilde{u}_1, \tilde{u}_2$ be two sets of inputs for the system with corresponding outputs e_1, e_2 and $\tilde{e}_1, \tilde{e}_2$. From (4.56) it follows that

$$|e_1 - \tilde{e}_1| \le |u_1 - \tilde{u}_1| + |N_2(u_2 + N_1(e_1)) - N_2(\tilde{u}_2 - N_1(\tilde{e}_1))|$$

$$\le |u_1 - \tilde{u}_1| + k_2|u_2 - \tilde{u}_2| + k_1 k_2 |e_1 - \tilde{e}_1|,$$

so that we have

$$|e_1 - \tilde{e}_1| \le \frac{|u_1 - \tilde{u}_1| + k_2|u_2 - \tilde{u}_2|}{1 - k_1 k_2}.$$

Analogously, similar bounds can be obtained for $|e_2 - \tilde{e}_2|, |y_1 - \tilde{y}_1|, |y_2 - \tilde{y}_2|$. This completes the proof. $\Box$

The small gain theorems, while simple in statement, are extremely useful in proving several interesting results:

4.3   Applications of the Small Gain Theorems

4.3.1   Robustness of Feedback Stability

Consider the unity feedback gain feedback control system shown in Figure 4.8. Assume that the closed loop system is incremental finite gain stable from $L_{pe}^{n_i} \mapsto L_{pe}^{n_i}$. From the results of Problem 4.13, it follows that there exists k such that

$$|(I + G)^{-1}u| \le k|u|$$

Now let us assume that the plant is perturbed to one of the form $G + \Delta$ with $\Delta : L_{pe}^{n_i} \mapsto L_{pe}^{n_i}$, an incremental finite gain operator as shown in the Figure 4.8. For the perturbed system the relation between u and e, abbreviated H_{eu} (even though it is not really a transfer function), is

$$H_{eu} = [I + G + \Delta]^{-1}$$

$$= \{(I + \Delta(I + G)^{-1})(I + G)\}^{-1}$$

$$= (I + G)^{-1}[I + \Delta(I + G)^{-1}]^{-1}.$$

Note that the preceding calculations involve nonlinear operators rather than matrices. Now, consider the small gain condition that the product of the (incremental)

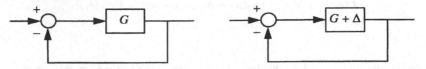

FIGURE 4.8. Unperturbed and perturbed unity feedback systems

gain of $(I + G)^{-1}$ and that of Δ is less than 1, namely,

$$k_\Delta < \frac{1}{k_{(I+G)^{-1}}}.$$

Under these conditions it follows that H_{eu} is also (incremental) finite gain stable.[3] In fact, the reader should verify that the incremental finite gain of H_{eu} may be bounded above by

$$k_{(I+G)^{-1}} \frac{1}{1 - k_\Delta k_{(I+G)^{-1}}}.$$

If the unperturbed system has open loop incremental gain k_G less than 1, then it follows that it is stable by the small gain theorem. Further, we can get a bound on $k_{(I+G)^{-1}}$ as follows:

$$(I + G)u = y \Rightarrow u = y - G(u) \Rightarrow |u| \le |y| + k_G|u|.$$

This, in turn, yields

$$|u| \le \frac{|y|}{1 - k_G} \Rightarrow k_{(I+G)^{-1}} \le \frac{1}{1 - k_G}.$$

In this case, a robustness result is that the perturbed closed loop system is stable if

$$k_\Delta < 1 - k_G.$$

This result is weaker (more conservative) than the one above, since

$$1 - k_G \le \frac{1}{k_{(I+G)^{-1}}}.$$

The small gain condition has an interpretation in terms of redrawing the perturbed feedback loop as shown in Figure 4.9.

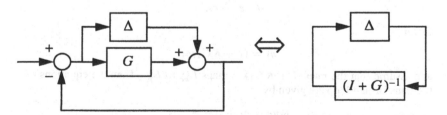

FIGURE 4.9. Redrawing the perturbed feedback loop to illustrate the small gain condition

---

[3]There is a version of this result that also holds for the case that G is only finite gain stable and not incremental finite gain stable; also Δ, with the added proviso that $\Delta(0) = 0$. The interested reader should work this out.

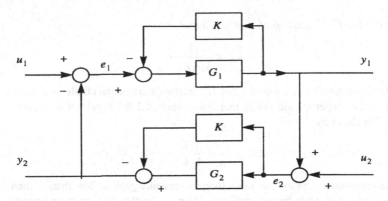

FIGURE 4.10. A loop transformation

4.3.2 Loop Transformation Theorem

Consider the feedback loop of Figure 4.7. In some applications, it is useful to transform the loop and redraw it as shown in Figure 4.10 with a *linear* operator K with finite gain $k_K < \infty$ added in both the upper and lower feedback loops. We will explain this transformation in terms of the loop equations we start with

$$u_1 = e_1 + G_2 e_2,$$
$$u_2 = e_2 - G_1 e_1. \tag{4.57}$$

Subtracting K times the second equation of (4.57) from the first and using (by the assumption that K is linear)

$$K(e_2 - G_1 e_1) = K e_2 - K G_1 e_1$$

we get

$$u_1 - K u_2 = (I + K G_1) e_1 + (G_2 - K) e_2,$$
$$u_2 = -G_1 e_1 + e_2. \tag{4.58}$$

Define

$$\eta_1 = (I + K G_1) e_1$$

and assume that the map $(I + K G_1)^{-1}$ maps $L^{n_i}_{pe}$ to $L^{n_i}_{pe}$. Then the equations of the transformed loop are given by

$$u_1 - K u_2 = \eta_1 + (G_2 - K) e_2,$$
$$u_2 = (I + K G_1)^{-1} \eta_1 + e_2. \tag{4.59}$$

Thus, we have shown that if K is linear and $(I + K G_1)^{-1}$ is *well defined as an operator from $L^{n_i}_{pe} \mapsto L^{n_i}_{pe}$* then the transformed feedback loop is equivalent to the given feedback loop. The restriction on the linearity of K can be relaxed if $u_2 = 0$.

The application of the small gain theorem is as follows:

Theorem 4.18 Small Gain and Loop Transformation.   *Consider the feedback loop of Figure 4.7 and its transformed version in Figure 4.10 with K linear and $(I + KG_1)^{-1}$ well posed as an operator from $L_{pe}^{n_i} \mapsto L_{pe}^{n_i}$. Then, if the operators $G_1(I + KG_1)^{-1}, G_2 - K$ are incremental finite gain (L_p) stable and*

$$k_{G_1(I+KG_1)^{-1}} k_{G_2 - K} < 1,$$

it follows that the transformed system $S_t : (u_1 - K u_2, u_2) \mapsto (\eta_1, e_2)$ is incremental finite gain (L_p) stable. In turn, this implies that $S : (u_1, u_2) \mapsto (e_1, e_2)$ is also incremental finite gain (L_p) stable.

Proof: The first part of the theorem is a straight forward application of the small gain theorem. To show the stability of the untransformed loop, we first note that $u_1 - K u_2, u_2 \in L_p \Rightarrow u_1, u_2 \in L_p$, since K is finite gain stable. Further note that

$$I - KG_1(I + KG_1)^{-1} = (I + KG_1)^{-1}.$$

Thus, the finite gain stability of K, $G_1(I + KG_1)^{-1}$ implies the finite gain stability of $(I + KG_1)^{-1}$. Since

$$e_1 = (I + KG_1)^{-1}\eta_1,$$

it follows that $\eta_1 \in L_p \Rightarrow e_1 \in L_p$.   □

4.4   Passive Nonlinear Systems

In this section we will consider nonlinear systems with the same number of inputs and outputs, that is $n_i = n_o$. Also, the nonlinear systems will need to be operators on inner product spaces, thus we will restrict ourselves to $L_{2e}^{n_i}$. To fix ideas physically, consider a nonlinear circuit as the system under consideration with inputs being a set of independent current and voltage sources and outputs being the corresponding (dependent) voltages and currents, respectively, as shown in Figure 4.11. Then, the quantity $y^T(t)u(t)$ has the interpretation of being the instantaneous power input into the circuit, with the reference directions chosen as shown in the figure. The circuit is said to be passive if the amount of input energy (the integral of the

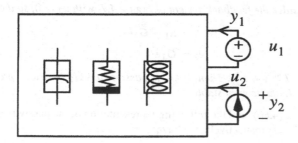

FIGURE 4.11. Nonlinear n port circuit

power) does not exceed the change in the amount of energy stored in the circuit. If the nonlinear circuit starts with all its storage elements, namely inductors and capacitors, uncharged, it follows that the circuit is passive if

$$\int_0^T y^T(t)u(t)\,dt \geq 0$$

for all T. This thinking is generalized as follows:

Definition 4.19 Passive Systems. *Let* $N : L_{2e}^{n_i} \mapsto L_{2e}^{n_i}$. *Then we say*

1. N *is passive if there exists* $\beta \in \mathbb{R}$ *such that for all* $T, u \in L_{2e}^{n_i}$,

$$\int_0^T (Nu)^T(t)u(t)\,dt \geq \beta.$$

2. N *is strictly passive if there exists* $\alpha > 0$, $\beta \in \mathbb{R}$ *such that for all* $T, u \in L_{2e}^{n_i}$

$$\int_0^T (Nu)^T(t)u(t)\,dt \geq \alpha \int_0^T u^T(t)u(t)\,dt + \beta.$$

3. N *is said to be* incrementally passive *if for all* $T, u_1, u_2 \in L_{2e}^{n_i}$, *we have*

$$\int_0^T (Nu_1 - Nu_2)^T(t)(u_1 - u_2)(t)\,dt \geq 0.$$

4. N *is said to be* incrementally strictly passive *if there exists a* δ *such that for all* $T, u_1, u_2 \in L_{2e}^{n_i}$, *we have*

$$\int_0^T (Nu_1 - Nu_2)^T(t)(u_1 - u_2)(t)\,dt \geq \delta \int_0^T (u_1 - u_2)^T(t)(u_1 - u_2)(t)\,dt.$$

In the sequel, we will use the notation $\langle \cdot, \cdot \rangle_T$ to denote the integral up to time T between two functions:

$$\langle u_1, u_2 \rangle_T = \int_0^T u_1^T(t)u_2(t)\,dt.$$

The first passivity theorem is as follows:

Theorem 4.20 Passivity Theorem for Passive + Strictly Passive Feedback Loop. *Consider the feedback system of Figure 4.7, with* $u_2 = 0$, *so that*

$$e_1 = u_1 - G_2 e_2, \tag{4.60}$$
$$e_2 = G_1 e_1.$$

with $G_1, G_2 : L_{2e}^{n_i} \mapsto L_{2e}^{n_i}$. *Then, if* G_1 *is passive and* G_2 *is strictly passive the closed loop is* L_2 *finite gain stable.*

Proof: Using (4.60), and the following representation of the passivity and strict passivity of G_1, G_2 respectively, we have

$$\langle e_1, G_1 e_1 \rangle_T \geq 0 \quad \langle e_2, G_2 e_2 \rangle_T \geq \delta_2 |e_2|_T^2 + \beta_2.$$

It follows that

$$\langle u_1, G_1 e_1 \rangle_T = \langle e_1, G_1 e_1 \rangle_T + \langle G_2 e_2, e_2 \rangle_T$$

$$\geq \delta_2 |e_2|_T^2 + \beta_2.$$

Using the Schwarz inequality, it follows that

$$|u_1|_T |G_1 e_1|_T \geq \delta_2 |G_1 e_1|_T^2 + \beta_2.$$

Since $\delta_2 > 0$, we may complete squares to get

$$\delta_2 \left(|G_1 e_1|_T - \frac{1}{\delta_2} |u_1|_T \right)^2 \leq \frac{1}{\delta_2} |u_1|_T^2 - \beta_2.$$

From this it follows that for some k, β, we have

$$|G_1 e_1|_T \leq k |e_1|_T + \beta. \qquad \square$$

Remark: It may be verified that this theorem holds when the conditions on G_1 and G_2 are interchanged. Thus, the first passivity theorem says that if one of the systems in the feedback loop is passive and the other strictly passive that the closed loop system is L_2 finite gain stable. The next theorem allows for one of the blocks in the closed loop to not be passive, provided the other is passive enough to compensate for it.

Theorem 4.21 Passivity Theorem for Feedback Loop When Neither System is Necessarily Passive. *Consider the feedback system of Figure 4.7. Assume that the two systems satisfy the following condition:*

$$\langle e_1, G_1 e_1 \rangle_T \geq \epsilon_1 |e_1|_T^2 + \delta_1 |G_1 e_1|_T^2,$$

$$\langle e_2, G_2 e_2 \rangle_T \geq \epsilon_2 |e_2|_T^2 + \delta_2 |G_2 e_2|_T^2.$$

(4.61)

If $\delta_1 + \epsilon_2 > 0$ and $\delta_2 + \epsilon_1 > 0$, the system is L_2 finite gain stable.

Proof: The feedback loop equations are $e_1 = u_1 - G_2 e_2$, $e_2 = u_2 + G_1 e_1$, so that

$$\langle e_1, G_1 e_1 \rangle_T + \langle e_2, G_2 e_2 \rangle_T = \langle u_2, e_1 \rangle_T + \langle u_1, e_2 \rangle_T.$$

Using (4.61), we have

$$\langle e_1, G_1 e_1 \rangle_T + \langle e_2, G_2 e_2 \rangle_T \geq \epsilon_1 |e_1|_T^2 + \delta_1 |G_1 e_1|_T^2 + \epsilon_2 |e_2|_T^2 + \delta_2 |G_2 e_2|_T^2.$$

Using this bound and the Schwartz inequality in the first bound, we have

$$\epsilon_1 |e_1|_T^2 + \delta_1 |G_1 e_1|_T^2 + \epsilon_2 |e_2|_T^2 + \delta_2 |G_2 e_2|_T^2 \leq |u_1|_T |e_2|_T + |u_2|_T |e_1|_T.$$

The loop equations can be used for $G_1 e_1 = e_2 - u_2$, $G_2 e_2 = u_1 - e_1$ to get

$$(\epsilon_1 + \delta_2) |e_1|_T^2 + (\epsilon_2 + \delta_1) |e_2|_T^2 - 2\delta_2 \langle e_1, u_1 \rangle_T - 2\delta_1 \langle e_2, u_2 \rangle_T$$

$$\leq -\delta_2 |u_1|_T^2 - \delta_1 |u_2|_T^2 + |u_1|_T |e_2|_T + |u_2|_T |e_1|_T.$$

Using the hypothesis that $\delta_1 + \epsilon_2 =: \mu_1 > 0$ and $\delta_2 + \epsilon_1 =: \mu_2 > 0$, we have

$$\mu_1 \left(|e_1|_T - \frac{\delta_2 + 1}{\mu_1} |u_1|_T \right)^2 + \mu_2 \left(|e_2|_T - \frac{\delta_1 + 1}{\mu_2} |u_2|_T \right)^2$$

$$\leq \left(\frac{(\delta_2 + 1)^2}{\mu_1} - \delta_2 \right) |u_1|_T^2 + \left(\frac{(\delta_1 + 1)^2}{\mu_2} - \delta_1 \right) |u_2|_T^2.$$

This bound can be used to get that

$$|e_1|_T \leq k_{11} |u_1|_T + k_{12} |u_2|_T + \beta_1,$$

$$|e_2|_T \leq k_{21} |u_1|_T + k_{22} |u_2|_T + \beta_2. \qquad \square$$

An interesting corollary of this passivity theorem is the following proposition

Proposition 4.22. *Consider the feedback system of Figure 4.7. Assume that the two systems satisfy the following condition:*

$$\langle e_1, G_1 e_1 \rangle_T \geq \epsilon |e_1|_T^2,$$

$$\langle e_2, G_2 e_2 \rangle_T \geq \delta_2 |G_2 e_2|_T^2, \qquad (4.62)$$

$$|G_2 e_2|_{T2} \leq \gamma |e_2|_T.$$

Under these conditions the closed loop system is L_2 finite gain stable.

Proof: Pick an $\alpha \in [0, \epsilon]$ and note that by using (4.62) we have that

$$\langle e_1, G_1 e_1 \rangle_T \geq (\epsilon - \alpha) |e_1|_T^2 + \alpha |e_1|_T^2,$$

$$\geq (\epsilon - \alpha) |e_1|_T^2 + \frac{\alpha}{\gamma^2} |G_1 e_1|_T^2. \qquad (4.63)$$

We use the last estimate of (4.63) with the estimate on G_2 of (4.62) as hypothesis for the application of Theorem 4.21 with $\epsilon_2 = 0, \epsilon_1 = \epsilon - \alpha, \delta_1 = \frac{\alpha}{\gamma^2}$. Now, we have that

$$\delta_1 + \epsilon_2 = \frac{\alpha}{\gamma^2} > 0, \qquad \delta_2 + \epsilon_1 = \epsilon + \delta - \alpha.$$

With α chosen small enough, the second term is greater than 0, enabling us to apply Theorem 4.21 to conclude the proposition. $\qquad \square$

Finally, there is an interesting generalization of the concept of passivity to what is known as dissipativity by Moylan and Hill [220] which is described in the Exercises (Problems 4.22 and 4.23).

4.5   Input–Output Stability of Linear Systems

We will specialize the framework of input–output stability to linear systems. We will take into account both lumped and distributed parameter linear systems in our treatment here. We first introduce the **algebra** of transfer functions:

Definition 4.23 Algebra of Transfer Functions. *The set of generalized functions (distributions) $f(\cdot)$ such that $f(t) \equiv 0$ for $t < 0$ of the form*

$$f(t) = \sum_{i=1}^{\infty} f_i \delta(t - t_i) + f_a(t) \tag{4.64}$$

where $f_i \in \mathbb{R}$, $t_i \in \mathbb{R}_+$, $\delta(\cdot)$ is the unit delta distribution, and $f_a(\cdot) \in L_1([0, \infty[)$, is called the algebra of stable convolution functions A, *provided that*

$$\sum_{i=1}^{\infty} |f_i| < \infty.$$

The norm of this function is given by

$$|f|_A := \sum_{i=1}^{\infty} |f_i| + \int_0^{\infty} |f_a(t)| dt.$$

The set A with the norm $|\cdot|_A$ is a Banach space (see Problem 4.16). It is made into an algebra by defining the binary convolution operation between two functions f, g in A as

$$f * g(t) := \int_{-\infty}^{t} f(t - \tau) g(\tau) d\tau.$$

It may be verified that convolution makes A into an algebra (Problem 4.16). The Laplace transform of a function in A belongs to the set $\hat{A}$. Thus $\hat{f}(s) \in \hat{A}$ is analytic in $\mathbb{C}_+$.

Remarks:

1. The set $\hat{A}$ includes the space of proper rational transfer functions (proper means that the degree of the numerator is less than or equal to that of the denominator), which are analytic in $\mathbb{C}_+$ and have no repeated poles on the $j\omega$ axis. On the other hand, functions in A of the kind

$$f(t) = \sum_{i=1}^{\infty} \frac{1}{(i + 1)^2} \delta(t - i)$$

   have Laplace transforms that belong to $\hat{A}$ but are not rational transfer functions.
2. For sequences $a(\cdot)$, we can define the analogue of the set A to be the space of l_1 sequences.
3. The *extension of the set A* is defined to be A_e, the set of all functions $f(\cdot)$ such that for all finite truncations T, $f_T(\cdot) \in A$.

Given a function $h(\cdot) \in A_e$, we define a zero initial condition single–input single–output (SISO) linear system with input $u \in L_{pe}$ to be $y(\cdot)$ defined by

$$y(t) = H(u) := \int_0^t h(t - \tau) u(\tau) d\tau := h * u \tag{4.65}$$

The following theorem gives stability conditions for SISO linear systems. Note that for linear systems finite gain stability and incremental finite gain stability are the same concept. We will use the term L_p stable for linear systems to mean finite gain or incremental finite gain L_p stable.

Theorem 4.24 Input Output Stability of SISO Linear Time Invariant Systems. *Consider the SISO linear time-invariant system H described by equation (4.65) with the convolution kernel $h(\cdot) \in A_e$. Then, we have that that for all $p = 1, 2, \ldots, \infty$, H is L_{pe} finite gain stable if and only if $h \in A$. Moreover in this case we have that*

$$|y|_p \le |h|_A |u|_p, \quad p = 1, 2, \ldots, \infty. \tag{4.66}$$

Proof: We begin with the proof that $h \in A$ implies L_{pe} bounded input bounded output stability. We start with $p = 1, 2, \ldots < \infty$. The $p = \infty$ case needs to be done separately. Let $u(\cdot) \in L_p$. Then, defining as before the integer q such that

$$\frac{1}{p} + \frac{1}{q} = 1,$$

we see that

$$
\begin{aligned}
|y(t)| &\le \int_0^t |h(t-\tau)||u(\tau)|d\tau \\
&= \int_0^t |h(t-\tau)|^{1/q}|h(t-\tau)|^{1/p}|u(\tau)|d\tau \\
&= \left[\int_0^t |h(t-\tau)|d\tau\right]^{1/q}\left[\int_0^t |h(t-\tau)||u(\tau)|^p d\tau\right]^{1/p}
\end{aligned} \tag{4.67}
$$

using the Hölder inequality. Taking the p-th power of both sides we get that

$$|y(t)|^p \le |h|_A^{p/q} \int_0^t |h(t-\tau)||u(\tau)|^p d\tau \tag{4.68}$$

Integrating both sides from 0 to ∞ and interchanging order of integration yields

$$
\begin{aligned}
\int_0^\infty |y(t)|^p dt &\le |h|_A^{p/q} \int_0^\infty \int_0^t |h(t-\tau)||u(\tau)|^p d\tau dt \\
&= |h|_A^{p/q} \int_0^\infty \int_\tau^\infty |h(t-\tau)||u(\tau)|^p dt d\tau \\
&= |h|_A^{p/q} \int_0^\infty \left[\int_\tau^\infty |h(t-\tau)|dt\right]|u(\tau)|^p d\tau \\
&= |h|_A^{p/q} |h|_A |f|_p^p.
\end{aligned} \tag{4.69}
$$

The last equation establishes that

$$|y(\cdot)|_p \le |h|_A |u(\cdot)|_p.$$

For $p = \infty$, we use the explicit form

$$h(t) = \sum_{i=1}^\infty h_i \delta(t - t_i) + h_a(t),$$

so that

$$y(t) = \sum_{i=1}^{\infty} h_i u(t - t_i) + \int_0^t h_a(t - \tau)u(\tau)d\tau,$$

$$|y(t)| \le \sup|u(t)| \left[\sum_{i=1}^{\infty} |h_i| + \int_0^{\infty} |h_a(\tau)|d\tau \right]$$

$$= |h|_A |u(\cdot)|_\infty.$$

The last equation implies that

$$|y(\cdot)|_\infty \le |h|_A |u(\cdot)|_\infty$$

For the converse, that is, to show that L_p stability implies that $h \in A$, we note that if the linear system H is stable for all L_p, it is in particular L_1 stable. Also the linear system H is a continuous linear operator from L_1 to L_1 (the finite gain is actually the Lipschitz constant of continuity!). Since every distribution in A can be obtained as a limit (in the sense of distributions or in the A norm) of functions in L_1, it follows that H also maps functions in A to functions in A. In particular, this implies that $H(\delta(t)) = h(t) \in A$, completing the proof. $\square$

Remarks and Extensions:

1. The preceding theorem can be generalized to time-varying SISO linear systems (see Problem 4.17).
2. The discrete time version of the theorem is also a straightforward extension with $h \in l_1$.
3. The bound (4.66) is not tight for $p \ne 1, \infty$.[4] Thus in general, it cannot be said that the induced norm of the linear operator $H : L_p \mapsto L_p$ is $|h|_A$. In fact, for the case of L_2 stability, Parseval's equality can be used to give a better bound than that given in (4.66).

Proposition 4.25 Frequency Domain Bounds for L_2 Stability.   *Consider the SISO time invariant linear operator H of equation (4.65) from $L_{2e} \mapsto L_{2e}$, with $h \in A$. Then H is L_2 finite gain stable and*

$$|y(\cdot)|_2 \le \sup_{\omega} |\hat{h}(j\omega)| |u(\cdot)|_2. \tag{4.70}$$

Proof: Since $y = h * u$ it follows that $\hat{y}(s) = \hat{h}(s)\hat{u}(s)$. Parseval's equality states that for functions $f \in L_2$,

$$|f|_2^2 = \frac{1}{2\pi} \int_{-\infty}^{\infty} |\hat{f}(j\omega)|^2 d\omega.$$

---

[4]The proof of this is involved; see [316].

Using this on $y(\cdot)$ we have

$$|y|_2^2 = \frac{1}{2\pi} \int_{-\infty}^{\infty} |\hat{h}(j\omega)|^2 |\hat{u}(j\omega)|^2 d\omega.$$
$$= \leq \sup_{\omega} |\hat{h}(j\omega)|^2 |u|_2^2. \qquad\qquad \square$$

The bound (4.70 is a better bound than (4.66), since it can be shown that $\sup_{\omega} |\hat{h}(j\omega)$ is the L_2 induced norm for the linear operator H. It has the physical interpretation of being the maximum amplitude of the frequency domain transfer function $\hat{h}(j\omega)$. The idea of the proof is to follow this intuition. The proof would be completely straightforward if sinusoidal signals belonged to L_2, but since they do not, the idea is to approximate a sinusoidal signal at the frequency ω_{sup} at which the supremum is achieved by

$$e^{-\epsilon t} \sin(\omega_{\text{sup}} t)$$

for progressively smaller ϵ (see [83] for details).

The multiple–input multiple–output (MIMO) generalization of this theorem to systems of the form $H : L_{pe}^{n_i} \mapsto L_{pe}^{n_o}$ described by

$$y(t) = \int_0^t h(t - \tau) u(\tau) d\tau \qquad\qquad (4.71)$$

where $u(t) \in \mathbb{R}^{n_i}$, $y(t) \in \mathbb{R}^{n_o}$, $h(t) \in \mathbb{R}^{n_o \times n_i}$ is as follows. The proof is quite procedural and is left to the exercises.

Theorem 4.26 Input–Output Stability of MIMO Linear Time Invariant Systems. *Consider the MIMO linear time invariant system H described by equation (4.71) with the convolution kernel $h(\cdot) \in A_e^{n_o \times n_i}$. Then, we have that that for all $p = 1, 2, \ldots, \infty$, H is L_{pe} stable if and only if $h \in A^{n_o \times n_i}$. Moreover in this case we have that*

$$|y|_p \leq |h|_A |u|_p \quad p = 1, 2, \ldots, \infty. \qquad\qquad (4.72)$$

The counterpart of the L_2 stability bound of Proposition 4.25 for MIMO systems is

$$|y|_2 \leq \sup_{\omega} |\hat{h}(j\omega)|_{i2} |u|_2, \qquad\qquad (4.73)$$

where the matrix norm is the l_2 induced norm on the transfer function matrix $|\hat{h}(j\omega)|$.

4.6   Input–Output Stability Analysis of Feedback Systems

In this section we combine the results from the small gain theorem and passivity based theorems to give conditions for the finite gain stability of nonlinear feedback systems, with some linear systems in the feedback loop. The results that we derive frequently have some very nice graphical interpretations. We begin with a lemma known as the *Paley–Wiener* lemma. It's proof while straightforward is not in the flow of this section, and we refer the interested reader to Rudin [246].

Lemma 4.27 Paley–Wiener Lemma.   *Let* $\hat{f} \in \hat{A}$. *Then, the function* $1/\hat{f} \in \hat{A}$ *if and only if*

$$\inf_{\mathrm{Re}\, s \geq 0} |\hat{f}(s)| > 0. \tag{4.74}$$

In the preceding equation, the notation Re refers to the real part of a complex number.

Remarks:

1. The elements of the algebra $\hat{A}$ that satisfy the Paley–Wiener condition (4.74) are the units of the algebra.
2. As a consequence of the Paley–Wiener lemma, it is easy to see (Problem 4.18) for a matrix $\hat{F}(s) \in \hat{A}^{n \times n}$, $\hat{F}^{-1}(s) \in \hat{A}^{n \times n}$ if and only if

$$\inf_{\mathrm{Re}\, s \geq 0} |\det \hat{F}(s)| > 0.$$

Now, consider the feedback interconnection of two linear time invariant MIMO systems as shown in Figure 4.12, which is essentially the same as Figure 4.7 with the nonlinear operators N_1, N_2 replaced by linear operators G_1, G_2. The inputs to the system are $u_1(\cdot) \in L_{pe}^{n_1}$ and $u_2(\cdot) \in L_{pe}^{n_2}$, and the outputs $y_1(\cdot) \in L_{pe}^{n_2}$ and $y_2 \in L_{pe}^{n_1}$. The loop equations are given by

$$
\begin{aligned}
e_1 &= u_1 - y_2, \\
e_2 &= u_2 + y_1, \\
y_1 &= \hat{G}_1(e_1), \\
y_2 &= \hat{G}_2(e_2).
\end{aligned}
\tag{4.75}
$$

We will assume that $\hat{G}_1 \in \hat{A}_e^{n_2 \times n_1}$, $\hat{G}_2 \in \hat{A}_e^{n_1 \times n_2}$.

Theorem 4.28 Input Output Stability of LTI Feedback Systems.   *Assume that the linear systems* G_1, G_2 *are open-loop stable, that is to say that* $\hat{G}_1 \in \hat{A}^{n_2 \times n_1}$, $\hat{G}_2 \in \hat{A}^{n_1 \times n_2}$. *Then the closed loop system is* L_p *finite gain stable if*

$$\inf_{\mathrm{Re}\, s \geq 0} (\det(I + \hat{G}_1(s)\hat{G}_2(s))) > 0.$$

Proof: Simplifying the loop equations of (4.75) yields

$$
\begin{bmatrix} \hat{y}_1 \\ \hat{y}_2 \end{bmatrix} =
\begin{bmatrix}
\hat{G}_1(I + \hat{G}_2\hat{G}_1)^{-1} & -\hat{G}_1\hat{G}_2(I + \hat{G}_1\hat{G}_2)^{-1} \\
\hat{G}_2\hat{G}_1(I + \hat{G}_2\hat{G}_1)^{-1} & \hat{G}_2(I + \hat{G}_1\hat{G}_2)^{-1}
\end{bmatrix}.
\tag{4.76}
$$

The hypothesis of the theorem guarantees that $(I + \hat{G}_1\hat{G}_2)^{-1} \in \hat{A}$. Also, since det $(I + \hat{G}_1\hat{G}_2)) = \det(I + \hat{G}_2\hat{G}_1)$ we are also guaranteed that $(I + \hat{G}_2\hat{G}_1)^{-1} \in \hat{A}$.

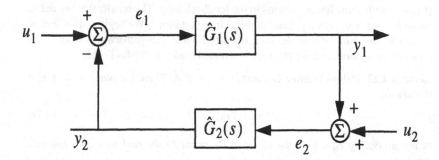

FIGURE 4.12. An interconnection of two linear MIMO systems

Since $\hat{A}$ is an algebra, it follows that all the entries of the matrix in (4.76) are in $\hat{A}$, establishing the theorem.   □

Remarks: While the preceding theorem is very elegant, it is a theorem about the interconnection of stable components to give a stable system. To be able to discuss the interconnection of unstable multi-input multi-output linear systems requires the development of more algebraic machinery of coprime factorizations and the ring of fractions on the algebra $\hat{A}$. This theory is very interesting and also has a nonlinear input output counterpart, but we will not begin a discussion of this direction here. Instead, we refer the reader to [317].

4.6.1   The Lur'e Problem

In this section we discuss the interconnection of a memoryless, but possibly time varying, nonlinearity with a linear time invariant system known as the Lur'e system, shown in Figure 4.13. The forward path has a SISO linear time invariant system with transfer function $\hat{g}(s)$ and the feedback path has a memoryless nonlinearity

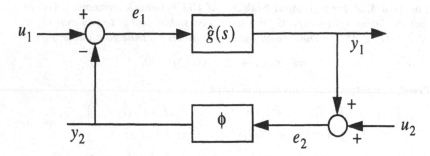

FIGURE 4.13. The Lur'e feedback interconnection problem

$\Phi(\cdot)$ on L_{2e} given by

$$\Phi(e_2)(t) := \phi(e_2(t), t).$$

The nonlinearity is said to *belong to the sector* $[a, b]$, if $\phi(0, t) \equiv 0$ and

$$a\sigma^2 \leq \sigma\phi(\sigma, t) \leq b\sigma^2 \quad \forall \sigma.$$

The following proposition is an easy consequence of the incremental small gain theorem (Theorem 4.17) and the loop transformation theorem (Theorem 4.18) using the formula for the L_2 gain of $\hat{g}$ (Proposition 4.25). It is attributed to Sandberg [256], [255] and Zames [335], [336].

Theorem 4.29 Circle Criterion. *Consider the Lur'e system of Figure 4.13, with $\hat{g} \in \hat{A}$ and ϕ in the sector $[-r, r]$. Then, the closed loop system is finite gain L_2 stable if*

$$\sup_{\omega \in \mathbb{R}} |\hat{g}(j\omega)| < r^{-1}. \tag{4.77}$$

In the instance that ϕ belongs to the sector $[a, b]$ for arbitrary a, b, define $k = \frac{a+b}{2}$ and $r = \frac{b-a}{2}$, and the transfer function

$$\hat{g}_c = \frac{\hat{g}}{1 + k\hat{g}}.$$

Then the closed loop system is L_2 stable if $\hat{g}_c \in \hat{A}$ and

$$\sup_{\omega \in \mathbb{R}} |\hat{g}_c(j\omega)| < r^{-1}. \tag{4.78}$$

Proof: The proof of the first part of the theorem is a simple application of the incremental small gain theorem (Theorem 4.17). The proof of the second part of the theorem uses the loop transformation theorem (Theorem 4.18) to transform the linear time invariant system $\hat{g}$ and nonlinearity Φ to

$$\hat{g}_c = \frac{\hat{g}}{1 + k\hat{g}}, \quad \Phi_c(\sigma, t) = \Phi(\sigma, t) - k\sigma.$$

Note that Φ_c lies in the sector $[-r, r]$ now and the first part of the proposition can be applied. The details are left to the exercises. $\square$

Remarks: The foregoing theorem is called the circle criterion because of its graphical interpretation. If $\hat{g}$ is a rational transfer function, then the Nyquist criterion may be used to give a graphical interpretation of the conditions of the theorem:

1. Condition (4.77) requires that the Nyquist locus of the open loop transfer function $\hat{g}(j\omega)$ lie inside a disk of radius r^{-1}, as shown in Figure 4.14.
2. Condition (4.78) has the following three cases, shown in Figure 4.15. We define $D(a, b)$ to be the disk in the complex plane passing through the points $-1/a$, $-1/b$ on the real axis.

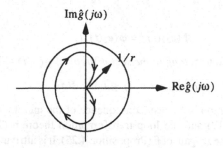

FIGURE 4.14. Interpretation of the circle criterion for symmetric sector bounded nonlinearities

a. $ab > 0$. With the $j\omega$ axis indented around the purely imaginary open loop poles of $\hat{g}(s)$ the Nyquist locus of the open loop transfer function $\hat{g}(j\omega)$ is bounded away from the disk $D(a, b)$ and encircles it counterclockwise ν_+ times as ω goes from $-\infty$ to ∞, where ν_+ is the number of $\mathbb{C}_+$ poles of the open loop transfer function $\hat{g}(s)$.

b. $0 = a < b$. Here $\hat{g}(s)$ has no $\mathbb{C}_+$ poles and the Nyquist locus of $\hat{g}(j\omega)$ satisfies

$$\inf_{\omega \in \mathbb{R}} \operatorname{Re} \hat{g}(j\omega) > -\frac{1}{b}$$

c. $a < 0 < b$. Here $\hat{g}(s)$ has no $\mathbb{C}_+$ poles and the Nyquist locus of $\hat{g}(j\omega)$ is contained inside the disk $D(a, b)$ bounded away from the circumference of the disk.

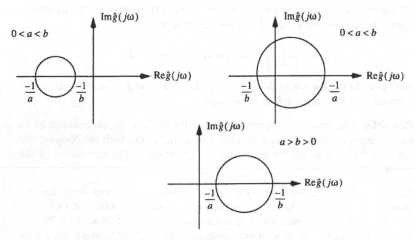

FIGURE 4.15. Graphical interpretation of the circle criterion

We will revisit the circle criterion in Chapter 6, where we will derive the result using Lyapunov techniques. It is worth noting that the statement of the theorem actually applies to some cases which are not covered in the Lyapunov techniques of the next two chapters, since it allows for the possibility of the linear system to be a distributed parameter or infinite dimensional linear system. Of course, in this case the extension of the graphical interpretation using the Nyquist criterion just described needs to be done with care.

A multivariable circle criterion can also be derived (see Problem 4.20), when $\hat{G}(s) \in \mathbb{R}^{n \times n}$ and the nonlinear Φ satisfies

$$a|\sigma|^2 \leq \sigma^T \phi(\sigma, t) \leq b|\sigma|^2. \tag{4.79}$$

The passivity theorem of the previous section can also be applied to the Lur'e system to give the following theorem:

Theorem 4.30 Popov Criterion. *Consider the Lur'e feedback system of Figure 4.13. Assume in addition that the nonlinearity is time invariant, i.e.,*

$$\Phi(e_2)(t) := \phi(e_2(t)),$$

and belongs to the sector $[0, k]$. Also assume that both $\hat{g}$, and $s\hat{g} \in \hat{A}$. Then, if there exist $q \in \mathbb{R}, \beta > 0$ such that

$$\inf_{\omega \in \mathbb{R}} \text{Re}\left[(1 + j\omega q)\hat{g}(j\omega)\right] + \frac{1}{k} := \beta > 0. \tag{4.80}$$

Then, the closed loop system is L_2 finite gain stable from the inputs $u_1, u_2, \dot{u}_2$ to the outputs y_1, y_2, that is there exists $\gamma < \infty$ such that

$$|y_i|_2 \leq \gamma(|u_1|_2 + |u|_2 + |\dot{u}_2|_2), \quad i = 1, 2.$$

Proof: The proof of the Popov criterion uses the following trick known as introducing "multipliers." We draw the Figure 4.13 as Figure 4.16 by introducing the "multiplier" $1 + qs$. In this process u_2 gets modified to $v_2 = u_2 + q\dot{u}_2$. Define $\hat{g}_1(s) = (1 + qs)\hat{g}(s)$, the corresponding linear operator G_1. and the nonlinear block G_2 as the concatenation of $1/(1 + qs)$ and the nonlinearity Φ. The input u_1 and the signals e_1, e_2, y_1, y_2 are unchanged, but they may not be present as the outputs or inputs of blocks as shown in the figure. The output of the G_1 block is labeled z_1 (its input is e_1), and the input to the G_2 block is labeled w_2 (its output is y_2). It is important to specify the initial condition of the block $1/(1 + qs)$ in G_2 as 0. Now,

$$\langle e_1, G_1 e_1 \rangle_T = \int_0^T e_1(t)(G_1 e_1)(t)dt$$

$$= \int_0^\infty e_{1T}(t)(G_1 e_{1T}(t)dt$$

where we have used the causality of G_1 in the last step to equate $(G_1(e_1))_T$ to $G_1(e_{1T})$. Now, by hypothesis, G_1 is a L_2 finite gain stable operator, hence $G_1(e_{1T})$

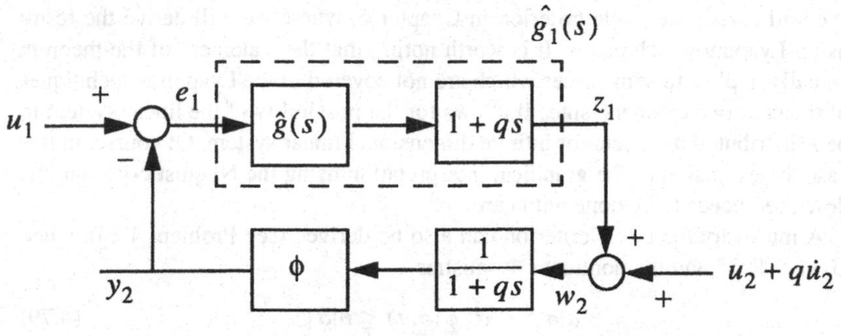

FIGURE 4.16. Showing the loop transformation required for the Popov criterion

belongs to L_2. Thus, using the Parseval theorem, we have that

$$\langle e_1, G_1(e_{1T}) \rangle = \frac{1}{2\pi} \mathrm{Re} \int_{-\infty}^{\infty} |\hat{e}_1(j\omega)^* \hat{g}_1(j\omega) e_1(j\omega) d\omega$$

$$= \frac{1}{2\pi} \int_{-\infty}^{\infty} \mathrm{Re}\, \hat{g}_1(j\omega) |\hat{e}_1(j\omega)|^2 d\omega$$

Using the bound of (4.80) it follows that if

$$\epsilon := \inf_{\omega \in \mathbb{R}} \mathrm{Re}(1 + j\omega q)\hat{g}(j\omega),$$

then

$$\langle e_1, G_1(e_{1T}) \rangle \geq \epsilon |e_{1T}|_2^2 = \epsilon |e_1|_T^2. \tag{4.81}$$

To complete the proof of the theorem, we will invoke the corollary of the passivity theorem 4.21 stated as Proposition 4.22. For this we need a bound on

$$\langle z_2, G_2 z_2 \rangle_T = \langle z_2, y_2 \rangle_T.$$

Note that $y_2 = \phi(e_2(t))$, $z_2 = e_2 + q\dot{e}_2$ Thus, it follows that

$$\langle z_2, y_2 \rangle_T = \int_0^T \phi(e_2(t)) e_2(t) dt + q \int_0^T \phi(e_2(t)) \dot{e}_2(t) dt.$$

The second term above can be shown to be positive, since

$$\int_0^T \phi(e_2(t)) \dot{e}_2(t) dt = \int_{e_2(0)}^{e_2(T)} \phi(\sigma) d\sigma = \int_0^{e_2(T)} \phi(\sigma) d\sigma \geq 0.$$

In turn, this implies that

$$\langle z_2, y_2 \rangle_T \geq \int_0^T \phi(e_2(t)) e_2(t) dt,$$

Using the sector bounds on $\phi(\cdot)$ it follows that

$$\phi(e_2(t)) e_2(t) \geq \frac{1}{k} (\phi(e_2(t)))^2.$$

Using this in the previous bound yields

$$\phi(e_2(t))e_2(t) \geq \frac{1}{k}|y_2|_T^2.$$

Combining this equation with equation (4.81) we see from Proposition 4.22 that the proposition holds.   □

Remarks:

1. The Popov criterion reduces to the circle criterion if the q in equation (4.80) is zero. Of course the circle criterion is more general in this instance, since it allows for arbitrary sector bounded nonlinearities and time varying nonlinearities.
2. The Popov criterion does not quite prove L_2 stability of the closed loop system since it requires also that $\dot{u}_2 \in L_2$.
3. In the instance that $\hat{g}, s\hat{g}$ are proper rational functions then there is a very pleasing graphical interpretation of the condition of (4.80) shown in Figure 4.17 which shows a plot of Re $\hat{g}(j\omega)$ against ωIm $\hat{g}(j\omega)$ as ω goes from 0 to ∞. The plot need not be drawn for negative ω since both Re$\hat{g}$ and ω Im $\hat{g}(j\omega)$ are even functions of ω. The inequality (4.80) means that there exists a straight line through $-1/k$ with slope $1/q$ which lies above the plot.

4.7 Volterra Input–Output Representations

In this section we will restrict our attention to single input single output (SISO) systems. The material in this section may be extended to multiple input multiple output systems with a considerable increase in notational complexity deriving from multilinear algebra in many variables. In an input-output context, linear time-invariant systems of a very general class may be represented by convolution operators of

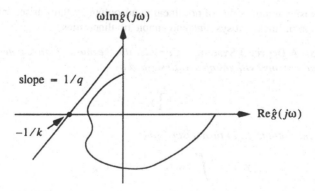

FIGURE 4.17. A graphical interpretation of the Popov criterion

the form

$$y(t) = \int_{-\infty}^{t} h(t - \tau)u(\tau)d\tau. \tag{4.82}$$

Here the fact that the integral has upper limit t models a *causal* linear system and the lower limit of $-\infty$ models the lack of an initial condition in the system description (hence, the entire past history of the system). In contrast to previous sections, where the dependence of the input output operator on the initial condition was explicit, here we will replace this dependence on the initial condition by having the limits of integration going from $-\infty$ rather than 0. In this section, we will explore the properties of a nonlinear generalization of (4.82) of the form

$$y(t) = \sum_{n=1}^{\infty} \int_{-\infty}^{t} \int_{-\infty}^{\tau_1} \cdots \int_{-\infty}^{\tau_{n-1}} h(t-\tau_1, \tau_1-\tau_2, \ldots, \tau_{n-1}-\tau_n)d\tau_1 \cdots d\tau_n. \tag{4.83}$$

Equation (4.83) is to be thought of as a polynomial or Taylor series expansion for the function $y(\cdot)$ in terms of the function $u(\cdot)$. Historically, Volterra [319] introduced the terminology "function of a function," or actually "function of lines," and defined the derivatives of such functions of functions or "functionals." Indeed, then, if F denotes the operator (functional) taking input functions $u(\cdot)$ to output functions $y(\cdot)$, then the terms listed above correspond to the summation of the n-th term of the power series for F. The first use of the Volterra representation in nonlinear system theory was by Wiener [327] and hence representations of the form of (4.83) are referred to as Volterra–Wiener series. Our development follows that of Boyd, Chua and Desoer [37], and Rugh [248], which have a nice extended treatment of the subject. For other details and many examples, see also Schetzen [267].

4.7.1 Homogeneous, Polynomial and Volterra Systems in the Time Domain

One of the most common kinds of nonlinear input-output systems arises from the interconnection of linear subsystems and simple nonlinearities.

Example 4.31 A Degree 3 System. *Consider the the multiplicative connection of three linear systems, with outputs y_i described by*

$$y = \prod_{i=1}^{3} y_i$$

with each of the outputs y_i, in turn described by

$$y_i(t) = \int_{-\infty}^{t} h_i(t - \tau_i)u(\tau_i)d\tau_i$$

(we will assume that $h_i(\cdot) \in L_1$, so that the integral above is well defined as a map from $L_\infty \mapsto L_\infty$). The overall input output system is a cubic or degree 3

homogeneous system described by

$$y(t) = \int_{-\infty}^{t} \int_{-\infty}^{t} \int_{-\infty}^{t} \prod_{i=1}^{3} h_i(t - \sigma_i) u(\sigma_i) d\sigma_1 d\sigma_2 d\sigma_3.$$

More generally, we can define a homogeneous degree n and a polynomial degree n nonlinear system as follows:

Definition 4.32 Homogeneous and Polynomial Time Invariant Nonlinear Systems. *An input-output nonlinear system is said to be* homogeneous of degree n *if it can be represented in input-output form as*

$$y(t) = \int_{-\infty}^{t} \cdots \int_{-\infty}^{t} h_n(t, \sigma_1, \sigma_2, \ldots, \sigma_n) u(\sigma_1) u(\sigma_2) \cdots u(\sigma_n) d\sigma_1 \cdots d\sigma_n$$

$$(4.84)$$

with the assumption that $h_n(\cdots) \in L_1$, that is,

$$\int_{-\infty}^{\infty} |h_n(t, \sigma_1, \cdots, \sigma_n)| \, dt \, d\sigma_1 \cdots d\sigma_n := k < \infty. \qquad (4.85)$$

We will also assume that the system of (4.84) is time-invariant, so that the kernel $h_n(t, \sigma_1, \sigma_2, \ldots, \sigma_n)$ is actually a function of $t - \sigma_1, \sigma_1 - \sigma_2, \ldots$. Thus, while we could continue the development of what follows for general time-varying systems, we will restrict attention in this section to time-invariant Volterra descriptions, and will emphasize the dependence of h_n on the difference of time arguments. An input output system is said to be polynomial of degree N *if it can be written as*

$$y(t) = \sum_{i=1}^{N} \left[\int_{-\infty}^{t} \cdots \int_{-\infty}^{t} h_i(t - \sigma_1, \ldots, \sigma_{i-1} - \sigma_i) d\sigma_1 \ldots d\sigma_i \right] \qquad (4.86)$$

with the assumption that $h_i(\cdots) \in L_1$.

It is easy to verify that if $u(\cdot) \in L_\infty$ with $|u(\cdot)|_\infty := M$ and if $h(\cdots) \in L_1$ as in equation (4.85), then the output of the homogeneous nonlinear system of degree n defined in (4.84) is in L_∞ with

$$|y(\cdot)|_\infty \le k|u(\cdot)|_\infty^n = kM^n.$$

This motivates the following definition:

Definition 4.33 Volterra System. *A nonlinear input-output system is said to be a* Volterra system *if its input-output description is given by*

$$y(t) = N(u)(t)$$

$$:= \sum_{n=1}^{\infty} \int_{-\infty}^{t} \cdots \int_{-\infty}^{t} h_n(t - \sigma_1, \sigma_1 - \sigma_2, \ldots, \sigma_{n-1} - \sigma_n) u(\sigma_1) \qquad (4.87)$$

$$\times u(\sigma_2) \cdots u(\sigma_n) d\sigma_1 d\sigma_2 \cdots d\sigma_n$$

with the assumption that each of the $h_n(\cdots)$ is in L_1 with $|h_n(\cdots)|_1 := k_n$. The Volterra system is well defined as a map from $L_\infty \mapsto L_\infty$ for those inputs $u(\cdot) \in$

L_∞ *with norm* $|u(\cdot)|_\infty := M$ *for which the following series converges:*

$$\sum_{n=1}^{\infty} k_n M^n.$$

The largest M satisfying this condition is referred to as the radius of convergence *of the Volterra series.*

Remarks:

1. A Volterra system with the series of (4.87) truncated to length k is a good approximation of the full Volterra system for those inputs $u(\cdot)$ for which

$$\sum_{n=k+1}^{\infty} k_n |u(\cdot)|_\infty^n$$

is small.

2. A Volterra system is incrementally finite gain L_∞ stable (see Problem 4.26 for a proof).

3. While Volterra series operators are usually viewed as maps on L_∞, we can also view them as operators on L_p spaces provided we exert care about their domain of definition (which may not include open subsets of L_p). However, if for some $u(\cdot) \in L_p$ we have that $N(u)(\cdot) \in L_p$ we can derive incremental finite gain bounds of the kind derived on L_∞ (see Problem 4.27).

4.7.2 Volterra Representations from Differential Equations

The main technique for obtaining Volterra series expansions from differential equation descriptions of nonlinear systems is *Peano–Baker* series expansion. We first develop the Peano–Baker Volterra series for *bilinear systems*:

Proposition 4.34 Peano–Baker–Volterra Series for Bilinear Systems. *Consider the bilinear system*

$$\dot{x} = A(t)x + D(t)xu + b(t)u, \quad x(0) = x_0,$$

$$y = c(t)x. \tag{4.88}$$

Here $A(t), D(t) \in \mathbb{R}^{n \times n}$, $b(t) \in \mathbb{R}^n$, $c^T(t) \in \mathbb{R}^n$ and the input $u(t)$, and output $y(t)$ are scalars. Then

$$y(t) = c(t)\Phi(t, 0)x_0$$

$$+ \sum_{k=1}^{\infty} \int_0^t \int_0^{\sigma_1} \cdots \int_0^{\sigma_{k-1}} c(t)\Phi(t, \sigma_1)D(\sigma_1)\Phi(\sigma_1, \sigma_2)D(\sigma_2)$$

$$\cdots D(\sigma_k)\Phi(\sigma_k, 0)x_0 u(\sigma_1) \cdots u(\sigma_k)d\sigma_k \cdots d\sigma_1 \tag{4.89}$$

$$+ \sum_{k=1}^{\infty} \int_0^t \int_0^{\sigma_1} \cdots \int_0^{\sigma_{k-1}} c(t)\Phi(t, \sigma_1)D(\sigma_1)\Phi(\sigma_1, \sigma_2)D(\sigma_2)$$

$$\cdots D(\sigma_k)\Phi(\sigma_k, 0)b(\sigma_k)u(\sigma_1) \cdots u(\sigma_k)d\sigma_k \cdots d\sigma_1.$$

Here $\Phi(t, \tau)$ stands for the state transition matrix of the unforced system $\dot{x} = A(t)x(t)$.

Proof: The proof follows along the lines of the Peano–Baker series of Problem 3.11. More precisely, define $z(t) = \Phi(t, 0)^{-1}x(t)$ and verify that

$$
\begin{aligned}
\dot{z} &= \tilde{D}(t)z(t)u + \tilde{b}(t)u(t), \quad z(t_0) = z_0, \\
y &= \tilde{c}(t)z.
\end{aligned}
\tag{4.90}
$$

where

$$
\tilde{D}(t) = \Phi^{-1}(t, 0)D(t)\Phi(t, 0), \quad \tilde{b}(t) = \Phi^{-1}(t, 0)b(t), \quad \tilde{c}(t) = c(t)\Phi(t, 0).
\tag{4.91}
$$

The Peano Baker series for (4.90) follows from the recursive formula

$$
z^i(t) = z_0 + \int_0^t \tilde{D}(\sigma_1)z^{i-1}(\sigma_1)u(\sigma_1) + \tilde{b}(\sigma_1)u(\sigma_1)d\sigma_1
$$

for $i = 0, 1, \ldots$, starting with $z^0(t) \equiv z_0$. It is easy to verify that the formula for $z^i(t)$ is like the formula (4.89) with the summation running from $k = 1$ to $k = i$, namely,

$$
\begin{aligned}
z^i(t) = z_0 &+ \sum_{k=1}^{i} \int_0^t \int_0^{\sigma_1} \cdot \int_0^{\sigma_{k-1}} \tilde{D}(\sigma_1)\tilde{D}(\sigma_2) \\
&\cdots \tilde{D}(\sigma_k)z_0 u(\sigma_1)\cdots u(\sigma_k)d\sigma_k\cdots d\sigma_1 \\
&+ \sum_{k=1}^{i} \int_0^t \int_0^{\sigma_1} \cdot \int_0^{\sigma_{k-1}} \tilde{D}(\sigma_1)\tilde{D}(\sigma_2) \\
&\cdots \tilde{D}(\sigma_k)\tilde{b}(\sigma_k)u(\sigma_1)\cdots u(\sigma_k)d\sigma_k\cdots d\sigma_1.
\end{aligned}
\tag{4.92}
$$

Now use the fact that $\Phi(t, 0)$ is bounded for given t, along with the boundedness of $D(t), b(t), c(t)$, to prove that as $i \to \infty$ the series of (4.92) is uniformly convergent for bounded $u(\cdot)$. The fact that the limit $z^\infty(\cdot)$ is the solution of the differential equation (4.90) follows by differentiating the limit of equation (4.92) and verifying the initial condition (by the existence uniqueness theorem of differential equations). The desired formula (4.89) follows now by substituting for $\tilde{D}(t), \tilde{b}(t), \tilde{c}(t), z_0 = \Phi^{-1}(t, 0)x_0$. $\square$

Note that the Volterra series expansion (4.89) of the preceding proposition actually does depend on the initial condition at time 0. As a consequence, the integrals run from 0 to t rather than from $-\infty$ to t. For more general nonlinear control systems of the form

$$
\begin{aligned}
\dot{x} &= f(x) + g(x)u, \\
y &= h(x)
\end{aligned}
\tag{4.93}
$$

the derivation of the Volterra series can be involved. It is traditional in this approach to append an extra state $x_{n+1} = y(t)$ to the dynamics of $x(t)$ so as to be able to assume that the output is a linear function of the (augmented) states. This having been done, the typical existence theorem for a Volterra series expansion uses the variational technique of considering the solution of the driven control system as a variation of the unforced system and proving that for small enough $|u(\cdot)|_\infty$ the Volterra series converges (see Chapter 3 of [248]). However, such proofs are not constructive. An interesting approach is the Carleman (bi)linearization of control systems discussed in Chapter 3. Indeed, recall from Problem 3.16 that the state trajectory of an arbitrary linear analytic system of the form

$$\dot{x} = \sum_{i=1}^{\infty} A_i x^{(i)} + \sum_{i=0}^{\infty} B_i x^{(i)} u,$$

$$y = \sum_{i=1}^{\infty} C_i x^{(i)}.$$

$$(4.94)$$

may be approximated by a bilinear control system with state $x^{\oplus} \in \mathbb{R}^{n+n^2+\cdots+n^N}$:

$$\frac{dx^{\oplus}}{dt} = \begin{bmatrix} A_{11} & A_{12} & \cdots & A_{1N} \\ 0 & A_{22} & \cdots & A_{2N} \\ 0 & 0 & \cdots & A_{3N} \\ \vdots & \vdots & \vdots & \vdots \\ 0 & 0 & \cdots & A_{N1} \end{bmatrix} x^{\oplus}$$

$$+ \begin{bmatrix} B_{11} & B_{12} & \cdots & B_{1N} \\ 0 & B_{22} & \cdots & B_{2N} \\ 0 & 0 & \cdots & B_{3N} \\ \vdots & \vdots & \vdots & \vdots \\ 0 & 0 & \cdots & B_{N1} \end{bmatrix} x^{\oplus} u + \begin{bmatrix} B_{10} \\ 0 \\ 0 \\ \vdots \\ 0 \end{bmatrix} u,$$

$$(4.95)$$

$$y = \begin{bmatrix} C_1 & C_2 & \cdots & C_N \end{bmatrix} x^{\oplus}$$

for suitably defined A_{ij}, B_{ij}. In principle, the Volterra series of (4.95) may be computed using Proposition 4.34. We say "in principle" because the dimension of the system grows rather dramatically with the parameter N of the approximation. Because of the upper triangular form of the Carleman bilinear approximation there is some simplification in the form of the Volterra kernels. Also, the Volterra series expansion of the approximate system of (4.95) agrees exactly with the first N terms of the Volterra series of the original system (that is, the terms that are polynomial of order less than or equal to N). The proof of this statement, while straightforward is notationally involved.

4.7.3 Frequency Domain Representation of Volterra Input Output Expansions

As with linear systems, it is very useful to use the Laplace transforms of the Volterra kernels defined by

$$\hat{h}_n(s_1, \ldots, s_n) = \int_{-\infty}^{\infty} \cdots \int_{-\infty}^{\infty} h_n(\tau_1, \tau_2, \ldots, \tau_n) d\tau_1 d\tau_2 \cdots d\tau_n. \qquad (4.96)$$

Since $h_n(\cdots) \in L_1$, it follows that the domain of definition of $\hat{h}_n(\cdots)$ includes $\mathbb{C}_+$. Thus, the Fourier transform of the Volterra kernels is also well defined, and is written as $\hat{h}_n(j\omega_1, \ldots, j\omega_n)$. It is clear that if the input to a Volterra system is a sinusoid at frequency ω, say $u(t) = \alpha e^{j\omega t} + \alpha^* e^{-j\omega t}$, then the output is periodic with period $2\pi/\omega$ and of the form

$$y(t) = \sum_{n=-\infty}^{\infty} \left(\alpha \hat{h}_n(j\omega, \ldots, j\omega) e^{jn\omega t} + \alpha^* \hat{h}_n(-j\omega, \ldots, -j\omega) \right).$$

In particular this implies that Volterra systems generate only harmonics of the input signal (and not sub-harmonics).

Proposition 4.35 Frequency Domain Formula for Multitone Signals. *Consider the Volterra System of (4.87) and let the input $u(t)$ be a* multitone signal *of the form*

$$u(t) = \sum_{k=-M}^{M} \alpha_k e^{j\omega k t}.$$

Then the output $y(t)$ is a periodic signal of the form

$$y(t) = \sum_{m=-\infty}^{\infty} \beta_m e^{j\omega m t}$$

with

$$\beta_m = \sum_{n=1}^{\infty} \left(\sum_{k_1 + \cdots + k_n = m} \alpha_{k_1} \cdots \alpha_{k_n} \hat{h}_n(j\omega k_1, \ldots, j\omega k_n) \right). \qquad (4.97)$$

Proof: Using the form of the multitone signal in the formula for the output $y(t)$ yields

$$y(t) = \sum_{n=1}^{\infty} h_n(\tau_1, \tau_2, \ldots, \tau_n) \prod_{i=1}^{n} \sum_{k=-M}^{M} \alpha_k e^{j\omega k(t-\tau_i)} d\tau_i$$

$$= \sum_{n=1}^{\infty} \left(\sum_{-M \le k_1 \cdots k_n \le M} \right) \alpha_{k_1} \cdots \alpha_{k_n} \qquad (4.98)$$

$$\times \hat{h}_n(j\omega k_1, \ldots, j\omega k_n) e^{j(\omega k_1 + \cdots + \omega k_n)t}.$$

The term $\alpha_{k_1} \cdots \alpha_{k_n} \hat{h}_n(j\omega k_1, \ldots, j\omega k_n) e^{j(\omega k_1 + \cdots + \omega k_n)t}$ is often called an n-th order *intermodulation product*, and $\hat{h}_n(j\omega k_1, \ldots, j\omega k_n)$ is the *intermodulation distor-*

tion measure.[5] Since the terms in equation (4.98) are L_1 it follows that we may evaluate the m th Fourier coefficient of $y(t)$

$$\beta_m := \frac{\omega}{2\pi} \int_0^{2\pi\omega^{-1}} y(t)e^{-j\omega mt} dt$$

$$= \sum_{n=1}^{\infty} \left\{ \sum_{k_1 + \cdots k_n = m} \right\} \alpha_{k_1} \cdots \alpha_{k_n} \hat{h}_n(j\omega k_1, \ldots, j\omega k_n). \qquad \square$$

Remarks: The fundamental formula of (4.97) may not be absolutely convergent, that is,

$$\sum_{n=1}^{\infty} \left(\sum_{k_1 + \cdots + k_n = m} |\alpha_{k_1}| \cdots |\alpha_{k_n}| |\hat{h}_n(j\omega k_1, \ldots, j\omega k_n)| \right)$$

may equal ∞ even in some very elementary situations. Thus, to extend formula (4.97) to more general periodic inputs and almost periodic inputs requires care. In [37] some sufficient conditions are given to establish the absolute convergence of the series above for periodic inputs: For example, if

$$\hat{h}_n(j\omega k_1, \ldots, j\omega k_n) = O\left(\frac{1}{k_1 \cdots k_n} \right)$$

then a periodic input $u(\cdot)$ results in a periodic output $y(\cdot)$, and if

$$u(t) = \sum_k \hat{\alpha}_k e^{j\omega kt},$$

then

$$y(t) = \sum_k \hat{\beta}_k e^{j\omega kt}$$

with

$$\hat{\beta}_m = \sum_{n=1}^{\infty} \left(\sum_{k_1 + \cdots + k_n = m} \right) \alpha_{k_1} \cdots \alpha_{k_n} \hat{h}_n(j\omega k_1, \ldots, j\omega k_n).$$

For more general stationary input signals of the kind discussed earlier in this chapter, frequency domain formulas for the output of a Volterra system remain an open problem.

4.8 Summary

This chapter gave an introduction to some frequency domain techniques for the study of nonlinear systems. Several kinds of techniques have been presented:

---

[5]This term is frequently used to rate high quality amplifiers and loudspeakers.

approximate techniques such as describing functions with a proof of when they predict oscillations in a nonlinear system. To this day, describing function techniques remain an important practical tool of engineers: several engineering applications are in [109], [206], and [69]. Likewise, Volterra series with their frequency domain and state space realization theory remain an extremely popular tool for the analysis of nonlinear systems, especially slightly nonlinear ones ([112], [267], [248] and [39] are excellent references for more details on this topic). There is also a large literature on realization theory for bilinear systems, see for example Isidori and Ruberti [151] and Evans [95].

Finally, input-output methods for the study of stability of nonlinear systems, their applications to passivity and to the circle, and Popov criterion, robustness analysis of linear and nonlinear control systems represent a large body of literature. A comprehensive treatment is in the book of Desoer and Vidyasagar [83], see also Vidysagar [317], Willems [330], and Safonov [249] for the application to robustness of control schemes.

4.9 Exercises

Problem 4.1. Consider $u(t) = \cos t + \cos(\pi t)$. Find the autocovariance of $u(t)$. Is $u(t)$ periodic? Is it *almost periodic* in any sense? A function $u(t)$ is said to be almost periodic if given any $\epsilon > 0$, there exists $T(\epsilon)$ such that

$$|u(t) - u(t + T)| < \epsilon \quad \text{for all } t.$$

Actually, there is a characterization of almost periodic functions as functions of the form

$$u(t) = \sum_{i=1}^{\infty} c_i \cos \omega_i t$$

with the $\{c_i\}$ being an l_2 sequence, that is to say,

$$\sum_{i=1}^{\infty} |c_i|^2 < \infty$$

What is the autocovariance of this function? Now, let $u(t) = \cos t + \cos(\pi t) + f(t)$, with $f(t)$ some continuous function that tends to zero as $t \to \infty$. Compute the autocovariance of $u(t)$. What can you say if $f(t)$ is bounded and is absolutely integrable? Compute the power spectral densities of the functions in this problem.

Problem 4.2. Prove that the autocovariance of a stationary signal $u(\cdot)$ is positive definite, that is given $t_1, \ldots t_k \in \mathbb{R}$ and $c_1, \ldots, c_k \in \mathbb{R}$, then

$$\sum_{i,j=1}^{n} c_i c_j R_u(t_i - t_j) \geq 0.$$

Problem 4.3. Prove that the autocovariance of a function is an even function of time, that is,

$$R_u(t) = R_u(-t).$$

For the cross covariance between u and y, prove that

$$R_{yu}(t) = R_{uy}(-t).$$

What implications do these properties have for the Fourier transforms $S_u(\omega)$ and $S_{yu}(\omega)$.

Problem 4.4. Prove Proposition 4.5.

Problem 4.5. Prove the formula of equation (4.22).

Problem 4.6. Prove Proposition 4.6 using the formula of (4.22).

Problem 4.7. Use harmonic balance to predict the existence (and amplitudes, frequencies) in the functional differential equation

$$\dot{x} + \alpha x(t - T) + Bx^3(t - T).$$

In this equation $T > 0$, $B > 0$, and it models the variation of orders in the shipbuilding industry (crudely).

Problem 4.8 Calcium release in cells [206]. The following equations model the release of calcium in certain cells

$$\dot{x}_1 = \frac{K}{1 + x_n} - b_1 x_1,$$

$$\dot{x}_j = x_{j-1} - b_j x_j, \quad j = 2, 3, \ldots, n.$$

Use the describing function to predict limit cycles say for $n = 10$. Verify your result using a simulation package.

Problem 4.9 Friction Controlled Backlash. Find the sinusoidal input describing function for friction controlled backlash, say in gears that do not quite mesh. This is shown in Figure 4.18 along with the hysteretic input-output characteristic.

Problem 4.10. Prove that if $f \in L_1 \cap L_\infty$, then $f \in L_p$ for $p \in [1, \infty]$.

Problem 4.11. Show that if $A : L_{pe}^n \mapsto L_{pe}^m$ is a causal operator and for two functions $u_1, u_2 \in L_{pe}^n$ we have that $(u_1)_T = (u_2)_T$ then it follows that

$$(A(u_1))_T = (A(u_2))_T.$$

Problem 4.12. Consider two nonlinear finite gain operators G_1, G_2 from $L_{pe}^{n_1} \mapsto L_{pe}^{n_2}$ and $L_{pe}^{n_2} \mapsto L_{pe}^{n_3}$, respectively. Prove that $G_2 \circ G_1$ is a finite gain operator as well. Repeat this also for the case that G_1, G_2 are both incremental finite gain stable.

Problem 4.13 Feedback interconnection of two systems. Consider the feedback interconnection of two nonlinear operators G_1, G_2 shown in Figure 4.19 with

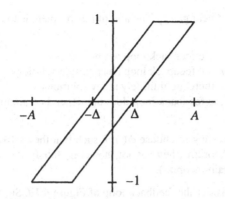

FIGURE 4.18. Friction controlled backlash

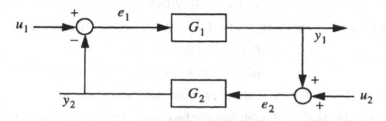

FIGURE 4.19. Feedback loop

inputs u_1, u_2 and two sets of outputs y_1, y_2 and e_1, e_2. Assume that the operators G_1, G_2 are defined from $L_{pe}^{n_i} \mapsto L_{pe}^{n_o}$ and $L_{pe}^{n_o} \mapsto L_{pe}^{n_i}$. Now show that the closed loop system from the inputs u_1, u_2 to the outputs y_1, y_2 is finite gain L_p stable iff the closed loop system defined from the inputs u_1, u_2 to y_1, y_2 is. Specialize this problem to the instance that $u_2 = 0$ to prove that

$$e_1 = (I + G_2 G_1)^{-1} u_1, \quad y_1 = G_1 G_2 (I + G_1 G_2)^{-1} u_1.$$

Problem 4.14 Simple memoryless feedback system [83].   Consider the simple single loop system shown in Figure 4.20 consisting of a memoryless nonlinearity

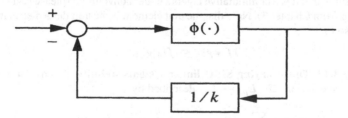

FIGURE 4.20. Memoryless feedback loop

$\phi : \mathbb{R} \mapsto \mathbb{R}$ and the linear gain $\frac{1}{k}$ The aim of this problem is to choose a $\phi(\cdot)$ and k such that

1. For some input u, the feedback loop has no solution.
2. For some input u, the feedback loop has multiple solutions.
3. For some input u, there are infinitely many solutions.
4. The output y of the feedback loop does not depend continuously on the input u.

It is especially interesting to choose $\phi(\cdot)$, k such that they satisfy the hypothesis of the small gain theorem (they cannot, however, satisfy the hypothesis of the incremental small gain theorem).

Problem 4.15. Consider the feedback loop of Figure 4.19. Set the input $u_2 \equiv 0$. Assume that $e_1 \in L_{pe}$ and define u_1 by

$$u_1 = e_1 + G_2(G_1(e_1))$$

Suppose that there exist constants $k_{21}, k_1, \beta_{21}, \beta_1$ with $k_{21}, k_1 \geq 0$ such that

$$|(G_2 G_1(e_1))_T| \leq k_{21}|e_1|_T + \beta_{21}$$

$$|(G_1(e_1)_T| \leq k_1 |e_1|_T + \beta_1$$

Prove that if $k_{21} < 1$, then the system has finite gain from u_1 to e_1, y_1.

Problem 4.16 Algebra of stable transfer functions. Prove that the space A of functions of the form

$$f(\cdot) = \sum_{i=1}^{\infty} f_i \delta(t - t_i) + f_a(\cdot)$$

with

$$\sum_{i=1}^{\infty} |f_i| < \infty, \qquad f_a \in L_1.$$

is a Banach space under the norm

$$|f(\cdot)|_A = \sum_{i=1}^{\infty} |f_i| + |f_a(\cdot)|_1.$$

Further show that it is a commutative algebra under convolution (please review the definition from Chapter 3). Note that the unit element is the unit delta distribution. Show also that

$$|f * g|_A \leq |f|_A |g|_A.$$

Problem 4.17 Time-varying SISO linear systems stability. Consider a time varying linear system $G : L_{pe} \mapsto L_{pe}$ described by

$$y(t) = \sum_{i=1}^{\infty} g_i(t) u(t - t_i) + \int_0^t g_a(t, \tau) u(\tau) d\tau, \qquad (4.99)$$

Since $u(t) = 0$ for $t \le 0$, it follows that if $I(t) := \{i : t_i \le t\}$, then the preceding equation can be written as

$$y(t) = \sum_{i \in I(t)} g_i(t)u(t - t_i) + \int_0^t g_a(t, \tau)u(\tau)d\tau.$$

Now assume that

$$c_\infty := \sup_t \left[\sum_{i \in I(t)} |g_i(t)| + \int_0^t |g_a(t, \tau)d\tau \right] < \infty,$$

$$c_1 := \sup_\tau \left[\sum_{i=1}^\infty |g_i(\tau + t_i)| + \int_\tau^\infty |g_a(t, \tau)dt \right] < \infty.$$

Show that this implies that the linear system described by equation (4.99) is L_{pe} finite gain stable if $c_1, c_\infty < \infty$. Relate this to Theorem 4.24.

Problem 4.18 Multivariable Paley–Wiener Condition. Prove the multivariable Paley–Wiener condition for the invertibility of a square transfer function $\hat{H}(s) \in \hat{A}^{n \times n}$.

Problem 4.19 Proof of the circle criterion. Prove Theorem 4.29.

Problem 4.20 Multivariable circle criterion. Derive a multivariable circle criterion for square linear systems (i.e., number of inputs equals the number of outputs) and a nonlinearity that satisfies the generalized sector condition of equation (4.79.

Problem 4.21 MIMO finite gain stability conditions. Derive the necessary and sufficient conditions for L_p finite gain stability of MIMO linear systems described by (4.71), given in Theorem 4.26.

Problem 4.22 Q, S, R dissipativity [220]. A nonlinear operator $G : L_{2e}^{n_i} \mapsto L_{2e}^{n_o}$ is said to be Q, S, R *dissipative* with $Q \in \mathbb{R}^{n_o \times n_o}, R \in \mathbb{R}^{n_i \times n_i}, S \in \mathbb{R}^{n_o \times n_i}$ if for all $T > 0$

$$\langle y, Qy \rangle_T + 2\langle y, Su \rangle_T + \langle u, Ru \rangle_T \ge 0. \tag{4.100}$$

This is a generalization of the definition of passivity and strict passivity. Now prove that if Q is negative definite, then the system is L_2 finite gain stable, that is there exists k, β such that

$$|Gu|_{T2} \le k|u|_{T2} + \beta.$$

Problem 4.23 Stability of interconnected systems [220]. Consider a collection of N nonlinear operators $G_j : L_{2e}^{n_{ij}} \mapsto L_{2e}^{n_{oj}}$ that are respectively Q_j, S_j, R_j dissipative. Now define the input vector $u = (u_1, u_2, \ldots, u_N)^T$ and the output vector $y = (y_1, y_2 \ldots, y_N)^T$ and the interconnection of the systems by

$$u = v - Hy \tag{4.101}$$

for some $H \in \mathbb{R}^{(n_{i1}+\cdots+n_{iN}) \times (n_{o1}+\cdots+n_{oN})}$ and an exogenous input v in $L_{2e}^{(n_{i1}+\cdots+n_{iN})}$. Also define $Q = \mathrm{diag}(Q_1, \ldots, Q_N)$, $S = \mathrm{diag}(S_1, \ldots S_N)$, and $R = \mathrm{diag}(R_1, \ldots R_N)$. Now prove that if

$$\bar{Q} = Q + H^T R H - S H H^T S^T$$

is negative definite, then the closed loop system is finite gain L_2 stable. Verify that this theorem generalizes the passivity theorem 4.21.

Problem 4.24 Discrete-time input-output theorems. Verify that the small gain theorem, incremental small gain theorem and the passivity theorems go through unchanged for discrete time nonlinear operators. The corresponding input and output spaces are spaces of sequences, denoted by l_{pe}.

Problem 4.25 Linear discrete time systems. Unlike in the continuous time case the space of discrete time convolution kernels does not need to belong to a space other than l_{1e}. Derive the input output stability conditions for discrete time linear time invariant systems. Use these theorems to derive analogues of the circle and Popov criterion for discrete time systems.

Problem 4.26 Incremental finite gain for Volterra systems. Consider the Volterra System of (4.87). Let two inputs $u(\cdot)$, $u(\cdot) + v(\cdot)$ both be in the domain of convergence of the Volterra series. Define the function $f : \mathbb{R}_+ \mapsto \mathbb{R}_+$ by

$$f(|u|) = \sum_{n=1}^{\infty} h_n |u|^n. \tag{4.102}$$

Now show that

$$|N(u+v)(\cdot) - N(u)(\cdot)|_\infty \le f(|u(\cdot)|_\infty + |v(\cdot)|) - f(|u(\cdot)|_\infty) \le f'(|u|_\infty + |v|_\infty)|v|_\infty$$

where f' stands for the derivative of the function f. Use this to show that $N(\cdot)$ is an incremental finite gain operator in its radius of convergence R.

Problem 4.27 Volterra series on L_p. Assume that for some $u(\cdot) \in L_p$ we have $Nu(\cdot) \in L_p$. Prove that

$$|N(u)(\cdot)|_p \le |u(\cdot)|_p \frac{f(|u|_\infty)}{|u|_\infty},$$

where $f(\cdot)$ is the gain function of (4.102) as defined in the preceding problem. Use this gain bound to establish that $N(u)$ is also incrementally finite gain stable on L_p.

Problem 4.28 Volterra series computations [248].

1. The equations of a *voltage controlled oscillator* in an FM transmitter are given by

$$\ddot{y}(t) + (\omega^2 + u(t))y(t) = 0.$$

Here $y(t)$ is the output waveform, $u(t)$ is the input which influences the frequency of the oscillator. Using $y(0) = 0$, $\dot{y}(0) = 1$ find the Volterra series of $y(t)$ as a function of $u(t)$.

2. The equations of a *dc motor* rotating at speed $\omega(t)$ (which is the output $y(t)$) when excited by the field current (which is the input $u(t)$) is modeled by writing an electrical circuit equation in the armature with state variable $i(t)$ and a mechanical equation for the variation of the speed $\omega(t)$ (the units in the equation below are all normalized, and we have assumed that the load varies linearly with ω) as

$$\dot{i} = -c_1 i - c_2 u(t)\omega + c_3,$$

$$\dot{\omega} = c_4 i - c_5\omega, \qquad\qquad (4.103)$$

$$y(t) = \omega(t).$$

Write the Volterra series for this system.

5
Lyapunov Stability Theory

5.1 Introduction

The study of the stability of dynamical systems has a very rich history. Many famous mathematicians, physicists, and astronomers worked on axiomatizing the concepts of stability. A problem, which attracted a great deal of early interest was the problem of stability of the solar system, generalized under the title "the N-body stability problem." One of the first to state formally what he called the principle of "least total energy" was Torricelli (1608–1647), who said that a system of bodies was at a stable equilibrium point if it was a point of (locally) minimal total energy. In the middle of the eighteenth century, Laplace and Lagrange took the Torricelli principle one step further: They showed that if the system is conservative (that is, it conserves total energy—kinetic plus potential), then a state corresponding to zero kinetic energy and minimum potential energy is a stable equilibrium point. In turn, several others showed that Torricelli's principle also holds when the systems are dissipative, i.e., total energy decreases along trajectories of the system. However, the abstract definition of stability for a dynamical system not necessarily derived for a conservative or dissipative system and a characterization of stability were not made till 1892 by a Russian mathematician/engineer, Lyapunov, in response to certain open problems in determining stable configurations of rotating bodies of fluids posed by Poincaré. The original paper of Lyapunov of 1892, was translated into French very shortly there after, but its English translation appeared only recently in [193]. The interested reader may consult this reference for many interesting details, as well as the historical and biographical introduction in this issue of the *International Journal of Control* by A. T. Fuller. There is another interesting survey paper about the impact of Lyapunov's stability theorem on feedback control by Michel [208].

At heart, the theorems of Lyapunov are in the spirit of Torricelli's principle. They give a precise characterization of those functions that qualify as "valid energy functions" in the vicinity of equilibrium points and the notion that these "energy functions" decrease along the trajectories of the dynamical systems in question. These precise concepts were combined with careful definitions of different notions of stability to give some very powerful theorems. The exposition of these theorems is the main goal of this chapter.

5.2 Definitions

This chapter is concerned with general differential equations of the form

$$\dot{x} = f(x, t), \quad x(t_0) = x_0, \tag{5.1}$$

where $x \in \mathbb{R}^n$ and $t \geq 0$. The system defined by (5.1) is said to be *autonomous* or *time-invariant* if f does not depend explicitly on t. It is said to be *linear* if $f(x, t) = A(t)x$ for some $A(\cdot) : \mathbb{R}_+ \mapsto \mathbb{R}^{n \times n}$ and nonlinear otherwise. In this chapter, we will assume that $f(x, t)$ is *piecewise continuous* with respect to t, that is, there are only finitely many discontinuity points in any compact set. The notation B_h will be short-hand for $B(0, h)$, the ball of radius h centered at 0. Properties will be said to be true

- *locally* if they are true for all x_0 in some ball B_h.
- *globally* if they are true for all $x_0 \in \mathbb{R}^n$.
- *semi-globally* if they are true for all $x_0 \in B_h$ with h arbitrary.
- *uniformly* if they are true for all $t_0 \geq 0$.

In the development that follows, unless explicitly specified, properties are true locally. The first few definitions and estimates are repeated from Chapter 3.

5.2.1 The Lipschitz Condition and Consequences

Definition 5.1 Lipschitz Continuity. *The function f is said to be* locally Lipschitz continuous *in x if for some $h > 0$ there exists $l \geq 0$ such that*

$$|f(x_1, t) - f(x_2, t)| \leq l|x_1 - x_2| \tag{5.2}$$

for all $x_1, x_2 \in B_h$, $t \geq 0$. The constant l is called the Lipschitz *constant. A definition for globally Lipschitz continuous functions follows by requiring equation (5.2) to hold for $x_1, x_2 \in \mathbb{R}^n$. The definition of semi-globally Lipschitz continuous functions holds as well by requiring that equation (5.2) hold in B_h for arbitrary h but with l possibly a function of h. The Lipschitz property is by default assumed to be uniform in t.*

If f is Lipschitz continuous in x, it is continuous in x. On the other hand, if f has bounded partial derivatives in x, then it is Lipschitz. Formally, if

$$D_1 f(x, t) := \left[\frac{\partial f_i}{\partial x_j} \right]$$

denotes the partial derivative matrix of f with respect to x (the subscript 1 stands for the first argument of $f(x, t)$), then $|D_1 f(x, t)| \leq l$ implies that f is Lipschitz continuous with Lipschitz constant l (again locally, globally or semi-globally depending on the region in x that the bound on $|D_2 f(x, t)|$ is valid). We have seen in Chapter 3 that if f is locally bounded and Lipschitz continuous in x, then the differential equation (5.1) has a unique solution on some time interval (so long as $x \in B_h$).

Definition 5.2 Equilibrium Point.   *x^* is said to be an* equilibrium point *of (5.1) if $f(x^*, t) \equiv 0$ for all $t \geq 0$.*

If $f(x, t)$ is Lipschitz continuous in x, then the solution $x(t) \equiv x^*$ for all t is called an *equilibrium solution*. By translating the origin to the equilibrium point x^* we can make 0 an equilibrium point. Since this is of great notational help, we will henceforth assume that 0 is an equilibrium point of (5.1). One of the most important consequences of the Lipschitz continuity hypothesis is that it gives bounds on the rate of convergence or divergence of solutions from the origin:

Proposition 5.3 Rate of Growth/Decay.   *If $x = 0$ is an equilibrium point of (5.1) f is Lipschitz in x with Lipschitz constant l and piecewise constant with respect to t, then the solution $x(t)$ satisfies*

$$|x_0| e^{l(t-t_0)} \geq |x(t)| \geq |x_0| e^{-l(t-t_0)} \tag{5.3}$$

as long as $x(t)$ remains in B_h.

Proof: Since $|x|^2 = x^T x$, it follows that

$$\left| \frac{d}{dt} |x|^2 \right| = 2 |x| \left| \frac{d}{dt} |x| \right|$$
$$= 2 \left| x^T \frac{d}{dt} x \right| \leq 2 |x| \left| \frac{d}{dt} x \right|, \tag{5.4}$$

so that

$$\left| \frac{d}{dt} |x| \right| \leq \left| \frac{d}{dt} x \right|.$$

Since $f(x, t)$ is Lipschitz continuous, and $f(x, 0) = 0$ it follows that

$$-l|x| \leq \frac{d}{dt} |x| \leq l|x| \tag{5.5}$$

Using the Bellman Gronwall lemma twice (refer back to Chapter 3 and work out the details for yourself) yields equation (5.3), provided that the trajectory stays in the ball B_h where the Lipschitz condition holds. $\square$

The preceding proposition implies that solutions starting inside B_h will stay in B_h for at least a finite time. Also, if $f(x, t)$ is globally Lipschitz, it guarantees that the solution has no finite escape time, that is, it is finite at every finite instant. The proposition also establishes that solutions $x(t)$ cannot converge to zero faster than exponentially.

We are now ready to make the stability definitions. Informally $x = 0$ is *stable* equilibrium point if trajectories $x(t)$ of (5.1) remain close to the origin if the initial condition x_0 is close to the origin. More precisely, we have the following definition.

Definition 5.4 Stability in the sense of Lyapunov. *The equilibrium point $x = 0$ is called a* stable *equilibrium point of (5.1) if for all $t_0 \geq 0$ and $\epsilon > 0$, there exists $\delta(t_0, \epsilon)$ such that*

$$|x_0| < \delta(t_0, \epsilon) \Rightarrow |x(t)| < \epsilon \quad \forall \; t \geq t_0, \tag{5.6}$$

where $x(t)$ is the solution of (5.1) starting from x_0 at t_0.

The definition is illustrated in Figure 5.1, showing the trajectories starting in a ball B_δ and not leaving the ball B_ϵ. Sometimes this definition is also called *stability in the sense of Lyapunov (i.s.L.) at time t_0.*

Definition 5.5 Uniform Stability. *The equilibrium point $x = 0$ is called a* uniformly stable *equilibrium point of (5.1) if in the preceding definition δ can be chosen independent of t_0.*

Intuitively, the definition of uniform stability captures the notion that the equilibrium point is not getting progressively less stable with time. Thus, in particular, it prevents a situation in which given an $\epsilon > 0$, the ball of initial conditions of radius $\delta(t_0, \epsilon)$ in the definition of stability required to hold trajectories in the ϵ ball tends to zero as $t_0 \to \infty$. The notion of stability is weak in that it does not require that trajectories starting close to the origin to tend to the origin asymptotically. That property is included in a definition of asymptotic stability:

Definition 5.6 Asymptotic Stability. *The equilibrium point $x = 0$ is an* asymptotically stable *equilibrium point of (5.1) if*

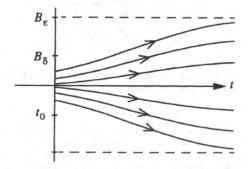

FIGURE 5.1. Illustrating the definition of stability

- $x = 0$ is a stable equilibrium point of (5.1),
- $x = 0$ is attractive, that is for all $t_0 \geq 0$ there exists a $\delta(t_0)$ such that

$$|x_0| < \delta \Rightarrow \lim_{t \to \infty} |x(t)| = 0.$$

An interesting feature of this definition is that it requires two separate conditions: one, that the equilibrium point be stable and two that trajectories tend to the equilibrium point as $t \to \infty$. Though, it may superficially appear to be the case, the requirement that trajectories converge to the origin does not imply the stability of the equilibrium point. To illustrate this, we consider the following example:

$$\dot{x}_1 = x_1^2 - x_2^2,$$
$$\dot{x}_2 = 2x_1 x_2. \qquad (5.7)$$

The phase portrait of this system (also presented as an example of an equilibrium point with index 2 in Chapter 2) is as shown in Figure 5.2. All trajectories tend to the origin as $t \to \infty$, except for the trajectory that follows the positive x_1 axis to $+\infty$. By assuming that this "point at infinity" is the same as the point at infinity at $x_1 = -\infty$ we may assert that all trajectories tend to the origin.[1] However, the equilibrium point at the origin is not stable in the sense of Lyapunov: Given any $\epsilon > 0$, no matter how small a δ we choose for the ball of initial condition, there are always some initial conditions close to the x_1 axis which will exit the ϵ ball before converging to the origin. The trajectory starting right on the x_1 axis gives a hint to this behavior.

Definition 5.7 Uniform Asymptotic Stability. *The equilibrium point $x = 0$ is a uniformly asymptotically stable equilibrium point of (5.1) if*

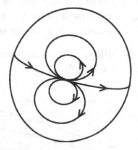

FIGURE 5.2. An equilibrium point, which is not stable, but for which all trajectories tend to the origin if one includes the point at infinity

---

[1]This procedure is referred to in mathematical terms as the Alexandroff or one-point compactification of $\mathbb{R}^2$. Another way of conceptualizing this procedure is to imagine that the plane is mapped onto a sphere with the origin corresponding to the north pole and the "point at infinity" to the south pole; that is, the state space of the system is really the sphere S^2.

- $x = 0$ is a uniformly stable equilibrium point of (5.1).
- The trajectory $x(t)$ converges uniformly to 0, that is, there exists $\delta > 0$ and a function $\gamma(\tau, x_0) : \mathbb{R}_+ \times \mathbb{R}^n \mapsto \mathbb{R}_+$ such that $\lim_{\tau \to \infty} \gamma(\tau, x_0) = 0$ for all $x_0 \in B_\delta$ and

$$|x_0| < \delta \Rightarrow |x(t)| \le \gamma(t - t_0, x_0) \quad \forall \ t \ge t_0.$$

The previous definitions are *local*, since they concern neighborhoods of the equilibrium point. *Global* asymptotic stability and global uniform asymptotic stability are defined as follows:

Definition 5.8 Global Asymptotic Stability. *The equilibrium point $x = 0$ is a globally asymptotically stable equilibrium point of (5.1) if it is stable and $\lim_{t \to \infty} x(t) = 0$ for all $x_0 \in \mathbb{R}^n$.*

Definition 5.9 Global Uniform Asymptotic Stability. *The equilibrium point $x = 0$ is a globally, uniformly, asymptotically stable equilibrium point of (5.1) if it is globally asymptotically stable and if in addition, the convergence to the origin of trajectories is uniform in time, that is to say that there is a function $\gamma : \mathbb{R}^n \times \mathbb{R}_+ \mapsto \mathbb{R}$ such that*

$$|x(t)| \le \gamma(x_0, t - t_0) \quad \forall \ t \ge 0.$$

It is instructive to note that the definitions of asymptotic stability do not quantify the speed of convergence of trajectories to the origin. For time-invariant linear systems, the speed of convergence of trajectories either to or from the origin is exponential, but for time varying and nonlinear systems, the rate of convergence can be of many different types, for example as

$$\frac{1}{t}, \quad \frac{1}{\sqrt{t}}.$$

It is a good exercise to write down examples of this behavior. There is a strong form of stability which demands an exponential rate of convergence:

Definition 5.10 Exponential Stability, Rate of Convergence. *The equilibrium point $x = 0$ is an exponentially stable equilibrium point of (5.1) if there exist $m, \alpha > 0$ such that*

$$|x(t)| \le me^{-\alpha(t-t_0)}|x_0| \tag{5.8}$$

for all $x_0 \in B_h, t \ge t_0 \ge 0$. The constant α is called (an estimate of) the rate of convergence.

Global exponential stability is defined by requiring equation (5.8) to hold for all $x_0 \in \mathbb{R}^n$. Semi-global exponential stability is also defined analogously except that m, α are allowed to be functions of h. For linear (possibly time-varying) systems it will be shown that uniform asymptotic stability is equivalent to exponential stability, but in general, exponential stability is stronger than asymptotic stability.

5.3   Basic Stability Theorems of Lyapunov

The so-called second method of Lyapunov enables one to determine the stability properties of a system (5.1) without explicitly integrating the differential equation. The method is a generalization of the basic notion that some measure of "energy dissipation" in a system enables us to conclude stability. To make this precise we need to define exactly what one means by a "measure of energy," that is, energy functions. This needs the following preliminary definitions (due, in the current form to Hahn [124]):

5.3.1   Energy-Like Functions

Definition 5.11 Class K, KR Functions.   *A function $\alpha(\cdot) : \mathbb{R}_+ \mapsto \mathbb{R}_+$ belongs to class K (denoted by $\alpha(\cdot) \in K$) if it is continuous, strictly increasing and $\alpha(0) = 0$. The function $\alpha(\cdot)$ is said to belong to class KR if α is of class K and in addition, $\alpha(p) \to \infty$ as $p \to \infty$.*

Definition 5.12 Locally Positive Definite Functions.   *A continuous function $v(x, t) : \mathbb{R}^n \times \mathbb{R}_+ \mapsto \mathbb{R}_+$ is called a* locally positive definite function (l.p.d.f) *if, for some $h > 0$ and some $\alpha(\cdot)$ of class K,*

$$v(0, t) = 0 \quad and \quad v(x, t) \geq \alpha(|x|) \quad \forall x \in B_h, \quad t \geq 0. \tag{5.9}$$

An l.p.d.f. is locally like an "energy function." Functions which are globally like "energy functions" are called positive definite functions (p.d.f.s) and are defined as follows:

Definition 5.13 Positive Definite Functions.   *A continuous function $v(x, t) : \mathbb{R}^n \times \mathbb{R}_+ \mapsto \mathbb{R}_+$ is called a* positive definite function (p.d.f.) *if for some $\alpha(\cdot)$ of class KR,*

$$v(0, t) = 0 \quad and \quad v(x, t) \geq \alpha(|x|) \quad \forall x \in \mathbb{R}^n, \quad t \geq 0 \tag{5.10}$$

and, in addition, $\alpha(p) \to \infty$ as $p \to \infty$.

In the preceding definitions of l.p.d.f.s and p.d.f.s, the energy was not bounded above as t varied. This is the topic of the next definition

Definition 5.14 Decrescent Functions.   *A continuous function $v(x, t) : \mathbb{R}^n \times \mathbb{R}_+ \mapsto \mathbb{R}_+$ is called a* decrescent *function if, there exists a function $\beta(\cdot)$ of class K, such that*

$$v(x, t) \leq \beta(|x|) \quad \forall x \in B_h, \quad t \geq 0. \tag{5.11}$$

Example 5.15 Examples of Energy-like Functions.   *Here are some examples of energy like functions and their membership in the various classes introduced above. It is an interesting exercise to check the appropriate functions of class K and KR that need to be used to verify these properties.*

1. $v(x, t) = |x|^2$: *p.d.f., decrescent.*
2. $v(x, t) = x^T P x$, *with* $P \in \mathbb{R}^{n \times n} > 0$: *p.d.f., decrescent.*
3. $v(x, t) = (t + 1)|x|^2$: *p.d.f.*
4. $v(x, t) = e^{-t}|x|^2$: *decrescent.*
5. $v(x, t) = \sin^2(|x|^2)$: *l.p.d.f., decrescent.*
6. $v(x, t) = e^t x^T P x$ *with* P *not positive definite: not in any of the above classes.*
7. $v(x, t)$ *not explicitly depending on time* t: *decrescent.*

5.3.2 Basic Theorems

Generally speaking, the basic theorem of Lyapunov states that when $v(x, t)$ is a p.d.f. or an l.p.d.f. and $dv(x, t)/dt \leq 0$ then we can conclude stability of the equilibrium point. The time derivative is taken along the trajectories of (5.1), i.e.,

$$\left.\frac{dv(x, t)}{dt}\right|_{(5.1)} = \frac{\partial v(x, t)}{\partial t} + \frac{\partial v(x, t)}{\partial x} f(x, t). \tag{5.12}$$

The rate of change of $v(x, t)$ along the trajectories of the vector field (5.1) is also called the *Lie derivative of* $v(x, t)$ *along* $f(x, t)$. In the statement of the following theorem recall that we have translated the origin to lie at the equilibrium point under consideration.

Theorem 5.16 Basic Lyapunov Theorems.

	Conditions on $v(x, t)$	Conditions on $-\dot{v}(x, t)$	Conclusions
1.	l.p.d.f.	≥ 0 locally	stable
2.	l.p.d.f., decrescent	≥ 0 locally	uniformly stable
3.	l.p.d.f., decrescent	l.p.d.f.	uniformly asymptotically stable
4.	p.d.f., decrescent	p.d.f.	globally unif. asymp. stable

Proof:

1. Since v is an l.p.d.f., we have that for some $\alpha(\cdot) \in K$,

$$v(x, t) \geq \alpha(|x|) \quad \forall \quad x \in B_s. \tag{5.13}$$

Also, the hypothesis is that

$$\dot{v}(x, t) \leq 0, \quad \forall t \geq t_0, \quad \forall x \in B_r \tag{5.14}$$

Given $\epsilon > 0$, define $\epsilon_1 = \min(\epsilon, r, s)$. Choose $\delta > 0$ such that

$$\beta(t_0, \delta) := \sup_{|x| \leq \delta} v(x, t_0) < \alpha(\epsilon_1).$$

Such a δ always exists, since $\beta(t_0, \delta)$ is a continuous function of δ and $\alpha(\epsilon_1) > 0$. We now claim that $|x(t_0)| \leq \delta$ implies that $|x(t)| < \epsilon_1 \; \forall t \geq t_0$. The proof is by contradiction. Clearly since

$$\alpha(|x(t_0)|) \leq v(x(t_0), t_0) < \alpha(\epsilon_1),$$

it follows that $|x(t_0)| < \epsilon_1$. Now, if it is not true that $|x(t)| < \epsilon_1$ for all t, let $t_1 > t_0$ be the first instant such that $|x(t)| \geq \epsilon_1$. Then

$$v(x(t_1), t_1) \geq \alpha(\epsilon_1) > v(x(t_0), t_0). \tag{5.15}$$

But this is a contradiction, since $\dot{v}(x(t), t) \leq 0$ for all $|x| < \epsilon_1$. Thus,

$$|x(t)| < \epsilon_1 \quad \forall\, t \geq t_0.$$

2. Since v is decrescent,

$$\beta(\delta) = \sup_{|x| \leq \delta} \sup_{t \geq t_0} v(x, t) \tag{5.16}$$

is nondecreasing and satisfies for some d

$$\beta(\delta) < \infty \quad \text{for} \quad 0 \leq \delta \leq d.$$

Now choose δ such that $\beta(\delta) < \alpha(\epsilon_1)$.

3. If $-\dot{v}(x, t)$ is an l.p.d.f., then $\dot{v}(x, t)$ satisfies the conditions of the previous proof so that 0 is a uniformly stable equilibrium point. We need to show the existence of a $\delta_1 > 0$ such that for $\epsilon > 0$ there exists $T(\epsilon) < \infty$ such that

$$|x_0| < \delta_1 \Rightarrow |\phi(t_1 + t, x_0, t_1)| < \epsilon \quad \text{when} \quad t > T(\epsilon)$$

The hypotheses guarantee that there exist functions $\alpha(\cdot), \beta(\cdot), \gamma(\cdot) \in K$ such that $\forall t \geq t_0, \forall x \in B_r$ such that

$$\alpha(|x|) \leq v(x, t) \leq \beta(|x|),$$

$$\dot{v}(x, t) \leq -\gamma(|x|).$$

Given $\epsilon > 0$, define δ_1, δ_2 and T by

$$\beta(\delta_1) < \alpha(r),$$

$$\beta(\delta_2) < \min(\alpha(\epsilon), \beta(\delta_1)),$$

$$T = \alpha(r)/\gamma(\delta_2).$$

This choice is explained in Figure 5.3. We now show that there exists at least one instant $t_2 \in [t_1, t_1 + T]$ when $|x_0| < \delta_2$. The proof is by contradiction. Recall the notation that $\phi(t, x_0, t_0)$ stands for the trajectory of (5.1) starting from x_0 at time t_0. Indeed, if

$$|\phi(t, x_0, t_1)| \geq \delta_2 \quad \forall t \in [t_1, t_1 + T],$$

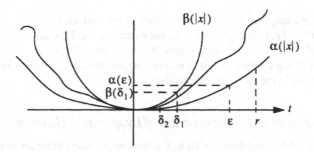

FIGURE 5.3. The choice of constants in the proof of Lyapunov's Theorem.

then it follows that

$$0 \leq \alpha(\delta_2) \leq v(s(t_1 + T, x_0, t_1), t_1 + T)$$

$$= v(t_1, x_0) + \int_{t_1}^{t_1 + T} \dot{v}(\tau, \phi(\tau, x_0, t_1)) d\tau$$

$$\leq \beta(\delta_1) - T\gamma(\delta_2)$$

$$\leq \beta(\delta_1) - \alpha(r)$$

$$< 0,$$

establishing the contradiction (compare the ends of the preceding chain of inequalities to see this). Now, if $t \geq t_1 + T$, then

$$\alpha(|\phi(t, x_0, t_1)|) \leq v(t, \phi(t, x_0, t_1))$$

$$\leq v(t_2, \phi(t_2, x_0, t_1)),$$

since $\dot{v}(x, t) \leq 0$. (Actually, the definition of δ_1 guarantees that the trajectory stays in B_r so that $\dot{v}(x, t) \leq 0$.) Thus,

$$\alpha(|\phi(t, x_0, t_1)|) \leq v(t_2, \phi(t_2, x_0, t_1)) \leq \beta(|\phi(t_2, x_0, t_1)|)$$

$$\leq \beta(\delta_2)$$

$$< \alpha(\epsilon)$$

so that $|\phi(t_2, x_0, t_1)| < \epsilon$ for $t \geq t_1 + T$.   □

Remarks:

1. The tabular version of Lyapunov's theorem is meant to highlight the following correlations between the assumptions on $v(x, t)$, $\dot{v}(x, t)$ and the conclusions:

   a. Decrescence of $v(x, t)$ is associated with uniform stability and the local positive definite character of $\dot{v}(x, t)$ being associated with asymptotic stability.

   b. $-\dot{v}(x, t)$ is required to be an l.p.d.f. for asymptotic stability,

   c. $v(x, t)$ being a p.d.f. is associated with global stability.

However, we emphasize that this correlation is not perfect, since $v(x, t)$ being l.p.d.f. and $-\dot{v}(x, t)$ being l.p.d.f. does not guarantee local asymptotic stability. (See Problem 5.1 for a counterexample from Massera [227]).

2. The proof of the theorem, while seemingly straightforward, is subtle in that it is an exercise in the use of contrapositives.

5.3.3   Examples of the Application of Lyapunov's Theorem

1. Consider the following model of an RLC circuit with a linear inductor, nonlinear capacitor, and inductor as shown in the Figure 5.4. This is also a model for a mechanical system with a mass coupled to a nonlinear spring and nonlinear damper as shown in Figure 5.4. Using as state variables x_1, the charge on the capacitor (respectively, the position of the block) and x_2, the current through the inductor (respectively, the velocity of the block) the equations describing the system are

$$\dot{x}_1 = x_2,$$
$$\dot{x}_2 = -f(x_2) - g(x_1).$$
(5.17)

Here $f(\cdot)$ is a continuous function modeling the resistor current–voltage characteristic, and $g(\cdot)$ the capacitor charge–voltage characteristic (respectively the friction and restoring force models in the mechanical analog). We will assume that f, g both model locally passive elements, i.e., there exists a σ_0 such that

$$\sigma f(\sigma) \geq 0 \ \forall \sigma \in [-\sigma_0, \sigma_0],$$
$$\sigma g(\sigma) \geq 0 \ \forall \sigma \in [-\sigma_0, \sigma_0].$$

The Lyapunov function candidate is the total energy of the system, namely,

$$v(x) = \frac{x_2^2}{2} + \int_0^{x_1} g(\sigma)d\sigma.$$

The first term is the energy stored in the inductor (kinetic energy of the body) and the second term the energy stored in the capacitor (potential energy stored in

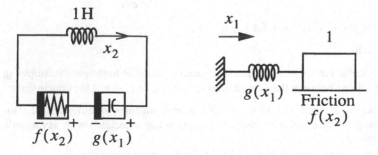

FIGURE 5.4. An RLC circuit and its mechanical analogue

the spring). The function $v(x)$ is an l.p.d.f., provided that $g(x_1)$ is not identically zero on any interval (verify that this follows from the passivity of g). Also,

$$\dot{v}(x) = x_2[-f(x_2) - g(x_1)] + g(x_1)x_2$$
$$= -x_2 f(x_2) \leq 0$$

when $|x_2|$ is less than σ_0. This establishes the stability but not asymptotic stability of the origin. In point of fact, the origin is actually asymptotically stable, but this needs the LaSalle principle, which is deferred to a later section.

2. *Swing Equation*
The dynamics of a single synchronous generator coupled to an infinite bus is given by

$$\dot{\theta} = \omega,$$
$$\dot{\omega} = -M^{-1}D\omega - M^{-1}(P - B\sin(\theta)). \qquad (5.18)$$

Here θ is the angle of the rotor of the generator measured relative to a synchronously spinning reference frame and its time derivative is ω. Also M is the moment of inertia of the generator and D its damping both in normalized units; P is the exogenous power input to the generator from the turbine and B the susceptance of the line connecting the generator to the rest of the network, modeled as an infinite bus (see Figure 5.5) A choice of Lyapunov function is

$$v(\theta, \omega) = \frac{1}{2}M\omega^2 + P\theta + B\cos(\theta).$$

The equilibrium point is $\theta = \sin^{-1}(\frac{P}{B})$, $\omega = 0$. By translating the origin to this equilibrium point it may be verified that $v(\theta, \omega) - v(\theta_0, 0)$ is an l.p.d.f. around it. Further it follows that

$$\dot{v}(\theta, \omega) = -D\omega^2,$$

yielding the stability of the equilibrium point. As in the previous example, one cannot conclude asymptotic stability of the equilibrium point from this analysis.

3. *Damped Mathieu Equation*
This equation models the dynamics of a pendulum with sinusoidally varying

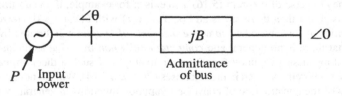

FIGURE 5.5. A generator coupled to an infinite bus

length; this is a good model for instance of a child pumping a swing.

$$\dot{x}_1 = x_2,$$
$$\dot{x}_2 = -x_2 - (2 + \sin t)x_1. \tag{5.19}$$

The total energy is again a candidate for a Lyapunov function, namely,

$$v(x, t) = x_1^2 + \frac{x_2^2}{2 + \sin t}.$$

It is an l.p.d.f. since

$$x_1^2 + x_2^2 \geq x_1^2 + \frac{x_2^2}{2 + \sin t} \geq x_1^2 + \frac{x_2^2}{3}.$$

Further, a small computation yields

$$\dot{v}(x, t) = -\frac{x_2^2(4 + 2\sin t + \cos t)}{(2 + \sin t)^2} \leq 0.$$

Thus, the equilibrium point at the origin is uniformly stable. However, we cannot conclude that the origin is uniformly asymptotically stable.

4. *System with a Limit Cycle*

Consider the system

$$\dot{x}_1 = -x_2 + x_1(x_1^2 + x_2^2 - 1),$$
$$\dot{x}_2 = x_1 + x_2(x_1^2 + x_2^2 - 1).$$

A choice of $v(x) = x_1^2 + x_2^2$ yields for $\dot{v}(x) = 2(x_1^2 + x_2^2)(x_1^2 + x_2^2 - 1)$, which is the negative of an l.p.d.f. for $\{x : x_1^2 + x_2^2 < 1\}$. Thus, 0 is a locally asymptotically stable equilibrium (though it is not globally, asymptotically stable, since there is a limit cycle of radius 1, as may be verified by changing into polar coordinates, see also Chapter 2).

Comments.

1. The theorems of Lyapunov give sufficient conditions for the stability of the equilibrium point at the origin of (5.1). They do not, however, give a prescription for determining the Lyapunov function $v(x, t)$. Since the theorems are only sufficient conditions for stability the search for a Lyapunov function establishing stability of an equilibrium point could be long. However, it is a remarkable fact that the converse of theorem (5.16) also exists: for example, if an equilibrium point is stable then there exists an l.p.d.f. $v(x, t)$ with $\dot{v}(x, t) \leq 0$. However, the utility of this and other *converse theorems of Lyapunov* is limited by the fact that there is no general and *computationally non-intensive* technique for generating these Lyapunov functions (an example of such a theorem, called Zubov's theorem, is given in the Exercises (Problem 5.14). We will not give the details of the construction of converse Lyapunov functions in general, but we focus on exponentially stable equilibria in the next subsection. However, the

method of construction of the Lyapunov function in Section 5.3.4 is prototypical of other converse theorems.

2. If the system has multiple equilibrium points, then by translating the equilibria in turn to the origin one may individually ascertain their stability.

5.3.4   Exponential Stability Theorems

The basic theorem of Lyapunov, Theorem 5.16, stops short of giving explicit rates of convergence of solutions to the equilibria. It may however be modified to do so in the instance of exponentially stable equilibria. We will pay special attention to exponentially stable equilibria, since they are robust to perturbation and are consequently desirable from the viewpoint of applications. We will now state necessary and sufficient conditions for the existence of an exponentially stable equilibrium point, that is, we state a converse theorem for exponentially stable systems:

Theorem 5.17 Exponential Stability Theorem and Its Converse.   *Assume that $f(x, t) : \mathbb{R}_+ \times \mathbb{R}^n \mapsto \mathbb{R}^n$ has continuous first partial derivatives in x and is piecewise continuous in t. Then the two statements below are equivalent:*

1. *$x = 0$ is a locally exponentially stable equilibrium point of*

$$\dot{x} = f(x, t),$$

i.e., if $x \in B_h$ for h small enough, there exist $m, \alpha > 0$ such that

$$|\phi(\tau, x, t)| \leq m e^{-\alpha(\tau - t)}.$$

2. *There exists a function $v(x, t)$ and some constants $h, \alpha_1, \alpha_2, \alpha_3, \alpha_4 > 0$ such that for all $x \in B_h, t \geq 0$*

$$\alpha_1 |x|^2 \leq v(x, t) \leq \alpha_2 |x|^2,$$
$$\left. \frac{dv(x, t)}{dt} \right|_{(5.1)} \leq -\alpha_3 |x|^2, \tag{5.20}$$
$$\left| \frac{\partial v(x, t)}{\partial x} \right| \leq \alpha_4 |x|.$$

Proof:

$(1) \Rightarrow (2)$

We prove the three inequalities of (5.20) in turn, starting from the definition of $v(t, x)$:

- Denote by $\phi(\tau, x, t)$ the solution of (5.1) at time τ starting from x at time t, and define

$$v(x, t) := \int_t^{t+T} |\phi(\tau, x, t)|^2 d\tau, \tag{5.21}$$

where T will be defined later. From the exponential stability of the system at rate α and the lower bound on the rate of growth given by Proposition 5.3, we have

$$m|x|e^{-\alpha(\tau-t)} \geq |\phi(\tau, x, t)| \geq |x|e^{-l(\tau-t)} \tag{5.22}$$

for $x \in B_h$ for some h. Also l the Lipschitz constant of $f(x, t)$ exists because of the assumption that $f(x, t)$ has continuous first partial derivatives with respect to x. This, when used in (5.21), yields the first inequality of (5.20) for $x \in B_{h'}$ (where h' is chosen to be h/m) with

$$\alpha_1 := \frac{(1 - e^{-2lT})}{2l}, \qquad \alpha_2 := m^2 \frac{(1 - e^{-2\alpha T})}{2\alpha}. \tag{5.23}$$

- Differentiating (5.21) with respect to t yields

$$\begin{aligned}
\frac{dv(x, t)}{dt} &= |\phi(t + T, x, t)|^2 - |\phi(t, x, t)|^2 \\
&\quad + \int_t^{t+T} \frac{d}{dt}(|\phi(\tau, x(t), t)|^2)d\tau.
\end{aligned} \tag{5.24}$$

Note that d/dt is the derivative with respect to the initial time t along the trajectories of (5.1). However, since for all Δt the solution satisfies

$$\phi(\tau, x(t + \Delta t), t + \Delta t) = \phi(\tau, x(t), t),$$

we have that that $\frac{d}{dt}(|\phi(\tau, x(t), t)|^2) \equiv 0$. Using the fact that $\phi(t, x, t) = x$ and the exponential bound on the solution, we have that

$$\frac{dv(x, t)}{dt} \leq -(1 - m^2 e^{-2\alpha T})|x|^2.$$

The second inequality of (5.20) now follows, provided that $T > (1/\alpha) \ln m$ and

$$\alpha_3 := 1 - m^2 e^{-2\alpha T}.$$

- Differentiating (5.21) with respect to x_j, we have

$$\frac{\partial v(x, t)}{\partial x_i} = 2 \int_t^{t+T} \sum_{j=1}^n \phi_j(\tau, x, t) \frac{\partial \phi_j(\tau, x, t)}{\partial x_i} d\tau. \tag{5.25}$$

By way of notation define

$$Q_{ij}(\tau, x, t) := \frac{\partial \phi_j(\tau, x, t)}{\partial x_i}$$

and

$$A_{ij}(x, t) := \frac{\partial f_i(t, x)}{\partial x_j}.$$

Interchanging the order of differentiation by τ, with differentiation by x_j yields that

$$\frac{d}{d\tau}Q(\tau, x, t) = A(\phi(\tau, x, t), t).Q(\tau, x, t). \tag{5.26}$$

Thus $Q(\tau, x, t)$ is the state transition matrix associated with the matrix $A(\phi(\tau, x, t), t)$. By the assumption on boundedness of the partials of f with respect to x, it follows that $|A(\cdot, \cdot)| \le k$ for some k, so that

$$|Q(\tau, x, t)| \le e^{k(\tau - t)}.$$

using this and the bound for exponential convergence in (5.25) yields

$$\left| \frac{\partial v(t, x)}{\partial x} \right| \le 2 \int_t^{t+T} m|x|e^{(k-\alpha)(\tau - t)}d\tau,$$

which is the last equation of (5.20) if we define

$$\alpha_4 := \frac{2m(e^{(k-\alpha)T} - 1)}{(k - \alpha)}.$$

This completes the proof, but note that $v(x, t)$ is only defined for $x \in B_{h'}$ with $h' = h/m$, to guarantee that $\phi(\tau, x, t) \in B_h$ for all $\tau \ge t$ (convince yourself of this point).

$(2) \Rightarrow (1)$
This direction is straightforward, as may be verified by noting that equation (5.20) implies that

$$\dot{v}(x, t) \le -\frac{\alpha_3}{\alpha_2}v(x, t). \tag{5.27}$$

This in turn implies that

$$v(t, x(t)) \le v(t_0, x(t_0))e^{-\frac{\alpha_3}{\alpha_2}(t - t_0)}. \tag{5.28}$$

Using the lower bound for $v(t, x(t))$ and the upper bound for $v(t_0, x(t_0))$ we get

$$\alpha_1|x(t)|^2 \le \alpha_2|x(t_0)|^2 e^{-\frac{\alpha_3}{\alpha_2}(t - t_0)}. \tag{5.29}$$

Using the estimate of (5.29) it follows that

$$|x(t)| \le m|x(t_0)|e^{-\alpha(t - t_0)}.$$

with

$$m := \left(\frac{\alpha_2}{\alpha_1} \right)^{1/2}, \quad \alpha := \frac{\alpha_3}{2\alpha_2}. \qquad \square$$

5.4   LaSalle's Invariance Principle

LaSalle's invariance principle has two main applications:

1. It enables one to conclude asymptotic stability even when $-\dot{v}(x, t)$ is not an l.p.d.f.
2. It enables one to prove that trajectories of the differential equation starting in a given region converge to one of many equilibrium points in that region.

However, the principle applies primarily to autonomous or periodic systems, which are discussed in this section. Some generalizations to the non–autonomous case are also discussed in the next section. We will deal with the autonomous case first. We recall some definitions from Chapter 2:

Definition 5.18 ω limit set. *A set $S \subset \mathbb{R}^n$ is the ω limit set of a trajectory $\phi(\cdot, x_0, t_0)$ if for every $y \in S$, there exists a sequence of times $t_n \to \infty$ such that $\phi(t_n, x_0, t_0) \to y$.*

Definition 5.19 Invariant set.   *A set $M \subset \mathbb{R}^n$ is said to be an* invariant set *if whenever $y \in M$ and $t_0 \geq 0$, we have*

$$\phi(t, y, t_0) \in M \quad \forall t \geq t_0.$$

The following propositions establish some properties of ω limit sets and invariant sets.

Proposition 5.20. *If $\phi(\cdot, x_0, t_0)$ is a bounded trajectory, its ω limit set is compact. Further, $\phi(t, x_0, t_0)$ approaches its ω limit set as $t \to \infty$.*

Proof: As in Chapter 2, see pages 46–50 of [329]. The proof is no different in $\mathbb{R}^n$.

Proposition 5.21. *Assume that the system (5.1) is autonomous and let S be the ω-limit set of any trajectory. Then S is invariant.*

Proof: Let $y \in S$ and $t_1 \geq 0$ be arbitrary. We need to show that $\phi(t, y, t_1) \in S$ for all $t \geq t_1$. Now $y \in S \Rightarrow \exists t_n \to \infty$ such that $\phi(t_n, x_0, t_0) \to y$ as $n \to \infty$. Since trajectories are continuous in initial conditions, it follows that

$$\phi(t, y, t_1) = \lim_{n \to \infty} \phi(t, \phi(t_n, x_0, t_0), t_1)$$
$$= \lim_{n \to \infty} \phi(t + t_n - t_1, x_0, t_0),$$

since the system is autonomous. Now, $t_n \to \infty$ as $n \to \infty$ so that by Proposition 5.20 the right hand side converges to an element of S. $\square$

Proposition 5.22 LaSalle's Principle. *Let $v : \mathbb{R}^n \to \mathbb{R}$ be continuously differentiable and suppose that*

$$\Omega_c = \{x \in \mathbb{R}^n : v(x) \leq c\}$$

is bounded and that $\dot{v} \leq 0$ for all $x \in \Omega_c$. Define $S \subset \Omega_c$ by

$$S = \{x \in \Omega_c : \dot{v}(x) = 0\}$$

*and let M be the largest invariant set in S. Then, whenever $x_0 \in \Omega_c$, $\phi(t, x_0, 0)$
approaches M as $t \to \infty$.*

Proof: Let $x_0 \in \Omega_c$. Now, since $v(\phi(t, x_0, 0))$ is a nonincreasing function of
time we see that $\phi(t, x_0, 0) \in \Omega_c$ $\forall t$. Further, since Ω_c is bounded $v(\phi(t, x_0, 0))$
is also bounded below. Let

$$c_0 = \lim_{t \to \infty} v(\phi(t, x_0, 0))$$

and let L be the ω limit set of the trajectory. Then, $v(y) = c_0$ for $y \in L$. Since L
is invariant we have that $\dot{v}(y) = 0 \forall y \in L$ so that $L \subset S$. Since, M is the largest
invariant set inside S, we have that $L \subset M$. Since $s(t, x_0, 0)$ approaches L as
$t \to \infty$, we have that $s(t, x_0, 0)$ approaches M as $t \to \infty$. $\square$.

Theorem 5.23 LaSalle's Principle to Establish Asymptotic Stability. *Let $v :
\mathbb{R}^n \mapsto \mathbb{R}$ be such that on $\Omega_c = \{x \in \mathbb{R}^n : v(x) \leq c\}$, a compact set we have
$\dot{v}(x) \leq 0$. As in the previous proposition define*

$$S = \{x \in \Omega_c : \dot{v}(x) = 0\}.$$

Then, if S contains no trajectories other than $x = 0$ then 0 is asymptotically stable.

Proof: follows directly from the preceding lemma.

An application of LaSalle's principle is to prove global asymptotic stability is
as follows:

**Theorem 5.24 Application of LaSalle's Principle to prove Global Asymptotic
Stability.** *Let $v(x) : \mathbb{R}^n \mapsto \mathbb{R}$ be a p.d.f. and $\dot{v}(x) \leq 0$ for all $x \in \mathbb{R}^n$. Also, let
the set*

$$S = \{x \in \mathbb{R}^n : \dot{v}(x) = 0\}$$

contain no nontrivial trajectories. Then 0 is globally, asymptotically stable.

Examples:

1. *Spring–mass system with damper*
   This system is described by

$$\dot{x}_1 = x_2,$$
$$\dot{x}_2 = -f(x_2) - g(x_1).$$

$$(5.30)$$

If f, g are locally passive, i.e.,

$$\sigma f(\sigma) \geq 0 \ \forall \sigma \in [-\sigma_0, \sigma_0],$$

then it may be verified that a suitable Lyapunov function (l.p.d.f.) is

$$v(x_1, x_2) = \frac{x_2^2}{2} + \int_0^{x_1} g(\sigma) d\sigma.$$

Further, it is easy to see that

$$\dot{v}(x_1, x_2) = -x_2 f(x_2) \leq 0 \ \text{for} \ x_2 \in [-\sigma_0, \sigma_0].$$

Now choose

$$c = \min(v(-\sigma_0, 0), v(\sigma_0, 0)).$$

Then $\dot{v} \leq 0$ for $x \in \Omega_c = \{(x_1, x_2) : v(x_1, x_2) \leq c\}$. As a consequence of LaSalle's principle, the trajectory enters the largest invariant set in $\Omega_c \cap \{(x_1, x_2) : \dot{v} = 0\} = \Omega_c \cap \{x_1, 0\}$. To obtain the largest invariant set in this region note that $x_2(t) \equiv 0 \Rightarrow x_1(t) \equiv x_{10} \Rightarrow \dot{x}_1(t) = 0 = -f(0) - g(x_{10})$. Consequently, we have that $g(x_{10}) = 0 \Rightarrow x_{10} = 0$. Thus, the largest invariant set inside $\Omega_c \cap \{(x_1, x_2) : \dot{v} = 0\}$ is the origin. Thus, the origin is locally asymptotically stable.

The application of LaSalle's principle shows that one can give interesting conditions for the convergence of trajectories of the system of (5.30) even when $g(\cdot)$ is not passive. It is easy to see that the arguments given above can be easily modified to obtain convergence results for the system (5.30), provided that $\int_0^{x_1} g(\sigma)d\sigma$ is merely *bounded below*. The next examples are in this spirit.

2. *Single generator coupled to an infinite bus*

If θ is the angle of the rotor of a generator with respect to a synchronously rotating frame of reference and $\omega = \dot{\theta}$, then the equations of the dynamics of the synchronous generator are given by

$$M\ddot{\theta} + D\dot{\theta} + B \sin \theta = P_m - P_e. \tag{5.31}$$

Here, M, D stand for the moment of inertia and damping of the generator rotor; P_m stands for the exogenous power input, and P_e stands for the local power demand. This equation can be written in state space form with $x_1 = \theta$, $x_2 = \omega$. The equilibrium points of the system are at $\theta = \sin^{-1}(\frac{P_m - P_e}{B})$, $\omega = 0$. Note that the equilibria repeat every 2π radians. Consider a choice of

$$v(\theta, \omega) = \frac{M\omega^2}{2} + (P_e - P_m)\theta - B \cos(\theta) \tag{5.32}$$

with

$$\dot{v} = -D\omega^2.$$

Since $v(\theta, \omega)$ is not bounded below, $\Omega_c = \{(\theta, \omega) : v(\theta, \omega) \leq c\}$ is not compact. However, if we know *a priori* that a trajectory is bounded, then we know from LaSalle's principle that it converges to the largest invariant set in a compact region with $\omega = 0$. Reasoning exactly as in the previous example guarantees that the trajectory converges to one of the equilibrium points of the system. (These are the only invariant sets in the region where $\dot{v} = 0$.) This behavior is referred to by power systems engineers as "trajectories that skip only finitely many cycles (2π multiples in the θ variable) converge to an equilibrium" owing to the periodic dependence of the dynamics on θ. However, if we view the dynamics as resident on the state space $S^1 \times \mathbb{R}^1$, we can no longer use the function $v(\theta, \omega)$ to conclude convergence of the trajectories. (Why? See Problem 5.11 for details on this point.)

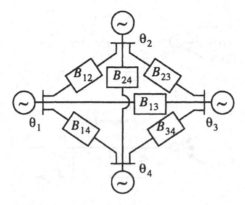

FIGURE 5.6. An interconnected power system

3. *Power system swing equations*

The dynamics of a power system modeled as shown in Figure 5.6 as a collection of generators interconnected by transmission lines of susceptance B_{ij} between generators i and j is a generalization of the swing equation for a single generator coupled to an infinite bus:

$$M_i \ddot{\theta}_i + D_i \dot{\theta}_i + \sum_{j=1, j \neq i}^{n} B_{ij} \sin(\theta_i - \theta_j) = P_{mi} - P_{ei} \qquad (5.33)$$

for $i = 1, \ldots, n$. Here M_i, D_i stand for the moment of inertia and damping of the ith generator. and P_{mi}, P_{ei} stand for the mechanical power input and electrical power output from the i-th generator. The total energy function for this system is given by

$$v(\theta, \omega) = \frac{1}{2} \dot{\theta}^T M \dot{\theta} - \sum_{i < j, j=1}^{n} B_{ij} \cos(\theta_i - \theta_j) + \theta^T (P_e - P_m), \qquad (5.34)$$

with its derivative given by

$$\dot{v} = -\dot{\theta}^T D \dot{\theta}.$$

As before the function $v(\theta, \omega)$ is not bounded below so that we cannot guarantee that Ω_c is compact for any c. However, we may establish as before that all *a priori bounded* trajectories converge. The same comments about convergence of finite–cycle–skipping trajectories 2π variations in one or more of the θ_i hold here as well. Also $v(\theta, \omega)$ is no longer a valid function to consider when the dynamics of the power system are considered on the state space $T^n \times \mathbb{R}^n$ (see Problem 5.11 for details).

4. *Hopfield Neural Networks [143]*

Figure 5.7 shows a circuit diagram of a so-called Hopfield network consisting of N operational amplifiers interconnected by an RC network. The operational amplifier input voltages are labeled u_i and the outputs x_i. The relationship between

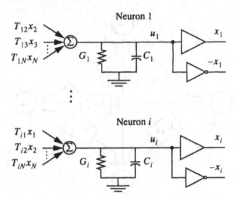

FIGURE 5.7. A Hopfield network

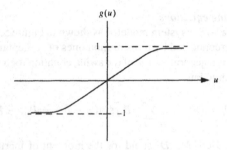

FIGURE 5.8. A sigmoidal nonlinearity

u_i and x_i for each i is given by a single, so-called *sigmoidal nonlinearity*: i.e., a monotone increasing odd-symmetric nonlinearity of the form shown in Figure 5.8 asymptotic to ± 1 as its argument tends to $\pm\infty$, respectively. Examples of sigmoidal functions $g(u)$ are

$$g(u) = \frac{2}{\pi} \tan^{-1} \frac{\lambda \pi u}{2},$$

$$g(u) = \frac{e^{\lambda u} - e^{-\lambda u}}{e^{\lambda u} + e^{-\lambda u}}.$$

(5.35)

The equations of the Hopfield network are given by

$$C_i \frac{du_i}{dt} = \sum_{j=1}^{N} T_{ij} x_j - G_i u_i + I_i,$$

$$x_i = g(u_i).$$

(5.36)

Here $C_i > 0$ is the capacitance of the i th capacitor, and $G_i > 0$ the conductance of the i-th amplifier, T_{ij} stands for the strength of the interconnection between the x_j and the i-th capacitor. I_i stands for the current injection at the i-th node. Using the relationship between u_i and x_i, we get the equations of the Hopfield network:

$$\dot{x}_i = h_i(x_i) \left[\sum_{j=1}^{N} T_{ij}x_j - G_i g^{-1}(x_i) + I_i \right], \qquad (5.37)$$

where

$$h_i(x) = \frac{1}{C_i} \frac{dg}{du}\bigg|_{u=g^{-1}(x)} > 0 \quad \forall x.$$

Consider the function

$$v(x) = -\frac{1}{2} x^T T x + \sum_{i=1}^{N} G_i \int_0^{x_i} g^{-1}(y)dy - \sum_{i=1}^{N} I_i x_i. \qquad (5.38)$$

It is now easy to verify that equation (5.37) may be rewritten as

$$\dot{x}_i = -h_i(x_i) \frac{\partial v}{\partial x_i}. \qquad (5.39)$$

Thus, it follows that the Hopfield network is a gradient flow of v (along the "metric" given by the $h_i(\cdot)$) and

$$\dot{v}|_{(5.37)} = -\sum_{i=1}^{N} h_i(x_i) \left(\frac{\partial v}{\partial x_i}\right)^2 \le 0.$$

Further, it follows that

$$\dot{v} = 0 \quad \Rightarrow \quad \frac{\partial v}{\partial x_i} = 0 \; \forall i \quad \Rightarrow \quad \dot{x} = 0.$$

Thus, to apply LaSalle's principle to conclude that trajectories converge to one of the equilibrium points, it is enough to find a region which is either invariant or on which v is bounded. Consider the set

$$\Omega_\epsilon = \{x \in \mathbb{R}^n : |x_i| \le 1 - \epsilon\}.$$

We will prove that the region Ω_ϵ is invariant for ϵ small enough. Note that

$$\frac{dx_i^2}{dt} = 2x_i h_i(x_i) \left(\sum_{j=1}^{N} T_{ij}x_j - g^{-1}(x_i) + I_i \right).$$

Since $g^{-1}(x_i) \to \pm\infty$ as $x_i \to \pm 1$, it follows that for $\epsilon > 0$ small enough,

$$\frac{dx_i^2}{dt} < 0 \quad \text{for} \quad 1 - \epsilon \le |x_i| < 1,$$

thus establishing that Ω_ϵ is invariant. In turn, this implies that $v(x)$ is bounded below on Ω_ϵ, so that initial conditions beginning inside Ω_ϵ tend to the equilibrium points inside Ω_ϵ. Thus, for symmetric interconnections T_{ij} the Hopfield network does not oscillate, and all trajectories converge to one of the equilibria. It is of great interest to estimate the number of equilibria of a Hopfield network (see [143] and a very large literature on neural networks for more details).

There is a globalization of LaSalle's principle, which is as follows:

Theorem 5.25 Global LaSalle's Principle. *Consider the system of (5.1). Let $v(x)$ be a p.d.f. with $\dot{v} \le 0 \forall x \in \mathbb{R}^n$. If the set*

$$s = \{x \in \mathbb{R}^n, \quad \dot{v}(x) = 0\}$$

contains no invariant sets other than the origin, the origin is globally asymptotically stable.

There is also a version of LaSalle's theorem that holds for periodic systems as well.

Theorem 5.26 LaSalle's Principle for Periodic Systems. *Assume that the system of (5.1) is periodic, i.e.,*

$$f(x, t) = f(x, t + T), \quad \forall t \; \forall x \in \mathbb{R}^n.$$

Further, let $v(x, t)$ be a p.d.f. which is periodic in t also with period T. Define

$$S = \{x \in \mathbb{R}^n : \dot{v}(x, t) = 0, \forall t \ge 0\}$$

Then if $\dot{v}(x, t) \le 0 \; \forall t \ge 0, \quad \forall x \in \mathbb{R}^n$ and the largest invariant set in S is the origin, then the origin is globally (uniformly) asymptotically stable.

5.5   Generalizations of LaSalle's Principle

LaSalle's invariance principle is restricted in applications because it holds only for time-invariant and periodic systems. For extending the result to arbitrary time-varying systems, two difficulties arise:

1. $\{x : \dot{v}(x, t) = 0\}$ may be a time-varying set.
2. The ω limit set of a trajectory is itself not invariant.

However, if we have the hypothesis that

$$\dot{v}(x, t) \le -w(x) \le 0,$$

then the set S may be defined to be

$$\{x : w(x) = 0\},$$

and we may state the following generalization of LaSalle's theorem:

Theorem 5.27 Generalization of LaSalle's Theorem. *Assume that the vector field $f(x, t)$ of (5.1) is locally Lipschitz continuous in x, uniformly in t, in a ball of radius r. Let $v(x, t)$ satisfy for functions α_1, α_2 of class K*

$$\alpha_1(|x|) \le v(x, t) \le \alpha_2(|x|). \tag{5.40}$$

Further, for some non-negative function $w(x)$, assume that

$$\dot{v}(x, t) = \frac{\partial v}{\partial t} + \frac{\partial v}{\partial x} f(x, t) \le -w(x) \le 0. \tag{5.41}$$

Then for all $|x(t_0)| \leq \alpha_2^{-1}(\alpha_1(r))$, *the trajectories* $x(\cdot)$ *are bounded and*

$$\lim_{t \to \infty} w(x(t)) = 0. \tag{5.42}$$

Proof: The proof of this theorem needs a fact from analysis called Barbalat's lemma (which we have left as an exercise: Problem 5.9; see also [259]), which states that if $\phi(\cdot) : \mathbb{R} \mapsto \mathbb{R}$ is a *uniformly continuous* integrable function with

$$\int_0^\infty \phi(t) < \infty,$$

then

$$\lim_{t \to \infty} \phi(t) = 0.$$

The requirement of uniform continuity of $\phi(\cdot)$ is necessary for this lemma, as easy counterexamples will show. We will use this lemma in what follows.

First, note that a simple contradiction argument shows that for any $\rho < r$,

$$|x(t_0)| \leq \alpha_2^{-1}(\alpha_1(\rho)) \Rightarrow |x(t)| \leq \rho \quad \forall \ t \geq t_0.$$

Thus $|x(t)| < r$ for all $t \geq t_0$, so that $v(x(t), t)$ is monotone decreasing. This yields that

$$\int_{t_0}^t w(x(\tau)) d\tau \leq -\int_t^{t_0} \dot{v}(x(\tau), \tau) d\tau \tag{5.43}$$

$$= v(x(t_0), t_0) - v(x(t), t).$$

Since $v(x, t)$ is bounded below by 0, it follows that

$$\int_{t_0}^\infty w(x(\tau)) d\tau < \infty.$$

By the continuity of $f(x, t)$ (Lipschitz in x, uniformly in t) and the boundedness of $x(t)$, it follows that $x(t)$ is uniformly continuous, and so is $w(x(t))$. Using Barbalat's lemma it follows that

$$\lim_{t \to \infty} w(x(t)) = 0. \qquad \square$$

Remarks: The preceding theorem implies that $x(t)$ approaches a set E defined by

$$E := \{x \in B_r : w(x) = 0\}.$$

However, the set E is not guaranteed to be an invariant set; so that one cannot, for example, assert that $x(t)$ tends to the largest invariant set inside E. It is difficult, in general, to show that the set E is invariant. However, this can be shown to be the case when $f(x, t)$ is autonomous, T-periodic, or asymptotically autonomous. Another generalization of LaSalle's theorem is given in the exercises (Problem 5.10).

5.6   Instability Theorems

Lyapunov's theorem presented in the previous section gives a sufficient condition for establishing the stability of an equilibrium point. In this section we will give some sufficient conditions for establishing the instability of an equilibrium point.

Definition 5.28 Unstable Equilibrium.   *The equilibrium point* 0 *is* unstable *at* t_0, *if it is not stable.*

This definition is more subtle than it seems: we may parse the definition, by systematically negating the definition of stability: *There exists* an $\epsilon > 0$, such that *for all* δ balls of initial conditions (no matter how small the ball), there exists at least one initial condition, such that the trajectory is not confined to the ϵ ball; that is to say,

$$\forall \delta \;\; \exists x_0 \in B_\delta$$

such that $\exists t_\delta$ with

$$|x_{t_\delta}| \geq \epsilon.$$

Instability is of necessity a local concept. One seldom has a definition for uniform instability. Note that the definition of instability does not require every initial condition starting arbitrarily near the origin to be expelled from a neighborhood of the origin, it just requires one from each arbitrarily small neighborhood of the origin to be expelled away. As we will see in the context of linear time-invariant systems in the next section, the linear time-invariant system

$$\dot{x} = Ax$$

is unstable, if just one eigenvalue of A lies in $\mathbb{C}^\circ_+$! The instability theorems have the same flavor: They insist on $\dot{v}$ being an l.p.d.f. so as to have a mechanism for the increase of v. However, since we do not need to guarantee that every initial condition close to the origin is repelled from the origin, we do not need to assume that v is an l.p.d.f.

We state and prove two examples of instability theorems:

Theorem 5.29 Instability Theorem.   *The equilibrium point* 0 *is unstable at time* t_0 *if there exists a decrescent function* $v : \mathbb{R}^n \times \mathbb{R}_+ \mapsto \mathbb{R}$ *such that*

1. $\dot{v}(x, t)$ *is an l.p.d.f.*
2. $v(0, t) = 0$ *and there exist points* x *arbitrarily close to* 0 *such that* $v(x, t_0) > 0$.

Proof: We are given that there exists a function $v(x, t)$ such that

$$v(x, t) \leq \beta(|x|) \quad x \in B_r,$$
$$\dot{v}(x, t) \geq \alpha(|x|) \quad x \in B_s.$$

We need to show that for some $\epsilon > 0$, there is no δ such that

$$|x_0| < \delta \Rightarrow |x(t)| < \epsilon \quad \forall t \geq t_0.$$

Now choose $\epsilon = \min(r, s)$. Given $\delta > 0$ choose x_0 with $|x_0| < \delta$ and $v(x_0, t_0) > 0$. Such a choice is possible by the hypothesis on $v(x, t_0)$. So long as $\phi(t, x_0, t_0)$ lies in B_ϵ we have $\dot{v}(x(t), t) \geq 0$, which shows that

$$v(x(t), t) \geq v(x_0, t_0) > 0$$

This implies that $|x(t)|$ is bounded away from 0. Thus $\dot{v}(x(t), t)$ is bounded away from zero. Thus, $v(x(t), t)$ will exceed $\beta(\epsilon)$ in finite time. In turn this will guarantee that $|x(t)|$ will exceed ϵ in finite time. □

Theorem 5.30 Chetaev's theorem.   *The equilibrium point 0 is unstable at time t_0 if there is a decrescent function $v : \mathbb{R}_+ \times \mathbb{R}^n \mapsto \mathbb{R}$ such that*

1. $\dot{v}(x, t) = \lambda v(x, t) + v_1(x, t)$, where $\lambda > 0$ and $v_1(x, t) \geq 0$ $\forall t \geq 0$ $\forall x \in B_r$.
2. $v(0, t) = 0$ and there exist points x arbitrarily close to 0 such that $v(x, t_0) > 0$.

   Proof: Choose $\epsilon = r$ and given $\delta > 0$ pick x_0 such that $|x_0| < \delta$ and $v(x_0, t_0) > 0$. When $|x(t)| \leq r$, we have

$$\dot{v}(x, t) = \lambda v(x, t) + v_1(x, t) \geq \lambda v(x, t).$$

If we multiply the inequality above by integrating factor $e^{-\lambda t}$, it follows that

$$\frac{d v(x, t) e^{-\lambda t}}{dt} \geq 0.$$

Integrating this inequality from t_0 to t yields

$$v(x(t), t) \geq e^{\lambda(t - t_0)} v(x_0, t_0).$$

Thus, $v(x(t), t)$ grows without bound. Since $v(x, t)$ is decrescent,

$$v(x, t) \geq \beta(|x|)$$

for some function of class K, so that for some t_δ, $v(x(t), t) > \beta(\epsilon)$, establishing that $|x(t_\delta)| > \epsilon$. □

5.7   Stability of Linear Time-Varying Systems

Stability theory for linear time varying systems is very easy and does not really need the application of the theorems of Lyapunov. However, Lyapunov theory provides valuable help in understanding the construction of Lyapunov functions for nonlinear systems. Thus, in this section we will consider linear time varying systems of the form

$$\dot{x} = A(t)x, \quad x(t_0) = x_0. \tag{5.44}$$

where $A(t) \in \mathbb{R}^{n \times n}$ is a piecewise continuous bounded function. As a consequence the system of (5.44) satisfies the conditions for the existence and uniqueness of solutions.

Definition 5.31 State Transition Matrix. *The state transition matrix* $\Phi(t, t_0) \in$
$\mathbb{R}^{n \times n}$ *associated with* $A(t)$ *is by definition the unique solution of the matrix
differential equation*

$$\dot{\Phi}(t, t_0) = A(t)\Phi(t, t_0), \quad \Phi(t_0, t_0) = I. \tag{5.45}$$

The flow of (5.44) can be expressed in terms of the solutions to (5.45) as

$$x(t) = \Phi(t, t_0)x(t_0). \tag{5.46}$$

In particular, this expression shows that the trajectories are proportional to the
size of the initial conditions, so that local and global properties are identical. Two
important properties of the state transition matrix are as follows:

1. *Group Property*

$$\Phi(t, t_0) = \Phi(t, \tau)\Phi(\tau, t_0) \ \forall \ t, \tau, t_0 \tag{5.47}$$

2. *Inverse*

$$\Phi(t, t_0)^{-1} = \Phi(t_0, t) \ \forall t, t_0. \tag{5.48}$$

Recalling the formula for differentiating the inverse of a matrix:

$$\frac{d}{dt}[M(t)]^{-1} = M^{-1}(t)\frac{d}{dt}M(t)M^{-1}(t),$$

we have

$$\begin{aligned}
\frac{d}{dt_0}\Phi(t, t_0) &= \frac{d}{dt_0}\left[(\Phi(t_0, t))^{-1}\right] \\
&= -\Phi(t_0, t)^{-1}A(t_0)\Phi(t_0, t)\Phi(t_0, t)^{-1} \\
&= -\Phi(t, t_0)A(t_0).
\end{aligned}$$

The following theorems are elementary characterizations of stability, uniform
stability, and asymptotic stability of linear systems. Uniform asymptotic stability
is characterized in the next theorem. Recall that for linear systems local and global
stability notions are identical.

Theorem 5.32 Stability of Linear Systems. *The right-hand-side of the following table gives the stability conclusions of the equilibrium point* 0 *of the linear
time-varying system (5.44).*

Conditions on $\Phi(t, t_0)$	Conclusions
$\sup_{t \geq t_0} \|\Phi(t, t_0)\| := m(t_0) < \infty$	Stable at t_0
$\sup_{t_0 \geq 0} \sup_{t \geq t_0} \|\Phi(t, t_0)\| < \infty$	Uniformly Stable
$\lim_{t \to \infty} \|\Phi(t, t_0)\| = 0$	Asymptotically stable

Theorem 5.33 Exponential and Uniform Asymptotic Stability. *The equilibrium point 0 of (5.44) is* uniformly asymptotically stable *if it is uniformly stable and* $|\Phi(t, t_0)| \to 0$ *as* $t \to \infty$, *uniformly in* t_0. *The point* $x = 0$ *is a uniform asymptotically stable equilibrium point of (5.44) iff* $x = 0$ *is an exponentially stable equilibrium point of (5.44)*.

Proof: The first part of the theorem is a statement of the definition of uniform asymptotic stability for linear time varying systems. For the second half, the fact that exponential stability implies uniform asymptotic stability follows from the definition. For the converse implication let us assume that the equilibrium is uniformly asymptotically stable, From the first part of the theorem, we have that uniform stability implies that $\forall t_1$, there exist m_0, T such that

$$|\Phi(t, t_1)| \le m_0, \quad t \ge t_1.$$

Uniform convergence of $\Phi(t, t_0$ to 0 implies that

$$|\Phi(t, t_1)| \le \frac{1}{2}, \quad \forall t \ge t_1 + T.$$

Now given any t, t_0 pick k such that

$$t_0 + kT \le t \le t_0 + (k + 1)T.$$

Since

$$\Phi(t, t_0) = \Phi(t, t_0 + kT) \prod_{j=1}^{k} \Phi(t_0 + jT, t_0 + jT - T),$$

it follows that

$$|\Phi(t, t_0)| \le |\Phi(t, t_0 + kT)| \prod_{j=1}^{k} |\Phi(t_0 + jT, t_0 + jT - T)|$$

$$\le m_0 2^{-k}$$

$$\le 2m_0 2^{-(t-t_0)/T}$$

$$\le m e^{-\lambda(t-t_0)}$$

where $m = 2m_0$ and $\lambda = \log 2/T$. □

5.7.1 Autonomous Linear Systems

The results stated above can be specialized to linear time-invariant systems of the form

$$\dot{x} = Ax \tag{5.49}$$

with $x \in \mathbb{R}^n$.

Theorem 5.34 Stability of Linear Time-Invariant Systems.

1. *The equilibrium at the origin of (5.49) is stable iff all the eigenvalues of A are in $\mathbb{C}_-$ and those on the $j\omega$ axis are simple zeros of the minimal polynomial of A.*
2. *The equilibrium at the origin of (5.49) is asymptotically stable iff all the eigenvalues lie in $\mathbb{C}_-^\circ$.*

An alternative way of studying the stability of (5.49) is by using the Lyapunov function with symmetric, positive definite $P \in \mathbb{R}^{n \times n}$,

$$v(x) = x^T P x.$$

An easy calculation yields that

$$\dot{v}(x) = x^T (A^T P + P A) x. \tag{5.50}$$

If there exists a symmetric, positive definite $Q \in \mathbb{R}^{n \times n}$ such that

$$A^T P + P A = -Q, \tag{5.51}$$

then we see that $-\dot{v}(x)$ is a p.d.f. Equation (5.51), which is a linear equation of the form

$$\mathcal{L}(P) = Q$$

where $\mathcal{L}$ is a map from $\mathbb{R}^{n \times n} \mapsto \mathbb{R}^{n \times n}$ referred to as a *Lyapunov equation*. Given a symmetric $Q \in \mathbb{R}^{n \times n}$, it may be shown that (5.51) has a unique symmetric solution $P \in \mathbb{R}^{n \times n}$ when $\mathcal{L}$ is invertible, that is, iff all the n^2 eigenvalues of the linear map $\mathcal{L}$ are different from 0:

$$\lambda_i + \lambda_j^* \neq 0 \quad \forall \; \lambda_i, \lambda_j \in \sigma(A).$$

Here $\sigma(A)$ refers to the spectrum, or the set of eigenvalues of A.

Claim 5.35. *If A has all its eigenvalues in $\mathbb{C}_-^\circ$, then the solution of (5.51) is given by*

$$P = \int_0^\infty e^{A^T t} Q e^{At} dt.$$

Proof: Define

$$S(t) = \int_0^t e^{A^T \tau} Q e^{A\tau} d\tau,$$

Using a change of variables, we may rewrite this as

$$S(t) = \int_0^t e^{A^T (t-\tau)} Q e^{A(t-\tau)} d\tau.$$

Differentiating this expression using the Leibnitz rule yields

$$\dot{S}(t) = A^T S + S A + Q.$$

Also, since A has eigenvalues in $\mathbb{C}^\circ_-$ the eigenvalues of the linear operator

$$S \mapsto A^T S + SA$$

lie in $\mathbb{C}^\circ_-$ as well, so that $\dot{S} \to 0$ as $t \to \infty$, whence $P = S(\infty)$ exists, and satisfies

$$A^T S(\infty) + S(\infty)A + Q = 0.$$

Hence the P of the claim satisfies equation (5.51). $\square$

Theorem 5.36 Lyapunov Lemma.   *For $A \in \mathbb{R}^{n \times n}$ the following three statements are equivalent:*

1. *The eigenvalues of A are in $\mathbb{C}^\circ_-$.*
2. *There exists a symmetric $Q > 0$ such that the unique symmetric solution to (5.51) satisfies $P > 0$.*
3. *For all symmetric $Q > 0$ (5.51) the unique symmetric solution satisfies $P > 0$.*
4. *There exists $C \in \mathbb{R}^{m \times n}$ (with m arbitrary) with the pair A, C observable,[2] there exists a unique symmetric solution $P > 0$ of the Lyapunov equation with $Q = C^T C \geq 0$,*

$$A^T P + PA + C^T C = 0. \tag{5.52}$$

5. *For all $C \in \mathbb{R}^{m \times n}$ (with m arbitrary), such that the pair A, C is observable, there exists a unique solution $P > 0$ of (5.52).*

Proof: (iii) $\Rightarrow$ (ii) and (v) $\Rightarrow$ (iv) are obvious.
(ii) $\Rightarrow$ (i) Choose $v(x) = x^T Px$ a p.d.f. and note that $-\dot{v}(x) = x^T Qx$ is also a p.d.f.
(iv) $\Rightarrow$ (i) Choose $v(x) = x^T Px$ a p.d.f. and note that $-\dot{v}(x) = x^T C^T Cx$. By LaSalle's principle, the trajectories starting from arbitrary initial conditions converge to the largest invariant set in the null space of C, where $\dot{v}(x) = 0$. By the assumption of observability of the pair A, C it follows that the largest invariant set inside the null space of C is the origin.
(i) $\Rightarrow$ (iii) By the preceding claim it follows that the unique solution to (5.51) is

$$P = \int_0^\infty e^{A^T t} Q e^{At} dt.$$

That $P \geq 0$ is obvious. It only needs to be shown that $P > 0$. Indeed, if $x^T Px = 0$ for some $x \in \mathbb{R}^n$ then it follows with $Q = M^T M$ that

$$0 = x^T Px = \int_0^\infty x^T e^{A^T t} M^T M e^{At} x \, dt = \int_0^\infty |M e^{At} x|^2 dt$$

Thus we have that $M e^{At} x \equiv 0$. In particular, $Mx = 0$. But M is nonsingular since Q is. Hence $x = 0$, establishing the positive definiteness of P.

---

[2] Recall from linear systems theory that a pair A, C is said to be *observable* iff the largest A-invariant space in the null space of C is the origin.

(i) $\Rightarrow$ (v) By the preceding claim it follows that the unique solution to (5.51) is

$$P = \int_0^\infty e^{A^T t} C^T C e^{At} dt.$$

That $P \geq 0$ is obvious. It only needs to be shown that $P > 0$. Indeed, if $x^T P x = 0$ for some $x \in \mathbb{R}^n$ then it follows that $C e^{At} x \equiv 0$. In particular, by differentiating $C e^{At} x$ n-times at the origin, we get $Cx = CAx = \cdots = CA^{n-1}x = 0$. Since the pair A, C are observable, it follows that $x = 0$ establishing the positive definiteness of P.   □

The following generalization of the Lyapunov lemma, called the Taussky lemma, is useful when $\sigma(A) \not\subset \mathbb{C}^\circ_-$. We drop the proof since it is not central to the developments of this chapter (see [296]).

Lemma 5.37 Taussky Lemma. *For $A \in \mathbb{R}^{n \times n}$ and given $Q \in \mathbb{R}^{n \times n}$ positive definite, if $\sigma(A) \cap \{j\omega : \omega \in \,]-\infty, \infty[\} = \emptyset$ the unique symmetric solution P to the Lyapunov equation (5.51)*

$$A^T P + PA = -Q$$

has as many positive eigenvalues as the number of eigenvalues of A in $\mathbb{C}^\circ_-$ and as many negative eigenvalues as the number of eigenvalues of A in $\mathbb{C}^\circ_+$.

5.7.2   Quadratic Lyapunov Functions for Linear Time Varying Systems

The time varying counterpart to the claim of (5.35) may be derived after a few preliminary results.

Claim 5.38. *Consider the system of (5.44) with uniformly asymptotically stable equilibrium 0:*

$$\dot{x} = A(t)x, \quad x(t_0) = x_0.$$

Further, assume that $Q(.)$ is a continuous, bounded function. Then for $t \geq 0$ the matrix

$$P(t) = \int_t^\infty \Phi^T(\tau, t) Q(\tau) \Phi(\tau, t) d\tau \tag{5.53}$$

is well-defined. Further, $P(t)$ is bounded.

Proof: Uniform asymptotic stability of (5.44) guarantees exponential stability so that

$$|\Phi(\tau, t)| \leq m e^{-\lambda(\tau - t)} \quad \forall \tau \geq t.$$

This estimate may be used to show that $P(t)$ as defined in (5.53) is well defined and bounded.   □

Claim 5.39. *If $Q(t) > 0$ $\forall$ $t \geq 0$ and further,*

$$\alpha x^T x \leq x^T Q(t)x, \quad \forall x \in \mathbb{R}^n, \; t \geq 0,$$

and $A(t)$ is bounded, say by m, then $P(t)$ of (5.53) is uniformly positive definite for $t \geq 0$, i.e., there exists $\beta > 0$ such that

$$\beta x^T x \leq x^T P(t)x.$$

Proof:

$$x^T P(t)x = \int_t^\infty x^T \Phi^T(\tau, t) Q(\tau) \Phi(\tau, t)x \, d\tau$$

$$\geq \alpha \int_t^\infty |\Phi(\tau, t)x|^2 d\tau$$

$$\geq \alpha \int_t^\infty x^T x e^{-2k(t-\tau)} d\tau.$$

Here the last inequality follows from the Bellman Gronwall lemma which gives a lower bound on $|\Phi(\tau, t)x|$ from the norm bound $|A(t)| \leq k$, as

$$|\Phi(\tau, t)x| \geq e^{-k(\tau-t)}|x|.$$

From the last inequality, it also follows that

$$x^T P(t)x \geq \frac{\alpha}{2k} x^T x,$$

establishing the claim. $\qquad\qquad\qquad\qquad\qquad\qquad\qquad\qquad\qquad\square$

Theorem 5.40 Time-Varying Lyapunov Lemma. *Assume that $A(\cdot)$ is bounded. If for some $Q(t) \geq \alpha I$, $P(t)$ as defined by (5.53) is bounded, then the origin is the uniformly asymptotically stable equilibrium point of (5.44).*

Proof: By the preceding claim, $P(t)$ satisfying (5.53) is bounded below, so that

$$\gamma |x|^2 \geq x^T P(t)x \geq \beta |x|^2$$

whence $x^T P(t)x$ is a decrescent p.d.f. Further, it is easy to see that

$$\dot{v}(x, t) = x^T (\dot{P}(t) + A^T(t)P(t) + P(t)A(t))x$$

$$= -x^T Q(t)x$$

$$\leq -\alpha |x|^2.$$

This establishes exponential stability of the origin and as a consequence uniform asymptotic stability. $\qquad\qquad\qquad\qquad\qquad\qquad\qquad\qquad\qquad\square$

Uniform Complete Observability and Stability

A version of the time-varying Lyapunov lemma for systems that are uniformly completely observable may also be stated and proved (see Exercise 5.12 for the definition of uniform complete observability and the statement of the lemma). The key philosophical content of this result and the corresponding time invariant one stated extremely loosely is that if for some Lyapunov function candidate $v(x, t)$, the nonlinear system

$$\dot{x} = f(x, t)$$

with output map

$$y(t) = h(x,t) := \frac{\partial v}{\partial t} + \frac{\partial v}{\partial x} f(x,t)$$

is "locally uniformly observable" near the origin, then the origin is uniformly asymptotically stable. The reason this statement is loose is that the concept of "local uniform observability" needs to be made precise. This is straightforward for the linear time varying case that we are presently discussing, as discussed in Problem 5.12.

5.8   The Indirect Method of Lyapunov

Consider an autonomous version of the system of (5.1), namely,

$$\dot{x} = f(x)$$

with $f(0) = 0$, so that 0 is an equilibrium point of the system. Define $A = \left.\frac{\partial f}{\partial x}\right|_{x=0} \in \mathbb{R}^{n \times n}$ to be the Jacobian matrix of $f(x)$. The system

$$\dot{z} = Az$$

is referred to as the linearization of the system (5.1) *around the equilibrium point* 0. For non-autonomous systems, the development is similar: Consider

$$\dot{x} = f(x,t)$$

with $f(0,t) \equiv 0$ for all $t \geq 0$. With

$$A(t) = \left.\frac{\partial f(x,t)}{\partial x}\right|_{x=0}$$

it follows that the remainder

$$f_1(x,t) = f(x,t) - A(t)x$$

is $o(|x|)$ for each fixed $t \geq 0$. It may not, however, be true that

$$\lim_{|x| \to 0} \sup_{t \geq 0} \frac{|f_1(x,t)|}{|x|} = 0. \tag{5.54}$$

that is $f(x,t)$ is a function for which the remainder is *not uniformly* of order $o(|x|)$. For a scalar example of this consider the function

$$f(x,t) = -x + tx^2.$$

In the instance that the higher order terms are uniformly, $o(|x|)$, i.e., if (5.54) holds, then the system of (5.55)

$$\dot{z} = A(t)z \tag{5.55}$$

is referred to as the *linearization of (5.1)* about the origin. It is important to note that analysis of the linearization of (5.55) yields conclusions about the nonlinear system (5.1) *when the uniform higher-order condition of (5.54) holds.*

Theorem 5.41 Indirect Theorem of Lyapunov: Stability from Linearization.
Consider the system of (5.1) and let

$$\lim_{|x|\to 0} \sup_{t\geq 0} \frac{|f_1(x,t)|}{|x|} = 0.$$

Further, assume that $A(\cdot)$ is bounded. If 0 is uniformly asymptotically stable equilibrium point of (5.55), then it is a locally uniformly asymptotically stable equilibrium point of (5.1).

Proof: Since $A(\cdot)$ is bounded and 0 is a uniformly asymptotically stable equilibrium point of (5.55), it follows from Claims 5.38, 5.39 that $P(t)$ defined by

$$P(t) = \int_t^\infty \Phi^T(\tau, t)\Phi(\tau, t)d\tau$$

is bounded above and below, that is, it satisfies

$$\beta x^T x \geq x^T P(t)x \geq \alpha x^T x$$

for some $\alpha, \beta > 0$. Thus $v(x, t) = x^T P(t)x$ is a decrescent p.d.f. Also,

$$\dot{v}(x, t) = x^T[\dot{P}(t) + A^T(t)P(t) + P(t)A(t)]x + 2x^T P(t)f_1(x, t),$$
$$= -x^T x + 2x^T P(t)f_1(x, t).$$

Since (5.54) holds, there exists $r > 0$ such that

$$|f_1(x, t)| \leq \frac{1}{3\beta}|x| \quad \forall x \in B_r, \quad t \geq 0,$$

and

$$|2x^T P(t)f_1(x, t)| \leq \frac{2|x|^2}{3} \quad \forall x \in B_r,$$

so that

$$\dot{v}(x, t) \leq -\frac{x^T x}{3} \quad \forall x \in B_r.$$

Thus $-\dot{v}(x, t)$ is an l.p.d.f. so that 0 is a locally (since $-\dot{v}$ is an l.p.d.f. only in B_r) uniformly asymptotically stable equilibrium point of (5.1). $\quad\square$

Remarks:

1. The preceding theorem requires *uniform* asymptotic stability of the linearized system to prove uniform asymptotic stability of the nonlinear system. Counterexamples to the theorem exist if the linearized system is not uniformly asymptotically stable.
2. The converse of this theorem is also true (see Problem 5.18).
3. If the linearization is time-invariant, then $A(t) \equiv A$ and $\sigma(A) \subset \mathbb{C}^\circ_-$, then the nonlinear system is uniformly asymptotically stable.

4. This theorem proves that *global* uniform asymptotic stability of the linearization implies *local* uniform asymptotic stability of the original nonlinear system. The estimates of the theorem can be used to give a (conservative) bound on the domain of attraction of the origin. For example, the largest level set of $v(x, t)$ that is uniformly contained in B_r is such a bound and can be further estimated using the bounds on $P(t)$ as $\{x \in \mathbb{R}^n : |x| \leq \frac{r}{\beta}\}$ Also, the estimates on r in the preceding proof could be refined. Systematic techniques for estimating the bounds on the regions of attraction of equilibrium points of nonlinear systems is an important area of research and involves searching for the 'best' Lyapunov functions.

To prove the instability counterpart to the above theorem one needs to assume that the linearized system is autonomous even though the original system is non-autonomous.

Theorem 5.42 Instability from Linearization. *Consider the nonlinear system (5.1) and let the linearization satisfy the conditions of equation (5.54). Further, let the linearization be time invariant, i.e.,*

$$A(t) = \frac{\partial f(x, t)}{\partial x}|_{x=0} \equiv A_0.$$

If A_0 has at least one eigenvalue in $\mathbb{C}^\circ_+$, then the equilibrium 0 of the nonlinear system is unstable.

Proof: Consider the Lyapunov equation

$$A_0^T P + P A_0 = I$$

Assume, temporarily, that A_0 has no eigenvalues on the $j\omega$ axis. Then by the Taussky lemma the Lyapunov equation has a unique solution, and further more if A_0 has at least one eigenvalue in $\mathbb{C}^\circ_+$, then P has at least one positive eigenvalue. Then $v(x, t) = x^T P x$ takes on positive values arbitrarily close to the origin and is decrescent. Further,

$$\dot{v}(x, t) = x^T (A_0^T P + P A_0) x + 2x^T P f_1(x, t)$$
$$= x^T x + 2x^T P f_1(x, t).$$

Using the assumption of (5.54) it is easy to see that there exists r such that

$$\dot{v}(x, t) \geq \frac{x^T x}{3} \quad \forall x \in B_r,$$

so that it is an l.p.d.f. Thus, the basic instability theorem yields that 0 is unstable.

In the instance that A_0 has some eigenvalues on the $j\omega$ axis in addition to at least one in the open right half plane, the proof follows by continuity.   □

Remarks:

1. We have shown using Lyapunov's basic theorems that if the the linearization of a nonlinear system is time invariant then:

- Having all eigenvalues in the open left half plane guarantees local uniform asymptotic stability of the origin for the nonlinear system.
- If at least one of the eigenvalues of the linearization lies in the open right half plane, then the origin is unstable.

The only situation not accounted for is the question of stability or instability when the eigenvalues of the linearization lie in the closed left half plane and include at least one on the $j\omega$ axis. The study of stability in these cases is delicate and relies on higher order terms than those of the linearization. This is discussed in Chapter 7 as stability theorems on the center manifold.

2. The technique provides only *local* stability theorems. To show global stability for equilibria of nonlinear systems there are no short cuts to the basic theorems of Lyapunov.

5.9 Domains of Attraction

Once the local asymptotic stability of an equilibrium point has been established, it is of interest to determine the set of initial conditions that converges to the equilibrium point.

Definition 5.43 Domain of Attraction at t_0. *Consider the differential equation (5.1) with equilibrium point x_0 at t_0. The domain of attraction of x_0 at t_0 is the set of all initial conditions x at time t_0, denote $\phi(t, t_0, x)$ satisfies*

$$\lim_{t \to \infty} \phi(t, t_0, x) = x_0.$$

For autonomous systems,

$$\dot{x} = f(x) \tag{5.56}$$

the domain of attraction of the equilibrium point x_0 (assumed to be 0, without loss of generality) is a set that is independent of the initial time, since the flow only depends on the time difference $t - t_0$. For this case, we use the notation of Chapter 2, namely $\phi_t(x)$, to mean the state at time t starting from x at time 0. We may characterize this set as follows:

$$\Omega := \left\{ x : \lim_{t \to \infty} \phi_t(x) = 0 \right\}.$$

Proposition 5.44 Topological Properties of the Domain of Attraction. *Let 0 be an asymptotically stable equilibrium point of (5.56). Then, the domain of attraction, Ω of 0 is an open, invariant set. Moreover, the boundary of Ω is invariant as well.*

Proof: Let $x \in \Omega$. For the invariance of Ω we need to show that $\phi_s(x) \in \Omega$ for all s. To this end, note that

$$\phi_t(\phi_s(x)) = \phi_{t+s}(x),$$

so that for all s, we have

$$\lim_{t \to \infty} \phi_t(\phi_s(x)) = 0,$$

establishing that $\phi_s(x) \in \Omega$ for all s.

To show that Ω is open, we first note that a neighborhood of the origin is contained in Ω: Indeed, this follows from the definition of asymptotic stability. Let us denote this by B_a, a ball of radius a. Now let $x \in \Omega$ since $\phi_t(x) \to 0$ as $t \to \infty$, it follows that there exists T such that $|\phi_t(x)| < a/2$ for $t \geq T$. From the fact that solutions of the differential equation (5.56) depend continuously on initial conditions on compact time intervals, it follows that there exists $\delta > 0$ such that for $|y - x| < \delta$, we have that $|\phi_T(y) - \phi_T(x)| < a/2$. In turn, this implies that $\phi_T(y) \in B_a$, so that $\lim_{t \to \infty} \phi_t(y) = 0$. Hence a ball of radius δ around x is contained in Ω, and this proves that Ω is open.

To show that the boundary of Ω, labeled $\partial \Omega$ is invariant, let $x \in \partial \Omega$. Then $x_n \to x$ with $x_n \in \Omega$. Given any time t, then there exists a sequence x_n such that $\phi_t(x_n) \to \phi_t(x)$. Now, $\phi_t(x_n) \in \Omega$, since Ω is invariant. Further, $\phi_t(x) \notin \Omega$ since $x \in \partial \Omega$. Hence, $\phi_t(x) \in \partial \Omega$ for all t. Hence, $\partial \Omega$ is invariant. $\square$

The proof of the preceding proposition hints at a way of constructing domains of attraction of an asymptotically stable equilibrium point. One starts with a small ball of initial conditions close to the equilibrium point and integrates them backwards in time. More specifically, let B_a be the initial ball of attraction around the equilibrium point at the origin. Then $\phi_{-T}(B_a) := \{\phi_{-T}(x) : x \in B_a\}$ corresponds to integrating equation (5.56) backwards for T seconds and is contained in Ω. By choosing T large, we get a good estimate for the domain of attraction.

Example 5.45 Power Systems: Critical Clearing Times, Potential Energy Surfaces, Alert States.   *The dynamics of an interconnected power system with n buses may be modeled by the so-called swing equations, which we have seen somewhat earlier in this chapter:*

$$M_i \ddot{\theta}_i + D_i \dot{\theta}_i + \sum_{i=1}^{n} B_{ij} \sin(\theta_i - \theta_j) = P_{mi} - P_{ei}.$$

Here θ_i stands for the angle of the i-th generator bus, M_i the moment of inertia of the generator, D_i the generator damping, P_{mi}, P_{ei} the mechanical input-power and the electrical output-power at the i-th bus, respectively. During the normal operation of the power system, the system dynamics are located at an equilibrium point of the system, $\theta^e \in \mathbb{R}^n$. In the event of a system disturbance, usually in the form of a lightning strike on a bus, say, B_{ij}, protective equipment on the bus opens, interrupting power transfer from generator i to generator j. In this instance, the dynamics of the system move from θ^e, since θ^e is no longer an equilibrium point of the modified system dynamics. Of interest to power utility companies is the so-called reclosure *time for the bus hit by lightning. This is the maximum time before the trajectory of the system under the disturbance drifts out of the domain* *of attraction of the equilibrium point θ^e. It is of interest to find the maximum such time under a variety of different fault conditions so as to give the fault as long as possible to clear.*

In the past, utilities relied on lengthy off-line simulations to determine these critical clearing times. More recently, analytic techniques for estimating the boundaries of the domains of attraction of equilibria have been developed. In particular, the potential energy boundary surface (PEBS) method relies on the "potential energy" function [239].

$$W(\theta) = \sum_{i=1}^{n} - \sum_{j \geq i} B_{ij} \cos(\theta_i - \theta_j) + \sum_{i=1}^{n} (P_{ei} - P_{mi})\theta_i \qquad (5.57)$$

which is a modification of the function of equation (5.34) in Section 5.4 to exclude the "kinetic energy" terms. It may now be verified that local minima of $W(\theta)$ correspond to stable equilibria of the swing dynamics and that level sets of the function $W(\theta)$ about local minima, which do not intersect any saddles of the function give conservative estimates of the domain of attraction (see Problem 5.30). In particular, a great deal of effort is spent on computing the "nearest unstable" equilibrium of the swing dynamics (i.e., the saddle extremum or maximum of the function $W(\theta)$) closest to a given operating point θ_0. An operating point or equilibrium of the form $(\theta_0, 0)$ is called an alert state *when the nearest unstable equilibrium point is close (in Euclidean norm) to θ_0.*

We will now illustrate some qualitative considerations about the computation of domains of attraction of equilibria of power systems. This is an extremely important part of the operational considerations of most electrical utilities which spend a great deal of resources on stability assessment of operating points [314]. We begin with some general considerations and remarks. Our treatment follows Chiang, Wu, and Varaiya [62]. Let x_s be a locally asymptotically stable equilibrium point of the time-invariant differential equation $\dot{x} = f(x)$ and define the *domain of attraction* of x_s to be

$$A(x_s) := \left\{ x : \lim_{t \to \infty} \phi_t(x) = x_s \right\}.$$

As we have shown earlier in this section $A(x_s)$ is an open set containing x_s. We will be interested in characterizing the boundary of the set $A(x_s)$. It is intuitive that the boundary of $A(x_s)$ contains other unstable equilibrium points, labeled x_i. We define their stable and unstable manifolds (in analogy to Chapter 2) as

$$\begin{aligned} W_s(x_i) &= \left\{ x : \lim_{t \to \infty} \phi_t(x) = x_i \right\}, \\ W_u(x_i) &= \left\{ x : \lim_{t \to -\infty} \phi_t(x) = x_i \right\}. \end{aligned} \qquad (5.58)$$

In Chapter 7 we will state the generalizations of the Hartman–Grobman and stable unstable manifold theorems which guarantee that the sets W_s, W_u defined above are manifolds. We now make three assumptions about the system $\dot{x} = f(x)$:

A1. All equilibrium points of the system are *hyperbolic*. Actually, we only need the equilibrium points on the boundary of $A(x_s)$ to be hyperbolic.

A2. If the stable manifold $W_s(x_i)$ of one equilibrium point intersects the unstable manifold of another equilibrium point $W_u(x_j)$, they intersect transversally. The reader may wish to review the definition of transversal intersection from Chapter 3, namely, if $x \in W_s(x_i) \cap W_u(x_j)$, then

$$T_x(W_s(x_i)) + T_x(W_u(x_j)) = \mathbb{R}^n.$$

A3. There exists an energy like function $v(x)$ satisfying:

a. $\dot{v}(x) < 0 \; x \notin \{x : f(x) = 0\}$.

b. $\{x : \dot{v}(x) = 0\}$ is the set of equilibrium points.

c. $v(x)$ is a proper function, that is, the inverse image of compact sets is compact.

The first two hypotheses are "generically valid hypotheses." The last hypothesis is a strong "global" version of a converse Lyapunov hypothesis, which insists that all trajectories converge to one or the other of the equilibrium points and that there are no closed orbits or other complex ω limit sets. The following proposition can now be proved.

Proposition 5.46 Unstable Equilibrium Points on the Boundary of $A(x_s)$.
Consider the time-invariant nonlinear system satisfying assumptions A1–A3 above, with x_s an asymptotically stable equilibrium point. Then, an unstable equilibrium point x_i belongs to the boundary of $A(x_s)$ if and only if

$$W_u(x_i) - \{x_i\} \cap \bar{A}(x_s) \neq \varnothing. \tag{5.59}$$

Proof: The proof relies on the following lemma, which we leave as exercise to the reader (taking the help of [62] if necessary):

Lemma 5.47.

1. *An equilibrium point x_i belongs to $\partial A(x_i)$ if and only if*

$$W_u(x_i) - \{x_i\} \cap \bar{A}(x_s) \neq \varnothing.$$

2. *Let x_i, x_j be two equilibrium points. Then if the unstable manifold of x_i intersects the stable manifold of x_j, that is*

$$W_u(x_i) - \{x_i\} \cap W_s(x_j) - \{x_j\} \neq \varnothing,$$

then the dimension of $W_u(x_i)$ is greater than the dimension of $W_u(x_j)$.

3. *Assumption A3 implies that every trajectory starting on the boundary of $A(x_s)$ converges to one of the equilibrium points on the boundary of $A(x_s)$.*

4. *If for equilibria x_1, x_2, x_3, $W_s(x_1)$ intersects $W_u(x_2)$ transversely and $W_u(x_2)$ intersects $W_s(x_3)$ transversely, then $W_u(x_1)$ intersects $W_s(x_3)$ transversely.*

The necessity of condition (5.59) follows in straightforward fashion from the first statement of Lemma 5.47.

For the sufficiency, we first start with the case that x_i has an unstable manifold of dimension 1 on the boundary of $A(x_s)$. We would like to make sure that $W_u(x_i)$ is not entirely contained in $\partial A(x_s)$. Indeed, if this were true, since all points on the boundary of $A(x_s)$ converge to one of the equilibrium points on the boundary $A(x_s)$, say $\bar{x}$, then we have that

$$W_u(x_i) - \{x_i\} \cap W_s(\bar{x}) - \{\bar{x}\} \neq \varnothing,$$

and from the preceding lemma, we would have that

$$1 = \dim W_u(x_i) > \dim W_u(\bar{x}),$$

implying that the dimension of $W_u(\bar{x}) = 0$, or that $\bar{x}$ is a stable equilibrium point on the boundary of $A(x_s)$. This establishes the contradiction.

If x_i has an unstable manifold of dimension 2 on the boundary of $A(x_s)$, as before assuming for the sake of contradiction that $W_u(x_i)$ lies entirely in $\partial A(x_s)$, then there exists $\bar{x}$ such that

$$W_u(x_i) - \{x_i\} \cap W_s(\bar{x}) - \{\bar{x}\} \neq \varnothing.$$

In going through the steps of the previous argument, we see that $\bar{x}$ is either stable or has an unstable manifold of dimension 1. Since $\bar{x}$ cannot be stable, it should have a stable manifold of dimension 1. But then, by the last claim of Lemma 5.47, it follows that

$$W_u(x_i) - \{x_i\} \cap W_s(\bar{x}) - \{\bar{x}\} \neq \varnothing \quad \text{and} \quad W_u(\bar{x}) \cap A(x_s) \neq \varnothing$$

implying that

$$W_u(x_i) \cap A(x_s) \neq \varnothing.$$

By induction on the dimension of the unstable manifold of x_i, we finish the proof.
□

Using this proposition, we may state the following theorem

Theorem 5.48 Characterization of the Stability Boundary. *For the time invariant nonlinear system $\dot{x} = f(x)$ satisfying assumptions A1–A3 above, let x_i, $i = 1, \ldots, k$ be the unstable equilibrium points on the boundary of the domain of attraction $\partial A(x_s)$ of a stable equilibrium point. Then,*

$$\partial A(x_s) = \bigcup_{i=1}^{k} W_s(x_i). \tag{5.60}$$

Proof: By the third statement of Lemma 5.47 it follows that

$$\partial A(x_s) \subset \bigcup_{i=1}^{k} W_s(x_i).$$

To show the reverse inclusion, note from the previous proposition that

$$W_u(x_i) \cap A(x_s) \neq \varnothing.$$

Choose $p \in W_u(x_i) \cap A(x_s)$ and let $p \in D \subset A(x_s)$ be a disk of dimension $n - \dim W_u(x_i)$ transversal to $W_u(x_i)$ at p. It is intuitive that as $t \to -\infty$ the flow $\phi_t(D)$ tends to the stable manifold of x_i. The precise proof of this fact is referred to as the λ lemma of Smale ([277], [140]) and is beyond the scope of this book. This lemma along with the fact that the invariance of $A(x_s)$ guarantees that

$$\phi_t(D) \subset A(x_s),$$

and yields that

$$W_s(x_i) = \lim_{t \to -\infty} \phi_t(D) \subset \bar{A}(x_s).$$

Since $W_s(x_i) \cap A(x_s) = \varnothing$, it follows that

$$W_s(x_i) \subset \partial A(x_s).$$

Consequently, we have that

$$\bigcup_{i=1}^{k} W_s(x_i) \subset \partial A(x_s),$$

completing the proof. □

This theorem characterizing the boundary of the stability region can be combined with the Lyapunov function to establish the following theorem, whose proof we leave to the exercises (Problem 5.29).

Theorem 5.49 Lyapunov Functions and the Boundary of the Domain of Attraction. *Consider the time invariant nonlinear system $\dot{x} = f(x)$ satisfying assumptions A1–A3, with x_s a locally asymptotically stable equilibrium point and $x_i, i = 1, \ldots, k$ the equilibrium points on the boundary of its domain of attraction $A(x_s)$. Then we have*

1. *On the stable manifold $W_s(x_i)$ of an equilibrium point, the Lyapunov function $v(x)$ attains its minimum at the equilibrium point x_i.*
2. *On $\partial A(x_s)$ the minimum of $v(\cdot)$ is achieved at a type 1 equilibrium point, that is, one whose stable manifold has dimension $n - 1$.*
3. *If the domain of attraction $A(x_s)$ is bounded, the maximum of $v(\cdot)$ on the boundary is achieved at a source (that is, an equilibrium with a zero dimensional stable manifold).*

Remark: The preceding theorem is extremely geometrical. It gives a picture of a dynamical system satisfying assumptions A1–A3 as "gradient like" [277]. The stable equilibria lie at the base of valleys. The region of attraction of these equilibria are separated by "ridges" which are the stable manifolds of saddles or peaks. This is very much the landscape of the earth as we fly over it in an aircraft.

Numerical Methods of Computing Domains of Attraction

There are many interesting methods for computing domains of attraction. A detailed discussion of these is deferred to the literature: An interesting set of methods was proposed by Brayton and Tong in [41]. While we have shown that quadratic Lyapunov functions are the most natural for linear systems, there has been a recent resurgence in methods for obtaining piecewise quadratic Laypunov functions for nonlinear systems [156]. In this paper, the search for piecewise quadratic Lyapunov functions is formulated as a convex optimization problem using linear matrix inequalities. In turn, algorithms for optimization with linear matrix inequality constraints have made a great deal of progress in recent years (see, for example, the papers of Packard [238], [19] and the monograph by Boyd, Balakrishnan, El Ghaoui, and Feron [36]). We refer the reader to this literature for what is a very interesting new set of research directions.

5.10 Summary

This chapter was an introduction to the methods of Lyapunov in determining the stability and instability of equilibria of nonlinear systems. The reader will enjoy reading the original paper of Lyapunov to get a sense of how much was spelled out over a hundred years ago in this remarkable paper. There are several textbooks which give a more detailed version of the basic theory with theorems stated in greater generality. An old classic is the book of Hahn [124]. Other more modern textbooks with a nice treatment include Michel and Miller [209], Vidyasagar [317], and Khalil [162]. The recent book by Liu and Michel [210] contains some further details on the use of Lyapunov theory to study Hopfield neural networks and generalizations to classes of nonlinear systems with saturation nonlinearities.

5.11 Exercises

Problem 5.1 A p.d.f. function v with $-\dot{v}$ not a p.d.f..   Consider a function $g^2(t)$ with the form shown in Figure 5.9 and the scalar differential equation

$$\dot{x} = \frac{\dot{g}(t)}{g(t)}x.$$

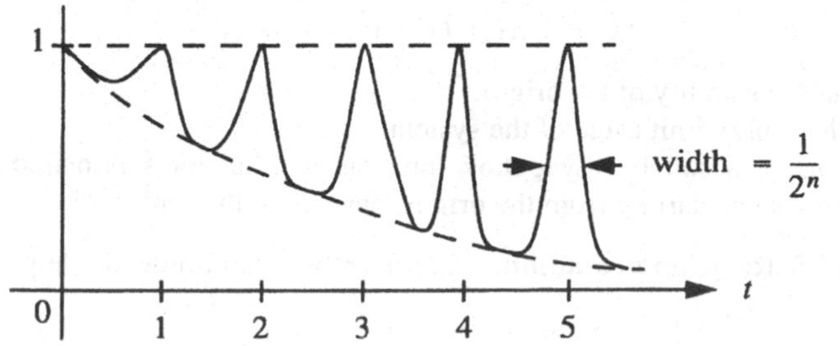

FIGURE 5.9. A bounded function $g^2(t)$ converging to zero but not uniformly.

Consider the Lyapunov function candidate

$$v(x, t) = \frac{x^2}{g^2(t)}\left[3 - \int_0^t g^2(\tau)d\tau\right].$$

Verify that $v(x, t)$ is a p.d.f. and so is $\dot{v}(t, x)$. Also, verify that the origin is stable but **not asymptotically stable**. What does this say about the missing statement in the table of Lyapunov's theorem in Theorem 5.16.

Problem 5.2 Krasovskii's theorem generalized.   (a) Consider the differential equation $\dot{x} = f(x)$ in $\mathbb{R}^n$. If there exist $P, Q \in \mathbb{R}^{n \times n}$, two constant, symmetric, positive definite matrices such that

$$P\frac{\partial f}{\partial x}(x) + \frac{\partial f}{\partial x}(x)^T P = -Q,$$

then prove that $x = 0$ is globally asymptotically stable. *Hint: Try both the Lyapunov function candidates $f(x)^T P f(x)$ and $x^T P x$. Be sure to justify why the former is a p.d.f. if you decide to use it!*

(b) Consider the differential equation $\dot{x} = A(x)x$ in $\mathbb{R}^n$. If there exist $P, Q \in \mathbb{R}^{n \times n}$, two constant, symmetric, positive definite matrices such that

$$PA(x) + A(x)^T P = -Q,$$

then prove that $x = 0$ is globally asymptotically stable.

Problem 5.3.   Consider the second-order nonlinear system:

$$\dot{x}_1 = -x_2 + \epsilon x_1(x_1^2 + x_2^2)\sin(x_1^2 + x_2^2),$$
$$\dot{x}_2 = x_1 + \epsilon x_2(x_1^2 + x_2^2)\sin(x_1^2 + x_2^2).$$

Show that the linearization is inconclusive in determining the stability of the origin. Use the direct method of Lyapunov and your own creative instincts to pick a V to study the stability of the origin for $-1 \le \epsilon \le 1$.

Problem 5.4.   Consider the second order nonlinear system:

$$\dot{x}_1 = x_2,$$
$$\dot{x}_2 = -x_1 + (1 - x_1^2 - x_2^2)x_2.$$

- Discuss the stability of the origin.
- Find the (only) limit cycle of the system.
- Prove using a suitable Lyapunov function and LaSalle's principle that all trajectories not starting from the origin converge to the limit cycle.

Problem 5.5 Rayleigh's equation.   Consider the second-order Rayleigh system:

$$\dot{x}_1 = -\epsilon(x_1^3/3 - x_1 + x_2),$$
$$\dot{x}_2 = -x_1.$$

This system is known to have one equilibrium and one limit cycle for $\epsilon \neq 0$. Show that the limit cycle does not lie inside the strip $-1 \leq x_1 \leq 1$. Show, using a suitable Lyapunov function, that it lies outside a circle of radius $\sqrt{3}$.

Problem 5.6 Discrete-time Lyapunov theorems.   Consider the discrete time nonlinear system

$$x(k+1) = f(x(k), k). \tag{5.61}$$

Check the definitions for stability, uniform stability, etc. and derive discrete time Lyapunov theorems for this system with suitable conditions on a Lyapunov function candidate $v(x, k)$ and on

$$\Delta v(x(k), k) := v(x(k+1), k+1) - v(x(k), k)$$
$$= v(f(x(k), k), k+1) - v(x(k), k).$$

Also, discuss theorems for exponential stability and instability. Is there a counterpart to LaSalle's principle for discrete time systems?

Problem 5.7 Discrete-time Lyapunov equation.   Prove that the following statements are equivalent for the discrete-time linear time invariant system

$$x(k+1) = Ax(k):$$

1. $x = 0$ is exponentially stable.
2. $|\lambda_i| < 1$ for all the eigenvalues of A.
3. Given any positive definite symmetric matrix $Q \in \mathbb{R}^{n \times n}$, there exists a unique, positive definite symmetric matrix $P \in \mathbb{R}^{n \times n}$ satisfying

$$A^T P A - P = -Q.$$

4. Given a matrix $C \in \mathbb{R}^{n_o \times n}$ such that A, C is observable, there exists a unique positive definite matrix P such that

$$A^T P A - P = -C^T C.$$

Problem 5.8 Discrete-time stability from the linearization.   Give a proof of the uniform local stability of the equilibrium at the origin of the discrete-time nonlinear system (5.61) when the linearization is uniformly stable.

Problem 5.9 Barbalat's lemma [259].   Assume that $\phi(\cdot) : \mathbb{R} \mapsto \mathbb{R}$ is **uniformly continuous** and integrable, that is,

$$\int_0^\infty \phi(t)dt < \infty.$$

Then show that

$$\lim_{t \to \infty} \phi(t) = 0.$$

Give a counterexample to show the necessity of uniform continuity.

Problem 5.10 Another generalization of LaSalle's principle. Consider a nonlinear time varying system Lipschitz continuous in x uniformly in t. Let $v; \mathbb{R} \times \mathbb{R}^n \mapsto \mathbb{R}$ be such that for $x \in B_r$, $\delta > 0$, and functions α_1, α_2 of class K,

$$\alpha_1(|x|) \le v(x, t) \le \alpha_2(|x|),$$

$$\dot{v}(t, x) \le 0, \tag{5.62}$$

$$\int_t^{t+\delta} \dot{v}(s(\tau, t, x), \tau)d\tau \le -\lambda v(x, t) \qquad 0 < \lambda < 1.$$

Now show that for $|x(t_0)| < \alpha_2^{-1}(\alpha_1(r))$ show that the trajectories converge uniformly, asymptotically to 0.

Problem 5.11 Power system swing dynamics on $T^n \times \mathbb{R}^n$ [9]. Consider the swing equations of the power system (5.33). Since the right-hand side is periodic in each of the θ_i with period 2π, we can assume that the state space of the system is $S^1 \times \cdots \times S^1 \times \mathbb{R}^n$ or $T^n \times \mathbb{R}^n$. Note that the function $v(\theta, \omega)$ of (5.34) is not a well defined function from $T^n \times \mathbb{R}^n \mapsto \mathbb{R}$, unless $P_{mi} - P_{ei} = 0$. What are the implications of the assertion in the text that bounded trajectories (on $\mathbb{R}^n \times \mathbb{R}^n$) converge to equilibrium points of the swing dynamics on the manifold $T^n \times \mathbb{R}^n$?

Prove that trajectories of the swing equations are bounded on the manifold $T^n \times \mathbb{R}^n$. Can you give counterexamples to the assertion that bounded trajectories on $T^n \times \mathbb{R}^n$ converge to equilibria (you may wish to review the Josephson junction circuit of Chapter 2 for hints in the case that $n = 1$)? Prove that all trajectories of the swing equation converge to equilibria for $P_{mi} - P_{ei}$ sufficiently small for $i = 1, \ldots, n$.

Problem 5.12 Generalization of the time-varying Lyapunov lemma. Consider the linear time varying system

$$\dot{x} = A(t)x.$$

Define it to be *uniformly completely observable* from the output $y(t) = C(t)x(t)$, if there exists a $\delta > 0$ such that the observability Gramian defined in (5.63) is uniformly positive definite.

$$W_0(t, t+\delta) := \int_t^{t+\delta} \Phi^T(\tau, t)C^T(\tau)C(\tau)\Phi(\tau, t)d\tau \ge kI \quad \forall t. \tag{5.63}$$

Now use Problem 5.10 to show that if there exists a uniformly positive definite bounded matrix $P(\cdot) : \mathbb{R} \mapsto \mathbb{R}^{n \times n}$ satisfying

$$k_1 I \le P(t) \le k_2 I,$$

$$-\dot{P}(t) = A^T(t)P(t) + P(t)A(t) + C^T(t)C(t). \tag{5.64}$$

with $A(t), C(t)$ a uniformly completely observable pair, then the origin is uniformly asymptotically stable and hence exponentially stable. Prove the converse as well.

Problem 5.13 Time-varying Lyapunov lemma generalized with output injection [259]. Prove that if a pair $(A(t), C(t))$ is uniformly completely observable, that is the observability Gramian of (5.63) is uniformly positive definite, then for any $K(t) \in \mathbb{R}^{n \times n}$ that $(A(t) + K(t)C(t), C(t))$ is also uniformly completely observable. You will need to do a perturbation analysis to compare trajectories of the two systems

$$\dot{x} = A(t)x(t),$$

$$\dot{w} = (A(t) + K(t)C(t))w(t),$$

with the same initial condition at time t_0 and give conditions on the terms of the second observability Gramian using a perturbation analysis. You may wish to refer to the cited reference for additional help.

Problem 5.14 Zubov's converse theorem.

1. For time-invariant nonlinear systems assume that there exists a function $v :$ $\Omega \mapsto [0, 1[$ with $v(x) = 0$ if and only if $x = 0$. Further, assume that if Ω is bounded, then as $x \to \partial\Omega$, $v(x) \to 1$. On the other hand, if Ω is unbounded assume that $|x| \to \infty \Rightarrow v(x) \to 1$. Assume also that there exists a p.d.f. $h(x)$, such that for $x \in \Omega$

$$\dot{v}(x) = -h(x)(1 - v(x)). \tag{5.65}$$

   Show that Ω is the domain of attraction of the asymptotically stable equilibrium point 0.

2. Let v, h be as in the previous part, but replace equation (5.65) by

$$\dot{v}(x) = -h(x)\left(1 - v(x)\right)(1 + |f(x)|^2)^{\frac{1}{2}}. \tag{5.66}$$

   Prove the same conclusion as in the previous part.

Problem 5.15 Modified exponential stability theorem. Prove the following modification of the exponential stability theorem (5.17) of the text: If there exists a function $v(x, t)$ and some constants $h, \alpha_1, \alpha_2, \alpha_3, \alpha_4$ such that for all $x \in B_h$, $t \geq 0$,

$$\alpha_1 |x|^2 \leq v(x, t) \leq \alpha_2 |x|^2,$$

$$\left.\frac{dv(x, t)}{dt}\right|_{(5.1)} \leq 0, \tag{5.67}$$

$$\int_t^{t+\delta} \left.\frac{dv(x, t)}{dt}\right|_{(5.1)} \leq -\alpha_3 |x(t)|^2,$$

then $x(t)$ converges exponentially to zero. This problem is very interesting in that it allows for a decrease in $v(x, t)$ proportional to the norm squared of $x(t)$ over a window of length δ rather than instantaneously.

Problem 5.16 Perturbations of exponentially stable systems. Consider the following system on $\mathbb{R}^n$

$$\dot{x} = f(x, t) + g(x, t). \tag{5.68}$$

Assume that $f(0, t) = g(0, t) \equiv 0$. Further, assume that:

1. 0 is an exponentially stable equilibrium point of

$$\dot{x} = f(x, t).$$

2.

$$|g(x, t)| \leq \mu|x| \quad \forall x \in \mathbb{R}^n.$$

Show that 0 is an exponentially stable equilibrium point of 5.68 for μ small enough. The moral of this exercise is that *exponential stability is robust*!

Problem 5.17 Bounded-input bounded-output or finite gain stability of non-linear systems. Another application of Problem 5.16 is to prove L_∞ bounded input bounded output stability of nonlinear systems (you may wish to review the definition from Chapter 4). Consider the nonlinear control system

$$\dot{x} = f(x, t) + g(x, t)u,$$
$$y = h(x). \tag{5.69}$$

Prove that if 0 is an exponentially stable equilibrium point of $f(x, t)$, $|g(x, t)| \leq \mu|x|$ and finally

$$h(0) = 0, \quad |h(x)| \leq \beta|x| \quad \forall x,$$

then the control system (5.69) is L_∞ finite gain stable. Give the bound on the L_∞ stability.

Problem 5.18 Local exponential stability implies exponential stability of the linearization. Prove the converse of Theorem 5.41, that is, if 0 is a locally exponentially stable equilibrium point of the nonlinear system (5.1) then it is a (globally) exponentially stable equilibrium point of the linearized system, provided that the linearization is uniform in the sense of Theorem 5.41. Use the exponential converse theorem of Lyapunov.

Problem 5.19 Interconnected systems. The methods of Problem 5.16 may be adapted to prove stability results for interconnected systems. Consider an interconnection of N systems on $\mathbb{R}^{n_i}$, $i = 1, \ldots N$, given by

$$\dot{x}_i = f_i(x_i, t) + \sum_{j=1}^{N} g_{ij}(x_1, \ldots, x_N, t) \tag{5.70}$$

with $f_i(x_i, t), g_{ij}(x_1, \ldots, x_N, t)$ vector fields on $\mathbb{R}^{n_i}$ for $i = 1, \ldots, N$. Now assume that in each of the decoupled systems

$$\dot{x}_i = f_i(x_i, t) \tag{5.71}$$

the origin is an exponentially stable equilibrium point, and that for i, $j = 1, \ldots, N$,

$$|g_{ij}(x_1, \ldots, x_N, t)| \le \sum_{j=1}^{N} \gamma_{ij}|x_j|.$$

Using the converse Lyapunov theorem for exponentially stable systems to generate Lyapunov functions $v_i(x_i, t)$ satisfying

$$\alpha_1^i|x_i|^2 \le v_i(x_i, t) \le \alpha_2^i|x_i|^2,$$
$$\frac{\partial v_i}{\partial t} + \frac{\partial v_i}{\partial x_i} f_i(x_i, t) \le -\alpha_3^i|x_i|^2, \qquad (5.72)$$
$$\left|\frac{\partial v_i}{\partial x_i}\right| \le \alpha_4^i|x_i|,$$

give conditions on the α_1^i, α_2^i, α_3^i, α_4^i, γ_{ij} to prove the exponential stability of the origin. *Hint: Consider the Lyapunov function*

$$\sum_{i=1}^{N} d_i v_i(x_i),$$

where $d_i > 0$ and $v_i(x_i)$ is the converse Lyapunov function for each of the systems (5.71).

See [162] for a generalization of this problem to the case when the stability conditions of equations (5.72) are modified to replace $|x_i|$ by $\phi_i(x_i)$ where $\phi_i(\cdot)$ is a positive definite function, and

$$|g_{ij}(x_1, \ldots, x_N)| \le \sum_{j=1}^{N} \gamma_{ij}\phi_j(x_j).$$

Problem 5.20 Hopfield networks with asymmetric interconnections. Give a counterexample to the lack of oscillations of the Hopfield network of (5.37) for asymmetric interconnections T_{ij}. Modify the method of Problem 5.19 to give conditions under which the Hopfield network with asymmetric interconnections does not have limit cycles.

Problem 5.21 Satellite stabilization. Consider the satellite rigid body B with an orthonormal body frame attached to its center of mass with axes aligned with the principal axes of inertia as shown in Figure 5.10. Let $b_0 \in \mathbb{R}^3$ be a unit vector representing the direction of an antenna on the satellite and $d_0 \in \mathbb{R}^3$ a fixed unit vector representing the direction of the antenna of the receiving station on the ground. Euler's equations of motion say that if $\omega \in \mathbb{R}^3$ is the angular velocity vector of the satellite about the body axes shown in Figure 5.10, then

$$I\dot{\omega} + \omega \times I\omega = \tau \qquad (5.73)$$

Here $I = \text{diag}\,(I_1, I_2, I_3)$. The aim of this problem is to find a control law to align b_0 with $-d_0$. In the body frame, d_0 is not fixed. It rotates with angular velocity $-\omega$

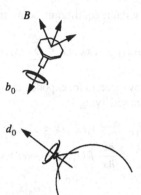

FIGURE 5.10. Satellite and earth station

so that

$$\dot{d}_0 = -\omega \times d_0 \qquad (5.74)$$

Now consider the control law

$$\tau = -\alpha\omega + d_0 \times b_0$$

applied to (5.73), (5.74). What are the equilibrium points of the resulting system in $\mathbb{R}^{3+3}$? Linearize the system about each equilibrium point and discuss the stability of each. Is it possible for a system in $\mathbb{R}^6$ to have this number and type of equilibria. Do the system dynamics actually live on a submanifold of $\mathbb{R}^6$? If so, find the state space of the system. Prove that the control law causes all trajectories to converge to one of the equilibrium points. It may help to use the Lyapunov function candidate

$$v(\omega, d_0) = \frac{1}{2}\omega^T I \omega + \frac{1}{2}|d_0 + b_0|^2.$$

Another approach to satellite stabilization is in a recent paper by D'Andrea Novel and Coron [77].

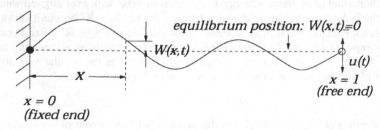

FIGURE 5.11. Boundary control stabilization of a vibrating string

Problem 5.22 Boundary control of a vibrating string. The equation of motion of a linear, frictionless, vibrating string of length 1 is given by the following partial differential equation:

$$m \frac{\partial^2 W(x,t)}{\partial t^2} - T \frac{\partial^2 W(x,t)}{\partial x^2} = 0, \qquad (5.75)$$

where $W(x,t)$ denotes the displacement of the string at location x at time t, with $x \in]0, 1[, t \geq 0$. The quantities m and T are the mass per unit length of the string and the tension in the string, respectively. Let the string be fixed at $x = 0$ as shown in Figure 5.11. Thus, $W(0,t) = 0$ for all $t \geq 0$. Let a vertical control force $u(\cdot)$ be applied to the free end of the string at $x = 1$. The balance of the forces in the vertical direction (at $x = 1$) yields:

$$u(t) = T \frac{\partial W(x,t)}{\partial x} \bigg|_{x=1} \qquad (5.76)$$

for all $t \geq 0$. The energy of the spring at time t is

$$E(t) = \frac{1}{2} \int_0^1 m \left(\frac{\partial W(x,t)}{\partial t} \right)^2 dx + \frac{1}{2} \int_0^1 T \left(\frac{\partial W(x,t)}{\partial x} \right)^2 dx \qquad (5.77)$$

which consists of the kinetic energy (first integral in the expression above) and the strain energy (second integral above). Show that if a *boundary control*

$$u(t) = -k \frac{\partial W(x,t)}{\partial t} \bigg|_{x=1} \qquad (5.78)$$

is applied, where $k > 0$, then $\dot{E}(t) \leq 0$ for all $t \geq 0$. See also [274].

Problem 5.23 Brockett's H dot equation [26], [132].

1. Consider the matrix differential equation ($H(t) \in \mathbb{R}^{n \times n}$

$$\dot{H} = [H, G(t)] \qquad (5.79)$$

Here $G(t) \in \mathbb{R}^{n \times n}$ is a skew-symmetric matrix. The notation $[A, B] := AB - BA$ refers to the Lie bracket, or commutator, between A and B. Show that if $H(0)$ is symmetric, then $H(t)$ is symmetric for all t. Further, show in this case that the eigenvalues of $H(t)$ are the same as the eigenvalues of $H(0)$. Equation (5.79) is said to generate an *iso-spectral* flow on the space of symmetric matrices $SS(n)$.

2. Consider the so-called double bracket equation

$$\dot{H} = [H, [H, N]] \qquad (5.80)$$

with $N \in \mathbb{R}^{n \times n}$ a constant matrix. The Lie bracket is defined as above. Show that if $H(0)$ and N are symmetric, then (5.80) generates an iso-spectral flow as well on $SS(n)$.

3. Now using the Lyapunov function candidate

$$v(H) = -\text{trace}(H^T N)$$

and a norm on $\mathbb{R}^{n \times n}$ given by $\|A\| = (\text{trace } A^T A)^{\frac{1}{2}}$, prove that

$$\dot{v}(H) = -\|[H, N]\|^2$$

Now use the fact that

$$|\text{trace } A^T B| \le \|A\| \, \|B\|$$

and LaSalle's principle to find the limits as $t \to \infty$ of $H(t)$.

4. Consider the double bracket equation with

$$N = \text{diag}(\lambda_1, \lambda_2, \dots, \lambda_n)$$

with $\lambda_1 > \lambda_2 > \cdots > \lambda_n > 0$. Use the iso-spectrality of (5.80) and the previous calculations to characterize the equilibrium points of (5.80). How many are there?

5. Linearize (5.80) about each equilibrium point and try to determine the stability from the linearization. What can you say if some of the eigenvalues of N, $H(0)$ are repeated?

Discuss how you might use the H dot equation to sort a given set of n numbers $\lambda_1, \cdots, \lambda_n$ in descending order. How about sorting them in ascending order?

Problem 5.24 Controlled Lagrangians for the satellite stabilization problem [28], [30].   Consider a satellite system with an internal rotor as shown in Figure 5.12 with equations given by

$$
\begin{aligned}
\lambda_1 \dot{\omega}_1 &= \lambda_2 \omega_2 \omega_3 - (\lambda_3 \omega_3 + J_3 \dot{\alpha}) \omega_2, \\
\lambda_2 \dot{\omega}_2 &= \lambda_1 \omega_1 \omega_3 - (\lambda_3 \omega_3 + J_3 \dot{\alpha}) \omega_1, \\
\lambda_3 \dot{\omega}_3 + J_3 \ddot{\alpha} &= (\lambda_1 - \lambda_2) \omega_1 \omega_2, \\
J_3 \dot{\omega}_3 + J_3 \ddot{\alpha} &= u,
\end{aligned}
\tag{5.81}
$$

where the body angular velocity is given by $(\omega_1, \omega_2, \omega_3)^T \in \mathbb{R}^3$, and I_1, I_2, I_3 are the satellite moments of inertia, $J_1 = J_2$, J_3 the rotor moments of intertia, and α the relative angle of the rotor. Furthermore, $\lambda_i = I_i + J_i$. It is easy to verify that regardless of the choice of u we have an equilibrium manifold $\omega_1 = \omega_3 = 0$ and $\omega_2 = m$ arbitrary. This corresponds to the satellite spinning freely about the intermediate axis. Prove that the control law

$$u = k(\lambda_1 - \lambda_2) \omega_1 \omega_2$$

stabilizes the satellite about each of these equilibria. Note that you are not required to prove anything about the variables α, $\dot{\alpha}$.

Problem 5.25 Rates of convergence estimates for the Lyapunov equation.
Consider the Lyapunov equation

$$A^T P + P A = -Q.$$

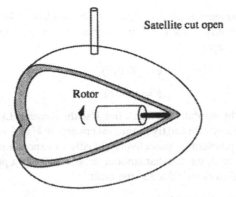

Satellite cut open

Rotor

FIGURE 5.12. Satellite with a spinning rotor aligned with its central principal axis as controller

Note that an estimate of the rate of convergence is given by

$$\mu_A(Q) := \frac{\lambda_{\min}(Q)}{2\lambda_{\max}(P)}.$$

Show the following three properties of $\mu_A(Q)$:

1. Show that $\mu_A(kQ) = \mu_A(Q)$ for all $k \in \mathbb{R}_+$.
2. Show that if $\lambda_{\min}(Q) = 1$, then $\mu_A(I) \geq \mu_A(Q)$.
3. Using the two previous facts, show that $\mu_A(I) \geq \mu_A(Q)$ for arbitrary Q.

Thus, if one were to use the Lyapunov equation to estimate the rate of convergence of a linear equation, it is best to use $Q = I$! Now, use this to study the stability of the perturbed linear system

$$\dot{x} = Ax + f(x, t). \tag{5.82}$$

Here, $A \in \mathbb{R}^{n \times n}$ has its eigenvalues in $\mathbb{C}_-^{\circ}$, and $|f(x, t)| \leq \gamma|x|$. Now prove that the perturbed linear system is stable if $\gamma < \mu_A(I)$.

Problem 5.26 Stability conditions based on the measure of a matrix. Recall the definition of the *measure of a matrix* $\mu(A)$ from Problem 3.2. Use this definition to show that if for some matrix measure $\mu(\cdot)$, there exists an $m > 0$, such that

$$\int_{t_0}^{t_0+T} \mu(A(\tau))d\tau < -m \quad \forall t \geq T \quad t_0 \geq 0,$$

then the origin of the linear time-varying system of (5.44) is exponentially stable. Apply this test to various kinds of matrix measures (induced by different norms).

Problem 5.27 Smooth stabilizability [279]. Assume that the origin $x = 0$ of the system

$$\dot{x} = f(x, u)$$

is stabilizable by smooth feedback, that is, there exists $u = k(x)$ such that 0 is a locally asymptotically stable equilibrium point of $\dot{x} = f(x, k(x))$. Now prove that the extended system

$$\dot{x} = f(x, z),$$
$$\dot{z} = h(x, z) + u$$

(5.83)

is also stabilizable by smooth feedback. *Hint:* Use the converse Lyapunov function associated with the x system and try to get z to converge to $k(x)$. Repeat the problem when the original problem is respectively globally or exponentially stabilizable. By setting $h(x, z) \equiv 0$, we see that smooth stabilizability is a property which is not lost when integrators are added to the input!

Problem 5.28. Prove Lemma 5.47.

Problem 5.29. Prove Theorem 5.49. Now apply this theorem to the following generalization of the power system swing dynamics

$$\dot{x}_1 = x_2.$$
$$\dot{x}_2 = -Dx_2 - \nabla w(x_1).$$

(5.84)

Here $x_1, x_2 \in \mathbb{R}^n$, $w(x_1)$ is a proper function with isolated non-critical stationary points and $D \in \mathbb{R}^{n \times n}$ is a symmetric positive definite matrix. Is this "second order" system gradient like? Characterize the domains of attraction of its stable equilibria. Show that all the equilibria of the system are either stable nodes, unstable nodes, or saddles (i.e., that the eigenvalues of their linearizations are real).

Problem 5.30 Potential energy boundary surfaces. Prove rigorously that the method sketched in the Example 5.45, a level set of $W(\theta)$ around a local minimum θ_0 just touching a saddle or local maximum of W, is a (conservative) estimate of the domain of attraction of the equilibrium θ_0 of the swing dynamics. You may wish to use the "Lyapunov" function of (5.34) for helping you along with the proof.

6

Applications of Lyapunov Theory

In this chapter we will give the reader an idea of the many different ways that Lyapunov theory can be utilized in applications. We start with some very classical examples of stabilization of nonlinear systems through their Jacobian linearization, and the circle and Popov criteria (which we have visited already in Chapter 4) for linear feedback systems with a nonlinearity in the feedback loop. A very considerable amount of effort has been spent in the control community in applying Lyapunov techniques to adaptive identification and control. We give the reader an introduction to adaptive identification techniques. Adaptive control is far too detailed an undertaking for this book, and we refer to Sastry and Bodson [259] for a more detailed treatment of this topic. Here, we study multiple time scale systems and singular perturbation and averaging from the Lyapunov standpoint here. A more geometric view of singular perturbation is given in the next chapter.

6.1 Feedback Stabilization

An application of the indirect methods of Lyapunov is to the stabilization of nonlinear autonomous control systems. The problem is described as follows: Consider an autonomous control system of the form

$$\dot{x} = f(x, u). \tag{6.1}$$

Here $f(x, u) : \mathbb{R}^n \times \mathbb{R}^{n_i} \mapsto \mathbb{R}^n$, and the objective is to find a *feedback control law*, i.e., a law of the form

$$u(t) = k(x(t)),$$

in such a way as to make the origin an asymptotically stable equilibrium point of the system

$$\dot{x} = f(x, k(x)). \tag{6.2}$$

We will make two assumptions:

1. $f(0, 0) = 0$, i.e., 0 is an equilibrium point of the undriven system.
2. If we define

$$A = \left[\frac{\partial f(x, u)}{\partial x}\right]_{x=0, u=0}, \quad B = \left[\frac{\partial f(x, u)}{\partial u}\right]_{x=0, u=0},$$

we assume that the pair A, B is completely controllable. [1] Thus we assume that the linearization of the control system (6.1) namely

$$\dot{z} = Az + Bu, \tag{6.3}$$

is completely controllable.

The following fact is elementary in the study of linear control systems and is referred to as the pole placement theorem.

Theorem 6.1 Pole Placement. *Consider the linear control system of (6.3). If the pair A, B is completely controllable, then given any constellation of eigenvalues $\lambda_1, \ldots, \lambda_n \in \mathbb{C}$ symmetric with respect to the real axis there exists a matrix $K \in \mathbb{R}^{n_i \times n}$ such that the eigenvalues of $A + BK$ are precisely $\lambda_1, \ldots, \lambda_n$.*

This theorem is easily extended to the nonlinear case as follows.

Theorem 6.2 Stabilization of Nonlinear Systems. *Consider the nonlinear system (6.1) with $f(0, 0) = 0$. Assume that its linearization given by (6.3) is completely controllable. Then with A, B defined as above there exists a matrix $K \in \mathbb{R}^{n \times n_i}$ such that $\sigma(A + BK) \subset \mathbb{C}_-^\circ$. Further, the linear control law*

$$u = Kx$$

stabilizes the nonlinear control system (6.1).

Proof: Consider the closed loop system

$$\dot{x} = f(x, Kx) := \tilde{f}(x). \tag{6.4}$$

We observe that $\tilde{f}(0) = 0$ and

$$\left[\frac{\partial \tilde{f}}{\partial x}\right]_{x=0} = A + BK.$$

---

[1] Recall that a pair A, B is completely controllable if the rank of the matrix $[B, AB, \ldots, A^{n-1}B]$ is n.

Since K was chosen so that the eigenvalues of $A + BK$ are in $\mathbb{C}_-^\circ$, we see that the linearization of $\tilde{f}$ at 0 is stable establishing the theorem.   □

Remarks:

1. The *linear* stabilizing control law also stabilizes the nonlinear system. Different choices of K may, however, result in larger or smaller domains of attraction.
2. The hypothesis of the previous theorem can be weakened somewhat as follows: Recall that a pair A, B is said to be *stabilizable* if the generalized eigenspace corresponding to the unstable eigenvalues, i.e., those in $\mathbb{C}_+$, are contained in the span of $[B, AB, \ldots, A^{n-1}B]$. Colloquially, this means that the unstable eigenmodes are controllable. In turn, this implies that there exists a $K \in \mathbb{R}^{n_i \times n}$ such that $\sigma(A + BK) \subset \mathbb{C}_-^\circ$. As in the preceding theorem the linear control law can be used to stabilize the nonlinear system (6.1).
3. To explore the necessity of the stabilizability of A, B for stabilizing the nonlinear system, we consider the instance that A, B is not stabilizable and has at least one unstable eigenmode in $\mathbb{C}_+^\circ$. In this instance no matter what the choice of K, there is at least one eigenvalue of $A + BK$ in $\mathbb{C}_+^\circ$. As a consequence, we may use the instability theorem of Lyapunov's indirect method to conclude that the nonlinear system of (6.1) cannot be stabilized regardless of the choice of $u(t) = k(x(t))$. (See Problem 6.2).
4. The only situation where we cannot determine whether it is possible to stabilize the nonlinear control system using state feedback is the one in that A, B is not stabilizable, but the only $\mathbb{C}_+$ modes which are not controllable are on the $j\omega$ axis. In this instance, linear feedback alone will not suffice, and we will take up this point again after we have studied center manifolds in Problem 7.26 of the next chapter (see Behtash and Sastry [21]).
5. When the linearization is time varying, then stabilization of the resulting time varying linear system stabilizes the nonlinear time varying system. In turn, uniform complete controllability of the time varying linearization guarantees the stabilizability of the linearization. This point is explored in the Exercises, Problem 6.3.
6. When full state information is not available, in the linear time invariant case, an observer is constructed to reconstruct the state. In the Exercises (Problem 6.4) we show how converse theorems may be used to construct nonlinear observers for observer-based stabilization.

6.2   The Lur'e Problem, Circle and Popov Criteria

In this section, we study the stability of a class of feedback systems whose forward path contains a single input single output linear, time-invariant system and whose reverse path contains a memoryless (possibly time-varying) nonlinearity as shown in Figure 6.1. If $x \in \mathbb{R}^n$ represents the state of the linear system, $u(t) \in \mathbb{R}$ its

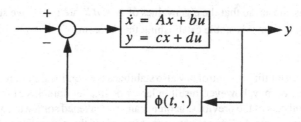

FIGURE 6.1. The system of Lur'e's problem

input, and $y(t) \in \mathbb{R}$ its output, it follows that the equations of the system are:

$$\dot{x} = Ax + bu,$$
$$y = cx + du, \qquad\qquad (6.5)$$
$$u = -\phi(y(t), t).$$

The nonlinearity $\phi(\cdot, \cdot)$ satisfies $\phi(0, t) = 0 \forall t \geq 0$ and

$$\sigma\phi(\sigma, t) \geq 0 \quad \forall \sigma \in \mathbb{R}^n \quad t \geq 0.$$

This is referred to as $\phi(\cdot, t)$ being a first and third quadrant nonlinearity. The Lur'e problem consists in finding conditions on A, b, c, d such that the equilibrium point 0 is a globally asymptotically stable equilibrium point of (6.5). This problem is also referred to as the *absolute stability problem* since we would like to prove global asymptotic stability of the system (6.5) for a whole class of feedback nonlinearities ϕ. In this section, we focus attention on the single-input single-output Popov and circle criteria. Generalizations of this to multi-input multi-output systems are left to the Exercises (see Problem 6.7). We place the following assumptions on the system:

1. A, b, c is minimal, that is A, b is completely controllable and A, c is completely observable.
2. A is a matrix whose eigenvalues lie in $\mathbb{C}^0_-$.
3. $\phi(\cdot, t)$ is *said to belong to the sector* $[k_1, k_2]$ for $k_2 \geq k_1 \geq 0$, as shown in Figure (6.2), i.e.,

$$k_1\sigma^2 \leq \sigma\phi(\sigma, t) \leq k_2\sigma^2.$$

Several researchers conjectured solutions to this absolute stability problem. We discuss two of these conjectures here:

Aizerman's Conjecture

Suppose that $d = 0$ and that for each $k \in [k_1, k_2]$ the matrix $A - bkc$ is Hurwitz (i.e., eigenvalues lie in $\mathbb{C}^0_-$). Then the nonlinear system is globally asymptotically stable. The conjecture here was that the stability of all the linear systems contained in the class of nonlinearities admissible by the sector bound determined the global asymptotic stability of the nonlinear system.

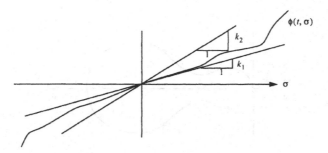

FIGURE 6.2. The sector condition for $\phi(t, \cdot)$

Kalman's Conjecture

Let $\phi(\sigma, t)$ satisfy for all σ

$$k_3 \le \frac{\partial \phi(\sigma, t)}{\partial \sigma} \le k_4$$

Now suppose that $d = 0$ and that for each $k \in [k_3, k_4]$ the matrix $A - bkc$ is Hurwitz (i.e., its eigenvalues lie in $\mathbb{C}^\circ_-$). Then the nonlinear system is globally asymptotically stable. The conjecture here is that the stability of the linearization determines the global asymptotic stability of the nonlinear system.

Both Aizerman's conjecture and Kalman's conjecture are false. However, they stimulated a great deal of activity in nonlinear stability theory. The criteria that can be derived both rely on the following generalization of a lemma called the Kalman–Yakubovich–Popov lemma, due to Lefschetz. Its generalization to multiple input multiple output systems is called the positive real lemma and is discussed in Problem 6.7.

Theorem 6.3 Kalman–Yacoubovich–Popov (KYP) Lemma. *Given a Hurwitz matrix $A \in \mathbb{R}^{n \times n}$, and $b \in \mathbb{R}^n$, $c^T \in \mathbb{R}^n$, $d \in \mathbb{R}$ such that the pair A, b is controllable, then there exist positive definite matrices $P, Q \in \mathbb{R}^{n \times n}$, a vector $q \in \mathbb{R}^n$, and scalar $\epsilon > 0$ satisfying*

$$A^T P + P A = -qq^T - \epsilon Q$$
$$Pb - \frac{1}{2}c^T = \sqrt{d}q \tag{6.6}$$

iff the scalar transfer function

$$\hat{h}(s) = d + c(sI - A)^{-1}b$$

is strictly positive real, i.e., $\hat{h}(s)$ is analytic in $\mathbb{C}_+$ and

$$Re\, \hat{h}(j\omega) > 0 \quad \forall \omega \in \mathbb{R}.$$

Proof: The proof of this lemma requires techniques from optimal control that are not central to this section so we omit it. See [5] for a proof, as well as the generalization of the lemma to multi-input multi-output systems due to Anderson.

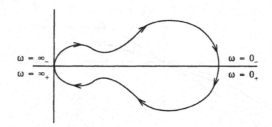

FIGURE 6.3. Nyquist locus of an SPR transfer function

Remarks:

1. Strictly positive real, or SPR, transfer functions are so called because they are positive, or strictly positive respectively, on the imaginary axis. In particular, since $\hat{h}(\infty) \geq 0$, we must have $d \geq 0$. In addition, since it is required that they are analytic in $\mathbb{C}_+$ it follows that

$$\text{Re } \hat{h}(s) > 0 \quad \forall s \in \mathbb{C}_+.$$

2. From the definition it follows that the Nyquist locus of a strictly positive real transfer function lies exclusively in the first and fourth quadrant as shown in Figure 6.3. Thus the phase of an SPR transfer function is less than or equal in magnitude than $\pi/2$ radians. Thus, if $d = 0$, the asymptotic phase at $\omega = \infty$ is $\pi/2$ so that the relative degree of $\hat{h}(s)$ is 1. Thus, SPR transfer functions have a relative degree of either 0 ($d > 0$) or 1.

3. SPR transfer functions model what are referred to as *passive* linear systems. In particular, network functions of linear networks with positive resistors, inductors and capacitors are strictly positive real.

4. SPR transfer functions are minimum phase (i.e., they have no zeros in the $\mathbb{C}_+^\circ$). (the proof is left as an exercise).

6.2.1   The Circle Criterion

For conditions guaranteeing absolute stability of the closed loop nonlinear system, we start with the case that $\phi(\cdot, t)$ lies in the sector $[0, k]$.

Proposition 6.4 Circle Criterion.   *Consider the feedback system of Figure 6.1 described by the equations of (6.5). Let the assumptions on minimality of A, b, c and stability of A be in effect. Then the origin of (6.5) is globally, asymptotically stable if*

$$Re \, (1 + k\hat{h}(j\omega)) > 0, \quad \forall \omega \tag{6.7}$$

This condition is visualized graphically in Figure 6.4.

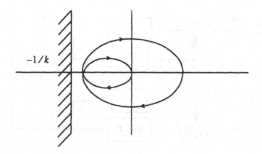

FIGURE 6.4. Visualization of the stability condition of Proposition 6.4

Proof: Consider the Lyapunov function $v(x) = x^T P x$, with $P > 0$ to be defined later. Now,

$$\dot{v}(x) = \dot{x}^T P x + x^T P \dot{x} = x^T (A^T P + P A)x + 2x^T P b u. \tag{6.8}$$

Using (6.5) and multiplying both sides of the equality $y = cx + du$ by ku and adding and subtracting u^2 it follows that

$$0 = kucx + (1 + kd)u^2 - u(ky + u). \tag{6.9}$$

Subtracting equation (6.9) from (6.8) yields

$$\dot{v}(x) = x^T (A^T P + P A)x + 2ux^T (Pb - 1/2kc^T) \\ - (1 + kd)u^2 + u(ky + u). \tag{6.10}$$

Now, since ϕ belongs to the sector $[0, k]$

$$0 \leq -uy \leq ky^2 \Rightarrow y(ky + u) \geq 0,$$

Since u and y have opposite signs it follows that $u(ky + u) \leq 0$. This yields

$$\dot{v}(x) \leq x^T (A^T P + P A)x + 2ux^T (Pb - 1/2kc^T) - (1 + kd)u^2. \tag{6.11}$$

Now, if the transfer function $(1 + kd) + kc(sI - A)^{-1}b$ is positive real, then by the KYP lemma it follows that $\exists P, Q > 0$ and $q \in \mathbb{R}^n$ such that

$$A^T P + P A = -qq^T - \epsilon Q, \\ Pb - \frac{1}{2}kc^T = (1 + kd)^{1/2}q. \tag{6.12}$$

Then, using (6.12) in (6.11) yields

$$\dot{v}(x) \leq -\epsilon x^T Q x - x^T qq^T x + 2ux^T (1 + kd)^{1/2}q - (1 + kd)u^2 \\ \leq -\epsilon x^T Q x - (x^T q - (1 + kd)^{1/2}u)^2 \\ \leq -\epsilon x^T Q x.$$

Thus, we have that $-\dot{v}$ is a p.d.f., and the origin is globally asymptotically stable. The condition assumed to generate the Lyapunov function is that $1 + k\hat{h}(s)$ is SPR. This, in turn, is equivalent to insisting that Re $\hat{h}(j\omega) \geq -\frac{1}{k}$, since k is positive.

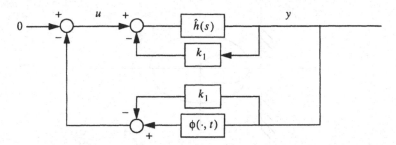

FIGURE 6.5. Showing loop transformations on the feedback loop

Note that the requirement of stability of the closed loop system for all possible $\phi(\cdot, t)$ in the sector $[0, k]$ forces the open loop system to be stable (since $\phi(\cdot, t)$ may be chosen to be $\equiv 0$). $\qquad\qquad\qquad\qquad\qquad\qquad\qquad\qquad\square$

This proposition can now be generalized to cover the case that $\phi(\cdot, t)$ lies in a sector $[k_1, k_2]$ for $k_2 \geq k_1 \geq 0$. Figure 6.5 shows how the feedback loop may be transformed to yield a nonlinearity $\psi(\cdot, t)$ in the sector $[0, k_2 - k_1]$.

The transformed loop has the same form as the loop we started with, with the difference that the linear system in the feed forward path is

$$\hat{g}(s) = \frac{\hat{h}(s)}{1 + k_1 \hat{h}(s)}$$

and the nonlinearity in the feedback path is

$$\psi(\sigma, t) = \phi(\sigma, t) - k_1 \sigma.$$

To apply the preceding theorem, we need to guarantee that the system in the forward path is stable. A necessary and sufficient condition for the stability of $\hat{g}(s)$ is given by the Nyquist criterion: If $\hat{h}(s)$ has no poles on the $j\omega$ axis and ν poles in the right half plane; then Nyquist's criterion states that $\hat{g}(s)$ is a stable transfer function iff the Nyquist contour of $\hat{h}(j\omega)$ does not intersect $-\frac{1}{k_1}$ and encircles it counterclockwise ν times. If $\hat{h}(s)$ has poles on the $j\omega$ axis, the Nyquist contour is indented to the left (or right) of the pole and the criterion applied in the limit that the indentation goes to zero. If the indentation is made to the right of a pole, the number of encirclements does not include the count of that pole, if it is made to the left it includes the count of that pole. Using this, along with the estimate of equation (6.7) gives the following condition, called the circle criterion:

Theorem 6.5 Circle Criterion.   *Let $\hat{h}(s)$ have ν poles in $\mathbb{C}_+$ and let the non-linearity $\phi(\cdot, t)$ be in the sector $[k_1, k_2]$. Then the origin of the system of (6.5) is globally asymptotically stable iff*

1. *The Nyquist locus of $h(s)$ does not touch $-\frac{1}{k_1}$ and encircles it ν times in a counterclockwise sense.*

2.

$$Re\left[\frac{1 + k_2\hat{h}(j\omega)}{1 + k_1\hat{h}(j\omega)}\right] > 0 \quad \forall \omega \in \mathbb{R}. \tag{6.13}$$

Remark: The preceding theorem is referred to as the circle criterion because the condition of (6.13) has a very nice graphical interpretation. We consider the following three cases:

1. $k_1 > 0$. If $\hat{h}(j\omega) = u(j\omega) + jv(j\omega)$, we may rewrite equation (6.13) as

$$\frac{(1 + k_2u)(1 + k_1u) + k_2k_1v^2}{(1 + k_1u)^2 + k_1^2v^2} > 0. \tag{6.14}$$

The denominator of the expression above is never zero, since the Nyquist locus of $\hat{h}(j\omega)$ does not touch the point $-\frac{1}{k_1}$. Hence, (6.14) may be written as

$$(1 + k_2u)(1 + k_1u) + k_2k_1v^2 > 0,$$

or

$$\left(u + \frac{1}{k_1}\right)\left(u + \frac{1}{k_2}\right) + v^2 > 0. \tag{6.15}$$

The inequality of equation (6.15) says that the Nyquist locus of $\hat{h}(j\omega)$ does not enter a disk in the complex plane passing through $-\frac{1}{k_1}$ and $-\frac{1}{k_2}$ as shown in figure (6.6). Combining this with the encirclement condition, we may restate the conditions of the preceding theorem as follows: The Nyquist locus of $\hat{h}(j\omega)$ encircles the disk $D(k_1, k_2)$ exactly ν times without touching it.

2. $k_1 < 0 < k_2$. A necessary condition for absolute stability is that A have all its eigenvalues in the open left half plane since, $\psi(\cdot, t) \equiv 0$ is an admissible nonlinearity. Proceeding as before from

$$(1 + k_2u)(1 + k_1u) + k_2k_1v^2 > 0$$

yields

$$\left(u + \frac{1}{k_1}\right)\left(u + \frac{1}{k_2}\right) + v^2 < 0, \tag{6.16}$$

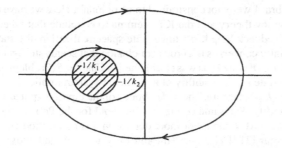

FIGURE 6.6. Graphical interpretation of the circle criterion when $k_2 \geq k_1 \geq 0$

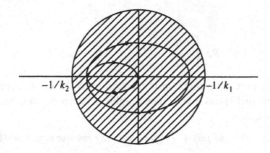

FIGURE 6.7. The graphical interpretation of the circle criterion when $k_2 \geq 0 > k_1$

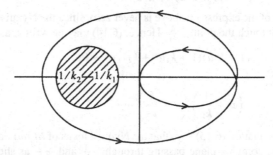

FIGURE 6.8. Graphical interpretation of the circle criterion when $k_1 < k_2 < 0$

since $k_1 k_2 < 0$. Inequality (6.16) specifies that the Nyquist locus does not touch or encircle $-\frac{1}{k_1}$ (since $\nu = 0$) and lies inside the disk $D(k_1, k_2)$, as shown in Figure 6.7.

3. $k_1 < k_2 < 0$. The results of the first case can be made to apply by replacing $\hat{h}$ by $-\hat{h}$, k_1 by $-k_1$, and k_2 by $-k_2$. The results are displayed graphically in Figure 6.8.

The circle criterion of this section should be compared with the L_2 input–output stability theorem, also called the circle criterion in Chapter 4. There we proved the input–output stability for a larger class of linear systems in the feedback loop (elements of the algebra A were not constrained to be rational). Here we have used the methods of Lyapunov theory and the KYP lemma to conclude that when the linear system in the feedback loop has a finite state space realization, the state of the system exponentially converges to zero, a conclusion that seems stronger than input output L_2 stability. Of course, as was established in Chapter 5 (Problem 5.17), exponential internal (state space) stability of a nonlinear system implies bounded input bounded output (L_∞) stability. The added bonus of the input–output analysis is the proof of L_2 stability. The same comments also hold for the Popov criterion which we derive in the next section. For more details on this connection we refer the reader to Vidyasagar [317] (see also Hill and Moylan [139] and Vidyasagar and Vannelli [318]).

6.2.2   The Popov Criterion

The circle criterion for absolute stability for a family of systems studied in (6.5) involves the use of a quadratic Lyapunov function $x^T P x$. When, the nonlinearity in the feedback loop $\phi(\cdot, t)$ is time invariant and $d = 0$, another choice of Lyapunov function involves the sum of $x^T P x$ and $\int_0^y \phi(\sigma) d\sigma$. As in the case of the circle criterion, we start with the case that $\phi(\cdot, t)$ lies in the sector $[0, k]$ with $k > 0$.

Proposition 6.6 Popov Criterion.   *Consider the Lur'e system of (6.5) with the nonlinearity $\phi(\cdot)$ in the sector $[0, k]$. The equilibrium at the origin is globally asymptotically (exponentially) stable, provided that there exists $r > 0$ such that the transfer function*

$$1 + k(1 + rs)\hat{h}(s)$$

is strictly positive real, where $\hat{h} = c(sI - A)^{-1}b$.

Proof: Consider the Lyapunov function

$$v(x) = x^T P x + r \int_0^y k\phi(\sigma) d\sigma.$$

Here P is a symmetric positive definite matrix to be determined in the proof. The sector bounds on y guarantee that v is a p.d.f. We have that

$$\dot{v} = x^T (A^T P + P A)x - 2x^T P b\phi(y) + rk\phi(y)\dot{y}. \tag{6.17}$$

Using the formula $\dot{y} = c\dot{x} = cAx - cb\phi(y)$, we have that

$$\dot{v} = x^T (A^T P + P A)x - x^T (2Pb - r A^T c^T k)\phi(y) - 2rkcb\phi^2(y). \tag{6.18}$$

From the sector bound on ϕ, it follows that $2\phi(y)(\phi(y) - ky) \leq 0$ so that we have

$$\dot{v} \leq x^T (A^T P + P A)x - x^T (2Pb - r A^T c^T k - kc^T)\phi(y) - (rkcb+1)\phi^2(y). \tag{6.19}$$

Choose r small enough that $1 + rkcb > 0$. Then consider the conditions for the transfer function

$$(1 + rkcb) + (kc + rkcA)(sI - A)^{-1}b$$

to be strictly positive real. Then by the KYP Lemma, there exists $P, Q > 0$ and $q \in \mathbb{R}^n$ such that

$$A^T P + P A = -Q - qq^T,$$

$$Pb - \frac{1}{2}(kc + rkcA) = \sqrt{1 + rkcb}\, q.$$

Using this, we obtain

$$\dot{v} \leq -x^T Q x - (q^T x - \sqrt{1 + rkcb}\, \phi(y))^2.$$

Thus, $-\dot{v}$ is a p.d.f., and we have proven the exponential stability of the equilibrium at the origin. The strict positive realness condition can be rewritten as the strict positive realness of

$$1 + kc(sI - A)^{-1}b + rkc(IA(sI - A)^{-1})b = 1 + k(1 + rs)\hat{h}(s).$$

This establishes the conditions of the theorem.   □

Remarks:

1. The graphical test for checking Popov's condition is based on whether there exists an $r > 0$ such that

$$\text{Re } (1 + rj\omega)\hat{h}(j\omega) \geq -\frac{1}{k}. \qquad (6.20)$$

To check this graphically, plot the real part of $\hat{h}(j\omega)$ against $\omega \times \text{Im } \hat{h}(j\omega)$. On this plot, called the Popov plot, inequality (6.20) is satisfied if there exists a straight line of slope $1/r$ passing through $-1/k$ on the real axis, so that the Popov plot lies underneath the line, as shown in Figure 6.9. Since the slope of the line is variable (with the only constraint that it be positive, since we are required to check for the existence of an r), the test boils down to finding a line of positive slope through $-1/k$ on the real axis so above the Popov plot of the open loop transfer function.

2. When the nonlinearity $\phi(\cdot)$ lies in a sector $[k_1, k_2]$ one proceeds as in the case of the circle criterion to transform the system so that the nonlinearity $\phi(y) - k_1 y$ lies in the sector $[0, k_2 - k_1]$. The Popov condition then reduces to requiring that the transfer function

$$\hat{g}(s) = \frac{\hat{h}(s)}{1 + k_1 \hat{h}(s)}$$

be such that for some choice of $r > 0$ the transfer function

$$\frac{1}{k_2 - k_1} + (1 + rs)\hat{g}(s)$$

is SPR.

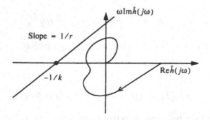

FIGURE 6.9. Graphical verification of the Popov criterion

6.3   Singular Perturbation

In a wide variety of practical applications, state variables of a nonlinear system evolve at different time-scales, i.e., at fast and slow time scales. This phenomenon is caused by a variety of different factors, the most common of which is the existence of small, or *parasitic*, elements in the system dynamics. This situation is modeled by the following set of differential equations

$$\left. \begin{array}{l} \dot{x} = f(x(t), y(t)) \\ \epsilon \dot{y} = g(x(t), y(t)) \end{array} \right\} \quad \Sigma_\epsilon, \tag{6.21}$$

where $x \in \mathbb{R}^n$, $y \in \mathbb{R}^m$. In equation (6.22) above, the small parameter ϵ models the fact that the y variable evolves at a far more rapid rate than x. For $\epsilon \neq 0$ the system of (6.21) has a state space of dimension $n + m$. However at $\epsilon = 0$ the equations degenerate to

$$\left. \begin{array}{l} \dot{x} = f(x(t), y(t)) \\ 0 = g(x(t), y(t)) \end{array} \right\} \quad \Sigma. \tag{6.22}$$

The system (6.22) consists of m algebraic equations and n differential equations. A more complete geometric description of them is given by noting that if 0 is a *regular* value of $g : \mathbb{R}^n \times \mathbb{R}^m \mapsto \mathbb{R}^m$, then the set $M = \{(x, y) : g(x, y) = 0\}$ is an n-dimensional manifold. The equations of (6.22) need to be interpreted. The simplest interpretation is that the system of (6.22) represents a dynamical system on M in the following way: If the second equation $g(x, y) = 0$ could be "solved" for y locally uniquely as a function of x, say as $y = \psi(x)$ (by the implicit function theorem, this is, in turn, possible if $D_2 g(x, y)$ is invertible for $(x, y) \in M$) then the manifold M may be locally parametrized by x, and the dynamical system on M may be described in terms of

$$\dot{x} = f(x, \psi(x)). \tag{6.23}$$

Notice that equation (6.23) represents an n-dimensional equation, which is referred to as the *reduced* system. It is of obvious importance to determine when the trajectories of the original system of (6.21) converge to those of the reduced order system. The behavior of the y variable of the reduced order system may also be described by a differential equation as

$$\dot{y} = \frac{\partial \psi(x)}{\partial x} f(x, \psi(x)). \tag{6.24}$$

The reduced order system is also referred to as the slow time scale system. The *fast* time scale system is the system obtained by rescaling time t to $\tau = t/\epsilon$ to get

$$\frac{\partial x}{\partial \tau} = \epsilon f(x, y),$$
$$\frac{\partial y}{\partial \tau} = g(x, y). \tag{6.25}$$

As $\epsilon \downarrow 0$ the equations become

$$\left.\begin{aligned}\frac{\partial x}{\partial \tau} &= 0, \\[2mm] \frac{\partial y}{\partial \tau} &= g(x, y).\end{aligned}\right\} \quad \Upsilon \qquad (6.26)$$

The dynamics of the sped-up system Υ in the fast time scale τ shows that the x variable is "frozen." The set of equilibria of the speeded up system Υ is precisely M, the configuration space manifold. The fast dynamics of

$$\frac{\partial y}{\partial \tau} = g(x, y)$$

describe whether or not the manifold M is attracting or repulsive to the parasitic dynamics. Now consider linear, singularly perturbed systems of the form

$$\begin{aligned}\dot{x} &= A_{11}x + A_{12}y, \\[2mm] \epsilon\dot{y} &= A_{21}x + A_{22}y.\end{aligned} \qquad (6.27)$$

By setting $\epsilon = 0$ we get the linear singularly perturbed system

$$\begin{aligned}\dot{x} &= A_{11}x + A_{12}y \\[2mm] 0 &= A_{21}x + A_{22}y\end{aligned} \qquad (6.28)$$

If A_{22} is nonsingular, then the linear algebraic equation can be solved to yield the reduced system

$$\dot{x} = (A_{11} - A_{12}A_{22}^{-1}A_{21})x \qquad (6.29)$$

along with

$$y = -A_{22}^{-1}A_{21}x.$$

The configuration space manifold is $M = \{(x, y) : y = -A_{22}^{-1}A_{21}x\}$. Also, the fast time scale system in the fast time scale is

$$\frac{\partial y}{\partial \tau} = A_{21}x + A_{22}y. \qquad (6.30)$$

Thus the stability of the matrix A_{22} determines whether or not the manifold M is attracting to the sped up dynamics. We generalize this notion to nonlinear systems:
Define

$$M_1 = \{(x, y) : g(x, y) = 0, \det D_2 g(x, y) \neq 0\}.$$

Let $(x_0, y_0) \in M_1$. Then by the implicit function theorem there is a unique integral curve of (6.22), $\gamma(t) = (x(t), y(t))$ through (x_0, y_0) defined on $[0, \alpha[$ for some $\alpha > 0$. Some trajectories of the original system Σ_ϵ originating from arbitrary $(x(0), y(0)) \in \mathbb{R}^{n+m}$ for all $\epsilon > 0$ tend uniformly (*consistently*) to $\gamma(t)$ as $\epsilon \downarrow 0$ on all closed intervals of $]0, \alpha[$ as follows (see also Figure 6.10). We state these theorems without proof, local versions of them were proved by Levin, Levinson

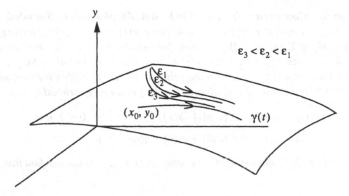

FIGURE 6.10. The consistency requirement

[179], Miscenko, Pontryagin and Boltyanskii [212] for bounded time intervals, and Hoppensteadt [144] for infinite time intervals. A good version of the proofs is given in Khalil [162]; see also [163]. Finally, a modern geometric version of these proofs using the Center Manifold theorem of the next Chapter 7 was given by Fenichel [97].

Theorem 6.7 Consistency over Bounded Time Intervals. *Given that a unique solution curve* $\gamma(t) = (x(t), y(t))$ *starting from* $(x_0, y_0) \in M_1$ *of* Σ *exists on* $[0, \alpha[$, *then* $\exists \delta > 0$ *such that solutions of* Σ_ϵ *starting from* $x(0), y(0)$ *with* $|x(0) - x_0| + |y(0) - y_0| < \delta$ *converge uniformly to* $\gamma(t)$ *on all closed subintervals of* $]0, \alpha[$ *as* $\epsilon \downarrow 0$, *provided that the spectrum of* $D_2g(x(t), y(t))$ *over the trajectory* $\gamma(t)$ *of* Σ *lies in the left half complex plane and is bounded away from the* $j\omega$ *axis.*

Remarks:

1. If the pointwise consistency condition $D_2g(x_0, y_0) \subset \mathbb{C}_-^\circ$ holds for $(x_0, y_0) \in M$ it holds in a neighborhood of (x_0, y_0). Thus, we label the subset M_a of M given by

$$M_a = \{(x, y) : g(x, y) = 0, \sigma(D_2g(x, y)) \subset \mathbb{C}_-^\circ\}$$

   as the "attracting" portion of the manifold M.

2. Some weaker consistency conditions hold if $(x_0, y_0) \in M_1$ but $\sigma(D_2g(x, y)) \cap \mathbb{C}_+ \neq \varnothing$. First, some notation: Let $P(t) \in \mathbb{R}^{m \times m}$ $(0 \leq t < \alpha)$ be a family of nonsingular matrices such that $(x(t), y(t)) = \gamma(t)$:

$$P^{-1}(t)D_2g(x(t), y(t))P(t) = \begin{pmatrix} B(t) & 0 \\ 0 & C(t) \end{pmatrix} \tag{6.31}$$

   for $0 \leq t < \alpha$ where $B(t) \in \mathbb{R}^{m_1 \times m_1}$ has spectrum in $\mathbb{C}_-^\circ$ bounded away from the $j\omega$ axis and $C(t) \in \mathbb{R}^{m-m_1 \times m-m_1}$ has spectrum in $\mathbb{C}_+^\circ$ bounded away from the $j\omega$ axis.

Theorem 6.8 Consistency from Stable Initial Manifolds over Bounded Time Intervals. *Given that a unique solution curve $\gamma(t) = (x(t), y(t))$ starting from $(x_0, y_0) \in M_1$ of Σ exists on $[0, \alpha[$, then there exists an m_1 dimensional manifold $S(\epsilon) \subset \mathbb{R}^m$ depending on ϵ for $0 < \epsilon \leq \epsilon_0$ such that if the initial vector for Σ_ϵ is $(x_0, y_\epsilon(0))$ with $y_\epsilon(0) \in S(\epsilon)$, then the solution of (6.21) satisfies the inequalities (with $b \in \mathbb{R}^{m_1}$ representing coordinates for $S(\epsilon)$ in a neighborhood of y_0).*

$$|x_\epsilon(t) - x(t)| \leq K(\epsilon|b| + \omega(\epsilon)) \qquad\qquad 0 \leq t < \alpha$$

$$|y_\epsilon(t) - y(t)| \leq K(\epsilon|b| + \omega(\epsilon) + |b|e^{-\sigma t/\epsilon})$$

where $K, \sigma \in \mathbb{R}_+$ independent of ϵ and $\omega(\epsilon)$ is a continuous function with $\omega(0) = 0$.

1. $\lim_{\epsilon \to 0} S(\epsilon)$ exists and is an m_1 dimensional manifold that is the local stable manifold of the equilibrium y_0 of the "frozen" boundary layer system Υ. This is depicted in Figure 6.11. In the sequel, we refer to the limit $S(0)$ as $S_{y_0}^{x_0}$ to show its dependence on the equilibrium y_0 with x_0 frozen.
2. The preceding theorems were stated as 'local' –they can be made global in a certain sense which we will discuss in Chapter 7. Actually the parameterized center manifold theorem of Chapter 7 has been used by Fenichel to derive 'global' versions of both Theorems 6.7 and 6.8.

It is of obvious importance to understand when trajectories of the full system (6.21) converge to those of the singularly perturbed system (6.22). In particular, it is important to understand which of the solutions of $g(x, y) = 0$ are chosen for a given value of x by the system.

6.3.1   Nonsingular Points, Solution Concepts, and Jump Behavior

We may aggregate the geometric picture obtained from Theorems 6.7 and 6.8: For a hyperbolic equilibrium point y_0 of the frozen boundary layer system Υ, we attach as a fiber the stable manifold $S_{y_0}^{x_0}$. When the fiber $S_{y_0}^{x_0}$ does not have full co-dimension (i.e., m), then disturbances and noise will not cause the "state" (x, y) of Σ to slip off M. If, however, the fiber is of less than full co-dimension, the state can

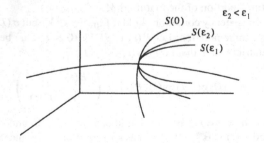

FIGURE 6.11. Stable initial manifolds for the augmented system

slip off M and transit infinitely rapidly (jump) in the y variable to another portion of M along the singularly perturbed (infinitely fast) limit of the parasitic dynamics. Thus, a valid definition of a solution concept for the singularly perturbed limit of (6.22) viewed as the limit of the augmented system of (6.21) has to allow for the possibility of jumps (in the y variable, the x variable is constrained to be at least an absolutely continuous function of time from the differential equation in (6.22)) from those parts of M that are not attractive, that is, for which

$$\sigma(D_2g(x_0, y_0)) \cap \mathbb{C}_+ \neq \varnothing.$$

However to make this intuitive picture correct, we will need to be sure that the frozen boundary layer system is "gradient–like" in the sense of the previous chapter (cf. Section 5.9), that is to say, for each x_0, each trajectory of

$$\frac{dy}{d\tau} = g(x_0, y), \quad y(0) = \tilde{y} \tag{6.32}$$

converges to one of the equilibrium points of the system (6.32). In particular this is guaranteed if $g(x, y)$ is indeed the gradient of some scalar function, that is,

$$g(x_0, y) = D_2^T S(x, y)$$

for some $S(x_0, y) : \mathbb{R}^m \mapsto \mathbb{R}$ that is a proper function for each x_0. The construction that we have described may be made more mathematically precise as a singular foliation of $\mathbb{R}^{n+m}$, with "fibers" of varying dimension attached to a "base space" M. The slow dynamics live on the base space and the infinitely fast dynamics on the fiber space. Now, define M_h to be the set of points for which the equilibria of the frozen boundary layer system are hyperbolic. That is

$$M_h = \{(x, y) \in M : g(x, y) = 0, \ \sigma(D_2g(x, y)) \cap]-j\infty, j\infty[= \varnothing\}.$$

It is clear that the stability properties of the base points changes on M_h. We refer to the points in M_h as the *non-singular* points. Of course, as we just pointed out, if the non-singular points are not stable, there could be jumps off them onto other segments of M.

At this point, the reader will find it instructive to return to the degenerate van der Pol example of Chapter 2 (equation 2.25) and relate the construction given here to that system. It follows in particular that any trajectories starting in the middle section of the curved resistor characteristic are liable to slip off the manifold M and transition to one or the other of the two outer legs of the resistor characteristic (cf. Figure 2.31). In re-examining this example, it appears that the truly interesting and amazing feature of the behavior of singularly perturbed nonlinear systems comes from the changing nature of the stability of points on M as x changes, i.e., at points $x \in M_h$ (cf. the "impasse" points in Figure 2.30); that is to say, x acts as a bifurcation parameter for the dynamics of the frozen boundary layer system. Thus, a solution concept for the singularly perturbed limit has to allow for jumps in the y variable as (x, y) transitions through M_h. We will take up this point in the next chapter, in Section 7.10.2, after we have studied bifurcations.

6.4   Dynamics of Nonlinear Systems and Parasitics

The dynamics of a large class of nonlinear systems are described by nonlinear differential equations with algebraic constraints. Thus, the model is of a form that we have called singularly perturbed (6.22) in the preceding section. If the algebraic constraints can be "solved" for the unknown state variables, the dynamics can be converted into normal form equations. However, if there are many solutions or changes in the number of solutions of the algebraic equation, one has to interpret solutions as being the singularly perturbed limits of a larger system of differential equations of the form (6.21). In this section, we give two examples: one of nonlinear circuits and the other of the power system dynamics and loads, where we can give physical meaning to the singular perturbation assumption.

6.4.1   Dynamics of Nonlinear Circuits

The treatment of this section follows Sastry and Desoer [263]. Consider the dynamics of a fairly general class of time-invariant nonlinear circuits represented as shown in Figure 6.12. We view the network as a resistive n-port terminated by capacitors and inductors. Let $z \in \mathbb{R}^{n_c + n_l}$ be the vector of charges on the capacitors ($z_1 \in \mathbb{R}^{n_c}$) and fluxes on the inductors ($z_2 \in \mathbb{R}^{n_l}$). Let $x \in \mathbb{R}^{n_c + n_l}$ be the vector of capacitor voltages ($x_1 \in \mathbb{R}^{n_c}$) and inductor currents ($x_2 \in \mathbb{R}^{n_l}$). Further, we will assume that the capacitors and inductors are reciprocal in the sense that they can be characterized by

$$x = h(z) = D^T H(z). \tag{6.33}$$

Here $H(z)$ is called the *stored energy* of the capacitor and inductor and $h(\cdot)$ is called the *constitutive equation* of the capacitors and inductors. We will assume that h is a diffeomorphism, or else continuous changes in the charge and flux may result in discontinuous changes in the voltage across the capacitor and current across the inductors.

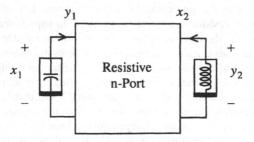

FIGURE 6.12. A nonlinear circuit as an n-port terminated by capacitors and inductors

The resistive $n = n_c + n_l$ port is characterized by the algebraic constitutive equation

$$g(x, y) = 0. \tag{6.34}$$

Here $y \in \mathbb{R}^n$ is the vector of capacitor port currents (y_1) and inductor port voltages (y_2), with reference directions chosen such that $x^T y$ represents the power into the n-port. Now, Coulomb's law and Faraday's law yield the equation

$$\dot{z} = -y. \tag{6.35}$$

This is to be combined with

$$x = h(z) = D^T H(z), \quad g(x, y) = 0 \tag{6.36}$$

to yield the circuit equation

$$\dot{x} = -D^T h(h^{-1}(x))y, \tag{6.37}$$
$$0 = g(x, y).$$

If the second of these two equations is solvable for y as a smooth function of x, that is

$$g(x, y) = 0 \Leftrightarrow y = \psi(x),$$

then the circuit equations of 6.37 can be solved to give normal form equations of the form

$$\dot{x} = -D^T h(h^{-1}(x))\psi(x).$$

However, if the circuit equations admit some hybrid representation, of the form

$$x_A = f_A(y_A, x_B), \tag{6.38}$$
$$y_B = f_B(y_A, x_B),$$

where A, B are a partition of $\{1, 2, \ldots, n\}$, equivalent to $g(x, y) = 0$, then the circuit equations are given by

$$\dot{x} = -D^T h(h^{-1}(x)) \begin{bmatrix} y_A \\ f_B(y_A, x_B) \end{bmatrix}, \tag{6.39}$$
$$0 = x_A - f_A(y_A, x_B).$$

These equations (6.39) represent singularly perturbed dynamics of the nonlinear circuit. The dynamics evolve on the n dimensional manifold embedded in $\mathbb{R}^{n+n_A}$ given by

$$M = \{x \in \mathbb{R}^n, y_A \in \mathbb{R}^{n_A} : x_A - f_A(y_A, x_B) = 0\},$$

provided that 0 is a regular value of the map f_A. Since it is not a priori clear which solution y_A of the algebraic equation is "chosen" by the circuit for given

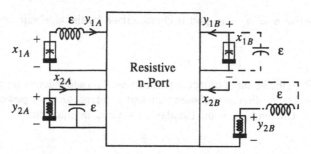

FIGURE 6.13. The introduction of parasitic capacitors and inductors to normalize the nonlinear circuit

$x \in \mathbb{R}^n$ (if multiple solutions exist), the understanding is that the second of these two equations is the singularly perturbed limit of

$$\epsilon \dot{y}_A = x_A - f_A(y_A, x_B). \tag{6.40}$$

This understanding has the physical interpretations of introducing small parasitic linear inductances and capacitances at the A ports of the circuit (with the capacitors in parallel and the inductors in series) as shown in Figure 6.13. The physical basis for introducing these ϵ parasitics is based on the notion that a *current controlled* resistor is envisioned as the singularly perturbed limit as $\epsilon \downarrow 0$ of the resistor in series with a small linear parasitic inductor (since current is the controlling variable). The dual holds for a *voltage controlled* resistor. From this standpoint, it should appear that we add parasitics at both the A and B ports. However, at the B ports the parasitics are paired with elements of the same kind and so are swamped (shown in dotted lines in the figure). At the B ports the parasitics are paired with elements of the opposite kind and cannot be neglected.

Note that as in the preceding section, it is not immediately obvious that there are limit trajectories to the augmented system consisting of (6.40) along with the first equation of (6.39). However, if the boundary layer system with x frozen is gradient like in the sense of Section 5.9, then there exists an energy like function $v(x, y)$ satisfying

1. $\dot{v}(x, y_A) = \frac{\partial v}{\partial y_A}(x_A - f_A(y_A, x_B)) \le 0$ for $y_A \notin \{y_A : x_A = f_A(y_A, x_B)\}$.
2. $\{y_A : \dot{v}(x, y_A) = 0\}$ is the set of equilibrium points.
3. $v(x, y_A)$ is a proper function of y_A, for each value of x, that is the inverse image of compact sets is compact.

Again, as in the preceding section, the attracting portion of the manifold M is precisely

$$M_a\{(x, y_A) : x_A = f_A(y_A, x_B), \ \sigma(D_1 f_A(y_A, x_B) \subset \mathbb{C}_+^\circ\}.$$

Note that the statement above has $\mathbb{C}_+^\circ$ only because the boundary layer system (6.40) has $-f_A$ in its right-hand side.

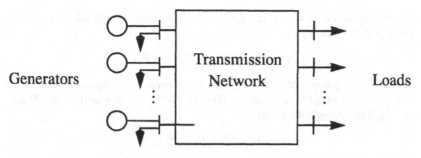

FIGURE 6.14. An interconnected power network

6.4.2 Dynamics of Power Systems

In the preceding chapter we had a model of a power system which we called the power system swing equations (5.33). However, this model did not take into account loads at some nodes without generation. To this end consider the model shown in Figure 6.14 with so-called PV loads, that is the voltage is kept constant and there is a fixed amount of power demanded. The equations of motion for the generator buses are given by

$$M_i \dot{\omega} + D_i \omega_i = P_i^m - P_i^g,$$
$$\dot{\delta}_i = \omega_i,$$

$$(6.41)$$

where for $i = 1, \ldots, g$, $M_i(D_i)$ represent the moment of inertia and damping constant, ω_i is the departure of generator frequency from synchronous frequency, and δ_i is the generator rotor angle measured relative to a synchronously rotating reference. Moreover, P_i^m represents the exogenously specified mechanical input power, and P_i^g is the electrical power output at the i-th node. Now let θ_i for $i = 1, \ldots, l$ be the angle of a set of l nodes which have no generation, and let P_i^l be the power demanded at these nodes. Now define B_{ij}^{gg} be the admittance of the line joining nodes i and j for the generator nodes, and B_{ij}^{gl}, B_{ij}^{ll} be similarly defined. If two buses are not connected then $B_{ij} = 0$. Then the so-called *load flow equations* are given by

$$P_i^g = f_i^g(\delta, \theta) = \sum_j Y_{ij}^{gg} \sin(\theta_i - \theta_j) + \sum_k Y_{ik}^{gl} \sin(\delta_i - \theta_k), \quad i = 1, \ldots, g,$$

$$P_i^l = f_i^l(\delta, \theta) = \sum_j Y_{ji}^{gl} \sin(\theta_i - \delta_j) + \sum_k Y_{ik}^{ll} \sin(\theta_i - \theta_k), \quad i = 1, \ldots, l.$$

$$(6.42)$$

Combining, equations (6.41) and (6.42), we get

$$\dot{\delta} = \omega,$$
$$\dot{\omega} = -M^{-1} D\omega - M^{-1} P^m - M^{-1} f^g(\delta, \theta),$$
$$0 = f^l(\delta, \theta) - P^l.$$

$$(6.43)$$

The state space of the system is a $2g$ dimensional manifold embedded in a $\mathbb{R}^{2g+l}$ space (δ, ω, θ), given by

$$M = \{(\omega, \delta, \theta) : f^l(\delta, \theta) = P^l\}.$$

To resolve the question of which solution θ is chosen, we recognize that the load P_i^l is not constant but is really of the form $P_i^l - \epsilon \theta_i$, so that equations (6.43) are the singularly perturbed limit of

$$\epsilon \dot{\theta} = f^l(\delta, \theta) - P^l.$$

It is left to the Exercises to verify that the boundary layer system is a gradient system (Problem 6.19). Our treatment in this section follows Sastry and Varaiya [265]; the model (6.43) first introduced there has found wide spread acceptance in the power systems community as the *structure preserving model* of power systems swing equations (see [23], also [314]).

6.5   Adaptive Identification of Single-Input Single-Output Linear Time-Invariant Systems

The design of adaptive identifiers serves as an interesting example of the application of the theory of the previous chapter. In this section, we explore the use of the Lyapunov theory to design schemes for estimating the parameters of an unknown system. Adaptive control is studied in the next section. Our treatment follows very closely the treatment of Chapter 2 of [259]. We will describe techniques for identifying linear systems and nonlinear systems: Linear systems will be identified exclusively using input-output signals, while the study of nonlinear systems uses measurements of the state variables. We will address this in the context of the adaptive control of nonlinear systems in Chapter 9. See also Problem 6.15 for an exercise in nonlinear adaptive identification.

Consider a single-input single-output linear time invariant system with transfer function $\hat{p}$ a rational function in s. This is the Laplace transform of the impulse response of the linear system with zero initial condition. In this section we will be somewhat cavalier, at the outset, in disregarding the effect of initial conditions. However, we will remedy this later in the section. By way of notation, we will use hatted variables here to represent Laplace transformed variables.

Some terminology will be useful in discussing the parameterization of the plant: a polynomial in s is said to be *monic* if the coefficient of the leading power of s is 1 and *Hurwitz* if its roots lie on $\mathbb{C}_-$. The *relative degree* of a transfer function is the excess of poles over zeros. A transfer function is said to be *proper* if its relative degree is at least 0 and *strictly proper* if its relative degree at least 1. A transfer function is said to be *stable* if it is proper and has a Hurwitz denominator. It is said to be *minimum phase* if its numerator is Hurwitz. The identifier presented in this section is known as an *equation error identifier*. We will denote the plant transfer function explicitly by

$$\hat{P} = \frac{\alpha_n s^n + \cdots + \alpha_1}{s^n + \beta_n s^{n-1} + \cdots + \beta_1} = \frac{k_P \hat{n}_P}{\hat{d}_P} \tag{6.44}$$

where the $2n$ coefficients $\alpha_1, \ldots, \alpha_n$ and $\beta_1, \ldots, \beta_n$ are unknown. This expression is a *parameterization* of the plant, that is, a model for which only a finite number of parameters need to be determined. If $\hat{y}_P$ and $\hat{u}_P$ stand for the input and output of the plant and the initial condition at $t = 0$ is zero, then it follows that from algebraic manipulation,

$$s^n \hat{y}_P = (\alpha_n s^{n-1} + \cdots + \alpha_1)\hat{u}_P - (\beta_n s^{n-1} + \cdots + \beta_1)\hat{y}_P.$$

The foregoing is an expression that is linear in the parameters, but unfortunately involves differentiation of the signals, $\hat{u}_P$, $\hat{y}_P$, an inherently noisy operation. In order to get around this problem, we define a monic Hurwitz polynomial, $\hat{\lambda}(s) \in \mathbb{R}[s] := s^n + \lambda_n s^{n-1} + \cdots + \lambda_1$. Then we rewrite the previous equation as

$$\hat{\lambda}(s)\hat{y}_P(s) = k_P \hat{n}_P(s)\hat{u}_P(s) + (\hat{\lambda}(s) - \hat{d}_P(s))\hat{y}_P(s). \tag{6.45}$$

Dividing through by $\hat{\lambda}(s)$ and using the preceding parameterization yields

$$\hat{y}_P(s) = \frac{\alpha_n s^{n-1} + \cdots + \alpha_1}{\hat{\lambda}(s)} \hat{u}_P(s) + \frac{(\lambda_n - \beta_n)s^{n-1} + \cdots + (\lambda_1 - \beta_1)}{\hat{\lambda}(s)} \hat{y}_P(s).$$

$$\tag{6.46}$$

This expression is now a new parameterization of the plant: Indeed, if we define

$$\hat{a}^*(s) = \alpha_n s^{n-1} + \cdots + \alpha_1 = k_P \hat{n}_P(s),$$

$$\hat{b}^*(s) = (\lambda_n - \beta_n)s^{n-1} + \cdots + (\lambda_1 - \beta_1) = \hat{\lambda}(s) - \hat{d}_P(s),$$

then the new representation of the zero initial condition response of the plant can be written as

$$\hat{y}_P(s) = \frac{\hat{a}^*(s)}{\hat{\lambda}(s)} \hat{u}_P(s) + \frac{\hat{b}^*(s)}{\hat{\lambda}(s)} \hat{y}_P(s), \tag{6.47}$$

and also

$$\hat{P}(s) = \frac{\hat{a}^*(s)}{\hat{\lambda}(s) - \hat{b}^*(s)}.$$

The parameterization is *unique* if the given plant numerator and denominator are *coprime*. Indeed, if there exist two different polynomials $\hat{a}^* + \hat{\Delta}a(s)$, $\hat{b}^* + \hat{\Delta}b(s)$ of degree at most $n - 1$ each so as to yield the same plant transfer function, then an easy calculation shows that we would have

$$\frac{\Delta\hat{a}(s)}{\Delta\hat{b}(s)} = -\frac{k_P \hat{n}_P}{\hat{d}_P} = -\hat{P}(s).$$

This contradicts the coprimeness of $\hat{n}_P$, $\hat{d}_P$ since $\Delta\hat{b}(s)$ has degree at most $n - 1$. The basis for the identifier is a state space realization of equation (6.46). In particular, we choose $\Lambda \in \mathbb{R}^{n \times n}$, $b_\lambda \in \mathbb{R}^n$ in controllable canonical form as

$$
\Lambda = \begin{bmatrix}
0 & 1 & 0 & \cdots & 0 \\
0 & 0 & 1 & \cdots & 0 \\
\vdots & \vdots & \vdots & \cdots & \vdots \\
0 & 0 & 0 & \cdots & 1 \\
-\lambda_1 & -\lambda_2 & -\lambda_3 & \cdots & -\lambda_n
\end{bmatrix}, \quad
b_\lambda = \begin{bmatrix}
0 \\
0 \\
\vdots \\
0 \\
1
\end{bmatrix}
$$

so as to yield

$$
(sI - \Lambda)^{-1} b_\lambda = \frac{1}{\hat{\lambda}(s)} \begin{bmatrix}
1 \\
s \\
\vdots \\
s^{n-1}
\end{bmatrix}.
$$

Now define the two signal vectors w_P^1, $w_P^2 \in \mathbb{R}^n$ by

$$
\begin{aligned}
\dot{w}_P^1 &= \Lambda w_P^1 + b_\lambda u_P, \\
\dot{w}_P^2 &= \Lambda w_P^2 + b_\lambda y_P.
\end{aligned}
\tag{6.48}
$$

From the form of the transfer function for $(sI - \Lambda)^{-1} b_\lambda$ above, the names *dirty derivatives* or *smoothed derivatives* of the input and output for the signals w_P^1, w_P^2 become clear. Further, define the constant vectors a^*, $b^* \in \mathbb{R}^n$ by

$$
a^{*T} := (\alpha_1, \ldots, \alpha_n), \quad b^{*T} := (\lambda_1 - \beta_1, \ldots, \lambda_n - \beta_n),
\tag{6.49}
$$

so that equation (6.46) may be written as

$$
\hat{y}_P(s) = a^{*T} \hat{w}_P^1(s) + b^{*T} \hat{w}_P^2(s)
\tag{6.50}
$$

modulo contributions due to the initial conditions $w_P^1(0)$, $w_P^2(0)$. This equation may be written in the time domain as

$$
y_P(t) = \theta^{*T} w(t),
\tag{6.51}
$$

where

$$
\theta^* := (a^{*T}, b^{*T})^T \in \mathbb{R}^{2n},
$$
$$
w_P := (w_P^{1T}, w_P^{2T})^T \in \mathbb{R}^{2n}.
$$

and as before, there may be contributions due to the initial conditions $w_P(0)$. Equations (6.48)–(6.50) define a realization of the new parameterization of the plant. The realization is not minimal since the dimension of the state space of the realization of (6.48)–(6.48) is $2n$. In fact, the realization is controllable but not observable, though the unobservable modes are at the zeros of $\hat{\lambda}(s)$ and are thus stable. The vector θ^* of unknown parameters is related affinely to the unknown

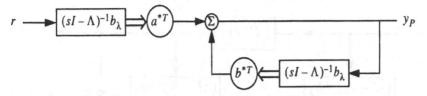

FIGURE 6.15. Plant parameterization

parameters of the plant. The plant parameterization is represented as in Figure (6.15). The purpose of the identifier is to produce a recursive estimate of the unknown parameter θ^*. Since u_P, y_P are available we define the *observer*,

$$\dot{w}^1 = \Lambda w^1 + b_\lambda u_P,$$
$$\dot{w}^2 = \Lambda w^2 + b_\lambda y_P. \tag{6.52}$$

to reconstruct the states of the plant. It is interesting to note that the error $w - w_P$ decays to zero regardless of the stability of the plant. Thus, the generalized state of the plant can be reconstructed without knowledge of the parameters of the plant. Thus, the (true) plant output may be written as

$$y_P(t) = \theta^{*T} w(t) + \epsilon(t),$$

where the signal $\epsilon(t)$ stands for additive exponentially decaying terms caused by initial conditions in the observer and also in the plant parameterization. We will first neglect these terms, but later we will show that they do not affect the properties of the identifier. In analogy with the expression of the plant output, we define the output of the identifier to be

$$y_i(t) = \theta^T(t)w(t). \tag{6.53}$$

We define the *parameter error* to be

$$\phi(t) := \theta(t) - \theta^* \in \mathbb{R}^{2n}$$

and the *identifier error* to be

$$e_i(t) := y_i(t) - y_P(t) = \phi^T w(t) + \epsilon(t). \tag{6.54}$$

The form of the identifier is as shown in Figure 6.16. The equation (6.54) is known as the *error equation* and shows a linear relationship between the parameter errors and the identifier error.

Equation (6.54) is the basis for the design of a number of identification algorithms, called *update laws* to reflect the fact that they update estimates of the parameter $\theta(t)$ based on measurements. Thus an identification algorithm is a differential equation of the form

$$\dot{\theta} = F(y_P, e_i, \theta, w),$$

where F is a function (or perhaps an operator) explicitly independent of θ^* that defines the evolution of the identifier parameter θ. The identifier update laws are

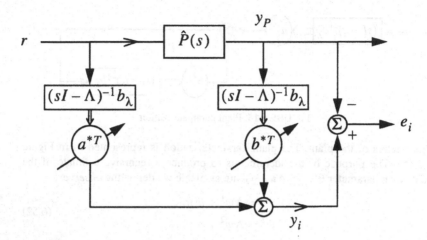

FIGURE 6.16. Identifier structure

sometimes written in terms of the parameter error as $\dot{\theta} = \dot{\phi}$, since $\phi = \theta - \theta^*$ with θ^* constant. Two classes of algorithms are commonly utilized for identification: gradient algorithms and least-squares algorithms.

Gradient Algorithms

The update law

$$\dot{\theta} = \dot{\phi} = -ge_i w, \quad g \in \mathbb{R} > 0, \tag{6.55}$$

is the so-called *standard gradient algorithm*, so named because the right hand side is the negative of gradient of the identifier output error squared, i.e., $e_i^2(\theta)$ with respect to θ, with θ assumed frozen. The parameter g is a fixed positive gain called the adaptation gain, which enables a variation in the speed of the adaptation. A variation of this algorithm is the normalized gradient algorithm

$$\dot{\theta} = \dot{\phi} = -\frac{ge_i w}{1 + \gamma w^T w}, \quad g, \gamma \in \mathbb{R} > 0 \tag{6.56}$$

The advantage of the normalized gradient algorithm is the fact that it may be used even when the signals w are unbounded, as is the case with unstable plants.

Least-Squares Algorithms

These are derived from writing down the Kalman-Bucy filter for the state estimation problem for the state θ^* of a (fictitious) system representing the plant with state variables being the unknown parameters:

$$\dot{\theta}^*(t) = 0$$

with output $y_P(t) = w^T(t)\theta^*(t)$. Assuming that the right hand sides of these equations are perturbed by zero mean, white Gaussian noise of spectral intensities

$Q \in \mathbb{R}^{2n \times 2n}$ and $\frac{1}{g} \in \mathbb{R}$ we get the Kalman—Bucy filter

$$\dot{\theta} = -gPwe_i,$$
$$\dot{P} = Q - gPww^T, \quad P > 0, \ Q > 0, \ g > 0 \tag{6.57}$$

The matrix P is called the *covariance matrix*[2] and acts on the update law as a time varying *directional* adaptation gain. It is common to use a version of the least squares algorithm with $Q = 0$, since the parameter θ^* is assumed to be constant. In this case the preceding update law for the covariance matrix has the form

$$\frac{dP}{dt} = -gPww^TP \Leftrightarrow \frac{dP^{-1}}{dt} = gww^T \tag{6.58}$$

where $g > 0$ is a constant. This equation shows that $dP^{-1}/dt \geq 0$, so that P^{-1} may grow without bound. Thus, some directions of P will become arbitrarily small, and the update in those directions will be arbitrarily slow. This is referred to as the *covariance wind-up problem*. One way of preventing covariance wind-up is to use a least squares with forgetting factor algorithm, defined by

$$\frac{dP}{dt} = -g(-\lambda P + Pww^TP),$$
$$\frac{dP^{-1}}{dt} = g(-\lambda P^{-1} + ww^T) \quad g, \lambda > 0. \tag{6.59}$$

Yet another remedy is to use a least squares algorithm with covariance resetting, where P is reset to some predetermined positive definite value whenever $\lambda_{\min}(P)$ falls under some threshold. A *normalized least-squares algorithm* may be defined as

$$\dot{\theta} = -g\frac{Pwe_i}{1 + \gamma w^TPw}, \quad g, \gamma > 0,$$
$$\dot{P} = -g\frac{Pww^TP}{1 + \gamma w^TPw}, \tag{6.60}$$
$$\dot{P^{-1}} = g\frac{ww^T}{1 + \gamma w^T(P^{-1})^{-1}w}.$$

Again, g, γ are fixed parameters and $P(0) > 0$. The same modifications as before may be made to avoid covariance wind-up. We will assume that we will use resetting in what follows, that is,

$$P(0) = P(t_r^+) = k_0 I > 0,$$
$$t_r \in \{t : \lambda_{\min}(P(t)) \leq k_1 < k_0\}. \tag{6.61}$$

The least-squares algorithms are somewhat more complicated to implement than gradient algorithms but are found to have faster convergence properties.

---

[2]In the theory of Kalman—Bucy filtering, P has the interpretation of being the covariance of the error in the state estimate at time t.

6.5.1   Linear Identifier Stability

We start with some general properties of the gradient algorithm of (6.55),

$$\dot{\phi} = \dot{\theta} = -ge_i w, \quad g > 0,$$

and the normalized gradient algorithm of (6.56)

$$\dot{\phi} = \dot{\theta} = -g\frac{e_i w}{1 + \gamma w^T w}, \quad g, \gamma > 0,$$

with the linear error equation (6.54),

$$e_i = \phi^T w.$$

Theorem 6.9 Linear Error Equation with Gradient Algorithm. *Consider the linear error equation of (6.54), together with the gradient algorithm (6.55). Assume that $w : \mathbb{R}_+ \mapsto \mathbb{R}^{2n}$ is piecewise continuous. Then*

- $e_i \in L_2$.
- $\phi \in L_\infty$.

Proof: The differential equation describing the evolution of ϕ is

$$\dot{\phi} = -gww^T\phi.$$

With Lyapunov function $v(\phi) = \phi^T\phi$, the derivative along trajectories is given by

$$\dot{v} = -2g(\phi^T w)^2 = -2ge_i^2 \le 0.$$

Hence, $0 \le v(t) \le v(0)$ for all $t \ge 0$, so that $v, \phi \in L_\infty$. Further, since $v(\infty)$ is well defined, we have

$$v(\infty) - v(0) = \int_0^\infty \dot{v}dt = -2g\int_0^\infty e_i^2 dt \le \infty$$

thereby establishing that $e_i \in L_2$.   □

Theorem 6.10 Linear Error Equation with Normalized Gradient Algorithm. *Consider the linear error equation of (6.54), together with the gradient algorithm (6.56). Assume that $w : \mathbb{R}_+ \mapsto \mathbb{R}^{2n}$ is piecewise continuous. Then*

- $\dfrac{e_i}{\sqrt{1+\gamma w^T w}} \in L_2 \cap L_\infty$.
- $\phi \in L_\infty, \dot{\phi} \in L_2 \cap L_\infty$.

Proof: With the same Lyapunov function $v(\phi) = \phi^T\phi$, we get

$$\dot{v} = -\frac{2ge_i^2}{1 + \gamma w^T w} \le 0.$$

Hence, $0 \le v(t) \le v(0)$ for all $t \ge 0$, so that from the same logic as before we have $v, \phi, e_i/\sqrt{1 + \gamma w^T w} \in L_\infty$. From the formula

$$\dot{\phi} = -g\frac{ww^T}{1 + \gamma w^T w}\phi$$

it follows that $|\dot{\phi}| \leq (g/\gamma)|\phi|$ and thus $\dot{\phi} \in L_\infty$. Since

$$\int_0^\infty \dot{v}dt = -2g\int_0^\infty \frac{e_i}{1+\gamma w^T w}dt \leq \infty,$$

it follows that $e_i/\sqrt{1+\gamma w^T w} \in L_2$. Also

$$|\dot{\phi}|^2 = \left| \frac{g^2 w^T w e_i^2}{(1+\gamma w^T w)^2} \right| \leq \frac{g^2}{\gamma}\frac{e_i^2}{1+\gamma w^T w}$$

implies that $\dot{\phi} \in L_2$. □

Theorem 6.11 Linear Error Equation with Normalized Least Squares Algorithm and Covariance Resetting. *Consider the linear error equation of (6.54), together with the normalized least squares algorithm (6.60) with covariance resetting (6.61). Assume that $w : \mathbb{R}_+ \mapsto \mathbb{R}^{2n}$ is piecewise continuous. Then:*

1.

$$\frac{e_i}{\sqrt{1+gw^T Pw}} \in L_2 \cap L_\infty.$$

2.

$$\phi \in L_\infty \quad \dot{\phi} \in L_2 \cap L_\infty.$$

Proof: The covariance matrix $P(t)$ is a discontinuous function of time. Between discontinuities, the evolution is described by the differential equation (6.60). Since $d/dt\, P^{-1} \geq 0$ between resets, we have that

$$P^{-1}(t) \geq P^{-1}(t_r^+) = P^{-1}(0) = k_0^{-1} I.$$

On the other hand due to resetting, we have $P(t) \geq k_1 I$ for all $t \geq 0$, so that

$$k_0 I \geq P(t) \geq k_1 I \quad k_1^{-1}I \geq P^{-1}(t) \geq k_0^{-1}.I$$

In addition, the time between resets is bounded below since it is easy to estimate from (6.60) that

$$\frac{dP^{-1}}{dt} \leq \frac{g}{\gamma \lambda_{min} P(t)}.$$

Thus, the set of resets is of measure 0. Now, choose the decrescent, p.d.f. $v(\phi, t) = \phi^T P^{-1}(t)\phi$, and verify that between resets

$$\dot{v} = -g\frac{e_i^2}{1+\gamma w^T Pw} \leq 0.$$

Also, at resets which are points of discontinuity of v we have

$$v(t_r^+) - v(t_r) = \phi^T (P^{-1}(t_r^+) - P^{-1}(t_r))\phi \leq 0.$$

Thus, we have $0 \leq v(t) \leq v(0)$ and given the bounds on $P(t)$, it follows that $\phi, \dot{\phi} \in L_\infty$. From the bounds on $P(t)$, it also follows that

$$\frac{e_i}{1 + \gamma w^T P w} \in L_\infty.$$

Also,

$$-\int_0^\infty \dot{v} dt < \infty \Rightarrow \frac{e_i}{1 + \gamma w^T P w} \in L_2.$$

This establishes the first claim of the theorem. Now, writing

$$\dot{\phi} = -g \frac{e_i}{\sqrt{1 + \gamma w^T P w}} \frac{P w}{\sqrt{1 + \gamma w^T P w}},$$

we see that the first term on the right hand side is in L_2 and the second term is in L_∞, so that we have $\dot{\phi} \in L_2$.   □

Theorems 6.9–6.11 state general properties of the identifiers with gradient, normalized gradient and normalized least-squares respectively. They appear to be somewhat weak in that they do not guarantee that the parameter error ϕ approaches 0 as $t \to \infty$ or for that matter that the identifier error e_i tends to 0. However, the theorems are extremely general in that they do not assume either boundedness of the regressor signal $w(\cdot)$ or stability of the plant. The next theorem gives conditions under which it can be proven that $e_i(\cdot) \to 0$ as $t \to \infty$. This property is called the *stability of the identifier.*

Theorem 6.12 Stability of the Identifier.   *Consider the identifier with the error equation (6.54) with either the gradient update law of (6.55) or the normalized gradient update law of (6.56) or the normalized least squares of (6.60) with covariance resetting given by (6.61). Assume that the input reference signal $u_P(\cdot)$ is piecewise continuous and bounded. In addition, assume that the plant is either stable or is embedded in a stabilizing loop so that the output $y_p(\cdot)$ is bounded.*

Then the estimation error $e_i \in L_2 \cap L_\infty$, with $e_i(t)$ converging to 0 as $t \to \infty$. In addition, $\dot{\phi} \in L_\infty \cap L_2$, and $\dot{\phi}(t)$ converges to 0 as $t \to \infty$.

Proof: Since u_P, y_P are bounded, it follows from the stability of Λ in equation (6.52) that $w, \dot{w}$ are bounded. By Theorems 6.9–6.11 it follows that $\phi, \dot{\phi}$ are bounded, so that $e_i, \dot{e}_i$ are bounded. Since $e_i \in L_2$, it follows from Barbalat's lemma (see Problem 5.9 of Chapter 5) that $e_i \to 0$ as $t \to \infty$. Similar conclusions follow directly for $\dot{\phi}$.   □

Remark: The preceding theorem gives conditions (basically the boundedness of the input and the stability of the plant or the embedding of the plant in a stable closed loop system) under which the identifier error goes to zero as $t \to \infty$. Under the same conditions $\dot{\phi}$ also goes to zero. However, this does not imply that $\phi(t)$ converges at all (let alone to zero)! Problem 6.11 gives an example of a function $\phi(t)$ which meets the requirements of Theorem 6.12 but does not converge.

6.5.2 Parameter Error Convergence

In this subsection we prove conditions under which the parameter error converges (exponentially) for the gradient update law. Surprisingly, the same convergence condition applies for both the modified gradient and the least squares, normalized least squares and normalized least squares with resetting update laws (see Problems 6.12, 6.13).

When we combine the adaptive identifier equation of (6.54) with the parameter update law of (6.55) we get the linear time-varying system

$$\dot{\phi} = -gw(t)w^T(t)\phi = A(t)\phi. \tag{6.62}$$

The time-varying linear system of (6.62) has a negative semi-definite $A(t)$. The condition for parameter convergence of this update law (and the others as well) relies on the following definition of *persistency of excitation*:

Definition 6.13 Persistency of Excitation. *A vector-valued signal* $z(\cdot) : \mathbb{R}_+ \mapsto \mathbb{R}^N$ *is said to be* persistently exciting *if there exists* $\delta > 0, \alpha_2 \geq \alpha_1 \geq 0$ *such that*

$$\alpha_1 I \leq \int_{t_0}^{t_0+T} w(\tau)w^T(\tau)d\tau \leq \alpha_2 I \quad \forall t_0. \tag{6.63}$$

A vector $z(\cdot)$ is said to be persistently exciting if it wobbles around enough in every window of length T so that the integral of the dyad associated with it (rank 1 for each instant of time t) is positive definite. Using this we have the following theorem:

Theorem 6.14 Parameter Convergence for the Gradient Algorithm. *Consider the identifier with gradient update resulting in the composite system (6.62). Now assume that the plant input u_P is bounded and the plant is either stable or in a feedback loop where the signals are all bounded. If $w(t)$ is persistently exciting. The system (6.62) is exponentially stable, i.e., the parameter error $\phi(t)$ converges to zero exponentially.*

Proof: The persistency of excitation assumption on $w(t)$ guarantees that the (hypothetical) linear system

$$\dot{\phi} = 0,$$
$$y(t) = w(t)\phi(t). \tag{6.64}$$

is uniformly completely observable, since its observability Gramian is precisely

$$\int_{t_0}^{t_0+\delta} w(\tau)w^T(\tau)d\tau.$$

In addition, since u_P, y_P are bounded by the assumptions on u_P and either the stability of the plant or the loop that it is located in, the uniform complete observability of (6.64) guarantees the uniform complete observability of (another hypothetical linear system (by Problem 5.13)):

$$\dot{\phi} = -g w(t) w^T(t),$$
$$y(t) = w(t)\phi(t). \tag{6.65}$$

Now consider the system of (6.62), which is the same as the first equation in the conceptual system (6.65) above, with the Lyapunov function $v(\phi) = \phi^T \phi$. Then, we have $\dot{v} = -g(w^T(t)\phi(t))^2$ so that we have

$$\int_{t_0}^{t_0+T} \dot{v}(\tau) d\tau = -\int_{t_0}^{t_0+T} g(w^T(\tau)\phi(\tau))^2 d\tau = -\int_{t_0}^{t_0+T} y^2(\tau) d\tau.$$

By the uniform complete observability of the conceptual system (6.65) it follows that for some $\alpha_3 > 0$,

$$\int_{t_0}^{t_0+T} \dot{v}(\tau) d\tau \leq -\alpha_3 |\phi(t_0)|^2.$$

By the modified exponential stability theorem of Problem 5.15 it follows that $\phi(t) \to 0$ as $t \to \infty$. $\qquad \Box$

As mentioned at the beginning of this section, the persistency of excitation condition on $w(\cdot)$ is also a sufficient condition for exponential parameter convergence of the normalized gradient update law (Problem 6.12). Persistency of excitation is also a sufficient condition for parameter convergence of the least-squares update law (6.57) and the normalized least-squares update law (6.60), though the convergence may not be exponential unless there is covariance resetting (see Problem 6.13).

Finally the methods of generalized harmonic analysis introduced in Chapter 4 can be used to prove that when the plant is stable and the input $u_P(\cdot)$ is stationary (cf. Definition 4.1), the regressor vector $w(\cdot)$ is persistently exciting, if and only if the input $u_p(\cdot)$ has at least as many sinusoids as there are unknown parameters, namely $2n$: that is the spectral measure or power density function of $u_P(\cdot)$, $S_{u_p}(\omega) \in \mathbb{R}_+$, has a support of at least $2n$ points (i.e., $S_{u_p}(\omega) \neq 0$ at at least $2n$ points). This statement, which is remarkable for its intuitive content that the system needs to be excited by a *sufficiently rich* input signal, was first proven in [38]. The reader is led through the details of the proof in Problem 6.14.

6.6   Averaging

Averaging is a method of approximating the dynamics of a (slowly) time-varying system by the dynamics of a time invariant system. More precisely, consider the system

$$\dot{x} = \epsilon f(x, t, \epsilon), \quad x(0) = x_0. \tag{6.66}$$

Here $\epsilon > 0$ is a small parameter which models the fact that the dynamics of x are slowly varying with respect to the time variation of the right hand side of (6.66).

The **averaged** system is described by

$$\dot{x}_{av} = \epsilon f_{av}(x_{av}), \quad x(0) = x_0, \tag{6.67}$$

where

$$f_{av}(x) := \lim_{T \to \infty} \frac{1}{T} \int_{t_0}^{t_0+T} f(x, \tau, 0)d\tau, \tag{6.68}$$

assuming that the limit exists. The method was proposed initially by Bogoliuboff and Mitropolsky [33], developed by several authors including Volosov and Sethna [271], and Hale [125], and stated in geometric form by Arnold [12] and Guckenheimer and Holmes [122]. Our treatment closely follows Chapter 4 of the book by Sastry and Bodson [259].

Definition 6.15 Mean Value of a Function, Convergence Function. *The function $f(x, t, 0)$ is said to have* mean value $f_{av}(x)$ *if there exists a continuous function* $\gamma(T) : \mathbb{R}_+ \mapsto \mathbb{R}_+$, *strictly decreasing and tending to zero as $T \to \infty$ such that*

$$\left| \frac{1}{T} \int_{t_0}^{t_0+T} f(x, \tau, 0)d\tau - f_{av}(x) \right| \le \gamma(T) \tag{6.69}$$

for all $t_0 \ge 0$, $T \ge 0$, $x \in B_h$. The function $\gamma(T)$ is called the convergence function.

The function $f(x, t, 0)$ has mean value $f_{av}(x)$ if and only if the function

$$d(x, t) = f(x, t, 0) - f_{av}(x)$$

has zero mean value. If $f(x, t, 0)$ is periodic, then $d(x, t)$ is periodic and integrable. If $d(x, t)$ is bounded in addition, $\gamma(T)$ in (6.69) is of the form a/T for some a. If $f(x, t, 0)$ is almost periodic in t (review the definition from Problem 4.1) and $d(x, t)$ is bounded, it may be verified (see [125]) that $d(x, t)$ has zero mean value. However, $\gamma(T)$ may not be of order $1/T$ (it could for example be of order $1/\sqrt{T}$).

We now address the following two issues:

1. The closeness of the trajectories of the unaveraged system (6.66) and the averaged system (6.68) on intervals of the form $[0, T/\epsilon]$.
2. The relationship between the stability properties of the two systems.

In order to prove these theorems we need a preliminary lemma:

Lemma 6.16 Smooth Approximate Integral of a Zero-Mean Function. *Assume that $d(x, t) : B_h \times \mathbb{R}_+ \mapsto \mathbb{R}^n$ is piecewise continuous with respect to t and has bounded and continuous first partial derivatives with respect to x, and assume also that $d(0, t) \equiv 0$. Further, assume that $d(x, t)$ has zero mean value, with convergence function $\gamma(T)|x|$, and $\partial d(x, t)/\partial x$ has zero mean value with convergence function $\gamma(T)$.*

Then there exists a function $w_\epsilon(x, t) : B_h \times \mathbb{R}_+ \mapsto \mathbb{R}^n$ and a function $\xi(\epsilon)$ of Class K such that[3]

$$|\epsilon w_\epsilon(x, t)| \le \xi(\epsilon)|x|,$$

$$\left| \frac{\partial w_\epsilon(x, t)}{\partial t} - d(x, t) \right| \le \xi(\epsilon)|x|,$$

(6.70)

$$\left| \frac{\partial w_\epsilon(x, t)}{\partial x} \right| \le \xi(\epsilon),$$

$$w_\epsilon(x, 0) = 0.$$

for all $t \in \mathbb{R}_+, x \in B_h$. Moreover, if $\gamma(T) = a/T^r$ for some $a \ge 0, r \in]0, 1]$, then $\xi(\epsilon) = 2a\epsilon^r$.

Remarks:

1. The main point of this lemma is that although the exact integral with respect to t of $d(x, t)$ may be an unbounded function of time, there exists a function $w_\epsilon(x, t)$ whose first partial derivative with respect to t is arbitrarily close to $d(x, t)$.

2. The preceding lemma may be proved with weaker hypotheses (in particular, without the hypotheses on $\partial w(x, t)/\partial x$ if no estimate on the derivative of the approximate integral $w_\epsilon(x, t)$ with respect to x is needed (see [259] for details). We choose the stronger hypotheses to be able to determine stability of the unaveraged system from the stability of the averaged system.

Proof: Define

$$w_\epsilon(x, t) := \int_0^t d(x, \tau)e^{-\epsilon(t-\tau)}d\tau.$$

Note that

$$w_0(x, t) := \int_0^t d(x, \tau)d\tau,$$

Integrating the expression for $w_\epsilon(x, t)$ by parts yields

$$w_\epsilon(x, t) = w_0(x, t) - \epsilon \int_0^t e^{-\epsilon(t-\tau)}w_0(x, \tau)d\tau.$$

Integrating by parts we get

$$\int_0^t e^{-\epsilon(t-\tau)}w_0(x, \tau)d\tau = w_0(x, t) - w_0(x, t)e^{-\epsilon t}$$

so that we have

$$w_\epsilon(x, t) = w_0(x, t)e^{-\epsilon t} - \epsilon \int_0^t e^{-\epsilon(t-\tau)}(w_0(x, t) - w_0(x, \tau))d\tau.$$

---

[3]Recall the definition of Class K functions from Section 5.3.

Using the convergence bounds (6.69) we get that

$$|w_\epsilon(x,t)| \leq \gamma(t)te^{-\epsilon t} + \epsilon \int_0^t e^{-\epsilon(t-\tau)}(t-\tau)\gamma(t-\tau)d\tau.$$

Multiplying both sides by ϵ and rescaling $s = \epsilon t$, we get

$$|\epsilon w_\epsilon(x,t)| \leq \sup_{s \geq 0} \gamma\left(\frac{s}{\epsilon}\right)se^{-s} + \int_0^\infty s\gamma\left(\frac{s}{\epsilon}\right)e^{-s}ds. \qquad (6.71)$$

Using the fact that if $d(x,t)$ is bounded say by β then $\gamma(\cdot)$ is also bounded by β a simple calculation yields[4]

$$|\epsilon w_\epsilon(x,t)| := \xi(\epsilon)$$

for some $\xi(\epsilon)$ of Class K. In the instance that $\gamma(T) = a/T^r$, equation (6.71) can be integrated explicitly to give

$$|\epsilon w_\epsilon(x,t)| \leq a\epsilon^r + a\epsilon^r \int_0^\infty \tau^{1-r}e^{-\tau}d\tau = a\epsilon^r + a\epsilon^r\Gamma(2-r),$$

where Γ is the standard gamma function. Since the $|\Gamma(2-r)| < 1$ when $r \in]0, 1]$, it follows that we may choose $\xi(\epsilon) = 2a\epsilon^r$. This gives the first estimate of (6.70). By construction, we may now check that

$$\frac{\partial w_\epsilon(x,t)}{\partial t} - d(x,t) = -\epsilon w_\epsilon(x,t),$$

so that the second estimate of (6.70) is verified. Using the hypotheses on

$$\frac{\partial d(x,t)}{\partial x}$$

yields the third estimate of (6.70). The last estimate follows simply because $d(x,t)$ is bounded. $\qquad \square$

We are ready to state the conditions for the basic averaging theorem. The assumptions are as follows, for some $h > 0$, $\epsilon_0 > 0$

A1. $x = 0$ is an equilibrium point of the unaveraged system, i.e., $f(0, t, 0) \equiv 0$. Actually, note that the conditions on f are only required to hold at $\epsilon = 0$. The function $f(x, t, \epsilon)$ is Lipschitz continuous in $x \in B_h$ uniformly in $t, \epsilon \in [0, \epsilon_0]$, i.e.,

$$|f(x_1, t, \epsilon) - f(x_2, t, \epsilon)| \leq l_1|x_1 - x_2|;$$

$f(x, t, \epsilon)$ is Lipschitz continuous in ϵ linearly in x, i.e.,

$$|f(x, t, \epsilon_1) - f(x, t, \epsilon_2)| \leq l_2|x||\epsilon_1 - \epsilon_2|.$$

A2. $f_{av}(0) = 0$, and $f_{av}(x)$ is Lipschitz continuous in $x \in B_h$.

---

[4] The interval $[0, \infty[$ needs to be broken into the intervals $[0, \sqrt{\epsilon}]$ and $]\sqrt{\epsilon}, \infty[$ to produce this estimate.

A3. The function $d(x, t) = f(x, t, 0) - f_{av}(x)$ satisfies the conditions of Lemma 6.16.

A4. $|x_0|$ is chosen small enough so that the trajectories of (6.66) and (6.67) lie in B_h for $t \in [0, T/\epsilon_0]$ (such a choice is possible because of the Lipschitz continuity assumptions).

Theorem 6.17 Basic Averaging Theorem. *If the original unaveraged system of (6.66) and the averaged system of (6.67) satisfy assumptions A1–A4 above, then there exist a function $\psi(\epsilon)$ of Class K, such that given $T > 0$ there exists $b_T > 0$, $\epsilon_T > 0$ such that for $t \in [0, T/\epsilon]$,*

$$|x(t) - x_{av}(t)| \leq \psi(\epsilon) b_T. \tag{6.72}$$

Proof: Consider the change of coordinates for (6.66) given by

$$x = z + \epsilon w_\epsilon(z, t)$$

where $w_\epsilon(z, t)$ is from Lemma 6.16. From the estimates of (6.70), it follows that there exists an $\epsilon_T > 0$ for which this transformation is a local diffeomorphism, i.e., that

$$I + \frac{\partial w_\epsilon(z, t)}{\partial z}$$

is invertible. Applying this transformation to 6.66 yields

$$\left[I + \epsilon \frac{\partial w(z, t)}{\partial z} \right] \dot{z} = -\frac{\partial w(z, t)}{\partial t} + \epsilon f(z + \epsilon w_\epsilon(z, t)), t, \epsilon). \tag{6.73}$$

The right-hand side of this equation may be written to group terms as follows:

$$\epsilon f_{av}(z) + \epsilon \left[f(z, t, 0) - f_a v(z) - \frac{\partial w_\epsilon}{\partial t} \right]$$

$$+ [f(z + \epsilon w_\epsilon, t, \epsilon) - f(z, t, \epsilon)]$$

$$+ \epsilon [f(z, t, \epsilon) - f(z, t, 0)].$$

Denoting this by $\epsilon f_{av}(z) + \epsilon \tilde{p}(z, t, \epsilon)$, it follows using Assumptions A1, A3 and Lemma 6.16 that

$$|\tilde{p}(z, t, \epsilon)| \leq \xi(\epsilon)|z| + \xi(\epsilon) l_1 |z| + \epsilon l_2 |z|.$$

Continuing the calculation of equation (6.73) yields

$$\dot{z} = \left[I + \epsilon \frac{\partial w_\epsilon(z, t)}{\partial z} \right]^{-1} \left[\epsilon f_{av}(z) + \epsilon \tilde{p}(z, t, \epsilon) \right]$$

$$= \epsilon f_{av}(z) + \epsilon p(z, t, \epsilon), \tag{6.74}$$

$$z(0) = x_0,$$

where

$$p(z, t, \epsilon) = \left[I + \epsilon \frac{\partial w_\epsilon(z, t)}{\partial z} \right]^{-1} \left[\tilde{p}(z, t, \epsilon) - \epsilon \frac{\partial w_\epsilon}{\partial z} f_{av}(z) \right],$$

so that we have the estimate

$$|p(z, t, \epsilon) \leq \frac{1}{1 - \xi(\epsilon_T)} \left[\xi(\epsilon) + \xi(\epsilon)l_1 + \epsilon l_2 + \xi(\epsilon)l_{av}\right] |z|$$

$$:= \psi(\epsilon)|z|. \tag{6.75}$$

Note that $\psi(\epsilon) \geq \xi(\epsilon)$. Now we compare (6.74) with the averaged system (6.67). We repeat the equations for convenience:

$$\dot{x}_{av} = f_{av}(x_{av}), \qquad\qquad x(0) = x_0,$$

$$\dot{z} = f_{av}(z) + \epsilon p(z, t, \epsilon), \quad z(0) = x_0.$$

Subtracting the two equations, we have

$$\dot{z} - \dot{x}_{av} = \epsilon \left[f_{av}(z) - f_{av}(x_{av})\right] + \epsilon p(z, t, \epsilon).$$

Taking norms and the estimate of $|p(z, t, \epsilon)|$, we have

$$|z(t) - x_{av}(t)| \leq \epsilon l_{av} \int_0^t |z(\tau) - x_{av}(\tau)d\tau + \epsilon \psi(\epsilon) \int_0^t |z(\tau)d\tau. \tag{6.76}$$

To use the bounds above we need $z(\cdot), x_{av}(\cdot) \in B_h$; this is guaranteed by A4 above by choosing ϵ small enough. Using the Bellman Gronwall lemma on equation (6.76) we get

$$|z(t) - x_{av}(t)| \leq \epsilon \psi(\epsilon) \int_0^t |z(\tau)|e^{\epsilon l_{av}(t-\tau)} d\tau. \tag{6.77}$$

Integrating this estimate with the bound $|z(t)| \leq h$, we get

$$|z(t) - x_{av}(t)| \leq \psi(\epsilon)h \frac{e^{\epsilon l_{av}T} - 1}{l_{av}} := \psi(\epsilon)a_T.$$

On the other hand, we have from the estimate on the transformation that

$$|x(\cdot) - z(\cdot)| \leq \psi(\epsilon)|z|.$$

Combining these two estimates, we get

$$|x(t) - x_{av}(t)| \leq |x(t) - z(t)| + |z(t) - x_{av}(t)|$$

$$\leq \psi(\epsilon)|x_{av}(t)| + (1 + \psi(\epsilon))|z(t) - x_{av}(t)|$$

$$\leq \psi(\epsilon)(h + (1 + \psi(\epsilon))a_T)$$

$$:= b_T.$$

Remarks: Theorem 6.17 establishes that the trajectories of the unaveraged system and averaged system remain close on time intervals of the form $[0, T/\epsilon]$ for ϵ small enough. How small they are ($O(\psi(\epsilon))$) is related to the convergence function $\gamma(t)$. It is important to remember that the theorem while applying to unbounded intervals as ϵ becomes small, *does not* state that the trajectories of the averaged and unaveraged system are close for *all time* $t \geq 0$. The relationship between the averaged and unaveraged system over *infinite time intervals* is investigated in the next theorem:

Theorem 6.18 Exponential Stability Theorems for Averaging. *Assume that the unaveraged system (6.66) and the averaged system (6.67) satisfy assumptions A1–A4 above. Further assume that f_{av} is a continuously differentiable vector field with a locally exponentially stable equilibrium point at 0. Then 0 is a locally exponentially stable equilibrium point of the unaveraged system (6.66).*

Proof: From the converse Lyapunov theorem for exponential stability (Theorem 5.17) it follows that there exists a converse Lyapunov function $v(x_{av})$ for the averaged system (6.67) satisfying

$$\alpha_1 |x_{av}|^2 \leq v(x_{av}) \leq \alpha_2 |x_{av}|^2,$$

$$\dot{v}(x_{av}) \leq -\epsilon \alpha_3 |x_{av}|^2, \tag{6.78}$$

$$\left| \frac{\partial v}{\partial x_{av}} \right| \leq \alpha_4 |x_{av}|.$$

Note the presence of the ϵ in $\dot{v}(x_{av})$ to account for the multiplication by ϵ in the right hand side of (6.67). We will now apply this Lyapunov function to study the stability of the unaveraged system transformed into the form of the second equation of (6.74), namely,

$$\dot{z} = \epsilon f_{av}(z) + \epsilon p(z, t, \epsilon). \tag{6.79}$$

Using the bounds on $p(z, t, \epsilon)$, we get that

$$\dot{v}|_{(6.79)}(z) = \dot{v}|_{(6.67)}(z) + \frac{\partial v}{\partial z} \epsilon p(z, t, \epsilon) \tag{6.80}$$

$$\leq -\epsilon \alpha_3 |z|^2 + \epsilon \alpha_4 \psi(\epsilon)|z|^2.$$

Define

$$\alpha(\epsilon) := \frac{\alpha_3 - \psi(\epsilon))\alpha_4}{\alpha_2}.$$

We may choose ϵ_1 such that $\alpha(\epsilon) > 0$ for $\epsilon \leq \epsilon_1$. Then, using the methods of Section 5.3.4, it may be verified that

$$|z(t)| \leq \sqrt{\frac{\alpha_2}{\alpha_1}} e^{-\epsilon \alpha(\epsilon)(t-t_0)} |z(t_0)|.$$

Using the estimate for $|x(t)|$ in terms of $|z(t)|$ from the Jacobian of the transformation

$$x = z + \epsilon w_\epsilon(z, t, \epsilon),$$

we get that

$$|x(t)| \leq \frac{1 + \xi(\epsilon)}{1 - \xi(\epsilon)} \sqrt{\frac{\alpha_2}{\alpha_1}} e^{-\epsilon \alpha(\epsilon)(t-t_0)} |x_0|. \tag{6.81}$$

$\square$

Remarks:

1. The preceding theorem is a local exponential stability theorem. It is easy to see that if the hypotheses of Lemma 6.16 are satisfied globally and the averaged system is exponentially stable, then the unaveraged system is also globally exponentially stable.
2. The proof of Theorem 6.18 gives useful bounds on the rate of convergence of the unaveraged system. The rates of convergence of the averaged and unaveraged systems are, respectively,

$$\lambda_{av} = \epsilon \frac{\alpha_3}{2\alpha_2}, \quad \lambda_\epsilon = \epsilon \frac{\alpha_3 - \psi(\epsilon)\alpha_4}{2\alpha_2}.$$

Since the averaged system is autonomous it is often easy to find Lyapunov functions for it and use them in turn to get estimates for the convergence of the unaveraged system (see Problem 6.16 for an example of this).
3. The conclusions of Theorem 6.18 are quite different from the conclusions of Theorem 6.17. Since both $x(t)$ and $x_{av}(t)$ go to zero exponentially, it follows that the error $x(t) - x_{av}(t)$ also goes to zero exponentially. Theorem 6.18 does not, however, relate this error bound to ϵ. However, it is possible to combine the two theorems to give a composite bound on the approximation error for all time $[0, \infty[$ of the form of equation (6.72).

6.7   Adaptive Control

In this section we give a brief introduction to the use of Lyapunov methods in adaptive control of linear systems. Adaptive control of nonlinear systems is discussed in the exercises of Chapter 9. Consider at first the first order single input single output linear time invariant plant with transfer function

$$\hat{P}(s) = \frac{k_p}{s + a_p},$$

where k_p, a_p are unknown. Let the input to the plant be called $u_p(\cdot)$ and the output $y_p(\cdot)$. We are also given a *reference model* $\hat{M}(s)$ with input $r(\cdot)$ and output $y_m(\cdot)$. We would like to design an adaptive controller for the plant, that is a prescription for designing $u_p(\cdot)$ such that $y_p(t)$ asymptotically tracks $y_m(\cdot)$, with all generated signals remaining bounded. The model is a single-input single-output linear time invariant system of the same order as the plant, with transfer function

$$\hat{M}(s) = \frac{k_m}{s + a_m},$$

where $k_m, a_m > 0$ are arbitrarily chosen by the designer. In the time domain we have the following equations:

$$\dot{y}_p = -a_p y_p + k_p u_p,$$
$$\dot{y}_m = -a_m y_m + k_m r. \tag{6.82}$$

Choose the control input to be of the form

$$u_p(t) = c_0(t)r(t) + d_0(t)y_p(t).$$

If the plant parameters k_p, a_p were known, we would choose

$$c_0^* = \frac{k_m}{k_p}, \quad d_0^* = \frac{a_p - a_m}{k_p}$$

such that the closed loop system for the first equation in (6.82) becomes identical to the second equation in (6.82). Denote the error variables

$$e_0 := y_p - y_m, \quad \phi_1 = c_0(t) - c_0^*, \quad \phi_2 = d_0(t) - d_0^*$$

Then, subtracting the second equation from the first yields

$$\dot{e}_0 = -a_m e_0 + k_p(\phi_1 e_0 + \phi_2 y_p). \tag{6.83}$$

Now, the rest of the adaptive control is to give the formula for the update of $c_0(t)$, $d_0(t)$ so as to guarantee the convergence of the output error e_0 to 0 as $t \to \infty$. In turn, we have that

$$\dot{c}_0 = \dot{\phi}_1, \quad \dot{d}_0 = \dot{\phi}_2.$$

Choosing as Lyapunov function

$$v(e_0, \phi) = e_0^2 + \phi_1^2 + \phi_2^2,$$

we see that with the choice of

$$\begin{aligned}
\dot{c}_0 = \dot{\phi}_1 &= -g e_0 r, \\
\dot{d}_0 = \dot{\phi}_2 &= -g e_0 y_p
\end{aligned} \tag{6.84}$$

and the sign of the gain g equal to that of k_p, we have that

$$\dot{v}(e_0, \phi) = -2|g k_p| e_0^2 \leq 0,$$

establishing that $e_0(\cdot)$, $\phi_1(\cdot)$, $\phi_2(\cdot)$ are bounded. In addition, if we assume that r is bounded, then y_m is bounded in turn, implying that $y_p = y_m + e_0$ is bounded. Now we may use Barbalat's lemma and the generalization of LaSalle's principle (Theorem 5.27) to see that $e_0 \to 0$ as $t \to \infty$.

This approach, called the *output-error-based adaptive control*, while simple and elegant, is not very straightforward to extend to the case when the relative degree of the plant is greater than 1. An alternative is a so-called input error approach. The key to this approach lies in the algebraic manipulation which establishes that

$$\hat{M}(s) = c_0^* \hat{P}(s) + d_0^* \hat{M}(s)\hat{P}(s).$$

This equality, in turn, implies that

$$\hat{M}(u_p) = c_0^* y_P + d_0^* \hat{M}(y_p)$$

and motivates the definition of the *input error*

$$\begin{aligned}
e_{in} &:= c_0(t)y_p(t) + d_0(t)\hat{M}(y_p) - \hat{M}(u_p) \\
&= \phi_1 y_p + \phi_2 \hat{M}(y_p).
\end{aligned} \tag{6.85}$$

Now note that the second error equation is of the same kind encountered in the context of identification in Section 6.5, and any one of the set of identification laws in this section, such as the gradient law or the least squares law or any of its variants, may be applied. For example, a gradient law is given by

$$\dot{c}_0 = \dot{\phi}_1 = -g\,e_{in}\,y_p,$$
$$\dot{d}_0 = \dot{\phi}_2 = -g\,e_{in}\,\hat{M}(y_p). \tag{6.86}$$

However, the proof of stability is now more involved, since after e_{in} and ϕ have been shown to be bounded, the proof that y_p is bounded requires a form of *"bounded-output bounded-input" stability* for the equation (6.85).

The two algorithms discussed here are *direct algorithms*, since they concentrate on driving the output error e_0 to zero. An alternate approach is the *indirect approach*, where any of the identification procedures discussed in Section 6.5 are used to provide estimates of the plant parameters k_p, a_p. These estimates are, in turn, used to compute the controller parameters. This is a very intuitive approach but may run into problems if the plant is unstable. The generalization of the preceding discussion of both input-error and output-error adaptive control of general linear systems and also multi-input multi-output linear systems is an interesting and detailed topic in itself and is the subject of the monograph by Sastry and Bodson [259]. In this monograph it is also shown how to apply the averaging theorems of the preceding section to analyze the stability and robustness of adaptive control schemes. Output error schemes were first proven by Morse and Feuer [101], Narendra, Valavani, and Annaswamy [228], and input error schemes were introduced by Goodwin and Mayne [116]. See also the discrete time treatment in Goodwin and Sin [117] and Astrom and Wittenmark [14].

6.8 Back-stepping Approach to Stabilization

Recently, there has been proposed an interesting new way for designing stabilizing controllers for nonlinear systems in so-called strict feedback form (see the book of Krstić, Kanellakopoulos, and Kokotović [171] for details):

$$\dot{x}_1 = x_2 + f_1(x_1),$$
$$\dot{x}_2 = x_3 + f_2(x_1, x_2),$$
$$\vdots$$
$$\dot{x}_i = x_{i+1} + f_i(x_1, x_2, \ldots, x_i), \tag{6.87}$$
$$\vdots$$
$$\dot{x}_n = f_n(x_1, x_2, \ldots, x_n) + u.$$

Note that the state equations for $\dot{x}_i$ depend only only on $x_1, x_2, \ldots, x_i$ and affinely on x_{i+1}. Conditions for converting general single input nonlinear systems into the form (6.87) are given in Chapter 9. The idea behind back-stepping is to consider the state x_2 as a sort of "pseudo-control" for x_1. Thus, if it were possible to make $x_2 = -x_1 - f_1(x_1)$, the x_1 state would be stabilized (as would be verified by the simplest Lyapunov function $v_1(x_1) = (1/2)x_1^2$. Since x_2 is not available, we define

$$z_1 = x_1,$$

$$z_2 = x_2 - \alpha_1(x_1),$$

with $\alpha_1(x_1) = -x_1 - f_1(x_1)$. Including a partial Lyapunov function $v_1(z_1) = (1/2)z_1^2$, we get that

$$\dot{z}_1 = -z_1 + z_2,$$

$$\dot{z}_2 = x_3 + f_2(x_1, x_2) - \frac{\partial \alpha_1}{\partial x_1}(x_2 + f_1(x_1)) := x_3 + \bar{f}_2(z_1, z_2),$$

$$\dot{v}_1 = -z_1^2 + z_1 z_2.$$

Proceeding recursively, define

$$z_3 = x_3 - \alpha_2(z_1, z_2),$$

$$v_2 = v_1 + \frac{1}{2}z_2^2.$$

To derive the formula for $\alpha_2(z_1, z_2)$, we verify that

$$\dot{z}_2 = z_3 + \alpha_2(z_1, z_2) + \bar{f}_2(z_1, z_2),$$

$$\dot{v}_2 = -z_1^2 + z_2(z_1 + z_3 + \alpha_2(z_1, z_2) + \bar{f}_2(z_1, z_2).$$

By choosing $\alpha_2(z_1, z_2) = -z_1 - z_2 - \bar{f}_2(z_1, z_2)$ we get that

$$\dot{z}_1 = -z_1 + z_2,$$

$$\dot{z}_2 = -z_1 - z_2 + z_3,$$

$$\dot{v}_2 = -z_1^2 - z_2^2 + z_2 z_3.$$

Proceeding recursively, we have at the i-th step that we define

$$z_{i+1} = x_{i+1} - \alpha_i(z_1, \ldots, z_i),$$

$$v_i = \frac{1}{2}(z_1^2 + z_2^2 + \cdots + z_i^2)$$

to get

$$\dot{z}_i = z_{i+1} + \alpha_i(z_1, \ldots, z_i) + \bar{f}_i(z_1, \ldots, z_i),$$

$$\dot{v}_i = -z_1^2 - \cdots - z_{i-1}^2 + z_{i-1}z_i + z_i(z_{i+1} + \alpha_i(z_1, \ldots, z_i) + \bar{f}_i(z_1, \ldots, z_i)).$$

Using $\alpha_i(z_1, \ldots, z_i) = -z_{i-1} - z_i - \bar{f}_i(z_1, \ldots, z_i)$, it follows that

$$\dot{z}_i = -z_{i_1} - z_i + z_{i+1},$$

$$\dot{v}_i = -z_1^2 - \cdots - z_i^2 + z_i z_{i+1}.$$

At the last step, we have that

$$\dot{z}_n = \bar{f}_n(z_1, \ldots, z_n) + u.$$

By defining

$$u = \alpha_n(z_1, \ldots, z_n) = -z_{n-1} - z_n - \bar{f}_n(z_1, \ldots, z_n)$$

and the composite Lyapunov function $v_n = \frac{1}{2}(z_1^2 + \cdots + z_n^2)$ we see that

$$\dot{z}_n = -z_{n-1} - z_n,$$
$$\dot{v}_n = -z_1^2 - z_2^2 - \cdots - z_n^2.$$

The foregoing construction calculates a diffeomorphism of the x coordinates into the z coordinates in which the stability is demonstrated on a simple quadratic Lyapunov function. Also, the dynamics are linear in the transformed z coordinates. However, one of the interesting flexibilities of the back-stepping method is that the calculation of the α_i is merely to produce negative definite $\dot{v}_i$ (up to terms in z_i) rather than to cancel all the nonlinearities the way we have described it thus far. This is referred to by Krstić, Kanellakopoulos, and Kokotović as *nonlinear damping*. Also, there is some latitude in the choice of the Lyapunov function v_i. This is illustrated in Problem 6.20. An adaptive version of back-stepping is discussed in Problem 6.21. A criticism that is sometimes made of back-stepping methods is that there is an explosion of the number of terms that follow as the iteration steps increase and there is a need to differentiate the initial functions many times (this was somewhat obscured by our definition of the functions $\bar{f}_i(z_1, z_2, \ldots, z_i)$). An alternative that has been proposed called dynamic surface control [292], is given in Chapter 9 as Problem 9.25.

6.9 Summary

In this chapter we have given the reader a snapshot of a number of very interesting and exciting areas of application of Lyapunov theory: stabilization of nonlinear systems, the Circle and Popov criteria, adaptive identification, averaging, singular perturbations. We gave a very brief introduction to a very large and interesting body of literature in adaptive control as well. For adaptive control there is a great deal more to be studied in terms of proofs of convergence (see, for example, [259]). In recent years there has been a resurgence of interest in new and streamlined proofs of stability using some unified concepts of detectability and stabilizability (see, for example, Morse [219], Ljung and Glad [186], and Hespanha and Morse [138]).

There is a very interesting new body of literature on the use of controlled Lyapunov functions; we gave the reader a sense of the kind of results that are obtained in Section 6.8. However, there is a lot more on tracking of nonlinear systems using Lyapunov based approaches, of which back-stepping is one constructive example. Of special interest is an approach to control of uncertain nonlinear systems developed by Corless and Leitmann [70]. For more details on this topic we refer to two excellent survey paper one by Teel and Praly [297], the other by Coron [71], and the recent books of Krstić, Kanellakopoulos, and Kokotović [171] and Freeman and Kokotović [106]. The use of passivity based techniques for stabilizing nonlinear systems has also attracted considerable interest in the literature: for more on this we refer the reader to the work of Ortega and co-workers [235; 236], Lozano, Brogliato, and Landau [189], and the recent book by Sepulchre, Janković, and Kokotović [269]. An elegant geometric point of view to the use of passivity based techniques in the controls of robots and exercise machines is in the work of Li and Horowitz [180; 181].

6.10   Exercises

Problem 6.1 Stabilizing using the linear quadratic regulator.   Consider the linear time-invariant control system

$$\dot{x} = Ax + Bu \tag{6.88}$$

with (A, B) completely controllable. Choose the feedback law

$$u = -R^{-1}B^T Px$$

with $P \in \mathbb{R}^{n \times n}$ being the unique positive definite solution of the steady state Riccati equation (with A, C completely observable)

$$PA + A^T P + C^T C - PBR^{-1}B^T P. \tag{6.89}$$

Prove that the closed loop system is exponentially stable using the Lyapunov function $x^T Px$. What does this imply for stabilization of nonlinear systems of the form

$$\dot{x} = f(x, u)?$$

Problem 6.2 Stabilization of linearly uncontrollable systems.   Consider a nonlinear control system

$$\dot{x} = f(x, u).$$

Show that if the linearization has some unstable modes that are not controllable, i.e., values of $\lambda \in \mathbb{C}^\circ_+$ at which the rank of the matrix $[\lambda I - A, B]$ is less than n, that no **nonlinear feedback** can stabilize the origin. Also show that if the linearization is **stabilizable**, i.e there are no eigenvalues $\lambda \in \mathbb{C}_+$ for which rank $[\lambda I - A, B]$ is less than n, then the nonlinear system is stabilizable by linear state feedback.

Finally, for the case in which there are some $j\omega$ axis eigenvalues that are not controllable give an example of a control system that is not stabilizable by linear feedback but is stabilizable by nonlinear feedback. Look ahead to the next chapter for hints on this part of the problem.

Problem 6.3 Time-varying stabilization, see [61], [322].   Consider a linear time varying system

$$\dot{x} = A(t)x(t) + B(t)u(t)$$

with $x \in \mathbb{R}^n, u \in \mathbb{R}^{n_i}$. Assume that the following Gramian is uniformly positive definite for some $\delta > 0$

$$W(t, t + \delta) = \int_t^{t+\delta} e^{4\alpha(t-\tau)} \Phi(t, \tau) B(\tau) B(\tau)^T \Phi(t, \tau)^T d\tau.$$

Here $\alpha > 0$. Define $P(t) := W(t, t + \delta)^{-1}$. Show that the feedback control law

$$u = -B^T(t)P(t)x$$

exponentially stabilizes the linear system. Give an estimate of the rate of convergence. Now apply the same feedback law to stabilize a control system

$$\dot{x} = f(x, u)$$

about a given nominal trajectory $u(\cdot), x^0(\cdot)$, that is, to drive the tracking error

$$x(\cdot) - x^0(\cdot)$$

to zero. Give the appropriate conditions on the linearization and rate of convergence of the nonlinear system to its nominal trajectory.

Problem 6.4 Observed based stabilization.   Consider the control system with $x \in \mathbb{R}^n, u \in \mathbb{R}^{n_i}, y \in \mathbb{R}^{n_o}$

$$\dot{x} = f(x, u), \tag{6.90}$$
$$y = g(x).$$

Assume that the *state feedback control law*

$$u = h(x)$$

locally exponentially stabilizes the closed-loop system. Assume further that the system is *exponentially detectable*, i.e., to say that there exists an "observer"

$$\dot{z} = r(z, u, y) \tag{6.91}$$

such that $r(0, 0, 0) = 0$ and there exists a function $W : \mathbb{R}^n \times \mathbb{R}^n \mapsto \mathbb{R}$ satisfying, for $x, z \in B_\rho, u \in B_\rho$,

- $\beta_1 |x - z|^2 \le W(x, z) \le \beta_2 |x - z|^2$,
- $D_1 W(x, z) f(x, u) + D_2 W(x, z) r(z, u, g(x)) \le -\beta_3 |x - z|^2$.

Show that the system of (6.90) with the control law

$$u = h(z)$$

results in the closed loop system having x, z converging to zero locally exponentially, provided that the observer is "fast enough," i.e., β_3 is large enough. You can assume that $f(\cdot, \cdot), h(\cdot)$ are Lipschitz continuous in their arguments.

Problem 6.5. Use the KYP Lemma to show that the zeros of an SPR transfer function lie in the open left half plane ($\mathbb{C}^\circ_-$).

Problem 6.6 Circle and Popov Criterion examples.   Using either the circle criterion or the Popov criterion determine ranges of sector-bounded nonlinearities in the feedback loop that can be tolerated by the following linear time invariant systems:

1. $\frac{1}{s(s+1)^2}$.

2. $\frac{1}{s(s+1)}$.

3. $\frac{1}{(s-1)(s+2)^2}$. Note that the open loop system is unstable here.

Problem 6.7 Multivariable circle and Popov criteria [162].   Consider the multi-input multi-output generalization of the Lur'e system

$$\dot{x} = Ax + Bu,$$
$$y = Cx + Du, \tag{6.92}$$
$$u = -\phi(y, t).$$

Here $x \in \mathbb{R}^n$, $u \in \mathbb{R}^p$, $y \in \mathbb{R}^p$, the matrices A, B, C, D are dimensioned compatibly and the nonlinearity $\phi(y, t) : \mathbb{R}^p \times \mathbb{R}_+ \mapsto \mathbb{R}^p$ is "diagonal," that is, ϕ_i is a function of y_i, t only and each of the ϕ_i satisfies a sector bound of the form $\alpha_i |y_i|^2 \leq y_i \phi(y_i, t) \leq \beta_i |y_i|^2$. If K_m (respectively K_M) $\in \mathbb{R}^{p \times p}$ is defined to be a diagonal matrix of all the lower bounds α_i (respectively β_i), then the bounds can be summarized as

$$[\phi(y, t) - K_m y]^T [\phi(y, t) - K_M y] \leq 0 \tag{6.93}$$

The multivariable generalization of the KYP lemma called the PR lemma [5] states that a transfer function $\hat{Z}(s) \in \mathbb{R}^{p \times p}$ with minimal realization $C(sI - A)^{-1}B + D$ is strictly positive real (that is, it is analytic in $\mathbb{C}$, satisfies $\hat{Z}^*(s) = \hat{Z}(s^*)$,[5] and for $s \in \mathbb{C}_+ > 0$, $Z^T(s^*) + Z(s)$ is positive definite), if and only if there exist P, Q symmetric positive definite matrices, $W \in \mathbb{R}^{p \times p}$, $L \in \mathbb{R}^{p \times n}$, satisfying

$$A^T P + PA = -L^T L - \epsilon Q,$$
$$PB - C^T = -L^T W, \tag{6.94}$$
$$D + D^T = W^T W.$$

Prove the *multivariable circle criterion*: The origin is exponentially stable for all nonlinearities (possibly time-varying) $\phi(\cdot, y)$ in the sector bound given by (6.93) if

---

[5]The notation $\hat{Z}^*$ refers to the complex conjugate transpose of the matrix, and the notation $\hat{Z}^T(s)$ stands for the ordinary transpose.

1. $\hat{G}(s) = \hat{H}(s)[I + K_m \hat{H}(s)]^{-1}$ is a stable transfer function (all poles in $\mathbb{C}_-$),

2. the transfer function $[I + K_M \hat{H}(s)][I + K_m \hat{H}(s)]^{-1}$ is strictly positive real,

where $\hat{H}(s) = C(sI - A)^{-1}B + D$.

Also prove the *multivariable Popov criterion* for the case that $\phi(y)$ is time invariant and $K_m = 0$: the origin is exponentially stable if there exists an $r > 0$ such that the transfer function $I + (1 + rs)K_M \hat{G}(s)$ is strictly positive real.

Problem 6.8 Further generalizations of the multivariable circle and Popov criteria. For the case that the nonlinearity of (6.92) is not diagonal but satisfies the bounds of (6.93) with K_m, K_M non diagonal but positive definite matrices give and prove generalizations of the circle and Popov criteria above. In particular pay close attention to how you define the Lyapunov function for the Popov criteria. *Hint: Do you know conditions under which a given nonlinear map $\phi : \mathbb{R}^p \times \mathbb{R}^p$ can be written as the gradient of a scalar function $W : \mathbb{R}^p \mapsto \mathbb{R}$?*

Problem 6.9 Brockett's necessary condition for asymptotic stability [47].

1. Consider the autonomous nonlinear differential equation

$$\dot{x} = f(x). \tag{6.95}$$

Let 0 be a locally asymptotically stable equilibrium point. Prove that $f : \mathbb{R}^n \mapsto \mathbb{R}^n$ maps a neighborhood of the origin onto a neighborhood of the origin. Do this by constructing a homotopy from $f(x)$ to $-x$, defined by

$$H(t, x) := \begin{cases} f(x), & t = 0, \\ -x, & t = 1, \\ \dfrac{1}{t}\left[\phi\left(\dfrac{t}{1-t}, x\right) - x\right], & 0 < t < 1, \end{cases}$$

where $\phi(s, x)$ is the flow of the differential equation (6.95) at time s starting from x at time 0. Verify that it is a homotopy, by checking the continuity with respect to t. Now use degree theory to derive the result. This method of proof is given in Sontag [279].

2. Now consider the control system

$$\dot{x} = f(x, u)$$

with $f(0, 0) = 0$. If $u = \alpha(x)$, use the previous result to give necessary conditions for smooth stabilizability of the control system. Is the following control system stabilizable by smooth feedback:

$$\dot{x}_1 = u_1,$$

$$\dot{x}_2 = u_2,$$

$$\dot{x}_3 = x_2 u_1?$$

Problem 6.10. For the identifier structure of this chapter, we chose Λ, b_λ to be in controllable canonical form so as to get w^1, w^2 to be *dirty derivatives* of the input and output. Verify that other choices of Λ, b_λ would also give unique parameterizations of the parameters a^*, b^*, for example,

$$(sI - \Lambda)^{-1}b_\lambda = \begin{bmatrix} \dfrac{1}{s+a} \\ \vdots \\ \dfrac{1}{(s+a)^n} \end{bmatrix}, \quad a > 0,$$

or

$$(sI - \Lambda)^{-1}b_\lambda = \begin{bmatrix} \dfrac{b_1}{s+a_1} \\ \vdots \\ \dfrac{b_n}{s+a_n} \end{bmatrix}, \quad a_i \neq a_j > 0, \quad b_i \neq 0.$$

More specifically, show that in each instance there are unique vectors $a^*, b^* \in \mathbb{R}^n$ such that

$$y_P = (a^*)^T (sI - \Lambda)^{-1}b_\lambda u_P + (b^*)^T (sI - \Lambda)^{-1}b_\lambda y_P,$$

provided that the plant numerator and denominator are coprime.

Problem 6.11. Consider $\phi(t) = [\cos(\log(1+t)), \sin(\log(1+t))]^T \in \mathbb{R}^2$. Show that it satisfies the conclusions of Theorem 6.12 on $d\phi/dt, \phi$ but do not converge. Can you contrive an example plant $\hat{P}$ and input $u_P(\cdot)$ for which this is in fact the parameter error vector?

Problem 6.12 Parameter convergence for the normalized gradient update law. Prove that if the input u_P is bounded and either the plant is stable or is embedded in a feedback loop so as to stablize the output, and the regressor $w(t)$ is *persistently exciting*, then the parameter error converges to zero exponentially. How do you expect the rate of convergence of the normalized gradient update law to compare with the rate of convergence of the gradient update law.

Problem 6.13 Parameter convergence for the least squares update law.   Consider first the least squares update law of (6.57) with $Q = 0$ applied to the identifier of (6.54). Assume that the input u_P is bounded and either the plant is stable or is embedded in a feedback loop so as to stablize the output, and the regressor $w(t)$ is *persistently exciting*. Now consider the Lyapunov function

$$v(\phi, t) = \phi^T P^{-1}(t)\phi.$$

For the system

$$\begin{aligned} \dot{\phi} &= -gP(t)w(t)w^T(t)\phi \\ \dot{P} &= -gPw(t)w^T(t)P \end{aligned} \tag{6.96}$$

it follows that $v(\phi, t)$ is a p.d.f. but is not decrescent when w is persistently exciting (verify this!). Now, show that

$$\dot{v}|_{(6.96)} = -g(w^T(t)\phi(t))^2 \le 0.$$

Use the existence of

$$\lim_{t \to \infty} v(\phi, t) \ge 0$$

and the persistency of excitation condition to prove that $\phi(t) \to 0$ as $t \to \infty$. Verify that the convergence is not, in general, exponential but may be as slow as $1/t$! Now show that if there is covariance resetting (using the same methods as in Theorem 6.11), then the parameter convergence is exponential. Repeat the analysis for the normalized least squares algorithm of (6.60).

Problem 6.14 Persistency of excitation and sufficient richness [38].   Assume that the input u_P to the identifier of a linear time invariant system is *stationary*. That is

$$R_{u_P}(\tau) := \lim_{T \to \infty} \int_t^{t+T} u_P(s)u_P(s + \tau)ds$$

exists uniformly in t. Prove that if the plant is stable, then the regressor $w(\cdot)$ is also stationary. You may wish to use the linear filter lemma 4.3 to prove this and establish the connection between the power spectral density of u_p and w.

Now, show that the persistency of excitation condition for a stationary regressor w is equivalent to the statement that its autocovariance at 0, $R_w(0) \in \mathbb{R}^{2n \times 2n}$, is positive definite. Use the inverse Fourier transform formula

$$R_w(0) = \int_{-\infty}^{\infty} S_w(\omega)d\omega$$

and the formula for $S_w(\omega)$ that you have derived above to prove that w is persistently exciting if and only if $S_{u_P}(\omega) \neq 0$ at at least $2n$ points (i.e., that the input signal is *sufficiently rich*). You will need to use the coprimeness of the numerator and denominator polynomials of the plant transfer function in an essential way.

Problem 6.15 Identification of nonlinear systems [158].   Consider a nonlinear multiple input ($u \in \mathbb{R}^r$) system depending linearly on unknown parameters $\theta \in \mathbb{R}^p$ given by

$$\dot{x} = \sum_{i=1}^{p} \left(f_i(x)\theta_i^* + \sum_{j=1}^{r} g_{ij}(x)\theta_i^* u_j \right). \tag{6.97}$$

For notational convenience this system may be written in the form

$$\dot{x} = w(x, u)\theta^*, \tag{6.98}$$

where the regressor is given by

$$w^T(x, u) = [f_1(x) + g_1(x)u, \ldots, f_p(x) + g_p(x)u].$$

Consider the following observer based on *the use of the full state x* given by

$$\dot{\hat{x}} = A(\hat{x} - x) + w^T(x, u)\hat{\theta},$$

$$\dot{\hat{\theta}} = -w(x, u)P(\hat{x} - x). \tag{6.99}$$

Here $A \in \mathbb{R}^{n \times n}$ has eigenvalues in $\mathbb{C}^{\circ}_{-}$, and $P \in \mathbb{R}^{n \times n}$ is chosen to satisfy the Lyapunov equation for some symmetric $Q \in \mathbb{R}^{n \times n}$,

$$A^T P + P A = -Q.$$

Give conditions under which this nonlinear observer cum identifier (first proposed in [172], [168]) has parameter convergence as well as $\hat{x} \to x$ as $t \to \infty$. Apply this to a model of the induction motor [192] with $x = [i_d, i_q, \phi_d, \phi_q]^T \in \mathbb{R}^4$, where i_d, ϕ_d stand for the direct axis stator current and flux, and i_q, ϕ_q for the quadrature axis stator current and flux.

$$\alpha = \frac{R_s}{\sigma L_s}, \quad \beta = \frac{R_r}{\sigma L_r}, \quad \sigma = 1 - \frac{M^2}{L_s L_r},$$

with R_s, L_s standing for stator resistance and inductance and R_r, L_r for rotor resistance and inductance. M stands for the mutual inductance.

$$f(x) = \begin{bmatrix} -(\alpha + \beta) & 0 & \dfrac{\beta}{L_s} & \dfrac{\omega}{\sigma L_s} \\ 0 & -(\alpha + \beta) & -\dfrac{\omega}{\sigma L_s} & \dfrac{\beta}{L_s} \\ -\alpha\sigma L_s & 0 & 0 & \omega \\ 0 & -\alpha\sigma L_s & -\omega & 0 \end{bmatrix} x,$$

$$g_1(x) = \begin{bmatrix} \dfrac{1}{\sigma L_s} \\ 0 \\ 1 \\ 0 \end{bmatrix}, \quad g_2(x) = \begin{bmatrix} 0 \\ \dfrac{1}{\sigma L_s} \\ 0 \\ 1 \end{bmatrix}, \quad g_3(x) = \begin{bmatrix} x_2 \\ -x_1 \\ x_4 \\ -x_3 \end{bmatrix}.$$

Assume that α, β are unknown parameters. Simulate the system with parameters $\alpha = 27.23, \beta = 17.697, \sigma = 0.064, L_s = 0.179$. How many input sinusoids did you need to guarantee parameter convergence.

Problem 6.16 Averaging applied to adaptive identification (Chapter 4 of [259]). Consider the adaptive identifier of (6.54) with the gradient update law of (6.55). Now assume that the update is slow, i.e., $g = \epsilon$ is small. Assume that the plant is stable, u_P is stationary (recall the definition 4.1 from Chapter 4). Prove that this implies that the regressor $w(t)$ is also stationary. Find the averaged version of the gradient update law. Express the convergence in terms of the eigenvalues of the autocovariance of $w(\cdot)$.

Repeat this exercise for the normalized gradient update law (6.56), and the least-squares (6.57) with $Q = 0$, and normalized least-squares update laws (6.60).

Compare simulations of the unaveraged and averaged systems for simple choices of the plant and the other parameters of the identifier. You may be surprised at how close they are for even moderately large values of ϵ.

Problem 6.17 Discrete time averaging [15].   Develop a theory of averaging for discrete time nonlinear systems with a slow update, that is, for the case that

$$x_{k+1} = x_k + \lambda f(x_k, k)$$

with λ small. Give conditions under which the evolution of the system may be approximated by a differential equation.

Problem 6.18 Circuit dynamics.   Sometimes resistors need not be either voltage or current controlled. Consider the series interconnection of a 1 F capacitor with a resistor whose i-v characteristic is as shown in Figure 6.17, characterized by a function $\psi(i, v) = 0$. Discuss how you would obtain a solution concept for the circuit equations for this simple circuit, that is.

$$\dot{v} = i,$$

$$0 = \psi(i, v).$$

Problem 6.19.   Consider the swing equation for the power system with PV loads of (6.43). Prove that the fast frozen system is a gradient system. Characterize the stable and unstable equilibria of this system in the limit $\epsilon \to 0$.

Problem 6.20.   Stabilize the nonlinear system

$$\dot{x}_1 = x_2 - x_1^3,$$

$$\dot{x}_2 = u.$$

Use back-stepping but note that it may not be necessary to cancel the term x_1^3.

Problem 6.21 Adaptive back-stepping.   Consider the back-stepping controller for systems in strict feedback form as described in Section 6.8. Assume that the nonlinearities $f_i(x_1, x_2, \ldots, x_i)$ may be expressed as $w_i^T(x_1, x_2, \ldots, x_i)\theta$, where

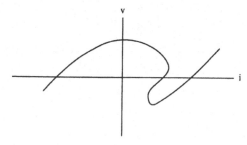

FIGURE 6.17. The $i - v$ characteristic of a non-voltage or non-current controlled resistor

$w_i(x_1, x_2, \ldots, x_i)$, $\theta^* \in \mathbb{R}^p$. Here the functions w_i are known, and the parameters $\theta^* \in \mathbb{R}^p$ are unknown. In the back-stepping controller replace θ^* by its estimate $\theta(t)$ and derive the equations for the z_i in terms of the parameter errors $\phi = \theta - \theta^*$. Give a parameter update law for the estimates $\dot{\theta}$ to guarantee adaptive stabilization. Use the usual quadratic Lyapunov function.

7

Dynamical Systems and Bifurcations

7.1 Qualitative Theory

In this section we develop the rudiments of the modern theory of dynamical systems. To give the reader a bird's eye view of this exciting field, we will for the most part skip the proofs. Some results are identical in $\mathbb{R}^n$ to those stated in $\mathbb{R}^2$ in Chapter 2, but the center manifold theorem and bifurcation calculations are new in this chapter. We will also use machinery that we have built up enroute to this chapter to sharpen statements made in Chapter 2. Readers who wish to consult more detailed books are referred to Guckenheimer and Holmes [122], Ruelle [247], or Wiggins [329]. Consider a standard autonomous, nonlinear time-invariant differential equation on $\mathbb{R}^n$,

$$\dot{x} = f(x), \quad x \in \mathbb{R}^n, \quad x(0) = x_0 \tag{7.1}$$

with $f : \mathbb{R}^n \mapsto \mathbb{R}^n$ smooth, with associated flow $\phi_t : \mathbb{R}^n \mapsto \mathbb{R}^n$ defined by $\phi(t, x_0) = x(t, x_0)$. By standard existence and uniqueness theorems the flow ϕ_t is well defined and a diffeomorphism at least for $t \in \ ]-c, c[$. Actually, we will assume that the flow is globally well defined, i.e., ϕ_t is well defined for all t. The set $\{\phi_t(x) : t \in \mathbb{R}_+\}$ is referred to as a *positive orbit* of (7.1). It is worth remembering that this usage means that we think of $\phi_t(\cdot)$ not as a one-parameter family of diffeomorphisms indexed by t.

Proposition 7.1. *If f is C^r then ϕ_t is a C^r diffeomorphism for each t.*

The zeros of f are the fixed points of the flow ϕ_t. They are also referred to as *equilibria* or *stationary solutions*. Let $\bar{x}$ be one such equilibrium and define the linearization of (7.1) as

$$\dot{z} = Df(\hat{x})z. \tag{7.2}$$

Proposition 7.2. *The linearization of the flow of (7.1) at $\bar{x}$ is the flow of the linearization of (7.2), i.e.,*

$$D\phi_t(\bar{x})z = e^{tDf(\bar{x})}z.$$

Proof: See exercises.

A fundamental question regarding the flow of (7.1) is that of what we can say about it based on the flow of (7.2). This is resolved by the *Hartman Grobman theorem*. The proof of this theorem is a little too technical and detracts from the flow of the chapter, hence we omit it (see, however, [128]). The theorem is no different on $\mathbb{R}^n$ than it was in $\mathbb{R}^2$ in Chapter 2.

Theorem 7.3 Hartman Grobman Theorem.   *Consider the system of (7.1) with the equilibrium point $\bar{x}$. If $Df(\bar{x})$ has no zero or purely imaginary eigenvalues, there is a homeomorphism h defined on a neighborhood U of $\bar{x}$ taking orbits of the flow ϕ_t to those of the linear flow $e^{tDf(\bar{x})}$ of (7.2). The homeomorphism preserves the sense of the orbits and is chosen to preserve parameterization by time.*

Remarks:

1. The homeomorphism h cannot be made a diffeomorphism (i.e., a *differentiable homeomorphism*) unless some further non-resonance conditions[1] on the eigenvalues of $Df(\bar{x})$ hold.
2. When $\bar{x}$ has no $j\omega$ axis eigenvalues, $\bar{x}$ is called a *hyperbolic* or *non-degenerate* fixed point, and the asymptotic behavior of solutions near it and its consequent stability type are determined by the linearization. When any one of the eigenvalues has zero real part, stability cannot be determined by linearization.
3. We first encountered the Hartman–Grobman theorem in Chapter 2, when we were studying planar dynamical systems. The version of the theorem stated then for planar systems is exactly identical to the one stated here for systems of arbitrary dimension.

Example 7.4. *Consider the system described by the second order equation*

$$\ddot{x} + \epsilon x^2 \dot{x} + x = 0.$$

The linearization has eigenvalues $\pm j$ for all ϵ. However, the origin is attracting for $\epsilon > 0$ and repelling for $\epsilon < 0$. (Prove this using the Lyapunov function $v(x, \dot{x}) = x^2 + \dot{x}^2$.)

The next result builds on the theme of the Hartman–Grobman theorem. We need the definitions of *local insets* and *local outsets* of an equilibrium point $\bar{x}$:

---

[1]Roughly speaking, that no integer linear combinations of the eigenvalues sum to zero.

Definition 7.5 Local Insets and Outsets.   *Consider an equilibrium point $\bar{x}$ of (7.1). Let $U \subset \mathbb{R}^n$ be a neighborhood of x. We define the* local inset *of $\bar{x}$ to be the set of initial conditions converging to $\bar{x}$ as follows:*

$$W_{loc}^s(\bar{x}) = \left\{ x \in U : \lim_{t \to \infty} \phi_t(x) \to \bar{x} \; \text{ and } \; \phi_t(x) \in U \; \forall t \geq 0 \right\}.$$

The local outset *is similarly defined as*

$$W_{loc}^u(\bar{x}) = \left\{ x \in U : \lim_{t \to -\infty} \phi_t(x) \to \bar{x} \; \text{ and } \; \phi_t(x) \in U \; \forall t \leq 0 \right\}.$$

Remark: The insets and outsets of an equilibrium point may not be manifolds when the equilibrium point is not hyperbolic. Indeed, consider the system

$$\dot{x}_1 = -x_1,$$
$$\dot{x}_2 = x_2^2.$$

The insets and outsets of this system are as shown in Figure (7.1). Note that the inset is a closed half space (not a manifold) and the outset a closed half line, the positive x_2 axis (also not a manifold).

Yet another example pictured in Figure 7.1 is the system

$$\dot{x}_1 = x_1^2 - x_2^2,$$
$$\dot{x}_2 = 2x_1 x_2.$$

Here the inset is the plane minus the positive x axis, but including the origin (not a manifold) and the outset is the positive x axis, also including the origin (also, not a manifold).

In the instance that the equilibrium point $\bar{x}$ is hyperbolic, the following theorem establishes that they are manifolds. First define $E^s \subset \mathbb{R}^n$, $E^u \subset \mathbb{R}^n$ to be the stable and unstable eigenspaces of the linear system (7.2), that is, E^s, E^u are the spans of the eigenvectors and generalized eigenvectors corresponding to the eigenvalues in $\mathbb{C}_-^\circ$, $\mathbb{C}_+^\circ$ respectively. Further, let their dimensions be n_s, n_u respectively.

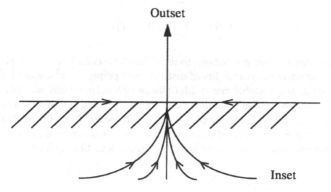

FIGURE 7.1. Insets and outsets are not manifolds

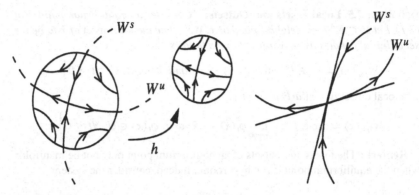

FIGURE 7.2. Visualizing the Hartman–Grobman and stable–unstable manifold theorems

Theorem 7.6 Stable–Unstable Manifold Theorem. *Let $\bar{x}$ be a* hyperbolic *equilibrium point of the system (7.1). Then the inset* $W^s_{loc}(\bar{x})$ *and the outset* $W^u_{loc}(\bar{x})$ *are manifolds referred to as the* local stable *and* local unstable manifolds *of $\bar{x}$ respectively. They are of the same dimensions as* E^s, E^u, *namely* n_s, n_u, *respectively. Further they are tangent to* E^s, E^u, *respectively, at $\bar{x}$.*

Remarks:

1. The Hartman–Grobman theorem and the stable–unstable manifold theorem are depicted in Figure 7.2.
2. The manifolds W^s_{loc}, W^u_{loc} are both *invariant* (think about why this may be true). They are, indeed, the nonlinear analogues of the stable and unstable eigenspaces, E^s, E^u, of the linearization.

The local invariant manifolds $W^s_{loc}(\bar{x})$, $W^u_{loc}(\bar{x})$ have global analogues $W^s(\bar{x})$, $W^u(\bar{x})$ which may be defined as follows

$$W^s(\bar{x}) = \bigcup_{t \leq 0} \phi_t(W^s_{loc}(\bar{x})),$$

$$W^u(\bar{x}) = \bigcup_{t \geq 0} \phi_t(W^u_{loc}(\bar{x})).$$

The existence of unique solutions to the differential equation (7.1) implies that two stable (or unstable) manifolds of distinct fixed points $\bar{x}^1$, $\bar{x}^2$ cannot intersect. However, stable and unstable manifolds of the same fixed point can intersect and are in fact a source of a lot of the complex behavior encountered in dynamical systems. It is, in general, difficult to compute W^s, W^u in closed form for all but some textbook examples; for others it is an experimental procedure on the computer. At any rate, here is an example where we can do the calculation by hand

$$\dot{x}_1 = x_1,$$
$$\dot{x}_2 = -x_2 + x_1^2.$$

The linear eigenspaces are respectively

$$E^s = \begin{pmatrix} 0 \\ 1 \end{pmatrix}, \quad E^u = \begin{pmatrix} 1 \\ 0 \end{pmatrix}.$$

Eliminating t from the differential equation yields

$$\frac{dx_2}{dx_1} = -\frac{x_2}{x_1} + x, \quad x_1 \neq 0,$$

which can be integrated to give the integral curves

$$x_2(x_1) = \frac{x_1^2}{3} + \frac{c}{x_1},$$

where c is a constant determined by initial conditions. By Hartman–Grobman $W_{loc}^u(0)$ is representable by a graph $y = h(x)$ with $h(0) = 0$. Further, by the stable–unstable manifold theorem $Dh(0) = 0$, since $W_{loc}^u(0)$ is tangent to E^u. Consequently, the constant c above has to be set to be zero to yield the (global) unstable manifold to be the set

$$W^u(0) = \left\{ \left(x_1, \frac{x_1^2}{3} \right) ; x \in \mathbb{R} \right\}.$$

Finally, note that if $x_1(0) = 0$, then $x_1(t) \equiv 0$, so that $W^s(0) = E^s$. Nonlinear systems possess limit sets other than fixed points, for example limit cycles. It is easy to study their stability by first studying nonlinear maps.

7.2 Nonlinear Maps

In this section we study the time invariant discrete time dynamical system

$$x_{n+1} = G(x_n). \tag{7.3}$$

Here $G : \mathbb{R}^n \mapsto \mathbb{R}^n$ is a smooth diffeomorphism. The reason for choosing G to be a diffeomorphism is so that we can consider flows in both forward and reverse time.

Example 7.7. *Consider the time-1 map of the differential equation (7.1). Note that*

$$x_{n+1} = \psi_1(x_n)$$

is a discrete time dynamical system of the form (7.3).

A point $\bar{x}$ is a *fixed point* of the map G iff

$$G(\bar{x}) = \bar{x}.$$

Define the linearization of the system (7.3) about a fixed point $\bar{x}$ by

$$z_{n+1} = DG(\bar{x})z_n. \tag{7.4}$$

Let us pause to study orbits of linear maps analogous to (7.4):

$$z_{n+1} = Bz_n.$$

We define the stable and unstable eigenspaces for B having no eigenvalues of modulus 1 as follows:

$$E^s = \text{Span}\{n_s \text{ generalized eigenvectors of modulus} < 1\},$$
$$E^u = \text{Span}\{n_u \text{ generalized eigenvectors of modulus} > 1\}.$$

If there are no multiple eigenvalues then the rates of contraction (expansion) in $E^s (E^u)$ are bounded by geometric series, i.e., $\exists c > 0, \alpha < 1$ such that

$$
\begin{aligned}
|x_n| \le c\alpha^n |x_0| \quad &\text{if} \quad x_0 \in E^s, \\
|x_{-n}| \le c\alpha^n |x_0| \quad &\text{if} \quad x_0 \in E^u.
\end{aligned}
\tag{7.5}
$$

If $DG(\bar{x})$ has no eigenvalues of unit modulus, its eigenvalues determine the stability of the fixed point $\bar{x}$ of G. The linearization theorem of Hartman–Grobman and the stable–unstable manifold theorem results apply to maps just as they do for flows.

Theorem 7.8 Hartman–Grobman Theorem for Maps.   *Let G be a diffeomorphism with hyperbolic fixed point $\bar{x}$ (i.e., no eigenvalues of $DG(\bar{x})$ have modulus 1). Then there exists a homeomorphism h defined on a neighborhood U of $\bar{x}$ such that $h(G(z)) = DG(\bar{x})h(z)$.*

It is important to remember that flows and maps differ crucially in that, while the *orbit* or *trajectory* of a flow $\phi_t(x)$ is a curve in $\mathbb{R}^n$, the orbit $G^n(x)$ is a sequence of points. (The notation $G^n(x)$ refers to the n-th iterate of x under G, that is, $G^n = G \circ \cdots \circ G(x)$ n times.) This is illustrated in Figure 7.3.

The insets and outsets of a fixed point are defined as before:

Definition 7.9 Local Insets and Outsets.   *Consider an equilibrium point $\bar{x}$ of (7.1). Let $U \subset \mathbb{R}^n$ be a neighborhood of x. We define the* local inset *of $\bar{x}$ to be the set of initial conditions converging to $\bar{x}$ as follows:*

$$W^s_{loc}(\bar{x}) = \left\{ x \in U : \lim_{n \to \infty} G^n(x) \to \bar{x} \text{ and } G^n(x) \in U \ \forall n \ge 0 \right\}$$

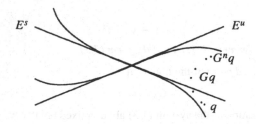

FIGURE 7.3. An orbit of a discrete time system with a saddle fixed point

and the local outset *to be*

$$W_{loc}^u(\bar{x}) = \left\{ x \in U : \lim_{n \to -\infty} G^n(x) \to \bar{x} \text{ and } G^n(x) \in U \ \forall n \leq 0 \right\}$$

As before, the inset and outset are not necessarily manifolds unless the fixed point is hyperbolic.

Theorem 7.10 Local Stable–Unstable Manifolds for Maps. *Let* $G : \mathbb{R}^n \mapsto \mathbb{R}^n$ *be a diffeomorphism with a hyperbolic fixed point* $\bar{x}$. *Then there are stable and unstable manifolds* $W_{loc}^s(\bar{x})$, $W_{loc}^u(\bar{x})$ *tangent to the eigenspaces* E^s, E^u *of* $DG(\bar{x})$ *and of corresponding dimension.*

Remark: As for flows, the global stable and unstable manifolds $W^s(\bar{x})$, $W^u(\bar{x})$ are defined by

$$W^s(\bar{x}) = \bigcup_{n \leq 0} G^n(W_{loc}^s(\bar{x})),$$

$$W^u(\bar{x}) = \bigcup_{n \geq 0} G^n(W_{loc}^u(\bar{x})).$$

Definition 7.11 Period-k Points and Orbits. *If there is a cycle of k distinct points* $p_j = G^j(p_0)$ *for* $j = 1, \ldots, k - 1$ *such that* $G^k(p_0) = p_0$, *then* p_0 *is referred to as a* period-k *point (as are* $p_1, \ldots, p_{j-1}$ *as well). The set of k points* $p_0, \ldots, p_{k-1}$ *is referred to as a* period-k *orbit.*

Note that a period-k point is a fixed point of $G^k(x)$. The stability of such a period-k point p_0 is determined by the linearization $DG^k(p_0) \in \mathbb{R}^{n \times n}$. By the chain rule,

$$DG^k(p_0) = DG(G^{k-1}(p_0)) \cdots DG(G(p_0))DG(p_0).$$

An obvious question to ask is about the relationship between $DG^k(p_0)$ and $DG^k(p_j)$ for $j = 1, \ldots, k - 1$.

Proposition 7.12. *If* $p_0, \ldots, p_{k-1}$ *constitute an orbit of period-k, then the set of eigenvalues of each of the k matrices* $DG^k(p_0), DG^k(p_1), \ldots, DG^k(p_{k-1})$ *is the same.*

Proof: See exercises.

As a consequence of this proposition we see that the stability of any point p_j on a periodic orbit determines the stability of the entire orbit. In general, the stability type of the fixed point $\bar{x}$ is determined by the magnitude of the eigenvalues of $DG(\bar{x})$. If $|\lambda_j| < 1$ for all eigenvalues, then we have a sink; if $|\lambda_j| > 1$ for some eigenvalues and $|\lambda_j| < 1$ for the others: a saddle point and $|\lambda_j| > 1$ for all eigenvalues a source. If an even number of eigenvalues have negative real part the map $DG(\bar{x})$ is *orientation preserving*.

7.3   Closed Orbits, Poincaré Maps, and Forced Oscillations

The construction of the Poincaré map is an extremely useful technique to convert the study of the flow in the vicinity of a closed orbit or a periodic solution of a forced system into the study of a certain map known as the Poincaré map. Roughly speaking, the Poincaré map corresponds to the process of periodically strobing a phase portrait. The details follow:

7.3.1   The Poincaré Map and Closed Orbits

Let γ be a periodic orbit of period T of some flow $\phi_t(x)$ arising from a vector field of the form of (7.1). Let $\Sigma \subset \mathbb{R}^n$ be a hypersurface of dimension $n-1$ transverse to the flow at a point $p \in \gamma$, i.e., if $n(x)$ is the normal to Σ at x, i.e., $f^T(p)n(p) \neq 0$. Let U be a neighborhood of p on Σ. If γ has multiple intersections with Σ, shrink Σ till there is only one intersection. The *first return* or *Poincaré* map $P : U \mapsto \Sigma$ is defined for $q \in U$ by

$$P(q) = \phi_\tau(q),$$

where $\tau(q)$ is the time taken for the orbit $\phi_t(q)$ based at q to first return to Σ. The map is well defined for U sufficiently small as depicted in figure (7.4). Note that in general, $\tau(q) \neq T$ but $\lim_{q \to p} \tau(q) = T$. (For a proof that the Poincaré map is well defined see Problem 7.4.)

Now note that p is a fixed point of the Poincaré map $P : U \mapsto U$. It is not difficult to see that the stability of p for the discrete-time dynamical system or map P reflects the stability of γ for the flow ϕ_t. In particular if p is hyperbolic and $DP(p)$, the linearized map has n_s eigenvalues of modulus less than 1 and n_u eigenvalues of modulus greater than 1 ($n_s + n_u = n - 1$), then the dimension of $W^s(p)$ is n_s and the dimension of $W^u(p)$ is n_u. If we define the local insets and

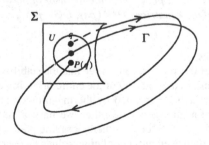

FIGURE 7.4. The Poincare or first return map

outsets of γ in $\bar{U}$ as

$$W_{loc}^s(\gamma) = \left\{ x \in \bar{U} : \lim_{t \to \infty} |\phi_t(x) - \gamma| = 0, \quad \phi_t(x) \in \bar{U}, \ t \geq 0 \right\},$$

$$W_{loc}^u(\gamma) = \left\{ x \in \bar{U} : \lim_{t \to -\infty} |\phi_t(x) - \gamma| = 0, \quad \phi_t(x) \in \bar{U}, \ t \leq 0 \right\}.$$

then it is intuitive that if p is a hyperbolic point, then W_{loc}^s and W_{loc}^u are both manifolds. Furthermore

$$W_{loc}^s(p) = W_{loc}^s(\gamma) \cap \Sigma,$$
$$W_{loc}^u(p) = W_{loc}^u(\gamma) \cap \Sigma.$$

This is depicted in Figure 7.5. The formal proof of this is somewhat involved but is in-effect that of the Hartman–Grobman theorem. In fact the Hartman Grobman theorem can be generalized to *arbitrary invariant manifolds* (rather than merely equilibrium points and closed orbits) and under a hyperbolicity assumption shown to result in stable and unstable manifolds (this is a celebrated result of Hirsch, Pugh, and Shub [140]).

Example 7.13. *We illustrate the Poincaré map on a planar dynamical system*

$$\dot{x}_1 = x_1 - x_2 - x_1(x_1^2 + x_2^2)$$
$$\dot{x}_2 = x_1 + x_2 - x_2(x_1^2 + x_2^2)$$

which in polar coordinates with $r \in \mathbb{R}_+, \theta \in S^1$ is

$$\dot{r} = r(1 - r^2),$$
$$\dot{\theta} = 1.$$

Choose as section $\Sigma = \{(r, \theta) \in \mathbb{R}_+ \times S^1 : r > 0, \theta = 0\}$. The equations above can be explicitly integrated to give the flow

$$\phi_t(r, \theta) = \left(\left(1 + \left(\frac{1}{r^2} - 1 \right) e^{-2t} \right)^{-1/2}, \ \theta + t \right)$$

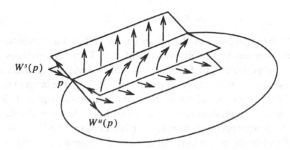

$W^s(p)$

p

$W^u(p)$

FIGURE 7.5. Stable and unstable manifolds of a limit cycle

Thus

$$P(r) = \left(1 + \left(\frac{1}{r^2} - 1 \right) e^{-4\pi} \right)^{-1/2}$$

with the first return time $\tau = 2\pi$ *for all* r. *Note that* $r = 1$ *is the location of the limit cycle and that* $DP(1) = e^{-4\pi} < 1$, *so that* γ, *a circular limit cycle of radius 1, is an attracting closed orbit.*

It is not, in general, easy to compute the Poincaré map, since it involves integrating the differential equation and computing the first return time experimentally. However, the eigenvalues of the linearization of the Poincaré map may be computed by linearizing the flow about the limit cycle γ. This is referred to as the *Floquet technique.* The linearization of (7.1) about the flow of γ is

$$\dot{\xi} = Df(\gamma(t))\xi. \tag{7.6}$$

Since $A(t) := Df(\gamma(t))$ is a T periodic matrix the equation (7.6) is a periodic linear system with the following property for its state transition map.

Proposition 7.14. *The state transition matrix of (7.6) may be written as*

$$\Phi(t, 0) = K(t)e^{Bt}. \tag{7.7}$$

where $K(t) = K(t + T) \in \mathbb{R}^{n \times n}$ *and* $K(0) = I$, *and* $B = \frac{1}{T} \log \Phi(T, 0)$.

Proof: Since $\Phi(t, 0)$ is the state transition matrix of a linear time varying system it is non singular for all t. Using the definition of B, define

$$K(t) := \Phi(t, 0)e^{-Bt}.$$

Now equation (7.7) follows from the definitions. To establish the periodicity of $K(\cdot)$ note that

$$\begin{aligned}
K(t + T) &= \Phi(t + T, 0)e^{-B(t+T)} \\
&= \Phi(t + T, T)\Phi(T, 0)e^{-B(t+T)} \\
&= \Phi(t + T, T)e^{-Bt} \\
&= \Phi(t, 0)e^{-Bt} \\
&= K(t). \qquad \square
\end{aligned}$$

It follows from the proposition that the behavior of the solutions in the neighborhood of γ is determined by the eigenvalues of e^{TB}. These eigenvalues $\lambda_1, \ldots, \lambda_n$ are called the *characteristic,* or *Floquet multipliers,* and the eigenvalues of B are referred to as the characteristic exponents of γ. If $v \in \mathbb{R}^n$ is the tangent to $\gamma(0)$, then v is the eigenvector corresponding to the characteristic multiplier 1. The moduli of the remaining $n - 1$ eigenvalues, if none are unity, determine the stability of γ. Choosing the basis appropriately to make the last column of $e^{BT} = (0, \ldots, 0, 1)^T$ the matrix $DP(p)$ of the linearized Poincaré map is simply the matrix belonging to $\mathbb{R}^{(n-1) \times (n-1)}$ obtained by deleting the n-th row and column of e^{TB}.

7.3.2   The Poincaré Map and Forced Oscillations

In this section we consider the dynamics of T-periodic forcing

$$\dot{x} = f(x, t), \quad f(x, t + T) = f(x, t) \tag{7.8}$$

The system (7.8) can be rewritten as an autonomous system

$$\begin{aligned} \dot{x} &= f(x, \theta), \\ \dot{\theta} &= 1 \quad (x, \theta) \in \mathbb{R} \times S^1. \end{aligned} \tag{7.9}$$

Here $\theta \in S^1$ reflects the T-periodicity of equation (7.9) with $S^1 \equiv \mathbb{R} \pmod{T}$. To study forced oscillations in (7.9) we define the section

$$\Sigma = \left\{ (x, \theta) \in \mathbb{R} \times S^1 : \theta = 0 \right\}.$$

All solutions of (7.9) cross Σ transversally, since $\dot{\theta} = 1$. The Poincaré map $P : \Sigma \mapsto \Sigma$ is given by

$$P(x) = \Pi \cdot \phi_t(x, 0)$$

where $\phi_t : \mathbb{R}^n \times S^1 \mapsto \mathbb{R}^n \times S^1$ is the flow of (7.8) and Π denotes projection onto the first (x) factor. Note that the time of flight is T for all x. This is shown in Figure 7.6. A fixed point p of P corresponds to a periodic orbit of period T for the flow. It is not immediate that every forced periodic system has a periodic orbit of period T. The following proposition is useful in understanding when certain kinds of periodic orbits appear:

Proposition 7.15 Closed Orbit Under Small Periodic Forcing. *Let $\bar{x}$ be a hyperbolic equilibrium point of an unforced system*

$$\dot{x} = f(x).$$

Then the forced periodic system

$$\dot{x} = f(x) + \epsilon g(x, t)$$

with $g(x, t + T) \equiv g(x, t)$ has a periodic orbit of period T for ϵ small enough. Moreover, the stability characteristics of the closed orbit are those of the equilibrium point $\bar{x}$. Thus for example, if $Df(\bar{x})$ has k eigenvalues in $\mathbb{C}^\circ_-$, then the

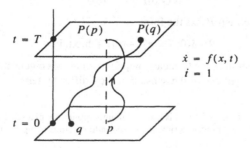

FIGURE 7.6. Construction of the Poincare map for forced systems

linearization of the Poincaré map of the periodic orbit, $DP(\bar{x})$, has k eigenvalues inside the unit disk.

Proof: We give only a formal proof: A more rigorous proof is not very involved; however, it needs an application of the implicit function theorem in function spaces. For the formal proof let us postulate that the solution which we are looking for is of order ϵ. If a periodic solution of the form $\bar{x} + \epsilon \gamma(t) +$ h. o. t. were to exist then using this in the right hand side of the equation above using the Taylor series expansion

$$f(\bar{x} + \epsilon \gamma(t)) = \epsilon Df(\bar{x})\gamma(t) + (h.o.t.)$$

and equating terms of order ϵ yields[2] the following linear equation for $\gamma(t)$

$$\dot{\gamma}(t) = Df(\bar{x})\gamma(t) + g(\bar{x}, t). \tag{7.10}$$

It remains to be shown that there is a periodic solution to equation (7.10). Using the notation $A := Df(\bar{x})$ and using the fact that a periodic solution satisfies $\gamma(0) = \gamma(T)$ yields

$$(I - e^{AT})\gamma(0) = \int_0^T e^{A(T-\tau)}g(\bar{x}, \tau)d\tau.$$

Since the equilibrium point $\bar{x}$ is hyperbolic, A has no purely imaginary eigenvalues. Consequently, none of the eigenvalues of e^{AT} is 1, and we may solve the equation for $\gamma(0)$ as

$$\gamma(0) = (I - e^{AT})^{-1} \int_0^T e^{A(T-\tau)}g(\bar{x}, \tau)d\tau.$$

To derive the Poincaré map choose initial conditions close to $\epsilon \gamma(0)$ at $t = 0$, say of the form $\epsilon(\gamma(0) + \delta(0))$. For this initial condition the solution at time T is found up to terms of higher order in ϵ by using equation (7.10) to be

$$\epsilon \left(e^{AT}(\gamma(0) + \delta(0)) + \int_0^T e^{A(T-\tau)}g(\bar{x}, \tau)d\tau \right).$$

By choice of $\gamma(0)$ this is equal, in turn, to

$$\epsilon(\gamma(0) + e^{AT}\delta(0)).$$

Thus the Poincaré map P has the form

$$P(\epsilon\delta(0)) = e^{AT}\epsilon\delta(0) + \text{h.o.t. in } \epsilon$$

Thus the linearization of the Poincaré map for ϵ small has eigenvalues close to the eigenvalues of e^{AT} (close in the sense that they differ by terms of order at most

---

[2]This is the part of the proof that is formal, though apparently plausible, since it involves equating terms of an asymptotic series. It can be made rigorous using the implicit function theorem.

ϵ). Thus the number of eigenvalues of the linearization of the Poincaré map inside the unit disk is equal to the number of eigenvalues of A in $\mathbb{C}^{\circ}_{-}$. The statement of the proposition thus follows.   □

The preceding result is a small perturbation result. The requirement that A be hyperbolic was only needed to guarantee that the matrix $I - e^{AT}$ was invertible. It is of interest to explore periodic forcing of linear systems with complex eigenvalues. This is a subject of a great deal of interest under the heading of *Forced Oscillation*. Of special interest is the scenario when a multiple of the forcing frequency (i.e., $(2\pi)/T$) is close to an imaginary eigenvalue of the system. To explore this scenario, we will need the following definition:

Definition 7.16 Subharmonic Oscillation.   *A period-k point of P (a point p such that $P^j(p) \neq p$ for $1 \leq j \leq k - 1$ and $P^k(p) = p$) corresponds to a subharmonic orbit of period kT.*

Figure 7.7 shows a forced system with a period-2 point. Let us pursue this point in some detail in the context of the forced oscillation of a planar system as in Problem 7.5 (the reader may wish to work his way through the problem now). We consider the case of $\delta = 0$ in equation (7.72). Our treatment follows [329] closely. As is pointed out in the problem, there is no difficulty with computing the Poincaré map explicitly in this linear case. The Poincaré map is well defined so long as $\omega \neq \omega_0$, that is, the frequency of forcing is different from the frequency of oscillation. In this case it should not be surprising that the solution breaks down into a superposition of solutions of frequency ω and ω_0. We will explore this case

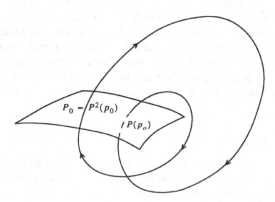

FIGURE 7.7. Showing a subharmonic of period 2T in a forced system

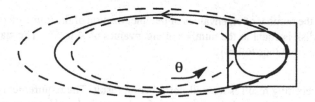

FIGURE 7.8. The Poincare map for a linear forced system

now. First verify from the Problem 7.5 that the Poincaré map is given by

$$P : \begin{pmatrix} x \\ y \end{pmatrix} \mapsto \begin{pmatrix} \cos\left(2\pi\dfrac{\omega_0}{\omega}\right) & \dfrac{1}{\omega_0}\sin\left(2\pi\dfrac{\omega_0}{\omega}\right) \\ -\omega_0\sin\left(2\pi\dfrac{\omega_0}{\omega}\right) & \cos\left(2\pi\dfrac{\omega_0}{\omega}\right) \end{pmatrix} \begin{pmatrix} x \\ y \end{pmatrix}$$

$$+ \begin{pmatrix} A\left(1 - \cos\left(2\pi\dfrac{\omega_0}{\omega}\right)\right) \\ \omega_0 A \sin\left(2\pi\dfrac{\omega_0}{\omega}\right) \end{pmatrix}.$$

(7.11)

where $A = (\epsilon)/(\omega_0^2 - \omega^2)$. A simple calculation shows that $(A, 0)^T$ is a fixed point of the Poincaré map. We will redraw the specialization of Figure 7.6 for our example as Figure 7.8. The periodic solution corresponding to the fixed point mentioned above is

$$x(t) = A\cos(\omega t),$$
$$y(t) = -A\omega\sin(\omega t).$$

In the figure, we have shown the projection of this orbit on the x-y plane at $t = 0$ as well as the rotation of this orbit through $(2\pi)/(\omega)$ seconds. The projection of the orbit on the x-y plane is a circle so that the rotation of the orbit through 2π seconds is a torus. A quick calculation with the Poincaré map of (7.11) reveals that the torus is invariant. As we saw in Chapter 3 a convenient way to draw the torus is as a rectangle with opposite edges identified. The orbit on the torus as shown in Figure 7.9 is a line joining diagonal edges. The two axes are labeled θ for the rotation in the x-y plane and ϕ for the revolution in the t direction. Both of these angles are normalized to 2π. We now discuss subtleties of the trajectories on this invariant torus:

Subharmonics of order m

Suppose that $\omega = m\omega_0$ for some $m > 1$, an integer. From an inspection of the Poincaré map of (7.11) it is clear that for $(x, y) \neq (A, 0)$ all points are period m points. Thus for m rotations in the θ direction we have one revolution in the ϕ direction. This trajectory is shown on the torus flattened out into a square in Figure 7.10.

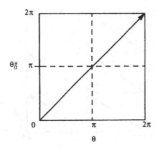

FIGURE 7.9. Trajectory of the forced oscillation on a flattened torus

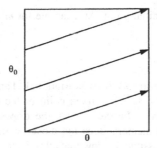

FIGURE 7.10. A subharmonic of order 3 on a flattened torus

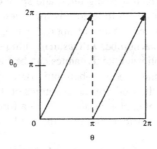

FIGURE 7.11. An ultraharmonic of order 2 on a flattened torus

Ultraharmonic of order n

Suppose that $n\omega = \omega_0$ for some $n > 1$. Now verify from equation (7.11) that every point is a fixed point of the Poincaré map. It is interesting to see that all trajectories complete n revolutions in the ϕ variable before they complete one rotation in the θ variable. An example for $n = 2$ is shown in Figure 7.11.

Ultrasubharmonic of order m, n

Suppose that $m\omega = n\omega_0$ for some relatively prime integers $m, n > 1$. Then all points except $(A, 0)^T$ are period m points with m rotations along with n revolutions

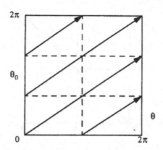

FIGURE 7.12. A subultraharmonic of order 2, 3

before returning to their initial point. An example for $m = 2, n = 3$ is shown in Figure 7.12.

Quasiperiodic response

Finally assume that the ratio $(\omega)/(\omega_0)$ is irrational. This corresponds to a case where only the point $(A < 0)^T$ is a fixed point of the Poincaré map. None of the other points are periodic. Further, when one draws the trajectories on the torus, one sees that any trajectory does not close on itself, it appears to wind around indefinitely. It is possible to show that all trajectories wind around *densely* on the surface of the torus. (This is an interesting exercise in analysis.) Such trajectories are called irrational winding lines, and since each trajectory is dense on the torus, the limit set of each trajectory is the whole torus. The trajectories are called quasiperiodic, since the act of approximating the irrational number arbitrarily closely by a rational number will result in a ultrasubharmonic oscillation.

We have seen how complicated the dynamics can be for a forced *linear system* when there is resonance and sub/ultra–harmonic periodic orbits. This is more generally true. Not only is the study of the dynamics of these systems subtle and complex, but the presence of subharmonic orbits in a forced nonlinear system is often a symptom that parametric variation in the model will cause complicated dynamics.

Since the definition of the Poincaré map for a forced system of the form of (7.8) relies on a knowledge of the flow of the forced system, it is not easy to compute unless the differential equation is easy to integrate. However, perturbation and averaging methods have often been used to approximate the map in several cases.

7.4   Structural Stability

We will try to make precise a "robustness" concept in dynamical systems: Under certain circumstances qualitative behavior of a dynamical system is unaffected by perturbation. Terms like closeness, small perturbations, genericity, and the like are often used loosely and to mean different things by different people. In what follows we will give a precise definition and then explain its limitations. The first notion to quantify is the closeness of two maps. Given a C^r map $F : \mathbb{R}^n \mapsto \mathbb{R}^n$; we say that a C^r map G is a C^k ϵ *perturbation of* F if there exists a compact set K such that $F(x) = G(x)$ for $x \in \mathbb{R}^n - K$ and for all $i_1, \ldots, i_n$ with $i_1 + \cdots + i_n = r \leq k$ we have

$$\left| \frac{\partial}{\partial x_1^{i_1} \ldots \partial x_n^{i_n}} (F - G) \right| < \epsilon.$$

Intuitively, G is a C^k ϵ perturbation of F if it and all its first k derivatives are ϵ close. We then have the following definition:

Definition 7.17 C^k Equivalence. *Two C^r maps F and G are C^k equivalent for $(k \leq r)$ if there exists a C^r diffeomorphism h such that*

$$h \circ F = G \circ h.$$

C^0 *equivalence is referred to as* topological equivalence.

The preceding definition implies that h takes orbits of F, namely $\{F^n(x)\}$; onto orbits of G, namely $\{G^n(x)\}$. For vector fields we define orbit equivalence as follows:

Definition 7.18 Orbit Equivalence. *Two C^r vector fields f and g are C^k orbit equivalent if there is a C^k diffeomorphism h that takes orbits $\phi_t^f(x)$ of f onto orbits $\phi_t^g(x)$ of g preserving sense but not necessarily parameterization by time. Thus we have that given t_1 there exists t_2 such that*

$$h(\phi_{t_1}^f(x)) = \phi_{t_2}^g(h(x)).$$

Also, C^0 *orbit equivalence is referred to as* topological orbit equivalence.

We may now define a robustness concept for both vector fields and flows, referred to as *structural stability*.

Definition 7.19 Structural Stability. *A C^r map $F : \mathbb{R}^n \mapsto \mathbb{R}^n$ (respectively, a C^r vector field f) is said to be* structurally stable *if there exists $\epsilon > 0$ such that all $C^1\epsilon$ perturbations of F (respectively f) are topologically (resp. topologically orbitally) equivalent to F (resp. f).*

At first sight it might appear that the use of topological equivalence is weak, and we may be tempted to use C^k equivalence with $k > 0$. It turns out that C^k equivalence is too strong: In particular, it would force the linearizations of the map (vector field) and its perturbation about their fixed (equilibrium) points to have exactly the same eigenvalue ratio. For example, the systems

$$\begin{pmatrix} \dot{x}_1 \\ \dot{x}_2 \end{pmatrix} = \begin{pmatrix} 1 & 0 \\ 0 & 1 \end{pmatrix} \begin{pmatrix} x_1 \\ x_2 \end{pmatrix} \quad \text{and} \quad \begin{pmatrix} \dot{x}_1 \\ \dot{x}_2 \end{pmatrix} = \begin{pmatrix} 1 & 0 \\ 0 & 1 + \epsilon \end{pmatrix} \begin{pmatrix} x_1 \\ x_2 \end{pmatrix}$$

are not C^k orbit equivalent for any $k \neq 1$, since $x_2 = c_1 x_1$ is not diffeomorphic to $x_2 = c_2 |x_1|^{1+\epsilon}$ at the origin. Note, however, that C^0 equivalence does not distinguish between *nodes, improper nodes and foci*, for example

$$\begin{bmatrix} -1 & 0 \\ 0 & -2 \end{bmatrix}, \quad \begin{bmatrix} -1 & 1 \\ 0 & -1 \end{bmatrix}, \quad \text{and} \quad \begin{bmatrix} -3 & -1 \\ 1 & -3 \end{bmatrix}$$

all have flows which are C^0 equivalent to the node

$$\begin{bmatrix} -1 & 0 \\ 0 & -1 \end{bmatrix}.$$

By an easy modification of the Hartman–Grobman theorem it is possible to show that the linear flow

$$\dot{x} = Ax$$

is structurally stable if A has no eigenvalues on the $j\omega$ axis. The map

$$x_{n+1} = Bx_n$$

is structurally stable if B has no eigenvalues on the unit disk. Thus, it follows that a map (or vector field) possessing even one non-hyperbolic fixed point (equilibrium point) cannot be structurally stable, since it can either be lost under perturbation or be turned into a hyperbolic sink, saddle, or source (see the section on bifurcations for an elaboration of this point). *Thus, a necessary condition for structural stability of maps and flows is that all fixed points and closed orbits be hyperbolic.*

As promised at the start of this section, we will discuss the limitations of the definition of structural stability using as an example a linear harmonic oscillator

$$\dot{x}_1 = -x_2,$$
$$\dot{x}_2 = x_1.$$

Elementary considerations yield that trajectories other than the origin are closed circular orbits. Let us now add small C^∞ perturbations to this nominal system:

Linear Dissipation. Consider the perturbed system

$$\dot{x}_1 = -x_2,$$
$$\dot{x}_2 = x_1 + \epsilon x_2.$$

It is easy to see that for $\epsilon \neq 0$ the origin is a hyperbolic equilibrium point that is stable for $\epsilon < 0$ and unstable for $\epsilon > 0$. The trajectories of this are not equivalent to those of the oscillator, but perhaps this is not surprising since the unperturbed system did not have a hyperbolic equilibrium point. However, if the perturbed system were Hamiltonian, then the presence of the dissipation term would not be an allowable perturbation.

Nonlinear Perturbation. Consider the perturbed system

$$\dot{x}_1 = -x_2,$$
$$\dot{x}_2 = x_1 + \epsilon x_1^2.$$

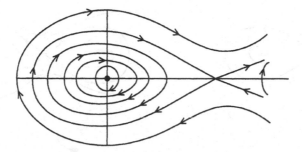

FIGURE 7.13. Phase portrait of a nonlinearly perturbed linear oscillator

This system now has two equilibria at $(0, 0)$ and at $(-\frac{1}{\epsilon}, 0)$. The origin is still a center, and the other equilibrium point is a saddle, which is far away for ϵ small. The perturbed system is still Hamiltonian with first integral given by $H(x, y) = x_2^2/2 + x_1^2 - \epsilon x_1^3/3$. The phase portrait of this system as shown in Figure 7.13 has a continuum of limit cycles around the origin merging into a saddle connection for the saddle at $(-\frac{1}{\epsilon}, 0)$. The point to this example is to show that though ϵ is small for large values of x_1, the perturbation is substantial and the phase portrait changes substantially.

The conclusions to be drawn from the preceding examples are the following:

1. It is important to specify the type of perturbations that are permissible in a discussion of generic properties of a family of systems. The structural stability definitions make no such distinction and allow for arbitrary perturbations. In applications one frequently considers perturbations of Hamiltonian systems or Volterra–Lotka equations where the perturbations are indeed structured. These scenarios are not covered by the definitions of structural stability.
2. In discussing small perturbations of vector fields in systems with unbounded state space, it is important to notice that perturbations which appear to be small for small values of the state variable are in fact large for large values of the state variable.
3. Small time-varying perturbations and periodic forcing of differential equations are not covered in the definitions. Indeed, they may cause tremendous changes in the nature of the phase portrait.

7.5 Structurally Stable Two Dimensional Flows

Nonwandering points are points whose neighborhood the flow visits infinitely often. More precisely, we have the following definition:

Definition 7.20 Nonwandering Points. *For either maps or flows, a point $x \in \mathbb{R}^n$ is said to be a nonwandering point if given any neighborhood U of x there exists a sequence of times $t_n \to \infty$ such that $\phi(t_n, x) \in U$ for all n.*

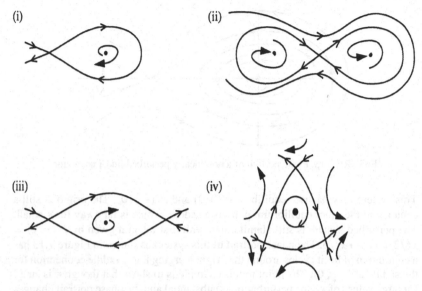

FIGURE 7.14. (i) A homoclinic orbit or saddle loop, (ii) a double saddle loop, (iii) a homoclinic cycle involving two saddles, and (iv) a homoclinic cycle involving three saddles.

It was shown by Andronov [122] that nonwandering points in $\mathbb{R}^2$ may be classified as:

1. Fixed points.
2. Closed orbits.
3. Unions of fixed points (saddles) and trajectories connecting them.

Sets of the last kind in the foregoing list are called *heteroclinic orbits* when they connect distinct saddles and *homoclinic orbits* when they connect a saddle to itself. Closed paths made up of heteroclinic orbits are called homoclinic cycles. Figure 7.14 shows four different kinds of homoclinic cycles. Dream up differential equations which generate these phase portraits (see Exercises). Since saddle connections are formed by the coincidence of stable and unstable manifolds of saddles, it is intuitive that they are not structurally stable. An important theorem of Peixoto characterizes structural stability in $\mathbb{R}^2$. In its original form it was stated for flows on compact two-dimensional manifolds. To apply it to $\mathbb{R}^2$ we assume that the flow points inwards on the boundary of a compact disk $D \subset \mathbb{R}^2$.

Theorem 7.21 Peixoto's Theorem.   *A vector field (or map) on $\mathbb{R}^2$ whose flow is eventually confined to a compact disk $D \subset \mathbb{R}^2$ is structurally stable iff*

1. *The number of fixed points and closed orbits is finite and they are hyperbolic.*
2. *There are no orbits connecting saddle points.*

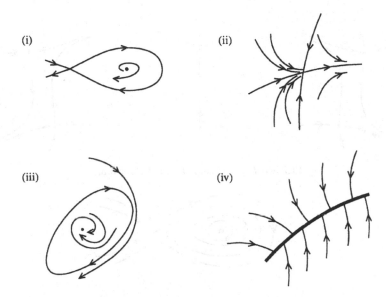

FIGURE 7.15. Structurally unstable flows in the plane: (i) a saddle loop, (ii) a nonhyperbolic fixed point, (iii) a nonhyperbolic limit cycle, and (iv) a continuum of equilibria.

Remark: By Andronov's and Bendixson's theorem we see that structurally stable flows in $\mathbb{R}^2$ have only fixed points and periodic orbits. Figure 7.15 visualizes some structurally unstable flows in $\mathbb{R}^2$.

Peixoto's theorem states that saddle connections are not structurally stable. We give three examples of how they break up under perturbation:

Example 7.22 Breakup of a Heteroclinic Orbit. *Consider the system*

$$\dot{x}_1 = \mu + x_1^2 - x_1 x_2,$$
$$\dot{x}_2 = x_2^2 - x_1^2 - 1. \tag{7.12}$$

When $\mu = 0$, the system has two saddles at $(0, \pm 1)$ and a heteroclinic orbit connecting the two saddles (see Figure 7.16). For $\mu \neq 0$ the saddles are perturbed to $(\pm \mu, \pm 1) + O(\mu^2)$ and the heteroclinic orbit disappears. It is useful to see that the system has an odd symmetry, that is, that the equations are unchanged if x_1, x_2 are replaced by $-x_1, -x_2$. Notice, however, the marked difference in the behavior of trajectories of initial conditions starting between the separatrices of the two saddles for $\mu < 0$ and $\mu > 0$.

Example 7.23 Appearance and Disappearance of Periodic Orbits from Homoclinic Connections. *Consider the system*

$$\dot{x}_1 = x_2,$$
$$\dot{x}_2 = -x_1 + x_1^2 + \mu x_2. \tag{7.13}$$

FIGURE 7.16. Breakup of a heteroclinic orbit

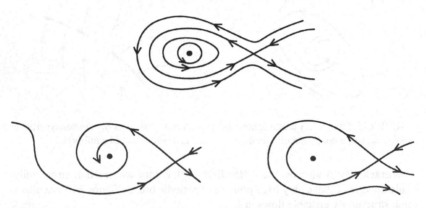

FIGURE 7.17. Appearance and disappearance of periodic orbits and a homoclinic connection

The phase portraits of Figure 7.17 show two equilibria at $(0, 0)$ and $(1, 0)$. For $\mu < 0$ the equilibrium at $(0, 1)$ is a stable focus and for $\mu < 0$ an unstable focus. At $\mu = 0$ the system is Hamiltonian, and the origin is a center surrounded by a continuum of limit cycles bounded by a homoclinic orbit of the saddle $(0, 0)$.

Example 7.24 Emergence of a Limit Cycle from a Homoclinic Orbit. *Figure 7.18 shows an example of a finite period limit cycle emerging from a saddle connection. An example of this was shown in Chapter 2 when a finite period limit cycle was born from a homoclinic orbit on $\mathbb{R} \times S^1$ (note that the saddles that are 2π radians apart are coincident on $\mathbb{R} \times S^1$).*

FIGURE 7.18. Emergence of limit cycle from a homoclinic orbit

7.6 Center Manifold Theorems

One of the main analytical techniques necessary for the analysis of dynamical systems that are *not structurally stable* is a set of center manifold theorems. The Hartman–Grobman theorem assures us that when an equilibrium point $\bar{x}$ of a continuous dynamical system is hyperbolic, its flow is locally approximated by its linearization. If the linearization has eigenvalues on the $j\omega$ axis the dynamics can be more complicated. For maps or discrete-time dynamical systems analogously, when the equilibrium point has eigenvalues of its linearization on the unit disk, the dynamics can be complex. To see this, work through several examples for the continuous time case:

1. $\dot{x} = x^2$. Compare the flow with that of the linearization.
2. Convert the following system to polar coordinates and study the trajectories of the system. Compare this with the trajectories of the linearization.

$$\dot{x}_1 = x_2 - x_1(x_1^2 + x_2^2),$$
$$\dot{x}_2 = -x_1 - x_2(x^2 + x_2^2).$$

3. For the following system show that the x_1, x_2 axes, and the lines $x_2 = \pm x_1$ are invariant and draw the phase portraits in these invariant sets.

$$\dot{x}_1 = x_1^3 - 3x_1x_2^2,$$
$$\dot{x}_2 = -3x_1^2x_2 + x_2^3.$$

To some analysts, a system truly manifests nonlinear behavior only when its linearization is not hyperbolic. For hyperbolic equilibria and maps, we had stable and unstable manifold theorems. In the next sections, we discuss their counterparts for non-hyperbolic equilibria.

7.6.1 Center Manifolds for Flows

We start with continuous-time dynamical systems. When the linearization of $Df(\bar{x})$ has eigenvalues on the $j\omega$ axis, define the center eigenspace E^c to be the span of the (generalized) eigenvectors corresponding to the eigenvalues on the $j\omega$ axis. In general, complicated behavior of non-hyperbolic equilibria is the asymptotic behavior on the *center manifold*—the invariant manifold tangent to E^c at $\bar{x}$. There are certain technical difficulties, however, with the *nonuniqueness* and *loss of smoothness* of the center manifold. We illustrate this with a celebrated example due to Kelley [122]:

$$\dot{x}_1 = x_1^2,$$
$$\dot{x}_2 = x_2.$$

It is easy to see that the solution curves of this equation satisfy

$$x_2(x_1) = ce^{\frac{1}{x_1}}, \tag{7.14}$$

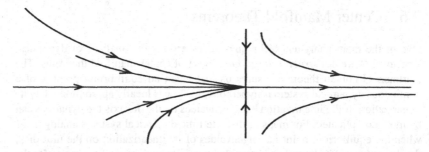

FIGURE 7.19. Nonuniqueness of the center manifold

as shown in Figure 7.19. Note that for $x_1 < 0$ the solution curves of (7.14) approach 0 "flatly," that is, all derivatives vanish at the origin. For $x_1 \geq 0$ the only solution curve approaching the origin is the x_1 axis. We can thus obtain a smooth manifold by piecing together any solution curve in the left half plane along with the positive half of the x_1 axis.

Theorem 7.25 Center Manifold Theorem for Flows [155]. *Let f be a C^k vector field with $f(0) = 0$ and $Df(0) = A$. Let $\sigma_s \subset \mathbb{C}^\circ_-$, $\sigma_u \subset \mathbb{C}^\circ_+$, and σ_c be a subset of the $j\omega$ axis be the disjoint partition of the spectrum of A with associated generalized eigenspaces E^s, E^u, E^c. Then there exist C^k stable and unstable invariant manifolds W^s, W^u tangent to E^s, E^u at the origin and a C^{k-1} center manifold W^c tangent to E^c at the origin. The manifolds W^s, W^u are unique, but W^c may not be unique.*

Let us study the implications of this theorem in the instance that A has eigenvalues only in $\mathbb{C}_-$, the closed left half plane. Further, we will assume that the equilibrium point $\bar{x}$ is at the origin. Then, one can use a change of coordinates to represent the nonlinear system in the form

$$\dot{y} = By + g(y, z),$$
$$\dot{z} = Cz + h(y, z). \tag{7.15}$$

Here $B \in \mathbb{R}^{n_s \times n_s}$ with eigenvalues in $\mathbb{C}^\circ_-$, $C \in \mathbb{R}^{n_0 \times n_0}$ with eigenvalues on the $j\omega$ axis, and g, h are smooth functions that along with their derivatives vanish at 0. In fact, the way that one would transform the original x system into the system of (7.15) is by choosing a change of coordinates of the form

$$\begin{bmatrix} y \\ z \end{bmatrix} = Tx$$

such that

$$TAT^{-1} = \begin{bmatrix} B & 0 \\ 0 & C \end{bmatrix}.$$

The preceding theorem guarantees the existence of a center manifold that is a manifold tangential to the $y = 0$ plane. Thus, a (any) center manifold is described by $y = \pi(z)$ with $\pi(0) = 0$ and

$$\frac{\partial \pi}{\partial z}(0) = 0.$$

Moreover, invariance of a center manifold guarantees that if $y(0) = \pi(z(0))$ then $y(t) = \pi(z(t))$. Differentiating the second equation and using the formulas for $\dot{y}, \dot{z}$ from (7.15) yields the following partial differential equation that $\pi(z)$ must satisfy

$$\frac{\partial \pi}{\partial z}(Cz + h(\pi(z), z)) = B\pi(z) + g(\pi(z), z) \qquad (7.16)$$

with the boundary conditions

$$\pi(0) = 0, \quad \frac{\partial \pi}{\partial z}(0) = 0.$$

The next theorem states that the dynamics on the center manifold, namely, the dynamics of

$$\dot{u} = Cu + h(\pi(u), u), \qquad (7.17)$$

determine the dynamics near the origin:

Theorem 7.26 Stability of the Center Manifold.   *Consider the system of equations (7.15) and its associated center manifold system (7.17). Suppose that the origin $u = 0$ is stable (asymptotically stable) (unstable) for (7.17). Then the origin $y = 0, z = 0$ is stable (asymptotically stable) (unstable) for (7.15). Further, if the origin of (7.17) is stable and $(y(t), z(t))$ is a solution of (7.15) starting with $(y(0), z(0))$ sufficiently small, then there is a solution $u(t)$ of (7.17) such that for t large enough,*

$$y(t) = u(t) + O(e^{-\gamma t}),$$

$$z(t) = \pi(u(t)) + O(e^{-\gamma t}),$$

where $\gamma > 0$ is a fixed constant.

The proof of this theorem is in [155]. The theorem is extremely powerful. It is a dimension reduction technique for determining the stability of systems with degenerate linearization, since the dimension of the system of (7.17) is n_0 which may be much less than $n_s + n_0$. The determination of the stability of (7.17) may not, however, be easy since it has a linearization with all eigenvalues on the $j\omega$ axis. In some sense, all of the "true" nonlinearity of the system is concentrated in (7.17). The obvious question that now arises is how to compute the center manifold. The solution of the partial differential equation (7.16) is not easy, but the following theorem gives a series approximation formula:

Theorem 7.27 Approximation of the Center Manifold [155].   *Let $\phi(z)$ be a function satisfying $\phi(0) = 0$ and $\frac{\partial \phi}{\partial z}(0) = 0$. Further, assume that ϕ satisfies* *equation (7.16) up to terms of order z^p, i.e.,*

$$\frac{\partial \phi}{\partial z}(Cz + h(\phi(z), z)) = B\phi(z) + g(\phi(z), z) + O(|z|^p).$$

Then it follows that as $|z| \to 0$,

$$\pi(z) = \phi(z) + O(|z|^p).$$

Thus, we can approximate $\pi(z)$ arbitrarily closely by series solutions. The convergence of the asymptotic series is not guaranteed, and further, even in the instance of convergence as in the Kelley example above, the convergent series is not guaranteed to converge to the correct value. However, this is not often the case, and the series approximation is valuable.

Example 7.28 Series Approximation of the Center Manifold.

$$\dot{y} = -y + \alpha z^2,$$

$$\dot{z} = zy.$$

Equation (7.16) specializes to

$$\pi'(z)(z\pi(z)) + \pi(z) - \alpha z^2 = 0.$$

Choose a Taylor series for $\pi(z) = az^2 + bz^3 + cz^4 + \cdots$. Then using this in the preceding equation and setting powers of z^2 to zero gives $a = \alpha$, and the order-2 solution is $\phi(z) = \alpha z^2$. Further, setting powers of $|z|^3$, $|z|^4$ to zero yields $b = 0, c = -2\alpha^2$ and the order-4 solution $\phi(z) = \alpha z^2 - 2\alpha^2 z^4$. Thus, the flow on the center manifold is given up to terms of order-3 by using the order-2 solution as

$$\dot{z} = \alpha z^3 + O(z^4).$$

The stability of the center manifold is determined then by the sign of α. The flows for the two cases are given in Figure 7.20.

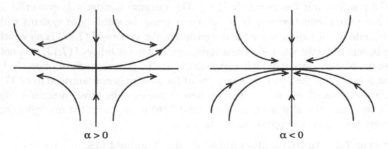

$$\alpha > 0 \qquad\qquad\qquad \alpha < 0$$

FIGURE 7.20. Center manifolds for α positive and α negative.

Including Linearly Unstable Directions

The system considered in (7.15) did not have any linearly unstable directions. We will remedy this shortcoming now: Most results carry forward with no problems. Consider

$$\dot{w} = Aw + f(w, y, z),$$

$$\dot{y} = By + g(w, y, z), \quad (w, y, z) \in \mathbb{R}^{n_u} \times \mathbb{R}^{n_s} \times \mathbb{R}^{n_0}, \qquad (7.18)$$

$$\dot{z} = Cz + h(w, y, z).$$

Here $A \in \mathbb{R}^{n_u \times n_u}$ has eigenvalues in $\mathbb{C}_+^\circ$, $B \in \mathbb{R}^{n_s \times n_s}$ has eigenvalues in $\mathbb{C}_-^\circ$ and $C \in \mathbb{R}^{n_0 \times n_0}$ has eigenvalues only on the $j\omega$ axis. The vector fields f, g, h are smooth and vanish along with their derivatives at the origin. A center manifold is guaranteed to exist as in the preceding. It goes through the origin and is tangential to the $x = 0$, $y = 0$ plane. Thus a (any) center manifold is described by $w = \pi_1(z)$, $y = \pi_2(z)$ with $\pi_1(0) = 0 < \pi_2(0) = 0$ and

$$\frac{\partial \pi_1}{\partial z}(0) = 0, \quad \frac{\partial \pi_2}{\partial z}(0) = 0.$$

As in the preceding, the invariance of the center manifold guarantees that the partial differential equations that π_1, π_2 satisfy are

$$\frac{\partial \pi_1}{\partial z}(Cz + h(\pi_1(z), \pi_2(z), z) = A\pi_1(z) + f(\pi_1(z), \pi_2(z), z),$$

$$\frac{\partial \pi_2}{\partial z}(Cz + h(\pi_1(z), \pi_2(z), z) = B\pi_2(z) + g(\pi_1(z), \pi_2(z), z). \qquad (7.19)$$

with the boundary conditions

$$\pi_1(0) = 0, \quad \pi_2(0) = 0.$$

and

$$\frac{\partial \pi_1}{\partial z}(0) = 0, \quad \frac{\partial \pi_2}{\partial z}(0) = 0.$$

The associated center manifold equation is

$$\dot{u} = Cu + h(\pi_1(u), \pi_2(u), u). \qquad (7.20)$$

The solution to the partial differential equations (7.19) can be mechanized as before by power series expansions. There is no counterpart to the stable center manifold theorem, since the system is unstable linearly to begin with in certain directions.

7.6.2 Center Manifolds for Flows Depending on Parameters

Consider the system of (7.15) with the added proviso that the equations depend linearly on a vector of parameters $\mu \in \mathbb{R}^p$. Thus we rewrite equation (7.15) as

$$\dot{y} = B(\mu)y + g(y, z, \mu),$$

$$\dot{z} = C(\mu)z + h(y, z, \mu), \qquad (7.21)$$

with $(y, z, \mu) \in \mathbb{R}^{n_0} \times \mathbb{R}^{n_u} \times \mathbb{R}^p$, where

$$g(0, 0, 0) = 0, \qquad Dg(0, 0, 0) = 0,$$
$$h(0, 0, 0) = 0 \qquad Dh(0, 0, 0) = 0.$$

The same assumptions as in the unparameterized case hold for B, C with the additional assumption that their eigenvalues lie on the $j\omega$ axis and in $\mathbb{C}^0_-$ at $\mu = 0$. The way to deal with this situation is to make μ a state variable as follows (in a technique known as **suspension**):

$$\dot{y} = B(\mu)y + g(y, z, \mu),$$
$$\dot{z} = C(\mu)z + h(y, z, \mu), \qquad (7.22)$$
$$\dot{\mu} = 0.$$

Consider the equilibrium point of this system at $y = 0$, $z = 0$, $\mu = 0$. There are $n_0 + p$ eigenvalues of the linearization on the $j\omega$ axis and n_s of them in $\mathbb{C}^0_-$. Thus, a center manifold is a graph from the z, μ space to the y space of the form $\pi(z, \mu)$ with the dynamics on the center manifold given by

$$\dot{u} = C(\mu)u + h(\pi(u, \mu), u, \mu),$$
$$\dot{\mu} = 0.$$

Also, as a consequence of the invariance of the center manifold the center manifold equation becomes

$$\frac{\partial \pi}{\partial z}(C(\mu)z + h(\pi(z, \mu), z, \mu)) = B(\mu)\pi(z, \mu) + g(\pi(z, \mu), z, \mu) \qquad (7.23)$$

As in the case of the unparameterized center manifold, the partial differential equation (7.23) can be solved by equating like powers in z, μ. It is important to note that product terms involving z, μ are now also second-order terms.

7.6.3   Center Manifolds for Maps

Center manifold theory for flows can be modified to apply to nonlinear maps. Analogous to Theorem 7.25, we have the following theorem:

Theorem 7.29 Center Manifold Theorem for Maps [155].   *Let f be a C^k map, with fixed point at the origin. That is, $f(0) = 0$ and define $Df(0) = A$. Let $\sigma_s, \sigma_u, \sigma_c$ be the disjoint partition of the spectrum of A with associated generalized eigenspaces E^s, E^u, E^c, corresponding to eigenvalues inside the open unit disk, outside the open unit disk, and on the unit disk. Then there exist C^k stable and unstable invariant manifolds W^s, W^u tangent to E^S, E^u at the origin and a C^{k-1} center manifold W^c tangent to E^c at 0. W^s, W^u are unique, but W^c may not be unique.*

Continuing with the analogue to the system of (7.15), consider the discrete time system

$$y_{n+1} = By_n + g(y_n, z_n),$$
$$z_{n+1} = Cz_n + h(y_n, z_n). \tag{7.24}$$

Here $B \in \mathbb{R}^{n_s \times n_s}$ with eigenvalues inside the open unit disk in the complex plane, $C \in \mathbb{R}^{n_c \times n_c}$ has eigenvalues on the boundary of the unit disk, and g, h are smooth functions that along with their derivatives vanish at the origin. As in the continuous-time case, the form (7.24) arises from a linear change of coordinates applied to a general discrete time nonlinear system. From Theorem 7.29, it follows that any center manifold is described by $y = \pi(z)$, which satisfies the equation:

$$\pi(Cz + h(\pi(z), z)) = B\pi(z) + g(\pi(z), z). \tag{7.25}$$

The dynamics on the center manifold are given by

$$u_{n+1} = Cu_n + h(\pi(u_n), u_n). \tag{7.26}$$

The counterparts of Theorems 7.26 and 7.27 now follow, verbatim, with the obvious changes in the notation for decaying exponentials of time. For example, the discrete time counterpart of Theorem 7.26 is the following:

Theorem 7.30 Stability of the Center Manifold for Maps. *Consider the system of (7.24) and its associated center manifold system of (7.26). Suppose that the fixed point at the origin $u = 0$, of the system of (7.26) is stable (asymptotically stable) (unstable). Then the origin $y = 0, z = 0$ is stable (asymptotically stable) (unstable) for (7.24). Further, if the origin of (7.26) is stable and (y_n, z_n) is a solution of (7.24) starting with y_0, z_0 sufficiently small, then there is a solution u_n of (7.26) such that for n large enough,*

$$|y_n - u_n| \le k\beta^n,$$
$$|z_n - h(u_n)| \le k\beta^n$$

for some $\beta < 1, k > 0$.

The inclusion of linearly unstable directions as well as the dependence of the center manifold for flows on parameters also carries through in verbatim fashion.

7.7 Bifurcation of Vector Fields: An Introduction

The term *bifurcation* was originally used by Poincaré to describe branching of equilibrium solutions in a family of differential equations.

$$\dot{x} = f(x, \mu), \quad x \in \mathbb{R}^n, \quad \mu \in \mathbb{R}^k. \tag{7.27}$$

The equilibrium solutions satisfy

$$f(x, \mu) = 0, \tag{7.28}$$

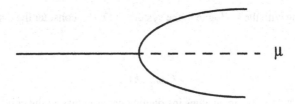

FIGURE 7.21. Bifurcation diagram of a cubic equation—the solid line indicates sinks and the dashed line indicates sources

As μ is varied, the solutions $x^*(\mu)$ of (7.28) are smooth functions of μ so long as $Df(x^*(\mu), \mu)$ does not have a zero eigenvalue (by the implicit function theorem, convince yourself of this point). The graph of each $x^*(\mu)$ is a *branch* of equilibria of (7.27). At an equilibrium $(x^*(\mu_0), \mu_0)$ where Df_μ has a zero eigenvalue, several branches of equilibria may come together, and one says that $(x_0 := x^*(\mu_0), \mu_0)$ is a point of bifurcation. Consider, for example,

$$f(x, \mu) = \mu x - x^3, \quad x \in \mathbb{R}, \quad \mu \in \mathbb{R}.$$

Here $D(x, \mu) = \mu - 3x^2$, and the only bifurcation point is $(0, 0)$. It is easy to verify that the unique fixed point is $x = 0$ for $\mu \leq 0$ and that $x = 0$ becomes unstable for $\mu > 0$. We draw the solution branches in Figure 7.21. The diagram of Figure 7.21 is referred to as a *bifurcation diagram*. Bifurcations of equilibria *usually* produce changes in the topological type of the flow, but there are many other changes that occur in the topological equivalence class of flows: We include all of these in our use of the term *bifurcation*. We give the following general definition for bifurcations of (7.21): A value μ_0 for which the flow of (7.21) is *not structurally stable* is a *bifurcation value of μ*.

Definition 7.31 Bifurcations. *A one-parameter family of vector fields of the form of (7.29) also satisfying the conditions $f(0, 0) = 0$ and $D_1 f(0, 0) = 0$ is said to undergo a bifurcation at $\mu = 0$ if the flow of the equation at $\mu = 0$ is not C^0 equivalent (not qualitatively similar) to the flow for μ near zero.*

The definition is not very satisfactory, since it impels one to study the detailed fine structure of flows, for which complete descriptions do not exist, namely, the structurally unstable flows. Consequently, construction of a systematic bifurcation theory leads to very difficult technical questions. However, we do wish to study more than merely the bifurcation of equilibria (so called *static bifurcations*). Another point to note is that a point of bifurcation need not actually represent a change in the topological equivalence class of a flow. For example, the system $\dot{x} = -(\mu^2 x + x^3)$ has a bifurcation value $\mu = 0$, but all of the flows in the family have an attracting equilibrium at $x = 0$. The changes in the topological type of the topological characteristics of the flow are represented as loci in the (x, μ) space of the invariant sets of (7.21), for example fixed points, periodic orbits. These are referred to as *bifurcation diagrams*.

7.8 Bifurcations of Equilibria of Vector Fields

The simplest kinds of bifurcations are the changes in the number and type of equilibrium points of dynamical systems or fixed points of maps. We will consider the parameterized family of vector fields of (7.27) and assume that the family of equilibrium points is of the form $x_0(\mu)$. The first question that comes to mind is that of the stability of the family of equilibrium points. For this purpose we consider the linearized system described by

$$\dot{\xi} = D_1 f(x_0(\mu_0), \mu_0)\xi.$$

The stability of this equation determines the stability of the equilibrium point except when the linearization has eigenvalues on the $j\omega$ axis. Further, if μ_0 corresponds to a hyperbolic equilibrium point, then since no eigenvalue of $D_1 f(x_0, \mu_0)$ lies at the origin, it follows from the implicit function theorem that there is a unique continuation of solutions $x(\mu)$ in a neighborhood of μ_0. The point of particular interest to us in this section is the case when the equilibrium point $x_0(\mu_0)$ is not a hyperbolic equilibrium point, i.e., it has eigenvalues that are on the $j\omega$ axis. To get started we begin with the simplest possible situation that the equilibrium point $x_0(\mu_0)$ can fail to be hyperbolic, namely a *single, simple zero eigenvalue* of the linearization. The question to be answered then is: What is the structure of the equilibrium points in a neighborhood of μ_0.

7.8.1 Single, Simple Zero Eigenvalue

Consider equation (7.27) and assume that at μ_0 the linearization has a simple zero eigenvalue. Apply the parameterized center manifold theorem to this situation. The following equation gives the dynamics on the center manifold in this instance:

$$\dot{z} = g(z, \mu), \tag{7.29}$$

where $z \in \mathbb{R}^1$. Also, $g(z_0, \mu_0) = 0$. Note that the application of the parameterized center manifold theorem results in a reduction of the dimension of the system from n to 1. Further, as a consequence of the fact that $C = 0$ the additional condition that g satisfies is

$$\frac{\partial g}{\partial z}(z_0, \mu_0) = 0.$$

To keep the notation simple, we will assume without loss of generality in what follows that $z_0 = 0$, $\mu_0 = 0$. In Chapter 2 we began a study of several systems that satisfied these conditions. We list them here again:

1. *Fold Bifurcation*

$$\dot{z} = \mu - z^2.$$

This gives rise to the so called fold bifurcation shown in Figure 7.22 with two equilibrium points for $\mu \geq 0$ and none for $\mu < 0$. Thus, $\mu = 0$ is a point at which two equilibrium points coalesce, producing none for $\mu < 0$.

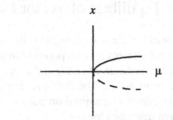

FIGURE 7.22. Fold bifurcation

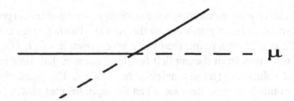

FIGURE 7.23. Transcritical bifurcation

2. *Transcritical or Exchange of Stability*

$$\dot{z} = \mu z - z^2.$$

The equilibrium points in this case are given by $z = 0$ and $z = \mu$. The equilibrium points trade stability type at $\mu = 0$ as depicted in Figure 7.23. There is no branching in solutions at $\mu = 0$.

3. *Pitchfork Bifurcation*

$$\dot{z} = \mu z - z^3.$$

This system has three equilibrium points for $\mu > 0$, namely $0, \pm\sqrt{\mu}$ and one for $\mu \leq 0$, namely 0. The stability type of the equilibrium points is shown in Figure 7.24. It is called a pitchfork because of its obvious resemblance to that barnyard implement.

4. *No bifurcation*

$$\dot{z} = \mu - z^3.$$

In this case there is one equilibrium for all values of μ, and in addition, it is stable for all μ.

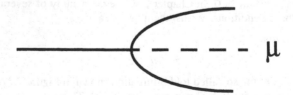

FIGURE 7.24. Pitchfork bifurcation

We now proceed to derive conditions on one parameter families of one dimensional vector fields of the form of (7.29), with $\mu \in \mathbb{R}^1$, that exhibit the behavior of the three examples described above. When $\mu \in \mathbb{R}^k$ for $k > 1$ the bifurcation generically occurs on a $(k - 1)$ dimensional surface in the space of parameters: $\kappa(\mu) = 0$ for some $\kappa : \mathbb{R}^k \mapsto \mathbb{R}$. The determination of this surface is straightforward, but notationally involved.

1. **Fold or Saddle Node Bifurcation** This is the situation where for $\mu > 0$ there are two equilibrium points and there are none for $\mu < 0$. Locally this curve of fixed points may be described by $\mu(z)$ with

$$\frac{d\mu}{dz}(0) = 0, \qquad \frac{\partial^2 \mu}{\partial z^2}(0) \neq 0.$$

If the second term is positive, we have the situation of Figure 7.22, with two equilibria for $\mu > 0$; if it is negative, we have two equilibria for $\mu < 0$, which would also be a saddle node or fold bifurcation. Further, requiring that

$$\frac{\partial g}{\partial \mu}(0, 0) \neq 0$$

implies the existence of $\mu(z)$ near $z = 0$, $\mu = 0$. We also have that

$$g(0, 0) = 0, \qquad \frac{\partial g}{\partial z}(0, 0) = 0. \tag{7.30}$$

Differentiating the equation

$$g(z, \mu(z)) = 0 \tag{7.31}$$

yields (see Problem 7.14)

$$\frac{d\mu}{dz}(0) = -\frac{\frac{\partial g}{\partial z}(0, 0)}{\frac{\partial g}{\partial \mu}(0, 0)} \tag{7.32}$$

and

$$\frac{\partial^2 \mu}{\partial z^2}(0) = -\frac{\frac{\partial^2 g}{\partial z^2}(0, 0)}{\frac{\partial g}{\partial \mu}(0, 0)}. \tag{7.33}$$

The term in (7.32) is zero from the first equation in (7.30). Thus, the system of (7.29) goes through a fold bifurcation if it is nonhyperbolic (satisfies (7.30)) and in addition,

$$\frac{\partial g}{\partial \mu}(0, 0) \neq 0,$$
$$\frac{\partial^2 g}{\partial z^2}(0, 0) \neq 0. \tag{7.34}$$

If one were to write the Taylor series of $g(z, \mu)$ in the form

$$g(z, \mu) = a_0 \mu + a_1 z^2 + a_2 \mu z + a_3 \mu^2 + \text{terms of order 3}, \tag{7.35}$$

then the conditions (7.34) imply that $a_0, a_1 \neq 0$, so that the *normal form*[3] for the system (7.29) is

$$\dot{z} = \mu \pm z^2,$$

As far as the original system of (7.27) is concerned, with $\mu \in \mathbb{R}$, if the parameterized center manifold system $g(z, \mu)$ satisfied the conditions of (7.34), the original system would have a *saddle node bifurcation*, meaning to say that a saddle and a node (equilibrium points of index $+1$ and -1) fuse together and annihilate each other.

2. *Transcritical Bifurcation*

In order to get an unbifurcated branch of equilibria at $z = 0$, we will need to explicitly assume that

$$g(z, \mu) = zG(z, \mu),$$

where

$$G(z, \mu) = \begin{cases} \dfrac{g(z, \mu)}{z}, & z \neq 0, \\[2mm] \dfrac{\partial g}{\partial z}(0, \mu) & z = 0. \end{cases}$$

From this definition it follows that $G(0, 0) = 0$, and

$$\frac{\partial G}{\partial z}(0, 0) = \frac{\partial^2 g}{\partial z^2}(0, 0),$$

$$\frac{\partial^2 G}{\partial z^2}(0, 0) = \frac{\partial^3 G}{\partial z^3}(0, 0), \tag{7.36}$$

$$\frac{\partial G}{\partial \mu}(0, 0) = \frac{\partial^2 g}{\partial z \, \partial \mu}(0, 0).$$

If the last expression is not equal to zero, there exists a function $\mu(z)$ defined in a neighborhood of 0 such that

$$G(z, \mu(z)) = 0.$$

To guarantee that $\mu(z) \not\equiv 0$ we require that

$$\frac{d\mu}{dz}(0) \neq 0.$$

In Problem 7.15 you are required to verify that this is equivalent to

$$\frac{d\mu}{dz}(0) = -\frac{\frac{\partial^2 g}{\partial z^2}(0, 0)}{\frac{\partial^2 g}{\partial z \partial \mu}(0, 0)}. \tag{7.37}$$

---

[3]This usage is formal: There exists a transformation that will transform the system of (7.29) into the normal form, but the details of this transformation are beyond the scope of this book.

Thus, the system of (7.29) goes through a transcritical bifurcation if it is nonhyperbolic (satisfies (7.30), and in addition

$$\frac{\partial g}{\partial \mu}(0, 0) = 0,$$

$$\frac{\partial^2 g}{\partial z \partial \mu}(0, 0) \neq 0, \tag{7.38}$$

$$\frac{\partial^2 g}{\partial z^2}(0, 0) \neq 0.$$

In terms of the Taylor series of $g(z, \mu)$ of (7.35), the conditions (7.38) imply that $a_0 = 0$, $a_1, a_2 \neq 0$, so that the *normal form* for the system (7.29) is

$$\dot{z} = \mu z \pm z^2,$$

As far as the original system of (7.27) is concerned, with $\mu \in \mathbb{R}$, if the parameterized center manifold system $g(z, \mu)$ satisfied the conditions of (7.38), the original system would have a *transcritical bifurcation*, meaning to say that a saddle and a node exchange stability type.

3. *Pitchfork Bifurcation*

In order to get an unbifurcated branch at $z = 0$ we will need to assume as in the transcritical bifurcation case above that

$$g(z, \mu) = zG(z, \mu)$$

with the same definition of $G(z, \mu)$ as above. To guarantee the pitchfork we will need to give conditions on $G(z, \mu)$ to make sure that it has a *fold bifurcation*, since the pitchfork is a combination of an unbifurcated branch and a fold. We leave the reader to work out the details in Problem 7.16 that the conditions for a pitchfork are the nonhyperbolicity conditions of (7.30) along with

$$\frac{\partial g}{\partial \mu}(0, 0) = 0,$$

$$\frac{\partial^2 g}{\partial z^2}(0, 0) = 0,$$

$$\frac{\partial^2 g}{\partial z \partial \mu}(0, 0) \neq 0, \tag{7.39}$$

$$\frac{\partial^3 g}{\partial z^3}(0, 0) \neq 0.$$

In terms of the Taylor series of $g(z, \mu)$ of (7.35), the conditions (7.39) imply that that the *normal form* for the system (7.29) is

$$\dot{z} = \mu z \pm z^3.$$

As far as the original system of (7.27) is concerned, with $\mu \in \mathbb{R}$, if the parameterized center manifold system $g(z, \mu)$ satisfied the conditions of (7.39), the original system would have a *pitch fork bifurcation*, meaning to say that either two saddles and a node coalesce into a saddle, or two nodes and a saddle coalesce into a node.

7.8.2 Pure Imaginary Pair of Eigenvalues: Poincaré–Andronov–Hopf Bifurcation

We now turn to the next most simple way that an equilibrium point of a vector field can become non-hyperbolic, namely that two of the eigenvalues of the linearization of the vector field are located on the $j\omega$ axis. More precisely, consider the case that

$$\dot{x} = f(x, \mu)$$

and at $\mu = \mu_0, x = x_0(\mu_0)$ there are two eigenvalues of $D_x f(x_0(\mu_0), \mu_0)$ on the $j\omega$ axis. Using the parameterized center manifold theorem, the flow on the two-dimensional center manifold can be written as

$$\begin{bmatrix} \dot{z}_1 \\ \dot{z}_2 \end{bmatrix} = \begin{bmatrix} \alpha(\mu) & -\omega(\mu) \\ \omega(\mu) & \alpha(\mu) \end{bmatrix} \begin{bmatrix} z_1 \\ z_2 \end{bmatrix} + \begin{bmatrix} g_1(z_1, z_2, \mu) \\ g_2(z_1, z_2, \mu) \end{bmatrix} \tag{7.40}$$

with $\alpha(\mu_0) = 0$. As in the previous section, to keep the notation simple we will (without loss of generality) set $\mu_0 = 0$. To be able to reduce this system to a normal form, it is convenient to conceptualize the system of (7.40) on the complex plane $z = z_1 + jz_2$. It can then be shown (again, the details of this are beyond the scope of this book, the interested reader may wish to get the details from [329]) that a normal form for the system is of the form

$$\dot{z}_1 = \alpha(\mu)z_1 - \omega(\mu)z_2 + (a(\mu)z_1 - b(\mu)z_2)(z_1^2 + z_2^2) + O(|z_1|^5, |z_2|^5),$$
$$\dot{z}_2 = \omega(\mu)z_1 + \alpha(\mu)z_2 + (b(\mu)z_1 + a(\mu)z_2)(z_1^2 + z_2^2) + O(|z_1|^5, |z_2|^5)$$

$$\tag{7.41}$$

for appropriately chosen $a(\mu), b(\mu)$. In polar coordinates $r = \sqrt{z_1^2 + z_2^2}, \theta = \tan^{-1}(z_2/z_1)$, this has the form

$$\dot{r} = \alpha(\mu)r + a(\mu)r^3 + O(r^5),$$
$$\dot{\theta} = \omega(\mu) + b(\mu)r^2 + O(r^4). \tag{7.42}$$

Close to $\mu = 0$, using the leading terms in the Taylor series for $\alpha(\mu), a(\mu), b(\mu)$, equation (7.42) has the form

$$\dot{r} = \alpha'(0)\mu r + a(0)r^3 + O(\mu^2 r, \mu r^3, r^5),$$
$$\dot{\theta} = \omega(0) + \omega'(0)\mu + b(0)r^2 + O(\mu^2, \mu r^2, r^4). \tag{7.43}$$

where α' refers to differentiation of α with respect to μ. Neglecting the higher-order terms gives the system

$$\dot{r} = d\mu r + ar^3,$$
$$\dot{\theta} = \omega + c\mu + br^2,$$ (7.44)

where $d = \alpha'(0)$, $a = a(0)$, $\omega = \omega(0)$, $c = \omega'(0)$, $b = b(0)$. It follows readily that for $\frac{\mu d}{a} < 0$ and μ small, the system of (7.44) has a circular periodic orbit of radius

$$\sqrt{\frac{-\mu d}{a}}$$

and frequency $\omega + (c - \frac{bd}{a}\mu)$. Furthermore, the periodic orbit is stable for $a < 0$ and unstable for $a > 0$. Indeed, this is easy to see from the pitchfork for the dynamics of r embedded in (7.44). We will now classify the four cases depending on the signs of a, d. We will need to assume them both to be non-zero. If either of them is zero, the bifurcation is more complex and is not referred to as the Hopf bifurcation.

1. **Supercritical Hopf Bifurcation**. Supercritical Hopf (more correctly Poincaré–Andronov–Hopf) bifurcations refer to the case in which a stable periodic orbit is present close to $\mu = 0$.

   a. $d > 0, a < 0$. The origin is stable for $\mu < 0$, unstable for $\mu > 0$, and surrounded by a stable periodic orbit for $\mu > 0$. This situation is shown in Figure 7.25.
   b. $d < 0, a < 0$. The origin is unstable for $\mu < 0$ and stable for $\mu > 0$. The origin is surrounded by a stable periodic orbit for $\mu < 0$.

2. **Subcritical Hopf Bifurcation**. Subcritical Hopf bifurcations refer to the case in which an unstable periodic orbit is present close to $\mu = 0$.

   a. $d > 0, a > 0$. The origin is stable for $\mu < 0$ and unstable for $\mu > 0$. In addition, the origin is surrounded by an unstable periodic orbit for $\mu < 0$.
   b. $d < 0, a > 0$. The origin is unstable for $\mu < 0$, stable for $\mu > 0$, and surrounded by an unstable periodic orbit for $\mu > 0$. This is shown in Figure 7.26.

It may be verified using the Poincaré–Bendixson theorem that the conclusions derived on the basis of the system (7.44) also hold for the original system (7.43) with the higher order terms present for μ small. The signs of the two numbers a, d determine the type of bifurcation. The value $a = a(0)$ is referred to as the *curvature coefficient* (see Marsden and McCracken [201]) for a formula for a in terms of the original system $f(x, \mu)$). The value $a < 0$ corresponds to the supercritical Hopf bifurcation and a value $a > 0$ to a subcritical Hopf bifurcation. As noted before, when $a = 0$, the bifurcation may be more complicated. The value

$$d = \frac{d\alpha(\mu)}{d\mu}(0)$$

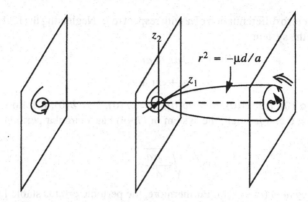

FIGURE 7.25. Supercritical Hopf bifurcation: a stable periodic orbit and an unstable equilibrium arising from a stable equilibrium point

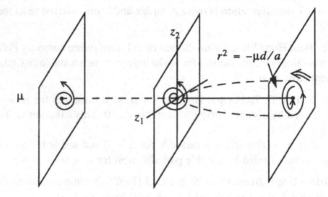

FIGURE 7.26. Subcritical Hopf bifurcation: a stable equilibrium and an unstable periodic orbit fusing into an unstable equilibrium point

represents the *rate at which the real part of the eigenvalue of the linearization crosses the $j\omega$ axis*. If $d > 0$ the eigenvalues cross from $\mathbb{C}_-$ to $\mathbb{C}_+$ as μ increases, and conversely for $d < 0$.

7.9   Bifurcations of Maps

Analogous to the study of the bifurcations of vector fields there is a theory of bifurcations of maps of parameterized families of discrete time systems of the form

$$x_{n+1} = G(x_n, \mu), \quad x \in \mathbb{R}^n, \quad \mu \in \mathbb{R}^k. \tag{7.45}$$

Suppose that (7.45) has a fixed point at $x(\mu)$. The associated linear map at μ_0 is given by

$$\xi \mapsto D_1 G(x(\mu_0), \mu_0)\xi.$$

If the eigenvalues of the matrix $D_1 G(x(\mu_0), \mu_0)$ do not lie on the boundary of the unit disk, the Hartman–Grobman theorem yields conclusions about the stable and unstable manifolds of the equilibrium. When there are eigenvalues of $D_1 G(x(\mu_0), \mu_0)$ on the boundary of the unit disk, the fixed point is not *hyperbolic* and the parameterized center manifold may be used to derive a reduced dimensional map on the center manifold

$$z_{n+1} = G(z_n, \mu). \tag{7.46}$$

In the remainder of this section we will assume $\mu \in \mathbb{R}$ and in addition discuss the simplest cases when eigenvalues of the linearization of (7.45) cross the boundary of the unit disk:

1. Exactly one eigenvalue of $D_1 G(x(\mu_0), \mu_0)$ is equal to 1, and the rest do not have modulus 1.
2. Exactly one eigenvalue of $D_1 G(x(\mu_0), \mu_0)$ is equal to -1, and the rest do not have modulus 1.
3. Exactly two eigenvalues of $D_1 G(x(\mu_0), \mu_0)$ are complex and have modulus 1, and the rest do not have modulus 1.

7.9.1 Single Eigenvalue 1: Saddle Node, Transcritical and Pitchfork

Here we consider systems of the form

$$z_{n+1} = G(z_n, \mu) \tag{7.47}$$

with $z, \mu \in \mathbb{R}$. We will assume without loss of generality that at $\mu_0 = 0$, the linearization of this system at the fixed point $z = 0$ has a single eigenvalue at 1; thus, we have

$$G(0,0) = 0, \qquad \frac{\partial G}{\partial z}(0,0) = 1. \tag{7.48}$$

Saddle Node Bifurcation

The prototype here is a map of the form

$$G(z, \mu) = z + \mu \pm z^2. \tag{7.49}$$

This system has two fixed points for either $\mu < 0$ or $\mu > 0$ and none for $\mu < 0$, $\mu > 0$, respectively, depending on the sign of z^2 in (7.49). For a general system (7.47), the same techniques as the continuous time (vector field) case of the previous section can be used to verify that (7.47) undergoes a saddle node bifurcation if

(7.48) holds and in addition

$$\frac{\partial G}{\partial \mu}(0, 0) \neq 0,$$

$$\frac{\partial^2 G}{\partial z^2}(0, 0) \neq 0. \tag{7.50}$$

The details of verifying this are left as an exercise (Problem 7.17). The sign of the ratio of the two non-zero terms in (7.50) determines whether the two fixed points are present for $\mu < 0$ or for $\mu > 0$.

Transcritical Bifurcation

The prototype map here is of the form

$$G(z, \mu) = z + \mu z \pm z^2. \tag{7.51}$$

There are two curves of fixed points passing through the bifurcation point namely $z = 0$ and $z = \pm \mu$. The more general system of (7.47) undergoes a transcritical bifurcation, if

$$\frac{\partial G}{\partial \mu}(0, 0) = 0,$$

$$\frac{\partial^2 G}{\partial z \, \partial \mu}(0, 0) \neq 0, \tag{7.52}$$

$$\frac{\partial^2 G}{\partial z^2}(0, 0) \neq 0.$$

The proof of this is delegated to Problem 7.18.

Pitchfork Bifurcation

Consider the family of maps given by

$$G(z, \mu) = z + \mu z \pm z^3. \tag{7.53}$$

There are two curves of fixed points, namely $z = 0$ and $\mu = \pm z^2$. The system of (7.47) undergoes a pitch fork bifurcation if in addition to (7.48, we have (see Problem 7.19):

$$\frac{\partial G}{\partial \mu}(0, 0) = 0,$$

$$\frac{\partial^2 G}{\partial z^2}(0, 0) = 0,$$

$$\frac{\partial^2 G}{\partial z \, \partial \mu}(0, 0) \neq 0, \tag{7.54}$$

$$\frac{\partial^3 G}{\partial z^3}(0, 0) \neq 0.$$

7.9.2   Single Eigenvalue -1: Period Doubling

Up to this point, bifurcations of one parameter families of maps have been very much the same as analogous cases for vector fields. The case of an eigenvalue at -1 has no analogue with one dimensional continuous time dynamics. The prototype example here is the family of one dimensional maps given by

$$G(z, \mu) = -z - \mu z + z^3, \tag{7.55}$$

This system has two curves of fixed points $z = 0$ and $z^2 = 2 + \mu$. It is easy to verify that the fixed point at $z = 0$ is unstable for $\mu \leq -2$, stable for $-2 \leq \mu \leq 0$, and unstable for $\mu > 0$. The two fixed points at $z = \pm\sqrt{2 + \mu}$ are unstable for $\mu \geq -2$ and do not exist for $\mu < -2$ (pitchfork bifurcation at $\mu = -2$). Thus for $\mu > 0$ there are *three unstable fixed points*. From examining the second iterate $G^2(z, \mu)$ given by

$$G^2(z, \mu) = z + \mu(2 + \mu)z - 2z^3 + O(4) \tag{7.56}$$

we see that the map $G^2(z, \mu)$ has a pitchfork at $\mu = 0$, with three fixed points for $\mu > 0$ at $z = 0, z = \pm\sqrt{\mu(2 + \mu)/2}$. Fixed points of G^2 are *period-two points of the original system* of (7.55). This situation is shown in Figure 7.27. Returning to the general one parameter family of maps (7.47), the conditions for having an eigenvalue at -1 are that

$$G(0, 0) = 0, \quad \frac{\partial G}{\partial z}(0, 0) = -1. \tag{7.57}$$

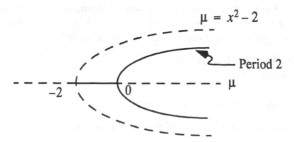

FIGURE 7.27. Showing the period doubling bifurcation superimposed on a pitch fork bifurcation

and in addition,

$$\frac{\partial G^2}{\partial \mu}(0,0) = 0,$$

$$\frac{\partial G^2}{\partial z^2}(0,0) = 0,$$

$$\frac{\partial^2 G^2}{\partial z \partial \mu}(0,0) \neq 0,$$

$$\frac{\partial^3 G^2}{\partial z^3}(0,0) \neq 0.$$

(7.58)

Note that the four conditions of (7.58) are precisely the pitchfork conditions of (7.54) applied to G^2. The sign of the ratio of the two nonzero terms in equation (7.58) tells us the side of $\mu = 0$ on which we have period-two points.

7.9.3   Pair of Complex Eigenvalues of Modulus 1: Naimark–Sacker Bifurcation

The Naimark–Sacker bifurcation is the discrete-time counterpart of the Poincaré Andronov Hopf bifurcation. As in the case of the Hopf bifurcation, the derivation of the normal form for this bifurcation is beyond the scope of this book (see, however Chapters 2 and 3 of Wiggins [329]). The normal form map $G(z, \mu) : \mathbb{R}^2 \mapsto \mathbb{R}^2$ with $\mu \in \mathbb{R}$ is given in polar coordinates by

$$r \mapsto r + (d\mu + ar^2)r,$$

$$\theta \mapsto \theta + \phi_0 + \phi_1 \mu + br^2.$$

(7.59)

The conditions for the existence of this normal form are a little more subtle than in the continuous time case, in that the pair of eigenvalues at the boundary of the unit disk at $\mu = 0$ are each required not to be ± 1 or any of the (complex) square roots, cube roots or fourth roots of 1. This is called avoiding *strong resonances*. The amazing intricacy of what happens if there are strong resonances is discussed in Arnold [12]. Strong resonances occur when a (continuous) system is forced at frequencies that are commensurate in ratios of $1/1$, $1/2$, $1/3$, $1/4$ to the natural frequency of the system. The fixed point of (7.59) at the origin is

- Asymptotically stable for $d\mu < 0$.
- Unstable for $d\mu > 0$.
- Unstable for $\mu = 0$, $a > 0$.
- Asymptotically stable for $\mu = 0$, $a < 0$.

As in the case of the Hopf bifurcation, the circle of radius

$$\sqrt{\frac{-\mu d}{a}}$$

is invariant under the dynamics of (7.59). This invariant circle is said to be asymptotically stable if initial conditions "close to it" converge to it as the forward iterations tend to ∞. Then it is easy to see that the invariant circle is asymptotically stable for $a < 0$ and unstable for $a > 0$.

As in the case of the Hopf bifurcation, the four cases corresponding to the signs of a, d give the supercritical and subcritical Naimark–Sacker bifurcation. We will not repeat them here. However, we would like to stress the differences between the situation for maps and vector fields: In the Hopf bifurcation the bifurcating invariant set (a periodic orbit) was a *single orbit*, while in the Naimark–Sacker bifurcation the bifurcating set consists of *an invariant circle* containing many orbits. We may study the dynamics of (7.59) restricted to the invariant circle and conclude that it is described by the circle map

$$\theta \mapsto \theta + \phi_0 + \left(\phi_1 - \frac{d}{a}\right)\mu. \tag{7.60}$$

It follows that the orbits on the invariant circle are periodic if $\phi_0 + (\phi_1 - \frac{d}{a}\mu$ is rational, and if it is irrational then the orbits are dense on the invariant circle. Thus, as μ is varied the orbits on the invariant circle are alternately periodic and quasi-periodic.

7.10 More Complex Bifurcations of Vector Fields and Maps

7.10.1 Bifurcations of Equilibria and Fixed Points: Catastrophe Theory

The preceding sections have given us a snapshot into an elementary set of bifurcations of equilibria of vector fields and fixed points of maps. However, these sections perhaps raise more questions than they answer. For example, in the context of bifurcations of equilibria of vector fields, we may ask several questions:

1. When a single eigenvalue of the linearization passes through the origin, what other bifurcations can one expect other than the fold, transcritical, and pitchfork? Is it possible to have more sophisticated branching of the number of equilibria?
2. In Chapter 2 we saw that the transcritical and the pitchfork may be perturbed or unfolded into folds and an unbifurcated branches. Is there a theory of unfolding of the bifurcations and a taxonomy of them?
3. What happens when more than one eigenvalue of the linearization crosses the origin?

The answers to this set of questions fall into the domain of singularity theory. The initial work in this area done by Thom [301] drew a great deal of press under the title of "catastrophe theory" and also strong scientific criticism [13]. Our short treatment follows the book by Poston and Stewart [244].

The Seven Elementary Catastrophes of Thom

The seven elementary catastrophes are studied as bifurcations of the zero set of canonical functions

$$f(x, \mu) = 0.$$

Here $x \in \mathbb{R}^1(\mathbb{R}^2)$ and $\mu \in \mathbb{R}^k$ and $f : \mathbb{R}(\mathbb{R}^2) \times \mathbb{R}^k \mapsto \mathbb{R}(\mathbb{R}^2)$. The first set of four catastrophes is in $\mathbb{R}$ and are referred to as the fold, cusp, swallowtail, and butterfly, respectively. The second set of three are referred to as the umbilics: elliptic, hyperbolic, and parabolic. One might ask oneself why these first seven catastrophes are singled out for special attention. A detailed and accurate discussion of this is beyond the scope of this book, but the rough answer is contained in the search for what is known as the *unfolding*, or *codimension* of a singular point. For example, if for $x \in \mathbb{R}$, $f(x, 0) = x^2$, then $x = 0$ is a singularity point. It may be shown that the complete unfolding of this singularity is obtained by using a single parameter $\mu \in \mathbb{R}$, of the form $f(x, \mu) = x^2 + \mu$ (which shows two solutions for $\mu < 0$, one solution for $\mu = 0$ and none for $\mu > 0$, in the sense that for all other parameterized families of functions $f(x, \mu)$ with $f(x, 0) = 0$, $\mu \in \mathbb{R}^k$ for arbitrary k has an equivalence of qualitative behavior, 2 solutions for an open subset of the μ space, 0 solutions for another open set of the μ space and 1 solution on the boundary. This *one parameter unfolding*, or *codimension* 1 singularity is called the fold. The program of Thom and Zeeman was to classify these unfoldings. This program is not yet completed and the first seven catastrophes are all the singularities of *codimension* 4 *or lower*. In this section we list the seven elementary catastrophes, with special attention to the first four, and in the next section we describe how to use these catastrophes in applications.

1. **Canonical Fold Catastrophe**
   Consider the function $f(x, a) = x^2 + a$ from $\mathbb{R} \mapsto \mathbb{R}$. Note that this equation has two roots for $a < 0$, one for $a = 0$ and none for $a > 0$. The fold is a building block for other catastrophes. The codimension of the fold catastrophe is 1.
2. **Canonical Cusp Catastrophe**
   Consider the function $f(x, a, b) = x^3 + ax + b$. From elementary algebra (7.22) it can be shown that this equation has three real roots when the discriminant $4a^3 + 27b^2$ is negative, two roots for $4a^3 + 27b^2 = 0$ and one root for $4a^3 + 27b^2$ is positive. The codimension of the cusp catastrophe is 2. This is illustrated in Figure 7.28.
3. **Canonical Swallowtail Catastrophe**
   Consider the function $f(x, a, b, c) = x^4 + ax^2 + bx + c$. The bifurcation set is shown to the left in Figure 7.29. The origin at $x = 0$, called the swallow tail point, lies in the middle of a line of cusps (two sets of cusps below and one above). In turn, this line of cusps runs through a plane of folds. The figure to the

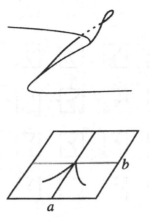

FIGURE 7.28. Two folded-over sheets of solutions of the canonical cusp catastrophe meeting at the "cusp point" at the origin

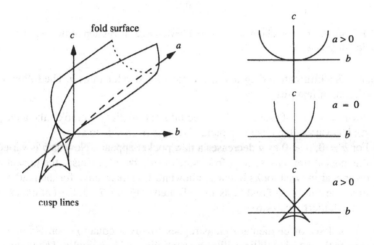

FIGURE 7.29. Lines of cusps and fold comprising the swallowtail bifurcation (on the left) and projections of the bifurcation surfaces on the b-c plane for $a > 0$, $a = 0$, and $a < 0$.

right shows the projection of the catastrophe set into the $b - c$ plane for differing values of a showing the number of real solutions of $x^4 + ax^2 + bx + c = 0$ in the different regions. The derivation of these regions is relegated to the exercises (Problem 7.23). The codimension of the swallowtail is 3.

4. **Canonical Butterfly Catastrophe**

Consider the function $f(x, a, b, c, d) = x^5 + ax^3 + bx^2 + cx + d$. Figure 7.30 shows the bifurcation diagram of the butterfly catastrophe. The figure shows a succession of bifurcation diagrams for various values of a, b. Each of the inset

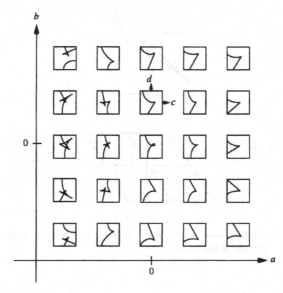

FIGURE 7.30. Bifurcation diagram of the swallowtail as a succession of two-dimensional bifurcation diagrams

squares is a bifurcation diagram in c, d for a fixed value of a, b. The following is a listing of features:

- For $b = 0$, $a > 0$ the section of the bifurcation diagram looks like a cusp. As b is varied, the cusp like picture swings from side to side.
- For $b = 0$, $a < 0$ as a decreases a new pocket appears. Now as b is varied, this pocket not only swings from side to side but also causes the pocket to disappear in what looks like a swallowtail. It is instructive for the reader to fill in the numbers of real roots are indicated in Figure 7.30. The codimension of the butterfly catastrophe is 4.

The second set of elementary catastrophes involves equations on $\mathbb{R}^2$ and is referred to as the set of umbilics: elliptic, parabolic, and hyperbolic. The *umbilics* are relevant to the singularity set of equations from $\mathbb{R}^2 \mapsto \mathbb{R}^2$.

1. **Canonical Ellipitic Umbilic** The canonical equations for the elliptic umbilic, which is a catastrophe of codimension 3, are

$$f_1(x_1, x_2) = 3x_1^2 - 3x_2^2 + 2ax_1 + b,$$
$$f_2(x_1, x_2) = -6x_1x_2 + 2ax_2 + c. \tag{7.61}$$

The singularity set for the elliptic umbilic is shown in Figure 7.31. The two conical regions are primarily fold regions except for the cusp lines embedded in them (on the right side there are lines of dual cusps). The central organizing point at $a = b = c = 0$ of the singularity is referred to as a monkey saddle. It

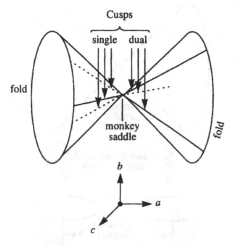

FIGURE 7.31. Singularity set of a canonical elliptic umbilic

is instructive for the reader to determine numbers of solutions to the equations (7.61) in the different regions (Problem 7.24).

2. **Canonical Hyperbolic Umbilic** The canonical form for the hyperbolic umbilic, another codimension 3 catastrophe, is given by

$$f_1(x_1, x_2) = 3x_1^2 + ax_2 + b,$$
$$f_2(x_1, x_2) = 3x_2^2 + ax_1 + c. \tag{7.62}$$

The singularity set is shown in Figure 7.32; it has, like the elliptic umbilic, two cones of fold surfaces with an embedded line of cusps on the left and double cusps on the right. There is nothing particularly striking about the organizing point of the hyperbolic umbilic. The discussion of the number of solutions is left to Problem 7.25.

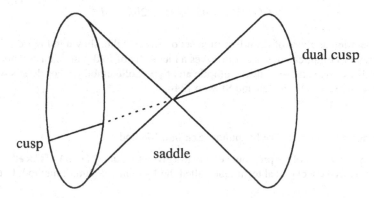

FIGURE 7.32. Singularity set of a canonical hyperbolic umbilic

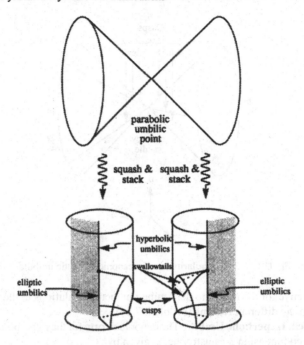

FIGURE 7.33. Singularity set of a canonical parabolic umbilic

3. Canonical Parabolic Umbilic

The canonical parabolic umbilic, the most sophisticated of the codimension 4 catastrophes, is given by

$$f_1(x_1, x_2) = 2x_1x_2 + 2ax_1 + c$$
$$f_2(x_1, x_2) = 4x_2^3 + x_1^2 + 2bx_2 + d \tag{7.63}$$

A detailed analysis of the bifurcation set of this umbilic, shown in Figure 7.33 is rather complicated, since it involves a hierarchy of fold points, cusp points, swallowtails and even lines of elliptic and hyperbolic umbilics. For details we refer the reader to Poston and Stewart [244].

Bifurcation Functions: The Lyapunov–Schmidt Method

To apply the theory of the previous section to a given function $f(x, \mu) : \mathbb{R}^n \times \mathbb{R}^k \mapsto \mathbb{R}^n$, we describe a classical technique called the Lyapunov Schmidt method. Let

$$f(x_0, \mu_0) = 0, \quad \det D_1 f(x_0, \mu_0) = 0.$$

If p is the rank defect of $D_1 f(x_0, \mu_0)$, choose nonsingular matrices $[P_u \mid U]$ and $[P_v \mid V]$ with $U, V \in \mathbb{R}^{n \times p}$ such that

$$\det P_v^T D_1 f(x_0, \mu_0) P_u \neq 0,$$
$$V^T D_1 f(x_0, \mu_0) = 0, \tag{7.64}$$
$$D_1 f(x_0, \mu_0) U = 0.$$

Decomposing $x = P_u w + U q$ with $w \in \mathbb{R}^{n-p}, q \in \mathbb{R}^p$, we can rewrite the original equation as

$$P_v^T f(P_u w + U q, \mu) = 0,$$
$$V^T f(P_u w + U q, \mu) = 0. \tag{7.65}$$

If w_0, q_0 correspond to x_0, it follows that in a neighborhood of w_0, q_0, μ_0, the first of the equations above can be solved by the implicit function theorem to give $w = w^*(q, \mu)$. Substituting this solution in the second equation above gives a set of equations from $\mathbb{R}^p \times \mathbb{R}^k \mapsto \mathbb{R}^p$ described by

$$N(q, \mu) := V^T f(P_u w^*(q, \mu) + U q, \mu) = 0. \tag{7.66}$$

Since $V^T D_1 f(x_0, \mu_0) = 0$, it follows that

$$D_1 N(q_0, \mu_0) = 0.$$

Since both N and its first derivative vanish at q_0, μ_0, N is referred to as *the bifurcation function*. The bifurcation function is a function on a lower dimensional space ($\mathbb{R}^p$), and the preceding procedure is referred to as the Lyapunov–Schmidt procedure (not unlike one you saw in Section 4.1.3). The Lyapunov–Schmidt procedure is also a generalization of the splitting lemma of Problem 3.34 in Chapter 3.

The bifurcation functions are now analyzed by transforming them into one of the "canonical" catastrophes described in the previous section for the case that $p = 1, 2$. Whether or not the full unfolding is observed in a given application depends on the nature of the parametric dependence of the bifurcation function on μ. However, as a consequence of the completeness of the unfolding presented earlier, if the first non-vanishing terms in the Taylor series of the bifurcation function match one of the canonical catastrophes described in the previous section, the bifurcation pattern that will be observed is a section of the fold, cusp, swallowtail, butterfly, or elliptic, hyperbolic or parabolic umbilic.

7.10.2 Singular Perturbations and Jump Behavior of Systems

We return to a topic that we began discussing in Chapter 6 when we were discussing solution concepts for implicitly defined differential equations of the form

$$\dot{x} = f(x, y),$$
$$0 = g(x, y), \tag{7.67}$$

with $x \in \mathbb{R}^n$, $y \in \mathbb{R}^m$. Of special interest to us will be the singular or so-called *impasse* points that is those x, y on the constraint manifold $g(x, y) = 0$ for which the matrix $D_2 g(x, y) \in \mathbb{R}^{m \times m}$ is singular. At the impasse points it may not be possible to solve for y continuously as a function of x. In the bifurcation terminology, x is a bifurcation parameter of the algebraic equation $g(x, y) : \mathbb{R}^m \mapsto \mathbb{R}^m$. The singular points are the points of bifurcation of the algebraic equation. Thus, a solution concept for the implicitly defined system (7.67) should allow for instantaneous jumps in the y variable (x is constrained to be at least an absolutely continuous function of time because of the first equation in (7.67) at points of bifurcation of the algebraic equation $g(x, y) = 0$). The question arises as to which possible values can y take after a discontinuous jump. Actually, without further information there is no reason to expect y to be a continuous function of time even away from singular points of the algebraic equation, since presumably y could take any of the multiple solutions of the equation

$$g(x, y) = 0,$$

switching arbitrarily frequently between different solution branches. To be able to insist on a "maximally smooth" variation of y with time, we regularize the system (7.67) as in Chapter 6 to be

$$\dot{x} = f(x, y),$$
$$\epsilon \dot{y} = g(x, y), \tag{7.68}$$

where $\epsilon \in \mathbb{R}_+$ models the parasitic dynamics associated with the system which were singularly perturbed away to get (7.67). One sees that if one were to define solutions of (7.67) as the limit of trajectories of (7.68), then so long as the values (x, y) represent stable equilibria of the frozen or fast boundary layer system (x is the frozen variable),

$$\frac{dy}{d\tau} = g(x_0, y). \tag{7.69}$$

As in the previous chapter, we will assume that the frozen boundary layer system is either a gradient system or a gradient-like system. Then the limit trajectories as $\epsilon \downarrow 0$ of (7.68) will have y varying continuously as a smooth function of time. However, once a point of singularity has been reached, the continuous variation of y is no longer guaranteed. By assumption, we have that the frozen boundary layer system (7.69) is gradient-like, that is, there exist finitely many equilibrium points for the system for each x_0 and further more all trajectories converge to one of the equilibrium points. This assumption is satisfied if, for example, there exists a proper function $S(x, y) : \mathbb{R}^n \times \mathbb{R}^m \mapsto \mathbb{R}$ such that $g(x, y) = D_2 S(x, y)$, and $S(x, y)$ has finitely many stationary points, i.e., where $g(x, y) = 0$ for each $x \in \mathbb{R}^n$. We will denote by $S_{y_i}^{x_0}$ the inset of the equilibrium point $y_i(x_0)$ of the frozen boundary layer system (7.69) with x_0 frozen. The index of the $p(x_0)$ equilibria is $i = 0, \ldots, p(x_0)$. Motivated by the analysis of stable sets of initial conditions in singularly perturbed systems as discussed in Section 6.3 of the preceding chapter, we will define the system to allow for *a jump* from y_0 if for all neighborhoods V

of (x_0, y_0) in $\{x_0\} \times \mathbb{R}^m$,

$$V \not\subset S_{y_0}^{x_0}.$$

Further, if all sufficiently small neighborhoods V of (x_0, y_0) in $\{x_0\} \times \mathbb{R}^m$ can be decomposed as

$$V = (V \cap S_{y_0}^{x_0}) \cup (V \cap S_{y_1}^{x_0}) \cup \cdots \cup (V \cap S_{y_q}^{x_0})$$

with $V \cap S_{y_i}^{x_0} \neq \varnothing$, for $i = 1, \ldots, q$, then the system (7.67) allows for jumps from (x_0, y_0) to one of the following points: $(x_0, y_1), \ldots, (x_0, y_q)$. It is useful to see that from this definition jump is optional from non- singular points if they are not stable equilibrium points of the frozen boundary layer system. Also the definition does not allow for jumps from stable equilibria of the boundary layer system.

With this definition of jump behavior it is easy to see how the degenerate van der Pol oscillator of Chapter 2 fits the definition. The reader may wish to review the example of (2.24) and use the foregoing discussion to explain the phase portrait of Figure 2.31. Other examples of jump behavior in circuits are given in the Exercises (Problems 7.27 and 7.28). These problems illustrate that jump behavior is the mechanism by which circuits with continuous elements produce digital outputs. In general terms it is of interest to derive the bifurcation functions associated with different kinds of bifurcations of the equilibria of the frozen boundary layer system and classify the flow of the systems (7.67) and (7.68) close to the singularity boundaries. For a start to this classification, the interested reader should look up Sastry and Desoer [263] (see also Takens [293]). The drawbacks to our definition of jumps are:

1. It is somewhat involved to prove conditions under which trajectories of the augmented system (7.68) converge to those defined above as $\epsilon \downarrow 0$. The chief difficulty is with the behavior around saddle or unstable equilibria of the frozen boundary layer system. A probabilistic definition is more likely to be productive (see [262]) but is beyond the scope of this book.

2. The assumption of "gradient like" frozen boundary layer dynamics is needed to make sure that there are no dynamic bifurcations, such as the Hopf bifurcation discussed above and others discussed in the next section, since there are no limits as $\epsilon \downarrow 0$ for closed orbits in the boundary layer system dynamics (again a probabilistic approach is productive here, see [262]).

3. It is difficult to prove an existence theorem for a solution concept described above, because of the possibility of an accumulation point of jump times, referred to as Zeno behavior. In this case a Fillipov-like solution concept (see Chapter 3) needs to be used.

7.10.3 Dynamic Bifurcations: A Zoo

Dynamic bifurcations are bifurcations involving not only equilibrium points of vector fields and fixed points of maps, but other invariant sets. Let us begin by considering bifurcations of periodic orbits of vector fields. Thus, let $\gamma(t, \mu) \in \mathbb{R}^n$,

$t \in [0, T(\mu)]$ with $\gamma(0) = \gamma(T(\mu))$ represent a periodic orbit with period $T(\mu) \in \mathbb{R}_+$ of the parameterized differential equation on $\mathbb{R}^n$ given by

$$\dot{x} = f(x, \mu). \tag{7.70}$$

As in Section 7.3 we may define the Poincaré map associated with the stability of this closed orbit. Denote the Poincaré map $P(\cdot, \mu) : \mathbb{R}^{n-1} \mapsto \mathbb{R}^{n-1}$ where $\mathbb{R}^{n-1}$ is a local parameterization of a transverse section Σ located at $\gamma(0)$. Without loss of generality map $\gamma(0)$ to the origin. Then, the stability of the fixed point at the origin of the discrete time system consisting of this map, namely

$$q_{k+1} = P(q_k, \mu), \tag{7.71}$$

determines the stability of the closed orbit. So long as the closed orbit is hyperbolic, that is the fixed point at the origin of (7.71) is hyperbolic, the fixed point changes continuously with the parameter μ, as does the periodic orbit $\gamma(\cdot, \mu)$ of the continuous time system. However, when eigenvalues of the Poincare map system (7.71) cross the boundary of the unit disk, the bifurcations of this system reflect into the dynamics of the continuous-time system:

1. *Eigenvalues of the Poincaré map cross the boundary of the unit disk at 1.* There are likely to be changes in the number of fixed points: for example, through folds, transcritical, or pitch fork bifurcations of the Poincaré system. For example, a saddle node in the Poincaré map system will be manifested as a saddle periodic orbit merging into a stable (or unstable) periodic orbit of (7.70). This is referred to in Abraham and Marsden [1] as *dynamic creation* or *dual suicide* depending on whether the two periodic orbits are created or disappear. Dynamic pitchforks consisting of the merger of three periodic orbits into one are also possible.

2. *Eigenvalues of the Poincaré map cross the boundary of the unit disk at −1.* A period doubling bifurcation in the Poincaré map system (7.71) may be manifested as an attractive closed orbit of the system (7.70) becoming a saddle closed orbit as one of the eigenvalues of the associated Poincaré map goes outside the unit disk, and a new attractive closed orbit of twice the period is emitted from it (see Figure 7.34). The reader should be able to fill in the details when the closed orbit is initially a saddle orbit as well. This bifurcation is referred to as *subharmonic resonance, flip bifurcation or subtle division [1]*.

3. *Eigenvalues of the Poincaré map cross the boundary of the unit disk at complex locations.* A Naimark–Sacker bifurcation of the system 7.71 results in the creation of an invariant torus T^2 in the original system (7.70). Thus, as the eigenvalues of the Poincaré map cross from the inside of the unit disk to the outside, the stable closed orbit becomes a saddle closed orbit surrounded by a stable invariant torus (see Figure 7.35). The dynamics on the stable invariant torus will consist of a finite number of closed orbits some attracting. This bifurcation is referred to as *Naimark excitation*.

One other class of dynamic bifurcations is the *saddle connection* bifurcation which we first saw in Chapter 2 in the context of the Josephson junction, when a saddle connection bifurcated into a finite period closed orbit on the cylinder

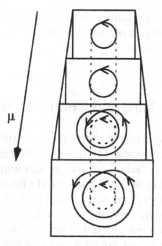

FIGURE 7.34. The flip bifurcation of a stable closed orbit into a saddle closed orbit (shown dotted) surrounded by a period-two stable closed orbit

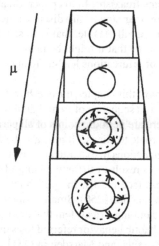

FIGURE 7.35. Naimark excitation showing the emergence of a stable invariant torus surrounding a closed orbit changing from being stable to a saddle

$\mathbb{R} \times S^1$. An example of saddle switching where the outset of a saddle moved from the domain of one attractor to another was presented earlier in this chapter. These bifurcations are *global*, in that they are difficult to predict by any linearization or characteristic values.

One can now conceive of other more complex bifurcations: One could, for example, conceive of a generalization of the Naimark excitation where a torus T^k gets excited to a torus T^{k+1} (called a *Takens excitation*). However, the list of all the possible bifurcations of dynamical systems is in its infancy: The examples that we have given are just a start in beginning to dive deeply into the many mysteries of nonlinear dynamical systems.

7.11   Routes to Chaos and Complex Dynamics

The qualitative analysis of dynamical systems is certainly the heart of nonlinear systems, and the names of Poincaré, Lyapunov, and others come to mind as early pioneers in the field. The advances, of course, also went hand in hand with developments in classical mechanics. Abraham and Marsden [1] give a good sense of the synergy and excitement, as does Arnold [12]. However, in the last thirty years, starting from the publication of a program laid out by Smale in [277], there has been a tremendous amount of interest, not only in an understanding of how dynamical systems come to have complex aperiodic behavior of the kind that we saw earlier in this book in Chapter 1. Hardy and Littlewood and van der Pol had already noticed this behavior (and by some accounts thought of this complex behavior of dynamical systems freakish). In his paper, Smale showed that complex behavior in dynamical systems arising from a discrete time nonlinear system called the Smale horseshoe is structurally stable, that is to say, that it could not be perturbed away. He also suggested that it might be possible to understand "routes to chaos," meaning a pattern of bifurcations leading to aperiodic complex or chaotic dynamics.

In light of the discussions of this chapter, the one-hump map example of Chapter 1 may now be understood as a sequence of period doublings leading up to a finite parameter value where there are periodic points of all periods, and the determination of a "period doubling route to chaos." It may also be determined that periodic perturbations of saddle connections in dynamical systems under certain conditions (made precise by Melnikov) result in the embedding of the Smale horseshoe in the dynamics of the system (see [328]). In applications this results in complex dynamics in the Josephson junction [250]. Small periodic perturbations or the addition of small amount of damping to a system that has invariant torii results in an even more subtle form of chaotic behavior referred to as the Arnold diffusion. This was discussed by Salam, Varaiya, and Marsden in [251] and [252], along with its implications for complex domains of attraction of equilibria of the power system swing dynamics.

The literature in complex or chaotic dynamics, its control, and bifurcation routes leading to it is immense and growing. There are several well known chaotic systems besides the one-hump map, the Smale horseshoe, the Arnold diffusion, such as the Lorenz attractor [187] and the Chua circuit [332]. The control of bifurcations and chaos is another very active area of the literature. We refer the reader to the work of Abed and coworkers for control of bifurcation [185], some connections between period doubling bifurcations and the harmonic balance method of Chapter 4 [300], The control of chaos has also been very a topic of great excitement, initially in the physics literature with the work of Ott, Grebogi and Yorke, but more recently in the control literature see [324; 230; 188].

7.12 Exercises

Problem 7.1. Give a proof of Proposition 7.2.

Problem 7.2. Give several examples of systems that have equilibrium points whose stable and unstable manifolds intersect.

Problem 7.3. Use the chain rule to prove Proposition 7.12.

Problem 7.4. Use the fact that solutions of a differential equation depend smoothly on their initial conditions over a compact time interval to prove that the Poincaré map is well defined for closed orbits.

Problem 7.5 Forced oscillation.   Consider the planar second-order system

$$\ddot{x} + \delta \dot{x} + \omega_0^2 x = \epsilon \cos(\omega t) \tag{7.72}$$

Derive the conditions for a stable forced oscillation and the Poincaré map for the case that $\omega_0 \neq \omega$. Now consider the case where $\omega = \omega_0$. Show that there is no difficulty when $\delta \neq 0$.

Problem 7.6 Recurrent points and minimal sets.   In Chapter 2, Problem 2.14 you showed that in $\mathbb{R}^2$, all recurrent points are either equilibrium points or lie on closed orbits. Give a few examples of why this is false in higher dimensions. Also, in Chapter 2, Problem 2.15, you showed that in $\mathbb{R}^2$ all unbounded minimal sets consists of a single trajectory with empty α, ω limit sets. Give examples of the falseness of this in higher dimensions.

Problem 7.7. Compute the center manifold for the following systems and study their stability

1. from [122]

$$\dot{y} = -y - \alpha(y + z)^2 + \beta(yz + y^2),$$
$$\dot{z} = \alpha(y + z)^2 - \beta(yz + z^2).$$

2. from [122]

$$\dot{x}_1 = -x_2^2,$$
$$\dot{x}_2 = -x_2 + x_1^2 + x_1 x_2.$$

3. from [122] with $a \neq 0$:

$$\dot{x}_1 = a x_1^2 - x_2^2,$$
$$\dot{x}_2 = -x_2 + x_1^2 + x_1 x_2.$$

4.

$$\dot{x}_1 = x_1^2 x_2 - x_1^5,$$
$$\dot{x}_2 = -x_2 + x_1^2.$$

5. from [155]

$$\dot{x}_1 = x_1 x_2 + a x_1^3 + b x_1 x_2^2,$$
$$\dot{x}_2 = -x_2 + c x_1^2 + d x_1^2 x_2,$$

in the following cases:

a. $a + c > 0$.
b. $a + c < 0$.
c. $a + c = 0$ and $cd = bc^2$ respectively less than 0, greater than 0 and equal to 0.

6.

$$\dot{x}_1 = -x_1 + a x_1 x_2 + b x_2^2,$$
$$\dot{x}_2 = c x_1 x_2 - x_2^3,$$

for different values of a, b, c.

Problem 7.8 Stablization on the center manifold.   Consider the control system (adapted from [149])

$$\dot{x}_1 = a x_2 + u,$$
$$\dot{x}_2 = -x_2^3 + b x_1 x_1^m. \tag{7.73}$$

Verify that the linearization of the control system has an uncontrollable eigenvalue at 0. Verify that it satisfies the necessary condition for stabilization from Problem 6.9 of Chapter 6. Now, consider the state feedback law:

$$u = -k x_1 + \psi(x_1),$$

where $\psi(0), \psi'(0) = 0$. Show that the origin is a stable (asymptotically stable, unstable) equilibrium of the center manifold in the following cases:

1. $m = 0$, $ab < 0$ asymptotically stable center manifold for all values of k.
2. $m = 1$ always unstable.
3. $m = 2$ asymptotically stable for $k > \max(0, ab)$.
4. $m \geq 3$ asymptotically stable for all $k > 0$.

Problem 7.9 Wind-induced oscillation system [122].   Consider the two dimensional system

$$\dot{x}_1 = -\mu_1 x_1 - \mu_2 x_2 + x_1 x_2,$$
$$\dot{x}_2 = \mu_2 x_1 - \mu_1 x_2 + \frac{1}{2}(x_1^2 + x_2^2),$$

where μ_1 represents a damping factor and μ_2 a de-tuning factor. Note that the system is Hamiltonian for $\mu_1 = 0$. Draw the phase portrait to explicitly indicate heteroclinic orbits. Is there resemblance to any of the portraits shown in Figure 7.15. What happens for $\mu_1 \neq 0$?

Problem 7.10 Models for surge in jet engine compressors [234]. A simple second order model for explaining surge in jet engine compressors (which can cause loss of efficiency and sometimes stall) is given by

$$
\begin{aligned}
\dot{x} &= B(C(x) - y), \\
\dot{y} &= \frac{1}{B}(x - f_\alpha^{-1}(y)).
\end{aligned}
\tag{7.74}
$$

Here x stands for the non-dimensional compressor mass flow, y represents the plenum pressure in the compressor unit, B stands for the non-dimensional compressor speed, and α, the throttle area. For definiteness, we will assume the functional form of the compressor characteristic $C(x) = -x^3 + (3/2)(b+a)x^2 - 3abx + (2c + 3ab^2 - b^3)/3$, and the functional form of the throttle characteristic $f_\alpha(x) = (x^2/\alpha^2)\mathrm{sgn}(x)$. Here, $a, b, c \in \mathbb{R}_+$ are constants determining the qualitative shape of the compressor characteristic. The goal of this problem is to study the bifurcations of this system for fixed a, b, c with bifurcation parameters α, B.

1. Linearize the system about its equilibria and find the stability number and type of the equilibria. Classify saddle nodes and possible Hopf bifurcations in the positive quadrant where $x, y > 0$. In particular show that an equilibrium at (x^*, y^*) is asymptotically stable provided that $x^* \notin \,]a, b[$. For $x^* \in \,]a, b[$ determine the stability as a function of the parameters B, α.
2. Use theorems and methods from Chapter 2, including the determination of an attractive positively invariant region and the Bendixson theorem to establish regions in the positive quadrant where the closed orbits lie.
3. Study what happens to the dynamics of (7.74) in the limit that the compressor speed $B \to \infty$.

Problem 7.11. Use the center manifold stability theorem to establish the local asymptotic stability of the origin of the system

$$
\begin{aligned}
\dot{z} &= f(z, y), \\
\dot{y} &= Ay + p(z, y),
\end{aligned}
\tag{7.75}
$$

given that the eigenvalues of $A \in \mathbb{R}^{n \times n}$ are in $\mathbb{C}_-^\circ$, $z = 0$ is an asymptotically stable equilibrium point of $\dot{z} = f(z, 0)$, $p(z, 0) \equiv 0$ for all z near 0, and further that

$$
\frac{\partial p}{\partial y}(0, 0) = 0.
$$

Problem 7.12 Center manifolds for maps. Compute the approximation of the center manifold for the following examples (from [329]):

1.

$$x_{1n+1} = 2x_{1n} + 3x_{2n},$$
$$x_{2n+1} = x_{1n} + x_{1_n}^2 + x_{1n}x_{2_n}^2.$$

2.

$$x_{1n+1} = x_{1n} + x_{3_n}^4,$$
$$x_{2n+1} = -x_{1n} - 2x_{2n} - x_{1_n}^3,$$
$$x_{3n+1} = x_{2n} - \frac{1}{2}x_{3n} + x_{2_n}^2.$$

3.

$$x_{1n+1} = x_{1n} - x_{3_n}^3,$$
$$x_{2n+1} = 2x_{2n} - x_{3n},$$
$$x_{3n+1} = x_{1n} + \frac{1}{2}x_{3n} + x_{1_n}^3.$$

Problem 7.13 Failure of series approximation of center manifolds. Find the family of center manifolds for the system

$$\dot{y} = -y + z^2,$$
$$\dot{z} = -z^3,$$

by explicitly solving the center manifold PDE. Now evaluate its Taylor series at the origin. Can you see what would happen if one were to try to obtain the Taylor series approximation?

Problem 7.14 Derivation of fold conditions. By differentiating the equation (7.31) prove the formulas given in equations (7.32, 7.33).

Problem 7.15 Derivation of transcritical conditions. Derive the conditions (7.37) and (7.38) for the transcritical bifurcation by following the steps for the fold bifurcation except applied to the function $G(x, \mu)$. Use the relationships (7.36) to obtain conditions on $g(x, \mu)$.

Problem 7.16 Derivation of pitchfork bifurcation. Derive the conditions (7.39) for the pitch fork bifurcation by applying the steps of the fold bifurcation to the function $G(x, \mu)$.

Problem 7.17 Derivation of the discrete-time fold. Derive the conditions for the discrete-time fold or saddle-node bifurcation, that is, equations (7.50).

Problem 7.18 Derivation of the discrete-time transcritical. Derive the conditions for the discrete-time transcritical bifurcation, that is, equations (7.52).

Problem 7.19 Derivation of the discrete-time pitchfork. Derive the conditions for the discrete-time pitchfork bifurcation, that is the equations (7.54).

Problem 7.20 Period doubling in the one hump map. Use the conditions for period doubling (7.58) to check the successive period doubling of the one hump map (1.3)

$$x_{n+1} = hx_n(1 - x_n)$$

of Chapter 1 at $h = 3, h = 1 = \sqrt{6}$, etc.

Problem 7.21 Double-zero eigenvalue bifurcations [329]. Consider the two-dimensional system with two parameters μ_1, μ_2 given by

$$\dot{x}_1 = x_2,$$
$$\dot{x}_2 = \mu_1 + \mu_2 x_2 + x_1^2 + x_1 x_2. \tag{7.76}$$

This is actually a normal form for a double-zero eigenvalue (with nontrivial Jordan structure) bifurcation at $\mu_1 = \mu_2 = 0$. Draw the bifurcation diagram in the μ_1, μ_2 plane taking care to draw bifurcation surfaces and identifying the various bifurcations. Also, repeat this problem with the last entry of equation (7.76) being $-x_1 x_2$.

Problem 7.22 Bifurcation diagram of a cusp. Derive the bifurcation diagram for the cubic equation $x^3 + ax + b = 0$. The key is to figure out when two out of the three real roots coincide and annihilate. You can do this by determining values of x at which both $x^3 + ax + b = 0$ and $\frac{d}{dx}x^3 + ax + b = 0$. Of course, you can also solve the cubic equation in closed form and get the answer that way. Be sure to derive Figure 7.28.

Problem 7.23 Bifurcation diagram of a swallowtail. Derive the bifurcation diagram of a swallowtail (Figure 7.29). Use the method of the previous problem to determine the regions where the number of solutions transition from 4 to 2 to 0.

Problem 7.24 Bifurcation diagram of an elliptic umbilic. Derive the bifurcation diagram of an elliptic umbilic as shown in Figure 7.31. Use computer tools judiciously to help you fill in the number and type of solutions in the different regions. It is useful to note that the elliptic umbilic of equation (7.61) is the gradient of the function $v(x_1, x_2) = x_1^3 - 3x_1 x_2^2 + a(x_1^2 + x_2^2) + bx_1 + cx_2$.

Problem 7.25 Bifurcation diagram of a hyperbolic umbilic. Derive the bifurcation diagram of an elliptic umbilic as shown in Figure 7.32. Use computer tools and elimination of one of the variables x_1, x_2 judiciously to help you fill in the number and type of solutions in the different regions. It is useful to note that the hyperbolic umbilic of equation (7.62) is the gradient of the function $v(x_1, x_2) = x_1^3 + x_2^3 + ax_1 x_2 + bx_1 + cx_2$.

Problem 7.26 Stabilization of systems with uncontrollable linearization [21]. Consider the problem of stabilizing a nonlinear, single input system by state feedback, to simplify matters assume that the control enters linearly: we consider systems of the form

$$\dot{x} = f(x) + bu$$

with $x \in \mathbb{R}^n$ and $f(0) = 0$. In Chapter 6 we established that the system is not stabilizable by smooth feedback if the linearization is not stabilizable for some $s \in \mathbb{C}_+^o$, rank$[sI - A \mid b] < n$. We also showed how a linear feedback locally asymptotically stabilizes the system if rank $[sI - A \mid b] = n$ for all $s \in \mathbb{C}$. In this exercise, we study how it may be possible to use some nonlinear (smooth) feedback law to asymptotically stabilize systems which may have some $j\omega$ axis uncontrollable modes, that is, for some ω, rank $[j\omega I - A \mid b] < n$. The current problem focuses on a very specific system; for the more general theory refer to Behtash and Sastry [21]. Consider the system

$$\dot{x}_1 = x_2 - x_1^3 + x_1 x_2^2 - 2x_2 x_3,$$

$$\dot{x}_2 = x_1^3 + x_1 x_3, \tag{7.77}$$

$$\dot{x}_3 = -5x_3 + u.$$

Check that there is a double-zero eigenvalue of the linearization that is not linearly controllable. Now, choose

$$u = \alpha x_1^2 + \beta x_1 x_2 + \gamma x_2^2.$$

Write the dynamics on the center manifold for this system. Determine α, β, γ to have an asymptotically stable center manifold system. You may wish to use quadratic or quartic Lyapunov functions to help you along.

To study the robustness of the control law that you derived above, perturb the first equation of (7.77) by ϵx_1. Then, for all positive ϵ, for all smooth feedback the origin is unstable. However, verify that the control law that you derived earlier actually stabilizes the x_1, x_2 variables to a neighborhood of the origin. Can you actually determine the ω limit set of trajectories by doing a bifurcation analysis of the perturbed system (7.77) with the perturbation parameter ϵ as bifurcation parameter?

Problem 7.27 Two tunnel diodes [66]. Consider the circuit of Figure 7.36 with two tunnel diodes in series with a 1 Henry inductor. The constitutive equations of the two tunnel diodes are also given in the figure along with the composite characteristic. By introducing ϵ parasitic capacitors around the two voltage controlled resistors (as suggested in Section 6.4) as shown in the figure, give the jump dynamics of the circuit.

Problem 7.28 Astable multivibrator or clock circuit. Consider the astable multivibrator circuit of Figure 7.37 with two n-p-n transistors. Assume that the transistor characteristics are the usual exponentials associated with a bipolar junction

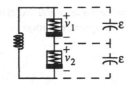

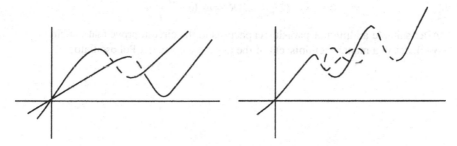

FIGURE 7.36. Two tunnel diode circuit with parasitics shown dotted

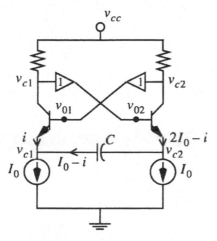

FIGURE 7.37. An astable multivibrator circuit

transistor as given in the equations (7.78) below. The circuit equations with the two transistors assumed identical are given by

$$\dot{v}_{e2} - \dot{v}_{e1} = \frac{1}{C}(I_0 - i),$$

$$v_{b2} = v_{c1} = v_{cc} - iR,$$

$$v_{b1} = v_{c2} = v_{cc} - (2I_0 - i)R,$$

$$v_{b1} - v_{e1} = v_T \log \frac{i}{I_S},$$

$$v_{b2} - v_{e2} = v_T \log \frac{2I_0 - i}{I_S}.$$

(7.78)

Here v_T, I_0, I_S are all constants; R, C stand for the resistance and capacitance values respectively. Simplify the equations to get an implicit equation for $v :=$ $v_{e2} - v_{e1}$:

$$\dot{v} = \frac{1}{C}(I_0 - i),$$

$$0 = v - (2I_0 - 2i)R - v_T \log \frac{2I_0 - i}{i}. \tag{7.79}$$

After suitable addition of parasitic capacitors to the circuit, prove that the circuit oscillates in a manner reminiscent of the degenerate van der Pol oscillator.

8
Basics of Differential Geometry

The material on matrix Lie groups in this chapter is based on notes written by Claire Tomlin, of Stanford University and Yi Ma.

In Chapter 3 we had an introduction to differential topology with definitions of manifolds, tangent spaces and their topological properties. In this chapter we study some elementary differential geometry which will provide useful background for the study of nonlinear control. In this chapter, we will study tangent spaces, vector fields, distributions, and codistributions from a differential geometric point of view. We will prove an important generalization of the existence–uniqueness theorem of differential equations for certain classes of distributions, called the Frobenius theorem. This theorem is fundamental to the study of control of nonlinear systems as we will see in the following chapters. Finally, we study matrix Lie groups, which are the simplest class of manifolds which are not vector spaces. The tangent space to Lie groups are especially simple and their study proves to be very instructive. In this context we will also study left invariant control systems on Lie groups. In some sense these are the simplest class of nonlinear control systems, behaving like linear systems with a Lie group as state space.

8.1 Tangent Spaces

In Chapter 3 we studied an intuitive definition of manifolds embedded in $\mathbb{R}^n$ and a geometric definition of their tangent spaces. However, for several formal reasons it is useful to have a more abstract definition of tangent spaces. In particular, it is useful to know what properties of a manifold are independent of the ambient space it is embedded in. Definitions that are independent of the embedding of a manifold in $\mathbb{R}^n$ are referred to as *intrinsic*. We first give an intuitive definition of tangent spaces. Let us denote by $C^\infty(p)$ the set of smooth real-valued functions defined in a neighborhood of a point p in $\mathbb{R}^n$ on the manifold.

Definition 8.1 Tangent Space. *Define the* tangent space *to M at p, denoted by* $T_p M$, *to be the vector space (over the reals) of mappings* $X_p : C^\infty(p) \mapsto \mathbb{R}$ *that satisfy the following two properties for all* $f, g \in C^\infty(p)$[1]:

- Linearity *For* $\alpha, \beta \in \mathbb{R}$,

$$X_p(\alpha f + \beta g) = \alpha X_p(f) + \beta X_p(g).$$

- Derivation

$$X_p(fg) = X_p(f)g(p) + f(p)X_p(g).$$

with the vector space operations in $T_p M$ *defined by*

$$(X_p + Y_p)(f) = X_p(f) + Y_p(f),$$
$$(\alpha X_p)(f) = \alpha X_p(f) \quad \alpha \in \mathbb{R}.$$

Elements of the tangent space *are called* tangent vectors.

This definition is somewhat abstract, but it will help the reader to think about tangent vectors as differential operators. For instance if $M = \mathbb{R}^m$, and the coordinate directions are r_i for $i = 1, \ldots, m$, examples of tangent vectors are the differential operators $E_{ia} = \partial/\partial r_i$ at every point $a \in \mathbb{R}^m$. They act on smooth functions by differentiation in the usual way as

$$E_{ia} f = \frac{\partial f}{\partial r_i}.$$

It is intuitively clear that the tangent space to $\mathbb{R}^m$ at any point can be identified with $\mathbb{R}^m$. We try to make this point clear by showing that the E_{ia} defined above for $i = 1, \ldots, m$ are a basis set for $T_a \mathbb{R}^m$.

It follows that every vector field in $\mathbb{R}^m$ can be described as

$$\alpha_i(x) E_{ia}$$

for some choice of smooth functions $\alpha_i(x)$. This construction can be used to develop an understanding of general vector fields on m dimensional manifolds. To this end, we develop a few definitions. If $F : M \mapsto N$ is a smooth map, we define a map called the *differential map* or *push forward map* as follows:

Definition 8.2 Push Forward Map or Differential Map. *Given a smooth map* $F : M \mapsto N$ *we define the* push forward, *or* differential, *map*

$$F_* : T_p M \mapsto T_{F(p)} N.$$

---

[1]Mappings satisfying these properties are referred to as *derivations*

at point $p \in M$ as follows: given $X_p \in T_p M$ and $f \in C^\infty(F(p))$ define

$$F_{*p} X_p(f) = X_p(f \circ F).$$

Indeed, it is easy to check that $F_{*p} X_p \in T_{F(p)} N$ and it is linear on the reals. As for checking whether $F_{*p} X_p$ is a derivation, note that

$$\begin{aligned}
F_{*p} X_p(fg) &= X_p((f \circ F).(g \circ F)) \\
&= X_p(f \circ F) g(F(p)) + f(F(p)) X_p(g \circ F) \\
&= F_{*p} X_p(f) g(F(p)) + f(F(p)) F_{*p} X_p(g).
\end{aligned}$$

The following properties are a consequence of this definition of *derivative maps*.

Proposition 8.3 Properties of Differential Maps.

1. *If $H = F \circ G$ is the composition of two smooth maps F, G then $H_{*p} = G_{*F(p)} \circ F_{*p}$.*
2. *If $id : M \mapsto M$ is the identity map, then $Id_{*p} : T_p M \mapsto T_p M$ is also the identity map.*
3. *For a diffeomorphism $F : M \mapsto N$ we have*

$$F_{*F(p)}^{-1} = (F_{*p})^{-1}.$$

Proof: See exercises.

The definition above can be used to develop an understanding of the tangent space of a general manifold M using the model of $\mathbb{R}^m$ discussed above: indeed, let U, ψ be a coordinate chart for the manifold M in a neighborhood of the point $p \subset M$. Then ψ^{-1} is a diffeomorphism from $\psi(U) \mapsto U$. The push forward of ψ^{-1} at $\psi(p)$ yields a basis set for $T_p M$ by Proposition 8.3 above

$$\psi_{*\psi(p)}^{-1} \frac{\partial}{\partial r_i}.$$

By abuse of notation, these basis elements are simply referred to as $\frac{\partial}{\partial x_i}|_p$ with the understanding that if $f \in C^\infty(p)$, then

$$\frac{\partial f}{\partial x_i}|_p(f) = \frac{\partial f \circ \psi^{-1}}{\partial r_i}.$$

Notice that the abuse of notation is consistent with the fact that $f \circ \psi^{-1}$ is a local representation of f. When the coordinate chart used to describe the neighborhood of the manifold changes the basis for $T_p M$ changes. For instance a given vector field $X_p \in T_p M$ may be represented differently with respect to the basis given by U, ψ namely $\frac{\partial}{\partial x_i}$ as

$$X_p = \sum_{i=1}^{m} \alpha_i(p) \frac{\partial}{\partial x_i},$$

or with respect to the basis given by v, ϕ namely $\frac{\partial}{\partial z_i}$ as

$$X_p = \sum_{i=1}^{m} \beta(p) \frac{\partial}{\partial z_i}.$$

Problem 3.29 gives the matrix transforming α into β. This discussion may be extended to a description of the tangent map to a smooth map $F : M \mapsto N$. Indeed, let U, x and V, z be coordinate charts for M, N in neighborhoods of $p \in M$ and $F(p) \in N$. Then, with respect to the bases $\frac{\partial}{\partial x_i}$ and $\frac{\partial}{\partial z_j}$ the representation of the tangent map is the Jacobian of the local representation of F, namely $\phi \circ F \circ \psi^{-1} = (F_1(x_1, \ldots, x_m), \ldots, F_m(x_1, \ldots, x_m))^T$. That is,

$$F_{*p} \left(\left. \frac{\partial}{\partial x_i} \right|_p \right) (z_j) = \frac{\partial}{\partial x_i} (z_j \circ F)|_p = \frac{\partial F_j}{\partial x_i}.$$

The *tangent bundle* of a smooth manifold is defined as

$$TM = \bigcup_{p \in M} T_p M. \tag{8.1}$$

There is a natural embedding of the manifold M in the tangent bundle TM. We refer to this embedding as the canonical projection $\pi : TM \mapsto M$. It is possible to endow the tangent bundle of a manifold with a manifold structure to obtain a manifold of dimension twice that of M. First note that if $M = \mathbb{R}^m$ then

$$T\mathbb{R}^m = \cup_{a \in \mathbb{R}^m} T_a \mathbb{R}^m \simeq \cup_{a \in \mathbb{R}^m} \mathbb{R}^m \simeq \mathbb{R}^m \times \mathbb{R}^m = \mathbb{R}^{2m}.$$

For a general manifold M with a coordinate chart U, ψ of the form $U, x_1, \ldots, x_m$ we define a coordinate chart $\tilde{U}, \tilde{\psi}$ on TM by setting

$$\tilde{U} = \pi^{-1}(U),$$

$$\tilde{\psi}(X_p) = (x_1(p), \ldots, x_m(p), x_{1*p} X_p, \ldots, x_{m*p} X_p),$$

Here $x_{i*p} X_p \in T_{x_i(p)} \mathbb{R} \simeq \mathbb{R}$ is the push forward of the coordinate map $x_i : M \mapsto \mathbb{R}$. These coordinate maps may be verified to be smooth and also to form an atlas for TM, establishing that TM is a $2m$ dimensional manifold. The choice of coordinates of $\tilde{\psi}$ is said to be *induced*.

Finally, given an injective smooth map $F : M \mapsto N$ there exists an extension to a *tangent map* $TF : TM \mapsto TN$ for $x \in M$ and $v \in T_p M$ defined by

$$TF(p, v) = (F(p), F_{*p} v).$$

8.1.1   Vector Fields, Lie Brackets, and Lie Algebras

Definition 8.4 Vector Field. *A smooth vector field X on a smooth manifold M is defined as a map*

$$X : M \mapsto TM$$

with the restriction that $\pi \circ X$ *is the identity on* M. *Here* π *is the natural projection from* TM *onto* M, *and* X *is referred to as a smooth* section *of the tangent bundle* TM.

The second condition in the above definition guarantees that there is exactly one tangent vector at each point $p \in M$. Further, the tangent vectors $X_p \in T_pM$ vary with respect to p in a smooth fashion. As a consequence of this definition, one can derive a local representation for a vector field consistent with a coordinate chart $U, \psi = U, x_1, \ldots, x_m$ for the manifold M. At a point $p \in M$ the vector X_p is equal to $\sum_{i=1}^{m} \alpha_i \frac{\partial}{\partial x_i}$. For other points in a neighborhood of the point p the formula for the vector field is given by

$$X_p = \sum_{i=1}^{m} \alpha_i(p) \frac{\partial}{\partial x_i} \tag{8.2}$$

where the $\alpha_i(p)$ are smooth functions of $p \in M$. Vector fields represent differential equations on manifolds. Indeed if $\sigma(t) :]a, b[\mapsto M$ is a curve on the manifold it is said to be an *integral curve* of the vector field X if

$$\dot{\sigma}(t) = X(\sigma(t)) \tag{8.3}$$

In local coordinates, these equations read

$$\dot{\sigma}_1(t) = \alpha_1(\sigma_1(t), \ldots, \sigma_m(t))$$
$$\vdots \tag{8.4}$$
$$\dot{\sigma}_m(t) = \alpha_n(\sigma_1(t), \ldots, \sigma_m(t))$$

By the existence and uniqueness theorem for ordinary differential equations, the existence of integral curves for a given smooth vector field is guaranteed. The vector field is said to be complete if the domain of definition of the integral curves can be chosen to be $]-\infty, \infty[$. In this case, the integral curves of a vector field define a one-parameter (parameterized by t) family of diffeomorphisms $\Phi_t(x) : M \mapsto M$ with the understanding that $\Phi_t(x)$ is the point on the integral curve starting from initial condition x at $t = 0$. (Make sure to convince yourself that the existence and uniqueness theorem for the solutions of a differential equation guarantees that the map $\Phi_t(x)$ is a diffeomorphism). This one parameter family of diffeomorphisms is referred to as the *flow of the vector field* X.

Vector fields transform naturally under diffeomorphisms. First, consider coordinate transformations: let $y = T(x)$ be a coordinate transformation from the local coordinates U, x to the coordinates V, y. Then, if the vector field X is represented as $\sum_{i=1}^{m} X_i(x) \frac{\partial}{\partial x_i}$ or $\sum_{i=1}^{m} Y_i(y) \frac{\partial}{\partial y_i}$ in the two coordinate systems, it follows that

$$\begin{bmatrix} Y_1(T(x)) \\ \vdots \\ Y_m(T(x)) \end{bmatrix} = \frac{\partial T}{\partial x}(x) \begin{bmatrix} X_1(x) \\ \vdots \\ X_m(x) \end{bmatrix}.$$

More generally, if $F : M \mapsto N$ is a diffeomorphism, it follows that we can induce a vector field Y on N from X defined on M using the push forward map, as follows:

$$Y_{F(p)} := F_{*p} X_p.$$

It is necessary either to have F be a diffeomorphism for the definition above to hold or to have a map satisfying

$$F(p) = F(q) \Rightarrow F_{*p} X_p = F_{*q} X_q \quad \forall p, q \in M.$$

In the sequel we will usually deal with the case that F is a diffeomorphism, in which case the vector field Y is said to be F *induced* and is denoted by $F_* X$. If $g : N \mapsto \mathbb{R}$ is a smooth function, it follows that

$$Y(g) = F_* X(g) = (X(g \circ F)) \circ F^{-1}.$$

We may now define two fundamental operations using vector fields:

Definition 8.5 Lie Derivative of a Function.   *Given a smooth vector field X : $M \mapsto TM$ and a smooth function $h : M \mapsto \mathbb{R}$ the* Lie derivative *of h with respect to the vector field X is a new function $L_X h : M \mapsto \mathbb{R}$ given by*

$$L_X h(p) = X(h)(p)$$

In local coordinates, if X is of the form $X = \sum_{i=1}^{m} X_i \frac{\partial}{\partial x_i}$, then we have

$$L_X h(p) = \sum_{i=1}^{m} \frac{\partial h}{\partial x_i}(x_1(p), \ldots, x_m(p)) X_i(x_1(p), \ldots, x_m(p)).$$

The Lie derivative of a function h with respect to a vector field X is the rate of change of h in the direction of X.

Definition 8.6 Lie Bracket of two Vector Fields.   *Given two smooth vector fields X, Y on M, we define a new vector field called the* Lie bracket $[X, Y]$ *by setting for given $h : M \mapsto \mathbb{R}$*

$$[X, Y]_p(h) = X_p Y(h) - Y_p X(h). \tag{8.5}$$

It is necessary to verify that the definition above does yield a vector field $[X, Y]$. Linearity over the reals is easy to verify for the definition of equation (8.5). To check the derivation property we choose two functions $g, h : M \mapsto \mathbb{R}$ and note that

$$
\begin{aligned}
[X, Y]_p(gh) &= X_p(Y(gh)) - Y_p(X(gh)) \\
&= X_p(Y(g)h + Y(h)g) - Y_p(X(g)h + X(h)g) \\
&= X_p(Y(g))h(p) + X_p(h)Y_p(g) + X_p(Y(h))g(p) + X_p(g)Y_p(h) \\
&\quad - Y_p(X(g))h(p) - Y_p(h)X_p(g) - Y_p(X(h))g(p) - Y_p(g)X_p(h) \\
&= [X, Y]_p(g)h(p) + g(p)[X, Y]_p(h).
\end{aligned}
$$

We may now derive the expression in local coordinates for $[X, Y]$. Indeed if in local coordinates $(x_1, \ldots x_m)$ the vector fields have the representations

$$X = \sum_{i=1}^{m} X_i(x) \frac{\partial}{\partial x_i}, \quad Y = \sum_{i=1}^{m} Y_i(x) \frac{\partial}{\partial x_i}$$

Then, choosing a coordinate function x_j, we get

$$[X, Y]_p(x_j) = X_p(Y(x_j)) - Y_p(X(x_j))$$
$$= X_p(Y_j(x)) - Y_p(X_j(x))$$
$$= \sum_{=1}^{m} \left(\frac{\partial Y_j}{\partial x_i} X_i - \frac{\partial X_j}{\partial x_i} Y_i \right).$$

Thus, we have that

$$[X, Y] = \sum_{j=1}^{m} \left(\sum_{i=1}^{m} \frac{\partial Y_j}{\partial x_i} X_i - \frac{\partial X_j}{\partial x_i} Y_i \right) \frac{\partial}{\partial x_j} \tag{8.6}$$

To get a feel for this formula consider the following example:

Example 8.7 Lie Brackets of Linear Vector Fields. *Consider two linear vector fields with coordinates given by $X = Ax, Y = Bx$ on $\mathbb{R}^m$, that is,*

$$X_i(x) = \sum_{j=1}^{m} a_{ij} x_j,$$

$$Y_i(x) = \sum_{j=1}^{m} b_{ij} x_j.$$

Then, applying the formula above yields

$$[X, Y](x) = (AB - BA)x.$$

in coordinates, as an easy calculation will show. Thus, the Lie bracket is another linear vector field, and its matrix is the commutator between the two matrices A, B.

The following properties of Lie brackets follow from the definitions:

Proposition 8.8 Properties of Lie Brackets. *Given vector fields X, Y, Z on a manifold M and smooth functions $g, h : M \mapsto \mathbb{R}$, we have*

1. Skew Symmetry

$$[X, Y] = -[Y, X].$$

2. Jacobi Identity

$$[[X, Y], Z] + [[Y, Z], X] + [[Z, X], Y] = 0.$$

3.

$$[fX, gY] = fg[X, Y] + fL_X(g)Y - gL_Y(f)X.$$

Proof: The proofs are by direct calculation (See Exercises).

It follows from the skew symmetry that the Lie bracket of a vector field with itself is zero. The Jacobi identity is a more subtle identity relating three vector fields. Its implications will become clear when we study controllability in Chapter 11. The preceding definition of Lie bracket is a little formal and it obscures a dynamic interpretation of $[X, Y]$ which we will endeavor to provide next. First, a proposition relating push forward maps and Lie brackets.

Proposition 8.9 Lie brackets and Push Forwards commute. *Consider two vector fields X, Y on a manifold M and a smooth map $F : M \mapsto N$ between manifolds M, N. Then, we have*

$$F_*[X, Y] = [F_*X, F_*Y]. \qquad (8.7)$$

Proof: Let $g : N \mapsto \mathbb{R}$ be a smooth function. Then we have

$$[F_*X, F_*Y]g = F_*X(F_*Y(g)) - F_*Y(F_*X(g))$$
$$= F_*X(Y(g \circ F)). \qquad \square$$

8.2   Distributions and Codistributions

In this section we will work in local coordinates so as to keep the exposition concrete. The methods of the preceding section may be used to give a more abstract description. This, of course is a valuable exercise for the reader. Thus, we will work on an open set $U \subset \mathbb{R}^n$ and represent vector fields as $X(x) = \sum_{i=1}^n f_i(x)\frac{\partial}{\partial x_i}$ with the x_i representing the canonical coordinates. It is convenient to think about a vector field as a vector $f(x)$ in $\mathbb{R}^n$ possibly depending on x.

$$f(x) = \begin{pmatrix} f_1(x) \\ f_2(x) \\ \vdots \\ f_n(x) \end{pmatrix}.$$

If $\alpha(x) : \mathbb{R}^n \mapsto \mathbb{R}$ is a smooth function, then it is easy to see that $\alpha(x)f(x)$ is also a vector field. Further, the sum of two vector fields is also a vector field. Thus, the space of all vector fields is a *module* over the *ring* of smooth functions on $U \subset \mathbb{R}^n$, denoted by $C^\infty(U)$. (A module is a linear space for which the scalars are elements of a ring.) Also, the space of vector fields is a *vector space* over the *field* of reals. We now make the following definition:

Definition 8.10 Distributions. *Given a set of smooth vector fields $X_1, X_2, \ldots, X_m$, we define the distribution $\Delta(x)$ to be*

$$\Delta = \text{span}\{X_1, X_2, \ldots, X_m\}.$$

In the preceding definition, span is understood to mean over the ring of smooth functions, i.e., elements of Δ at the point x are of the form

$$\alpha_1(x)X_1(x) + \alpha_2(x)X_2(x) + \cdots + \alpha_m(x)X_m(x)$$

with the $\alpha_i(x)$ all smooth functions of x.

Remarks:

1. A distribution is a smooth assignment of a subspace of $\mathbb{R}^n$ to each point x.
2. As a consequence of the fact that at a fixed x, $\Delta(x)$ is a subspace of $\mathbb{R}^n$, we may check that if Δ_1, Δ_2 are two smooth distributions, then their intersection $\Delta_1(x) \cap \Delta_2(x)$ defined pointwise, i.e., at each x is also a distribution as also their sum (subspace sum) $\Delta_1(x) + \Delta_2(x)$. Also note that the union of two distributions is not necessarily a distribution (give a counterexample to convince yourself of this). The question of the smoothness of the sum distribution and the intersection distribution is answered in the Proposition 8.11 below.
3. If $X(x) \in \mathbb{R}^{m \times n}$ is the matrix obtained by stacking the $X_i(x)$ next to each other, then

$$\Delta(x) = \text{Image } X(x).$$

We may now define the **rank** of the distribution at x to be the rank of $X(x)$. This rank, denoted by $m(x)$ is not, in general, constant as a function of x. If the rank is locally constant, i.e., it is constant in a neighborhood of x, then x is said to be a *regular point* of the distribution. If every point of the distribution is regular, the distribution is said to be **regular**.

Proposition 8.11 Smoothness of the Sum and Intersection Distributions. *Let $\Delta_1(x)$, $\Delta_2(x)$ be two smooth distributions. Then their sum $\Delta_1(x) + \Delta_2(x)$ is also a smooth distribution. Let x_0 be a regular point of the distributions Δ_1, Δ_2, $\Delta_1 \cap \Delta_2$. Then there exists a neighborhood of x_0 such that $\Delta_1 \cap \Delta_2$ is a smooth distribution in that neighborhood.*

Proof: Let

$$\Delta_1 = \text{span}\{X_1, \ldots, X_{d_1}\}, \quad \Delta_1 = \text{span}\{Y_1, \ldots, Y_{d_2}\}.$$

It follows that the sum distribution

$$\Delta_1 + \Delta_2 = \text{span}\{X_1(x), \ldots, X_{d_1}(x), Y_1(x), \ldots, Y_{d_2}(x)\}$$

is also a smooth distribution. As for the intersection, vector fields in the intersection are obtained by solving the following equation expressing equality of vectors in Δ_1 and Δ_2 as follows

$$\sum_{i=1}^{d_1} a_i(x)X_i(x) - \sum_{j=1}^{d_2} b_j(x)Y_j(x) = 0,$$

in the unknowns $a_i(x), b_j(x)$. If the intersection has constant dimension d in a neighborhood of x_0 it follows that there are d sets of $a_i(x).b_j(x)$ which solve the foregoing equation. If d_1, d_2, d are constant then they will be smooth functions of x establishing the smoothness of $\Delta_1 \cap \Delta_2$. $\square$

Remark: The preceding proposition says that if x_0 is a regular point of Δ_1, Δ_2, $\Delta_1 \cap \Delta_2$ then $\Delta_1 \cap \Delta_2$ is a smooth distribution. However, it is not true, in general, that the intersection of two smooth distributions is a smooth distribution. Indeed, consider

$$\Delta_1 = \text{span} \begin{bmatrix} 1 \\ 1 \end{bmatrix}, \quad \Delta_2 = \text{span} \begin{bmatrix} 1 + x_1 \\ 1 \end{bmatrix},$$

then $\Delta_1 \cap \Delta_2 = \{0\}$ if $x_1 \neq 0$ and $\Delta_1 \cap \Delta_2 = \Delta_1 = \Delta_2$ if $x_1 = 0$. Since we cannot find a smooth distribution that vanishes precisely at $x_1 = 0$, it follows that $\Delta_1 \cap \Delta_2(x)$ is not a smooth distribution.

Definition 8.12 Involutive Distributions. *A distribution Δ is called* involutive *if for any two vector fields τ_1, $\tau_2 \in \Delta(x)$ their Lie bracket $[\tau_1, \tau_2] \in \Delta$.*

Remarks:

1. Any one dimensional distribution is involutive, since if $\Delta = \text{span}(f(x))$ then

$$[\alpha_1(x)f(x), \alpha_2(x)f(x)] \in \text{span}(f(x))$$

for any $\alpha_1(x)$, $\alpha_2(x) \in C^\infty(\mathbb{R}^n)$.
2. A basis set is said to *commute* if $[X_i, X_j] = 0$. A distribution with a commutative basis set is involutive.
3. If two distributions Δ_1, Δ_2 are involutive and contain Δ, then their intersection $\Delta_1 \cap \Delta_2$ is also involutive (though it may not be smooth!). As a consequence we may define the *involutive closure* of a given distribution Δ to be the intersection of all involutive distributions containing it. It is non-empty since $\mathbb{R}^n$ is an involutive distribution containing Δ.

The dual of a distribution is a *codistribution*. Since we have chosen to represent elements of $\mathbb{R}^n$ by column vectors, we represent elements of the dual space of $\mathbb{R}^n$, denoted by $\mathbb{R}^{n*}$, by row vectors. (Recall that the dual space of a vector space is the space of all linear operators on the space. A row vector $v \in \mathbb{R}^{n*}$ may be thought of as linear operators on $\mathbb{R}^n$ by defining $v(x) = vx$.) Just as a vector field on $\mathbb{R}^n$ is a smooth assignment of a column vector $f(x)$ to each point $x \in \mathbb{R}^n$, we define a *covector field* to be a smooth assignment of a covector $\omega(x)$ to each point $x \in \mathbb{R}^n$. Now we may define codistributions as follows:

Definition 8.13 Codistributions. *Given a set of k smooth covector fields $\omega_1, \ldots, \omega_k$, we define the* codistribution Ω *by*

$$\Omega = \text{span}\{\omega_1, \ldots, \omega_k\}.$$

Remarks:

1. The rank of a codistribution may be defined just like the rank of a distribution. Also, regular points for codistributions and regular codistributions are analogously defined.

2. A codistribution Ω is said to *annihilate* or be *perpendicular to* to a distribution Δ if for every $\omega \in \Omega$, $f \in \Delta$, we have

$$\omega f \equiv 0.$$

3. Given a distribution $\Delta(x)$ we may construct its *annihilator* to be the codistribution Ω of all covectors that annihilate Δ. This is abbreviated as $\Omega = \Delta^\perp$ or $\Delta = \Omega^\perp$. As in the instance of intersections of distributions care must be taken about smoothness. It is not always true that the annihilator of a smooth distribution (resp. codistribution) is a smooth codistribution (resp. distribution). See, however, Proposition 8.14 below.

Proposition 8.14 Smooth Annihilators.   *Let x_0 be a regular point of a smooth distribution $\Delta(x)$. Then x_0 is a regular point of the annihilator $\Delta^\perp$. Also, in a neighborhood of x_0, $\Delta^\perp$ is a smooth codistribution.*

Proof: See Exercises.

8.3   Frobenius Theorem

Consider a nonsingular distribution Δ, defined on an open set of $\mathbb{R}^n$ with dimension d. Thus, it may be represented as

$$\Delta = \text{span}\{X_1, \ldots, X_d(x)\}.$$

By Proposition 8.14 it follows that the codistribution $\Omega = \Delta^\perp$ is again smooth and nonsingular with dimension $n - d$. Thus, it is spanned by $n - d$ covector fields $\omega_1, \ldots, \omega_{n-d}$. Thus, we have that for $i = 1, \ldots, d$ and $j = 1, \ldots, n - d$,

$$\omega_j(x)X_i(x) \equiv 0.$$

The question we ask ourselves now is whether the covector fields $\omega_j(x)$ are *exact*, i.e., whether there exist smooth functions $\lambda_1, \ldots, \lambda_{n-d}$ such that

$$\omega_j = \frac{\partial \lambda_j}{\partial x} = d\lambda_j(x).$$

This question is equivalent to solving the d partial differential equations

$$\frac{\partial \lambda_j}{\partial x}[X_1(x) \ldots X_d(x)] \equiv 0$$

and finding $n - d$ independent equations. By "independent" is meant the linear independence of the covector fields that are the differentials of the λ_j. A distribution Δ is said to be *integrable* if there exist $n - d$ real valued functions $\lambda_1, \ldots, \lambda_{n-d}$ such that

$$\text{span}\{d\lambda_1, \ldots, d\lambda_{n-d}\} = \Delta^\perp.$$

The following celebrated theorem of Frobenius characterizes integrable distributions:

Theorem 8.15 Frobenius Theorem.   *A nonsingular distribution is completely integrable if and only if it is involutive.*

Proof: The necessity of the involutivity condition is easy to establish. If, $\Delta^{\perp} = \text{span}\{d\lambda_1, \ldots, d\lambda_{n-d}\}$ then it follows that

$$d\lambda_j X_i \equiv 0, \quad i = 1, \ldots, d.$$

In other words, we may say that

$$L_{X_i}\lambda_j(x) \equiv 0$$

for $i = 1, \ldots, d$. Now a simple calculation establishes that

$$L_{[X_i, X_k]}\lambda_j = L_{X_i}L_{X_k}\lambda_j - L_{X_k}L_{X_i}\lambda_j.$$

Hence, we have that

$$L_{[X_i, X_k]}\lambda_j \equiv 0.$$

Since, by assumption, the differentials $\{d\lambda_1, \ldots, d\lambda_{n-d}\}$ span $\Delta^{\perp}$, we deduce that the vector field $[X_i, X_k]$ is also in Δ. Since X_i, X_k are arbitrary, it follows that Δ is involutive.

The proof of the sufficiency of the involutivity condition is by construction: Denote by $\phi_{t_i}^{X_i}(x)$ the flow along the vector field X_i for t_i seconds starting from x at 0. Thus, $\phi_{t_i}^{X_i}(x)$ satisfies the differential equation

$$\frac{d}{dt}\phi_t^{X_i}(x) = X(\phi_t^{X_i}(x)), \quad \phi_0^{X_i}(x) = x.$$

We will also define $n - d$ additional vector fields in addition to the $f_1, \ldots, f_d \in \Delta$ which are complementary in the sense that

$$\text{span}\{X_1, \ldots, X_d, X_{d+1}, \ldots, X_n\} = \mathbb{R}^n.$$

Now consider the mapping from a neighborhood of the origin to $\mathbb{R}^n$ described by

$$F : (t_1, \ldots, t_n) \rightarrow \phi_{t_1}^{X_1} \circ \cdots \circ \phi_{t_n}^{X_n}.$$

We claim two facts about the mapping F:

1. F is defined for all $(t_1, \ldots, t_n)$ in a neighborhood of the origin and is a diffeomorphism onto its image.
2. The first d columns of the Jacobian of F are linearly independent vectors (actually in Δ).

Before we prove the two claims, let us show that the two claims enable us to conclude the proof of the theorem. Let x be a point in the range of F and define

$$\begin{bmatrix} \psi_1(x) \\ \vdots \\ \psi_n(x) \end{bmatrix} = F^{-1}(x);$$

where the $\psi_i(x)$ are real-valued functions. We claim that the last $n - d$ of these functions are independent solutions of the given partial differential equations. Indeed, from the identity

$$\left[\frac{\partial F^{-1}}{\partial x}\right]_{x=F(t_1,\dots,t_n)}\left[\frac{\partial F}{\partial t}\right] = I$$

it follows that the last $n - d$ rows of the Jacobian of F^{-1}, i.e., the differentials $d\psi_{d+1}, \dots, d\psi_n$ annihilate the first d columns of the Jacobian of F. But, by the second claim above, the first d columns of the Jacobian of F span the distribution Δ. Also, by construction the differentials $d\psi_{d+1}, \dots, d\psi_n$ are independent, completing the proof.

For the proof of claim (1) above, note that for all $x \in \mathbb{R}^n$ the flow $\phi_{t_i}^{f_i}$ is well defined for t_i small enough. Thus the map F is well defined for $t_1, \dots, t_n$ sufficiently small. To show that F is a local diffeomorphism, we need to evaluate the Jacobian of F. It is convenient at this point to introduce the notation for *push forwards* or differentials of maps $M(x) : \mathbb{R}^n \mapsto \mathbb{R}^n$ by the symbol

$$(M)_* = \frac{\partial M}{\partial x}$$

Now by the chain rule we have that

$$\frac{\partial F}{\partial t_i} = (\phi_{t_1}^{X_1})_* \cdots (\phi_{t_{i-1}}^{X_{i-1}})_* \frac{\partial}{\partial t_i}(\phi_{t_i}^{X_i} \circ \cdots \circ \phi_{t_n}^{X_n})$$

$$= (\phi_{t_1}^{X_1})_* \cdots (\phi_{t_{i-1}}^{X_{i-1}})_* f_i(\phi_{t_i}^{X_i} \circ \cdots \circ \phi_{t_n}^{X_n}).$$

In particular, at $t = 0$, since $F(0) = x_0$, it follows that

$$\frac{\partial F}{\partial t_i}(0) = X_I(x_0).$$

Since the tangent vectors, $X_1(x_0), \dots, X_n(x_0)$ are linearly independent the columns of $(F)_*$ are also linearly independent. Hence, the mapping F is a local diffeomorphism.

For the proof of claim (ii) above, we have to show that for all x in a neighborhood of x_0 the vectors of

$$\frac{\partial F}{\partial t_i} = (\phi_{t_1}^{X_1})_* \cdots (\phi_{t_{i-1}}^{X_{i-1}})_* f_i(\phi_{t_i}^{X_i} \circ \cdots \phi_{t_n}^{X_n})$$

are linearly dependent. To do this we will show that for small $|t|$ and two vector fields $\tau, \theta \in \Delta$

$$(\phi_t^\theta)_* \tau \circ \phi_{-t}^\theta(x) \in \Delta(x)$$

We will, in fact, specialize the choice of τ to be one of the X_i and define

$$V_i(t) = (\phi_{-t}^\theta)_* X_i \circ \phi_t^\theta(x)$$

for $i = 1, \dots, d$. Since

$$\frac{d}{dt}(\phi_{-t}^\theta)_* = -(\phi_{-t}^\theta)_* \frac{\partial \theta}{\partial x}$$

and

$$\frac{d}{dt}(f \circ \phi_t^\theta(x)) = \frac{\partial f}{\partial x}\theta \circ \phi_t^\theta(x),$$

it follows that the vector field $V_i(t)$ satisfies the differential equation

$$\dot{V}_i = (\phi_{-t}^\theta)_*[\theta, X_i] \circ \phi_t^\theta(x).$$

By hypothesis, Δ is involutive, so that we may write

$$[\theta, X_i] = \sum_{j=1}^d \mu_{ij}X_j,$$

so that we get the linear differential equation

$$\dot{V}_i = (\phi_{-t}^\theta)_* \sum_{j=1}^d \mu_{ij}f_j \circ \phi_t^\theta(x) = \sum_{j=1}^d \mu_{ij}(\phi_t^\theta(x))V_j(t).$$

We may solve this linear equation to get

$$[V_1(t), \ldots, V_d(t)] = [V_1(0), \ldots, V_d(0)]X(t),$$

where $X(t) \in \mathbb{R}^{d \times d}$ is the state transition matrix associated with μ_{ij}. Premultiplying this by $(\phi_t^\theta)_*$ yields

$$[X_1(\phi_t^\theta(x)), \ldots, X_d(\phi_t^\theta(x))] = [(\phi_{-t}^\theta)_*X_1(x), \ldots, (\phi_{-t}^\theta)_*X_d(x)]X(t).$$

Now substituting $\phi_{-t}^\theta(x)$ for x yields

$$[X_1(x), \ldots, X_d(x)] = [(\phi_{-t}^\theta)_*X_1(x) \circ \phi_t^\theta(x), \ldots, (\phi_{-t}^\theta)_*X_d(x) \circ \phi_t^\theta(x)]X(t).$$

By the non-singularity of the state transition matrix $X(t)$. it follows that for $i = 1, \ldots, d$,

$$(\phi_{-t}^\theta)_*X_i(x) \circ \phi_t^\theta(x) \in \text{span}\{X_1(x), \ldots, X_d(x)\}.$$

This completes the proof of the theorem.   □

Remarks: The constructive sufficiency proof of the theorem is quite interesting in that it shows that the solution of the integral manifolds can be reduced to the solution of n ordinary differential equations followed by the inversion of the mapping describing their flow. This may be exploited as in Problem 8.6.

8.4   Matrix Groups

In this section we will be defining similar constructions for each of the fields $\mathbb{R}$, $\mathbb{C}$, and the Hamiltonian field $\mathbb{H}$ (see Chapter 3). For ease of notation, we denote the set as $\mathbb{K} \in \{\mathbb{R}, \mathbb{C}, \mathbb{H}\}$. We write $\mathbb{K}^n$ as the set of all n-tuples whose elements are in $\mathbb{K}$. If we define $\psi : \mathbb{K}^n \mapsto \mathbb{K}^n$ as a linear map, then ψ has matrix representation $M_n(\mathbb{K}) \in \mathbb{K}^{n \times n}$. $\mathbb{K}^n$ and $M_n(\mathbb{K})$ are both vector spaces over $\mathbb{K}$. Some examples of groups whose elements are $n \times n$ matrices is now introduced:

1. The group of units of $M_n(\mathbb{K})$ is the set of matrices M for which $\det(M) \neq 0$, where 0 is the additive identity of $\mathbb{K}$. This group is called the *general linear group* and is denoted by $GL(n, \mathbb{K})$.

2. $SL(n, \mathbb{K}) \subset GL(n, \mathbb{K})$ is the subgroup of $GL(n, \mathbb{K})$ whose elements have determinant 1. $SL(n, \mathbb{K})$ is called the *special linear group*.

3. *Orthogonal Matrix Groups.* $O(n, \mathbb{K}) \subset GL(n, \mathbb{K})$ is the subgroup of $GL(n, \mathbb{K})$ whose element are matrices A which satisfy the orthogonality condition: $\bar{A}^T = A^{-1}$, where $\bar{A}^T$ is the complex conjugate transpose of A. Examples of orthogonal matrix groups are:

   a. $O(n) \equiv O(n, \mathbb{R})$ is called the *orthogonal group*
   b. $U(n) \equiv U(n, \mathbb{C})$ is called the *unitary group*
   c. $Sp(n) \equiv Sp(n, \mathbb{H})$ is called the *symplectic group*

   If, for $A \in GL(n, \mathbb{H})$, $\bar{A}$ denotes the complex conjugate of the quaternion, defined by conjugating each element using

   $$\overline{x + \imath y + \jmath z + kw} = x - \imath y - \jmath z - kw,$$

   then

   $$Sp(n) = \{B \in M_n(\mathbb{H}), \ \bar{B}^T B = I\}.$$

   An equivalent way of defining the symplectic group is as a subset of $GL(2n, \mathbb{C})$, such that

   $$Sp(n) = \{B \in GL(2n, \mathbb{C}) : B^T J B = J; \bar{B}^T = B^{-1}\},$$

   where the matrix J is called the *infinitesimal symplectic matrix* and is written as

   $$J = \begin{bmatrix} 0 & I_{n \times n} \\ -I_{n \times n} & 0 \end{bmatrix}.$$

   It can also be verified (see Exercises) that

   $$Sp(n) = U(2n) \cap Sp(2n, \mathbb{C}). \tag{8.8}$$

4. *Special Orthogonal Matrix Groups.* $SO(n) = O(n) \cap SL(n, \mathbb{R})$ is the set of all orthogonal matrices of determinant 1. It is called the *special orthogonal group*. $SU(n) = U(n) \cap SL(n, \mathbb{C})$, the set of all unitary matrices of determinant 1, is called the *special unitary group*.

5. *Euclidean Matrix Groups* The *Euclidean group* is the set of matrices $E(n) \subset \mathbb{R}^{(n+1) \times (n+1)}$ such that

   $$E(n) = \left\{ A \in \mathbb{R}^{(n+1) \times (n+1)} : A = \begin{bmatrix} R & p \\ 0 & 1 \end{bmatrix}, R \in GL(n), p \in \mathbb{R}^n \right\}.$$

   The *special Euclidean group* is the set of matrices $SE(n)$ such that

   $$SE(n) = \left\{ A \in \mathbb{R}^{(n+1) \times (n+1)} : A = \begin{bmatrix} R & p \\ 0 & 1 \end{bmatrix}, R \in SO(n), p \in \mathbb{R}^n \right\}.$$

We will see in the next section that matrix groups are manifolds. Owing to their special structure, it is possible to characterize their tangent spaces. Here we will first give an algebraic definition of the tangent space. In the next section, we will see that it is straightforward to verify that the following algebraic definition of the tangent space to the matrix group is the same as the geometric definition given earlier in this chapter.

Proposition 8.16 The Tangent Space of a Matrix Group. *Let $G \subset M_n(\mathbb{K})$ be a matrix group. Let $\gamma : [a, b] \mapsto G$ be a curve with $0 \in (a, b)$ and $\gamma(0) = I$. Let T be the set of all tangent vectors $\gamma'(0)$ to curves γ. Then T is a real subspace of $M_n(\mathbb{K})$.*

Proof: If $\gamma(\cdot)$ and $\sigma(\cdot)$ are two curves in G, then $\gamma'(0)$ and $\sigma'(0)$ are in T. Also, $\gamma\sigma$ is a curve in G with $(\gamma\sigma)(0) = \gamma(0)\sigma(0) = I$.

$$\frac{d}{du}(\gamma(u)\sigma(u)) = \gamma'(u)\sigma(u) + \gamma(u)\sigma'(u),$$

$$(\gamma\sigma)'(0) = \gamma'(0)\sigma(0) + \gamma(0)\sigma'(0) = \gamma'(0) + \sigma'(0).$$

Since $\gamma\sigma$ is in G, $(\gamma\sigma)'(0)$ is in T. Therefore, $\gamma'(0)\sigma(0) + \gamma(0)\sigma'(0)$ is in T, and T is closed under vector addition. Also, if $\gamma'(0) \in T$ and $r \in \mathbb{R}$, and if we let $\sigma(u) = \gamma(ru)$, then $\sigma(0) = \gamma(0) = I$ and $\sigma'(0) = r\sigma'(0)$. Therefore, $r\sigma'(0) \in T$, and T is closed under scalar multiplication.   □

Definition 8.17 Dimension of a Matrix Group. *The dimension of the matrix group G is the dimension of the vector space T of tangent vectors to G at I.*

We now introduce a family of matrices that we will use to determine the dimensions of our matrix groups.

1. Let $so(n)$ denote the set of all *skew-symmetric matrices* in $M_n(\mathbb{R})$,

$$so(n) = \{A \in M_n(\mathbb{R}) : A^T + A = 0\}.$$

2. Let $su(n)$ denote the set of *skew-Hermitian matrices*

$$su(n) = \{A \in M_n(\mathbb{C}) : \bar{A}^T + A = 0\}.$$

3. Let $sp(n)$ denote the set of *skew-symplectic matrices*:

$$sp(n) = \{A \in M_n(\mathbb{H}) : \bar{A}^T + A = 0\}.$$

4. Let $sl(n)$ denote the set of *zero trace matrices*:

$$sl(n) = \{A \in M_n(\mathbb{R}) : \text{trace}(A) = 0\}.$$

5. Let $se(n)$ denote the set of *generalized twist matrices*

$$se(n) = \left\{ A \in \mathbb{R}^{(n+1)\times(n+1)} : A = \begin{bmatrix} \hat{w} & p \\ 0 & 0 \end{bmatrix}, \hat{w} \in so(n), p \in \mathbb{R}^n \right\}.$$

Now consider the orthogonal matrix group. Let $\gamma : [a, b] \mapsto O(n)$, such that $\gamma(u) = A(u)$, where $A(u) \in O(n)$. Therefore, $A^T(u)A(u) = I$. Taking the derivative of this identity with respect to u, we have:

$$A'^T(u)A(u) + A^T(u)A'(u) = 0.$$

Since $A(0) = I$,

$$A'^T(0) + A'(0) = 0.$$

Thus, the vector space $T_I O(n)$ of tangent vectors to $O(n)$ at I is a subset of the set of skew-symmetric matrices, $so(n)$:

$$T_I O(n) \subset so(n).$$

Similarly, we can derive

$$T_I U(n) \subset su(n),$$
$$T_I Sp(n) \subset sp(n).$$

From our definition of the dimension of a matrix group, it follows that

$$\dim O(n) \leq \dim so(n),$$
$$\dim U(n) \leq \dim su(n),$$
$$\dim Sp(n) \leq \dim sp(n).$$

We will show that these inequalities are actually equalities.

Definition 8.18 Exponential and Logarithm.   *The matrix exponential function,* $\exp : M_n(\mathbb{K}) \mapsto M_n(\mathbb{K})$, *is defined in terms of the Taylor series expansion of the exponential:*

$$\exp(A) := e^A = I + A + \frac{A^2}{2!} + \frac{A^3}{3!} + \cdots.$$

The matrix logarithm $\log : M_n(\mathbb{K}) \mapsto M_n(\mathbb{K})$ *is defined only for matrices close to the identity matrix I:*

$$\log X = (X - I) - \frac{(X - I)^2}{2} + \frac{(X - I)^3}{3} - + \cdots.$$

Proposition 8.19. $A \in so(n)$ *implies* $e^A \in SO(n)$.

Proof: $(e^A)^T = e^{A^T} = e^{A^{-1}} = (e^A)^{-1}$. Therefore, $e^A \in O(n)$. Using $\det(e^A) = e^{\text{trace}(A)}$, we have $\det(e^A) = e^0 = 1$. □

Similarly (see Problem 8.7), we have

$$A \in su(n) \Rightarrow e^A \in U(n),$$
$$A \in sp(n) \Rightarrow e^A \in Sp(n),$$
$$A \in sl(n) \Rightarrow e^A \in SL(n),$$
$$A \in se(n) \Rightarrow e^A \in SE(n).$$

Proposition 8.20. $X \in SO(n)$ *implies* $\log(X) \in so(n)$.

Proof: Noting that $\log(XY) = \log(X) + \log(Y)$ iff $XY = YX$, we take the logarithm on both sides of the equation: $XX^T = X^TX = I$. Thus, $\log(X) + \log(X^T) = 0$, so $\log(X) \in so(n)$.   □

It may also be verified that (see Problem 8.7)

$$X \in U(n) \Rightarrow \log(X) \in su(n),$$
$$X \in Sp(n) \Rightarrow \log(X) \in sp(n),$$
$$X \in SL(n) \Rightarrow \log(X) \in sl(n),$$
$$X \in SE(n) \Rightarrow \log(X) \in se(n).$$

The logarithm and exponential thus define maps that operate between the matrix group G to its tangent space at the identity $T_I G$.

Definition 8.21 One-Parameter Subgroup. *A one-parameter subgroup of a matrix group G is the image of a smooth homomorphism* $\gamma : \mathbb{R} \mapsto G$.

The group operation in $\mathbb{R}$ is addition; thus, $\gamma(u + v) = \gamma(u) \cdot \gamma(v)$. Since $\mathbb{R}$ is an Abelian group under addition, we have that $\gamma(u + v) = \gamma(v + u) = \gamma(u) \cdot \gamma(v) = \gamma(v) \cdot \gamma(u)$. Note that by defining γ on some small neighborhood U of $0 \in \mathbb{R}$, γ is defined over all $\mathbb{R}$, since for any $x \in \mathbb{R}$, some $\frac{1}{n}x \in U$ and $\gamma(x) = (\gamma(\frac{1}{n}x))^n$.

Proposition 8.22. *If* $A \in M_n(\mathbb{K})$, *then* e^{Au} *for* $u \in \mathbb{R}$ *is a one-parameter subgroup.*

Proof: Noting that $e^{X+Y} = e^X e^Y$ iff $XY = YX$, we have

$$e^{A(u+v)} = e^{Au+Av} = e^{Au}e^{Av},$$

since A commutes with itself.   □

Proposition 8.23. *Let* γ *be a one-parameter subgroup of* $M_n(\mathbb{K})$. *Then there exists* $A \in M_n(\mathbb{K})$ *such that* $\gamma(u) = e^{Au}$.

Proof: Define $A = \sigma'(0)$, where $\sigma(u) = \log \gamma(u)$, (i.e., $\gamma(u) = e^{\sigma(u)}$). We need to show that $\sigma(u) = Au$, a line through 0 in $M_n(\mathbb{K})$.

$$
\begin{aligned}
\sigma'(u) &= \lim_{v \to 0} \frac{\sigma(u+v) - \sigma(u)}{v} = \lim_{v \to 0} \frac{\log \gamma(u+v) - \log \gamma(u)}{v} \\
&= \lim_{v \to 0} \frac{\log \gamma(u)\gamma(v) - \log \gamma(u)}{v} = \lim_{v \to 0} \frac{\log \gamma(v)}{v} = \sigma'(0) \\
&= A.
\end{aligned}
$$

Therefore, $\sigma(u) = Au$.   □

Thus, given any element in the tangent space of G at I, its exponential belongs to G.

Proposition 8.24. *Let* $A \in T_I O(n, \mathbb{K})$, *the tangent space at I to* $O(n, \mathbb{K})$. *Then there exists a unique one parameter subgroup* γ *in* $O(n, \mathbb{K})$ *with* $\gamma'(0) = A$.

Proof: $\gamma(u) = e^{Au}$ is a one parameter subgroup of $GL(n, \mathbb{K})$, and γ lies in $O(n, \mathbb{K})$, since $\gamma(u)^T \gamma(u) = (e^{Au})^T e^{Au} = I$. The uniqueness follows from the preceding proposition. □

Thus,

$$\dim O(n, \mathbb{K}) \geq \dim so(n, \mathbb{K}).$$

But we have shown using our definition of the dimension of a matrix group that

$$\dim O(n, \mathbb{K}) \leq \dim so(n, \mathbb{K}).$$

Therefore,

$$\dim O(n, \mathbb{K}) = \dim so(n, \mathbb{K}),$$

and the tangent space at I to $O(n, \mathbb{K})$ is exactly the set of skew-symmetric matrices. The dimension of $so(n, \mathbb{R})$ is easily computable: We simply find a basis for $so(n, \mathbb{R})$. Let E_{ij} be the matrix whose entries are all zero except the ij-th entry, which is 1, and the ji-th entry, which is -1. Then E_{ij}, for $i < j$, form a basis for $so(n)$. There are $(n(n - 1))/2$ of these basis elements. Therefore, $\dim O(n) = (n(n - 1))/2$.

Similarly, (see Problem 8.8), one may compute that $\dim SO(n) = (n(n-1))/2$, $\dim U(n) = n^2$, $\dim SU(n) = n^2 - 1$, and $\dim Sp(n) = n(2n + 1)$.

8.4.1 Matrix Lie Groups and Their Lie Algebras

Definition 8.25 Lie Group. *A Lie group is a group G that is also a differentiable manifold such that, for any $a, b \in G$, the multiplication $(a, b) \mapsto ab$ and inverse $a \mapsto a^{-1}$ are smooth maps.*

It is a theorem (beyond the scope of this book) that all compact finite-dimensional Lie groups may be represented as matrix groups. This was the reason that we introduced matrix groups first in the preceding section. For example, since the function det: $\mathbb{R}^{n^2} \mapsto \mathbb{R}$ is continuous, the matrix group $GL(n, \mathbb{R}) = \det^{-1}(\mathbb{R}-\{0\})$ can be given a differentiable structure which makes it an open submanifold of $\mathbb{R}^{n^2}$. Multiplication of matrices in $GL(n, \mathbb{R})$ is continuous, and smoothness of the inverse map follows from Cramer's rule. Thus, $GL(n, \mathbb{R})$ is a Lie group. Similarly, $O(n)$, $SO(n)$, $E(n)$, and $SE(n)$ are Lie groups.

In order to study the algebras associated with matrix Lie groups, the concepts of differential maps and left translations are first introduced. Let G be a Lie group with identity I, and let X_I be a tangent vector to G at I. We may construct a vector field defined on all of G in the following way. For any $g \in G$, define the *left translation* by g to be a map $L_g : G \mapsto G$ such that $L_g(x) = gx$, where $x \in G$. Since G is a Lie group, L_g is a diffeomorphism of G for each g. Taking the differential of L_g at I results in a map from the tangent space of G at I to the tangent space of G at g,

$$dL_g : T_I G \mapsto T_g G,$$

such that

$$X_g = dL_g(X_I).$$

Note that dL_g is the push forward map $(L_g)_*$. The vector field formed by assigning $X_g \in T_g G$ for each $g \in G$ is called a *left-invariant* vector field.

Proposition 8.26. *If X and Y are left-invariant vector fields on G, then so is* $[X, Y]$.

Proof: It follows by applying the definitions, that

$$[dL_g X, dL_g Y] = dL_g [X, Y] \qquad \Box$$

Also, if X and Y are left-invariant vector fields, then $X + Y$ and $rX, r \in \mathbb{R}$ are also left-invariant vector fields on G. Thus, the left-invariant vector fields of G form an algebra under $[\cdot, \cdot]$, which is called the *Lie algebra* of G and denoted $\mathcal{L}(G)$. The Lie algebra $\mathcal{L}(G)$ is actually a subalgebra of the Lie algebra of all smooth vector fields on G. With this notion of a Lie group's associated Lie algebra, we can now look at the Lie algebras associated with some of our matrix Lie groups. We first look at three examples, and then, in the next section, study the general map from a Lie algebra to its associated Lie group.

Examples:

- The Lie algebra of $GL(n, \mathbb{R})$ is denoted by $gl(n, \mathbb{R})$, the set of all $n \times n$ real matrices. The tangent space of $GL(n, \mathbb{R})$ at the identity can be identified with $\mathbb{R}^{n^2}$, since $GL(n, \mathbb{R})$ is an open submanifold of $\mathbb{R}^{n^2}$. The Lie bracket operation is simply $[A, B] = AB - BA$, matrix multiplication.
- The special orthogonal group $SO(n)$ is a submanifold of $GL(n, \mathbb{R})$, so $T_I SO(n)$ is a subspace of $T_I GL(n, \mathbb{R})$. The Lie algebra of $SO(n)$, denoted $so(n)$, may thus be identified with a certain subspace of $\mathbb{R}^{n^2}$. We have shown in the previous section that the tangent space at I to $SO(n)$ is the set of skew-symmetric matrices; it turns out that we may identify $so(n)$ with this set. For example, for $SO(3)$, the Lie algebra is

$$so(3) = \left\{ \hat{w} \equiv \begin{bmatrix} 0 & -w_3 & w_2 \\ w_3 & 0 & -w_1 \\ -w_2 & w_1 & 0 \end{bmatrix}, \ w = \begin{bmatrix} w_1 \\ w_2 \\ w_3 \end{bmatrix} \right\}.$$

The Lie bracket on $so(n)$ is defined as $[\hat{w}_a, \hat{w}_b] := \hat{w}_a \hat{w}_b - \hat{w}_b \hat{w}_a = (\widehat{w_a \times w_b})$, the skew-symmetric matrix form of the vector cross product.
- The Lie algebra of $SE(3)$, called $se(3)$, is defined as follows:

$$se(3) = \left\{ \hat{\xi} = \begin{bmatrix} \hat{w} & v \\ 0 & 0 \end{bmatrix}, \ w, v \in \mathbb{R}^3 \right\}.$$

The Lie bracket on $se(3)$ is defined as

$$[\hat{\xi}_1, \hat{\xi}_2] = \hat{\xi}_1\hat{\xi}_2 - \hat{\xi}_2\hat{\xi}_1 = \begin{bmatrix} \widehat{(w_1 \times w_2)} & w_1 \times v_2 - w_2 \times v_1 \\ 0 & 0 \end{bmatrix}.$$

8.4.2   The Exponential Map

In computing the dimension of $O(n, \mathbb{K})$ in Section 8.4, we showed that for each matrix A in $T_I O(n, \mathbb{K})$ there is a unique one-parameter subgroup γ in $O(n, \mathbb{K})$, with $\gamma(u) = e^{Au}$, such that $\gamma'(0) = A$. In this section we introduce a function

$$\exp : T_I G \mapsto G$$

for a general Lie group G. This map is called the *exponential map* of the Lie algebra $\mathcal{L}(G)$ into G. We then apply this exponential map to the Lie algebras of the matrix Lie groups discussed in the previous section. Of course, all compact finite dimensional Lie groups may be represented as matrix Lie groups as we have remarked earlier.

Consider a general Lie group G with identity I. For every $\xi \in T_I G$, let $\phi_\xi :$ $\mathbb{R} \mapsto G$ denote the integral curve of the left invariant vector field X_ξ, associated to ξ passing through I at $t = 0$. Thus,

$$\phi_\xi(0) = I$$

and

$$\frac{d}{dt}\phi_\xi(t) = X_\xi(\phi_\xi(t)).$$

One can show that $\phi_\xi(t)$ is a one-parameter subgroup of G. Now the *exponential map* of the Lie algebra $\mathcal{L}(G)$ into G is defined as $\exp : T_I G \mapsto G$ such that for $s \in \mathbb{R}$,

$$\exp(\xi s) = \phi_\xi(s).$$

Thus, a line ξs in $\mathcal{L}(G)$ is mapped to a one parameter subgroup $\phi_\xi(s)$ of G.

We differentiate the map $\exp(\xi s) = \phi_\xi(s)$ with respect to s at $s = 0$ to obtain $d(\exp) : T_I G \mapsto T_I G$ such that:

$$d(\exp)(\xi) = \phi_\xi'(0) = \xi.$$

Thus, $d(\exp)$ is the identity map on $T_I G$. By the inverse function theorem,

$$\exp : \mathcal{L}(G) \mapsto G$$

is a local diffeomorphism from a neighborhood of zero in $\mathcal{L}(G)$ onto a neighborhood of I in G, denoted by U. The connected component of G which contains the identity I is denoted by G_0, called the *identity component* of G. Then clearly $U \subset G_0$. We now discuss the conditions under which the exponential map is surjective onto the Lie group. U is path connected by construction: The one-parameter subgroup $\exp(\xi s) = \phi_\xi(s)$ defines a path between any two elements in U.

Proposition 8.27. *If G is a path connected Lie group, ie $G = G_0$, and H is a subgroup which contains an open neighborhood U of I in G, then $H = G$.*

Proof: See Curtis [75].

We may thus conclude that if G is a path connected Lie group, then $\exp : \mathcal{L}(G) \mapsto G$ is surjective. If G is not path connected, $\exp(\mathcal{L}(G))$ is the identity component G_0 of G.

For matrix Lie groups, the exponential map is just the matrix exponential function, $e^A := \exp(A)$, where A is a matrix in the associated Lie algebra.

- For $G = SO(3)$, the exponential map $\exp \hat{w}$, $\hat{w} \in so(3)$, is given by

$$e^{\hat{w}} = I + \hat{w} + \frac{\hat{w}^2}{2!} + \frac{\hat{w}^3}{3!} + \cdots,$$

which can be written in closed form solution as:

$$e^{\hat{w}} = I + \frac{\hat{w}}{|w|} \sin |w| + \frac{\hat{w}^2}{|w|^2}(1 - \cos |w|).$$

This is known as the *Rodrigues formula*.

- For $G = SE(3)$, the exponential map $\exp \hat{\xi}$, $\hat{\xi} \in se(3)$ is given by

$$e^{\hat{\xi}} = \begin{bmatrix} I & v \\ 0 & 1 \end{bmatrix}$$

for $w = 0$ and

$$e^{\hat{\xi}} = \begin{bmatrix} e^{\hat{w}} & Av \\ 0 & 1 \end{bmatrix}$$

for $w \neq 0$, where

$$A = I + \frac{\hat{w}}{|w|^2}(1 - \cos |w|) + \frac{\hat{w}^2}{|w|^3}(|w| - \sin |w|).$$

8.4.3   Canonical Coordinates on Matrix Lie Groups

Let $\{X_1, X_2, \ldots, X_n\}$ be a basis for the Lie algebra $\mathcal{L}(G)$. Since

$$\exp : \mathcal{L}(G) \mapsto G$$

is a local diffeomorphism, the mapping $\sigma : \mathbb{R}^n \mapsto G$ defined by

$$g = \exp(\sigma_1 X_1 + \cdots + \sigma_n X_n)$$

is a local diffeomorphism between $\sigma \in \mathbb{R}^n$ and $g \in G$ for g in a neighborhood of the identity I of G. Therefore, $\sigma : U \mapsto \mathbb{R}^n$, where $U \subset G$ is a neighborhood of I, may be considered a coordinate mapping with coordinate chart (σ, U). Using the left translation L_g, we can construct an atlas for the Lie group G from this single coordinate chart. The functions σ_i are called the *Lie–Cartan coordinates of the first kind* relative to the basis $\{X_1, X_2, \ldots, X_n\}$. A different way of writing coordinates on a Lie group using the same basis is to define $\theta : \mathbb{R}^n \mapsto G$ by

$$g = \exp(X_1\theta_1)\exp(X_1\theta_2)\cdots\exp(X_n\theta_n)$$

for g in a neighborhood of I. The functions $(\theta_1, \theta_2, \ldots, \theta_n)$ are called the *Lie–Cartan coordinates of the second kind.*

An example of a parameterization of $SO(3)$ using the Lie–Cartan coordinates of the second kind is just the product of exponentials formula:

$$R = e^{\hat{z}\theta_1} e^{\hat{y}\theta_2} e^{\hat{x}\theta_3}$$

$$= \begin{bmatrix} \cos(\theta_1) & -\sin(\theta_1) & 0 \\ \sin(\theta_1) & \cos(\theta_1) & 0 \\ 0 & 0 & 1 \end{bmatrix} \begin{bmatrix} \cos(\theta_2) & 0 & \sin(\theta_2) \\ 0 & 1 & 0 \\ -\sin(\theta_2) & 0 & \cos(\theta_2) \end{bmatrix}$$

$$\times \begin{bmatrix} 1 & 0 & 0 \\ 0 & \cos(\theta_3) & -\sin(\theta_3) \\ 0 & \sin(\theta_3) & \cos(\theta_3) \end{bmatrix},$$

where $R \in SO(3)$ and

$$\hat{x} = \begin{bmatrix} 0 & 0 & 0 \\ 0 & 0 & -1 \\ 0 & 1 & 0 \end{bmatrix}, \quad \hat{y} = \begin{bmatrix} 0 & 0 & 1 \\ 0 & 0 & 0 \\ -1 & 0 & 0 \end{bmatrix}, \quad \hat{z} = \begin{bmatrix} 0 & -1 & 0 \\ 1 & 0 & 0 \\ 0 & 0 & 0 \end{bmatrix}.$$

This is known as the ZYX Euler angle parameterization. Similar parameterizations are the YZX Euler angles and the ZYZ Euler angles.

A *singular configuration* of a parameterization is one in which there does not exist a solution to the problem of calculating the Lie–Cartan coordinates from the matrix element of the Lie group. For example, the ZYX Euler angle parameterization for $SO(3)$ is singular when $\theta_2 = -\pi/2$. The ZYZ Euler angle parameterization is singular when $\theta_1 = -\theta_3$ and $\theta_2 = 0$, in which case $R = I$, illustrating that there are infinitely many representations of the identity rotation in this parameterization.

8.4.4 The Campbell–Baker–Hausdorff Formula

The exponential map may be used to relate the algebraic structure of the Lie algebra $\mathcal{L}(G)$ of a Lie group G with the group structure of G. The relationship is described through the Campbell–Baker–Hausdorff (CBH) formula, which is introduced in this section.

Definition 8.28 Actions of a Lie Group. *If M is a differentiable manifold and G is a Lie group, we define a left action of G on M as a smooth map $\Phi : G \times M \mapsto M$ such that:*

1. $\Phi(I, x) = x$ for all $x \in M$.

2. For every $g, h \in G$ and $x \in M$, $\Phi(g, \Phi(h, x)) = \Phi(gh, x)$.

The action of G on itself defined by $C_g : G \mapsto G$:

$$C_g(h) = ghg^{-1} = R_{g^{-1}}L_g h$$

is called the conjugation map associated with g. The derivative of the conjugation map at the identity I is called the Adjoint map, defined by $Ad_g : \mathcal{L}(G) \mapsto \mathcal{L}(G)$ such that for $\xi \in \mathcal{L}(G), g \in G$,

$$Ad_g(\xi) = d(C_g)(\xi) = d(R_{g^{-1}}L_g)(\xi).$$

If $G \subset GL(n, \mathbb{C})$, then $Ad_g(\xi) = g\xi g^{-1}$.

If we view the Lie algebra $\mathcal{L}(G)$ as a vector space, Ad_g is an element in the Lie group Aut $\mathcal{L}(G)$, the Lie group of bijective linear maps from $\mathcal{L}(G)$ into itself, called *automorphisms*. Then the differential, or the Lie algebra, of the Lie group Aut $\mathcal{L}(G)$ is the space End $\mathcal{L}(G)$, the space of all linear maps, called *endomorphisms*, from $\mathcal{L}(G)$ into itself.

The *adjoint map* (note that the word "adjoint" is written lower case here) $ad_\xi : \mathcal{L}(G) \mapsto \mathcal{L}(G)$ is defined for any $\xi \in \mathcal{L}(G)$ as

$$ad_\xi(\eta) = [\xi, \eta].$$

Clearly $ad_\xi \in$ End $\mathcal{L}(G)$. ad can then be viewed as the differential of Ad in the following sense. Let $\gamma(t) \in G$ is a curve in G with $\gamma(0) = I$ and $\gamma'(0) = \xi$. It is direct to check that:

$$\frac{d}{dt} Ad_{\gamma(t)}(\eta) = ad_\xi(\eta).$$

We know that an element $g \in G$ can usually be expressed as the exponential $\exp(X)$ for some $X \in \mathcal{L}(G)$. Then $Ad_g \in$ Aut $\mathcal{L}(G)$ and $ad_X \in$ End $\mathcal{L}(G)$ should also be related through the exponential map:

Lemma 8.29 Exponential Formula. *For any elements* $X, Y \in \mathcal{L}(G)$,

$$Ad_{\exp(X)}Y = \exp(X)Y\exp(-X)$$

$$= Y + [X, Y] + \frac{1}{2!}[X, [X, Y]] + \frac{1}{3!}[X[X, [X, Y]]] + \cdots$$

$$= Y + ad_X Y + \frac{1}{2!}ad_X^2 Y + \frac{1}{3!}ad_X^3 Y + \cdots$$

$$= \exp(ad_X)Y.$$

Proof: Both $Ad_{\exp(tX)}$ and $\exp(t\,ad_X)$ (note that the exponential maps $\exp(tX)$ and $\exp(t\,ad_X)$ are defined on different domains) give one-parameter subgroups of the Lie group Aut $\mathcal{L}(G)$ which have the same tangent vector ad_X for $t = 0$ and hence coincide for all t. □

The formula given in this lemma is a measure of how much X and Y fail to commute over the exponential: If $[X, Y] = 0$, then $Ad_{\exp(X)}Y = Y$.

If $\{X_1, X_2, \ldots, X_n\}$ is a basis for the Lie algebra $\mathcal{L}(G)$, the *structure constants* of $\mathcal{L}(G)$ with respect to $\{X_1, X_2, \ldots, X_n\}$ are the values $c_{ij}^k \in \mathbb{R}$ defined by

$$[X_i, X_j] = \sum_k c_{ij}^k X_k.$$

Structure constants may be used to prove the following useful lemma:

Lemma 8.30. *Consider the matrix Lie algebra $\mathcal{L}(G)$ with basis $\{X_1, X_2, \ldots, X_n\}$ and structure constants c_{ij}^k with respect to this basis. Then*

$$\prod_{j=1}^{r} \exp(p_j X_j) X_i \prod_{j=r}^{1} \exp(-p_j X_j) = \sum_{k=1}^{n} \xi_{ki} X_k,$$

where $p_j \in \mathbb{R}$ and $\xi_{ki} \in \mathbb{R}$.

Proof: We first prove the lemma for $r = 1$. Using the formula given in Lemma 8.29, write

$$\exp(p_1 X_1) X_i \exp(-p_1 X_1) = X_i + \sum_{k=1}^{\infty} \frac{ad_{X_1}^k X_i}{k!} p_1^k.$$

The terms $ad_{X_1}^k X_i$ are calculated using the structure constants:

$$ad_{X_1} X_i = \sum_{n_1=1}^{n} c_{1i}^{n_1} X_{n_1}$$

$$ad_{X_1}^2 X_i = \sum_{n_1=1}^{n} \sum_{n_2=1}^{n} c_{1i}^{n_1} c_{1n_1}^{n_2} X_{n_2}$$

$$\vdots$$

$$ad_{X_1}^k X_i = \sum_{n_1=1}^{n} \sum_{n_2=1}^{n} \cdots \sum_{n_k=1}^{n} c_{1i}^{n_1} c_{1n_1}^{n_2} c_{1n_2}^{n_3} \cdots c_{1n_{k-1}}^{n_k} X_{n_k}.$$

Using the foregoing formulas for $ad_{X_1}^k X_i$ into

$$X_i + \sum_{k=1}^{\infty} \frac{ad_{X_1}^k X_i}{k!} p_1^k,$$

note that since each of the c_{ij}^k is finite, the infinite sum is bounded. The ξ_{ki} are consequently bounded and are functions of c_{ij}^k, $k!$, and p_1^k. The proof is similar for $r > 1$. □

The *Campbell–Baker–Hausdorff formula* is another important measure of the commutativity of two Lie algebra elements X and Y over their exponential. It can be stated as follows:

Proposition 8.31 Campbell–Baker–Hausdorff (CBH) Formula. *Consider a neighborhood U of the identity element I of a Lie group G, such that every element* *in U can be represented as $\exp(X)$ for some $X \in \mathcal{L}(G)$. Then, for any two elements $\exp(X)$ and $\exp(Y)$ in U there exists an element $Z \in \mathcal{L}(G)$ such that:*

$$\exp(Z) = \exp(X) \cdot \exp(Y).$$

Formally we denote $Z = \log(\exp(X)\exp(Y))$. The element Z can be further explicitly expressed in the Dynkin form as (see [288]):

$$Z = X + Y + \frac{1}{2}[X, Y] + \frac{1}{12}[X, [X, Y]] + \frac{1}{12}[Y, [Y, X]] + \cdots$$

$$= \sum_{m=1}^{\infty} \sum \frac{(-1)^{m-1} ad_Y^{q_m} ad_X^{p_m} \cdots ad_Y^{q_1} ad_X^{p_1}}{m(\sum_{i=1}^{m}(p_i + q_i)) \prod_{i=1}^{m}(p_i! q_i!)} \qquad (8.9)$$

where the inner sum is over all m-tuples of pairs of nonnegative integers (p_i, q_i) such that $p_i + q_i > 0$, and in order to simplify the notation, we use the convention $ad_X = X$ or $ad_Y = Y$ for the final terms in the multiplication.

Proof: A combinatorial proof of the CBH formula can be found in [243].

Using the definition of the adjoint map ad, each term in the summation of the Dynkin's formula is in fact:

$$ad_Y^{q_m} ad_X^{p_m} \cdots ad_Y^{q_1} ad_X^{p_1}$$

$$= [\underbrace{Y, \ldots [Y}_{q_m \text{ times}}, \underbrace{[X, \ldots [X}_{p_m \text{ times}}, \ldots \underbrace{[Y, \ldots [Y}_{q_1 \text{ times}}, \underbrace{[X, \ldots [X}_{p_1 \text{ times}}, [X, X]\ldots].$$

Note that if X and Y are commutative, i.e., $[X, Y] = 0$, then the CBH formula simply becomes $\log(\exp(X)\exp(Y)) = X + Y$. An important element of the CBH formula is that the element Z is a formal power series in elements of the Lie algebra generated by X and Y. Another place that we have seen such a power series are the Peano–Baker–Volterra series for bilinear time-varying system (Proposition 4.34). We will also encounter such a series in the context of the Chen–Fliess series for affine nonlinear system in Chapter 11 (Theorem 11.30).

Another useful way of writing the CBH formula is in term of degrees (powers) in X or Y:

$$Z = Y + \frac{ad_Y}{\exp(ad_Y) - 1} X + \cdots,$$

$$Z = X + \frac{-ad_X}{\exp(-ad_X) - 1} Y + \cdots,$$

where the dots represent terms of at least second-degree in X or Y respectively. One can directly check that (see [243]) up to linear terms in X, the Campbell–Baker–Hausdorff formula has the terms:

$$\frac{ad_Y}{\exp(ad_Y) - 1} X = \sum_{n=0}^{\infty} \frac{B_n}{n!} ad_Y^n X$$

where the B_n are the so-called *Bernoulli numbers*. In fact, the recursive definition of Bernoulli numbers is simply the simplification of the exponential formulas given above.

8.5 Left-Invariant Control Systems on Matrix Lie Groups

This section uses the mathematics developed in the previous section to describe control systems with left-invariant vector fields on matrix Lie groups. For an n-dimensional Lie group G, the type of system described in this section has state variable that can be represented as an element $g \in G$. The differential equation which describes the evolution of g can be written as:

$$\dot{g} = g \left(\sum_{i=1}^{n} X_i u_i \right),$$

where the u_i are the inputs and the X_i are a basis for the Lie algebra $\mathcal{L}(g)$. In the above equation, $g X_i$ is the notation for the left-invariant vector field associated with X_i. The equation represents a *driftless* system, since if $u_i = 0$ for all i, then $\dot{g} = 0$.

In the next subsection, the state equation describing the motion of a rigid body on $SE(3)$ is developed. The following subsection develops a transformation, called the Wei–Norman formula, between the inputs u_i, i.e., the Lie-Cartan coordinates of the first kind and the Lie-Cartan coordinates of the second kind. Applications to steering a control system on $SO(3)$ are given in the Exercises (8.11).

8.5.1 Frenet–Serret Equations: A Control System on $SE(3)$

In this section, arc-length parameterization of a curve describing the path of a rigid body in $\mathbb{R}^3$ is used to derive the state equation of the motion of this left invariant system.

Consider a curve

$$\alpha(s) : [0, 1] \mapsto \mathbb{R}^3$$

representing the motion of a rigid body in $\mathbb{R}^3$. Represent the tangent to the curve as

$$t(s) = \alpha'(s).$$

Constrain the tangent to have unity norm, $|t(s)| = 1$, ie the inner product

$$\langle t(s), t(s) \rangle = 1.$$

Now taking the derivative of the above with respect to s, we have

$$\langle t'(s), t(s) \rangle + \langle t(s), t'(s) \rangle = 0,$$

so that $t'(s) \perp t(s)$. Denote the norm of $t'(s)$ as

$$|t'(s)| = \kappa(s),$$

where $\kappa(s)$ is called the *curvature* of the motion: It measures how quickly the curve is pulling away from the tangent. Let us assume $\kappa > 0$. Denoting the unit normal vector to the curve $\alpha(s)$ as $n(s)$, we have that

$$t'(s) = \kappa(s)n(s),$$

and also the inner product

$$\langle n(s), n(s) \rangle = 1,$$

so that $n'(s) \perp n(s)$.

The *binormal* to the curve at s is denoted by $b(s)$, where

$$b(s) = t(s) \times n(s),$$

or equivalently,

$$n(s) = b(s) \times t(s).$$

Let

$$n'(s) = \tau(s)b(s),$$

where $\tau(s)$, called the *torsion* of the motion, measures how quickly the curve is pulling out of the plane defined by $n(s)$ and $t(s)$. Thus

$$\begin{aligned} b'(s) &= t'(s) \times n(s) + t(s) \times n'(s) \\ &= \kappa(s)n(s) \times n(s) + t(s) \times \tau(s)b(s) \\ &= \tau(s)n(s), \end{aligned}$$

since $n(s) \times n(s) = 0$.

Similarly,

$$\begin{aligned} n'(s) &= b'(s) \times t(s) + b(s) \times t'(s) \\ &= -\tau(s)b(s) - \kappa(s)t(s). \end{aligned}$$

We thus have

$$\alpha'(s) = t(s),$$
$$t'(s) = \kappa(s)n(s),$$
$$n'(s) = -\tau(s)b(s) - \kappa(s)t(s),$$
$$b'(s) = \tau(s)n(s).$$

Since $t(s)$, $n(s)$, and $b(s)$ are all orthogonal to each other, the matrix with these vectors as its columns is an element of $SO(3)$:

$$[t(s), \quad n(s), \quad b(s)] \in SO(3).$$

Thus,

$$g(s) := \left[\begin{array}{ccc|c} t(s) & n(s) & b(s) & \alpha(s) \\ \hline 0 & 0 & 0 & 1 \end{array} \right] \in SE(3),$$

and

$$\frac{d}{ds}g(s) = g(s) \left[\begin{array}{ccc|c} 0 & -\kappa(s) & 0 & 1 \\ \kappa(s) & 0 & \tau(s) & 0 \\ 0 & -\tau(s) & 0 & 0 \\ \hline 0 & 0 & 0 & 0 \end{array} \right].$$

These are known at the *Frenet–Serret* equations of a curve. The evolution of the Frenet–Serret frame in $\mathbb{R}^3$ is given by

$$\dot{g} = gX,$$

where $g \in SE(3)$ and X is an element of the Lie algebra $se(3)$. We may regard the curvature $\kappa(s)$ and the torsion $\tau(s)$ as inputs to the system, so that if

$$u_1 = \kappa(s),$$
$$u_2 = -\tau(s),$$

then

$$\dot{g} = g \left[\begin{array}{ccc|c} 0 & -u_1 & 0 & 1 \\ u_1 & 0 & -u_2 & 0 \\ 0 & u_2 & 0 & 0 \\ \hline 0 & 0 & 0 & 0 \end{array} \right],$$

which is a special case of the general form describing the state evolution of a left invariant control system in $SE(3)$.

An example of the general form of a left invariant control system in $SE(3)$ is given by an aircraft flying in $\mathbb{R}^3$:

$$\dot{g} = g \left[\begin{array}{ccc|c} 0 & -u_3 & u_2 & u_4 \\ u_3 & 0 & -u_1 & 0 \\ -u_2 & u_1 & 0 & 0 \\ \hline 0 & 0 & 0 & 0 \end{array} \right].$$

The inputs u_1, u_2, and u_3 control the *roll, pitch,* and *yaw* of the aircraft, and the input u_4 controls the *forward velocity.*

Specializing the above to $SE(2)$, we have the example of the unicycle rolling on the plane with speed u_1:

$$\dot{g} = g \left[\begin{array}{cc|c} 0 & -u_2 & u_1 \\ u_2 & 0 & 0 \\ \hline 0 & 0 & 0 \end{array} \right]. \tag{8.10}$$

In this case, the input u_2 controls the angle of the wheel.

The previous formulation describes *kinematic* steering problems since it is assumed that we have direct control of the velocities of the rigid bodies. In the control of physical systems, though, we generally have access only to the forces and torques which drive the motion. A more realistic approach would therefore be to formulate the steering problem with a *dynamic* model of the rigid body, which uses these forces and torques as inputs. Dynamic models are more complex than their kinematic counterparts, and the control problem is harder to solve.

8.5.2   The Wei–Norman Formula

In this section we derive the Wei–Norman formula, which describes a relationship between the open loop inputs to a system and the Lie–Cartan coordinates used to parameterize the system. Consider the state equation of a left-invariant control system on a Lie group G with state $g \in G$:

$$\dot{g} = g \left(\sum_{i=1}^{n} X_i u_i \right),$$

where the u_i are inputs and the X_i are a basis of the Lie algebra $\mathcal{L}(g)$. We may express g in terms of its Lie–Cartan coordinates of the second kind:

$$g(t) = \exp(\gamma_1(t)X_1) \exp(\gamma_2(t)X_2) \cdots \exp(\gamma_n(t)X_n).$$

Thus,

$$\dot{g} = \sum_{i=1}^{n} \gamma_i'(t) \prod_{j=1}^{i-1} \exp(\gamma_j X_j) X_i \prod_{j=i}^{n} \exp(\gamma_j X_j)$$

$$= g \sum_{i=1}^{n} \gamma_i'(t) \left(\prod_{j=1}^{n} \exp(\gamma_j X_j) \right)^{-1} \left(Ad_{\prod_{j=1}^{i-1} \exp(\gamma_j X_j)} \right) X_i \left(\prod_{j=1}^{n} \exp(\gamma_j X_j) \right)$$

$$= g \sum_{i=1}^{n} \gamma_i'(t) Ad_{(\prod_{j=1}^{n} \exp(\gamma_j X_j))^{-1}} \left(\left(Ad_{\prod_{j=1}^{i-1} \exp(\gamma_j X_j)} \right) X_i \right)$$

$$= g \sum_{i=1}^{n} \gamma_i'(t) \sum_{k=1}^{n} \xi_{ki}(\gamma) X_k,$$

where the last equation results from Lemma 8.30. If we compare this equation with the state equation, we may generate a formula for the inputs to the system in terms of the Lie–Cartan coordinates:

$$\begin{bmatrix} u_1(t) \\ u_2(t) \\ \cdot \\ \cdot \\ u_n(t) \end{bmatrix} = \begin{bmatrix} & & \\ & \xi_{ij}(\gamma) & \\ & & \end{bmatrix} \begin{bmatrix} \gamma_1'(t) \\ \gamma_2'(t) \\ \cdot \\ \cdot \\ \gamma_n'(t) \end{bmatrix}$$

so that

$$
\begin{bmatrix} \gamma_1'(t) \\ \gamma_2'(t) \\ \cdot \\ \cdot \\ \cdot \\ \gamma_n'(t) \end{bmatrix} = \begin{bmatrix} & & \\ & \xi_{ij}(\gamma) & \\ & & \end{bmatrix}^{-1} \begin{bmatrix} u_1(t) \\ u_2(t) \\ \cdot \\ \cdot \\ \cdot \\ u_n(t) \end{bmatrix}.
$$

The above is known as the *Wei-Norman formula*. It transforms the problem from a differential equation on Lie groups to one in $\mathbb{R}^n$: Steering from an initial configuration g_i to a final configuration g_f is converted into steering from $\gamma(0)$ to $\gamma(1)$, both vectors in $\mathbb{R}^n$.

8.6 Summary

This chapter has given the reader an abbreviated vista into differential geometry. The treatment is rather abbreviated especially in topics of interest to Riemannian geometry where an inner product structure is placed on the tangent space. There are some excellent textbooks in differential geometry including Boothby [34] and [325]. For a great deal of the recent work on the intersection of differential geometry, classical mechanics and nonlinear control theory, see for example Murray [224], and Kang and Marsden [159].

For more on Lie groups we refer the reader to Curtis [75]. More advanced material is to be found in [312]. The literature on control of systems on Lie groups is very extensive with early contributions by Brockett [43], [44], and Baillieul [16]. The use of Lie groups in kinematics and control of robots introduced by Brockett was developed in Murray, Li and Sastry in [225]. Steering of systems on Lie groups was studied by Walsh, Sarti and Sastry in [320], [257]. More recently the use of the Wei Norman formula and averaging was studied by Leonard and Krishnaprasad [178] and Bullo, Murray, and Sarti [51]. Interpolation problems for systems on Lie groups can be very interesting in applications such as the "landing tower problem" posed by Crouch, see for example Crouch and Silva-Leite [73] and Camarinha et al [58]. Steering of systems on some other classical Lie groups is very important in quantum and bond-selective chemistry and is a growing area of research, see especially the work of Tarn and coworkers [295; 294] and Dahleh, Rabitz, and co-workers [76]. Optimal control of systems on Lie groups is the subject of a recent book by Jurdjevic [157], see also Montgomery [218].

8.7 Exercises

Problem 8.1. Prove Proposition 8.3.

Problem 8.2. Prove that the tangent bundle TM of a manifold is also a manifold of dimension twice the dimension of M.

Problem 8.3. Find a geometric description of the tangent bundle of a circle S^1, a torus $S^1 \times S^1$ and a cylinder $S^1 \times \mathbb{R}$. More generally what can you say about the tangent bundle of the product of two manifolds $M_1 \times M_2$.

Problem 8.4. Prove that the definition of the involutivity of a distribution is basis independent. More precisely, show that if for

$$\Delta(x) = \text{span}\{f_1(x), \ldots, f_m(x)\}$$

it has been verified that $[f_i, f_j] \in \Delta$, then for another basis

$$\Delta(x) = \text{span}\{g_1, \ldots, g_m(x)\}$$

it follows that $[g_i, g_j] \in \Delta$.

Problem 8.5. Prove Proposition 8.14.

Problem 8.6 Constructions using the Frobenius theorem. The sufficiency part of the proof of the Frobenius theorem is interesting in that if one can explicitly construct the diffeomorphism F and its inverse, its last $n - d$ rows are the desired annihilator of the distribution. This is particularly useful when $n - d$ is a small integer, for example, one or two. Pursue this program for the following two examples:

1. Consider $n = 2$ and $d = 1$ with

$$\Delta = \text{span}\left\{ \begin{bmatrix} e^{x_2} \\ 1 \end{bmatrix} \right\}.$$

Find the function $\lambda(x)$ such that $d\lambda = \Delta^\perp$.

2. For $n = 3$ and $d = 2$ consider

$$\Delta = \text{span}\left\{ \begin{bmatrix} 2x_3 \\ -1 \\ 0 \end{bmatrix}, \begin{bmatrix} -x_1 \\ -2x_2 \\ x_3 \end{bmatrix} \right\}.$$

Find the function $\lambda(x)$ such that $d\lambda = \Delta^\perp$.

Problem 8.7. Prove that the exponential map maps the Lie algebras $o(n)$, $su(n)$, $sp(n)$, $se(n)$ into their corresponding Lie groups $O(n)$, $SU(n)$, $Sp(n)$, $SE(n)$, respectively, Also prove that the logarithm map maps the same Lie groups into their respective Lie algebras.

Problem 8.8. Find the dimensions of the Lie Groups $O(n)$, $SU(n)$, $Sp(n)$, $SE(n)$, $SL(n)$.

Problem 8.9 Explicit exponential formulas [225]. Compute the exponential formula explicitly for $SO(3)$, $SE(3)$. The formula for $SO(3)$ is called the Rodrigues formula and that for $SE(3)$ is called the *Chasles formula*.

Problem 8.10 Product of exponentials formula for robot manipulators [225].
In [225] (Chapter 2) it is shown that the kinematics describing the position and
orientation of a frame attached to the end of the robot denoted $g \in SE(3)$ for a
rigid robot with serial connected links may be related to the joint angles of the
links labeled $\theta_i, i = 1, \ldots, n$, using Lie–Cartan coordinates of the second kind as

$$g(\theta) = g(0) \exp^{\xi_1 \theta_1} \exp^{\xi_2 \theta_2} \cdots \exp^{\xi_n \theta_n}, \tag{8.11}$$

where $g(0) \in SE(3)$ stands for the position and orientation of the robot in some
reference (home) configuration and $\xi_i \in se(3)$ are elements of the Lie algebra
representing the so-called twist axes of the robot links. Now assume that the joint
angles are functions of time t. Derive a formula (called the manipulator Jacobian)
relating the velocity of the manipulator defined by $g(\theta(t))^{-1}\dot{g}(\theta(t))$ to the joint
velocities $\dot{\theta}_i$. For a typical robot manipulator $n = 6$. Can you think of how you
may be able to invert the formula (8.11), that is given $g \in SE(3)$ find the θ_i which
satisfy the formula.

Problem 8.11 Satellite Control (see [320], [257]).   A satellite is a rigid body
floating in space. There are *rotors* or *momentum wheels* attached to its body which
create linearly independent momentum fields which rotate the satellite to any
configuration in $SO(3)$. The satellite may be modeled as a drift free system on
$SO(3)$:

$$\dot{g} = g\hat{b}_1 u_1 + g\hat{b}_2 u_2 + g\hat{b}_3 u_3,$$

where $g \in SO(3)$, $\hat{b}_i \in so(3)$, and the $u_i \in \mathbb{R}$ are scalars. The vector $b_i \in \mathbb{R}^3$
describes the direction and magnitude of the momentum field created by the ith
momentum wheel on the satellite. Given an initial state g_i and a desired final state
g_f, we wish to find control inputs $u_1(t), u_2(t), u_3(t)$ which will steer the system
from g_i to g_f in finite time T.

1. Consider the case in which b_i are linearly independent for $i \in \{1, 2, 3\}$. If we
   assume that the inputs are constant (u_1, u_2, u_3) and applied over one second,
   the solution to the state equation is

   $$g_f = g_i \exp(\hat{b}_1 u_1 + \hat{b}_2 u_2 + \hat{b}_3 u_3).$$

   Use the results of this chapter to find the constant inputs to
2. The second case that we consider is the two input system in which $b_3 = 0$, but
   b_1, b_2 are linearly independent. If the input vectors b_1 and b_2 are not orthogonal,
   the first step is to orthogonalize them using the Gram-Schmidt algorithm to
   make $b_1 = \hat{x}, b_2 = \hat{y}$. Give the details of this transformation. Now, consider
   an input consisting of having $u_1 = 1, u_2 = 0$ for a_1 seconds followed by
   $u_1 = 0, u_2 = 1$ for a_2 seconds followed by the input $u_1 = 1, u_2 = 0$ for a_3
   seconds. Then, we have

   $$g_f = g_i \exp(\hat{x}a_1) \exp(\hat{y}a_2) \exp(\hat{x}a_3). \tag{8.12}$$

   Solve this equation for a_1, a_2, a_3 given g_i, g_f. Are the solutions unique? Note
   that you have provided a strategy for steering from g_i to g_f in $a_1 + a_2 + a_3$

seconds. Of course, you can also steer in an arbitrary time T seconds. Equation (8.12) is a roll-pitch-roll coordinatization of $SO(3)$ or a generalization of the *Lie Cartan coordinates of the second kind* discussed in this chapter. Can you think of other strategies for steering the satellite from g_i to g_f?

3. Consider the two input situation of the previous example, with the input directions normalized. Consider the inputs $u_1 = a_0 + a_i \sin 2\pi t$ and $u_2 = b_1 + b_2 \cos 2\pi t$. Can you find values of a_0, a_1, b_0, b_1 to steer the system from g_i to g_f?

4. Consider the case that $b_3 = 0$ and b_1, b_2 are linearly independent, but $u_2 \equiv 1$. This corresponds to the case that the satellite is drifting. Derive conditions under which you can steer from g_i to g_f in 1 second.

Problem 8.12 Steering the Unicycle.   Consider the unicycle of equation (8.10). Steer it from an initial to a final configuration in $SE(2)$ using piecewise constant inputs u_1, u_2.

Problem 8.13 Stiefel Manifolds.   Consider the space of all matrices $U \in \mathbb{R}^{n \times m}$ with $m \leq n$, such that $UU^T = I \in \mathbb{R}^{m \times m}$. Prove that the space of these matrices is a manifold. Find its dimension. Characterize the tangent space of this manifold. What can you say about this class of manifolds, called *Stiefel manifolds* when $m = n$ and $m = n - 1$.

Problem 8.14 Grassmann Manifolds.   Consider the space of all m dimensional subspaces in $\mathbb{R}^n$ ($m \leq n$). Prove that this is a smooth manifold (the so called Grassmann manifold), usually denoted by $G(n, m)$. Characterize the tangent space of this manifold and find its dimension. Consider all the orthonormal bases for each m-dimensional subspace in $G(n, m)$. Convince yourself that $G(n, m)$ is the quotient space $V(n, m)/O(m)$, where $V(n, m)$ is the Stiefel manifold defined in the previous problem. Check that

$$\dim(G(n, m)) = \dim(V(n, m)) - \dim(O(m)).$$

Problem 8.15 Several Representations of $SO(3)$.   Prove that the following Lie groups can be naturally identified with each other:

1. $SO(3)$;
2. the quotient $SU(2)/(\pm I)$ of the group $SU(2)$ by its center (an element in the center of a group is one which commutes with all elements in the group);
3. the quotient $S^3/(\pm 1)$ of unitary quaternions (see Chapter 3 for the definition) by its center.

Prove that, as manifolds, each of them is diffeomorphic to all the following manifolds, and therefore induces the same group structure on them:

1. $V(3, 2)$ the Stiefel manifold of all 3×2 orthogonal matrices in $\mathbb{R}^{3 \times 2}$;
2. $T_1 S^2$ the set of unit tangent vectors to the unit sphere in $\mathbb{R}^3$.
3. $\mathbb{R}P^3$ the set of all lines through the origin in $\mathbb{R}^4$ (this is also the Grassmann manifold $G(4, 1)$, see Problem 8.14).

Problem 8.16 Essential Manifold [197; 107]. This exercise gives a geometric characterization of one of the most important manifolds used in computer vision, the so-called *essential manifold*:

$$E = \{R\hat{p} \mid R \in SO(3), p \in \mathbb{R}^3, |p|^2 = 1\}$$

where $\hat{p}$ is the skew symmetric matrix

$$\hat{p} = \begin{pmatrix} 0 & -p_3 & p_2 \\ p_3 & 0 & -p_1 \\ -p_2 & p_1 & 0 \end{pmatrix} \in \mathbb{R}^{3\times3}$$

associated to $p = (p_1, p_2, p_3)^T \in \mathbb{R}^3$. Show that this is a smooth manifold. Following the steps given below establish the fact that the unit tangent bundle of $SO(3)$, i.e., the unit ball of tangent vectors at each point on the manifold, is a double covering of the essential manifold E:

1. Show that $R\hat{p}$ is in $so(3)$ for $R \in SO(3)$ if and only if $R = I_{3\times3}$ or $R = e^{\hat{p}\pi}$.
2. Show that given an essential matrix $X \in E$, there are exactly two pairs (R_1, p_1) and (R_2, p_2) such that $R_i \hat{p}_i = X, i = 1, 2$ (Can you find an explicit formula for R_i, $\hat{p}_i$? Consider the singular value decomposition of X).
3. Show that the unit tangent bundle of $SO(3)$ (using the metric $\frac{1}{2}$ trace$(A^T A)$ for $A \in so(3)$) is a double covering of the essential manifold E.
4. Conclude that E is 5 dimensional compact connected manifold.

Problem 8.17 Conjugate Group of $SO(3)$. In this exercise, we study the conjugate group of $SO(3)$ in $SL(3)$ which plays a fundamental role in camera self-calibration in computer vision [196]. Given a matrix $A \in SL(3)$, a conjugate class of $SO(3)$ is given by:

$$G = \{ARA^{-1} \mid R \in SO(3)\}.$$

Show that G is also a Lie group and determine its Lie algebra? Follow the steps given below to establish necessary and sufficient conditions for determining the matrix A from given matrices in the conjugate group G:

1. Show that the matrix $A \in SL(3)$ can only be determined up to a rotation matrix, i.e., A can only be recovered as an element in the quotient space $SL(3)/SO(3)$ from given matrices in G.
2. Show that, for a given $C \in G$ with $C \neq I$, the real symmetric kernel of the linear map:

$$L : S \mapsto S - CSC^T$$

is exactly two dimensional.
3. Show that, given $C_i = AR_i A^{-1} \in G, i = 1, \ldots, n$, the symmetric form $S = AA^T$ is uniquely determined if and only if at least two axes of the rotation matrices R_i's are linearly independent.

9
Linearization by State Feedback

9.1 Introduction

In this chapter we begin with a study of the modern geometric theory of nonlinear control. The theory began with early attempts to extend results from linear control theory to the nonlinear case, such as results on controllability and observability. This work was pioneered by Brockett, Hermann, Krener, Fliess, Sussmann and others in the 1970s. Later, in the 1980s in a seminal paper by Isidori, Krener, Gori-Giorgi, and Monaco [150] it was shown that not only could the results on controllability and observability be extended but that large amounts of the linear geometric control theory, as represented, say, in Wonham [331] had a nonlinear counterpart. This paper, in turn, spurred a tremendous growth of results in nonlinear control in the 1980s. On a parallel course with this one, was a program begun by Brockett and Fliess on embedding linear systems in nonlinear ones. This program can be thought of as one for linearizing systems by state feedback and change of coordinates. Several breakthroughs, beginning with [45; 154; 148; 67], and continuing with the work of Byrnes and Isidori [56; 54; 55], the contents are summarized are in Isidori's book [149], yielded a fantastic set of new tools for designing control laws for large classes of nonlinear systems (see also a recent survey by Krener [170]). This theory is what we refer to in the chapter title. It does not refer to the usual Jacobian linearization of a nonlinear system, but rather the process of making *exactly linear* the input–output response of nonlinear systems. It is worthwhile noting that this material is presented not in the order of its discovery, but rather from the point of view of ease of development.

9.2 SISO Systems

A large class of SISO (Single-Input Single-Output) nonlinear systems can be made to have linear input-output behavior through a choice of *nonlinear state feedback control law*.

9.2.1 Input–Output Linearization

Consider, at first, the single-input single-output system

$$\dot{x} = f(x) + g(x)u,$$
$$y = h(x). \tag{9.1}$$

with $x \in \mathbb{R}^n$, f, g smooth vector fields on $\mathbb{R}^n$ and h a smooth nonlinear function. *Smooth* will mean an infinitely differentiable function.[1] For the more advanced results in geometric nonlinear control of Chapter 11, we will actually need to assume that f, g are analytic vector fields, that is to say that not only are they truly infinitely differentiable, they have convergent Taylor series. The following calculations will be made for $x \in U$, an open subset of $\mathbb{R}^n$. Typically U is an open set containing an equilibrium point x_0 of the undriven system, that is a point at which $f(x_0) = 0$. Differentiating y with respect to time, one obtains

$$\dot{y} = \frac{\partial h}{\partial x} f(x) + \frac{\partial h}{\partial x} g(x)u$$
$$:= L_f h(x) + L_g h(x)u. \tag{9.2}$$

Here, $L_f h(x) : \mathbb{R}^n \mapsto \mathbb{R}$ and $L_g h(x) : \mathbb{R}^n \mapsto \mathbb{R}$ stand for the *Lie derivatives* of h with respect to f and g respectively. Thus, $L_f h(x)$ is a function giving the rate of change of h along the flow of the vector field f; similarly[2] for $L_g h(x)$. If $L_g h(x)$ is bounded away from zero[3] for all $x \in U$, the state feedback law given by

$$u = \frac{1}{L_g h(x)} (-L_f h(x) + v) \tag{9.3}$$

yields the first-order linear system from the (new) input v to the output y:

$$\dot{y} = v. \tag{9.4}$$

Thus, we see that there exist functions $\alpha(x)$, $\beta(x)$ such that the state feedback law $u = \alpha(x) + \beta(x)v$ renders the system input-output linear. If the original system (9.1) were minimal, i.e., both controllable and observable (here we use these terms

---

[1] Infinitely differentiable, or C^∞, is shorthand for saying that we are too lazy to count the number of times the function needs to be continuously differentiable in order to use all of the formulas in the ensuing development.

[2] At this point, the reader may find it useful to rethink the definitions of Lyapunov stability of Chapter 5 in the language of Lie derivatives.

[3] Note that this is a stronger requirement than $L_g h(x) \neq 0$ for all $x \in U$, so as to make sure that the control law does not become unbounded.

without giving the precise definitions, the reader will need to wait till Chapter 11 to get precise definitions of these notions for nonlinear systems), then the control law (9.3) has the effect of rendering $(n - 1)$ of the states of the system (9.1) unobservable through state feedback. (Convince yourself that this is what is going on in the linear case, where you know precise definitions and characterizations of minimality. That is, in the case that $f(x) = Ax$, $g(x) = b$, convince yourself that the observability rather than controllability of the original system (9.1) is being changed.)

In the instance that $L_g h(x) \equiv 0$, meaning that $L_g h(x) = 0 \ \forall \ x \in U$, we differentiate (9.2) to get

$$\ddot{y} = \frac{\partial L_f h}{\partial x} f(x) + \frac{\partial L_f h}{\partial x} g(x) u,$$

$$:= L_f^2 h(x) + L_g L_f h(x) u.$$

(9.5)

In (9.5) above, $L_f^2 h(x)$ stands for $L_f(L_f h)(x)$ and $L_g L_f h(x) = L_g(L_f h(x))$. Now, if $L_g L_f h(x)$ is bounded away from zero for all $x \in U$, the control law given by

$$u = \frac{1}{L_g L_f h(x)} (-L_f^2 h(x) + v)$$

(9.6)

yields the linear second-order system from input v to output y:

$$\ddot{y} = v.$$

(9.7)

More generally, if γ is the *smallest* integer for which $L_g L_f^i h(x) \equiv 0$ on U for $i = 0, \dots, \gamma - 2$ and $L_g L_f^{\gamma-1} h(x)$ is bounded away from zero on U, then the control law given by

$$u = \frac{1}{L_g L_f^{\gamma-1} h(x)} (-L_f^\gamma h(x) + v)$$

(9.8)

yields the γ-th order linear system from input v to output y:

$$y^\gamma = v.$$

(9.9)

The procedure described above terminates at some finite γ, under some specific conditions to be given shortly. The theory is considerably more complicated if $L_g L_f^i h(x)$ is not identically zero on U, but zero for some values of $x \in U$. To set the preceding discussions on a firm analytical footing, we will make the following definition of *strict relative degree* of a nonlinear system:

Definition 9.1 Strict Relative Degree. *The SISO nonlinear system (9.1) is said to have* strict relative degree γ at $x_0 \in U$ if

$$L_g L_f^i h(x) \equiv 0 \quad \forall x \in U, \ i = 0, \dots, \gamma - 2,$$

$$L_g L_f^{\gamma-1} h(x_0) \neq 0.$$

(9.10)

Remarks:

1. This definition is compatible with the usual definition of relative degree for linear systems (as being the excess of poles over zeros). Indeed, consider the linear system described by

$$\dot{x} = Ax + bu,$$
$$y = cx.$$

(9.11)

Now note that the Laurent expansion of the transfer function is given by

$$c(sI - A)^{-1}b = \frac{cb}{s} + \frac{cAb}{s^2} + \frac{cA^2b}{s^3} + \cdots,$$

valid for $|s| > \max_i |\lambda_i(A)|$, so that the first nonzero term gives the relative degree of the system. Thus, if the system has relative degree is γ then $cb = cAb = \cdots = cA^{\gamma-2}b = 0$ and $cA^{\gamma-1}b \neq 0$. This is consistent with the preceding definition, since for $i = 0, 1, 2, \ldots$, we have that

$$L_g L_f^i h(x) = cA^i b.$$

2. It is worthwhile noting that if $L_g L_f^{\gamma-1} h(x_0) \neq 0$, then there exists a neighborhood U of x_0 such that $L_g L_f^{\gamma-1} h(x)$ is bounded away from zero.

3. The relative degree of some nonlinear systems *may not be defined* at some points $x_0 \in U \subset \mathbb{R}^n$. Indeed, it may happen that for some γ, $L_g L_f^{\gamma-1} h(x_0) = 0$, but $L_g L_f^{\gamma-1} h(x) \neq 0$ for x near x_0. For example, it may be the case that $L_g L_f^{\gamma-1} h(x)$ changes sign on a surface going through x_0 as shown in Figure 9.1.

Example 9.2. *Consider the nonlinear system given in state space form by*

$$\dot{x} = \begin{bmatrix} x_2 \\ -x_2 - x_1^3 - x_1 \end{bmatrix} + \begin{bmatrix} 0 \\ 1 \end{bmatrix} u.$$

If the output function is

$$y = h(x) = x_1,$$

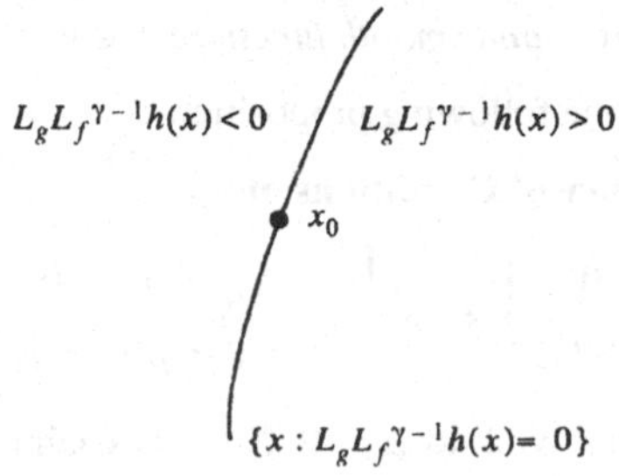

FIGURE 9.1. Lack of existence of relative degree in the state space

it follows that

$$L_g h(x) = \frac{\partial h}{\partial x} g(x) = 0,$$

and that

$$L_f h(x) = \frac{\partial h}{\partial x} f(x) = x_2,$$

so that we may obtain

$$L_g L_f h(x) = 1.$$

Thus with this choice of output the system has relative degree 2 at each value of x_0. However, if the output function is $y = x_2$, it is easy to see that $L_g h(x) = 1$ so that the system has relative degree 1. Also, if the output is $y = \cos x_2$ it may be verified that $L_g h(x) = -\sin x_2$. Note that the system has relative degree 1 except at points at which $x_2 = \pm n\pi$, where it is undefined.

"Normal" Form for SISO Nonlinear Systems

If a SISO nonlinear system has a relative degree $\gamma \le n$ at some point x_0, it is possible to transform the nonlinear system into a "normal" form as follows. In order to do this, we will recall from the previous chapter. that the Lie bracket of two vector fields f, g denoted $[f, g]$ is given in coordinates by

$$[f, g] = \frac{\partial g}{\partial x} f(x) - \frac{\partial f}{\partial x} g(x). \tag{9.12}$$

Denote by ad_f the operator from vector fields to vector fields defined by

$$ad_f g = [f, g]. \tag{9.13}$$

We define successive applications of ad_f by

$$ad_f^k g := [f, ad_f^{k-1} g] \quad \text{for } k \ge 0 \tag{9.14}$$

with the convention that $ad_f^0 g := g$.

Claim 9.3. *By straightforward calculation it follows that*

$$L_{[f,g]}\lambda = L_f L_g \lambda - L_g L_f \lambda \tag{9.15}$$

for all vector fields $f(x)$, $g(x)$ and smooth functions $\lambda \in C^\infty(\mathbb{R}^n)$.

Using this we may state the following proposition:

Proposition 9.4 Equivalence of Conditions on h.

$$\left\{ \begin{array}{c} L_g L_f^k h(x) \equiv 0 \\ 0 \le k \le \mu \forall x \in U \end{array} \right\} \Longleftrightarrow \left\{ \begin{array}{c} L_{ad_f^k g} h(x) \equiv 0 \\ 0 \le k \le \mu \forall x \in U \end{array} \right\}. \tag{9.16}$$

Proof: The proposition is trivial for $\mu = 0$. Now, by Claim 9.3 above,

$$L_{ad_f g} h(x) = L_f L_g h(x) - L_g L_f h(x). \tag{9.17}$$

Thus, if $L_g h(x) \equiv 0$, for $x \in U$, we have

$$L_{ad_f g} h(x) \equiv 0 \iff L_g L_f h(x) \equiv 0 \tag{9.18}$$

for $x \in U$. Proceeding further,

$$\begin{aligned} L_{ad_f^2 g} h &= L_f L_{ad_f g} h - L_{ad_f g} L_f h \\ &= L_f L_{ad_f g} h - L_f L_g L_f h + L_g L_f^2 h. \end{aligned} \tag{9.19}$$

Thus, if $L_g L_f h(x) \equiv 0 \Leftrightarrow L_{ad_f g} h(x) \equiv 0$ for $x \in U$, it follows that

$$\begin{aligned} L_{ad_f^2 g} h &= -L_{ad_f g} L_f h \\ &= -L_f L_g L_f h + L_g L_f^2 h(x). \end{aligned}$$

Using the fact that $L_g L_f h(x) \equiv 0$, it follows that

$$L_{ad_f^2 g} h(x) \equiv 0 \Leftrightarrow L_g L_f^2 h(x) \equiv 0 \tag{9.20}$$

for all $x \in U$. The rest of the proof proceeds by recursion. $\square$

Now define

$$\begin{aligned} \phi_1(x) &= h(x), \\ \phi_2(x) &= L_f h(x), \\ &\vdots \\ \phi_\gamma(x) &= L_f^{\gamma-1} h(x). \end{aligned} \tag{9.21}$$

In words, the $\phi_i(x)$ are the coordinates described by y and the first $\gamma - 1$ time derivatives of y. It is a consequence of the definition of relative degree that these $\gamma - 1$ time derivatives of y do not depend on u. We claim that these coordinates qualify as a partial (since $\gamma \leq n$) change of coordinates for the system. For this purpose, we need to check the following claim:

Claim 9.5 Partial Change of Coordinates. *The derivatives $d\phi_i(x)$ for $i = 1, \ldots, \gamma$ are linearly independent (over $C^\infty(\mathbb{R}^n)$) in U.*

Proof: We first claim that for all $i + j \leq \gamma - 2$ we have that for $x \in U$

$$L_{ad_f^i g(x)} L_f^j h(x) \equiv 0. \tag{9.22}$$

Indeed, for $i = 0$ the equation is a restatement of the definition of relative degree. For $i > 1$ we have that

$$L_{ad_f^i g(x)} L_f^j h(x) = L_{ad_f^{i-1} g(x)} L_f^{j+1} h(x) - L_f L_{ad_f^{i-1} g(x)} L_f^j h(x).$$

Thus, the proof of the equation (9.22) follows by recursion on the index i. Also continuing the iteration one step further yields that for $i + j = \gamma - 1$,

$$L_{ad_f^i g(x)} L_f^j h(x_0) = (-1)^{\gamma-1-j} L_g L_f^{\gamma-1} h(x_0) \neq 0,$$

Using this calculation we may verify that

$$
\begin{bmatrix} dh(x_0) \\ dL_f h(x_0) \\ \vdots \\ dL_f^{\gamma-1} h(x_0) \end{bmatrix}
\begin{bmatrix} g(x_0) & ad_f g(x_0) & \cdots & ad_f^{\gamma-1} g(x_0) \end{bmatrix}
$$

$$
= \begin{bmatrix} 0 & & \cdots & L_{ad_f^{\gamma-1} g} h(x_0) \\ 0 & & L_{ad_f^{\gamma-2} g} L_f h(x_0) & * \\ \cdots & & \cdots & * \\ L_g L_f^{\gamma-1} h(x_0) & & * & * \end{bmatrix}. \tag{9.23}
$$

The anti-diagonal entries are all equal to $\pm L_g L_f^{\gamma-1} h(x_0)$ and are hence nonzero, establishing that the ranks of the two matrices belonging to $\mathbb{R}^{\gamma \times n}$ and $\mathbb{R}^{n \times \gamma}$ on the left hand side of equation (9.23) are each γ (the product has rank γ, and consequently each matrix should have rank γ!). Thus, $d\phi_i(x_0)$ are linearly independent. Since these vectors are independent at x_0 they are also independent in a neighborhood of x_0. $\qquad \square$

As a by-product of the proof of the preceding claim we have the following claim which also establishes that if a SISO system has strict relative degree γ that $\gamma \leq n$.

Claim 9.6 Independence of $g, ad_f g, \ldots, ad_f^{\gamma-1} g$. *The vector fields* $g, ad_f g,$ $\ldots, ad_f^{\gamma-1} g$ *are linearly independent. Consequently* $\gamma \leq n$.

Further, since the one dimensional distribution $\Delta = \text{span}\{g(x)\}$ is involutive, there exists by the theorem of Frobenius (Theorem 8.15) $n - 1$ functions $\eta_1(x), \ldots, \eta_{n-1}(x)$ such that the matrix

$$
\begin{bmatrix} d\eta_1(x) \\ d\eta_2(x) \\ \vdots \\ d\eta_{n-1}(x) \end{bmatrix} \tag{9.24}
$$

has rank $n - 1$ at x_0 and

$$
d\eta_i(x) g(x) = 0 \quad \forall x \in U. \tag{9.25}
$$

Now, the matrix

$$
\begin{bmatrix}
dh(x) \\
dL_f h(x) \\
\vdots \\
dL_f^{\gamma-1} h(x) \\
d\eta_1(x) \\
\vdots \\
d\eta_{n-1}(x)
\end{bmatrix}
\qquad (9.26)
$$

has rank n at x_0. (Why? Consider the definition of strict relative degree.) Using this fact as well as the Claim 9.5 above, we may choose $n - \gamma$ of the $\eta_i(x)$; for notational convenience we will say the first $n - \gamma$ of them, so that the matrix

$$
\begin{bmatrix}
dh(x) \\
dL_f h(x) \\
\vdots \\
dL_f^{\gamma-1} h(x) \\
d\eta_1(x) \\
\vdots \\
d\eta_{n-\gamma}(x)
\end{bmatrix}
\qquad (9.27)
$$

has rank n at x_0, and the nonlinear coordinate transformation defined by

$$
\Phi : x \mapsto
\begin{pmatrix}
h(x) \\
L_f h(x) \\
\vdots \\
L_f^{\gamma-1} h(x) \\
\eta
\end{pmatrix}
\qquad (9.28)
$$

is a local diffeomorphism. We refer to the γ new variables $h, L_f h, \ldots, L_f^{\gamma-1} h$ as ξ. In the ξ, η coordinates, the system equations (9.1) read

$$\dot{\xi}_1 = \xi_2,$$
$$\dot{\xi}_2 = \xi_3,$$
$$\vdots$$
$$\dot{\xi}_\gamma = b(\xi, \eta) + a(\xi, \eta)u,$$
$$\dot{\eta} = q(\xi, \eta),$$
$$y = \xi_1.$$

(9.29)

Here, $b(\xi, \eta), a(\xi, \eta)$ stand for $L_f^\gamma h(x), L_g L_f^{\gamma-1} h(x)$ in the (ξ, η) coordinates and the $q_i(\xi, \eta)$ stand for $L_f \eta_i$ in the ξ, η coordinates. Thus, for example,

$$b(\xi, \eta) = L_f^\gamma h(\Phi^{-1}(\xi, \eta)). \tag{9.30}$$

Note the lack of input terms in the differential equations for η: This is a consequence of (9.25). The system description (9.29) is called the **"normal" form** for the system (9.1)

Example 9.7. *Consider the control system*

$$\dot{x} = \begin{bmatrix} -x_1^3 \\ \cos x_1 \cos x_2 \\ x_2 \end{bmatrix} + \begin{bmatrix} \cos x_2 \\ 1 \\ 0 \end{bmatrix} u,$$

$$y = h(x) = x_3.$$

This system has relative degree 2, as may be verified by the following calculation

$$L_g h(x) = 0, \quad L_f h(x) = x_2,$$
$$L_g L_f h(x) = 1, \quad L_f^2 h(x) = \cos x_1 \cos x_2.$$

To obtain the normal form coordinates we choose

$$\xi_1 = \phi_1(x) = h(x) \quad = x_3,$$
$$\xi_2 = \phi_2(x) = L_f h(x) = x_2,$$

and we choose the last coordinate $\eta(x)$ such that

$$\frac{\partial \eta}{\partial x} g(x) = \frac{\partial \eta}{\partial x_1} \cos x_2 + \frac{\partial \eta}{\partial x_2} = 0.$$

Note that a solution to the foregoing partial differential equation is

$$\eta(x) = x_1 - \sin x_2 = \phi_3(x).$$

Of course, other solutions are easily obtained by adding constants to $\eta(x)$. The Jacobian matrix of the transformation is

$$\frac{\partial \phi}{\partial x} = \begin{bmatrix} 0 & 0 & 1 \\ 0 & 1 & 0 \\ 1 & -\cos x_2 & 0 \end{bmatrix}$$

is easily seen to be nonsingular, and in fact, it is a global diffeomorphism with inverse given by

$$x_1 = \eta + \cos \xi_2,$$

$$x_2 = \xi_2,$$

$$x_3 = \xi_1.$$

In the new coordinates, the equations are described by

$$\dot{\xi}_1 = \xi_2,$$

$$\dot{\xi}_2 = \cos(\eta + \sin \xi_2) \cos \xi_2 + u,$$

$$\dot{\eta} = (\eta + \sin \xi_2)^3 - \cos^2 \xi_2 \cos(\eta + \sin \xi_2).$$

Full State Linearization by State Feedback

It is instructive to consider the case that the system (9.1) has relative degree *exactly* n. Then, the normal form of the system reads

$$\dot{\xi}_1 = \xi_2,$$

$$\dot{\xi}_2 = \xi_3,$$

$$\vdots \qquad\qquad (9.31)$$

$$\dot{\xi}_n = b(\xi) + a(\xi)u,$$

$$y = \xi_1.$$

The feedback law given by

$$u = \frac{1}{a(\xi)}[-b(\xi) + v] \qquad (9.32)$$

yields a linear n-th order system. In the original x coordinates, the transformation of the system (9.1) into a linear system with transfer function $1/s^n$ consists of a change of coordinates given by

$$\Phi : x \mapsto \xi = (h, \ldots, L_f^{n-1}h)^T \qquad (9.33)$$

and a static (memoryless) feedback

$$u = \frac{1}{L_g L_f^{n-1} h(x)}[-L_f^n h(x) + v], \qquad (9.34)$$

both defined in a neighborhood of x_0. The reason that Φ is a diffeomorphism is the statement of the Claim 9.5, which implies that

$$dh, dL_f h, \ldots, dL_f^{n-1} h$$

are linearly independent in a neighborhood of x_0. The closed loop system, since it is controllable from the input v can now be made to have its poles placed at the zeros of a desired polynomial $s^n + \alpha_{n-1} s^{n-1} + \cdots + \alpha_0$ by choosing

$$v = -\alpha_0 \xi_1 - \alpha_1 \xi_2 - \cdots - \alpha_{n-1} \xi_n. \tag{9.35}$$

Thus, combining (9.34) and (9.35) in the "old" x coordinates yields

$$u = \frac{1}{L_g L_f^{n-1} h(x)} [-L_f^n h(x) - \alpha_0 h(x) - \cdots - \alpha_{n-1} L_f^{n-1} h(x)]. \tag{9.36}$$

Let us now ask ourselves the following question: Given the system of (9.1) with no output,

$$\dot{x} = f(x) + g(x) u \tag{9.37}$$

is it possible to find a $h(x)$ such that the relative degree of the resulting system of the form (9.1) is precisely n? There is a rather elegant solution to this problem due to Brockett [45] (see also Jacubcyzk and Respondek [154] and Hunt, Su and Meyer [148]).

The system (9.1) has relative degree n at x_0 iff

$$L_g h = L_g L_f h = \cdots = L_g L_f^{n-2} h \equiv 0 \forall\, x \tag{9.38}$$

and $L_g L_f^{n-1} h(x_0) \neq 0$. Using the results of Proposition 9.4 it follows that the system has relative degree n at x_0 iff for some neighborhood U of x_0:

$$L_g h = L_{ad_f g} h = \cdots = L_{ad_f^{n-2} g} h \equiv 0 \; \forall x \in U,$$
$$L_{ad_f^{n-1} g} h(x_0) \neq 0. \tag{9.39}$$

The set of equations (9.39) may be rewritten as

$$\frac{\partial h}{\partial x} [g(x)\, ad_f g(x) \cdots ad_f^{n-2} g(x)] = 0. \tag{9.40}$$

Note that this is a set of $n - 1$ first order linear partial differential equations. The last equation of (9.39) is not included, but since the results of Claim 9.5 guarantee that the vector fields

$$g(x), \ldots, ad_f^{n-2} g(x), ad_f^{n-1} g(x) \tag{9.41}$$

are linearly independent at x_0, it follows that any non trivial $h(x)$ which satisfies the conditions of equation (9.40) will automatically satisfy the last equation of (9.39). Now, also, the distribution spanned by the vector fields

$$g(x), \ldots, ad_f^{n-3} g(x), ad_f^{n-2} g(x) \tag{9.42}$$

is (locally) of constant rank $n - 1$ (this was one of the by-products of the proof of Claim (9.5). The Frobenius theorem (Theorem 8.15) guarantees that the equations (9.40) have a solution iff the set of vector fields in (9.42) is *involutive* in a neighborhood of x_0.

Remarks and Interpretation: The conditions on (9.41), (9.42) are the necessary and sufficient conditions for the solution of (9.39). We may interpret these conditions as necessary and sufficient for the choice of an output variable $h(x)$ so that the system is fully state linearizable by state feedback and a change of coordinates. It may be verified (see Exercises) that if $x_0 = 0$ and $f(x_0) = 0$ and

$$A = \frac{\partial f}{\partial x}(x_0), \quad b = g(x_0). \tag{9.43}$$

then the condition of (9.41) reduces to

$$\text{rank}[b, Ab, \dots, A^{n-1}b] = n. \tag{9.44}$$

Thus the rank n condition on (9.41) is a sort of controllability condition on the original system (9.1), as we will see more precisely in Chapter 11, and is satisfied generically in the sense that it holds for almost all pairs of vector fields f, g. The second condition of involutivity (9.42) is more subtle and is not generically satisfied.

We have now developed the ingredients of the machinery required to address the following problem: Consider the system of (9.37) with no output described by

$$\dot{x} = f(x) + g(x)u$$

and a point $x_0 \in \mathbb{R}^n$. Assuming that the system is controllable, in the sense that the equation (9.41) holds, solve the following *state space, exact linearization problem*: Find, if possible, a feedback of the form

$$u = \alpha(x) + \beta(x)v$$

and a coordinate transformation $z = \Phi(x)$ so as to transform the system into the form

$$\dot{z} = Az + bv, \tag{9.45}$$

which is linear and controllable. In other words, we ask for the existence of functions of $x, \alpha(x), \beta(x), \Phi(x)$, such that

$$\left(\frac{\partial \phi}{\partial x}(f(x) + g(x)\alpha(x)) \right)_{x = \phi^{-1}(z)} = Az$$

$$\left(\frac{\partial \phi}{\partial x}(g(x)\beta(x)) \right)_{x = \phi^{-1}(z)} = b$$

where $A \in \mathbb{R}^{n \times n}$, $b \in \mathbb{R}^n$ satisfy the controllability condition

$$\text{rank}(b, Ab, A^2b, \dots, A^{n-1}b) = n.$$

Theorem 9.8 State Space Exact Linearization. *The state space exact linearization problem is solvable (that is, there exist functions $\alpha(x)$, $\beta(x)$, $\Phi(x)$) such that the system of equation (9.37) can be transformed into a linear system of the form of (9.45) iff there exists a function $\lambda(x)$ such that the system*

$$\dot{x} = f(x) + g(x)u,$$
$$y = \lambda(x)$$

has relative degree n at x_0.

Proof: In the light of the preceding discussion, the proof of the sufficiency of the condition is easy. Indeed, if the output $\lambda(x)$ results in a relative degree of n, then we may use

$$\Phi(x) = \begin{pmatrix} \lambda(x) \\ L_f\lambda(x) \\ L_f^2\lambda(x) \\ \cdots \\ L_f^{n-1}\lambda(x) \end{pmatrix} \tag{9.46}$$

and the control law

$$u(x) = \frac{1}{L_g L_f^{n-1}\lambda(x)}(-L_f^n\lambda(x) + v) \tag{9.47}$$

to yield the linear system

$$\dot{z} = \begin{bmatrix} 0 & 1 & 0 & \cdots & 0 \\ 0 & 0 & 1 & \cdots & 0 \\ \vdots & \vdots & \vdots & \ddots & \vdots \\ 0 & 0 & 0 & \cdots & 1 \\ 0 & 0 & 0 & \cdots & 0 \end{bmatrix} z + \begin{bmatrix} 0 \\ 0 \\ \vdots \\ 0 \\ 1 \end{bmatrix} u. \tag{9.48}$$

For the proof of the necessity, we first note that relative degree is invariant under changes of coordinates; that is, if

$$\bar{f}(z) = \left(\frac{\partial \Phi}{\partial x} f(x)\right)_{x=\Phi^{-1}(z)}, \qquad \bar{g}(z) = \left(\frac{\partial \Phi}{\partial x} g(x)\right)_{x=\Phi^{-1}(z)},$$

and

$$\bar{h}(z) = h(\Phi^{-1}(z)),$$

then, an easy calculation verifies that

$$L_{\bar{f}}\bar{h}(z) = L_f h(x)|_{x=\Phi^{-1}(z)}.$$

Iterating on this calculation yields

$$L_{\tilde{g}}L_{\tilde{f}}^k\bar{h}(z) = \left[L_g L_f^k h(x)\right]_{x=\phi^{-1}(z)}.$$

Relative degree is also invariant under state feedback of the form

$$u = \alpha(x) + \beta(x)v$$

with $\beta(x)$ bounded away from 0. Indeed, first note that

$$L_{f+g\alpha}^k h(x) = L_f^k h(x) \quad \forall \ 0 \le k \le \gamma - 1. \tag{9.49}$$

The equality above is trivially true for $k = 0$. For the induction proof, assume that it is true for some k. Then, we have that

$$\begin{aligned}
L_{f+g\alpha}^{k+1} h(x) &= L_{f+g\alpha} L_f^k h(x) && \text{by the induction hypothesis} \\
&= L_f^{k+1} h(x) + L_g L_f^k h(x)\alpha(x) \\
&= L_f^{k+1} h(x).
\end{aligned}$$

The last step uses the fact that $L_g L_f^k h(x) \equiv 0$ for $k < \gamma - 1$. From the equation (9.49) it follows that

$$L_{g\beta}L_{f+g\alpha}^k h(x) = 0 \quad \forall \ 0 \le k \le \gamma - 1,$$

and further if $\beta(x_0) \ne 0$, then

$$L_{g\beta}L_{f+g\alpha}^{\gamma-1} h(x_0) = \beta(x_0)L_g L_f^{\gamma-1} h(x_0) \ne 0.$$

We have shown that relative degree is invariant under state feedback and coordinate transformation. Thus, if the given nonlinear system can be fully linearized by state feedback and change of coordinates to the form of equation (9.34), then since the pair A, b is completely controllable, we may assume using further linear state feedback, if necessary, that they are in the form of equation (9.48), that is, the controllable canonical form. To these equations append the output $y = z_1$. The resulting system has relative degree n, as an easy calculation shows. Since the relative degree is unaffected by state feedback and coordinate change, as we have just established, it must follow that the output $y = z_1$ in the original x coordinates labeled $y = \lambda(x)$ has relative degree n. $\qquad\square$

Example 9.9. *Consider the system*

$$\dot{x} = \begin{bmatrix} 0 \\ x_1 + x_2^2 \\ x_1 - x_2 \end{bmatrix} + \begin{bmatrix} e^{x_2} \\ e^{x_2} \\ 0 \end{bmatrix} u$$

with the output $y = x_3$. For this system

$$\begin{array}{ll}
L_f h(x) = x_1 - x_2 & L_g h(x) = 0, \\
L_f^2 h(x) = -x_1 - x_2^2 & L_g L_f h(x) = 0, \\
L_f^3 h(x) = -2x_2(x_1 + x_2^2) & L_g L_f^2 h(x) = -(1 + 2x_2)e^{x_2}.
\end{array}$$

We see that the system has relative degree 3 at each point, and using the feedback control

$$u = \frac{-2x_2(x_1 + x_2^2) - v}{(1 + 2x_2)e^{x_2}}$$

and the change of coordinates

$$\xi_1 = h(x) \qquad = x_3,$$
$$\xi_2(x) = L_f h(x) = x_1 - x_2,$$
$$\xi_3(x) = L_f^2 h(x) = -x_1 - x_2^2.$$

yields

$$\dot{\xi} = \begin{bmatrix} 0 & 1 & 0 \\ 0 & 0 & 1 \\ 0 & 0 & 0 \end{bmatrix} \xi + \begin{bmatrix} 0 \\ 0 \\ 1 \end{bmatrix} v.$$

9.2.2 Zero Dynamics for SISO Systems

The exact input–output linearizing control law of (9.8) renders the closed loop system linear with transfer function from v to y to be $1/s^\gamma$. Thus, $n - \gamma$ of the state variables are rendered unobservable by state feedback. To see this more clearly, consider the linear case, i.e., $f(x) = Ax$, $g(x) = b$, $h(x) = cx$. Further, assume that the linear system described by A, b, c is minimal. Then, the condition that the system has relative degree γ is equivalent to

$$cb = cAb = cA^2b = \cdots = cA^{\gamma-2}b = 0, \quad \text{and} \quad cA^{\gamma-1}b \neq 0, \qquad (9.50)$$

and the control law of (9.8) yields the closed loop system

$$\dot{x} = \left[I - \frac{bcA^{\gamma-1}}{cA^{\gamma-1}b} \right] Ax + \frac{b}{cA^{\gamma-1}b} v,$$
$$y = cx,$$

with transfer function $1/s^\gamma$ from v to y. Since the original linear system is minimal, it follows that $n - \gamma$ of the eigenvalues of

$$(I - bcA^{\gamma-1}/cA^{\gamma-1}b)A$$

are located at the zeros of the original system and the remaining at the origin. Thus, we have placed $n - \gamma$ of the closed loop poles at the zeros of the original system. The exact input–output linearizing control law may be thought of as the nonlinear counterpart of this *zero-canceling* control law.

We will now attempt to think about the nonlinear version of zero dynamics. We have not, as yet, discussed questions of minimality for nonlinear systems. In particular, since no assumptions will be made about the observability properties of the system (9.1) to begin with, some of the variables which appear in our discussion of zero dynamics may have been unobservable prior to the application of state feedback.[4] We can use the intuition of linear systems to define the *zero dynamics* of SISO nonlinear systems. We first make a slight detour to solve the so-called output-zeroing problem:

Output-Zeroing Problem

Find, if possible, an initial state $\overline{x} \in U$ and an input $\overline{u}(t)$ such that the output $y(t)$ is identically zero for all $t \geq 0$.

The answer to this problem may be easily given, using the normal form of (9.29). First, note that by differentiation, we have

$$y(t) \equiv 0 \implies \xi_1 = \xi_2 = \cdots = \xi_\gamma \equiv 0. \tag{9.51}$$

Thus, in order to keep $\xi_\gamma \equiv 0$ it is necessary that

$$u(t) = -\frac{b(0, \eta(t))}{a(0, \eta(t))}, \tag{9.52}$$

where $\eta(t)$ is any solution of

$$\dot{\eta} = q(0, \eta), \quad \eta(0) \text{ arbitrary}. \tag{9.53}$$

Thus, the output can be held identically zero, provided that the $\xi_i(0) = 0$ for $i = 1, \ldots, \gamma$. in other words, $\overline{x}$ is chosen to lie on the manifold

$$M = \{x : h(x) = L_f h(x) = \cdots = L_f^{\gamma-1} h(x) = 0\}. \tag{9.54}$$

Recall from Chapter 3 that a sufficient condition that for M as defined above to be a manifold is that 0 is a regular value of the map

$$(h, L_f h, \ldots, L_f^{\gamma-1} h) : \mathbb{R}^n \mapsto \mathbb{R}^\gamma.$$

In Claim 9.5 we have shown that the derivatives

$$dh(x), dL_f h(x), \ldots, dL_f^{\gamma-1} h(x)$$

are linearly independent in a neighborhood of x_0 if the system (9.1) has relative degree γ at x_0. Thus, M is a manifold of dimension $n - \gamma$ in a neighborhood of x_0 and $\overline{u}(t)$ the state feedback law as given in (9.52). In the original x coordinates it is easy to see that the law is precisely (9.8) with v set to zero. We will, by abuse of notation refer to this control law in the original x coordinates as $\overline{u}(x)$. This feedback law renders M invariant, i.e, when the initial condition of (9.1) lie on M, the entire trajectory lies in M. These dynamics are precisely the dynamics of

$$\dot{x} = f(x) + g(x)\overline{u}(x) \tag{9.55}$$

---

[4] Recall that state feedback does not affect controllability of a system but may affect observability.

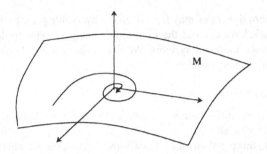

FIGURE 9.2. The manifold M corresponding to the zero dynamics

restricted to M, as shown in Figure 9.2. These are referred to as the *zero dynamics* of the system (9.1). M is parameterized by η; thus, the dynamics of

$$\dot{\eta} = q(0, \eta) \tag{9.56}$$

are the internal dynamics consistent with the constraint that $y(t) \equiv 0$ and hence are referred to as the *zero dynamics*. The description of the zero dynamics in terms of (9.56), a system evolving in $\mathbb{R}^{(n-\gamma)}$, allows for a certain greater economy than (9.55), which is a system evolving in $\mathbb{R}^n$. A similar interpretation holds for the case that the output is not required to be identically zero, but is required to follow some prescribed function $\bar{y}(t)$ (see Problem 9.4). In order to define the notion of minimum phase nonlinear systems we will need to stipulate that $\eta = 0$ is an equilibrium point of (9.56). This, in turn, is equivalent to stipulating that the point x_0, which is an equilibrium point of the undriven system, i.e., $f(x_0) = 0$ also corresponds to zero output. Further, the map Φ is chosen to map x_0 into $\xi = 0, \eta = 0$.

Definition 9.10 Minimum Phase.   *The nonlinear system (9.1) is said to be lo-cally asymptotically (exponentially) minimum phase at x_0 if the equilibrium point $\eta = 0$ of (9.56) is locally asymptotically (exponentially) stable.*

Remarks: It is important to note that the minimum phase property of a nonlinear system depends on the equilibrium point x_0 of the undriven system under consid-eration (and also whether or not the output $h(x)$ vanishes at these points). Thus, a nonlinear system may be minimum phase at some equilibrium points and be nonminimum phase at some others. Also, some equilibrium points of the undriven system may not be equilibrium points for the zero dynamics.

It is easy to characterize the stability of the zero dynamics in terms of the stability of the origin, which is an equilibrium point of (9.56). In particular, if the eigenvalues of the linearization of (9.56) lie in the left half plane the system is *locally exponentially minimum phase*. Thus, if the eigenvalues of

$$\frac{\partial q}{\partial \eta}(0, 0)$$

are in $\mathbb{C}^\circ_-$ then the system is exponentially minimum phase (locally). It is also easy to see that if even one eigenvalue of the linearization above is in $\mathbb{C}^\circ_+$ that the system is nonminimum phase. Finally, the system may be only *asymptotically* but not exponentially minimum phase if it has all the eigenvalues of its linearization in $\mathbb{C}_-$ but some of them are on the $j\omega$ axis. Problem (9.8) establishes the connection between the eigenvalues of the linearization and the zeros of the Jacobian linearization of the nonlinear system. It is of interest to note that if some of the eigenvalues of

$$\frac{\partial q}{\partial \eta}(0,0)$$

are on the $j\omega$ axis, then the linearization is inconclusive in determining the locally, asymptotic minimum phase characteristic of the equilibrium point. A detailed Lyapunov or center manifold analysis is needed for this (in this case, there is no hope of the system being exponentially minimum phase. Why?).

Example 9.11. *Consider the system*

$$\dot{x} = \begin{bmatrix} x_3 - x_2^3 \\ -x_2 \\ x_1^2 - x_3 \end{bmatrix} + \begin{bmatrix} 0 \\ -1 \\ 1 \end{bmatrix} u$$

with output $y = x_1$. *Then the system has relative degree 2 for all x. the change of coordinates to get the system in "normal form" is* $\xi_1 = x_1, \xi_2 = x_3 - x_2^3$. *To choose* η, *note that the two covector fields orthogonal to* $g(x)$ *are*

$$[1\ 0\ 0], \quad [0\ 1\ 1].$$

Since the differentials $d\xi_1, d\xi_2$ *are given by*

$$\begin{bmatrix} 1 & 0 & 0 \\ 0 & -3x_2^2 & 1 \end{bmatrix},$$

it follows that a choice of η *such that it is independent of* ξ_1, ξ_2 *and also orthogonal to* g *is* $\eta(x) = x_2 + x_3$. *In these coordinates, the system looks like*

$$\dot{\xi}_1 = \xi_2,$$
$$\dot{\xi}_2 = b(\xi, \eta) + a(\xi, \eta)u,$$
$$\dot{\eta} = \xi_1^2 - \eta.$$

Thus, the zero dynamics at the point $x = 0$ *are given by*

$$\dot{\eta} = -\eta,$$

which means that the system is exponentially minimum phase.

Example 9.12. *Consider the system*

$$\dot{x}_1 = x_2,$$
$$\dot{x}_2 = \sin x_3 + x_1 + u,$$
$$\dot{x}_3 = x_4,$$
$$\dot{x}_4 = x_1 + u,$$

with output $y = x_1$. The system has relative degree 2, and it is easy to see that $\xi_1 = x_1$ and $\xi_2 = x_2$. Also, we may choose $\eta_1 = x_3$ and $\eta_2 = x_4 - x_2$. The system dynamics are given by

$$\dot{\xi}_1 = \xi_2$$
$$\dot{\xi}_2 = \sin \eta_1 + \xi_1 + u$$
$$\dot{\eta}_1 = \eta_2 + \xi_2$$
$$\dot{\eta}_2 = -\sin \eta_1$$

At the equilibrium points of the system $x_1 = 0, x_2 = 0, x_3 = n\pi, x_4 = 0$ the zero dynamics are given by

$$\dot{\eta}_1 = \eta_2,$$
$$\dot{\eta}_2 = -\sin \eta_1.$$

Depending on whether the integer n is even or odd, the linearization of the zero dynamics is either a center (eigenvalues on the $j\omega$ axis) or a saddle (one stable and the other unstable eigenvalue).

9.2.3 Inversion and Exact Tracking

The problem of inversion or exact tracking for systems of the form (9.1) is to determine the input and initial conditions required to make the output of the plant exactly track a given signal $y_d(t)$. It is easy to solve this problem using the normal form equations of the previous subsection. Indeed, if the output y is to be identically equal to y_d, then we have that

$$\xi(t) = \begin{bmatrix} y_d(t) \\ \dot{y}_d(t) \\ \vdots \\ y_d^{(\gamma-1)}(t) \end{bmatrix}.$$

Let us denote the right hand side of this equation, consisting of y_d and its first $\gamma - 1$ derivatives by ξ^d. Then it follows that the input $u_d(t)$ required to generate the output $y_d(t)$ is given by solving the equation

$$y_d^{(\gamma)} = b(\xi^d, \eta) + a(\xi^d, \eta)u_d \tag{9.57}$$

to yield

$$u_d = \frac{1}{a(\xi^d, \eta)}(y_d^{(\gamma)} - b(\xi^d, \eta)). \tag{9.58}$$

Since $a(0, 0)$ is guaranteed to be nonzero if the system has strict relative degree, it follows that $a(\xi^d, \eta) \neq 0$ for ξ^d, η small enough. The η variables appear to be unconstrained by the tracking objective and as a consequence we may use any initial condition as $\eta(0)$. As for the ξ variables, exact tracking demands that

$$\xi(0) = \xi^d(0).$$

Thus, the system generating the control law is described as

$$\dot{\eta} = q(\xi^d, \eta), \quad \eta(0) \text{ arbitrary} \tag{9.59}$$

with the tracking input given by u_d as given by (9.58) so long as $a(\xi^d, \eta)$ is non-zero, is referred to as the inverse system. The inverse system is shown in Figure (9.3) with a chain of γ differentiators at the output of the plant followed by the η dynamics which take values in $\mathbb{R}^{n-\gamma}$ and the read out map of (9.58) to generate the tracking input. It should come as no surprise that the dynamics of the inverse system are:

1. Non-proper, with as many differentiators as the relative degree γ of the system.
2. A copy of the zero dynamics of the plant of (9.1) driven by the desired output and its first $\gamma - 1$ derivatives.

The preceding discussion has established that nonlinear SISO systems that have strict relative degree are *uniquely invertible* (i.e., the u_d required for exact tracking is unique) at least in the instance that y_d and its first $\gamma - 1$ derivatives are small enough.

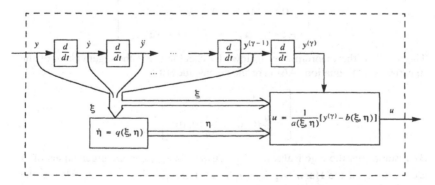

FIGURE 9.3. The inverse of a SISO nonlinear system

9.2.4 Asymptotic Stabilization and Tracking for SISO Systems

The preceding section dealt with the problem of exact tracking. The drawbacks of the approach of the previous section were that the generation of the exact tracking input needed the use of differentiators. However, with the concepts of zero dynamics in hand, we may also tackle the problems of asymptotic stabilization and tracking for nonlinear systems with well defined relative degree. We see that it is possible under the assumption of minimum phase to build controllers for asymptotic (rather than exact) tracking and stabilization. We start with stabilization:

Theorem 9.13 Stabilization of Minimum Phase Systems. *If the system (9.1) has relative degree γ is locally exponentially minimum phase and if the polynomial $s^\gamma + \alpha_{\gamma-1}s^{\gamma-1} + \cdots + \alpha_1 s + \alpha_0$ is Hurwitz, then the state feedback law*

$$u = \frac{1}{L_g L_f^{\gamma-1}h(x)}[-L_f^\gamma h(x) - \alpha_{\gamma-1}L_f^{\gamma-1}h(x) - \cdots - \alpha_1 L_f h(x) - \alpha_0 h(x)]$$

$$(9.60)$$

results in a (locally) exponentially stable system.

Proof: Consider the system in the normal form with the feedback control law of (9.60). The closed loop system has the form

$$\dot{\xi} = A\xi,$$
$$\dot{\eta} = q(\xi, \eta).$$
$$(9.61)$$

with

$$A = \begin{bmatrix} 0 & 1 & 0 & \cdots & 0 \\ 0 & 0 & 1 & \cdots & 0 \\ \vdots & \vdots & \vdots & \ddots & \vdots \\ 0 & 0 & 0 & \cdots & 1 \\ -\alpha_0 & -\alpha_1 & -\alpha_2 & \cdots & -\alpha_{\gamma-1} \end{bmatrix}.$$

The result of the theorem now follows by checking that the eigenvalues of the linearization of equation (9.61) are the eigenvalues of

$$\begin{bmatrix} A & 0 \\ \dfrac{\partial q}{\partial \xi}(0,0) & \dfrac{\partial q}{\partial \eta}(0,0) \end{bmatrix}.$$

By assumption, the eigenvalues of $\frac{\partial q}{\partial \eta}(0,0)$ are in $\mathbb{C}^\circ_-$, and the eigenvalues of A are the zeros of the Hurwitz polynomial $s^\gamma + c_{\gamma-1}s^{\gamma-1} + \cdots + c_0$. Hence, the linearization has eigenvalues in $\mathbb{C}^\circ_-$. This completes the proof.   □

The preceding theorem is not the strongest possible; using center manifold theorems of Chapter 7, we can extend the theorem to the case when the system is only asymptotically minimum phase (i.e., the eigenvalues of the linearization lie in $\mathbb{C}_-$, the closed left half plane possibly on the $j\omega$ axis) and still derive the conclusion that the closed loop system is asymptotically stable (see Problem 9.9).

It is also easy to modify the control law (9.60) in the instance that the control objective is tracking rather than stabilization. Consider the problem of getting the output $y(t)$ to track a given prespecified reference signal $y_m(t)$. Define the tracking error e_0 to be equal to $y - y_m$. The counterpart of the stabilizing control law of (9.60) is

$$u = \frac{1}{L_g L_f^{\gamma-1} h(x)} [y_m^\gamma(t) - L_f^\gamma h(x) - \alpha_{\gamma-1} e_0^{\gamma-1} - \cdots - \alpha_0 e_0]. \qquad (9.62)$$

In (9.62) above e_0^i stands for the ith derivative of the output error e_0. Also, though it is not immediately obvious, the feedback law of (9.62) is a state feedback law since

$$e_0^i(t) = L_f^i h(x) - y_m^i(t) \qquad (9.63)$$

for $i = 0, \ldots, \gamma - 1$ (convince yourself of this). The counterpart of Theorem 9.13 to tracking is the following:

Theorem 9.14 Bounded Tracking. *Consider the system of (9.1) with the assumptions of Theorem 9.13 in effect with the strengthening of the minimum phase assumption as follows:*

1. *The zero dynamics of (9.56) are globally defined.*
2. *The zero dynamics are globally exponential minimum phase.*
3. *The zero dynamics are Lipschitz continuous in ξ, η.*

Now, if the reference trajectory $y_m(t)$ and its first γ derivatives are bounded then the control law of (9.62) results in bounded tracking, i.e., e_0 and its first γ derivatives tend to zero asymptotically and the state x is bounded.

Proof: The first order of business is to write down the closed loop equations for the system using the control law of equation (9.62). For this purpose we define the error coordinates $e_{i+1} = e_0^i$ for $i = 0, \ldots, \gamma - 1$. These coordinates will be chosen to replace the ξ coordinates. The η coordinates will, however, be unchanged. It is easy to check that the form of the closed loop equation in the new coordinates is a great deal like that of equation (9.61). Thus, we have

$$\begin{aligned} \dot{e} &= Ae, \\ \dot{\eta} &= q(\xi, \eta). \end{aligned} \qquad (9.64)$$

where A is the companion form matrix defined by

$$
A = \begin{bmatrix}
0 & 1 & 0 & \cdots & 0 \\
0 & 0 & 1 & \cdots & 0 \\
\vdots & \vdots & \vdots & \ddots & \vdots \\
0 & 0 & 0 & \cdots & 1 \\
-\alpha_0 & -\alpha_1 & -\alpha_2 & \cdots & -\alpha_{\gamma-1}
\end{bmatrix},
$$

and it is understood that

$$
\xi = -e + \begin{pmatrix}
y_m(t) \\
\dot{y}_m(t) \\
\vdots \\
y_m^{(\gamma-1)}(t)
\end{pmatrix}.
$$

Now, since the zero dynamics of the nonlinear system described by (9.56) are exponentially stable and Lipschitz continuous it follows by a converse theorem of Lyapunov (Theorem 5.17) that there exists a Lyapunov function satisfying the following properties:

$$
\alpha_1 |\eta|^2 \le V(\eta) \le \alpha_2 |\eta|^2,
$$

$$
\left| \frac{\partial V}{\partial \eta} \right| \le \alpha_3 |\eta|,
$$

$$
\dot{V}(\eta)|_{(9.56)} = \frac{\partial V}{\partial \eta} q(0, \eta) \le -\alpha_4 |\eta|^2.
$$

Now, since the matrix A is Hurwitz, it follows that that $e \to 0$ as $t \to \infty$. Also, as a consequence of the assumption of the boundedness of y_m and its first $\gamma - 1$ derivatives, it follows that ξ is bounded, say by K. For the η variables consider the rate of change of $V(\eta)$ along trajectories of (9.64). This yields

$$
\begin{aligned}
\dot{V}(\eta) &= \frac{\partial V}{\partial \eta} q(\xi, \eta) \\
&= \frac{\partial V}{\partial \eta} q(0, \eta) + \frac{\partial V}{\partial \eta} (q(\xi, \eta) - q(0, \eta)) \\
&\le -\alpha_4 |\eta|^2 + \alpha_3 |\eta| L |\xi|.
\end{aligned}
$$

Here, L is the Lipschitz constant of q with respect to ξ, η. Since $|\xi|$ bounded by K, it follows that

$$
\dot{V} \le -\alpha_4 \left(|\eta| - \frac{\alpha_3 L K}{2\alpha_4} \right)^2 + \frac{\alpha_3^2 L^2 K^2}{4\alpha_4}.
$$

Thus $\dot{V}$ is negative for $|\eta| \geq ((1 + \epsilon)\alpha_3 LK)/\alpha_4$ for any $\epsilon > 0$. Using the bounds on V one can establish that all trajectories in η are confined to a ball of radius

$$\left(\frac{\alpha_2}{\alpha_1}\right)^{\frac{1}{2}} (1 + \epsilon)\frac{\alpha_3 LK}{\alpha_4}$$

(verify this). Thus, the η variables are bounded. Since the ξ variables were already established as being bounded and the diffeomorphism Φ is smooth, it follows that the entire state x is bounded.   □

Remarks:

1. If the nonlinear system is only locally exponentially stable, then the hypotheses of the theorem need to be modified to allow only for those tracking trajectories that keep the η variables in the region where the dynamics are exponentially stable.
2. The global Lipschitz hypothesis on the zero dynamics is needed in the proof in two places, first, to invoke the specific converse Lyapunov theorem and second, to obtain the bound on $\dot{V}$. This is a strong assumption, but unfortunately counterexamples exist to the statement of the theorem, if this is not satisfied.

9.3   MIMO Systems

For the Multi-Input Multi-Output (MIMO) case, we consider *square* systems (that is, systems with as many inputs as outputs) of the form

$$\begin{aligned}
\dot{x} &= f(x) + g_1(x)u_1 + \cdots + g_p(x)u_p, \\
y_1 &= h_1(x), \\
&\vdots \\
y_p &= h_p(x).
\end{aligned} \qquad (9.65)$$

Here $x \in \mathbb{R}^n, u \in \mathbb{R}^p, y \in \mathbb{R}^p$, and f, g_i are assumed to be smooth vector fields and h_j to be smooth functions. There is a class of MIMO systems for which the development very closely parallels that for the SISO case of the previous section. We start with this class:

9.3.1   MIMO Systems Linearizable by Static State Feedback

Start by differentiating the jth output y_j of the system (9.65) with respect to time to get

$$\dot{y}_j = L_f h_j + \sum_{i=1}^{p}(L_{g_i} h_j)u_i. \qquad (9.66)$$

In (9.66) above, note that if each of the $L_{g_i}h_j(x) \equiv 0$, then the inputs do not appear in the equation. Define γ_j to be the smallest integer such that at least one of the inputs appears in $y_j^{\gamma_j}$, that is,

$$y_j^{\gamma_j} = L_f^{\gamma_j}h_j + \sum_{i=1}^{p} L_{g_i}(L_f^{\gamma_j-1}h_j)u_i, \qquad (9.67)$$

with at least one of the $L_{g_i}(L_f^{\gamma_j-1}h_j) \neq 0$, for some x. Define the $p \times p$ matrix $A(x)$ as

$$A(x) := \begin{bmatrix} L_{g_1}L_f^{\gamma_1-1}h_1 & \cdots & L_{g_p}L_f^{\gamma_1-1}h_1 \\ \vdots & \ddots & \vdots \\ L_{g_1}L_f^{\gamma_p-1}h_p & \cdots & L_{g_p}L_f^{\gamma_p-1}h_p \end{bmatrix}. \qquad (9.68)$$

Using these definitions, we may give a definition of *relative degree* for MIMO systems:

Definition 9.15 Vector Relative Degree. *The system (9.65) is said to have vector relative degree* $\gamma_1, \gamma_2, \ldots, \gamma_p$ *at* x_0 *if*

$$L_{g_i}L_f^k h_i(x) \equiv 0, \quad 0 \leq k \leq \gamma_i - 2 \qquad (9.69)$$

for $i = 1, \ldots, p$ *and the matrix* $A(x_0)$ *is nonsingular.*

If a system has well defined vector relative degree, then (9.66) may be written as

$$\begin{bmatrix} y_1^{\gamma_1} \\ \vdots \\ y_p^{\gamma_p} \end{bmatrix} = \begin{bmatrix} L_f^{\gamma_1}h_1 \\ \vdots \\ L_f^{\gamma_p}h_p \end{bmatrix} + A(x) \begin{bmatrix} u_1 \\ \vdots \\ u_p \end{bmatrix}. \qquad (9.70)$$

Since $A(x_0)$ is non-singular, it follows that $A(x) \in \mathbb{R}^{p \times p}$ is bounded away from nonsingularity for $x \in U$ a neighborhood U of x_0, meaning that $A^{-1}(x)$ and has bounded norm on U. Then the state feedback control law

$$u = -A^{-1}(x) \begin{bmatrix} L_f^{\gamma_1}h_1 \\ \vdots \\ L_f^{\gamma_p}h_p \end{bmatrix} + A^{-1}(x)v. \qquad (9.71)$$

yields the *linear* closed loop system

$$\begin{bmatrix} y_1^{\gamma_1} \\ \vdots \\ y_p^{\gamma_p} \end{bmatrix} = \begin{bmatrix} v_1 \\ \vdots \\ v_p \end{bmatrix}. \qquad (9.72)$$

Note that the system of (9.72) is, in addition, *decoupled*. Thus, decoupling is a by product of linearization. A happy consequence of this is that a large number of results concerning SISO nonlinear systems can be easily extended to this class of MIMO nonlinear systems. Thus, further control objectives, such as model matching, pole placement, and tracking can be easily accommodated. The reader should work through the details of how both MIMO exact and asymptotic tracking control laws can be obtained for the case that the MIMO system has vector relative degree. The feedback law (9.71) is referred to as a *static, state feedback linearizing control law*.

MIMO "Normal" form

If a MIMO system has relative degree $\gamma = \gamma_1 + \cdots + \gamma_p < n$, then we can write a normal form for the equations (9.65) by choosing as coordinates

$$
\begin{aligned}
&\xi_1^1 = h_1(x), \quad \xi_2^1 = L_f h_1(x), \quad \ldots, \quad \xi_{\gamma_1}^1 = L_f^{\gamma_1 - 1} h_1(x), \\
&\xi_1^2 = h_2(x), \quad \xi_2^2 = L_f h_2(x), \quad \ldots, \quad \xi_{\gamma_2}^2 = L_f^{\gamma_2 - 1} h_2(x), \\
&\vdots \\
&\xi_1^p = h_p(x), \quad \xi_2^p = L_f h_p(x), \quad \ldots, \quad \xi_{\gamma_p}^p = L_f^{\gamma_p - 1} h_p(x).
\end{aligned}
\tag{9.73}
$$

These ξ_i^j qualify as a partial set of coordinates since the differentials

$$
dL_f^j h_i(x), \quad 0 \le j \le \gamma_i - 1, \quad 1 \le i \le p
\tag{9.74}
$$

are linearly independent. The proof of this is a small modification of the proof of Claim 9.5 (Problem 9.13). Now, complete the basis choosing $n - \gamma$ more functions $\eta_1(x), \eta_2(x), \ldots, \eta_{n-\gamma}(x)$. It is no longer possible as in the SISO case to guarantee that

$$
L_{g_j} \eta_i(x) \equiv 0, \quad 1 \le j \le p, \quad 1 \le i \le n - \gamma,
\tag{9.75}
$$

unless the distribution spanned by $\{g_1(x), \ldots, g_p(x)\}$ is involutive. In these ξ, η coordinates the system equations (9.65) have the following normal form

$$
\dot{\xi}_1^1 = \xi_2^1,
$$

$$
\vdots
$$

$$
\dot{\xi}_{\gamma_1}^1 = b_1(\xi, \eta) + \sum_{j=1}^p a_j^1(\xi, \eta) u_j,
$$

$$
\dot{\xi}_1^2 = \xi_2^2,
$$

$$
\vdots
$$

$$
\dot{\xi}_{\gamma_2}^2 = b_2(\xi, \eta) + \sum_{j=1}^p a_j^2(\xi, \eta) u_j,
$$

$$
\vdots
$$

$$
\tag{9.76}
$$

$$\dot{\xi}_1^p = \xi_2^p,$$

$$\vdots$$

$$\dot{\xi}_{\gamma_p}^p = b_p(\xi, \eta) + \sum_{j=1}^{p} a_j^p(\xi, \eta)u_j,$$

$$\dot{\eta} = q(\xi, \eta) + P(\xi, \eta)u,$$

$$y_1 = \xi_1^1,$$

$$\vdots$$

$$y_p = \xi_1^p.$$

Here

$$b_i(\xi, \eta) = L_f^{\gamma_i} h_i \circ \Phi^{-1}(\xi, \eta)$$

$$a_j^i(\xi, \eta) = L_{g_j} L_f^{\gamma_i-1} \circ \Phi^{-1}(\xi, \eta)$$

$$q_i(\xi, \eta) = L_f \eta_i \circ \Phi^{-1}(\xi, \eta)$$

$$P_{ij}(\xi, \eta) = L_{g_j} \eta_i \circ \Phi^{-1}(\xi, \eta)$$

where $\Phi : x \mapsto (\xi, \eta)$ is the diffeomorphism mapping from x into the normal form coordinates. Note that $P \in \mathbb{R}^{(n-\gamma) \times p}$, $q \in \mathbb{R}^{n-\gamma}$ respectively. As in the SISO case, the feedback law of (9.71) renders the η states unobservable. In the instance that the decoupling matrix $A(x)$ is nonsingular, the zero dynamics are easily defined in analogy to the SISO case by solving the output zeroing problem: Find, if possible, an initial state and an input $\bar{u}(t)$ so as to hold the outputs $y_i(t) \equiv 0$ for all t. Note that for $i = 1, \ldots, p$ we have that $y_i \equiv 0 \Rightarrow \xi_j^i \equiv 0$ for all j Thus from the form of the normal form equations(9.76) it follows that

$$\bar{u}(t) = -A^{-1}(\xi, \eta)b(\xi, \eta). \tag{9.77}$$

This is the decoupling control law of (9.72) with $v(t)$ set equal to zero. Also, the initial conditions on the ξ_i^j for $i = 1, \ldots, \gamma_j$ and $j = 1, \ldots, p$ have to be set to zero, and initial conditions on η are arbitrary. In the original x coordinates the initial conditions have to be chosen to belong to

$$M^* = \{x : h_1(x) = L_f h_1(x) = \cdots = L_f^{\gamma_1-1} h_1(x)$$

$$= h_2(x) = L_f h_2(x) = \cdots = L_f^{\gamma_2-1} h_2(x) \tag{9.78}$$

$$\cdots$$

$$= h_p(x) = L_f h_p(x) = \cdots = L_f^{\gamma_p-1} h_p(x) = 0\}.$$

The dynamics of the η variables, with the control law of $\bar{u}(t)$ in effect, are given by

$$\dot{\eta} = q(0, \eta) - P(0, \eta)A^{-1}(0, \eta)b(0, \eta). \tag{9.79}$$

If $f(x_0) = 0, h_1(x_0) = \cdots = h_p(x_0) = 0$ and Φ maps x_0 into $(0, 0)$, then it follows that $\eta = 0$ is an equilibrium point of the zero dynamics of (9.79).

The MIMO system (with vector relative degree) is said to be minimum phase, as before, if the equilibrium point 0 is asymptotically stable; similarly, exponentially minimum phase.

9.3.2   Full State Linearization of MIMO Systems

In this subsection we discuss the full state linearization of MIMO systems by static state feedback. This is to be contrasted with the results of the previous subsection, which have to do with input–output linearization of MIMO systems by static state feedback. In analogy to the SISO case, we will look for the existence of conditions under which there exist outputs $h_1, \ldots, h_p$ such that the MIMO system has vector relative degree and furthermore is such that $\gamma_1 + \cdots + \gamma_p = n$. In this case, the system of (9.65) can be converted into a controllable linear system by :

- The feedback law $u(x) = A^{-1}(x)[-b(x) + v]$.
- The coordinate transformation

$$\xi = (L_f^j h_i(x), \ 0 \le j \le \gamma_i - 1, \ 1 \le i \le p)^T. \tag{9.80}$$

Define the distributions

$$
\begin{aligned}
G_0(x) &= \mathrm{span}\{g_1(x), \ldots, g_p(x)\}, \\
G_1(x) &= \mathrm{span}\{g_1(x), \ldots, g_p(x), ad_f g_1, \ldots, ad_f g_p(x)\}, \\
&\cdots \\
G_i(x) &= \mathrm{span}\{ad_f^k g_j(x) : 0 \le k \le i; 1 \le j \le p\}.
\end{aligned}
\tag{9.81}
$$

for $i = 1, \ldots, n - 1$. Then we have the following result

Proposition 9.16 Full State MIMO Linearization. *Suppose that the matrix* $g(x_0)$ *has rank p. Then, there exist p functions $\lambda_1, \ldots, \lambda_p$ such that the system*

$$
\begin{aligned}
\dot{x} &= f(x) + g(x)u, \\
y &= \lambda(x),
\end{aligned}
$$

has vector relative degree $(\gamma_1, \ldots, \gamma_p)$ with

$$\gamma_1 + \gamma_2 + \cdots + \gamma_p = n$$

iff

1. *For each $0 \le i \le n - 1$ the distribution G_i has constant dimension in a neighborhood U of x_0.*
2. *The distribution G_{n-1} has dimension n.*
3. *For each $0 \le i \le n - 2$ the distribution G_i is involutive.*

Proof: Since G_{n-1} has dimension n, it follows that there exists some κ such that $\dim G_{\kappa-1} = n$ but $\dim G_{\kappa-2} < n$ with $\kappa \le n$. Set $\dim G_{\kappa-2} = n - p_1$. Since $G_{\kappa-2}$ is involutive, it follows that there exist p_1 functions $\lambda_i(x)$, $i = 1, \ldots, p_1$

such that

$$\text{span}\{d\lambda_i : 1 \le i \le p_1\} = G_{\kappa-2}^{\perp}$$

Hence

$$d\lambda_i(x)ad_f^k g_j(x) \equiv 0 \qquad (9.82)$$

for $0 \le k \le \kappa - 2, 1 \le j \le p, 1 \le i \le p_1$. Further, the $p_1 \times p$ matrix

$$A^1(x) = \{a_{ij}^1(x)\} = \{L_{g_j} L_f^{\kappa-1} \lambda_i(x)\}$$

has rank p_1 at x_0. Otherwise, we would have $c_1, \ldots, c_{p_1} \in \mathbb{R}$ such that for $1 \le j \le p$

$$\sum_{i=1}^{p_1} c_i L_{g_j} L_f^{\kappa-1} \lambda_i(x_0) = 0.$$

Using equation (9.82) and Proposition (9.4) we get for $1 \le j \le p$

$$\sum_{i=1}^{p_1} (-1)^{\kappa-1} c_i d\lambda_i(x) ad_f^{\kappa-1} g_j(x_0) = 0. \qquad (9.83)$$

Again, using equation (9.82) yields for $1 \le j \le p$ that

$$\sum_{i=1}^{p_1} c_i d\lambda_i(x) ad_f^k g_j(x_0) = 0 \quad \text{for all} \quad 0 \le k \le \kappa - 2.$$

Also, equation (9.83) yields that the foregoing equation is also valid for $k = \kappa - 1$. This shows that

$$\sum_{i=1}^{p_1} c_i d\lambda_i(x_0) \in G_{\kappa-1}^{\perp}(x_0).$$

But by assumption $G_{\kappa-1}$ has dimension n and since the row vectors $d\lambda_1(x_0), \ldots, d\lambda_{p_1}(x_0)$ are linearly independent, we must have that all the c_i are zero. Thus, the $\lambda_i(x)$ qualify as a good choice of output coordinates for our problem. Of course $p_1 \le p$ since rank $A_1(x) \in \mathbb{R}^{p_1 \times p} = p_1 \le \text{minimum}\ (p_1, p)$ for each x. If $p_1 = p$ we are done, since we have p outputs with the relative degrees each given by $\gamma_i = \kappa$. It is also clear (verify this) that in this case $n = p\kappa$. If $p_1 < p$, then one has to continue the search for additional outputs. To this end we try to find new functions in $G_{\kappa-3}^{\perp}$. We start by constructing a subset of $G_{\kappa-3}^{\perp}$ described by

$$\Omega_1 = \text{span}\{d\lambda_1, \ldots, d\lambda_{p_1}, dL_f\lambda_1, \ldots, \ldots, dL_f\lambda_{p_1}\}$$

To verify that Ω_1 is indeed in $G_{\kappa-3}^{\perp}$ first note that $d\lambda_i(x), 1 \le i \le p_1$ which are in $G_{\kappa-2}^{\perp}$ by construction are also in $G_{\kappa-3}^{\perp}$ since $G_{\kappa-3} \subset G_{\kappa-2}$. Also since

$$d\lambda_i(x) ad_f^k g_j(x) = 0$$

for $0 \le k \le \kappa - 2$ and $1 \le j \le p, 1 \le i \le p_1$ it follows that

$$dL_f\lambda_i(x)ad_f^k g_j(x) = 0$$

for $0 \le k \le \kappa - 3$ and $1 \le j \le p, 1 \le i \le p_1$. Thus, the $dL_f\lambda_i$ are in $G_{\kappa-3}^\perp$. We claim that the dimension of the codistribution Ω_1 is $2p_1$. Indeed, if not, there exist c_i, d_i such that

$$\sum_{i=1}^{p_1} c_i d\lambda_i(x_0) + d_i dL_f\lambda_i(x_0) = 0$$

Thus, we have that

$$\left(\sum_{i=1}^{p_1} c_i d\lambda_i(x_0) + d_i dL_f\lambda_i(x_0) \right) ad_f^{\kappa-2} g_j(x_0) = 0$$

for all $1 \le j \le m$. This, in turn, implies that

$$\left(\sum_{i=1}^{p_1} d_i d\lambda_i(x_0) \right) ad_f^{\kappa-1} g_j(x_0) = 0.$$

Using the linear independence of the rows of $A^1(x)$, we see that the foregoing equation implies that the d_i are zero. Similarly, the c_i are also zero. Thus the dimension of $G_{\kappa-3}$ is at least $2p_1$. If it is, in fact, strictly greater than $2p_1$ we define

$$p_2 = \dim G_{\kappa-3}^\perp - 2p_1.$$

Since by hypothesis, $G_{\kappa-3}$ is involutive, there exist $2p_1 + p_2$ exact differentials that span $G_{\kappa-3}^\perp$. Of these $2p_1$ can be chosen to be those spanning Ω_1 and the remaining m_2 can be chosen to be the differentials of $\lambda_i(x)$, $p_1 + 1 \le i \le p_1 + p_2$. By construction these functions satisfy

$$L_{g_j} L_f^k \lambda_i(x) = 0$$

for $0 \le k \le \kappa - 3, 1 \le j \le p, p_1 + 1 \le i \le p_1 + p_2$. In complete analogy to the foregoing, it follows that the matrix $A^2(x)$ defined by

$$a_{ij}^2(x) = \begin{cases} d\lambda_i(x) ad_f^{\kappa-1} g_j(x) & 1 \le i \le p_1 \\ d\lambda_i(x) ad_f^{\kappa-2} g_j(x) & p_1 + 1 \le i \le p_1 + p_2 \end{cases} \tag{9.84}$$

has rank $p_1 + p_2$ at x_0. Now, if $p_1 + p_2 = p$, we are done, since we will have found p output functions with relative degrees $\gamma_1 = \gamma_2 = \cdots = \gamma_{p_1} = \kappa$ and $\gamma_{p_1+1} = \gamma_{p_1+2} = \cdots = \gamma_m = \kappa - 1$. Further, $\gamma_1 + \cdots + \gamma_p = n$, because

$$n = \dim(G_{\kappa-2}) + p_1 \le p(\kappa - 1) + p_1 = p_1\kappa + p_2(\kappa - 1) \le n.$$

If $p_1 + p_2 < m$, we keep searching for functions whose differentials span $G_{\kappa-4}^\perp$, etc. We have to show that the procedure terminates at G_0 with p functions. Indeed, after $\kappa - 1$ iterations of the procedure we have $p_1 + \cdots + p_{\kappa-1}$ functions, with the property that the differentials

$$d\lambda_i, dL_f\lambda_i, \ldots, dL_f^{\kappa-2}\lambda_i, \quad 1 \leq i \leq p_1,$$
$$d\lambda_i, dL_f\lambda_i, \ldots, dL_f^{\kappa-3}\lambda_i, \quad p_1 + 1 \leq i \leq p_1 + p_2,$$

$$\cdots$$

$$d\lambda_i, dL_f\lambda_i, \quad p_1 + \cdots + p_{\kappa-3} + 1 \leq i \leq p_1 + \cdots + p_{\kappa-2},$$
$$d\lambda_i, \quad p_1 + \cdots + p_{\kappa-2} + 1 \leq i \leq p_1 + \cdots + p_{\kappa-1},$$

are a basis of $G_0^\perp$. Since G_0 has dimension p, by hypothesis, it follows that

$$n - p = (\kappa - 1)p_1 + (\kappa - 2)p_2 + \cdots + p_{\kappa-1}.$$

Also, as above, we may prove that the $\kappa p_1 + (\kappa - 1)p_2 + \cdots + 2p_{\kappa-1}$ differentials

$$d\lambda_i, dL_f\lambda_i, \ldots, dL_f^{\kappa-1}\lambda_i, \quad 1 \leq i \leq p_1,$$
$$d\lambda_i, dL_f\lambda_i, \ldots, dL_f^{\kappa-2}\lambda_i, \quad p_1 + 1 \leq i \leq p_1 + p_2,$$

$$\cdots$$

$$d\lambda_i, dL_f\lambda_i, dL_f^2\lambda_i, \quad p_1 + \cdots + p_{\kappa-3} + 1 \leq i \leq p_1 + \cdots + p_{\kappa-2},$$
$$d\lambda_i, dL_f\lambda_i, \quad p_1 + \cdots + p_{\kappa-2} + 1 \leq i \leq p_1 + \cdots + p_{\kappa-1},$$

are independent in a neighborhood of x_0. Thus, we must have that $n - (\kappa p_1 + (\kappa - 1)p_2 + \cdots 2p_{\kappa-1}) \geq 0$. If the inequality is strict, set

$$p_\kappa = n - (\kappa p_1 + (\kappa - 1)p_2 + \cdots + 2p_{\kappa-1}).$$

Using this in the formula for $n - p$ yields

$$p_1 + \cdots + p_\kappa = p,$$

so that we have generated p outputs with relative degrees

$$\gamma_i = \kappa, \qquad 1 \leq i \leq p_1,$$
$$\gamma_i = \kappa - 1, \quad p_1 + 1 \leq i \leq p_1 + p_2,$$

$$\cdots$$

$$\gamma_i = 2, \qquad p_1 + \cdots + p_{\kappa-2} + 1 \leq i \leq p_1 + \cdots + p_{\kappa-1},$$
$$\gamma_i = 1, \qquad p_1 + \cdots + p_{\kappa-1} + 1 \leq i \leq p_1 + \cdots + p_\kappa = p.$$

Moreover, $\kappa p_1 + \cdots + p_\kappa = n$, so that the proof of the sufficiency is now complete. □

9.3.3   Dynamic Extension for MIMO Systems

The conditions required for a MIMO system to have well-defined vector relative degree can fail in several ways. A MIMO system can fail to have vector relative degree, for example, when $A(x)$ is singular. If $A(x)$ has constant rank $r < p$, that is to say that the rank of $A(x)$ is r for all x near x_0, then we may be able to extend the system by adding integrators to certain input channels to obtain a system that does have vector relative degree. The following algorithm makes this precise. In what follows the variables with the tilde are the new state space variables which are being generated during the course of the extension algorithm.

Dynamic Extension Algorithm

Step 0 Set $\tilde{n} \leftarrow n$, $\tilde{x} \leftarrow x$, $\tilde{x}_0 \leftarrow x_0$, $\tilde{f} \leftarrow f$, $\tilde{g} \leftarrow g$, $\tilde{h} \leftarrow h$, and $\tilde{u} \leftarrow u$, $\tilde{p} \leftarrow p$.

Step 1 Calculate the decoupling matrix, $\tilde{A}(\tilde{x})$, valid on $\tilde{U}$, a neighborhood of $\tilde{x}_0$.

1. If the rank $\tilde{A}(\tilde{x}) = p$ on $\tilde{U}$, **stop**: The system $(\tilde{f}, \tilde{g}, \tilde{h})$ has a vector relative degree.
2. If rank $\tilde{A}(\tilde{x})$ is not constant in a neighborhood of $\tilde{x}_0$, **stop**: The system cannot be extended to a system with vector relative degree.
3. If rank $\tilde{A}(\tilde{x}) = r$, continue to **Step 2**.

Step 2 Calculate a smooth matrix $\beta(\tilde{x})$ of elementary column operations to *compress* the columns of $\tilde{A}(\tilde{x})$ so that the last $(\tilde{p} - \tilde{r})$ columns of

$$\tilde{A}(\tilde{x})\beta(\tilde{x})$$

are identically zero on $\tilde{U}$. This is possible, since the rank of $\tilde{A}(\tilde{x})$ is constant on $\tilde{U}$. Partition β as

$$\beta(\tilde{x}) = [\beta_1(\tilde{x}) \, \beta_2(\tilde{x})]$$

such that β_1 consists of the first $\tilde{r}$ columns of β.

Step 3 Extend the system by adding one integrator to each of the first r redefined input channels. Specifically, define $z_1 \in \mathbb{R}^{\tilde{r}}$, $w_2 \in \mathbb{R}^{\tilde{p}-\tilde{r}}$ by

$$\begin{bmatrix} z_1 \\ w_2 \end{bmatrix} := \beta(\tilde{x})^{-1}\tilde{u}.$$

Then, with $w_1 \in \mathbb{R}^{\tilde{r}}$ the extended system is given by

$$\begin{bmatrix} \dot{\tilde{x}} \\ \dot{z}_1 \end{bmatrix} = \begin{bmatrix} \tilde{f} + \tilde{g}\beta_1 z_1 \\ 0 \end{bmatrix} + \begin{bmatrix} 0 & \tilde{g}\beta_2 \\ I & 0 \end{bmatrix} \cdot \begin{bmatrix} w_1 \\ w_2 \end{bmatrix},$$
$$y = \tilde{h}(\tilde{x}).$$

The extended state is $\bar{x} = (\tilde{x}^T, z_1^T)^T$ and the new input $\bar{u} = (w_1^T, w_2^T)^T$. The output is unchanged, but the output function is defined to be $\bar{h}(\bar{x}) = \tilde{h}(\tilde{x})$.

Step 4 Rename $\bar{x}, \bar{u}, \bar{f}, \bar{g}$ and $\bar{h}$ as $\tilde{x}, \tilde{u}, \tilde{f}, \tilde{g}$, and $\tilde{h}$ respectively and return to Step 1.

It has been shown that this algorithm will be successful (terminating in a finite number of steps) if the system is invertible provided that the assumption that at each step of the algorithm, the decoupling matrices $\tilde{A}(\tilde{x})$ are of constant rank in a neighborhood of $\tilde{x}_0$. This assumption is referred to as a *regularity assumption* on the MIMO system. It is the MIMO counterpart of the assumption of strict relative degree for a SISO plant. Characterizing the regularity of nonlinear systems is a substantial undertaking involving somewhat more machinery than we have introduced thus far (for details, see [90; 87]). Here it will suffice to state that the convergence of the algorithm just presented along with the assumption of regularity is equivalent to the assumption of invertibility for the MIMO nonlinear system. Further, it may be shown that if the algorithm is successful it will terminate in at most n iterations. When the dynamic extension algorithm is successful, the extended system will have a well-defined vector relative degree, and the compensator will be a dynamical system of the form

$$\dot{z} = c(x, z) + d(x, z)w,$$
$$u = \alpha(x, z) + \beta(x, z)w. \tag{9.85}$$

Indeed, if only one iteration of the algorithm was necessary, the resulting extension would be

$$\dot{z}_1 = w_1,$$
$$u = \beta_1(x)z_1 + \beta_2(x)w. \tag{9.86}$$

Once dynamic extension has been performed successfully both exact and asymptotic tracking can be achieved on the decoupled systems. Here is an example:

Example 9.17. *Consider the planar vehicle shown in Figure 9.4 with* x_1, x_2 *representing the* x, y *coordinates of the vehicles position and* $\phi = x_3$, *the vehicle*

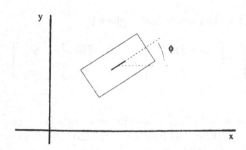

FIGURE 9.4. A steerable unicycle

heading. Assume that the inputs are the vehicle speed and turning rate. The system equations are

$$\begin{bmatrix} \dot{x}_1 \\ \dot{x}_2 \\ \dot{x}_3 \end{bmatrix} = \begin{bmatrix} \cos x_3 \\ \sin x_3 \\ 0 \end{bmatrix} u_1 + \begin{bmatrix} 0 \\ 0 \\ 1 \end{bmatrix} u_2,$$ (9.87)

$$y_1 = x_1,$$

$$y_2 = x_2.$$

Differentiating each output till the inputs appear, we get

$$\begin{bmatrix} \dot{y}_1 \\ \dot{y}_2 \end{bmatrix} = \begin{bmatrix} \cos x_3 & 0 \\ \sin x_3 & 0 \end{bmatrix} \begin{bmatrix} u_1 \\ u_2 \end{bmatrix}.$$

Since $A(x)$ has rank 1 and already has a zero second column we set $z_1 = u_1, w_2 = u_2$ and differentiate the outputs once more to get

$$\begin{bmatrix} \dot{x}_1 \\ \dot{x}_2 \\ \dot{x}_3 \\ \dot{z}_1 \end{bmatrix} = \begin{bmatrix} z_1 \cos x_3 \\ z_1 \sin x_3 \\ 0 \\ 0 \end{bmatrix} + \begin{bmatrix} 0 \\ 0 \\ 0 \\ 1 \end{bmatrix} w_1 + \begin{bmatrix} 0 \\ 0 \\ 1 \\ 0 \end{bmatrix} w_2.$$

Now differentiating the outputs till the inputs w_1, w_2 appear, we get

$$\begin{bmatrix} \ddot{y}_1 \\ \ddot{y}_2 \end{bmatrix} = \begin{bmatrix} \cos x_3 & -z_1 \sin x_3 \\ \sin x_3 & z_1 \cos x_3 \end{bmatrix} \begin{bmatrix} w_1 \\ w_2 \end{bmatrix}.$$

So long as z_1 is bounded away from zero, we can choose

$$\begin{bmatrix} w_1 \\ w_2 \end{bmatrix} = \begin{bmatrix} \cos x_3 & \sin x_3 \\ -\dfrac{\sin x_3}{z_1} & \dfrac{\cos x_3}{z_1} \end{bmatrix} \begin{bmatrix} v_1 \\ v_2 \end{bmatrix}.$$

to decouple the system. Recall also that $\dot{z}_1 = w_1, u_1 = z_1$.

The one question that we have not fully answered in this section is: Is it possible to input–output linearize a MIMO system without decoupling it? For this we refer the reader to [149].

9.4 Robust Linearization

It is clear that the process of linearization by state feedback involves the exact cancellation of nonlinearities. Consequently, it relies on a rather precise description of nonlinear functions. In this section we will ask ourselves the question of how robust linearization is to uncertainties in the description of the plant. We will begin by developing some machinery in the context of *disturbance rejection*. We start with the SISO case. The setup for disturbance rejection is as follows: Consider the system of (9.1) with an additive disturbance w,

$$
\begin{aligned}
\dot{x} &= f(x) + g(x)u + p(x)w, \\
y &= h(x).
\end{aligned}
\tag{9.88}
$$

Here $p(x)$ refers to the smooth vector field through which the disturbance enters the system. The aim is to find, if possible, a state feedback law

$$
u = \alpha(x) + \beta(x)v
\tag{9.89}
$$

such that w has no effect on the output. We will assume that the unperturbed system has relative degree γ.

Theorem 9.18 Disturbance Rejection. *The disturbance rejection problem for the system (9.88) is solvable iff*

$$
L_p L_f^k h(x) \equiv 0 \quad \text{for} \quad k \le \gamma - 1.
\tag{9.90}
$$

Proof: Writing the equations of (9.88) in the normal form given by the same coordinates as in (9.29) yields

$$
\begin{aligned}
\dot{\xi}_1 &= \xi_2 + L_p h(x)w, \\
\dot{\xi}_2 &= \xi_3 + L_p L_f h(x)w, \\
&\;\;\vdots \\
\dot{\xi}_\gamma &= b(\xi, \eta) + a(\xi, \eta)u + L_p L_f^{\gamma-1} h(x)w, \\
\dot{\eta} &= q(\xi, \eta) + P(\xi, \eta)w.
\end{aligned}
\tag{9.91}
$$

In (9.91) above, $P(\xi, \eta) \in \mathbb{R}^{n-\gamma}$ is the vector $(L_p \eta_1, \ldots, L_p \eta_{n-\gamma})^T$ expressed in the ξ, η coordinates. Using the condition of (9.90) yields that the first γ equations above do not contain w. Thus the usual linearizing control law

$$
u = \frac{1}{a(\xi, \eta)}[-b(\xi, \eta) + v]
\tag{9.92}
$$

yields the η states unobservable and consequently isolates y from w. This completes the "if" part of the proof.

For the "only if" part of the proof, we assume that there exists a control law of the form (9.89) that results in disturbance rejection. Using this law yields a closed loop system given by

$$
\begin{aligned}
\dot{x} &= f(x) + g(x)\alpha(x) + g(x)\beta(x)v + p(x)w, \\
y &= h(x).
\end{aligned}
\tag{9.93}
$$

With $v = 0$ the output needs to be independent of w. Thus,

$$
\dot{y} = L_{f+g\alpha}h + L_p h w
\tag{9.94}
$$

needs to be independent of w. Thus $L_p h(x) \equiv 0$. Differentiating (9.94) further we get that

$$L_p L_{f+g\alpha}^i h(x) \equiv 0, \quad 0 \leq i \leq \gamma - 1. \tag{9.95}$$

Now note that

$$L_{f+g\alpha} h = L_f h + \alpha L_g h. \tag{9.96}$$

Hence

$$L_g h \equiv 0 \implies L_{f+g\alpha} h = L_f h. \tag{9.97}$$

Proceeding recursively, we see that the conditions (9.95) are equivalent to (9.90) (convince yourself of this point). □

Remarks:

1. The conditions of Theorem (9.18) say that the relative degree of the disturbance to the output needs to be strictly greater than the relative degree of the unperturbed nonlinear system.

2. *Disturbance rejection with disturbance measurement*
   If measurements of the disturbance w were available, then the form of the feedback law could be changed to

$$u = \alpha(x) + \beta(x)v + \delta(x)w. \tag{9.98}$$

   Then an easy modification of the preceding theorem (cf. Exercise 9.10) is that disturbance rejection with disturbance measurement is possible iff

$$L_p L_f^i h(x) \equiv 0, \quad 0 \leq i \leq \gamma - 2. \tag{9.99}$$

   In words, this says that the problem of disturbance decoupling with disturbance measurement feedback is solvable if and only if the disturbance relative degree is greater than or equal to the relative degree of the unperturbed system. This problem appears to be somewhat artificial, since disturbances are seldom measurable. However, the problem is of conceptual value in solving the problem of *model reference control*, explained next.

3. Given a linear reference model of the form

$$\dot{z} = Az + br,$$
$$y_m = cz, \tag{9.100}$$

   with $z \in \mathbb{R}^{n_m}, r \in \mathbb{R}, y_m \in \mathbb{R}$ and A, b, c matrices of compatible dimensions, find, if possible, a feedback law that makes the output of the system (9.1) equal to y_m. Combining equations (9.1) and (9.100) yields

$$\begin{bmatrix} \dot{z} \\ \dot{x} \end{bmatrix} = \begin{bmatrix} Az \\ f(x) \end{bmatrix} + \begin{bmatrix} 0 \\ g(x) \end{bmatrix} u + \begin{bmatrix} b \\ 0 \end{bmatrix} r. \tag{9.101}$$

   r plays the role of the disturbance w and $e_0 := h(x) - cz$ the role of the (error) output. We are allowed a state feedback law which depends on x, z, r.

Application of the preceding problem yields that a solution is possible if and only if *the relative degree of the reference model of (9.100) is greater than or equal to that of the system (9.1)* (cf. Problem 9.11). Furthermore, in that instance, the feedback law is

$$u(x, z) = \frac{1}{L_g L_f^{\gamma-1} h(x)} [-L_f^{\gamma} h(x) + cA^{\gamma} z + cA^{\gamma-1} br]. \tag{9.102}$$

As part of Problem 9.11 you will establish that the control law of (9.102) causes only the γ-th derivatives of y and y_m to match. Thus, unless the initial conditions on the first $\gamma - 1$ derivatives of y and y_m match, they will diverge. Following the lines of Theorem 9.14 and the control law of (9.62), it is easy to modify the control law of (9.102) to obtain asymptotic tracking with arbitrary initial conditions.

We now return to the problem of linearization in the presence of uncertainty: For this purpose, we will assume that the uncertainties may be modeled as a perturbation of (9.1)

$$\dot{x} = f(x) + \Delta f(x) + g(x)u + \Delta g(x)u,$$
$$y = h(x). \tag{9.103}$$

Note that we have disallowed perturbations in the output function h. In the context of the discussion of this section we may treat this perturbation arising from $\Delta f(x)$, $\Delta g(x)$ as disturbance vector fields p_1, p_2. Applying Theorem 9.18 to each of these vector fields yields the following proposition.

Proposition 9.19 Robust Linearization. *Consider the uncertain nonlinear system (9.103). Assume that the nominal system (9.1) has relative degree γ. Then, if the perturbations Δf, Δg satisfy*

$$L_{\Delta f} L_f^i h \equiv 0 \quad \text{for} \quad 0 \le i \le \gamma - 1,$$
$$L_{\Delta g} L_f^i h \equiv 0 \quad \text{for} \quad 0 \le i \le \gamma - 1, \tag{9.104}$$

the linearizing control law for the nominal system also linearizes the perturbed system.

Remark: It is important to stress that the results of Proposition 9.19 only decouple the ξ variables of the normal form from the perturbations of Δf, Δg. The η variables associated with the zero dynamics may be perturbed. In particular, the zero dynamics may be destabilized by the perturbation. Further characterization of the perturbations is needed to study this aspect.

Since the disturbance functions associated with Δf, Δg are respectively 1, u it is possible to apply the techniques associated with the disturbance decoupling problem with measurement of disturbance. We apply the result for Δf alone since the disturbance associated with Δg is the control u that we seek. The solvability condition for this scenario is a generalization of a condition referred to in the literature as the *matching condition*. We refer to it as the *generalized matching condition*:

$$L_{\Delta f} L_f^i h \equiv 0 \quad \text{for} \quad 0 \le i \le \gamma - 2,$$
$$L_{\Delta g} L_f^i h \equiv 0 \quad \text{for} \quad 0 \le i \le \gamma - 1. \tag{9.105}$$

Compare equations (9.104) and (9.105) carefully and note that the latter conditions are weaker than the former. The generalized matching conditions are useful in devising a class of linearizing control laws called sliding mode control laws which are the topic of the next subsection.

Before we undertake a detailed discussion of sliding mode control laws, let us note the generalizations of the preceding conditions for MIMO systems that are linearizable (decouplable) by static state feedback, namely, systems with vector relative degree. Thus we consider systems of the form (9.65) with additive disturbance w,

$$\dot{x} = f(x) + g_1(x)u_1 + \cdots + g_p(x)u_p + p(x)w,$$
$$y_1 = h_1(x),$$
$$\vdots \tag{9.106}$$
$$y_p = h_p(x).$$

The analogue of Theorem 9.18 is the following:

Theorem 9.20 MIMO Disturbance Rejection. *Consider the MIMO system of (9.106). Further, assume that with the disturbance set equal to zero the system has vector relative degree $\gamma_1, \ldots, \gamma_p$. Then it is possible to solve the disturbance decoupling problem, namely, to find a state feedback of the form $u = \alpha(x) + \beta(x)v$ such that w has no effect on the output y iff*

$$L_p L_f^k h_i(x) \equiv 0, \quad \text{for} \ k \le \gamma_i - 1. \tag{9.107}$$

for $i = 1, \ldots, p$.

Proof: Since the system has vector relative degree, it is possible to choose the $dL_f^j h_i(x)$ for $0 \le j \le \gamma_i - 1$, $1 \le i \le p$ and the complementary η coordinates such that we obtain the following modification of the normal form coordinates of equation (9.76):

$$\dot{\xi}_1^1 = \xi_2^1 + L_p h_1(x)w,$$

$$\dot{\xi}_2^1 = \xi_3^1 + L_p L_f h_1(x)w,$$

$$\vdots$$

$$\dot{\xi}_{\gamma_1}^1 = b_1(\xi, \eta) + \sum_{j=1}^{p} a_j^1(\xi, \eta)u_j + L_p L_f^{\gamma_1 - 1} h_1(x)w,$$

$$\dot{\xi}_1^2 = \xi_2^2 + L_p h_2(x)w,$$

$$\dot{\xi}_2^2 = \xi_3^2 + L_p L_f h_2(x)w,$$

$$\vdots$$

$$\dot{\xi}_{\gamma_1}^2 = b_2(\xi, \eta) + \sum_{j=1}^{p} a_j^2(\xi, \eta)u_j + L_p L_f^{\gamma_2 - 1} h_2(x)w,$$

$$\vdots$$ (9.108)

$$\dot{\xi}_1^p = \xi_2^p + L_p h_p(x)w,$$

$$\dot{\xi}_2^p = \xi_3^p + L_p L_f h_p(x)w,$$

$$\vdots$$

$$\dot{\xi}_{\gamma_1}^p = b_p(\xi, \eta) + \sum_{j=1}^{p} a_j^p(\xi, \eta)u_j + L_p L_f^{\gamma_p - 1} h_p(x)w,$$

$$\dot{\eta} = q(\xi, \eta) + P(\xi, \eta)u + W(\xi, \eta)w,$$

$$y_1 = \xi_1^1,$$

$$\vdots$$

$$y_p = \xi_1^p.$$

In equation (9.108) above, $W(\xi, \eta) \in \mathbb{R}^{n-\gamma}$ stands for the vector $(L_p \eta_1, \ldots, L_p \eta_{n-\gamma})$ expressed in the ξ, η coordinates. Using the condition (9.107) and the usual linearizing control law

$$u = A^{-1}(\xi, \eta)[-b(\xi, \eta) + v]$$

yields the η states unobservable and consequently isolates y from w. This completes the "if" part of the proof.

The "only if" part of the proof proceeds as in the SISO case by showing that for each of the y^i to be independent of w we need to have that for given $u = \alpha(x) + \beta(x)v$ we have

$$L_p L_{f+g\alpha}^k h_i(x) \equiv 0, \quad 0 \le k \le \gamma_i.$$

Proceeding recursively, as before, yields the conditions of (9.107).   □

For the question of linearization in the presence of uncertainty, we return to the system of

$$\dot{x} = f(x) + \Delta f(x) + g_1(x)u_1 + \cdots + g_p(x)u_p$$
$$+ \Delta g_1(x)u_1 + \cdots + \Delta g_p(x)u_p,$$
$$y_1 = h_1(x), \tag{9.109}$$
$$\vdots$$
$$y_p = h_p(x).$$

Applying the preceding theorem to this system yields the following result on robust linearization.

Proposition 9.21 MIMO Robust Linearization. *Consider the uncertain non-linear system (9.109). Assume that the nominal system (9.65) has vector relative degree $\gamma_1, \ldots, \gamma_p$. Then, if the perturbations $\Delta f, \Delta g_j$ satisfy for $j = 1, \ldots, p$*

$$L_{\Delta f} L_f^i h_j \equiv 0 \quad \text{for} \quad 0 \le i \le \gamma_j - 1,$$
$$L_{\Delta g_k} L_f^i h_j \equiv 0 \quad \text{for} \quad 0 \le i \le \gamma_j - 1, \quad 1 \le k \le p. \tag{9.110}$$

the linearizing control law for the nominal system also linearizes the perturbed system.

The question of when disturbance rejection and robust linearization can be achieved when the MIMO system fails to have relative degree is an interesting one. We discuss it briefly here and leave the details to the Exercises (Problem 9.21). Consider that the nominal system has a singular decoupling matrix, and assume that the matching conditions for the perturbation terms $\Delta f(x)$, $\Delta g(x)$ of (9.110) hold. Continue with the steps in the dynamic extension algorithm of Section 9.3.3, with the definition of $\beta(\tilde{x}) \in \mathbb{R}^{m \times m}$, and the extended states z_1. The condition (9.110) will guarantee that the construction thus far does not depend on the perturbation vector fields $\Delta f(x)$, $\Delta g(x)$. Now, the *extended matching conditions* will require that when a new decoupling matrix is computed for the extended state consisting of $\tilde{x}$, z_1 it too does not depend on $\Delta f(x)$, $\Delta g(x)$, and the new drift terms also do not depend on these vector fields. We leave to Problem 9.21 the task of writing out the details.

9.5 Sliding Mode Control

9.5.1 SISO Sliding Mode Control

The set up is the same of (9.103), namely

$$\dot{x} = f(x) + \Delta f(x) + g(x)u + \Delta g(x)u,$$
$$y = h(x).$$

The assumption is that we would like to devise a tracking control law for tracking $y_m(t)$ with as few assumptions as possible on the perturbation vector fields $\Delta f(x)$, $\Delta g(x)$. The assumption in effect will be that the perturbation vector fields satisfy the *generalized matching conditions* of (9.105). Under these conditions the SISO normal form of the perturbed system can be written by choosing the coordinates to be the coordinates for the unperturbed system, namely $h(x), L_f h(x), \ldots, L_f^{\gamma-1} h(x), \eta(x)$ with $\eta(x)$ satisfying

$$L_g \eta(x) \equiv 0.$$

In these coordinates the system can be written as

$$\dot{\xi}_1 = \xi_2,$$
$$\dot{\xi}_2 = \xi_3,$$
$$\vdots \tag{9.111}$$
$$\dot{\xi}_\gamma = b(\xi, \eta) + \Delta b(\xi, \eta) + a(\xi, \eta)u,$$
$$\dot{\eta} = q(\xi, \eta) + \Delta q(\xi, \eta) + \Delta p(\xi, \eta)u.$$

Here $\Delta b(\xi, \eta) = L_{\Delta f} L_f^{\gamma-1} h$, $\Delta q(\xi, \eta) = L_{\Delta f} \eta$, $\Delta p(\xi, \eta) = L_{\Delta g} \eta$. Further, we will assume that there is a magnitude bound on the nonlinear perturbation described by

$$|\Delta b(\xi, \eta)| < K.$$

Now given the desired output tracking function $y_m(\cdot)$, we define the so-called *sliding surface* using $e_0(t) = y_m(t) - y(t)$

$$s(x, t) = e_0^{\gamma-1} + \alpha_{\gamma-2} e_0^{\gamma-2} + \cdots + \alpha_0 e_0$$
$$= y_m^{\gamma-1} - L_f^{\gamma-1} h + \alpha_{\gamma-2}(y_m^{\gamma-2} - L_f^{\gamma-2} h) + \cdots + \alpha_0(y_m - h(x)),$$

where the coefficients α_i are chosen so that the polynomial $s^{\gamma-1} + \alpha_{\gamma-2} s^{\gamma-2} + \cdots + \alpha_0$ is Hurwitz. Note that if the trajectory is confined to the sliding surface the error tends to zero exponentially according to the zeros of this polynomial. The sliding mode control law is a discontinuous control law chosen to make the time varying manifold described by $\{x : s(x, t) = 0\}$ a manifold which is attractive to initial conditions in its neighborhood in finite time. Indeed, consider the following sliding mode control law:

$$u = \frac{1}{L_g L_f^{\gamma-1} h(x)} [y_m^\gamma - L_f^\gamma h(x) + \alpha_{\gamma-2} e_0^{\gamma-1} + \cdots + \alpha_0 \dot{e}_0 - 1.1 K \, \mathrm{sgn} s(x, t)].$$

$$\tag{9.112}$$

This control law is an exact linearizing control law of the form

$$u = \frac{1}{L_g L_f^{\gamma-1} h(x)} [-L_f^\gamma h(x) + v]$$

with the choice of v as given in equation (9.112) above. The choice of 1.1 in the form of the control law is not magical, any number greater than 1 would suffice. The control law of (9.112) is chosen so that in using it in the normal form of (9.111) we get that

$$\dot{s}(x, t) = -1.1 K \operatorname{sgns}(x, t) + L_{\Delta} f L_f^{\gamma-1} h. \qquad (9.113)$$

The last term in the foregoing equation is $\Delta b(\xi, \eta)$. Thus, it follows that $s(x, t)$ tends to zero in finite time since when $s > 0$ we have that

$$\dot{s} \le -0.1 K,$$

and conversely, when $s < 0$, we have

$$\dot{s} \ge 0.1 K$$

Thus $s(x, t)$ goes to zero in finite time. Once the trajectory reaches the sliding surface, it follows that $e_0 \to 0$ asymptotically from the choice of the coefficients α_i. The foregoing is a rudimentary development of sliding mode control for SISO systems under the assumption that the perturbation vector fields result in a bounded perturbation in the normal form in addition to satisfying the *extended matching conditions*. More specifically, we demand that $L_{\Delta f} L_f^{\gamma-1} h(x)$ be globally bounded. The question of whether the η dynamics remain bounded under perturbation is an interesting one and needs further analysis.

9.5.2   MIMO Sliding Mode Control

As before, consider the uncertain nonlinear system (9.109). Assume that the nominal system (9.65) has vector relative degree $\gamma_1, \ldots, \gamma_p$. Then, if the perturbations $\Delta f, \Delta g_j$ for $j = 1, \ldots, p$ satisfy the *generalized matching condition*:

$$\begin{aligned} L_{\Delta f} L_f^i h_j &\equiv 0 \quad \text{for} \quad 0 \le i \le \gamma_j - 2, \\ L_{\Delta g_k} L_f^i h_j &\equiv 0 \quad \text{for} \quad 0 \le i \le \gamma_j - 1 \;\; 1 \le k \le p, \end{aligned} \qquad (9.114)$$

then one can follow the steps outlined for SISO sliding mode control to get a MIMO sliding mode control law. The details of this are left to Problem 9.22. Also when the system is decouplable only by dynamic extension, the derivation of the conditions for the existence of sliding mode control referred to as *extended generalized matching conditions* is left to Problem 9.24.

9.6   Tracking for Nonminimum Phase Systems

In the preceding sections we have seen methods for tracking (with bounded state trajectories) of nonlinear minimum phase systems. When the nonlinear system is nonminimum phase, two separate approaches may be followed. We discuss each in turn in this section.

9.6.1   The Method of Devasia, Chen, and Paden

A very interesting method for tracking of nonlinear nonminimum phase systems
was proposed by Devasia, Chen and Paden ([85] for the time invariant case and
[86] for the time varying case). Our own treatment follows a modified version of
their work presented in Tomlin and Sastry [307]. See also [147]. In this section, we
discuss the SISO case, the MIMO case is left to the Exercises. It applies to nonlinear
systems that have strict relative degree γ, so that the normal form of equation (9.29)
holds. The control objective is for $y(\cdot)$ to track a bounded reference input $y_D(\cdot)$.
At the outset, we will assume that $y_D(\cdot)$ is defined on the time interval $]-\infty, \infty[$.
We will revert to the familiar $[0, \infty[$ shortly. If $y(\cdot)$ is to track $y_d(\cdot)$, it follows that

$$\xi(\cdot) \equiv \xi_D(\cdot) = \left[y_D(\cdot) \, \dot{y}_D(\cdot) \cdots y_D^{(\gamma-1)}(\cdot) \right]^T \in \mathbb{R}^\gamma.$$

We will need to assume that $\xi_D(\cdot)$ is bounded, that is to say, $y_D, \dot{y}_D, \ldots, y_D^{(m-1)}$
are all bounded. It is assumed that the equilibrium point of the undriven system x_0
is mapped onto $\xi = 0$, $\eta = 0$. It remains to determine a *bounded* solution to the
driven zero dynamics equation

$$\dot{\eta} = q(\xi_D, \eta). \tag{9.115}$$

The Devasia–Chen–Paden scheme begins by assuming that the linearization of the
system of (9.115) about $\eta = 0$

$$Q(t) := \frac{\partial q}{\partial \eta}(\xi_D(t), 0), \tag{9.116}$$

is *kinematically hyperbolic* in the following strong sense: There exists a matrix
$T(t) \in \mathbb{R}^{(n-\gamma) \times (n-\gamma)}$ such that

$$\dot{T}(t) + T(t)Q(t)T^{-1}(t) = \begin{bmatrix} A_1(t) & 0 \\ 0 & A_2(t) \end{bmatrix}, \tag{9.117}$$

where $A_1(t) \in \mathbb{R}^{n_s \times n_s}$ represents an exponentially stable linear system, and the
state transition matrix of $A_2(t) \in \mathbb{R}^{n_u \times n_u}$ is exponentially unstable, that is to say
that there exist $m_1, m_2, \lambda_1, \lambda_2$ such that the state transition matrices of $A_1(t), A_2(t)$
denoted respectively by $\Phi_1(t, t_0), \Phi_2(t, t_0)$ satisfy

$$|\Phi_1(t, t_0)| \le m_1 e^{-\lambda_1(t-t_0)}, \quad t \ge t_0,$$
$$|\Phi_2(t, t_0)| \le m_2 e^{\lambda_2(t-t_0)}, \quad t \le t_0. \tag{9.118}$$

Conditions under which $Q(t)$ is kinematically equivalent to a time varying hyperbolic system of the form of (9.117) are discussed in [86]. Examples of conditions
that guarantee the existence of the transformation of (9.117) are:

- $|\eta_D(\cdot)|_\infty$ is small enough and the undriven zero dynamics are hyperbolic, that is,

$$\frac{\partial q}{\partial \eta}(0,0) \in \mathbb{R}^{(n-\gamma)\times(n-\gamma)}$$

is hyperbolic.
- $Q(t)$ in equation (9.116) is slowly time-varying.

It is easy to verify that the transition matrix

$$X(t) = T^{-1}(t) \begin{bmatrix} \Phi_1(t,t_0) & 0 \\ 0 & \Phi_2(t,t_0) \end{bmatrix} T(t) \tag{9.119}$$

is the Caratheodory solution of the differential equation

$$\dot{X} = Q(t)X, \quad X(\pm\infty) = 0, \quad X(0+) - X(0-) = I. \tag{9.120}$$

Note that the transition matrix of (9.119) is bounded on $]-\infty, \infty[$. Returning to the driven zero dynamics of (9.115), we write

$$q(\xi_D(t), \eta) = Q(t)\eta + r(\xi_D(t), \eta).$$

We will assume that the remainder $r(\xi_D(t), \eta)$ is Lipschitz continuous in both its arguments and in addition that it is *locally approximately linear* in η that is

$$|r(\xi_1, \eta_1) - r(\xi_2, \eta_2)| \le k_1|\xi_1 - \xi_2| + k_2|\eta_1 - \eta_2|, \tag{9.121}$$

with $k_2 \in \mathbb{R}_+$ small. We now claim that for k_2 small enough there is a unique bounded $\eta(\cdot)$ solving the driven zero dynamics equation (9.115):

$$\eta(t) = \int_{-\infty}^{\infty} X(t-\tau)r(\xi_D(\tau), \eta(\tau))d\tau. \tag{9.122}$$

The solution to this equation is obtained by applying the Picard–Lindelöf iteration scheme of Chapter 3 to this equation, that is,

$$\eta^m(t) = \int_{-\infty}^{\infty} X(t-\tau)r(\xi_D(\tau), \eta^{m-1}(\tau))d\tau,$$

starting from any initial guess $\eta^0(t)$ (convergence is in the L_∞ norm, and the proof is deferred to Problem 9.28 of the Exercises). In addition it may be verified that the solution denoted by $\eta_D(t)$ is bounded and satisfies

$$|\eta_D(t)| \le k_3 \sup_t |\xi_D(t)|.$$

The feedforward or exact tracking control law is now given by

$$u_{ff}(t) = \frac{y_D^\gamma(t) - b(\xi_D(t), \eta_D(t))}{a(\xi_D(t), \eta_D(t))}. \tag{9.123}$$

The main drawback of the control law (9.123) is that it is not causal, since the computation of $\eta_D(t)$ needs future values of $\xi_D(t)$. If $y_D(\cdot)$ is specified a-priori ahead of time, then this calculation can be done off-line. Another way that this is remedied is to use a preview of a certain duration, Δ, meaning to say that the control law at time t is actually the one that would have been the correct one at time $t - \Delta$ with the additional proviso that the integral in equation (9.122) runs from $-\infty$ to Δ. Also, when the desired trajectory is only specified on the interval $[0, \infty[$, the integral in (9.122) runs only from $[0, \Delta]$ corresponding to setting the desired output to be identically zero for $t \leq 0$. Once the feedforward control law has been chosen, the overall scheme needs to be stabilized. For this purpose the system in normal form is linearized about the nominal trajectory, i.e., $\xi(t) \equiv \xi_D(t)$, $\eta(t) \equiv \eta_D(t)$, $u(t) \equiv u_{ff}(t)$. Of course, the linearization could also be performed in the original x coordinates, since $x_D(t) = \Phi^{-1}(\xi_D(t), \eta_D(t))$, where Φ stands for the diffeomorphism of the zero dynamics (refer to equation (9.28)). In these original x coordinates define the error coordinates $e(t) = x(t) - x_D(t)$, $u_{fb} = u(t) - u_{ff}(t)$. Then the linearized system is of the form

$$\dot{e} = A(t)e(t) + b(t)u_{fb}(t), \tag{9.124}$$

where

$$A(t) = \frac{\partial f}{\partial x}(x_D(t)), \quad b(t) = g(x_D(t)).$$

Now the stabilization scheme of Problem 6.3 of Chapter 6 can be applied. More precisely, we will need to assume that the following controllability Gramian for the error system of (9.124) is uniformly positive definite for some $\delta > 0$

$$W(t, t + \delta) = \int_t^{t+\delta} \exp^{4\alpha(t-\tau)} \Psi(t, \tau)b(\tau)b(\tau)^T \Psi(t, \tau)^T d\tau.$$

Here $\alpha > 0$, and $\Psi(t, t_0)$ is the state transition matrix of $A(t)$. Define $P(t) := W(t, t + \delta)^{-1}$. Then we may show that the feedback control law

$$u_{fb} = -b^T(t)P(t)e$$

exponentially stabilizes the error system of (9.124). The composite control law used for asymptotic tracking of the nonlinear system is

$$u(t) = u_{ff}(t) + u_{fb}(t). \tag{9.125}$$

The subscript "ff" stands for feedforward or the exact tracking control law and the subscript "fb" stands for the stabilization of the overall system. It is useful to understand the philosophy of this approach. The computation of $u_{ff}(t)$ consisted in finding the unique bounded solution of the driven zero dynamics (9.115) to obtain the $\eta_D(t)$ and in turn the entire desired state trajectory $x_D(t)$. The stabilization was then applied not just to the ξ dynamics as in the case of input–output linearization by state feedback, but rather to the full state. The two drawbacks of the scheme that we have presented are:

1. The exact formula for $u_{ff}(t)$ is non-causal if $y_D(t)$ is not specified ahead of time and needs to be approximated.

2. The domain of validity of the scheme is usually restricted to track trajectories that are both small in amplitude and somewhat slowly varying (owing to the need to verify the strong hyperbolicity conditions needed for (9.117)). Further, since only the linearization is stabilized (9.124), the domain of validity of the overall control law is dependent on the domain of attraction of the stabilizing control law.

9.6.2   The Byrnes–Isidori Regulator

We will discuss the SISO case again here and leave the discussion of the MIMO case to the Exercises. Consider the SISO system of (9.1). It is required that $y(\cdot)$ asymptotically track $y_D(\cdot)$. In turn, it is assumed that $y_D(t)$ is generated by a so-called *exosystem* described by

$$\begin{aligned} \dot{w} &= s(w), \\ y_D &= r(w), \end{aligned} \tag{9.126}$$

with $w \in \mathbb{R}^e$. Thus, we are required to regulate the output e of the composite system described by

$$\begin{aligned} \dot{x} &= f(x) + g(x)u, \\ \dot{w} &= s(w), \\ e &= h(x) - r(w), \end{aligned} \tag{9.127}$$

to zero asymptotically, We will assume that $x = 0, w = 0$ are equilibria of the undriven system (9.1) and the exosystem (9.126), respectively and that $h(0) = 0$, $r(0) = 0$. We are asked to find a control law $u = \alpha(x, w)$ such that:

1. The equilibrium $x = 0$ of the system

$$\dot{x} = f(x) + g(x)\alpha(x, 0) \tag{9.128}$$

is locally exponentially stable.
2. There exists a neighborhood W of $(0, 0)$ in $\mathbb{R}^n \times \mathbb{R}^{n_e}$ such that for $x(0), w(0) \in W$

$$\lim_{t \to \infty} (h(x(t)) - r(w(t))) = 0.$$

We define the linearization matrices of the composite system about $x = 0, w = 0$ by

$$A = \frac{\partial f}{\partial x}(0) \in \mathbb{R}^{n \times n}, \quad S = \frac{\partial s}{\partial w}(0) \in \mathbb{R}^{n_e \times n_e}, \quad b \in \mathbb{R}^n = g(0),$$

$$K = \frac{\partial \alpha}{\partial x}(0, 0) \in \mathbb{R}^{1 \times n}, \quad L = \frac{\partial \alpha}{\partial w}(0, 0) \in \mathbb{R}^{1 \times n_e}. \tag{9.129}$$

Using these matrices, the closed loop system may be written as

$$\dot{x} = (A + bK)x + bLw + r_1(x, w),$$
$$\dot{w} = Sw + r_2(w),$$
(9.130)

where the remainders $r_1(x, w)$, $r_2(w)$ vanish along with their derivatives at $x = 0$, $w = 0$. We now make two key hypotheses about the system being regulated:

1. *Stabilizability.* We assume that the pair A, b is stabilizable, that is there exists at least one $K^T \in \mathbb{R}^n$ such that the eigenvalues of $A + bK$ lie in $\mathbb{C}_-^{\circ}$.
2. *Poisson stability of the equilibrium of the exosystem.* We assume that the equilibrium $w = 0$ of the exosystem of (9.126) is stable in the sense of Lyapunov. In addition, we require that it be Poisson stable, that is to say that all points in a neighborhood U_e of $w = 0$ are recurrent (for the definition see Problem 2.14 and Problem 7.6). In words, all trajectories starting close to $w = 0$ return infinitely often arbitrarily close to the initial condition. This hypothesis implies, in particular, that all the eigenvalues of S lie on the $j\omega$ axis and also that all trajectories starting in U_e are bounded.

We are now ready to state the solution of the feedback regulator problem:

Theorem 9.22 Solution of State Feedback Regulator Problem. *Under the two hypotheses listed above, namely stabilizability of the linearization and Poisson stability of the origin of the exosystem, there exists a control law $u = \alpha(x, w)$ that stabilizes the equilibrium at $x = 0$ of (9.128) and in addition yields for $(x, w) \in W$ that*

$$\lim_{t \to \infty} h(x(t)) - r(w(t)) = 0$$

iff there exist smooth maps $\pi : \mathbb{R}^{n_e} \mapsto \mathbb{R}^n$, $u_{ff} : \mathbb{R}^{n_e} \mapsto \mathbb{R}$ such that

$$\frac{\partial \pi}{\partial w} s(w) = f(\pi(w)) + g(\pi(w))u_{ff}(w),$$
$$h(\pi(w)) - r(w) = 0.$$
(9.131)

Furthermore, the control law which achieves these properties is given by

$$\alpha(x, w) = u_{ff}(w) + K(x - \pi(w)),$$
(9.132)

where $K \in \mathbb{R}^{1 \times n}$ is any stabilizing state feedback for the pair A, b of the linearization.

Proof: For the necessity, let us assume that there is a given feedback law $\alpha(x, w)$ that does the requisite two functions. With A, b, K, L as defined in (9.129), it follows that the eigenvalues of $A + bK$ lie in $\mathbb{C}_-^{\circ}$. Let $T \in \mathbb{R}^{n \times n_e}$ be the unique solution of the linear equation (the uniqueness follows from the fact that the spectrum of $A + bK$ is disjoint from that of S)

$$TS - (A + bK)T = bL.$$
(9.133)

Use the change of state variable $z = x + Tw$ to rewrite (9.130) as

$$\dot{z} = (A + bK)z + \bar{r}_1(z, w),$$
$$\dot{w} = Sw + r_2(w),$$
(9.134)

where $\bar{r}_1, r_2$ vanish along with their derivatives at the origin. By the center manifold theorem 7.25 of Chapter 7, it follows that the center manifold is defined by $z = \bar{\pi} : \mathbb{R}^{n_e} \mapsto \mathbb{R}^n$. In the original coordinates the center manifold is given by the graph of the mapping $x := \pi(w) = Tw + \bar{\pi}(w)$. It is easy to check that since this manifold is to be invariant, we have that

$$\frac{\partial \pi}{\partial w} s(w) = f(\pi(w)) + g(\pi(w))u_{ff}(w).$$

It remains to be shown that the output error is zero in this invariant manifold, i.e., that

$$h(\pi(w)) - r(w) = 0.$$

We use the Poisson stability of the exosystem for this purpose: For the sake of contradiction assume that there exists w_0 close to zero such that $|h(\pi(w_0)) - r(w_0)| := \epsilon > 0$. Then there exists a neighborhood V of w_0 such that for all $w \in V$ we have that $|h(\pi(w) - r(w)| > \epsilon/2$. Now, we are guaranteed that the trajectory starting from $\pi(w_0), w_0$ results in the output error going to zero as $t \to \infty$. Thus, there exists a $T \in \mathbb{R}_+$ such that for $t \geq T$ we have that $|h(\pi(w(t))) - r(w(t))| < \epsilon/2$. By the assumption of Poisson stability of the exosystem, w_0 is recurrent; hence there exists a $t > T$ such that $w(t) \in V$. But inside V we have that $|h(\pi(w(t)) - r(w(t))| > \epsilon/2$. This establishes the contradiction.

For the sufficiency, if we have a solution to the partial differential equation (9.131). we choose K such that the eigenvalues of $A + bK$ lie in $\mathbb{C}_-^\circ$ (such a choice always exists by the stabilizability hypothesis) and define

$$\alpha(x, w) = u_{ff}(w) + K(x - \pi(w)).$$

We claim that $x - \pi(w) = 0$ is an invariant manifold under this control law. Indeed, on this manifold,

$$\frac{d}{dt}(x - \pi(w)) = f(\pi(w)) + g(\pi(w))(u_{ff}(w) + K(x - \pi(w))) - \frac{\partial \pi}{\partial w} s(w).$$

The right hand side is zero on the set $x - \pi(w) = 0$ by the first equation of (9.131). This equation also yields that the invariant manifold is locally attracting since $g(0) = b$ and the eigenvalues of $A = bK$ are in $\mathbb{C}_-^\circ$. On this invariant manifold we have by the second equation of (9.131) that

$$h(\pi(w)) - r(w) = 0,$$

so that the output error is zero on this invariant manifold. Consequently the manifold $\{(x, w) : x - \pi(w)\}$ is referred to as the *stable, invariant, output zeroing manifold* when equations (9.131) hold and the control law of (9.132) is used. $\square$

Remarks:

1. The Byrnes–Isidori regulator guarantees that if we stabilize the closed loop system, then there exists an invariant manifold $x = \pi(w)$ which is a center manifold of the composite system, and furthermore the center manifold can be chosen so as to have the output error be identically zero on it.

2. The draw back of the Byrnes–Isidori regulator is that it does not give explicit solutions for the partial differential equation (9.131). However, it does not assume the existence of relative degree, or minimum phase on the part of the open loop system. Indeed, the first of the equations of (9.131) is a statement of the existence of an invariant manifold of the form $x = \pi(w)$ for the composite system

$$\dot{x} = f(x) + g(x)u_{ff}(w),$$
$$\dot{w} = s(w), \tag{9.135}$$
$$e = h(x) - r(w),$$

for a suitably chosen $u_{ff}(w)$. The second equation asserts that this invariant manifold is contained in the output zeroing manifold

$$h(\pi(w)) - r(w) = 0.$$

The stabilizing term $K(x - \pi(w))$ is designed to make this invariant manifold contained in the output zeroing manifold attractive. Center manifold theorems may be used to obtain solutions to the partial differential equation (9.131). It is instructive to note that in the linear case, that is $f(x) = Ax$, $g(x) = b$, $s(w) = Sw$, $h(x) = cx$, $r(w) = Qw$, the equation (9.131) has a solution of the form $x = \Pi w$, $u_{ff}(w) = \Gamma(w)$ with $\Pi \in \mathbb{R}^{n \times n_e}$ satisfying the matrix equation

$$\Pi S = A\Pi + B\Gamma,$$
$$0 = C\Pi + Q. \tag{9.136}$$

If the original system A, B, C is minimal, the linear regulator theory of Francis and Wonham [105] yields that equations (9.136) have a solution if no eigenvalue of S is a zero of transmission of the open loop system. The nonlinear counterpart of this condition needs the augmented system of (9.135) to have a well defined zero-dynamics manifold M^a in the neighborhood of the origin. Then, from the preceding theorem it follows that the zero-dynamics manifold M^a contains within it the manifold

$$M_s := \{(x, w) : x = \pi(w)\}$$

(with the dynamics on M_s a replica of the dynamics of the exosystem). Furthermore, the tangent space to the zero-dynamics manifold M^a of the augmented system at the origin $T_0 M^a$ can be decomposed as

$$T_0 M^a = T_0 M_s \oplus T_0\{M^a \cap \{(x, 0) : x \in \mathbb{R}^n\}. \tag{9.137}$$

For the proof of this and other interesting details about the solvability by series of the equations (9.131) we refer the reader to the recent book of Byrnes, Delli Priscoli, and Isidori [57].

3. In Problem 9.31 we study calculation of $\pi(w)$, $u_{ff}(w)$ when the open loop system has relative degree and has the normal form of (9.29).

4. Like the Devasia–Chen–Paden scheme the overall scheme has two steps, one that produces a feedforward input $u_{ff}(w)$ and the other that stabilizes the closed loop system $u_{fb}(x, w) = K(x - \pi(w))$.

9.7 Observers with Linear Error Dynamics

We seek the dual of the results presented in the section on linearization by state feedback. For simplicity, we will first consider systems without input and a single output of the form

$$\dot{x} = f(x),$$
$$y = h(x). \tag{9.138}$$

Now, if there exists a change of coordinates of the form $z = \phi(x)$ such that the vector field f and the output function h become

$$\left[\frac{\partial \phi}{\partial x} f(x)\right]_{x=\phi^{-1}(z)} = Az + k(Cz),$$
$$h(\phi^{-1}(z)) = Cz \tag{9.139}$$

with the pair A, C observable and $k : \mathbb{R} \mapsto \mathbb{R}^n$ then it would be possible to build an observer of the form

$$\dot{\hat{z}} = A\hat{z} + k(y) + L(\hat{y} - y),$$
$$\hat{y} = C\hat{z}. \tag{9.140}$$

Further, the observation error $e = \hat{z} - z$ satisfies the linear differential equation

$$\dot{e} = (A + LC)e.$$

If the pair A, C is observable, then it follows that one can choose $L \in \mathbb{R}^{n \times n_o}$ so as to place the eigenvalues of $A + LC$ in any desired locations. Motivated by this consideration, we now seek to given conditions on the f, h of system (9.138) so as to be able to transform it into the form given by equation (9.139). This problem is referred to as the *observer linearization problem*. Necessary conditions for the solvability of this problem, that is the existence of the transformation ϕ are given in the following proposition:

Proposition 9.23 Observer Linearization: Necessary Conditions. *The ob-server linearization problem is solvable only if*

$$\dim(\operatorname{span}\{dh(x_0), dL_f h(x_0), \ldots, dL_f^{n-1} h(x_0)\}) = n. \tag{9.141}$$

Proof: If the observer linearization problem could be solved, and the pair A, C were observable, since a further linear transformation of coordinates and linear output injection could convert the pair into observable canonical form, i.e., there exist $T \in \mathbb{R}^{n \times n}, L \in \mathbb{R}^{n \times n_o}$ such that

$$T(A + LC)T^{-1} = \begin{bmatrix} 0 & 0 & \cdots & 0 & 0 \\ 1 & 0 & \cdots & 0 & 0 \\ \cdots & \cdots & \cdots & \cdots & \cdots \\ 0 & 0 & \cdots & 1 & 0 \end{bmatrix}, \quad CT^{-1} = [0\ 0\ \cdots\ 0\ 1],$$

we could assume without loss of generality that A, C are already in this desired form, i.e.,

$$A = \begin{bmatrix} 0 & 0 & \cdots & 0 & 0 \\ 1 & 0 & \cdots & 0 & 0 \\ \cdots & \cdots & \cdots & \cdots & \cdots \\ 0 & 0 & \cdots & 1 & 0 \end{bmatrix}, \quad C = [0\ 0\ \cdots\ 0\ 1].$$

Then, it follows that there exists ϕ such that

$$\left[\frac{\partial \phi}{\partial x} f(x) \right]_{x = \phi^{-1}(z)} = \begin{bmatrix} 0 & 0 & \cdots & 0 & 0 \\ 1 & 0 & \cdots & 0 & 0 \\ & & \cdots & & \\ 0 & 0 & \cdots & 1 & 0 \end{bmatrix} z + k([0\ 0\ \cdots\ 0\ 1] z) \quad (9.142)$$

and

$$h(\phi^{-1}(z)) = [0\ 0\ \cdots\ 0\ 1] z. \quad (9.143)$$

If these equations hold then we have

$$h(x) = z_n(x),$$

$$\frac{\partial z_1}{\partial x} f(x) = k_1(z_n(x)),$$

$$\frac{\partial z_2}{\partial x} f(x) = z_1(x) + k_2(z_n(x)),$$

$$\cdots$$

$$\frac{\partial z_n}{\partial x} f(x) = z_{n-1}(x) + k_n(z_n(x)).$$

Thus observe that

$$L_f h(x) = \frac{\partial z_n}{\partial x} f(x) = z_{n-1}(x) + k_n(z_n(x)),$$

$$L_f^2 h(x) = \frac{\partial z_{n-1}}{\partial x} f(x) + \frac{\partial k_n}{\partial z_n} \frac{\partial z_n}{\partial x} f(x),$$

$$= z_{n-2} + \frac{\partial k_n}{\partial z_n} \frac{\partial z_n}{\partial x} f(x) + k_{n-1}(z_n(x)),$$

$$= z_{n-2} + \frac{\partial k_n}{\partial z_n}(z_{n-1} + k_n(z_n)),$$

$$= z_{n-2} + \tilde{k}_{n-1}(z_n(x), z_{n-1}(x)).$$

Proceeding inductively, for $L_f^i h(x)$ for $1 \le i \le n-1$, we get

$$L_f^i h(x) = z_{n-i}(x) + \tilde{k}_{n-i+1}(z_n(x), \ldots, z_{n-i+1}(x)),$$

Putting these relations together we get

$$
\begin{bmatrix} \dfrac{\partial h}{\partial x} \\[2mm] \dfrac{\partial L_f h}{\partial x} \\[2mm] \cdots \\[2mm] \dfrac{\partial L_f^{n-1} h}{\partial x} \end{bmatrix}
=
\begin{bmatrix} \dfrac{\partial h}{\partial z} \\[2mm] \dfrac{\partial L_f h}{\partial z} \\[2mm] \cdots \\[2mm] \dfrac{\partial L_f^{n-1} h}{\partial z} \end{bmatrix}
\frac{\partial z}{\partial x}
=
\begin{bmatrix} 0 & 0 & \cdots & 0 & 1 \\ 0 & 0 & \cdots & 1 & * \\ & & \cdots & & \\ 1 & * & \cdots & * & * \end{bmatrix}
\frac{\partial z}{\partial x}.
$$

Since, $\frac{\partial z}{\partial x}$ is nonsingular, the claim of the proposition follows. □

From the fact that a necessary condition for the observer linearization problem is equation (9.141) it follows that it is possible to define a vector field $g(x)$ satisfying

$$L_g h(x) = L_g L_f h(x) = \cdots = L_g L_f^{n-2} h(x) = 0,$$
$$L_g L_f^{n-1} h(x) = 1.$$

More precisely, $g(x)$ satisfies

$$
\begin{bmatrix} dh(x) \\ dL_f h(x) \\ \vdots \\ dL_f^{n-2} h(x) \\ dL_f^{n-1} h(x) \end{bmatrix}
g(x) =
\begin{bmatrix} 0 \\ 0 \\ \vdots \\ 0 \\ 1 \end{bmatrix}.
\tag{9.144}
$$

The following proposition gives a necessary and sufficient condition for the solution of the *observer linearization problem*

Proposition 9.24 Observer Linearization: Necessary and Sufficient Conditions.   *The observer linearization problem is solvable iff*

1. $\dim(\mathrm{span}\{dh(x_0), dL_f h(x_0), \ldots, dL_f^{n-1} h(x_0)\}) = n$.
2. *There exists a mapping* $F : V \subset \mathbb{R}^n \mapsto U$ *a neighborhood of* x_0 *such that*

$$\frac{\partial F}{\partial z} = \left[g(x) - ad_f g(x) \cdots (-1)^{n-1} ad_f^{n-1} g(x) \right]_{x=F(z)}, \tag{9.145}$$

where $g(x)$ *is the unique solution of equation (9.144) above.*

Proof: We already know about the necessity of the first condition from the preceding proposition. Suppose that there is a diffeomorphism ϕ such that the system can be transformed into the linearized observer form. Now set $F(z) = \phi^{-1}(z)$. Also, define (actually, recall from Chapter 3) the push forward of a vector field $\theta(z)$ by a diffeomorphism $x = F(z)$ as

$$F_*\theta = \left(\frac{\partial F}{\partial z}\theta\right)_{z=F^{-1}(x)}.$$

Thus, the push-forward of a vector field is equivalent to transforming it into new coordinates. Now, define the constant vector fields

$$e_1 = \begin{bmatrix} 1 \\ 0 \\ \vdots \\ 0 \end{bmatrix}, \quad e_2 = \begin{bmatrix} 0 \\ 1 \\ \vdots \\ 0 \end{bmatrix}, \quad \text{etc..}$$

Then, define

$$\tau(x) = F_*e_1.$$

We claim that

$$ad_f^k\tau(x) = (-1)^k F_*e_{k+1}.$$

Indeed, we have that f is the push forward of $\tilde{f}(z)$ given as $Az + k(cz)$, that is

$$f(x) = F_*\tilde{f}(z) = F_* \begin{bmatrix} k_1(z_n) \\ z_1 + k_2(z_n) \\ \cdots \\ z_{n-1} + k_n(z_n) \end{bmatrix}.$$

The proof of the claim now follows by induction: The claim is clearly true for $k = 0$. If it is true for k then

$$ad_f^{k+1}\tau(x) = [f, ad_f^k\tau] = (-1)^k F_*[\tilde{f}(z), e_{k+1}].$$

Using the form of $\tilde{f}$ it follows that

$$[\tilde{f}(z), e_{k+1}] = -e_{k+2}.$$

To verify that $\tau = g$, note that

$$(-1)^k L_{ad_f^k\tau}h(x) = L_{e_{k+1}}h(F(z))|_{z=F^{-1}(x)}.$$

But, since $h(F(z)) = z_n$ it follows that

$$L_{ad_f^k\tau}h(x) = 0$$

for $0 \le k \le n-2$ and that

$$(-1)^{n-1} L_{ad_f^{n-1} \tau} h(x) = 1.$$

In turn this implies that τ satisfies the equations of (9.144). Since the solutions of equations (9.144) are unique, the result follows.

The proof of the sufficiency is as follows: Suppose that the conclusion (1) of the proposition holds and let τ be the solution of equation (9.144). Using the results of Proposition 9.4 it follows that the matrix

$$\begin{bmatrix} dh(x) \\ dL_f h \\ \vdots \\ dL_f^{n-1} h(x) \end{bmatrix} \begin{bmatrix} \tau(x) & ad_f \tau(x) & \cdots & ad_f^{n-1} \tau(x) \end{bmatrix}$$

has rank n near x_0. Thus, the vector fields $\tau, ad_f \tau, \ldots, ad_f^{n-1} \tau$ are linearly independent near x_0. Now assume that F is a solution of the partial differential equation (9.145). Set $\Phi = F^{-1}$ and

$$\tilde{f}(z) = \Phi_* f(x).$$

By definition, the mapping F is such that

$$\frac{\partial F}{\partial z_{k+1}}\bigg|_{z=F^{-1}(x)} = (-1)^k ad_f^k \tau(x),$$

so that we have that

$$\frac{\partial \Phi}{\partial x} ad_f^k \tau(x)\big|_{x=\Phi^{-1}(z)} = (-1)^k e_{k+1}$$

for $0 \le k \le n-1$. Using this we see that

$$\begin{aligned}
(-1)^{k+1} e_{k+2} &= \Phi_* ad_f^{k+1} \tau \\
&= \Phi_* [f, ad_f^k \tau] \\
&= [\tilde{f}, (-1)^k e_{k+1}] \\
&= (-1)^{k+1} \frac{\partial \tilde{f}}{\partial z_{k+1}}.
\end{aligned}$$

In turn, this implies that for $0 \le k \le n-2$, we have

$$\frac{\partial \tilde{f}_{k+2}}{\partial z_{k+1}} = 1, \quad \frac{\partial \tilde{f}_i}{\partial z_{k+1}} = 0, \quad \text{for } i \ne k+2.$$

Using these relations the functional form of the $\tilde{f}$ follows. Further, since we have that for $0 \le k < n-1$ that

$$L_{ad_f^k \tau} h(x) = 0$$

and $L_{ad_f^{n-1}\tau}h(x) = (-1)^{n-1}$ we see that that

$$h \circ F = \tilde{h}(z) = z_n,$$

completing the proof.   □

It is of interest to make the second condition of the preceding proposition more explicit, i.e., the existence of a diffeomorphism F such that

$$\frac{\partial F}{\partial z} = \left[g(x) - ad_f g(x) \cdots (-1)^{n-1} ad_f^{n-1} g(x) \right]_{x=F(z)}.$$

The following theorem (which has some independent interest as well) gives conditions on a given $\mathbb{R}^{n \times n}$ matrix that make it the Jacobian of a diffeomorphism:

Theorem 9.25. *Let $X_1(x), \ldots, X_n(x)$ be n given linearly independent (over the ring of smooth functions) vector fields. Then there exists a diffeomorphism ψ : $\mathbb{R}^n(x) \mapsto \mathbb{R}^n(z)$ satisfying*

$$\psi_* e_i = X_i$$

iff the X_i commute, i.e.,

$$[X_i, X_j] = 0.$$

Here as above, e_i stands for the constant vector field with all except the i-th component equal to zero and the i-th one equal to 1.

Proof: The necessity part of the proof is easy: Indeed, if the said diffeomorphism exists, then

$$[X_k, X_l] = \psi_*[e_k, e_l] = 0,$$

since $[e_k, e_l] = 0$. For the sufficiency, construct the n distributions

$$\Delta_i = \text{span}\{X_1, \ldots, X_{i-1}, X_{i+1}, \ldots, X_n\}.$$

Each distribution Δ_i is of constant dimension $n - 1$ and is involutive since the $[X_k, X_l] = 0$. Thus, by the Frobenius theorem (Theorem 8.15) there exists a function $\phi_i(z)$ whose differential spans $\Delta_i^\perp$, i.e.,

$$d\phi_i X_j = 0, \quad j \neq i.$$

Now define

$$c_i(x) = d\phi_i X_i.$$

Since the X_i are linearly independent it follows that $c_i(x) \neq 0$. Also, as a consequence, the $d\phi_i$ are linearly independent. Indeed, if the differentials $d\phi_i(x_0), \ldots, d\phi_n(x_0)$ were linearly dependent then it follows that there will exist some nonzero scalars p_i such that

$$\sum_{i=1}^{n} p_i d\phi_i(x_0) = 0.$$

Post-multiplying this expression by $X_i(x_0)$ yields

$$d\phi_i(x_0)X_i(x_0) = c_i(x_0) = 0.$$

Thus we have that

$$\frac{\partial\phi}{\partial x}\left[\begin{array}{ccc} X_i(x) & \cdots & X_n(x) \end{array}\right] = \left[\begin{array}{ccc} c_1(x) & \cdots & 0 \\ 0 & \ddots & 0 \\ 0 & \cdots & c_n(x) \end{array}\right].$$

In other words, if we define $\xi = \phi(x)$, it follows that

$$(\phi_i)_*X_i = \left[\begin{array}{c} 0 \\ \vdots \\ c_i(x) \\ \vdots \\ 0 \end{array}\right].$$

Since the X_i commute, it follows that $c_i \circ \Phi^{-1}(\xi)$ is a function of ξ_i alone. Thus, there exist functions μ_i such that

$$\frac{\partial\mu_i}{\partial\xi_i} = \frac{1}{c_i(\Phi^{-1}(\xi))}.$$

The transformation obtained by composing μ and ξ,

$$z = \psi(x) = (\mu_1(\xi_1), \ldots, \mu_n(\xi_n))^T,$$

is the required diffeomorphism. $\square$

9.8   Summary

This chapter has covered an introduction to input–output linearization by state feedback for continuous time nonlinear systems. Our treatment has been rather brief, but there are many other interesting subtleties of the theory which are discussed in Isidori [149] and Nijmeijer and van der Schaft [232]. Questions of partial feedback linearization, robust and adaptive linearization are discussed in the books by Marino and Tomei [199], Krstic, Kanellakopoulos and Kokotovic [171], and Sastry and Bodson [259] (see also the Exercises in the next section). The main omission from this chapter is the coverage of a large body of literature on techniques for the linearization of discrete-time and sampled-data nonlinear systems. One of the reasons for doing this is that it is not possible to do justice to this subject without a thorough rethinking of the ideas of relative degree and relative degree under sampling. The main reasons why the discrete-time and sampled-data theory are so much richer than their continuous counterparts is because even if discrete time systems start off affine in the input, a one step time delay already results in a non-affine dependence of the output on the input. For more details, we refer the reader to Monaco and Normand-Cyrot [213], [214] (and their monograph in-preparation), Grizzle [121], and Lee, Arapostathis, and Marcus [176]. Another active area of research is the elaboration of the construction of non-linear observers. We gave a reader just a flavor of the results in the last section. It is of interest for nonlinear control systems to not reconstruct the full-state but merely the part that is needed for the state feedback law ($\alpha(x)$, $\beta(x)$ rather all of x). For this and other interesting points, we refer the reader to a thriving literature in nonlinear control.

9.9   Exercises

Problem 9.1. Prove Claim 9.5.

Problem 9.2. Prove Claim 9.3.

Problem 9.3. Prove the statement (9.44).

Problem 9.4. Consider the *output reproduction problem* for the system (9.1): Find, if possible, a pair $\bar{x}$, $\bar{u}(t)$ such that the output $y(t)$ is identically equal to $\bar{y}(t)$. Find the internal dynamics consistent with this output and relate them to the zeros of the system. Prove that if $\bar{y}(t)$ and its derivatives are bounded, then the state x is bounded as well.

Problem 9.5 Poincaré linearization.   Consider the nonlinear equation

$$\dot{x} = Ax + v(x) \qquad (9.146)$$

with $x \in \mathbb{R}^n$, $A \in \mathbb{R}^{n \times n}$, and $v(x)$ a vector of polynomials in x of degree greater than or equal to 2. We wish to find a transformation of the form $x = y + h(y)$: $\mathbb{R}^n \mapsto \mathbb{R}^n$, with $h(0) = 0$, $Dh(0) = 0$ to transform (9.146) into

$$\dot{y} = Ay.$$

1. Find the equation that h should satisfy. Use the operator L_A taking smooth vector fields in $\mathbb{R}^n$ into smooth vector fields $\psi(x)$ in $\mathbb{R}^n$ given by

$$L_A \psi := \frac{\partial \psi}{\partial x} Ax - A\psi.$$

2. Let us study the operator in the case that A is diagonal with real eigenvalues $\lambda_1, \ldots, \lambda_n$. Show that for each choice of integers $m_1, m_2, \ldots, m_n$, the following

$\psi(x)$ are eigenvectors of the operator L_A:

$$\psi^i_{m_1 m_2 \cdots m_n}(x) = x_1^{m_1} x_2^{m_2} \ldots x_n^{m_n} \begin{bmatrix} 0 \\ \cdot \\ 1 \\ \cdot \\ 0 \end{bmatrix} \leftarrow i\text{-th entry}$$

with the 1 in the i-th location, Show that

$$L_A \psi^i_{m_1 m_2 \cdot m_n}(x) = \mu \psi^i_{m_1 m_2 \cdot m_n}(x)$$

What are the eigenvalues μ?

3. Show that if L_A is invertible, one can solve the equation of part (1) of this problem. What is the condition on $\lambda_1, \ldots, \lambda_n$ for invertibility?

4. Try to figure out what would happen if A is not diagonal: Consider the cases when A is diagonalizable and when A has non-trivial Jordan blocks.

Problem 9.6 Vehicle dynamics.   The dynamics (extremely simplified) of a wheeled vehicle on a flat road from engine force input F to the position of the vehicle center of mass x are described by

$$F - \rho(\dot{x})^2 \text{sgn}(\dot{x}) - d_m = m\ddot{x}.$$

In this equation ρ stands for the coefficient of wind drag, d_m the mechanical drag and m the mass of the vehicle. Further, the engine dynamics are modeled by the first order system

$$\dot{\xi} = -\frac{\xi}{\tau(\dot{x})} + \frac{u}{m\tau(\dot{x})}.$$

Here ξ is a state variable modeling all of the complex engine dynamics, u is the throttle input, and $\tau(\dot{x})$ is a velocity-dependent time constant. Also $F = m\xi$. Write these equations in state space form and examine the input–output linearizability of the engine from input u to the output x. What are the relative degree, zero dynamics of the system?

Problem 9.7 Control of satellites.   Consider the control of a rigid spacecraft by a single input about its center of mass. The equations of motion of this system are given by:

$$\dot{\omega} = J^{-1} S(\omega) J \omega + J^{-1} B u,$$
$$\dot{R} = S(\omega) R. \tag{9.147}$$

Here $R \in SO(3)$ is a 3×3 unitary matrix modeling the orientation of the spacecraft relative to the inertial frame, and $\omega \in \mathbb{R}^3$ models the instantaneous angular velocity. J stands for the moment of inertia of the spacecraft about a frame attached to its center of mass (if the frame lines up with the principal axes J is diagonal) and

$S(\omega)$, also sometimes called $\omega\times$, is the skew matrix representation $(S(\omega) \in so(3))$ of the cross-product operator, given by

$$
\begin{bmatrix}
0 & \omega_3 & -\omega_2 \\
-\omega_3 & 0 & \omega_1 \\
\omega_2 & -\omega_1 & 0
\end{bmatrix}.
$$

Further, $B \in \mathbb{R}^3$ is the input matrix. To define the scalar output y we define the roll–pitch–yaw (also known as XYZ Euler angles) representation of R as

$$
R = e^{S(\hat{x})\phi} e^{-S(\hat{y}\theta)} e^{S(\hat{z})\psi},
$$

where $\hat{x}, \hat{y}, \hat{z}$ stand for unit vectors along the coordinate directions. The angles ϕ, θ, ψ are referred to as roll, pitch, and yaw respectively. Be sure to compute R as a function of ϕ, θ, ψ. Now determine the relative degree of the system for choice of output $y = \phi$, $y = \theta$ and $y = \psi$. Assume that J is diagonal and $B = \hat{x}$. Think about what would happen if $B = \hat{y}$ or $B = \hat{z}$.

Problem 9.8 Zeros of the linearization. Let 0 be an equilibrium point of the undriven nonlinear system of (9.1) and $A = Df(0)$, $b = g(0)$, $c = Dh(0)$ its Jacobian linearization. Assume that the Jacobian linearization is minimal. Show that the eigenvalues of the linearization of the zero dynamics of (9.56) are the same as the zeros of the Jacobian linearized linear system. What can you say about the minimum phase or nonminimum phase properties of the nonlinear system when 0 is a zero of the Jacobian linearized system?

Problem 9.9. Using the center manifold methods of Chapter 7, show that the stabilizing control law of (9.60) stabilizes the nonlinear system when the zero dynamics are asymptotically stable (i.e., the center manifold, if it exists, is stable).

Problem 9.10. Show that the disturbance decoupling problem with measurement feedback is solvable iff the conditions of (9.99) hold.

Problem 9.11. Using Problem 9.10 show that the *Model Reference Adaptive Control* problem can be solved iff the relative degree of the reference model of (9.100) exceeds that of the system (9.1). Further, show that the control law is given by (9.102). Derive the equation satisfied by the output e_0. Modify the control law of (9.102) so that it can be made zero asymptotically for arbitrary initial conditions on x, z, and any input r.

Problem 9.12. Prove Proposition 9.19.

Problem 9.13. Show that the differentials

$$
dL_f^j h_i(x), \quad 0 \le j \le \gamma_i - 1, \quad 1 \le i \le p, \tag{9.148}
$$

are linearly independent if the system (9.65) has well defined relative degree. Hint: Use the nonsingularity of the decoupling matrix of (9.68).

Problem 9.14 Brunovsky Normal Form. Assume that the conditions for full state MIMO linearization have been satisfied, i.e., the conditions of Theorem 9.16

are met. Derive the normal form of the MIMO system under these conditions, referred to as the Brunovsky normal form. The indices $0 \leq \kappa_i \leq \kappa$ of the m outputs are referred to as the Kronecker indices.

Problem 9.15 Chained form conversion. In this chapter you have studied how to do full state linearization. Here we will give sufficient conditions for conversion of a *driftless system* into a different kind of canonical form, called *chained* or *Goursat normal form*. This is discussed in greater detail in Chapter 12. Consider

$$\dot{x} = g_1(x)u_1 + g_2(x)u_2. \tag{9.149}$$

Define the following three distributions:

$$\begin{aligned}
\Delta_0 &= \text{span}\{g_1, g_2, \text{ad}_{g_1}g_2, \text{ad}^2_{g_1}g_2, \ldots, \text{ad}^{n-2}_{g_1}g_2\}, \\
\Delta_1 &= \text{span}\{g_2, \text{ad}_{g_1}g_2, \text{ad}^2_{g_1}g_2, \ldots, \text{ad}^{n-2}_{g_1}g_2\}, \\
\Delta_2 &= \text{span}\{g_2, \text{ad}_{g_1}g_2, \text{ad}^2_{g_1}g_2, \ldots, \text{ad}^{n-3}_{g_1}g_2\}.
\end{aligned} \tag{9.150}$$

You are given that $\Delta_0(x) = \mathbb{R}^n$ and that $\Delta_1(x)$, $\Delta_2(x)$ are involutive for $x \in U$ an open set containing the origin. Also, you are told that you can find a function $h_1 : U \mapsto \mathbb{R}$ such that $dh_1 \Delta_1 = 0$ and $L_{g_1}h_1 = 1$. Prove that there exists a transformation $z = \phi(x)$ and a transformation of the inputs $u = \beta(x)v$ that transform (9.149) into the form

$$\begin{aligned}
\dot{z}_1 &= v_1, \\
\dot{z}_2 &= v_2, \\
\dot{z}_3 &= z_2 v_1, \\
\dot{z}_4 &= z_3 v_1 \\
&\vdots \\
\dot{z}_n &= z_{n-1} v_1.
\end{aligned} \tag{9.151}$$

Can you apply your results to the kinematics of a front wheel drive car:

$$\begin{aligned}
\dot{x}_1 &= u_1, \\
\dot{x}_2 &= \tan x_4 u_1, \\
\dot{x}_3 &= u_2, \\
\dot{x}_4 &= \frac{1}{\cos x_4} \tan x_3 u_1.
\end{aligned} \tag{9.152}$$

Problem 9.16 Rigid robot computed torque controller [225; 285]. The dynamics of an n link rigid robot manipulator with joint angles θ are given by

$$M(\theta)\ddot{\theta} + \tilde{C}(\theta, \dot{\theta}) = u. \tag{9.153}$$

Here $\theta \in \mathbb{R}^n$ is the vector of joint angles and $u \in \mathbb{R}^n$ is the vector of motor torques exerted by the motors at each joint. Further $M(\theta) \in \mathbb{R}^{n \times n}$ is a positive definite matrix called the moment of inertia matrix and $\tilde{C}(\theta, \dot{\theta}) \in \mathbb{R}^n$ stands for Coriolis, gravitation, and frictional forces. Using as outputs the angles θ linearize the system of (9.153). What are the zero dynamics of this system? When the joint angles are required to follow a desired trajectory $\theta_d(\cdot)$ give the asymptotic tracking control law. This is known as the *computed torque control law* for robot manipulators. (see [225], Chapter 4, or [285], Chapters 7–11 for details).

Problem 9.17 Rigid robot PD set point controller [225; 285; 20].   Consider the rigid robot dynamics of equation (9.153). From physical considerations it follows that the second term on the left-hand side of the dynamics actually has the form

$$\tilde{C}(\theta, \dot{\theta}) = C(\theta, \dot{\theta})\dot{\theta} + N(\theta) + K_f \dot{\theta}, \qquad (9.154)$$

Here $C(\theta, \dot{\theta}) \in \mathbb{R}^{n \times n}$ represents the contribution of the Coriolis and centripetal terms, $N(\theta) \in \mathbb{R}^n$ represents the gravitational forces on the robot links, and $K_f \in \mathbb{R}^{n \times n}$ is a positive definite matrix representing the friction terms in the robot joints. It is a basic fact [225] that for open-link robot manipulators

$$\dot{M}(\theta) - 2C(\theta, \dot{\theta})$$

is a skew symmetric matrix. For the case that $\theta_d(t) \equiv \theta_0$, consider the following *proportional derivative (PD) control law*:

$$u = N(\theta) - K_v \dot{\theta} - K_p(\theta - \theta_0) \qquad (9.155)$$

with $K_v, K_p \in \mathbb{R}^{n \times n}$. Show using LaSalle's principle that if K_p is positive definite and K_v is positive semidefinite, then $\theta(t) \to \theta_0$ as $t \to \infty$. *Hint: Use the Lyapunov function*

$$v(\theta, \dot{\theta}) = \frac{1}{2}\dot{\theta}^T M(\theta)\dot{\theta} + \frac{1}{2}\theta^T K_p \theta.$$

Problem 9.18 Modified PD control of rigid robots [225].   Consider once again the rigid robot equations of (9.153) with the form of $\tilde{C}$ as given in (9.154). The control law of the previous problem (9.155) needs to be modified to the form

$$u = M(\theta)\ddot{\theta}_d + C(\theta, \dot{\theta})\dot{\theta}_d + N(\theta, \dot{\theta}) + K_f \dot{\theta} - K_v(\theta - \dot{\theta}_D) - K_p(\theta - \theta_d), \quad (9.156)$$

when $\theta(\cdot)$ is required to track a non-constant $\theta_d(\cdot)$. Use this control law to prove that the tracking error $e(\cdot) = \theta(\cdot) - \theta_d(\cdot)$ converges *exponentially to zero. Use the Lyapunov function*

$$v(e, \dot{e}) = \frac{1}{2}\dot{e}^T M(\theta)\dot{e} + \frac{1}{2}e^T K_p e + \epsilon e^T M(\theta)\dot{e}$$

for suitably chosen ϵ.

Problem 9.19 Rigid robot with flexible joints [190].   In some commercially available robots there is compliance between the motors and the robot joints caused by the presence of harmonic drives and other drive train characteristics. The dynamics of (9.153) are then modified. By way of notation we refer to the vector of motor angles as $\theta_m \in \mathbb{R}^n$. The coupling between the motor and the robot joints is assumed to be through a stiff spring:

$$M(\theta)\ddot{\theta} + C(\theta, \dot{\theta}) + K(\theta - \theta_m) = 0,$$
$$M_m\ddot{\theta}_m + B_m\dot{\theta}_m = K(\theta - \theta_m) + u. \tag{9.157}$$

In the equation above M_m, B_m stand for the moment of inertia matrix and the damping matrix associated with the motors, K is the matrix of torsional spring constants between the motor and robot links (usually a diagonal matrix), and u is, as before, the vector of input torques. Linearize the dynamics of (9.157) using θ as the outputs.

Discuss what happens to the linearization in the limit that the torsional springs become stiff, i.e., $K \to \infty$. In particular, can you recover the linearizing control law of the previous problem (9.16)?

Problem 9.20. Use the linearizing control law of Problem 9.16 on the dynamics of (9.157). Show that if the control objective is for the joint angles θ to track a reference trajectory that the tracking error is of order $1/|K|$.

Problem 9.21 Extended Matching Conditions for MIMO Systems needing Dynamic Extension. Consider the uncertain nonlinear system of (9.109). Assume that the nominal nonlinear system fails to have vector relative degree. Following the discussion in the text augment the matching conditions of (9.110) to derive extended matching conditions for robust linearization. Apply the results to the unicycle model:

$$\dot{x}_1 = \cos x_3 u_1 + \Delta f_1(x)$$
$$\dot{x}_2 = \sin x_3 u_1 + \Delta f_2(x)$$
$$\dot{x}_3 = u_2 + \Delta f_3(x)$$
$$y_1 = x_1$$
$$y_2 = x_2$$

Problem 9.22 MIMO Sliding for decouplable systems [100]. Consider the uncertain nonlinear system of (9.109). Assume that the nominal nonlinear system of (9.65) has vector relative degree. Further assume that the uncertainty satisfies the generalized matching condition of (9.114). Construct multiple sliding surfaces and a sliding control law which gives robust tracking under certain conditions on the perturbation.

Problem 9.23 Sliding Mode Control of a Continuous Stirred Tank Reactor [100]. Consider the following (celebrated) equations for a continuous stirred tank reactor involving the concentrations of three reagents A, B, C whose fractions in the tank are given by $x_1, x_2, x_3 \in \mathbb{R}$. The sensed temperature is given by $x_4 \in \mathbb{R}$.

The dynamics are given by

$$\dot{x}_1 = -r_1 + (c_1 - x_1)u_1$$
$$\dot{x}_2 = r_1 - r_2 + (c_2 - x_2)u_1$$
$$\dot{x}_3 = r_2 + (c_3 - x_3)u_1 \tag{9.158}$$
$$\dot{x}_4 = \alpha_1 r_1 + \alpha_2 r_2 + (c_4 - x_4)u_1 + u_2$$

Here $c_1, c_2, c_3, c_4, \alpha_1, \alpha_2$ are constants and

$$r_1 = \exp^{(\beta_1 - \beta_2)x_1^2} \quad r_2 = \exp^{(\beta_3 - \beta_4)x_2}$$

for some constants $\beta_1, \beta_2, \beta_3, \beta_4$. Derive conditions on the kinds of uncertainties on f, g_1, g_2 that can be overcome by MIMO sliding mode control. For details on numeric values for these constants please refer to Fernandez and Hedrick [100] and the references there in.

Problem 9.24 MIMO Sliding for systems not decouplable by static state feedback. Consider the uncertain nonlinear system of (9.109). Assume that the nominal nonlinear system of (9.65) fails to have vector relative degree. Further assume that the uncertainty satisfies an *extended generalized matching condition* obtained from modifying the conditions of Problem 9.21. Derive these conditions and construct multiple sliding surfaces and a sliding control law which gives robust tracking under certain conditions on the perturbation.

Problem 9.25 Dynamic Surface Control [292]. Consider the following single input control system in strict feedback form with output $y = x_1$.

$$\dot{x}_1 = x_2 + f_1(x_1) + \Delta f_1(x_1)$$
$$\dot{x}_2 = x_3 + f_2(x_1, x_2) + \Delta f_2(x_1, x_2)$$
$$\vdots$$
$$\dot{x}_{n-1} = x_n + f_n(x_1, \ldots, x_{n-1}) + \Delta f_n(x_1, \ldots, x_{n-1})$$
$$\dot{x}_n = u$$

Here $\Delta_i f_i(\cdot)$ are unknown but are bounded with known Lipschitz constants. The control objective is for $y = x_1$ to track $y_d(\cdot) := x_{1d}$. Now define the error surface $s_1 = x_1 - x_{1d}$. After differentiating we get

$$\dot{s}_1 = x_2 + f_1(x_1) - \dot{y}_d - f_1(x_1)$$

Define x_{id}, s_i for $i = 1, 2, \ldots, n - 1$ such that

$$s_i = x_i - x_{id}$$
$$\tau_i \dot{x}_{(i+1)d} + x_{(i+1)d} = \dot{x}_{1d} - k_i s_i - f_i(x_1, \ldots, x_i)$$

for some positive constants τ_i, k_i to be determined later. This results in

$$\dot{s}_i = x_{i+1} + f_i(x_1, \ldots, x_i) - \dot{x}_{id} + \Delta f_i(x_1, \ldots, x_i)$$

for $i = 1, \ldots, n - 1$. For $i = n$ choose

$$u = \dot{x}_{nd} - k_n s_n$$

Now prove using a succession of Lyapunov functions s_i^2 that given the Lipschitz constants of the $f_i, \Delta f_i$ there exist $k_1, \ldots, k_n, \tau_2, \ldots, \tau_n$ such that a bounded trajectory can be tracked asymptotically arbitrarily closely. The proof uses singular perturbation ideas and the reader may wish to consider the limit that the $\tau_i = 0$ to get the intuition involved in this scheme.

Problem 9.26 Poincaré like control system linearization. The aim of this problem is to give necessary conditions to convert a single input nonlinear system

$$\dot{x} = f(x) + g(x)u \tag{9.159}$$

into a single input controllable linear system

$$\dot{z} = Az + bu \tag{9.160}$$

using a change of coordinates $z = \phi(x)$ **alone** and **not** state feedback. Thus, define the push forward operator ϕ_* operating on $f(x), g(x)$ by

$$\phi_*(x) f(x) = d\phi(x) f(x)|_{x=\phi^{-1}(z)}$$

$$\phi_*(x) g(x) = d\phi(x) g(x)|_{x=\phi^{-1}(z)}$$

Now, we seek ϕ such that

$$\phi_* f = Az \quad \phi_* g = b$$

Verify that

$$\phi_*[f_1(x), f_2(x)] = [\phi_* f_1, \phi_* f_2]$$

In words this equation says that the push forward of the Lie brackets of f_1, f_2 is the Lie bracket of the push forward of f_1, f_2. Use this fact to give necessary conditions on $f, g, \mathrm{ad}_f g, \mathrm{ad}_f^2 g, \ldots$, to enable the conversion of the system of (9.159) into (9.160). Show that your conditions imply the conditions for full state linearization of (9.159). Can you say whether these conditions that you have derived are sufficient?

Problem 9.27 Adaptive Nonlinear Control. Consider the SISO nonlinear system

$$\dot{x} = f_0(x) + g_0(x)u + \sum_{i=1}^{p} \theta_i^* f_i(x) \tag{9.161}$$

$$y = h(x)$$

The functions $f_0, g_0, f_i(x)$ are considered known but the parameters $\theta^* \in \mathbb{R}^p$ are fixed and unknown. These are referred to as the *true* parameters. We will explore a control law for making y track a given **bounded** trajectory $y_d(t)$. Assume that $L_{g_0} h(x)$ is bounded away from zero and the control law is the following modification of the standard relative degree one asymptotic tracking control law

$$u = \frac{1}{L_{g_0}h}\left(-L_{f_0}h + \dot{y}_d + \alpha(y_d - y) - \sum_{i=1}^{p}\hat{\theta}_i(t)L_{f_i}h\right) \qquad (9.162)$$

Here $\hat{\theta}_i(t)$ is our estimate of θ_i^* at time t. Prove (using the simplest quadratic Lyapunov function of $y - y_d$, $\hat{\theta} - \theta^*$ that you can think of) that, under the parameter update law

$$\dot{\hat{\theta}} = -\begin{bmatrix} L_{f_1}h \\ \vdots \\ L_{f_p}h \end{bmatrix}(y - y_d) \qquad (9.163)$$

for all initial conditions $\hat{\theta}(0)$ and $x(0)$, $y(t) - y_d(t) \to 0$ as $t \to \infty$. Note that you are **not** required to prove anything about $\hat{\theta}(t)$!

Sketch the proof of the fact that if the true system of (9.161) is exponentially minimum phase, then $x(\cdot)$ is bounded as well. Can you modify the control law (9.162) and the parameter update law (9.163) for the case that the right hand side of the system (9.161) has terms of the form

$$\sum_{i=1}^{p}(f_i(x) + g_i(x)u)\theta_i^*$$

with f_i, g_i known and $\theta^* \in \mathbb{R}^p$ unknown. You can assume here that

$$L_{g_0}h + \sum_{i=1}^{p}\hat{\theta}_i(t)L_{g_i}h$$

is invertible.

Problem 9.28 Convergence of the Picard iteration for the Devasia Paden Chen method [85].   Prove that the Picard Lindelöf iteration applied to equation (9.122) converges by showing that the right hand side is a contraction map for k_2 small enough.

Problem 9.29 MIMO Devasia Chen Paden Scheme.   Verify that the Devasia Chen Paden scheme can be applied to nonminimum phase MIMO systems, provided that they are decouplable by (possibly dynamic) state feedback.

Problem 9.30 MIMO Byrnes Isidori Regulator.   Assume that the open loop system is a square MIMO system. Retrace the steps in the proof to give necessary and sufficient conditions for the existence of the Byrnes Isidori regulator.

Problem 9.31 Byrnes Isidori regulator for systems which have relative degree. First consider a SISO system with normal form given by (9.29). Note that in these coordinates $h(\xi, \eta) = \xi_1$. Use this to get a more explicit formula for $\pi(w)$, $u_{ff}(w)$ for the regulator. *Hint* $\xi_1 = r(w) := \pi_1(w)$, $\xi_2 = \frac{\partial r}{\partial w}s(w) := \pi_2(w)$, ...

For the MIMO case (assuming that the system is decouplable by static state feedback) repeat this calculation.

10

Design Examples Using Linearization

This chapter has been heavily based on the research work and joint papers with John Hauser, of the University of Colorado, George Meyer, of NASA Ames, Petar Kokotović of the University of California, Santa Barbara, and Claire Tomlin, of Stanford University.

10.1 Introduction

In this chapter we discuss the application of the theory of input–output linearization by state feedback to a few classes of applications. We have deliberately chosen two examples where the application of the theory is not completely straightforward. Several more straightforward applications of the theory are covered in the Exercises of this chapter and Chapter 9. Our motivation for choosing non-straightforward applications is twofold:

1. To show that whenever one applies theory there is always a need to make a number of practical "tweaks" to make the theory go.
2. To introduce some very important new ideas concerning the

   a. control of nonregular systems or nonlinear systems that fail to have relative degree,
   b. control of "slightly non-minimum phase systems" where the presence of high speed non-minimum phase dynamics makes the application of traditional input–output linearization infeasible.

The first example studied, the ball and beam system, is an example of a system which fails to have relative degree. We discuss how to approximately control it in Section 10.2. The generalization of this example to nonregular systems is given in Section 10.3. The case of flight control of a vertical take-off and landing aircraft, which is a canonical example of slightly non-minimum phase systems, is discussed in Section 10.4. The general theory for control of slightly nonminimum phase nonlinear systems is presented in Section 10.5. Section 10.6 gives some very interesting results about how the zero dynamics of nonlinear systems respond to perturbations, a subject which is the under-current of the preceding sections. In the Exercises we explore flight control of conventional aircraft and helicopters, automotive control, and other applications.

10.2   The Ball and Beam Example

We discuss an example of how to *approximately* linearize a system which fails to have relative degree. Before we launch into a discussion of the general theory let us begin with an example drawn from undergraduate control laboratories, the familiar ball and beam experiment depicted in Figure 10.1. In this setup, the beam is symmetric and is made to rotate in a vertical plane by applying a torque at the point of rotation (the center). Rather than have the ball roll on top of the beam as usual, we restrict the ball to frictionless sliding along the beam (as a bead along a wire). Note that this allows for complete rotations and arbitrary angular accelerations of the beam without the ball losing contact with the beam. We shall be interested in controlling the position of the ball along the beam. However, in contrast to the usual set-point problem, we would like the ball to track an arbitrary trajectory.

In this section we first derive the equations of motion for the ball and beam system. Then, we try to apply the techniques of *input–output linearization* and *full state linearization* to develop a control law for the system and demonstrate the shortcomings of these methods as they fail on this simple nonlinear system. At the outset we will closely follow the approach of Hauser, Sastry, and Kokotović [130]. Finally, we demonstrate a method of control law synthesis based on *approximate input-output linearization* and compare the performance of two control laws derived using differing approximations with that derived from the standard Jacobian approximation.

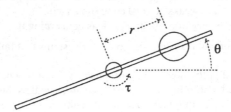

FIGURE 10.1. The ball and beam system

10.2.1 Dynamics

Consider the ball and beam system depicted in Figure 10.1. Let the moment of inertia of the beam be J, the mass of the ball be M, and the acceleration of gravity be G. Choose as generalized coordinates for this system the angle, θ, of the beam and the position, r, of the ball. Then the Lagrangian equations of motion are given by

$$
\begin{aligned}
0 &= \ddot{r} + G\sin\theta - r\dot{\theta}^2, \\
\tau &= (Mr^2 + J)\ddot{\theta} + 2Mr\dot{r}\dot{\theta} + MGr\cos\theta,
\end{aligned}
\tag{10.1}
$$

where τ is the torque applied to the beam and there is no force applied to the ball. Using the invertible transformation

$$
\tau = 2Mr\dot{r}\dot{\theta} + MGr\cos\theta + (Mr^2 + J)u
\tag{10.2}
$$

to define a new input u, the system can be written in state-space form as

$$
\begin{bmatrix} \dot{x}_1 \\ \dot{x}_2 \\ \dot{x}_3 \\ \dot{x}_4 \end{bmatrix} = \underbrace{\begin{bmatrix} x_2 \\ x_1 x_4^2 - G\sin x_3 \\ x_4 \\ 0 \end{bmatrix}}_{f(x)} + \underbrace{\begin{bmatrix} 0 \\ 0 \\ 0 \\ 1 \end{bmatrix}}_{g(x)} u,
$$

$$
y = \underbrace{x_1}_{h(x)},
\tag{10.3}
$$

where $x = (x_1, x_2, x_3, x_4)^T := (r, \dot{r}, \theta, \dot{\theta})^T$ is the state and $y = h(x) := r$ is the *output* of the system (i.e., the variable that we want to control). Note that (10.2) is a *nonlinear* input transformation.

10.2.2 Exact Input–Output Linearization

We are interested in making the system output, $y(t)$, track a specified trajectory, $y_d(t)$, i.e., $y(t) \to y_d(t)$ as $t \to \infty$. To this end, we might try to *exactly* linearize the input–output response of the system. Following the usual procedure, we differentiate the output until the input appears:

$$
\begin{aligned}
y &= x_1, \\
\dot{y} &= x_2, \\
\ddot{y} &= x_1 x_4^2 - G\sin x_3, \\
y^{(3)} &= \underbrace{x_2 x_4^2 - Gx_4 \cos x_3}_{b(x)} + \underbrace{2x_1 x_4}_{a(x)}\, u.
\end{aligned}
\tag{10.4}
$$

At this point if $a(x)$ were nonzero in the region of interest, we could use the control law

$$u = \frac{1}{a(x)} [-b(x) + v] \tag{10.5}$$

to yield a linear input–output system described by

$$y^{(3)} = v. \tag{10.6}$$

Unfortunately, for the ball and beam, the control coefficient $a(x)$ is zero whenever the angular velocity $x_4 = \dot{\theta}$ or ball position $x_1 = r$ are zero. Therefore, the *relative degree* of the ball and beam system *is not well defined*! This is due to the fact that neither is

$$L_g L_f^2 h(x) = 2x_1 x_4 \tag{10.7}$$

nonzero at $x = 0$ (an equilibrium point of the undriven system), nor is it identically zero on a neighborhood of $x = 0$. This is a characteristic unique to *nonlinear* systems. Thus, when the system has nonzero angular velocity and nonzero ball position, the input acts one integrator sooner than when the angular velocity is zero.

Perhaps we can learn something more about the system by exploring its *zero dynamics* i.e., the internal dynamics of the system consistent with the constraint that the output be held identically zero, provided that they are well-defined. For the ball and beam, it is interesting to look at the case when the output is held to a nonzero position, say $y \equiv K$ (strictly speaking this corresponds to modifying the output to $h(x) = x_1 - K$. To maintain this position, it is necessary that all higher order derivatives vanish:

$$
\begin{aligned}
y &= x_1 = K, \\
\dot{y} &= x_2 = 0, \\
\ddot{y} &= K x_4^2 - G \sin x_3 = 0,
\end{aligned}
\tag{10.8}
$$

so that the *zero dynamics manifold* is given by

$$M = \left\{ x \in \mathbb{R}^4 : x_1 = K, x_2 = 0, K x_4^2 - G \sin x_3 = 0 \right\}. \tag{10.9}$$

To find the dynamics that may evolve on this manifold, we note that

$$y^{(3)} = 2K x_4 \dot{x}_4 - G x_4 \cos x_3 = 0 \tag{10.10}$$

implies that

$$x_4 = 0 \quad \text{or} \quad \dot{x}_4 = \frac{G}{2K} \cos x_3. \tag{10.11}$$

Thus we have the possibility of three distinct behaviors for the zero dynamics (which are constrained to live on M):

$$(x_3, x_4) \equiv (0, 0), \qquad \text{(a)}$$

$$(x_3, x_4) \equiv (\pi, 0), \qquad \text{(b)} \qquad\qquad (10.12)$$

$$\dot{x}_4 = \frac{G}{2K} \cos x_3. \qquad \text{(c)}$$

Note that the third possibility does *not* contain any *equilibrium points*! The connection between this unusual zero dynamics behavior and the lack of relative degree bears some discussion. The two isolated points (10.12(a) and (b)) arise from the constraints (10.8), and $x_4 \equiv 0$ would correspond to a system with relative degree four so that the zero dynamics have dimension zero (i.e., consist of isolated point(s)). On the other hand, the one-dimensional dynamics of (10.12(c)) resemble those of a system with relative degree three. As it turns out, this last zero dynamics are possible only because the first term in the equation for $y^{(3)}$ (from (10.4)), namely

$$L_f^3 h(x) = \frac{-2M x_1^2 x_4}{M x_1^2 + J} (2 x_2 x_4 + G \cos x_3), \qquad (10.13)$$

vanishes on the same submanifold, $x_4 \equiv 0$, as $L_g L_f^2 h(x)$ does. If this were not true, then it would be possible for $y^{(3)}$ to take on a nonzero value at a point when the input had no effectiveness (i.e., when $x_4 = 0$). Thus, the ball and beam while failing to have relative degree is an interesting example of a system which lies "in between" a relative degree 3 and relative degree 4 system.

10.2.3   Full State Linearization

We will now try to find a set of coordinates and a feedback law such that the input-to-state behavior of the transformed system is linear, using the results of the preceding chapter. First we check the *controllability condition* of (9.41), which is the first necessary condition for controllability for:

$$\text{span} \left\{ g \; ad_f g \; \cdots \; ad_f^3 g \right\}, \qquad (10.14)$$

where $ad_f^i g$ denotes the iterated *Lie bracket* $[f, [f, \cdots [f, g] \cdots]]$. In turn this is given by

$$\left\{ \begin{array}{cccc} 0 & 0 & 2x_1 x_4 & 4x_2 x_4 + G \cos x_3 \\ 0 & -2x_1 x_4 & -2x_2 x_4 - G \cos x_3 & -4x_1 x_4^3 + 3G x_4 \cos x_3 \\ 0 & -1 & 0 & 0 \\ 1 & 0 & 0 & 0 \end{array} \right\}. \qquad (10.15)$$

This distribution has dimension 4 in a neighborhood of $x = 0$, establishing that that the ball and beam system satisfies the conditions (9.41). For the involutivity condition of (9.42), we see that

$$[g, ad_f^2 g] = (2x_1 \quad -2x_2 \quad 0 \quad 0)^T \tag{10.16}$$

does not lie within the span of the first three columns (vector fields) of (10.15). Failing this condition, we see that it is not possible to full-state linearize the ball and beam system.

10.2.4   Approximate Input–Output Linearization

In this section, we see that by appropriate choice of vector fields *close* to the system vector fields, we can design a feedback control law to achieve bounded error output tracking. The control law will, in fact, be the *exact* output tracking control law for an *approximate system* defined by these vector fields. Ideally, we would like to find a state feedback control law, $u(x) = \alpha(x) + \beta(x)v$, that would transform the ball and beam system into a linear system of the of the form $y^{(4)} = v$. Unfortunately, due to the presence of the centrifugal term $r\dot{\theta}^2 = x_1 x_4^2$, the input–output response of the ball and beam cannot be exactly linearized. Here we try to find an input–output linearizable system that is *close* to the true system. We present two such approximations for the ball and beam system. In each case, we will design a nonlinear change of (state) coordinates, $\xi = \phi(x)$ and a state dependent feedback $u(x, v) = \alpha(x) + \beta(x)v$ to make the system look like a chain of integrators perturbed by small higher order terms, $\psi(x, v)$, as depicted in Figure 10.2. We then build an approximate tracking control law by designing u so that

$$\begin{aligned} v = y_d^{(4)}(t) &+ \alpha_3(y_d^{(3)}(t) - \phi_4(x)) + \alpha_2(\ddot{y}_d(t) - \phi_3(x)) \\ &+ \alpha_1(\dot{y}_d(t) - \phi_2(x)) + \alpha_0(y_d(t) - \phi_1(x)), \end{aligned} \tag{10.17}$$

making the error system into an exponentially stable linear system perturbed by small nonlinear terms. We will compare the performance of these designs to a *linear* controller based on the standard Jacobian approximation to the system. For each approximation, we present simulation results depicting
(a) the output error, $y_d(t) - \phi_1(x(t))$,
(b) the neglected nonlinearity, $\psi(x, u)$,

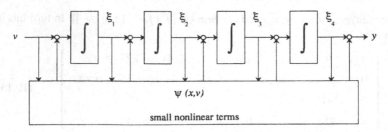

FIGURE 10.2. Approximate input–output linearization: a chain of intergrators perturbed by small nonlinear terms

(c) the angle of the beam, $\theta(t) = x_3(t)$, and

(d) the position of the ball, $r(t) = x_1(t)$, for a desired trajectory of $y_d(t) = R \cos \pi t / 30$, with $R = 5, 10$, and 15.

Approximation 1. Let $\xi_1 = \phi_1(x) = h(x)$. Then along the system trajectories, we have

$$
\begin{aligned}
\dot{\xi}_1 &= \underbrace{x_2}_{\xi_2 = \phi_2(x)}, \\
\dot{\xi}_2 &= \underbrace{-G \sin x_3}_{\xi_3 = \phi_3(x)} + \underbrace{x_1 x_4^2}_{\psi_2(x)}, \\
\dot{\xi}_3 &= \underbrace{-G x_4 \cos x_3}_{\xi_4 = \phi_4(x)}, \\
\dot{\xi}_4 &= \underbrace{G x_4^2 \sin x_3}_{b(x)} + \underbrace{(-G \cos x_3)}_{a(x)} u,
\end{aligned}
\qquad \text{or} \qquad
\begin{aligned}
\dot{\xi}_1 &= \xi_2, \\
\dot{\xi}_2 &= \xi_3 + \psi_2(x), \\
\dot{\xi}_3 &= \xi_4, \\
\dot{\xi}_4 &= b(x) + a(x)u =: v(x, u).
\end{aligned}
$$

$$(10.18)$$

In this case, the approximate system is obtained by a modification of the f vector field (neglecting $\psi_2(\cdot)$).

The simulation results in Figure 10.3 show that the closed loop system provides good tracking. Notice that the tracking error increases in a nonlinear fashion as the amplitude of the desired trajectory increases. This is expected since the approximation error term $\psi_2(x)$ is a nonlinear function of the state. A good a-priori estimate of the mismatch of the approximate system for a desired trajectory can be calculated using $\psi(\Phi^{-1}(y_d, \dot{y}_d, \ddot{y}_d, y_d^{(3)}))$, where $\Phi^{-1} : \xi \mapsto x$ is the inverse of the coordinate transformation. This may be a useful way to define a class of trajectories that the system can track with small error.

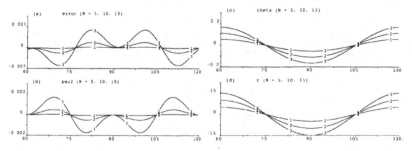

FIGURE 10.3. Simulation results for $y_d(t) = R \cos \pi t / 30$ using the first approximation ((a) $e = y_d - \phi_1$, (b) ψ_2, (c) θ, (d) r)

Approximation 2. Again, let $\xi_1 = \phi_1(x) = h(x)$. Then, along the system trajectories, we have

$$\dot{\xi}_1 = \underbrace{x_2}_{\xi_2 = \phi_2(x)},$$

$$\dot{\xi}_2 = \underbrace{-G \sin x_3 + x_1 x_4^2}_{\xi_3 = \phi_3(x)},$$

$$\dot{\xi}_3 = \underbrace{-G x_4 \cos x_3 + x_2 x_4^2}_{\xi_4 = \phi_4(x)} + \underbrace{2 x_2 x_4 u}_{\psi_3(x,u)},$$

$$\dot{\xi}_4 = \underbrace{x_1 x_4^4}_{b(x)} + \underbrace{(-G \cos x_3 + 2 x_2 x_4)}_{a(x)} u,$$

or

$$\dot{\xi}_1 = \xi_2,$$
$$\dot{\xi}_2 = \xi_3,$$
$$\dot{\xi}_3 = \xi_4 + \psi_3(x),$$
$$\dot{\xi}_4 = b(x) + a(x)u =: v(x, u).$$

$$(10.19)$$

This time the approximate system is obtained by modifying the g vector field. Pulling back the modified g vector field (obtained by neglecting $\psi_3(x, u)$) to the original x coordinates, we get

$$\underbrace{\begin{bmatrix} 0 \\ 0 \\ 0 \\ 1 \end{bmatrix}}_{g(x)} + \underbrace{\begin{bmatrix} 0 \\ 0 \\ \dfrac{2G x_1 x_4 \cos x_3 - 4 x_1 x_2 x_4^2}{G(G \cos^2 x_3 - 2 x_2 x_4 \cos x_3 - x_1 x_4^2 \sin x_3)} \\ \dfrac{2 x_1 x_4^2 \sin x_3}{G \cos^2 x_3 - 2 x_2 x_4 \cos x_3 - x_1 x_4^2 \sin x_3} \end{bmatrix}}_{\Delta g(x)}. \qquad (10.20)$$

The system with g modified in this manner is input–output linearizable and is an approximation to the original system since Δg is small for small angular velocity, $\dot{\theta} = x_4$.

The simulation results in Figure 10.4 show that the tracking error is substantially less than that obtained by the first design. Interestingly, this occurs despite the fact

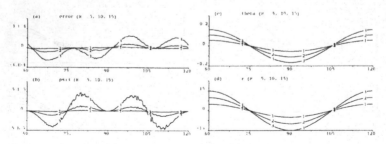

FIGURE 10.4. Simulation results for $y_d(t) = R \cos \pi t / 30$ using the second approximation ((a) $e = y_d - \phi_1$, (b) ψ_3, (c) θ, (d) r)

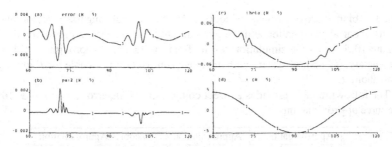

FIGURE 10.5. Simulation results for $y_d(t) = R \cos \pi t / 30$ using the Jacobian approximation ((a) $e = y_d - \phi_1$, (b) ψ_3, (c) θ, (d) r)

that $\psi_3(x, u)$ (approximation 2) is larger than $\psi_2(x)$ (approximation 1). This makes sense intuitively since ψ_3 enters one integrator further from the output than ψ_2 and therefore, roughly speaking, experiences more filtering.

Jacobian Approximation

To provide a basis for comparison, we will also calculate a linear control law based on the Jacobian approximation. Previously, we used the invertible nonlinear transformation of (10.2) to simplify the form of $\dot{x}_4$. Since we are only allowed *linear* functions in the control, we must work directly with the original input τ and the true angular acceleration $\ddot{\theta} = \dot{x}_4$ given by

$$\dot{x}_4 = \frac{-MGx_1 \cos x_3 - 2Mx_1x_2x_4}{Mx_1^2 + J} + \frac{1}{Mx_1^2 + J} \tau . \tag{10.21}$$

We will linearize about $x = 0$, $\tau = 0$.

Since the output is a *linear* function of the state, we begin with $\xi_1 = \phi_1(x) = h(x)$. Then, along the system trajectories, we have

$$\dot{\xi}_1 x = \underbrace{x_2}_{\xi_2 = \phi_2(x)} ,$$

$$\dot{\xi}_2 = \underbrace{-Gx_3}_{\xi_3 = \phi_3(x)} + \underbrace{x_1 x_4^2 + G(x_3 - \sin x_3)}_{\psi_2(x)},$$

$$\dot{\xi}_3 = \underbrace{-Gx_4}_{\xi_4 = \phi_4(x)} ,$$

$$\dot{\xi}_4 = \underbrace{\frac{MG^2}{J} x_1}_{b(x)} + \underbrace{\frac{-G}{J}}_{a(x)} \tau$$

$$+ \underbrace{\frac{MG^2x_1 \cos x_3 + 2MGx_1x_2x_4}{Mx_1^2 + J} - \frac{MG^2x_1}{J} + \left(\frac{G}{J} - \frac{G}{Mx_1^2 + J} \right) \tau}_{\psi_4(x, \tau)} .$$

$$\tag{10.22}$$

The Jacobian approximation is, of course, obtained by replacing the f vector field by its linear approximation and the g vector field by its constant approximation. Figure 10.5 shows the simulation results from the Jacobian approximation. Unfortunately, the control system with the linear controller is not stable for R greater than about 7.

The following table provides a direct comparison of the error $e = y_d - \phi_1$ for the three approximations:

R	Approximation 1	Approximation 2	Jacobian Approximation
5	$\pm 9.6 \cdot 10^{-5}$	$\pm 1.5 \cdot 10^{-5}$	$-4.7 \cdot 10^{-3} +3.0 \cdot 10^{-3}$
10	$\pm 7.5 \cdot 10^{-4}$	$\pm 6.5 \cdot 10^{-5}$	*unstable*
15	$\pm 2.5 \cdot 10^{-3}$	$\pm 1.9 \cdot 10^{-4}$	*unstable*

Note that approximation 2 provides better tracking for this class of inputs by about an order of magnitude over approximation 1. Due to the large excursions from the origin, the Jacobian approximation is no longer a good approximation so the system becomes unstable. Of course, the other approximations will eventually become unstable as R becomes large.

10.2.5 Switching Control of the Ball and Beam System

We now introduce a control scheme that switches between the approximate tracking law in a neighborhood of the singularity $x_1 x_4 = 0$, and an exact tracking law outside of this neighborhood. To do this, we partition the state space into the regions $M_0 = \{x \in \mathbb{R}^4 : |x_1 x_4| \leq \delta\}$, $M_- = \{x \in \mathbb{R}^4 : x_1 x_4 < -\delta\}$, $M_+ = \{x \in \mathbb{R}^4 : x_1 x_4 > \delta\}$ for small $\delta \in \mathbb{R}^+$. In M_0, we use the approximate control law u^{app} (10.27). In the regions M_+ and M_-, we define (as in the preceding subsection) the coordinate system $\Phi : x \mapsto (\xi, \eta)$, such that $\xi_1 = x_1 - K, \xi_2 = x_2, \xi_3 = Bx_1 x_4^2 - BG \sin x_3$, $\eta = x_4$, where the nonlinear coordinate transformation Φ is a local diffeomorphism away from $x_3 = \frac{\pi}{2}$. Using the calculations of (10.4), it follows that the system dynamics in these coordinates are

$$
\begin{aligned}
\dot{\xi}_1 &= \xi_2, \\
\dot{\xi}_2 &= \xi_3, \\
\dot{\xi}_3 &= Bx_2 x_4^2 - BGx_4 \cos x_3 + 2Bx_1 x_4 u, \\
\dot{\eta} &= u.
\end{aligned}
\tag{10.23}
$$

The *exact tracking law* with a singularity at $x_1 x_4 = 0$ is therefore given by

$$
u^{\text{ex}} = \frac{1}{2Bx_1 x_4}(-Bx_2 x_4^2 + BGx_4 \cos x_3 + v)
\tag{10.24}
$$

for an auxiliary input v, and the (one-dimensional) inverse dynamics, as discussed in the preceding subsection, are

$$\dot{x}_4 = \frac{1}{2Bx_1x_4}(-Bx_2x_4^2 + BGx_4 \cos x_3 + v). \tag{10.25}$$

Setting $v = 0$ gives the zero dynamics of the system (10.23) under the control law (10.24) which are of the type (c) of equation (10.12) of the preceding subsection, namely,

$$\dot{x}_4 = \frac{G \cos x_3}{2K}, \quad \text{where } Kx_4^2 = G \sin x_3. \tag{10.26}$$

These zero dynamics evolve on the one-dimensional manifold S^1, causing the beam to oscillate between $x_3 = 0$ and $x_3 = \pi$. Now consider the problem of tracking a desired output trajectory $y_D(t)$ corresponding to a periodic motion of the ball along the beam. State trajectories which implement $y_D(t)$ traverse the region in which $x_1x_4 = 0$, since x_4 must necessarily change sign. In M_0, the *approximate* tracking law

$$u^{\text{app}} = \frac{[-B(Bx_1x_4^2 - BG \sin x_3)x_4^2 - BGx_4^2 \sin x_3 + v]}{2Bx_2x_4 - BG \cos x_3} \tag{10.27}$$

may be used to stabilize the system to $y_D(t)$ with

$$v = y_D^{(4)} - \sum_{i=1}^{4} \alpha_i \left(\xi_i - y_D^{(i-1)}\right),$$

where the α_i are selected so that $s^4 + \alpha_4 s^3 + \alpha_3 s^2 + \alpha_2 s + \alpha_1$ is Hurwitz. We expect the system to stabilize nicely to this δ-region (around $x_3 = 0$) since the zero dynamics correspond to a single equilibrium. In the region $\{x : |x_1x_4| \geq \delta\}$, the *exact* tracking law of equation (10.24) may be used to stabilize the output of the system to $y_D(t)$:

$$v = y_D^{(3)} - \sum_{i=1}^{3} \alpha_i \left(\xi_i - y_D^{(i-1)}\right),$$

where $s^3 + \alpha_3 s^2 + \alpha_2 s + \alpha_1$ is Hurwitz. Because the zero dynamics in this region are constrained to lie on S^1, we cannot expect such a nice behavior of the inverse dynamics of the system in this region. Figure 10.6 shows the result of tracking $y_D(t) = 1.9 \sin(1.3t) + 3$ using control based on approximate linearization only, and Figure 10.7 shows the result of tracking the same trajectory using a control law that switches between approximate and exact tracking. The approximate tracking scheme is unstable, due to the neglect of the $2Bx_1x_4$ term away from the singularity where the magnitude of x_1 is large. The switched scheme is stable, yet the driven dynamics cause the beam to continually flip upside down and right-side up.

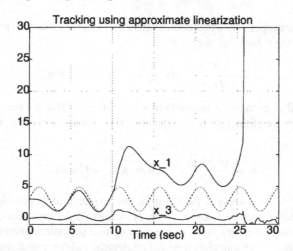

FIGURE 10.6. Tracking $y_D(t) = 1.9 \sin(1.3t) + 3$ (shown as a dashed curve) using a tracking law based on approximate linearization only. The parameters used in this simulation are $B = 0.7143$, $(\alpha_1, \alpha_2, \alpha_3, \alpha_4) = (0.6, 6, 4, 4)$.

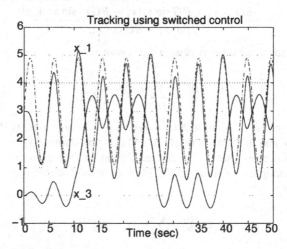

FIGURE 10.7. Tracking $y_D(t) = 1.9 \sin(1.3t) + 3$ (shown as a dashed curve) using the switched tracking law for the same parameters as in the previous figure, and $\delta = 3$. The system is stable, although the driven dynamics $x_3 = \theta$ flip between $\theta = 0$ and $\theta = \pi$, as expected.

10.3   Approximate Linearization for Nonregular SISO Systems

In this section, we will generalize the discussion of the ball and beam example to single-input single-output systems of the form

$$\dot{x} = f(x) + g(x)u,$$
$$y = h(x),$$
(10.28)

where $x \in \mathbb{R}^n$, $u, y \in \mathbb{R}$, f and g are smooth vector fields on $\mathbb{R}^n$, and h is a smooth function, which *fail to have relative degree*, referred to as nonregular. We will assume, as before, that $x = 0$ is an equilibrium point of the undriven system. Thus, $L_g L_f^{\gamma-1} h(x) = 0$ at $x = 0$ but is not identically zero in a neighborhood U of $x = 0$, i.e., $L_g L_f^{\gamma-1} h(x)$ is a function that is of order $O(x)$ rather than $O(1)$. Here and in the sequel, we will use the O notation. Recall that a function $\delta(x)$ is said to be $O(x)^n$ if

$$\lim_{|x| \to 0} \frac{|\delta(x)|}{|x|^n} \text{ exists and is } \neq 0.$$

Also, functions that are $O(x)^0$ are referred to as $O(1)$. By abuse of notation, we will also use the notation $O(x, u)^2$ to mean functions of x, u which are sums of terms of $O(x)^2$, $O(xu)$, and $O(u)^2$. Similarly for $O(x, u)^\rho$. Then the relative degree of the system is not well defined and the input–output linearizing control law cannot be derived in the usual fashion. Hence, we seek a set of functions of the state, $\phi_i(x)$, $i = 1, \ldots, \gamma$, that approximate the output and its derivatives in a special way. The integer γ will be determined during the approximation process. Since our control objective is tracking, the first function, $\phi_1(x)$, should approximate the output function, that is

$$h(x) = \phi_1(x) + \psi_0(x, u),$$
(10.29)

where $\psi_0(x, u)$ is $O(x, u)^2$ (actually, ψ_0 does not depend on u, but for consistency below we include it). Differentiating $\phi_1(x)$ along the system trajectories, we get

$$\dot{\phi}_1(x) = L_f h(x) + L_g h(x)u.$$
(10.30)

If $L_g h(x)$ is $O(x)$ or of higher order, we cannot effectively control the system at this level so we neglect it (and a small part of $L_f h(x)$ if we so desire) in our choice of $\phi_2(x)$:

$$L_{f+gu}\phi_1(x) = \phi_2(x) + \psi_1(x, u)$$
(10.31)

where $\psi(x, u)$ is $O(x, u)^2$. We continue this procedure with

$$L_{f+gu}\phi_i(x) = \phi_{i+1}(x) + \psi_i(x, u)$$
(10.32)

until at some step, say γ, the control term, $L_g \phi_\gamma(x)$, is $O(1)$, that is,

$$L_{f+gu}\phi_\gamma(x) = b(x) + a(x)u$$
(10.33)

where $a(x)$ is $O(1)$. Apparently, we have found an approximate system of relative degree γ. This motivates the following definition:

Definition 10.1 Robust Relative Degree. *We say that a nonlinear system (10.28) has a* robust relative degree *of γ about $x = 0$ if there exist smooth functions*

$\phi_i(x), i = 1, \ldots, \gamma$, *such that*

$$h(x) = \phi_1(x) + \psi_0(x, u),$$

$$L_{f+gu}\phi_i(x) = \phi_{i+1}(x) + \psi_i(x, u), \qquad i = 1, \ldots, \gamma - 1, \qquad (10.34)$$

$$L_{f+gu}\phi_\gamma(x) = b(x) + a(x)u + \psi_\gamma(x, u),$$

where the functions $\psi_i(x, u)$, $i = 0, \ldots, \gamma$, *are* $O(x, u)^2$ *and* $a(x)$ *is* $O(1)$.

Remarks:

- In equation (10.34) above, the ψ_i have the form

$$\psi_0(x, u) = \psi_0^1(x),$$

$$\psi_i(x, u) = \psi_i^1(x) + \psi_i^2(x)u, \qquad i = 1, \ldots, \gamma - 1, \qquad (10.35)$$

  where $\psi_i^1(x)$ is $O(x)^2$ and $\psi_i^2(x)$ is $O(x)$.
- There is considerable latitude in the definition of the $\phi_i(x)$, since each $\psi_i^1(x)$ may be chosen in a number of ways as long as it is $O(x)^2$.

The robust relative degree is easily characterized (see Problem 10.1 to be the relative degree of the Jacobian linearized version of the system (10.28) about $x = 0, u = 0$:

$$\dot{z} = Az + bu,$$

$$y = cz, \qquad (10.36)$$

with $A = Df(0)$, $b = g(0)$, and $c = dh(0)$.

An immediate corollary of this result is that the robust relative degree of a nonlinear system (10.28) is *invariant* under a state dependent change of control coordinates of the form

$$u(x, v) = \alpha(x) + \beta(x)v, \qquad (10.37)$$

where α and β are smooth functions and $\alpha(0) = 0$, while $\beta(0) \neq 0$. In order to show that this procedure produces an approximation of the true system, it is left to Problem 10.3 to prove that the functions $\phi_i(\cdot)$ can be used as part of a local nonlinear change of coordinates.

Proposition 10.2. *Suppose that the nonlinear system (10.28) has robust relative degree* γ. *Then the functions* $\phi_i(\cdot)$, $i = 1, \ldots, \gamma$, *are independent in a neighborhood of the origin.*

With γ independent functions $\phi_i(\cdot)$ in hand, we can, by the Frobenius theorem, complete the nonlinear change of coordinates with a set of functions, $\eta_i(x)$, $i = 1, \ldots, n - \gamma$, such that

$$L_g\eta_i(x) = 0, \quad x \in U. \qquad (10.38)$$

Choosing $n - \gamma$ of these independent of the $\phi_i(x)$ as before we can rewrite the true system (10.28) as

$$\dot{\xi}_1 = \xi_2 + \psi_1(x, u),$$

$$\vdots$$

$$\dot{\xi}_{\gamma-1} = \xi_\gamma + \psi_{\gamma-1}(x, u),$$

$$\dot{\xi}_\gamma = b(\xi, \eta) + a(\xi, \eta)u, \qquad (10.39)$$

$$\dot{\eta} = q(\xi, \eta),$$

$$y = \xi_1 + \psi_0(x, u),$$

where $q(\xi, \eta)$ is $L_f \eta$ expressed in (ξ, η) coordinates. Note that the form (10.39) is a generalization of the standard normal form of Chapter 9. with the extra terms $\psi_i(x, u), i = 0, \ldots, \gamma$ of $O(x, u)^2$. Thus the control law

$$u = \frac{1}{a(\xi, \eta)} [-b(\xi, \eta) + v] \qquad (10.40)$$

approximately linearizes the system (10.28) from the input v to the output y up to terms of $O(x, u)^2$. If the robust relative degree of the system (10.28) is $\gamma = n$, then the system (10.28) is almost completely linearizable from input to state as well (since there will be no η state variables). This situation was investigated by Krener [169], who showed that the system

$$\dot{x} = f(x) + g(x)u \qquad (10.41)$$

is full-state linearizable to terms of $O(x, u)^\rho$ if and only if the distribution

$$\text{span} \left\{ g \; ad_f g \; \cdots \; ad_f^{n-1} g \right\} \text{ has rank } n \qquad (10.42)$$

and the distribution

$$\text{span} \left\{ g \; ad_f g \; \cdots \; ad_f^{n-2} g \right\} \text{ is order } \rho \text{ involutive}, \qquad (10.43)$$

i.e., has a basis, up to terms of $O(x)^\rho$, that is involutive up to terms of $O(x)^\rho$. Equivalently, conditions (10.42) and (10.43) guarantee (through a version of the Frobenius theorem with remainder [169]) the existence of an *output* function $h(x)$ with respect to which the system (10.41) has *robust relative degree* n and further that the remainder functions $\psi_i(x, u)$ are $O(x, u)^\rho$. Our development differs somewhat from that in [169] in that we are given a specific output function $y = h(x)$ and a tracking objective for this output. However, there is a happy confluence of our results and those of Krener for the ball and beam example of the previous section, where it may be verified that the condition of (10.43) is satisfied for $\rho = 3$ and further more the desired output function $h(x)$ is, in fact, an order $\rho = 3$ integral manifold of the distribution of that equation. Consequently, the ball and beam can be input–output and full-state linearized up to terms of order 3.

As was remarked after Definition 10.1, there is a great deal of latitude in the choice of the functions $\psi_i^1(x), i = 0, \ldots, \gamma - 1$, so long as they are $O(x)^2$. To improve the quality of the approximation, one may insist on choosing these terms to be $O(x)^\rho$ for some $\rho \geq 2$. There is less latitude in the choice of the functions $\psi_i^2(x)$. They must be neglected if they are $O(x)$ or higher and not neglected if they are $O(1)$ (this determines γ). We cannot, in general, guarantee that an approximation of $O(x, u)^\rho$ for $\rho > 2$ can be found. At this level of generality, it is difficult to give analytically rigorous design guidelines for the choice of the functions $\psi_i^1(x)$. However, from the ball and beam example of Section 10.2, it would appear that it is advantageous to have the $\psi_i^1(x)$ be identically zero for as long (as large an i) as possible. *We conjecture that the larger the value of the first i at which either $\psi_i^1(x)$ or $\psi_i^2(x)$ are nonzero, the better the approximation.* It is also important to note the distinction between the *nonlinear* feedback control law (10.40) which approximately linearizes the system (10.39), and the *linear* feedback control law obtained from the Jacobian linearization of the original system (10.28) given by

$$u = \frac{1}{cA^{\gamma-1}b} \left[-cA^\gamma x + v \right]. \tag{10.44}$$

It may be verified (reader, please verify!) that they agree up to first order at $x = 0$ since $cA^{\gamma-1}b = a(0)$ and $cA^\gamma = dL_f\phi_\gamma(0) = dh(0)$. It is also useful to note that the control law (10.40) is the *exact* input-output linearizing control law for the *approximate system*

$$\dot{\xi}_1 = \xi_2,$$

$$\vdots$$

$$\dot{\xi}_{\gamma-1} = \xi_\gamma,$$

$$\dot{\xi}_\gamma = b(\xi, \eta) + a(\xi, \eta)u, \tag{10.45}$$

$$\dot{\eta} = q(\xi, \eta),$$

$$y = \xi_1.$$

In general, we can only guarantee the existence of control laws of the form (10.40) that approximately linearize the system up to terms of $O(x, u)^2$; the Jacobian law of (10.44) is such a law. In specific applications, we see that the control law (10.40) may produce better approximations (the ball and beam of Section 10.2 was linearized up to terms of $O(x, u)^3$). Furthermore, the resulting approximations may be valid on larger domains than the Jacobian linearization (also seen in the ball and beam example). We try to make this notion precise by studying the properties enjoyed by the approximately linearized system (10.28), (10.40) on a parameterized family of *operating envelopes*) defined as follows

Definition 10.3 Operating Envelopes.   *We call $U_\epsilon \subset \mathbb{R}^n$, $\epsilon > 0$, a family of operating envelopes provided that*

$$U_\delta \subset U_\epsilon \quad whenever \quad \delta < \epsilon \tag{10.46}$$

and

$$\sup\{\delta : B_\delta \subset U_\epsilon\} = \epsilon, \tag{10.47}$$

where B_δ is a ball of radius δ centered at the origin.

- It is not necessary that each U_ϵ be bounded (or compact), although this might be useful in some cases.
- Since the largest ball that fits in U_ϵ is B_ϵ, the set U_ϵ must get smaller in at least one direction as ϵ is decreased.

The functions $\psi_i(x, u)$ that are omitted in the approximation are of $O(x, u)^2$ in a neighborhood of the origin. However, if we are interested in extending the approximation to (larger) regions, say of the form of U_ϵ, we will need the following definition:

Definition 10.4. *A function $\psi : \mathbb{R}^n \times \mathbb{R} \mapsto \mathbb{R}$ is said to be of* uniformly higher order *on $U_\epsilon \times B_\sigma \subset \mathbb{R}^n \times \mathbb{R}$, $\epsilon > 0$, if for some $\sigma > 0$ there exists a monotone increasing function of ϵ, K_ϵ, such that*

$$|\psi(x, u)| \leq \epsilon K_\epsilon(|x| + |u|) \quad for \ \ x \in U_\epsilon, \ |u| \leq \sigma. \tag{10.48}$$

Remarks:

- If $\psi(x, u)$ is uniformly of higher order on $U_\epsilon \times B_\sigma$ then it is $O(x, u)^2$.
- This definition is a refinement of the condition that $\psi(x, u)$ be $O(x, u)^2$ inasmuch as it does not allow for terms of the form $O(u)^2$.

Now, if the approximate system (10.45) is exponentially minimum phase and the error term is uniformly of higher order on $U_\epsilon \times B_\sigma$, we may use the stable tracking control law for the approximate system given by

$$u = \frac{1}{a(\xi, \eta)} \left[-b(\xi, \eta) + y_d^{(\gamma)} + \alpha_{\gamma-1}(y_d^{(\gamma-1)} - \xi_\gamma) + \cdots + \alpha_0(y_d - \xi_1) \right],$$

$$\tag{10.49}$$

(with $s^\gamma + \alpha_{\gamma-1}s^{\gamma-1} + \cdots + \alpha_0$ a Hurwitz polynomial). We can now prove the following result:

Theorem 10.5. *Let U_ϵ, $\epsilon > 0$, be a family of operating envelopes and suppose that:*

- *The zero dynamics of the approximate system (10.45) (i.e., $\dot{\eta} = q(0, \eta)$) are exponentially stable and q is Lipschitz in ξ and η on $\Phi(U_\epsilon)$ for each ϵ.*
- *The functions $\psi_i(x, u)$ are uniformly higher order on $U_\epsilon \times B_\sigma$.*

Then for ϵ sufficiently small and desired trajectories with sufficiently small derivatives $(y_d, \dot{y}_d, \ldots, y^{(\gamma)})$, the states of the closed loop system (10.28), (10.49) will remain bounded, and the tracking error will be $O(\epsilon)$.

Proof: Define the trajectory error $e \in \mathbb{R}^\gamma$ to be

$$
\begin{bmatrix} e_1 \\ e_2 \\ \vdots \\ e_\gamma \end{bmatrix} = \begin{bmatrix} \xi_1 \\ \xi_2 \\ \vdots \\ \xi_\gamma \end{bmatrix} - \begin{bmatrix} y_d \\ \dot{y}_d \\ \vdots \\ y_d^{(\gamma-1)} \end{bmatrix}. \tag{10.50}
$$

Then the closed loop system (10.28), (10.49) (equivalently, (10.39), (10.49)) may be expressed as

$$
\begin{bmatrix} \dot{e}_1 \\ \vdots \\ \dot{e}_{\gamma-1} \\ \dot{e}_\gamma \end{bmatrix} = \begin{bmatrix} 0 & 1 & \cdots & 0 \\ \vdots & & \ddots & \vdots \\ 0 & & & 1 \\ -\alpha_0 & -\alpha_1 & \cdots & -\alpha_{\gamma-1} \end{bmatrix} \begin{bmatrix} e_1 \\ \vdots \\ e_{\gamma-1} \\ e_\gamma \end{bmatrix} + \begin{bmatrix} \psi_1(x, u(x, \bar{y}_d)) \\ \vdots \\ \psi_{\gamma-1}(x, u(x, \bar{y}_d)) \\ \psi_\gamma(x, u(x, \bar{y}_d)) \end{bmatrix},
$$

$$
\dot{\eta} = q(\xi, \eta), \tag{10.51}
$$

or, compactly,

$$
\dot{e} = Ae + \psi(x, u(x, \bar{y}_d)), \tag{10.52}
$$
$$
\dot{\eta} = q(\xi, \eta),
$$

where $\bar{y}_d := (y_d, \dot{y}_d, \ldots, y_d^{(\gamma)})$. Since the zero dynamics are exponentially stable, the converse Lyapunov theorem implies the existence of a Lyapunov function for the system

$$
\dot{\eta} = q(0, \eta), \tag{10.53}
$$

satisfying

$$
k_1 |\eta|^2 \leq v_2(\eta) \leq k_2 |\eta|^2,
$$
$$
\frac{\partial v_2}{\partial \eta} q(0, \eta) \leq -k_3 |\eta|^2,
$$
$$
\left| \frac{\partial v_2}{\partial \eta} \right| \leq k_4 |\eta|,
$$

for some positive constants k_1, k_2, k_3, and k_4. We first show that e and η are bounded. To this end, consider as Lyapunov function for the error system (10.52)

$$
v(e, \eta) = e^T P e + \mu v_2(\eta), \tag{10.54}
$$

where $P > 0$ is chosen so that

$$
A^T P + PA = -I \tag{10.55}
$$

(possible since A has eigenvalues in $\mathbb{C}^\circ_-$) and μ is a positive constant to be determined later. By assumption, y_d and its first γ derivatives are bounded:

$$|\xi| \leq |e| + b_d \text{ and } |y^{(\gamma)}| \leq b_d. \tag{10.56}$$

Further, the function, $q(\xi, \eta)$ is Lipschitz, i.e.,

$$|q(\xi^1, \eta^1) - q(\xi^2, \eta^2)| \leq l_q(|\xi^1 - \xi^2| + |\eta^1 - \eta^2|), \tag{10.57}$$

and the function, $\psi(x, u)$, is uniformly of higher order with respect to $U_\epsilon \times B_\sigma$ and $u(x, \bar{y}_d)$ locally Lipschitz in its arguments with $u(0, 0) = 0$:

$$|2P\psi(x, u(x, \bar{y}_d))| \leq \epsilon K_\epsilon l_u(|x| + b_d), \quad (x, u) \in U_\epsilon \times B_\sigma. \tag{10.58}$$

Since x is a local diffeomorphism of (ξ, η),

$$|x| \leq l_x(|\xi| + |\eta|). \tag{10.59}$$

Using these bounds and the properties of $v_2(\cdot)$, we have

$$\frac{\partial v_2}{\partial \eta} q(\xi, \eta) = \frac{\partial v_2}{\partial \eta} q(0, \eta) + \frac{\partial v_2}{\partial \eta}(q(\xi, \eta) - q(0, \eta)) \tag{10.60}$$
$$\leq -k_3|\eta|^2 + k_4 l_q|\eta|(|e| + b_d).$$

Taking the derivative of $v(\cdot, \cdot)$ along the trajectories of (10.52), we obtain, for $(x, u) \in U_\epsilon \times B_\sigma$,

$$\dot{v} = -|e|^2 + 2e^T P \psi(x, u(x, \bar{y}_d)) + \mu \frac{\partial v_2}{\partial \eta} q(\xi, \eta)$$

$$\leq -|e|^2 + \epsilon|e|K_\epsilon l_x(|e| + b_d + |\eta|) + \mu(-k_3|\eta|^2 + k_4 l_q|\eta|(|e| + b_d))$$

$$\leq -\left(\frac{|e|}{2} - \epsilon K_\epsilon l_x b_d\right)^2 + (\epsilon K_\epsilon l_x b_d)^2$$

$$- \left(\frac{|e|}{2} - (\epsilon K_\epsilon l_x + \mu k_4 l_q)|\eta|\right)^2 + (\epsilon K_\epsilon l_x + \mu k_4 l_q)^2|\eta|^2$$

$$- \mu k_3\left(\frac{|\eta|}{2} - \frac{k_4 l_q b_d}{k_3}\right)^2 + \mu \frac{(k_4 l_q b_d)^2}{k_3}$$

$$- \left(\frac{1}{2} - \epsilon K_\epsilon l_x\right)|e|^2 - \frac{3}{4}\mu k_3|\eta|^2$$

$$\leq -\left(\frac{1}{2} - \epsilon K_\epsilon l_x\right)|e|^2 - \left(\frac{3}{4}\mu k_3 - (\epsilon K_\epsilon l_x + \mu k_4 l_q)^2\right)|\eta|^2$$

$$+ (\epsilon K_\epsilon l_x b_d)^2 + \mu \frac{(k_4 l_q b_d)^2}{k_3}.$$

Define

$$\mu_0 = \frac{k_3}{4(K_\epsilon l_x + k_4 l_q)^2}. \tag{10.61}$$

Then for all $\mu \leq \mu_0$ and all $\epsilon \leq \min(\mu, \frac{1}{4K_\epsilon l_x})$, we have

$$\dot{v} \leq -\frac{|e|^2}{4} - \frac{\mu k_3 |\eta|^2}{2} + \frac{\mu(k_4 l_q b_d)^2}{k_3} + (\epsilon K_\epsilon l_x b_d)^2. \qquad (10.62)$$

Thus, $\dot{v} < 0$ whenever $|\eta|$ or $|e|$ is large which implies that $|\eta|$ and $|e|$ and, hence, $|\xi|$ and $|x|$, are bounded. The above analysis is valid for $(x, u) \in U_\epsilon \times B_\sigma$. Indeed, by choosing b_d sufficiently small and appropriate initial conditions, we can guarantee that the state remains in U_ϵ and the input is bounded by σ. Using this fact, we may abuse notation and write the function $\psi(x, u(x, \bar{y}_d))$ as $\epsilon \psi(t)$ and note that

$$\dot{e} = Ae + \epsilon \psi(t) \qquad (10.63)$$

is an exponentially stable linear system driven by an order ϵ input. Thus, we conclude that the tracking error will be $O(\epsilon)$.   □

10.3.1   Tracking for Nonregular Systems

Systems that fail to have relative degree are called "nonregular systems." In this section, we have discussed how we can regularize nonregular SISO nonlinear systems so that we may obtain approximate tracking results for these systems. One can also define notions of regularity and robust relative degree for MIMO nonlinear systems. We will not do this in greater detail here, but point the reader to two papers by Grizzle, Di Benedetto, and Lamnabhi-Lagarrique [87; 120] for a discussion of regularity for nonlinear systems. In particular, the latter paper shows that it is *necessary* for nonlinear systems to be regular if they are to show asymptotic tracking.

In some sense, nonregular systems are singular: They represent points of transition or bifurcation between different classes of regular systems. For example, the ball and beam system is on the boundary between relative degree 4 systems for small $x_1 x_4$ and relative degree 3 systems for large $x_1 x_4$. However, it may not be a good idea to switch between the relative degree 3 and 4 systems, as was shown in the ball and beam example. We refer the interested reader to Tomlin and Sastry [308] for a discussion of a bifurcation point of view to nonregular systems as well as switching control of nonregular systems; see also Problem 10.13.

10.4   Nonlinear Flight Control

The results of this section show the application of nonlinear control to flight control. It is interesting to see the subtleties that are revealed in the application of the basic theory. We follow the work of Hauser, Sastry, and Meyer [131]. In general, the complete dynamics of an aircraft, taking into account flexibility of the wings and fuselage, the (internal) dynamics of the engine and control surface actuators, and the multitude of changing variables, are quite complex and somewhat unmanageable for the purposes of control. A more useful approach is to consider the aircraft as a rigid body upon which a set of forces and moments act. Then, with $r \in \mathbb{R}^3$, $R \in SO(3)$, and $\omega \in \mathbb{R}^3$ being the aircraft position, orientation, and angular velocity, respectively, the equations of motion can be written as

$$m\ddot{r} = Rf_a + mg,$$
$$\dot{R} = \omega \times R, \tag{10.64}$$
$$J\dot{\omega}_a = \tau_a - \omega_a \times J\omega_a,$$

where f_a and τ_a are the force and moment acting on the aircraft expressed relative to the aircraft. The subscript "a" means that a quantity is expressed with respect to the aircraft reference frame. Depending on the aircraft and its mode of flight, the forces and moments can be generated by aerodynamics (lift, drag, and roll–pitch–yaw moments), by momentum exchange (gross thrust vectoring and reaction controls to generate moments), or a combination of the two. The flight envelope of the aircraft is the set of flight conditions for which the pilot and/or the control system can affect the forces and moments needed to remain in the envelope and achieve the desired task.

10.4.1 Force and Moment Generation

For the sake of presentation, we will focus our attention on a particular aircraft, the YAV-8B Harrier produced by McDonnell Douglas (now Boeing) Aircraft Company [204]. The Harrier is a single-seat trans-sonic light attack V/STOL (vertical/short takeoff and landing) aircraft powered by a single turbo-fan engine. Figure 10.8

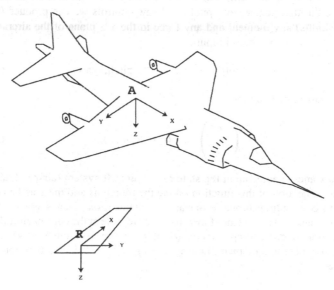

FIGURE 10.8. Aircraft coordinate systems

shows the aircraft with the coordinate frame A attached at the (nominal) center of mass. The x-axis is directed forward toward the nose of the aircraft and is also known as the *roll* axis since positive rotation about the x-axis coincides with rolling the aircraft to the right. The y-axis is directed toward the right wing and is called the *pitch* axis (positive rotation is a pitch up). The z-axis is directed downward and is also known as the *yaw* axis (we yaw to the right about this axis). Also shown in Figure 10.8 is the (inertial) *runway* coordinate frame R. The x-, y-, and z-axes of the runway frame are often oriented in the north, east, and down (N-E-D) directions, respectively. Four exhaust nozzles on the turbo-fan engine provide the gross thrust for the aircraft. These nozzles can be rotated from the aft position (used for conventional wing-borne flight) forward approximately 100 degrees allowing jet-borne flight and nozzle braking. The throttle and nozzle controls thus provide two degrees of freedom of thrust vectoring within the x-z plane of the aircraft. (If the line of action of the gross thrust does not pass through the object center of mass, then this thrust will also produce a net pitching moment.)

In addition to the conventional aerodynamic control surfaces (aileron, stabilator, and rudder for roll, pitch, and yaw moments, respectively), the Harrier also has a reaction control system (RCS) to provide moment generation during jet-borne and transition flight. Reaction valves in the nose, tail, and wingtips use bleed–air from the high-pressure compressor of the engine to produce thrust at these points and therefore moments (and forces) at the aircraft center of mass. The design of the aerodynamic and reaction controls provides complete (three degree of freedom) moment generation throughout the flight envelope of the aircraft. Since moments are often produced by applying a single force rather than a couple, a nonzero force (proportional to the moment) will usually be seen at the aircraft center of mass. Using the throttle, nozzle, roll, pitch, and yaw controls we can produce (within physical limits) any moment and any force in the x-z plane of the aircraft. The function $\mathcal{F}$, taking the control inputs

$$c = (\text{throttle, nozzle, roll, pitch, yaw})^T \qquad (10.65)$$

to the aircraft force and moment $(f_a^T, \tau_a^T)^T$,

$$\begin{pmatrix} f_a \\ \tau_a \end{pmatrix} = \mathcal{F}(r, \dot{r}, R, \omega, c), \qquad (10.66)$$

is complex and depends upon the state of the aircraft system (airspeed, altitude, etc.). The projection of this function taking the moments and the x and z components of force as outputs is one-to-one and hence invertible. That is, given a desired aircraft moment and (x-z plane) force that is achievable at the current aircraft state, there is a unique control input vector (throttle, nozzle, roll, pitch, yaw) that will produce that force and moment. Letting $u = (f_{ax}, f_{az}, \tau_a^T)^T$, this function can be written as

$$c = \mathcal{C}(r, \dot{r}, R, \omega, u), \qquad (10.67)$$

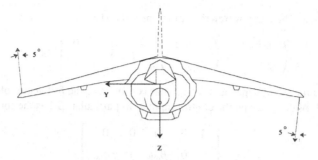

FIGURE 10.9. Reaction control system geometry

where r, $\dot{r}$, R, and ω compose the aircraft state. Given the function C, we are free to consider the desired moment and (x-z) force as the aircraft control input in place of the true control input. The idea of inverting the algebraic nonlinearities present in the system has been applied to real flight control problems [207]. It is now natural to ask what form the (state dependent) function taking u to f_{ay} will take. Since moments are produced by applying forces to the aircraft, one is hopeful that the resulting y-axis force at the aircraft center of mass will be a (state dependent) linear function of the moment acting on the aircraft. Note that this is not necessarily the case. For example, consider the generation of a right rolling moment (from the pilot's point of view) during jet-borne flight. Figure 10.9 shows the geometry of the wingtip reaction control valves. For small ranges of moment, the left reaction valve opens and blows downward, creating a net upward force. Once the left reaction valve is fully open, the right reaction valve opens and blows upward, which reduces the net upward force. In this case, there is a nonlinear coupling between the rolling moment and the force in the vertical (z-axis) direction on the aircraft. This case is easily reconciled, however, since we can directly affect the vertical (z-axis) force using the throttle and nozzle. Clearly, forces in the x-z plane and moments about the y-axis will not contribute to y-axis forces. Thus we consider the y-axis forces generated by rolling (x-axis) and yawing (z-axis) moments. Yawing moments are generated by applying a force at the tail of the aircraft (by aerodynamic or reaction control methods). As long as this force is effectively applied at the same point regardless of the magnitude of the moment, there will be a (state dependent) linear relationship between the z-axis moment and the resulting y-axis force. The coupling between rolling (x-axis) moments and y-axis forces is more subtle and is the result of the geometry of the reaction control system. As Figure 10.9 shows, the forces used to generate the rolling moment are not perpendicular to the y-axis of the aircraft. Thus, when a positive rolling moment is commanded, a negative force is generated in the y-axis direction (i.e., the airplane will initially accelerate to the left when it is commanded to go right). Also, as mentioned above, depending on the magnitude of the rolling moment, the right reaction valve could be actively blowing upward or be fully closed. Fortunately, the distance to and angle of the upward and downward reaction valve thrust vectors are equal. For this reason, the relationship between the rolling moment and y-axis force is linear. We can now rewrite equations (10.64) as

$$\begin{pmatrix} m\ddot{r} \\ J\dot{\omega}_a \end{pmatrix} = \begin{pmatrix} mg \\ -\omega_a \times J\omega_a \end{pmatrix} + \begin{bmatrix} R & 0 \\ 0 & I \end{bmatrix} Bu, \qquad (10.68)$$

where B is the (state dependent) 6×5 matrix providing the full vector of aircraft forces and moments given the control input u. In particular, B has the form

$$B = \begin{bmatrix} 1 & 0 & 0 & 0 & 0 \\ 0 & 0 & \beta_{\text{roll}} & 0 & \beta_{\text{yaw}} \\ 0 & 1 & 0 & 0 & 0 \\ 0 & 0 & 1 & 0 & 0 \\ 0 & 0 & 0 & 1 & 0 \\ 0 & 0 & 0 & 0 & 1 \end{bmatrix} \qquad (10.69)$$

where β_{roll} and β_{yaw} are the (scalar) functions giving the coupling between the roll and yaw moments and the y-axis force.

10.4.2   Simplification to a Planar Aircraft

It is particularly useful to consider a simple toy aircraft that has a minimum number of states and inputs but retains many of the features that must be considered in designing control laws for a real aircraft such as the Harrier. Figure 10.10 shows our prototype PVTOL (planar vertical takeoff and landing) aircraft. The aircraft state is simply the position x, y of the aircraft center of mass, the angle θ of the aircraft relative to the x-axis, and the corresponding velocities, $\dot{x}, \dot{y}, \dot{\theta}$. The control inputs u_1, u_2 are the thrust (directed out the bottom of the aircraft) and the rolling moment.

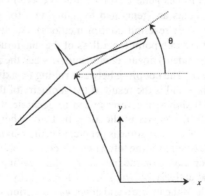

FIGURE 10.10. The planar vertical takeoff and landing (PVTOL) aircraft

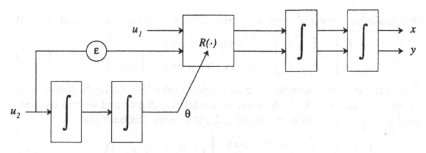

FIGURE 10.11. Block diagram of the PVTOL aircraft system

The equations of motion for our PVTOL aircraft are given by

$$\ddot{x} = -\sin\theta u_1 + \epsilon\cos\theta u_2,$$
$$\ddot{y} = \cos\theta u_1 + \epsilon\sin\theta u_2 - 1, \qquad (10.70)$$
$$\ddot{\theta} = u_2,$$

where '-1' is the gravitational acceleration and ϵ is the (small) coefficient giving the coupling between the rolling moment and the lateral acceleration of the aircraft. Note that $\epsilon > 0$ means that applying a (positive) moment to roll left produces an acceleration to the right (positive x). Figure 10.11 provides a block diagram representation of this dynamical system. The PVTOL aircraft system is the natural restriction of V/STOL aircraft to jet-borne operation (e.g., hover) in a vertical plane. Our study of this simple planar model is but the first step in an ongoing project to understand and develop robust methods for the control of highly maneuverable aircraft systems.

10.4.3 Exact Input–Output Linearization of the PVTOL Aircraft System

Consider the PVTOL aircraft system given by (10.70). Choosing x and y as the outputs to be controlled, we seek a (possibly dynamic) state feedback law of the form

$$u = a(z) + b(z)v \qquad (10.71)$$

such that for some $\gamma = (\gamma_1, \gamma_2)^T$,

$$x^{(\gamma_1)} = v_1,$$
$$y^{(\gamma_2)} = v_2. \qquad (10.72)$$

Here v is our new input, and z is used to denote the entire state of the system (including compensator states, if necessary). Proceeding in the usual way, we differentiate each output until at least one of the inputs appears. This occurs after differentiating twice and is given by (rewriting the first two equations of (10.70))

$$\begin{bmatrix} \ddot{x} \\ \ddot{y} \end{bmatrix} = \begin{bmatrix} 0 \\ -1 \end{bmatrix} + \begin{bmatrix} -\sin\theta & \epsilon\cos\theta \\ \cos\theta & \epsilon\sin\theta \end{bmatrix} \begin{bmatrix} u_1 \\ u_2 \end{bmatrix}. \qquad (10.73)$$

Since the matrix operating on u (the *decoupling* matrix) is nonsingular (barely—its determinant is $-\epsilon$!), the system has vector relative degree and we can linearize (and decouple) the system by choosing the static state feedback law

$$\begin{bmatrix} u_1 \\ u_2 \end{bmatrix} = \begin{bmatrix} -\sin\theta & \cos\theta \\ \dfrac{\cos\theta}{\epsilon} & \dfrac{\sin\theta}{\epsilon} \end{bmatrix} \left(\begin{bmatrix} 0 \\ 1 \end{bmatrix} + \begin{bmatrix} v_1 \\ v_2 \end{bmatrix} \right). \qquad (10.74)$$

The resulting system is

$$\ddot{x} = v_1,$$
$$\ddot{y} = v_2 \qquad (10.75)$$
$$\ddot{\theta} = \frac{1}{\epsilon}(\sin\theta + \cos\theta v_1 + \sin\theta v_2).$$

This feedback law makes our input–output map linear, but has the unfortunate side-effect of making the dynamics of θ unobservable. In order to guarantee the internal stability of the system, it is not sufficient to look at input–output stability, we must also show that all internal (unobservable) modes of the system are stable as well.

The first step in analyzing the internal stability of the system (10.75) is to look at the *zero dynamics* of the system. Constraining the outputs and derivatives to zero by setting $v_1 = v_2 = 0$ (and using appropriate initial conditions), we find the zero dynamics of (10.75) to be

$$\ddot{\theta} = \frac{1}{\epsilon}\sin\theta. \qquad (10.76)$$

Equation (10.76) is simply the equation of an undamped pendulum. Figure 10.12 shows the phase portrait ($\dot{\theta}$ vs. θ) of the pendulum (10.76) with $\epsilon = 1$. The phase portrait for $\epsilon < 0$ is simply a π-translate of Figure 10.12. Thus, for $\epsilon > 0$, the equilibrium point $(\theta, \dot{\theta}) = (0, 0)$ is unstable and the equilibrium point $(\pi, 0)$ is

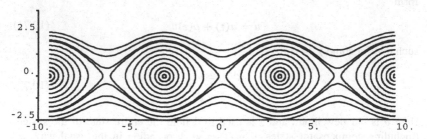

FIGURE 10.12. Phase portrait of an undamped pendulum ($\epsilon = 1$)

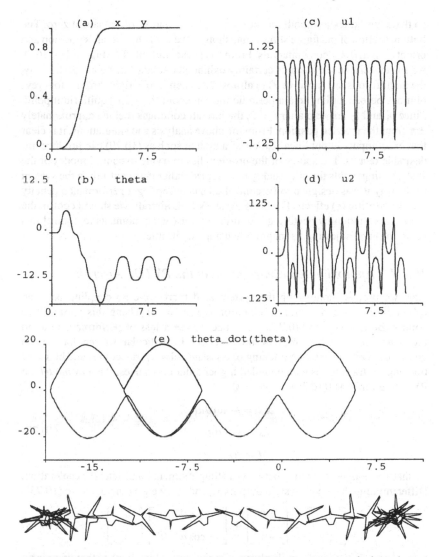

FIGURE 10.13. Response of nonminimum phase PVTOL system to smoothed *step* input

stable, but not asymptotically stable, and is surrounded by a family of closed orbits
with periods ranging from $2\pi\sqrt{\epsilon}$ up to ∞ (there is a heteroclinic orbit connecting
two saddles). Outside of these orbits is a family of unbounded trajectories. Thus,
depending on the initial condition, the aircraft will either rock from side to side
forever or roll continuously in one direction (except at the isolated equilibria).

Thus, this nonlinear system is *nonminimum phase*. Figure 10.13 shows the response of the system (10.75) when (v_1, v_2) is chosen (by a stable feedback law)

so that x will track a smooth trajectory from 0 to 1 and y will remain at zero. The bottom section of the figure shows snapshots of the PVTOL aircraft's position and orientation at 0.2 second intervals. From the phase portrait of θ (Figure 10.13(e)), we see that the zero dynamics certainly exhibit pendulum-like behavior. Initially, the aircraft rolls left (positive θ) to almost 2π. Then it rolls right through four revolutions before settling into a periodic motion about the -3π equilibrium point. Since v_1 and v_2 are zero after $t = 5$, the aircraft continues rocking approximately $\pm\pi$ from the inverted position. From the above analysis and simulations, it is clear that exact input–output linearization of a system such as (10.70) can produce undesirable results. The source of the problem lies in trying to control modes of the system using inputs that are weakly (ϵ) coupled rather than controlling the system in the way it was designed to be controlled and accepting a performance penalty for the parasitic (ϵ) effects. For our simple PVTOL aircraft, we should control the linear acceleration by vectoring the thrust vector (using moments to control this vectoring) and adjusting its magnitude using the throttle.

10.4.4   Approximate Linearization of the PVTOL Aircraft

Here we propose controlling the system as if there were no coupling between rolling moments and lateral acceleration (i.e., $\epsilon = 0$). Using this approach to control the true system (10.70), we expect to see a loss of performance due to the *unmodeled* dynamics present in the system. In particular, we see that we can guarantee stable asymptotic tracking of constant velocity trajectories and bounded tracking for trajectories with bounded higher order derivatives. We now *model* the PVTOL aircraft as ((10.70) with $\epsilon = 0$)

$$\ddot{x}_m = -\sin\theta u_1,$$
$$\ddot{y}_m = \cos\theta u_1 - 1, \tag{10.77}$$
$$\ddot{\theta} = u_2,$$

so that there is no coupling between rolling moments and lateral acceleration. Differentiating the model system outputs x_m and y_m, we get (analogous to (10.73))

$$\begin{bmatrix} \ddot{x}_m \\ \ddot{y}_m \end{bmatrix} = \begin{bmatrix} 0 \\ -1 \end{bmatrix} + \begin{bmatrix} -\sin\theta & 0 \\ \cos\theta & 0 \end{bmatrix} \begin{bmatrix} u_1 \\ u_2 \end{bmatrix}. \tag{10.78}$$

Now, however, the matrix multiplying u is singular, which implies that there is no *static* state feedback that will linearize (10.77). Since u_2 comes into the system (10.77) through $\ddot{\theta}$, we must differentiate (10.78) at least two more times. Let u_1 and $\dot{u}_1$ be states (in effect, placing two integrators before the u_1 input) and differentiate (10.78) twice, giving

$$\begin{bmatrix} x_m^{(4)} \\ y_m^{(4)} \end{bmatrix} = \begin{bmatrix} (\sin\theta)\dot{\theta}^2 u_1 - 2(\cos\theta)\dot{\theta}\dot{u}_1 \\ (-\cos\theta)\dot{\theta}^2 u_1 - 2(\sin\theta)\dot{\theta}\dot{u}_1 \end{bmatrix} + \begin{bmatrix} -\sin\theta & -\cos\theta u_1 \\ \cos\theta & -\sin\theta u_1 \end{bmatrix} \begin{bmatrix} \ddot{u}_1 \\ u_2 \end{bmatrix}. \tag{10.79}$$

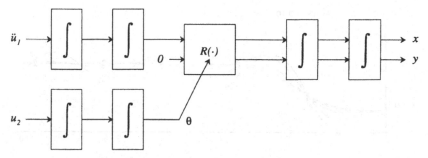

FIGURE 10.14. Block diagram of the augmented model PVTOL aircraft system

The matrix operating on our new inputs $(\ddot{u}_1, u_2)^T$ has determinant equal to u_1 and therefore is invertible as long as the thrust u_1 is nonzero. Figure 10.14 shows a block diagram of the model system with u_1 and $\dot{u}_1$ considered as states. Note that each input must go through four integrators to get to the output. Thus, we linearize (10.77) using the *dynamic* state feedback law

$$
\begin{bmatrix} \ddot{u}_1 \\ u_2 \end{bmatrix} = \begin{bmatrix} -\sin\theta & \cos\theta \\ \dfrac{\cos\theta}{u_1} & -\dfrac{\sin\theta}{u_1} \end{bmatrix} \left(\begin{bmatrix} -(\sin\theta)\dot{\theta}^2 u_1 + 2(\cos\theta)\dot{\theta}\dot{u}_1 \\ (\cos\theta)\dot{\theta}^2 u_1 + 2(\sin\theta)\dot{\theta}\dot{u}_1 \end{bmatrix} + \begin{bmatrix} v_1 \\ v_2 \end{bmatrix} \right)
$$

$$
= \begin{bmatrix} \dot{\theta}^2 u_1 \\ -\dfrac{2\dot{\theta}\dot{u}_1}{u_1} \end{bmatrix} + \begin{bmatrix} -\sin\theta & \cos\theta \\ \dfrac{\cos\theta}{u_1} & -\dfrac{\sin\theta}{u_1} \end{bmatrix} \begin{bmatrix} v_1 \\ v_2 \end{bmatrix}, \tag{10.80}
$$

resulting in

$$
\begin{bmatrix} x_m^{(4)} \\ y_m^{(4)} \end{bmatrix} = \begin{bmatrix} v_1 \\ v_2 \end{bmatrix}. \tag{10.81}
$$

Unlike the previous case (equation (10.75)), the linearized model system does not contain any unobservable (zero) dynamics. Thus, using a stable tracking law for v, we can track an arbitrary trajectory and guarantee that the (model) system will be stable.

Of course, the natural question that comes to mind is: Will a control law based on the model system (10.77) work well when applied to the true system (10.70)? In the next section we will show (in a more general setting) that, if ϵ is *small* enough, then the system will have reasonable properties (such as stability and bounded tracking).

How small is small enough? Figure 10.15 shows the response of the true system with epsilon ranging from 0 to 0.9 (0.01 is typical during jet-borne flight, i.e., hover, for the Harrier). As in Section 10.4.3, the desired trajectory is a smooth lateral motion from $x = 0$ to $x = 1$ with the altitude (y) held constant at 0. The figure also shows snapshots of the PVTOL aircraft's position orientation at 0.2 second intervals for $\epsilon = 0.0, 0.1$, and 0.3. Since the snapshots were taken

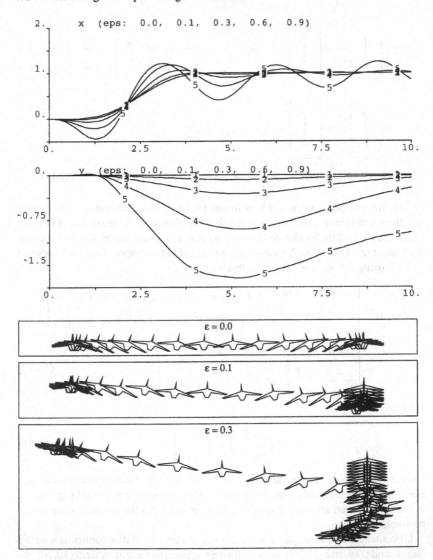

FIGURE 10.15. Response of the true PVTOL aircraft system under the approximate control

at uniform intervals, the spacing between successive pictures gives a clue to the aircraft's velocity and acceleration. The movie of the trajectories provides an even better sense of the system response. Interestingly, the x response is quite similar to the step response of a non-minimum phase linear system. Note that for ϵ less than approximately 0.6, the oscillations are reasonably damped. Although performance is certainly worse at higher values of ϵ, stability does not appear to be lost until

ϵ is in the neighborhood of 0.9. A value of 0.9 for ϵ means that the aircraft will experience almost $1g$ (the acceleration of gravity) in the wrong direction when a rolling acceleration of one radian per second per second is applied. For the range of ϵ values that will normally be expected, the performance penalty due to approximation is small, almost imperceptible.

Note that while the PVTOL aircraft system (10.70) with the approximate control (10.80) is stable for a large range of ϵ, this control allows the PVTOL aircraft to have a bounded but unacceptable altitude (y) deviation. Since the ground is hard and quite unforgiving and vertical takeoff and landing aircraft are designed to be maneuvered close to the ground, it is extremely desirable to find a control law that provides exact tracking of altitude if possible. Now, ϵ enters the system dynamics (10.70) in only one (state-dependent) direction. We therefore expect that one should be able to modify the system (by manipulating the inputs) so that the effects of the ϵ-coupling between rolling moments and aircraft lateral acceleration do not appear in the y output of the system.

Consider the decoupling matrix of the true PVTOL system (10.70) given in (10.73) as

$$
\begin{bmatrix} -\sin\theta & \epsilon\cos\theta \\ \cos\theta & \epsilon\sin\theta \end{bmatrix}. \tag{10.82}
$$

To make the y output independent of ϵ requires that the last row of this decoupling matrix be independent of ϵ. The only legal way to do this is by multiplication on the right (i.e., column operations) by a nonsingular matrix V which corresponds to multiplying the inputs by V^{-1}. In this case, we see that

$$
\begin{bmatrix} -\sin\theta & \epsilon\cos\theta \\ \cos\theta & \epsilon\sin\theta \end{bmatrix} \begin{bmatrix} 1 & -\epsilon\tan\theta \\ 0 & 1 \end{bmatrix} = \begin{bmatrix} -\sin\theta & \dfrac{\epsilon}{\cos\theta} \\ \cos\theta & 0 \end{bmatrix} \tag{10.83}
$$

is the desired transformation. Defining new inputs $\tilde{u}$ as

$$
\begin{bmatrix} \tilde{u}_1 \\ \tilde{u}_2 \end{bmatrix} = \begin{bmatrix} 1 & \epsilon\tan\theta \\ 0 & 1 \end{bmatrix} \begin{bmatrix} u_1 \\ u_2 \end{bmatrix}, \tag{10.84}
$$

we see that (10.73) becomes

$$
\begin{bmatrix} \ddot{x} \\ \ddot{y} \end{bmatrix} = \begin{bmatrix} 0 \\ -1 \end{bmatrix} + \begin{bmatrix} -\sin\theta & \dfrac{\epsilon}{\cos\theta} \\ \cos\theta & 0 \end{bmatrix} \begin{bmatrix} \tilde{u}_1 \\ \tilde{u}_2 \end{bmatrix}. \tag{10.85}
$$

Following the previous analysis, we set $\epsilon = 0$ and linearize the resulting approximate system using the dynamic feedback law

$$
\begin{bmatrix} \ddot{\tilde{u}}_1 \\ \tilde{u}_2 \end{bmatrix} = \begin{bmatrix} \dot{\theta}^2\tilde{u}_1 \\ -\dfrac{2\dot{\theta}\dot{\tilde{u}}_1}{\tilde{u}_1} \end{bmatrix} + \begin{bmatrix} -\sin\theta & \cos\theta \\ -\dfrac{\cos\theta}{\tilde{u}_1} & -\dfrac{\sin\theta}{\tilde{u}_1} \end{bmatrix} \begin{bmatrix} v_1 \\ v_2 \end{bmatrix}. \tag{10.86}
$$

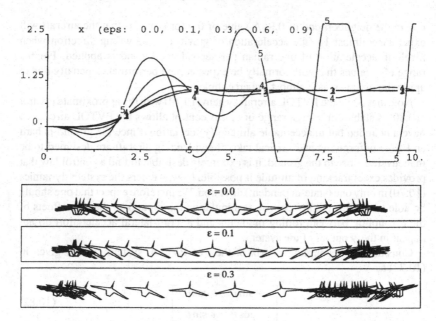

FIGURE 10.16. Response of the true PVTOL aircraft system under the approximate control with input transformation

The true system inputs are then calculated as

$$\begin{bmatrix} u_1 \\ u_2 \end{bmatrix} = \begin{bmatrix} 1 & -\epsilon \tan\theta \\ 0 & 1 \end{bmatrix} \begin{bmatrix} \tilde{u}_1 \\ \tilde{u}_2 \end{bmatrix}. \tag{10.87}$$

Figure 10.16 shows the response of the true system using the control law specified by equations (10.86) and (10.87) for the same desired trajectory. With this control law, our PVTOL aircraft maintains the altitude as desired and provides stable, bounded lateral (x) tracking for ϵ up to at least 0.6. Note, however, that the system is decidedly unstable for $\epsilon = 0.9$. Since we have forced the error into one direction (i.e., the x-channel), we expect the approximation to be more sensitive to the value of ϵ. In particular, compare the second column of the decoupling matrices of (10.73) and (10.85), i.e.,

$$\begin{bmatrix} \epsilon \cos\theta \\ \epsilon \sin\theta \end{bmatrix} \quad \text{and} \quad \begin{bmatrix} \dfrac{\epsilon}{\cos\theta} \\ 0 \end{bmatrix}. \tag{10.88}$$

Notice that the first is simply ϵ times a bounded function of θ, while the second contains ϵ times an unbounded function of θ (i.e., $1/\cos\theta$). Thus, for (10.85) with $\epsilon = 0$ to be a good approximation to (10.85) with nonzero ϵ requires that θ be bounded away from $\pm\pi/2$. This is not a completely unreasonable requirement, since most V/STOL aircraft do not have a large enough thrust to weight ratio to maintain level flight with a large roll angle. Since the physical limits of the aircraft usually place constraints on the achievable trajectories, a control law analogous to that defined by (10.86) and (10.87) can be used for systems with small ϵ on reasonable trajectories.

10.5 Control of Slightly Nonminimum Phase Systems

In this section we will take a more formal approach to the control of systems that are slightly nonminimum phase. Consider the class of square MIMO nonlinear systems of the form

$$\dot{x} = f(x) + g(x)u,$$
$$y = h(x),$$

$$\text{(10.89)}$$

where $x \in \mathbb{R}^n$, $u, y \in \mathbb{R}^m$, and $f : \mathbb{R}^n \mapsto \mathbb{R}^n$ and $g : \mathbb{R}^n \mapsto \mathbb{R}^{n \times m}$ are smooth vector fields and $h : \mathbb{R}^n \mapsto \mathbb{R}^m$ is a smooth function with $h(0) = 0$. We will also assume that the origin is an equilibrium point of (10.89), i.e., $f(0) = 0$, and will consider x in an open neighborhood U of the origin, i.e., the analysis will be *local*. All statements that we make, such as the existence of certain diffeomorphisms, will be assumed to merely hold in U. Also, when we say that a function is zero, it vanishes on U, and when we say that it is nonzero, we mean that it is bounded away from zero on U. While we will not precisely define *slightly nonminimum phase* systems, the concept is easy enough to explain.

10.5.1 Single Input Single Output (SISO) Case

Consider first the single-input single-output (SISO) case. Suppose that $L_g h(x) = \epsilon \psi(x)$ for some scalar function $\psi(x)$ with $\epsilon > 0$ small. In other words, the *relative degree* of the system is 1, but is very close to being greater than 1. As always, $L_g h(x)$ is the Lie derivative of $h(\cdot)$ along $g(\cdot)$ and is defined to be

$$L_g h(x) = \frac{\partial h(x)}{\partial x} g(x).$$

$$\text{(10.90)}$$

Now define two systems in normal form using the following two sets of local diffeomorphisms of $x \in \mathbb{R}^n$

$$(\xi^T, \eta^T)^T = (\xi_1 := h(x), \eta_1(x), \ldots, \eta_{n-1}(x))^T,$$

$$\text{(10.91)}$$

and

$$(\tilde{\xi}^T, \tilde{\eta}^T)^T = (\tilde{\xi}_1 := h(x), \tilde{\xi}_2(x) := L_f h(x), \tilde{\eta}_1(x), \ldots, \tilde{\eta}_{n-2}(x))^T,$$

$$\text{(10.92)}$$

with

$$\frac{\partial \eta_i}{\partial x} g(x) = 0, \quad i = 1, \ldots, n - 1,$$

$$\text{(10.93)}$$

and

$$\frac{\partial \bar{\eta}_i}{\partial x} g(x) = 0, \qquad i = 1, \ldots, n - 2. \tag{10.94}$$

System 1 (true system):

$$\dot{\xi}_1 = L_f h(x) + L_g h(x) u,$$
$$\dot{\eta} = q(\xi, \eta). \tag{10.95}$$

System 2 (approximate system):

$$\dot{\tilde{\xi}}_1 = \tilde{\xi}_2,$$
$$\dot{\tilde{\xi}}_2 = L_f^2 h(x) + L_g L_f h(x) u, \tag{10.96}$$
$$\dot{\tilde{\eta}} = \tilde{q}(\tilde{\xi}, \tilde{\eta}).$$

Note that the system (10.95) represents system (10.89) in normal form and the dynamics of $q(0, \eta)$ represent the zero dynamics of the system (10.89). System (10.96) does not represent the system (10.89), since in the $(\tilde{\xi}, \tilde{\eta})$ coordinates of (10.92), the dynamics of (10.89) are given by

$$\dot{\tilde{\xi}}_1 = \tilde{\xi}_2 + L_g h(x) u,$$
$$\dot{\tilde{\xi}}_2 = L_f^2 h(x) + L_g L_f h(x) u, \tag{10.97}$$
$$\dot{\tilde{\eta}} = \tilde{q}(\tilde{\xi}, \tilde{\eta}).$$

System (10.89) is said to be *slightly* non-minimum phase *if the true system (10.95) is nonminimum phase but the approximate system (10.96) is minimum phase.* Since $L_g h(x) = \epsilon \psi(x)$, we may think of the system (10.96) as a perturbation of the system (10.95) (or equivalently (10.97)). Of course, there are two difficulties with exact input–output linearization of (10.95):

- The input–output linearization requires a large control effort, since the linearizing control is

$$u^*(x) = \frac{1}{L_g h(x)}(-L_f h(x) + v) = \frac{-L_f h(x) + v}{\epsilon \psi(x)}. \tag{10.98}$$

  This could present difficulties in the instance that there is saturation at the control inputs.
- If (10.95) is non-minimum phase, a tracking control law producing a linear input-output response may result in unbounded η states.

Our prescription for the *approximate* input–output linearization of the system (10.95) is to use the input-output linearizing control law for the approximate system (10.96); namely

$$u_a^* = \frac{1}{L_g L_f h(x)}(-L_f^2 h(x) + v), \tag{10.99}$$

where v is chosen depending on the control task. For instance, if y is required to track y_d, we choose v as

$$v = \ddot{y}_d + \alpha_1(\dot{y}_d - \tilde{\xi}_2) + \alpha_2(y_d - \tilde{\xi}_1)$$
$$= \ddot{y}_d + \alpha_1(\dot{y}_d - L_f h(x)) + \alpha_2(y_d - h(x)). \tag{10.100}$$

Using (10.99) and (10.100) in (10.97) along with the definitions

$$e_1 = \tilde{\xi}_1 - y_d,$$
$$e_2 = \tilde{\xi}_2 - \dot{y}_d. \tag{10.101}$$

yields

$$\dot{e}_1 = e_2 + \epsilon \psi(x) u_a^*(x)$$
$$\dot{e}_2 = -\alpha_1 e_2 - \alpha_2 e_1 \tag{10.102}$$
$$\dot{\tilde{\eta}} = \tilde{q}(\tilde{\xi}, \tilde{\eta}).$$

The preceding discussion may be generalized to the case where the difference in the relative degrees between the true system and the approximate system is greater than 1. For example, if

$$L_g h(x) = \epsilon \psi_1(x),$$
$$L_g L_f h(x) = \epsilon \psi_2(x),$$
$$\vdots \tag{10.103}$$
$$L_g L_f^{\gamma-2} h(x) = \epsilon \psi_{\gamma-1}(x),$$

but $L_g L_f^{\gamma} h(x)$ is not of order ϵ, we define

$$(\tilde{\xi}^T, \tilde{\eta}^T) = (h(x), L_f h(x), \dots, L_f^{\gamma-1} h(x), \tilde{\eta}^T)^T \in \mathbb{R}^n \tag{10.104}$$

and note that the true system is

$$\dot{\tilde{\xi}}_1 = \tilde{\xi}_2 + \epsilon \psi_1(x) u,$$
$$\vdots$$
$$\dot{\tilde{\xi}}_{\gamma-1} = \tilde{\xi}_\gamma + \epsilon \psi_{\gamma-1}(x) u, \tag{10.105}$$
$$\dot{\tilde{\xi}}_\gamma = L_f^{\gamma} h(x) + L_g L_f^{\gamma-1} h(x) u,$$
$$\dot{\tilde{\eta}} = \tilde{q}(\tilde{\xi}, \tilde{\eta}).$$

The approximate (minimum phase) system (with $\epsilon = 0$) is given by

$$\dot{\tilde{\xi}}_1 = \tilde{\xi}_2,$$

$$\vdots$$

$$\dot{\tilde{\xi}}_{\gamma-1} = \tilde{\xi}_\gamma, \tag{10.106}$$

$$\dot{\tilde{\xi}}_\gamma = L_f^\gamma h(x) + L_g L_f^{\gamma-1} h(x) u,$$

$$\dot{\tilde{\eta}} = \tilde{q}(\tilde{\xi}, \tilde{\eta}).$$

The approximate tracking control law for (10.106) is

$$u_a = \frac{1}{L_g L_f^{\gamma-1} h(x)} \left(-L_f^\gamma h(x) + y_d^{(\gamma)} + \alpha_1 (y_d^{(\gamma-1)} - L_f^{\gamma-1} h(x)) \right.$$

$$\left. + \cdots + \alpha_\gamma (y_d - y) \right). \tag{10.107}$$

The following theorem provides a bound for the performance of this control when applied to the true system.

Theorem 10.6. *Assume that*

- *the desired trajectory and its first $\gamma - 1$ derivatives (i.e., y_d, $\dot{y}_d$, $\ldots$, $y^{(\gamma-1)}$) are* bounded,
- *the zero dynamics of the approximate system (10.106) are locally exponentially stable and $\tilde{q}$ is locally Lipschitz in $\tilde{\xi}$ and $\tilde{\eta}$, and*
- *the functions $\psi(x) u_a(x)$ are locally Lipschitz continuous.*

Then for ϵ sufficiently small, the states of the system (10.105) are bounded, and the tracking error e_1 satisfies

$$|e_1| := |\tilde{\xi}_1 - y_d| \le k\epsilon \tag{10.108}$$

for some $k < \infty$.

When the control objective is stabilization and the approximate system has no zero dynamics, we can do much better. In this case, one can show that the control law that stabilizes the approximate system also stabilizes the original system (see Problem 10.6).

10.5.2   Generalization to MIMO Systems

We now consider square MIMO systems of the form (10.89), which, for the sake of convenience, we rewrite as

$$\dot{x} = f(x) + g_1(x) u_1 + \cdots + g_m(x) u_m,$$

$$y_1 = h_1(x),$$

$$\vdots \tag{10.109}$$

$$y_m = h_m(x).$$

Assume that the system has vector relative degree $\gamma_1, \ldots, \gamma_m$, with decoupling matrix defined to be $A(x) \in \mathbb{R}^{m \times m}$ with

$$
A(x) = \begin{bmatrix} L_{g_1} L_f^{\gamma_1 - 1} h_1 & \cdots & L_{g_m} L_f^{\gamma_1 - 1} h_1 \\ \vdots & \ddots & \vdots \\ L_{g_1} L_f^{\gamma_m - 1} h_m & \cdots & L_{g_m} L_f^{\gamma_m - 1} h_m \end{bmatrix}, \tag{10.110}
$$

so that

$$
\begin{bmatrix} y_1^{(\gamma_1)} \\ \vdots \\ y_m^{(\gamma_m)} \end{bmatrix} = \begin{bmatrix} L_f^{\gamma_1} h_1 \\ \vdots \\ L_f^{\gamma_m} h_m \end{bmatrix} + A(x) \begin{bmatrix} u_1 \\ \vdots \\ u_m \end{bmatrix}. \tag{10.111}
$$

Since the decoupling matrix $A(x)$ is non-singular, the control law

$$
u(x) = A(x)^{-1} \left(- \begin{bmatrix} L_f^{\gamma_1} h_1 \\ \vdots \\ L_f^{\gamma_m} h_m \end{bmatrix} + v \right) \tag{10.112}
$$

with $v \in \mathbb{R}^m$ linearizes (and decouples) system (10.109) resulting in

$$
\begin{bmatrix} y_1^{(\gamma_1)} \\ \vdots \\ y_m^{(\gamma_m)} \end{bmatrix} = \begin{bmatrix} v_1 \\ \vdots \\ v_m \end{bmatrix}. \tag{10.113}
$$

To take up the ideas of Section 10.5.1, we first consider the case where $A(x)$ is non-singular but is *close to being singular*, that is, its smallest singular value is uniformly small for $x \in U$. Since $A(x)$ is close to being singular, i.e., it is close in norm to a matrix of rank $m - 1$, we may transform $A(x)$ using elementary column operations to get

$$
\bar{A}^0(x) = A(x) V^0(x) = \begin{bmatrix} \bar{a}_{.1}^0(x) & \cdots & \bar{a}_{.m-1}^0(x) & \epsilon \bar{a}_{.m}^0(x) \end{bmatrix}, \tag{10.114}
$$

where each $\bar{a}_{.i}^0$ is a column of $\bar{A}^0$. This corresponds to redefining the inputs to be

$$
\begin{bmatrix} \bar{u}_1^0 \\ \vdots \\ \bar{u}_m^0 \end{bmatrix} = (V^0(x))^{-1} \begin{bmatrix} u_1 \\ \vdots \\ u_m \end{bmatrix}. \tag{10.115}
$$

Now, the *normal* form of the system (10.109) is given by defining the following local diffeomorphism of $x \in \mathbb{R}^n$,

$$(\xi^T, \eta^T) = \xi_1^1 = h_1(x), \dots, \xi_{\gamma_1}^1 = L_f^{\gamma_1 - 1} h_1(x), \tag{10.116}$$

$$\xi_1^2 = h_2(x), \dots, \xi_{\gamma_2}^2 = L_f^{\gamma_2 - 1} h_2(x),$$

$$\vdots$$

$$\xi_1^m = h_m(x), \dots, \xi_{\gamma_m}^m = L_f^{\gamma_m - 1} h_m(x),$$

$$\eta^T)$$

and noting that

$$\dot{\xi}_1^1 = \xi_2^1,$$

$$\vdots$$

$$\dot{\xi}_{\gamma_1}^1 = b_1(\xi, \eta) + \sum_{j=1}^{m-1} \bar{a}_{1j}^0 \bar{u}_j^0 + \epsilon \bar{a}_{1m}^0 \bar{u}_m^0,$$

$$\dot{\xi}_1^2 = \xi_2^2, \tag{10.117}$$

$$\vdots$$

$$\dot{\xi}_{\gamma_m}^m = b_m(\xi, \eta) + \sum_{j=1}^{m-1} \bar{a}_{mj}^0 \bar{u}_j^0 + \epsilon \bar{a}_{mm}^0 \bar{u}_m^0,$$

$$\dot{\eta} = q(\xi, \eta) + P(\xi, \eta)\bar{u}^0.$$

where $b_i(\xi, \eta)$ is $L_f^{\gamma_i} h_i(x)$ for $i = 1, \dots, m$ in (ξ, η) coordinates. The zero dynamics of the system are the dynamics of the η coordinates in the subspace $\xi = 0$ with the linearizing control law of (10.112) (with $v = 0$) substituted, i.e.,

$$\dot{\eta} = q(0, \eta) - P(0, \eta)(\bar{A}^0(0, \eta))^{-1} b(0, \eta). \tag{10.118}$$

We will assume that (10.109) is nonminimum phase, that is to say that the origin of (10.118) is not stable.

Now, an approximation to the system is obtained by setting $\epsilon = 0$ in (10.117). The resultant decoupling matrix is singular, and the procedure for linearization by (dynamic) state feedback (the so-called *dynamic extension* process) proceeds by differentiating (10.117) and noting that

$$\dot{x} = f(x) + \bar{g}_1^0(x)\bar{u}_1^0 + \dots + \bar{g}_m^0(x)\bar{u}_m^0, \tag{10.119}$$

where

$$\begin{bmatrix} \bar{g}_1^0(x) & \cdots & \bar{g}_m^0(x) \end{bmatrix} = \begin{bmatrix} g_1(x) & \cdots & g_m(x) \end{bmatrix} V^0(x). \tag{10.120}$$

We then get

$$
\begin{bmatrix} y_1^{(\gamma_1+1)} \\ \vdots \\ y_{m-1}^{(\gamma_{m-1}+1)} \\ y_m^{(\gamma_m+1)} \end{bmatrix} = b^1(x, \bar{u}_1^0, \ldots, \bar{u}_{m-1}^0) + A^1(x, \bar{u}_1^0, \ldots, \bar{u}_{m-1}^0) \begin{bmatrix} \overset{\cdot 0}{\bar{u}}_1 \\ \vdots \\ \overset{\cdot 0}{\bar{u}}_{m-1} \\ \bar{u}_m^0 \end{bmatrix}
$$

$$
= b^1(x^1) + A^1(x^1)u^1, \tag{10.121}
$$

where

$$
u^1 = (\overset{\cdot 0}{\bar{u}}_1, \ldots, \overset{\cdot 0}{\bar{u}}_{m-1}, \bar{u}_m^0)^T \tag{10.122}
$$

is the *new* input and

$$
x^1 = (x^T, \bar{u}_1^0, \ldots, \bar{u}_{m-1}^0)^T \tag{10.123}
$$

is the *extended* state. Note the appearance of terms of the form $\overset{\cdot 0}{\bar{u}}_1, \ldots, \overset{\cdot 0}{\bar{u}}_{m-1}$ in (10.121). System (10.121) is linearizable (and decouplable) if $A^1(x^1)$ is nonsingular. We will assume that the singular values of A^1 are all of order 1 (i.e., A^1 is uniformly nonsingular) so that (10.121) is linearizable. The normal form for the approximate system is determined by obtaining a local diffeomorphism of the states $x, \bar{u}_1^0, \ldots, \bar{u}_{m-1}^0$ ($\in \mathbb{R}^{n+m-1}$) given by

$$
(\tilde{\xi}_1^1 = h_1(x), \ldots, \tilde{\xi}_{\gamma_1}^1 = L_f^{\gamma_1-1}h_1(x), \tilde{\xi}_{\gamma_1+1}^1 = L_f^{\gamma_1}h_1(x) + \sum_{j=1}^{m-1}\bar{a}_{1j}^0\bar{u}_j^0,
$$

$$
\tilde{\xi}_1^2 = h_2(x), \ldots, \tilde{\xi}_{\gamma_2}^2 = L_f^{\gamma_2-1}h_2(x), \tilde{\xi}_{\gamma_2+1}^2 = L_f^{\gamma_2}h_2(x) + \sum_{j=1}^{m-1}\bar{a}_{2j}^0\bar{u}_j^0,
$$

$$
\vdots
$$

$$
\tilde{\xi}_1^m = h_m(x), \ldots, \tilde{\xi}_{\gamma_m}^m = L_f^{\gamma_m-1}h_m(x), \tilde{\xi}_{\gamma_m+1}^m = L_f^{\gamma_m}h_m(x) + \sum_{j=1}^{m-1}\bar{a}_{mj}^0\bar{u}_j^0,
$$

$$
\eta^T). \tag{10.124}
$$

Note that $\tilde{\xi} \in \mathbb{R}^{\gamma_1+\cdots+\gamma_m+m}$ and $\tilde{\eta} \in \mathbb{R}^{n-\gamma_1-\cdots-\gamma_m-1}$, as compared with $\xi \in \mathbb{R}^{\gamma_1+\cdots+\gamma_m}$ and $\eta \in \mathbb{R}^{n-\gamma_1-\cdots-\gamma_m}$. With these coordinates, the *true* system (10.109) is given by

$$\dot{\bar{\xi}}_1^1 = \bar{\xi}_2^1,$$

$$\vdots$$

$$\dot{\bar{\xi}}_{\gamma_1}^1 = \bar{\xi}_{\gamma_1-1}^1 + \epsilon \bar{a}_{1m}^0 \bar{u}_m^1,$$

$$\dot{\bar{\xi}}_{\gamma_1+1}^1 = b_1^1(\bar{\xi}, \bar{\eta}) + a_{1.}^1(\bar{\xi}, \bar{\eta})u^1,$$

$$\dot{\bar{\xi}}_1^2 = \bar{\xi}_2^2, \qquad\qquad\qquad (10.125)$$

$$\vdots$$

$$\dot{\bar{\xi}}_{\gamma_m}^m = \bar{\xi}_{\gamma_m-1}^m + \epsilon \bar{a}_{mm}^0 \bar{u}_m^1,$$

$$\dot{\bar{\xi}}_{\gamma_m+1}^m = b_m^1(\bar{\xi}, \bar{\eta}) + a_{m.}^1(\bar{\xi}, \bar{\eta})u^1,$$

$$\dot{\bar{\eta}} = \bar{q}(\bar{\xi}, \bar{\eta}) + \bar{P}(\bar{\xi}, \bar{\eta})u^1.$$

In (10.125) above, $b_i^1(\bar{\xi}, \bar{\eta})$ and $a_{i.}^1(\bar{\xi}, \bar{\eta})$ are the ith element and row of b^1 and A^1, respectively, in (10.121) above (in the $\bar{\xi}$, $\bar{\eta}$ coordinates). The approximate system used for the design of the linearizing control is obtained from (10.125) by setting $\epsilon = 0$. The zero dynamics for the approximate system are obtained in the $\bar{\xi} = 0$ subspace by linearizing the approximate system using

$$u_*^1(\bar{\xi}, \bar{\eta}) = -(A^1(\bar{\xi}, \bar{\eta}))^{-1} \begin{bmatrix} b_1^1(\bar{\xi}, \bar{\eta}) \\ \vdots \\ b_m^1(\bar{\xi}, \bar{\eta}) \end{bmatrix}, \qquad (10.126)$$

to be

$$\dot{\bar{\eta}} = \bar{q}(0, \bar{\eta}) + \bar{P}(0, \bar{\eta})u_*^1(0, \bar{\eta}). \qquad (10.127)$$

Note that the dimension of $\bar{\eta}$ is *one less* than the dimension of η in (10.118). It would appear that we are actually determining the zero dynamics of the approximation to system (10.109) with dynamic extension—that is to say with integrators appended to the first $m - 1$ inputs $\bar{u}_1^0, \bar{u}_2^0, \ldots, \bar{u}_{m-1}^0$. While this is undoubtedly true, it is easy to see that the zero dynamics of systems are unchanged by dynamic extension (see also [54]). Thus, the zero dynamics of (10.127) are those of the approximation to system (10.109). The system (10.109) is said to be *slightly non-minimum phase* if the equilibrium $\eta = 0$ of (10.118) is not asymptotically stable, but the equilibrium $\bar{\eta} = 0$ of (10.127) is.

It is also easy to see that the preceding discussion may be iterated if it turns out that $A^1(\bar{\xi}, \bar{\eta})$ has some small singular values. At each stage of the *dynamic extension* process $m - 1$ *integrators* are added to the dynamics of the system, and the act of approximation reduces the dimension of the zero dynamics by 1. Also, if at any stage of this dynamic extension process there are two, three, ... singular values of order ϵ, the dynamic extension involves $m - 2$, $m - 3$, ... integrators.

If the objective is tracking, the approximate tracking control law is

$$
u_a^1(\bar{\xi}, \bar{\eta}) = (A^1(\bar{\xi}, \bar{\eta}))^{-1} \left(- \begin{bmatrix} b_1^1(\bar{\xi}, \bar{\eta}) \\ \vdots \\ b_m^1(\bar{\xi}, \bar{\eta}) \end{bmatrix} \right.
$$
$$
\left. + \begin{bmatrix} y_{d1}^{(\gamma_1 + 1)} + \alpha_1^1(y_{d1}^{(\gamma_1)} - \bar{\xi}_{\gamma_1 + 1}^1) + \cdots + \alpha_{\gamma_1 + 1}(y_{d1} - \bar{\xi}_1^1) \\ \vdots \\ y_{dm}^{(\gamma_m + 1)} + \alpha_1^m(y_{dm}^{(\gamma_m)} - \bar{\xi}_{\gamma_m + 1}^m) + \cdots + \alpha_{\gamma_m + 1}(y_{dm} - \bar{\xi}_1^m) \end{bmatrix} \right)
$$

$$(10.128)$$

with the polynomials

$$
s^{\gamma_i + 1} + \alpha_1^i s^{\gamma_i} + \cdots + \alpha_{\gamma_i + 1}^i, \quad i = 1, \ldots, m, \tag{10.129}
$$

chosen to be Hurwitz. The following theorem is the analog of Theorem 10.6 in terms of providing a bound for the system performance when the control law (10.128) is applied to the true system (10.109).

Theorem 10.7. *Assume that*

- *the desired trajectory y_d and the first $\gamma_i + 1$ derivatives of its ith component are* bounded,
- *the zero dynamics (10.127) of the approximate system are locally exponentially stable and $\bar{q} + \bar{P}u_*^1$ is locally Lipschitz in $\bar{\xi}$ and $\bar{\eta}$, and*
- *the functions $\bar{a}_{im}^0 u_m^1$ are locally Lipschitz continuous for $i = 1, \ldots, m$.*

Then for ϵ sufficiently small, the states of the system (10.125) are bounded and the tracking errors satisfy

$$
|e_1| = |\bar{\xi}_1^1 - y_{d1}| \leq k\epsilon,
$$
$$
|e_2| = |\bar{\xi}_1^2 - y_{d2}| \leq k\epsilon,
$$
$$
\vdots \tag{10.130}
$$
$$
|e_m| = |\bar{\xi}_1^m - y_{dm}| \leq k\epsilon,
$$

for some $k < \infty$.

As in the SISO case, stronger conclusions can be stated when the control objective is stabilization and the approximate system has no zero dynamics.

10.6   Singularly Perturbed Zero Dynamics for Regularly Perturbed Nonlinear Systems

In this section we will generalize the PVTOL and flight control example to discuss the perturbation of the zero dynamics of nonlinear systems caused by parametric variation in the parameters of the nonlinear system. Thus, we will be concerned with a family of systems depending on a parameter ϵ, described by equations of the form

$$\dot{x} = f(x, \epsilon) + g(x, \epsilon)u,$$
$$y = h(x, \epsilon), \tag{10.131}$$

where $f(x, \epsilon)$ and the columns of $g(x, \epsilon)$ are smooth vector fields and $h(x, \epsilon)$ is a smooth function defined in a neighborhood of $(x_0, 0)$ in $\mathbb{R}^n \times \mathbb{R}_+$. We will refer to the system of (10.131) with $\epsilon = 0$ as the *nominal* system and with $\epsilon \neq 0$ as the *perturbed* system. We will assume that $x = x_0$ is an equilibrium point for the nominal system, that is, $f(x_0, 0) = 0$, and without loss of generality we will assume that $h(x_0, 0) = 0$. Our treatment in this section follows Tomlin and Sastry [307]. See also [18].

10.6.1   SISO Singularly Perturbed Zero and Driven Dynamics

In the two papers [261; 152], it was shown that if the system (10.131) has relative degree $r(\epsilon) = r$ for $\epsilon \neq 0$, and relative degree $r(\epsilon) = r + d$ for $\epsilon = 0$, then there are *fast time scale zero dynamics* for the perturbed nonlinear system. This is, in itself, a rather surprising conclusion even for linear systems. Indeed if $f(x, \epsilon) = A(\epsilon)x$, $g(x, \epsilon) = b(\epsilon)$, $h(x, \epsilon) = c(\epsilon)x$ represent a controllable and observable family of linear systems, it may be verified that the poles of the transfer function

$$c(\epsilon)(sI - A(\epsilon))^{-1}b(\epsilon)$$

vary continuously (as a set) with ϵ, but the number of zeros and their locations vary in much more subtle fashion, since the relative degree, that is, the smallest integer $\gamma(\epsilon)$ such that

$$c(\epsilon)A^{\gamma(\epsilon)-1}b(\epsilon) \neq 0$$

may change at some values of ϵ (see Problem 10.9). We now discuss this for the nonlinear case. We will highlight the case where $\gamma(\epsilon) = r + d$ for $\epsilon \neq 0$ and $\gamma(0) = r$. As a consequence of the definition of relative degree we have that $\gamma(\epsilon) = r$ and $\gamma(0) = r + d$ implies that $\forall \epsilon \neq 0$,

$$L_g h(x, \epsilon) = L_g L_f h(x, \epsilon) = \cdots = L_g L_f^{r-2} h(x, \epsilon) = 0 \quad \forall x \text{ near } x_0 \quad (10.132)$$

and $L_g L_f^{r-1} h(x_0, \epsilon) \neq 0$ for $\epsilon = 0$,

$$L_g h(x, 0) = L_g L_f h(x, 0) = \cdots = L_g L_f^{r+d-2} h(x, 0) = 0 \quad \forall x \text{ near } x_0$$
$$L_g L_f^{r+d-1} h(x_0, 0) \neq 0. \tag{10.133}$$

To keep the singularly perturbed zero dynamics from demonstrating *multiple time scale* behavior,[1] we assume that for $0 \leq k \leq d$

$$L_g L_f^{r-1+k} h(x, \epsilon) = \epsilon^{d-k} \alpha_k(x, \epsilon) \qquad (10.134)$$

where each $\alpha_k(x, \epsilon)$ is a smooth function of (x, ϵ) in a neighborhood of $(x_0, 0)$. The choice of $L_g L_f^{r-1} h(x, \epsilon) = O(\epsilon^d)$ rather than $O(\epsilon)$ is made to avoid fractional powers of ϵ. What is critical about the assumption (10.134) is the dependence on decreasing powers of ϵ as k increases from 0 to d.

As before, we will denote by $\xi \in \mathbb{R}^{r+d}$ the vector corresponding to the output and first $r + d - 1$ derivatives of the system in (10.131), given by

$$\xi = \begin{pmatrix} h(x, \epsilon) \\ L_f h(x, \epsilon) \\ \vdots \\ L_f^{r-1} h(x, \epsilon) \\ \vdots \\ L_f^{r+d-1} h(x, \epsilon) \end{pmatrix}, \qquad (10.135)$$

where the first r coordinates correspond to the coordinates of the perturbed system, and the full set of $r + d$ coordinates, with $\epsilon = 0$, are the coordinates of the nominal system. It may be verified, that for small ϵ we have the following normal form (in the sense of Chapter 9):

$$\dot{\xi}_1 = \xi_2,$$
$$\dot{\xi}_2 = \xi_3,$$

$$\vdots$$

$$\dot{\xi}_r = \xi_{r+1} + \epsilon^d \alpha_0(\xi, \eta, \epsilon) u,$$
$$\dot{\xi}_{r+1} = \xi_{r+2} + \epsilon^{d-1} \alpha_1(\xi, \eta, \epsilon) u, \qquad (10.136)$$
$$\dot{\xi}_{r+2} = \xi_{r+3} + \epsilon^{d-2} \alpha_2(\xi, \eta, \epsilon) u,$$

$$\vdots$$

$$\dot{\xi}_{r+d} = b(\xi, \eta, \epsilon) + a(\xi, \eta, \epsilon) u,$$
$$\dot{\eta} = q(\xi, \eta, \epsilon).$$

We have introduced the smooth functions a, b, and q, we invite the reader to work out the the details of how a and b depend on f, g, and h. Using the change of

---

[1]This is an interesting case, and though it is no different conceptually, the notation is more involved. See Problem 10.10.

coordinates for the perturbed system given by

$$z_1 = \xi_{r+1}, \quad z_2 = \epsilon \xi_{r+2}, \quad \ldots, \quad z_d = \epsilon^{d-1} \xi_{r+d}, \tag{10.137}$$

it may be verified that the zero dynamics of the perturbed system have the form

$$\epsilon \dot{z}_1 = -\frac{\alpha_1}{\alpha_0} z_1 + z_2,$$

$$\epsilon \dot{z}_2 = -\frac{\alpha_2}{\alpha_0} z_1 + z_3,$$

$$\vdots \tag{10.138}$$

$$\epsilon \dot{z}_d = -\frac{a}{\alpha_0} z_1 + \epsilon^d b$$

$$\dot{\eta} = q(z, \eta, \epsilon).$$

Note that $\eta \in \mathbb{R}^{n-r-d}$, $z \in \mathbb{R}^d$. Also, we have abused notation for q from equation (10.136). Thus, the zero dynamics appear in singularly perturbed form, i.e.,

$$\epsilon \dot{z} = r(z, \eta, \epsilon),$$

$$\dot{\eta} = q(z, \eta, \epsilon), \tag{10.139}$$

with $n - r - d$ slow states (η) and d fast states (z). This is now consistent with the zero dynamics for the system at $\epsilon = 0$ given by

$$\dot{\eta} = q(0, \eta, 0), \tag{10.140}$$

Thus, the presence of small terms in $L_g L_f^{r-1+k} h(x, \epsilon)$ for $0 \le k \le d$ causes the presence of singularly perturbed zero dynamics. The Jacobian matrix evaluated at $z = 0$, $\epsilon = 0$ of the fast zero subsystem is obtained as

$$\begin{bmatrix} a_1(0, \eta, 0) & 1 & 0 & \cdots & 0 \\ a_2(0, \eta, 0) & 0 & 1 & \cdots & 0 \\ \vdots & \vdots & \vdots & \ddots & 0 \\ a_{d-1}(0, \eta, 0) & 0 & 0 & \cdots & 1 \\ a_d(0, \eta, 0) & 0 & 0 & \cdots & 0 \end{bmatrix}. \tag{10.141}$$

Here $a_i = -\frac{\alpha_i}{\alpha_0}(\xi, \eta, \epsilon)$ for $1 \le i < d$, and $a_d = -\frac{a}{\alpha_0}(\xi, \eta, \epsilon)$. It is clear that the perturbed system may be nonminimum phase either for positive ϵ, negative ϵ, or both positive and negative ϵ (according to whether the matrix in (10.141) has eigenvalues in $\mathbb{C}_-$, $\mathbb{C}_+$, or has indefinite inertia, respectively). If (10.140) has a stable equilibrium point at the origin (corresponding to the nominal system being minimum phase), but the origin of the system (10.138) is unstable, (corresponding to the perturbed system being nonminimum phase), we refer to these systems as *slightly nonminimum phase*.

We will need to generalize the preceding discussion of *zero dynamics* to the *driven dynamics* corresponding to the problem of tracking a desired output trajectory $y_D(t)$. If the output $y(t)$ is identically equal to $y_D(t)$, it follows that, for the perturbed system,

$$
\xi_D(t) = \begin{pmatrix} y_D(t) \\ \dot{y}_D(t) \\ \vdots \\ y_D^{(r-1)}(t) \end{pmatrix}.
$$

Then the driven dynamics of the system are given by (10.136) with the choice of error coordinates

$$
v_1 = \xi_{r+1} - y_D^{(r)}(t), \quad v_2 = \epsilon(\xi_{r+1} - y_D^{(r+2)}(t)), \quad \ldots,
$$
$$
v_d = \epsilon^{d-1}(\xi_{r+d} - y_D^{(r+d-1)}(t)) \tag{10.142}
$$

and the input

$$
u = \frac{-z_1^D}{\epsilon^d \alpha_0},
$$

namely,

$$
\epsilon \dot{v}_1 = a_1 v_1^D + v_2^D,
$$
$$
\epsilon \dot{v}_2 = a_2 v_1^D + v_3^D,
$$
$$
\vdots \tag{10.143}
$$
$$
\epsilon \dot{v}_d = a_d v_1^D + \epsilon^d (b - y_D^{(r+d)}(t)),
$$
$$
\dot{\eta} = q(\xi, \eta, \epsilon).
$$

10.6.2 MIMO Singularly Perturbed Zero and Driven Dynamics

To keep the notation in this section from becoming too complicated, we will restrict attention to two-input two-output systems of the form

$$
\dot{x} = f(x, \epsilon) + g_1(x, \epsilon)u_1 + g_2(x, \epsilon)u_2,
$$
$$
y_1 = h_1(x, \epsilon), \tag{10.144}
$$
$$
y_2 = h_2(x, \epsilon).
$$

We will need to distinguish in what follows between the following two cases:

1. Each perturbed system has vector relative degree at a point x_0, but as a function of ϵ the vector relative degree is not constant in a neighborhood of $\epsilon = 0$.
2. The perturbed system has vector relative degree, but the nominal system with $\epsilon = 0$ fails to have vector relative degree.

One could also consider as a third case, a scenario in which the perturbed system and nominal system fail to have vector relative degree, but they need different orders of dynamic extension for linearization. We will not consider this case here.

Case 1: Both the perturbed system and the nominal system have vector relative degree

Suppose that the system has vector relative degree $(r_1(\epsilon), r_2(\epsilon))$, with

$$r_1(\epsilon) = s \qquad \forall \epsilon,$$

$$r_2(\epsilon) = r \qquad \forall \epsilon \neq 0,$$

$$r_2(0) = r + d.$$

This implies that for the matrix

$$\begin{bmatrix} L_{g_1} L_f^i h_1(x, \epsilon) & L_{g_2} L_f^i h_1(x, \epsilon) \\ L_{g_1} L_f^j h_2(x, \epsilon) & L_{g_2} L_f^j h_2(x, \epsilon) \end{bmatrix} \tag{10.145}$$

1. The first row is identically zero for x near x_0 and all ϵ for $i < s - 1$ and nonzero at $(x_0, 0)$ for $i = s - 1$.
2. The second row is identically zero for x near x_0 when $\epsilon \neq 0$ for $j < r - 1$, and is identically zero when $\epsilon = 0$ for $j < r + d - 1$.
3. The matrix (10.145) is nonsingular at (x_0, ϵ) with $\epsilon \neq 0$ for $i = s-1$, $j = r-1$, and is nonsingular at $(x_0, 0)$ with $\epsilon = 0$ for $i = s - 1$, $j = r + d - 1$.

As in the SISO case, we will assume that there are only two time scales in the zero dynamics by assuming that for $0 \leq k \leq d$,

$$\begin{bmatrix} L_{g_1} L_f^{s-1} h_1(x, \epsilon) & L_{g_2} L_f^{s-1} h_1(x, \epsilon) \\ L_{g_1} L_f^{r-1+k} h_2(x, \epsilon) & L_{g_2} L_f^{r-1+k} h_2(x, \epsilon) \end{bmatrix} = \begin{bmatrix} 1 & 0 \\ 0 & \epsilon^{d-k} \end{bmatrix} A_k(x, \epsilon), \tag{10.146}$$

where $A_k(x, \epsilon)$ is a matrix of smooth functions, and both $A_0(x_0, 0)$ and $A_d(x_0, 0)$ are nonsingular. By using elementary column operations to modify g_1, g_2, we can assume that $A_0(x, \epsilon) = I \in \mathbb{R}^{2 \times 2}$. Since the first row of the left hand side of (10.146) does not change with k, we can assume $A_k(x, \epsilon)$ to be of the form

$$A_k(x, \epsilon) = \begin{bmatrix} 1 & 0 \\ \gamma_k(x, \epsilon) & \alpha_k(x, \epsilon) \end{bmatrix} \quad \text{with} \quad \alpha_d(x_0, 0) \neq 0. \tag{10.147}$$

We define the state variables

$$\xi_1^1 = h_1(x, \epsilon), \quad \dots \quad \xi_s^1 = L_f^{s-1} h_1(x, \epsilon),$$

$$\xi_1^2 = h_2(x, \epsilon), \quad \dots \quad \xi_{r+d}^2 = L_f^{r+d-1} h_2(x, \epsilon).$$

Then, using variables $\eta \in \mathbb{R}^{n-s-r-d}$ to complete the basis, the MIMO normal form of the system is given by

$$\dot{\xi}_1^1 = \xi_2^1,$$

$$\vdots$$

$$\dot{\xi}_s^1 = b_1(\xi, \eta, \epsilon) + u_1,$$

$$\dot{\xi}_1^2 = \xi_2^2,$$

$$\vdots$$

$$\dot{\xi}_r^2 = \xi_{r+1}^2 + \epsilon^d(\gamma_0(\xi, \eta, \epsilon)u_1 + \alpha_0(\xi, \eta, \epsilon)u_2),$$

$$\dot{\xi}_{r+1}^2 = \xi_{r+2}^2 + \epsilon^{d-1}(\gamma_1(\xi, \eta, \epsilon)u_1 + \alpha_1(\xi, \eta, \epsilon)u_2),$$

$$\vdots$$

$$\dot{\xi}_{r+d}^2 = b_2(\xi, \eta, \epsilon) + \gamma_d(\xi, \eta, \epsilon)u_1 + \alpha_d(\xi, \eta, \epsilon)u_2,$$

$$\dot{\eta} = q(\xi, \eta, \epsilon) + P(\xi, \eta, \epsilon)u. \tag{10.148}$$

Defining, in analogy to the SISO case,

$$z_1 = \xi_{r+1}^2, \quad z_2 = \epsilon\xi_{r+2}^2, \quad \cdots, \quad z_d = \epsilon^{d-1}\xi_{r+d}^2,$$

it may be verified, using the controls

$$u_1 = -b_1,$$

$$u_2 = \frac{\gamma_0}{\alpha_0}b_1 - \frac{1}{\epsilon^d}\frac{z_1}{\alpha_0},$$

that the zero dynamics are given by

$$\epsilon\dot{z}_1 = a_1 z_1 + z_2 + \epsilon^d(-\gamma_1 b_1 - a_1\gamma_0 b_1),$$

$$\epsilon\dot{z}_2 = a_2 z_1 + z_3 + \epsilon^d(-\gamma_2 b_1 - a_2\gamma_0 b_1),$$

$$\vdots$$

$$\epsilon\dot{z}_d = a_d z_1 + \epsilon^d(b_2 - \gamma_d b_1 - a_d\gamma_0 b_1),$$

$$\dot{\eta} = q\left(0, z_1, \frac{z_2}{\epsilon}, \ldots, \eta, \epsilon\right) + P = \left(0, z_1, \frac{z_2}{\epsilon}, \ldots, \eta, \epsilon\right)\begin{bmatrix} -b_1 \\ \frac{\gamma_0}{\alpha_0}b_1 - \frac{z_1}{\epsilon^d\alpha_0} \end{bmatrix},$$

$$\tag{10.149}$$

where the a_i are defined as before, with $a_d = -\frac{\alpha_d}{\alpha_0}$. The difference between the MIMO case and the SISO case is the presence of input terms in the right hand side of the η dynamics in (10.149). We will need to verify that no "bounded peaking" occurs in the dynamics of $\dot{\eta}$ in (10.149) when we estimate the magnitude of the variables z_i. Without this verification, it is possible that some of the η variables will also become fast time scale dynamics.

For the driven dynamics, one considers, as in the SISO case, new "error coordinates" given by

$$v_1 = \xi_{r+1}^2 - y_{D2}^{(r)}(t), \quad v_2 = \epsilon(\xi_{r+2}^2 - y_{D2}^{(r+1)}(t)), \quad \ldots,$$
$$v_d = \epsilon^{d-1}(\xi_{r+d}^2 - y_{D2}^{(r+d-1)}(t)). \tag{10.150}$$

In these coordinates

$$\epsilon \dot{v}_1 = a_1 v_1 + v_2 + \epsilon^d (y_{D1}^s - b_1)(\gamma_1 + a_1 \gamma_0),$$
$$\epsilon \dot{v}_2 = a_2 v_1 + v_3 + \epsilon^d (y_{D1}^s - b_1)(\gamma_2 + a_2 \gamma_0),$$

$$\vdots$$

$$\epsilon \dot{v}_d = x a_d v_1 + \epsilon^d \big(b_2 + (\gamma_d + a_d \gamma_0)(y_{D1}^{(s)} - b_1) - y_{D2}^{(r+d)}\big),$$
$$\dot{\eta} = q\left(\xi_D, v_1, \frac{v_2}{\epsilon}, \ldots, \eta, \epsilon, t\right)$$
$$+ P\left(\xi_D, v_1, \frac{v_2}{\epsilon}, \ldots, \eta, \epsilon, t\right)\begin{bmatrix} y_{D1}^{(s)} - b_1 \\ -\dfrac{\gamma_0}{\alpha_0}(y_{D1}^s - b_1) - \dfrac{v_1}{\epsilon^d \gamma_0} \end{bmatrix}. \tag{10.151}$$

Case 2: The perturbed system has vector relative degree, but the nominal system does not have vector relative degree. We will suppose that the matrix in (10.145) satisfies

1. The first row is identically zero for x near x_0 and all ϵ for $i < s - 1$ and nonzero at $(x_0, 0)$ for $i = s - 1$.
2. The second row is identically zero for x near x_0, for all ϵ for $j < r - 1$, and nonzero at $(x_0, 0)$ if $j = r - 1$.
3. The matrix is nonsingular at (x_0, ϵ) with $\epsilon \neq 0$, for $i = s - 1$, $j = r - 1$, but is singular at $(x_0, 0)$ with $\epsilon = 0$.

This implies that the system (10.144) has vector relative degree $(r_1(\epsilon), r_2(\epsilon)) = (s, r)$ for $\epsilon \neq 0$, but its relative degree cannot be defined at $\epsilon = 0$.

Under these assumptions and under dynamic extension of the system through u_1 and its first d derivatives, with $u_1^{(d)} = \bar{u}_1$, the system is decouplable. We will make the two time scale assumption, namely that the decoupling matrix at the k-th step of the dynamic extension algorithm (see Chapter 9 and [81]) has the form

$$\begin{bmatrix} \beta_k(x, u_1, \ldots, u_1^{(k-1)}, \epsilon) & \gamma_k(x, u_1, \ldots, u_1^{(k-1)}, \epsilon) \\ \delta_k(x, u_1, \ldots, u_1^{(k-1)}, \epsilon) & \alpha_k(x, u_1, \ldots, u_1^{(k-1)}, \epsilon) \end{bmatrix}\begin{bmatrix} 1 & 0 \\ 0 & \epsilon^{d-k} \end{bmatrix}. \tag{10.152}$$

We denote the first matrix in (10.152) above by $A_k(x, u_1, \ldots, u_1^{(k-1)}, \epsilon)$ for $0 \leq k \leq d$ and assume that $A_d(x, u_1, \ldots, u_1^{(k-1)}, \epsilon)$ is nonsingular. In fact, it may be verified that $\beta_k(x, u_1, \ldots, u_1^{(k-1)}, \epsilon) = \beta_0(x, \epsilon)$ and

$$\delta_k(x, u_1, \ldots, u_1^{(k-1)}, \epsilon) = \delta_0(x, \epsilon) \quad \text{for } 0 \leq k \leq d.$$

Now, define

$$\xi_1^1 = h_1(x, \epsilon),$$

$$\vdots$$

$$\xi_s^1 = L_f^{s-1} h_1(x, \epsilon),$$

$$\xi_{s+1}^1 = L_f^s h_1(x, \epsilon) + \beta_0(x, \epsilon) u_1,$$

$$\vdots$$

$$\xi_{s+d}^1 = L_f^{s+d-1} h_1(x, \epsilon) + \phi_1\left(x, u_1, \dot{u}_1, \ldots, u_1^{(d-2)}\right) + \beta_0(x, \epsilon) u_1^{(d-1)},$$

$$\xi_1^2 = h_2(x, \epsilon),$$

$$\vdots$$

$$\xi_r^2 = L_f^{r-1} h_2(x, \epsilon),$$

$$\xi_{r+1}^2 = L_f^r h_2(x, \epsilon) + \delta_0(x, \epsilon) u_1,$$

$$\vdots$$

$$\xi_{r+d}^2 = L_f^{r+d-1} h_2(x, \epsilon) + \phi_2\left(x, u_1, \dot{u}_1, \ldots, u_1^{(d-2)}\right) + \delta_0(x, \epsilon) u_1^{(d-1)}.$$

$$(10.153)$$

Using $\eta \in \mathbb{R}^{n-r-s-d}$ to complete the coordinate transformation from $\mathbb{R}^n \times \mathbb{R}^d \mapsto \mathbb{R}^n \times \mathbb{R}^d$, the MIMO normal form for Case 2 is given by

$$\dot{\xi}_1^1 = \xi_2^1,$$

$$\vdots$$

$$\dot{\xi}_s^1 = \xi_{s+1}^1,$$

$$\dot{\xi}_{s+1}^1 = \xi_{s+2}^1 + \epsilon^{d-1} \gamma_1(\xi, \eta, \epsilon) u_2,$$

$$\vdots$$

$$\dot{\xi}_{s+d}^1 = b_1(\xi, \eta, \epsilon) + \beta_0(\xi, \eta, \epsilon) \overline{u_1} + \gamma_d(\xi, \eta, \epsilon) u_2,$$

$$\dot{\xi}_1^2 = \xi_2^2, \qquad\qquad (10.154)$$

$$\vdots$$

$$\dot{\xi}_r^2 = \xi_{r+1}^2 + \epsilon^d \alpha_0(\xi, \eta, \epsilon) u_2,$$

$$\dot{\xi}_{r+1}^2 = \xi_{r+2}^2 + \epsilon^{d-1} \alpha_1(\xi, \eta, \epsilon) u_2,$$

$$\vdots$$

$$\dot{\xi}_{r+d}^2 = b_2(\xi, \eta, \epsilon) + \delta_0(\xi, \eta, \epsilon) \overline{u_1} + \alpha_d(\xi, \eta, \epsilon) u_2,$$

$$\dot{\eta} = q(\xi, \eta, \epsilon) + P(\xi, \eta, \epsilon) u.$$

Note that the variables $\xi^1_{s+1}, \ldots, \xi^1_{s+d}$ are affinely related to $u_1, \dot{u}_1, \ldots, u_1^{(d-1)}$ Defining, as before, the "fast" zero dynamics variables as

$$z_1 = \xi^2_{r+1}, \quad z_2 = \epsilon\xi^2_{r+2}, \quad \ldots, \quad z_d = \epsilon^{d-1}\xi^2_{r+d}$$

and using the control inputs

$$u_1 = -\frac{1}{\beta_0(x,\epsilon)}L^s_f h_1(x,\epsilon),$$

$$u_2 = -\frac{1}{\epsilon^d\alpha_0(x,\epsilon)}(L^r_f h_2(x,\epsilon) - \frac{\delta_0(x,\epsilon)}{\beta_0(x,\epsilon)}L^s_f h_1(x,\epsilon)),$$

the zero dynamics are given by

$$\epsilon\dot{z}_1 = a_1 z_1 + z_2,$$

$$\epsilon\dot{z}_2 = a_2 z_1 + z_3,$$

$$\vdots$$

$$\epsilon\dot{z}_d = a_d z_1 + \epsilon^d(b_2 + \delta_0\bar{u}_1),$$

$$\dot{\eta} = q\left(0, z_1, \frac{z_2}{\epsilon}, \ldots, \eta, \epsilon\right)$$

$$+ P\left(0, z_1, \frac{z_2}{\epsilon}, \ldots, \eta, \epsilon\right)\begin{bmatrix} -\frac{1}{\beta_0(x,\epsilon)}L^s_f h_1(x,\epsilon) \\ -\frac{1}{\epsilon^d\alpha_0(x,\epsilon)}(L^r_f h_2(x,\epsilon) - \frac{\delta_0(x,\epsilon)}{\beta_0(x,\epsilon)}L^s_f h_1(x,\epsilon)) \end{bmatrix}.$$

$$(10.155)$$

The driven dynamics are derived accordingly.

10.7   Summary

This chapter, along with the exercises that follow, gives the reader a sense of how to apply the theory of input–output linearization by state feedback to a number of applications. The main point that we made in the ball and beam example is that even if a system fails to have relative degree and is singular in a certain sense, it can be approximately input–output linearized for small signal tracking. For large signals a switched scheme of the form proposed in Tomlin and Sastry [308] may be useful. Further results inspired by the ball and beam example for MIMO systems which fail to have vector relative degree are in [120; 17; 142]. Another interesting approach to approximate feedback linearization using splines is due to Bortoff [35]. The application of the methods of input-output linearization to flight control featured approximate linearization for slightly nonminimum phase systems. While the chapter dealt with VTOL aircraft, slightly nonminimum phase systems are encountered in conventional take-off and landing aircraft [305; 129] as well as in helicopter flight dynamics [165] as illustrated in the exercises (Problems 10.7 and

10.15, respectively). Applications of feedback linearization to aerospace vehicle entering the atmosphere are presented in the work of Mease and co-workers [24; 167]. Other applications of feedback linearization are included in the Exercises here and in the preceding chapter for the control of robots [225; 231], stirred tank reactors [133] (see also the book by Henson and Seborg [134]), and automotive engine dynamics [64]. For geometric methods used to control electric motors see the recent survey by Bodson and Chiasson [32]. Nonlinear control methods applied to problems in robotics also have a very extensive literature, see for example the recent books of Sciavicco and Siciliano [268], Canudas de Wit, de Luca and Bastin [79], and underactuated robotic systems [191]. The control of electro-mechanical systems with friction is an especially important application area; we refer the reader to the survey paper of Armstrong-Helouvry, Dupont, and Canudas de Wit [10]. An especially interesting case study of different methods of nonlinear control for a system known as an Acrobot (which is a double inverted pendulum with one actuator, originally designed by Fearing at Berkeley to model an acrobat) is surveyed in [283]. Control of free floating robot systems in space are surveyed in the paper of Egeland and Pettersen [94], see also Egeland and Godhavn [93], and Wen and Kreutz-Delgado [326].

10.8   Exercises

Problem 10.1 Robust relative degree.   The robust relative degree of the nonlinear system (10.28) is equal to the relative degree of the Jacobian linearized system (10.36) and so is well defined.

Problem 10.2 Robust relative degree under feedback.   Prove that the robust relative degree of a nonlinear system (10.28) is *invariant* under a state dependent change of control coordinates of the form

$$u(x, v) = \alpha(x) + \beta(x)v,$$

where α and β are smooth functions and $\alpha(0) = 0$, while $\beta(0) \neq 0$.

Problem 10.3.   Prove Proposition 10.2.

Problem 10.4.   Provide a proof for Theorem 10.6.

Problem 10.5 Byrnes–Isidori regulator applied to the ball and beam system [298].   Apply the Byrnes–Isidori regulator of Chapter 9 to the ball and beam system when the signal to be tracked is a constant plus a sinusoid. Thus, the exo-system may be modeled by

$$
\begin{aligned}
\dot{w}_1 &= 0, \\
\dot{w}_2 &= -\omega w_3, \\
\dot{w}_3 &= \omega w_2, \\
y &= w_1 + w_2.
\end{aligned}
\tag{10.156}
$$

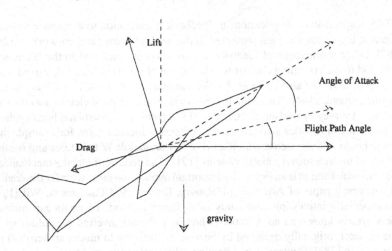

FIGURE 10.17. A two dimensional aircraft in flight with attached axes

Mimic the procedure of Problem 9.31 and approximate the solution of the PDE obtained by a polynomial in w. The paper [298] discusses the small domains of attraction that are obtained by the resulting controller. Do you also find that the domain of validity of your controller requires that $w_1(0)$, $w_2(0)$, $w_3(0)$, ω be very small? See also [299] for some hints on how to do better.

Problem 10.6. If a SISO nonlinear system is slightly nonminimum phase with relative degree $n - 1$, prove that the approximate control law for stabilizing the system also stabilizes the original system. Give the necessary Lipschitz conditions that are required for the neglected term.

Problem 10.7 Conventional aircraft (CTOL) control [305].

Consider the dynamics of an aircraft in normal aerodynamic flight in the longitudinal plane, as shown in Figure (10.17). The horizontal and vertical axes are respectively the x, y axes, and θ is the angle made by the aircraft axis with the x axis. The three inputs available to the aircraft are the horizontal thrust u_1 and the pitch moment u_2. To be able to write the simplest rigid body equations for the airframe we need to define the *flight path angle* γ as $tan^{-1}(\frac{\dot{y}}{\dot{x}})$ and the *angle of attack* $\alpha = \theta - \gamma$. The lift and drag forces L, D are given by the following with a_L, a_D being the lift and drag coefficients:

$$L = a_L(\dot{x}^2 + \dot{y}^2)(1 + c\alpha),$$
$$D = a_D(\dot{x}^2 + \dot{y}^2)\alpha^2. \tag{10.157}$$

The equations are given by

$$\begin{bmatrix} \ddot{x} \\ \ddot{y} \end{bmatrix} = R(\theta)\left[R^T(\alpha)\begin{bmatrix} -D \\ L \end{bmatrix} + \begin{bmatrix} u_1 \\ -\epsilon u_2 \end{bmatrix} \right] + \begin{bmatrix} 0 \\ -1 \end{bmatrix}, \tag{10.158}$$

and in addition we have that

$$\ddot{\theta} = u_2 \tag{10.159}$$

as a model for the generation of the pitching moment. The equations are highly stylized though accurate in form, with all masses and moments of inertia normalized away, and the gravity set to be one unit. Also in equation (10.158) above the matrices $R(\alpha)$, $R(\theta)$ are rotation matrices of the form

$$R(\theta) = \begin{bmatrix} \cos(\theta) & -\sin(\theta) \\ \sin(\theta) & \cos(\theta) \end{bmatrix}. \tag{10.160}$$

Physically, the presence of ϵ models the fact that the process of generaing pitching moment results in a small downward force. Use your analysis and the methods of this chapter to explain the fact that aircraft dip down first when commanded to rise in the vertical, y, direction.

Linearize the system from the inputs u_1, u_2 to the outputs $y_1 = \dot{x} - v$, $\dot{y}$ both for ϵ equal to zero and not equal to zero.[2] Here v is the horizontal "trim" velocity. For the case that $\epsilon \neq 0$ derive the zero dynamics. Plot the zero dynamics for some normalized representative numbers (for a DC 8) given by $\epsilon = 0.1$, $v = 0.18$, $a_L = 30$, $a_D = 2$, $c = 6$. Use the methods of this chapter to explain which control law you would use.

Problem 10.8 Automotive engine control (adapted from [64]). Consider the two state model of an automotive engine coupled to transmission and wheels:

$$\begin{aligned} \dot{m} &= c_1 T(\alpha) - c_2 \omega m, \\ \dot{\omega} &= \frac{1}{J}[T_i - T_f - T_d - T_r]. \end{aligned} \tag{10.161}$$

Here, we have

- m is the mass of air in the intake manifold of the car in kg.
- ω is the engine speed in rad/sec. The speed of the car is the *output* $y = 0.1289\omega$ meters per second.
- $T(\alpha)$ is the throttle characteristic of the car, with $\alpha = u$ being the *input* throttle angle in degrees.

$$T(\alpha) = 1 - \cos(1.14\alpha - 1.06)$$

for $\alpha \leq 79.46°$ and $T(\alpha) = 1$ for $\alpha > 79.46°$.
- J is the effective vehicle inertia reflected to the engine, $36.42 \ \text{kgm}^2$.
- T_i is the engine torque in Newton meters as $T_i = c_3 m$.
- T_f is the engine friction torque in Newton meters given by $T_f = 0.1056\omega + 15.1$.
- T_d is the wind drag given by $T_d = c_4\omega^2$ Newton meters.
- T_r is the rolling resistance torque given by $T_r = 21.5$ Newton meters.

---

[2]You will need a symbolic algebra package for the case that $\epsilon = 0$. Use D, L as long as you can in your calculations.

The nominal values for the constants are

$$c_1 = 0.6 \text{ kg/sec},$$

$$c_2 = 0.0952,$$

$$c_3 = 47469 \text{ N.m/sec},$$

$$c_4 = 0.0026 \text{ N.msec}^2.$$

We would like to build a controller to cause the output sped of the car to transition from $y = 20$ m/sec to $y = 30$ m/sec:

1. **Step 1** Build a second-order linear reference model such that the output of the reference model, y_M smoothly transitions from 20m/sec to 30m/sec in about 5 seconds. For example, use a linear model with transfer function

$$\frac{\omega_n^2}{s^2 + 2\zeta\omega_n s + \omega_n^2}$$

and design ω_n and ζ so that with a step reference input $r(t)$ going from 20 to 30, the output $y_M(t)$ has the desired characteristic.

2. **Step 2** Now build a feedback linearizing controller such that $y(t)$ tracks $y_M(t)$. Note that the input throttle angle α does not enter linearly in the dynamics (10.161). One strategy is to make α a state and define the input u to be

$$\dot{\alpha} = u. \tag{10.162}$$

Now the composite system (10.161, 10.162) is linear in the input u. Do you have any ideas about what you may be sacrificing for this easy fix? Another strategy is to invert the nonlinearity $T(\alpha)$, at least locally. Compare the two approaches.

3. **Step 3** Simulate the controller and see how sensitive the performance of the controller is to a 20% error in the values of the parameters c_1, c_2, c_3, c_4.

4. **Step 4** Use the techniques of Chapter 9 to build a **sliding mode** linearizing controller for the system with 20% error in the parameters.

5. Use the techniques of Problem 9.27 to build an adaptive linearizing controller for the system with a larger initial error in the parameter estimate of, say, 50%.

Problem 10.9 Zeros and poles of linear systems. Consider a family of SISO minimal linear systems with parameters $A(\epsilon) \in \mathbb{R}^{n \times n}, b(\epsilon) \in \mathbb{R}^n, c(\epsilon)^T \in \mathbb{R}^n$. Let $\gamma(\epsilon)$ be the relative degree of the linear system and assume that it jumps from r at $\epsilon = 0$ to $r + d$ at $\epsilon \neq 0$. Give asymptotic formulas for how d zeros of the linear system at $\epsilon \neq 0$ tend to ∞ as $\epsilon \to 0$. In other words, give the *zero locus* of the linear system. What implications does this have for the effect of numerical perturbations on the calculation of the zeros of a linear system, since a linear system "generically" has relative degree 1?

Problem 10.10 Multiple time scales in zero dynamics. In the case that $L_g L_f^{r-1+k}$ does not satisfy the one-time-scale assumption of (10.134) derive the multiple-time-scale zero dynamics. To be specific, let $d = 3$ and assume that

$$L_g L_f^{r-1} h(x, \epsilon) = \epsilon^3 \alpha_0(x, \epsilon), \quad L_g L_f^r h(x, \epsilon) = \epsilon \alpha_1(x, \epsilon),$$
$$L_g L_f^{r+1} h(x, \epsilon) = \epsilon \alpha_2(x, \epsilon), \quad L_g L_f^{r+2} h(x, \epsilon) = \alpha_3(x, \epsilon).$$

Show the existence of two time scales in the fast zero dynamics. What are the time scales?

Problem 10.11 Singularly perturbed zero dynamics for VTOL and CTOL aircraft [307]. While we have studied the zero dynamics of the PVTOL and CTOL longitudinal axis in detail in this chapter and in Problem 10.7 apply the methods of Section 10.6.2 to models of each, repeated here for your convenience, and with the powers of ϵ scaled:

- *PVTOL Dynamics*

$$\dot{x}_1 = x_2,$$
$$\dot{x}_2 = -\sin x_5 u_1 + \epsilon^2 \cos x_5 u_2,$$
$$\dot{x}_3 = x_4,$$
$$\dot{x}_4 = \cos x_5 u_1 + \epsilon^2 \sin x_5 u_2 - 1,$$
$$\dot{x}_5 = x_6,$$
$$\dot{x}_6 = u_2.$$

- *CTOL Dynamics*

$$\dot{x}_1 = x_2,$$
$$\dot{x}_2 = (-D + u_1) \cos x_5 - (L - \epsilon^2 u_2) \sin x_5,$$
$$\dot{x}_3 = x_4,$$
$$\dot{x}_4 = (-D + u_1) \sin x_5 + (L - \epsilon^2 u_2) \sin x_5 - 1,$$
$$\dot{x}_5 = x_6,$$
$$\dot{x}_6 = u_2,$$

with L, D the aerodynamic lift and drag given by

$$L = a_l (x_2^2 + x_4^2)(1 + c\alpha),$$
$$D = a_D (x_2^2 + x_4^2)(1 + b(1 + c\alpha)^2),$$

and α, the angle of attack is given by

$$\alpha = x_5 - \tan^{-1} \frac{x_4}{x_2}.$$

Problem 10.12 Approximate decoupling [114]. Give a "robust" version of the dynamic extension algorithm of the previous chapter for approximately decoupling a nonlinear system when there are some singular values of the decoupling matrix that are small relative to the others at some step in the algorithm. Discuss the zeros that are being singularly perturbed to ∞ as a result of any approximations that you make in the decoupling procedure.

Problem 10.13 "Zeros" of nonregular systems [308].   Consider a SISO non-linear system with $L_g L_f^{\gamma-2} h(x_0) = O(|x - x_0|^r)$ and $L_g L_f^{\gamma-1} h(x) = O(1)$. Thus, the nonlinear system has robust relative degree γ, but the exact zero dynamics are potentially in $\mathbb{R}^{n-\gamma+1}$. Since $f(x_0) = 0$, it follows that $L_f^{\gamma-1} h(x_0) = O(|x - x_0|^s)$. Classify what you think would happen if one used a switching control law for tracking $y_D(\cdot)$ in the three cases, a sort of classification of the "bifurcation" in the zero dynamics:

1. $s < r$ (No zero dynamics are defined)
2. $s = r$ (Zero dynamics are well-defined, and high-dimensional invariant sets are possible as in the case of the ball and beam.
3. $s > r$ (Zero dynamics are well-defined).

Problem 10.14 Bounded tracking for nonminimum phase nonlinear systems [307].   Combine the discussion of zeros for perturbed nonlinear systems and the discussion of the Devasia–Chen–Paden approach of Chapter 9 to come up with an approach for tracking for nonlinear control systems with singularly perturbed zero dynamics, of the form

$$\dot{x} = f(x, \epsilon) + g(x, \epsilon)u,$$

$$y = h(x, \epsilon).$$

In this chapter we simply neglected the high frequency zeros and assuming the rest of the system was minimum phase simply used the tracking control law associated with input–output linearization. Here try the following two step approach:

1. Step 1: Find the control law $u^0(\cdot)$ for tracking the nonlinear system with $\epsilon = 0$ and with relative degree $r + d$. This is done using the Devasia–Chen–Paden approach if the nominal system ($\epsilon = 0$) is nonminimum phase.
2. Step 2: Use the nominal control u^0 to get the nominal normal form coordinates ξ^0, η^0. Comparing the perturbed normal form of (10.136) with the standard normal form at $\epsilon = 0$, derive the formula

$$u = -\frac{\xi_{r+1} - \xi_{r+1}^0}{\epsilon^d \alpha_0(\xi, \eta, \epsilon)}. \tag{10.163}$$

Now define

$$v_i = \epsilon^{i-1}(\xi_{r+i} - \xi_{r+i}^0), \quad i = 1, \ldots, d$$
$$v_{d+i} = \eta_i - \eta_i^0, \qquad\quad i = 1, \ldots, n - r - d.$$

Subtracting the equation for the derivatives of $\dot{\xi}_i^0, \dot{\eta}^0$ from that for $\dot{\xi}, \dot{\eta}$ results in

$$v_1 = -\epsilon^d \alpha_0(\xi, \eta, \epsilon)u \tag{10.164}$$

and an error system

$$\epsilon\dot{v}_1 = v_2 + a_1(\xi, \eta, \epsilon)v_1,$$

$$\epsilon\dot{v}_2 = v_3 + a_2(\xi, \eta, \epsilon)v_1,$$

$$\vdots$$

$$\epsilon\dot{v}_d = a_d(\xi, \eta, \epsilon)v_1 + \epsilon^d(b(\xi, \eta, \epsilon) - b(\xi^0, \eta^0, 0) - a(\xi^0, \eta^0, 0)u^0),$$

$$\dot{v}_{d+1} = q_1(\xi, \eta, \epsilon) - q_1(\xi^0, \eta^0, 0),$$

$$\vdots$$

$$\dot{v}_{n-r} = q_{n-r-d}(\xi, \eta, \epsilon) - q_{n-r-d}(\xi^0, \eta^0, 0). \tag{10.165}$$

Now apply the Devasia–Chen–Paden algorithm to the system of (10.165) to find the value of $v_1(t)$ and hence the control $u(t, \epsilon)$ for exact tracking by (10.163).

The control law of (10.163) is bounded in t for each ϵ. Prove by analyzing the system of (10.165) that as ϵ tends to zero the control $u(t, \epsilon)$ is bounded.

Problem 10.15 Control of a helicopter [165].   The equations of motion of a helicopter can be written with respect to the body coordinate frame, attached to the center of mass of the model helicopter. The x axis is pointed to the body head, and the y axis goes to the right of the body. The z-axis is defined by the right-hand rule, as shown in Figure 10.18. The equations of motion for a rigid body subject to an external wrench $F^b = [f^b, \tau^b]^T$ applied at the center of mass and specified

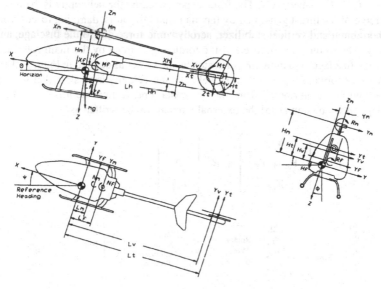

FIGURE 10.18. The free-body diagram of a helicopter in flight

with respect to the body coordinate frame is given by Newton-Euler equations, which can be written as

$$
\begin{bmatrix} mI & 0 \\ 0 & \mathcal{I} \end{bmatrix} \begin{bmatrix} \dot{v}^b \\ \dot{\omega}^b \end{bmatrix} + \begin{bmatrix} \omega^b \times mv^b \\ \omega^b \times \mathcal{I}\omega^b \end{bmatrix} = \begin{bmatrix} f^b \\ \tau^b \end{bmatrix}, \tag{10.166}
$$

where $\mathcal{I}$ is an inertia matrix $\in \mathbb{R}^{3 \times 3}$. Define the state vector $[p \ v^p \ R \ \omega^b]^T \in \mathbb{R}^3 \times \mathbb{R}^3 \times SO(3) \times \mathbb{R}^3$, where p and v^p are the position and velocity vectors of center of mass in spatial coordinates, R is the rotation matrix of the body axes relative to the spatial axes, and the body angular velocity vector is represented by ω^b. Using $\dot{p} = v^p = Rv^b$ and $\hat{\omega}^b = R^T \dot{R}$, we can rewrite the equations of motion of a rigid body.

$$
\begin{bmatrix} \dot{P} \\ \dot{v}^p \\ \dot{R} \\ \dot{\omega}^b \end{bmatrix} = \begin{bmatrix} v^p \\ \dfrac{1}{m} R f^b \\ R \hat{\omega}^b \\ \mathcal{I}^{-1}(\tau^b - \omega^b \times \mathcal{I}\omega^b) \end{bmatrix}. \tag{10.167}
$$

For $R \in SO(3)$, we parameterize R by ZYX Euler angles with ϕ, θ, and ψ about x, y, z-axes respectively, and hence $R = e^{\hat{z}\psi} e^{\hat{y}\theta} e^{\hat{x}\phi}$ The helicopter force and moment generation system consists of a main rotor, a tail rotor, a horizontal stabilizer, a vertical stabilizer, and fuselage, which are denoted by the subscripts M, T, H, V, F, respectively. The force experienced by the helicopter is the resultant force of the thrust generated by the main and tail rotors, damping forces from the horizontal and vertical stabilizer, aerodynamic force due to the fuselage, and gravity. The torque is composed of the torques generated by the main rotor, tail rotor and fuselage, and moment generated by the forces as defined in Figure 10.19. In hover or forward flight with slow velocity, the velocity is so slow that we can ignore the drag contributed from the horizontal and vertical stabilizers, and the fuselage. As shown in [165] the external wrench can be written as

$$
f^b = \begin{bmatrix} X_M \\ Y_M + Y_T \\ Z_M \end{bmatrix} + R^T \begin{bmatrix} 0 \\ 0 \\ mg \end{bmatrix},
$$

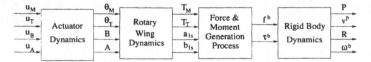

FIGURE 10.19. Helicopter dynamics

$$\tau^b = \begin{bmatrix} R_M \\ M_M + M_T \\ N_M \end{bmatrix} + \begin{bmatrix} Y_M h_M + Z_M y_M + Y_T h_T \\ -X_M h_M + Z_M l_M \\ -Y_M l_M - Y_T l_T \end{bmatrix}.$$

The forces and torques generated by the main rotor are controlled by T_M, a_{1s}, and b_{1s}, in which a_{1s} and b_{1s} are the longitudinal and lateral tilt of the tip path plane of the main rotor with respect to the shaft respectively. The tail rotor is considered as a source of pure lateral force Y_T and antitorque Q_T, which are controlled by T_T. The forces and torques can be expressed as

$$X_M = -T_M s a_{1s}, \qquad R_M \approx \frac{\partial R_M}{\partial b_{1s}} b_{1s} - Q_M s a_{1s},$$

$$Y_M = T_M s b_{1s}, \qquad M_M \approx \frac{\partial M_M}{\partial a_{1s}} a_{1s} + Q_M s b_{1s},$$

$$Z_M = -T_M c a_{1s} c b_{1s}, \qquad N_M \approx -Q_M c a_{1s} c b_{1s},$$

$$Y_T \quad -T_T \qquad M_T = -Q_T,$$

with $s a_{1s} = \sin a_{1s}$, $c a_{1s} = \cos a_{1s}$, $s b_{1s} = \sin b_{1s}$, $c b_{1s} = \cos b_{1s}$. The moments generated by the main and tail rotors can be calculated by using the constants $\{l_M, y_M, h_M, h_T, l_T\}$ as defined in Figure 10.18. We approximate the rotor torque equations by $Q_i \approx C_i^Q T_i^{1.5} + D_i^Q$ for $i = M, T$. Following are representative values of some of the constants above:

$$I_x = 0.142413, \qquad I_y = 0.271256, \qquad I_z = 0.271492,$$

$$l_M = -0.015, \qquad y_M = 0, \qquad h_M = 0.2943,$$

$$h_T = 0.1154, \qquad l_T = 0.8715, \qquad m = 4.9,$$

$$C_M^Q = 0.004452, \qquad D_M^Q = 0.6304, \qquad \frac{\partial R_M}{\partial b_{1s}} = 25.23,$$

$$C_T^Q = 0.005066, \qquad D_T^Q = 0.008488, \qquad \frac{\partial M_M}{\partial a_{1s}} = 25.23.$$

The operating region in radians for a_{1s}, b_{1s} and Newtons for T_M, T_T is described by: $|a_{1s}| \leq 0.4363$, $|b_{1s}| \leq 0.3491$, $-20.86 \leq T_M \leq 69.48$, $-5.26 \leq T_T \leq 5.26$.

To represent the system in an input–affine form, we assume that the inputs of the nonlinear system are the derivatives of T_M, T_T, a_{1s} and b_{1s}. Then, the system equations become

$$
\underbrace{\begin{bmatrix} \dot{p} \\ \dot{v}^P \\ \dot{\Theta} \\ \dot{\omega}^b \\ \dot{T}_M \\ \dot{T}_T \\ \dot{a}_{1s} \\ \dot{b}_{1s} \end{bmatrix}}_{\dot{x}} = \underbrace{\begin{bmatrix} v^P \\ \dfrac{1}{m} R f^b \\ \Psi \omega^b \\ \mathcal{I}^{-1}(\tau^b - \omega^b \times \mathcal{I}\omega^b) \\ w_1 \\ w_2 \\ w_3 \\ w_4 \end{bmatrix}}_{f(x)+\sum_{i=1}^{4} g_i w_i}, \tag{10.168}
$$

$$
\text{and} \quad y = \underbrace{[p_x,\ p_y,\ p_z,\ \phi,\ \theta,\ \psi]^T}_{h^T(x)}. \tag{10.169}
$$

where $w = [w_1, w_2, w_3, w_4]$ are defined as auxiliary inputs to the system. Here the states x are in $\mathbb{R}^{16}$ and the inputs $w \in \mathbb{R}^4$. We have chosen 6 output variables, however, since there are four control inputs the maximum number of outputs that can be input–output linearized is four. Since six variables, 3 positions and 3 Euler angles, are candidate outputs, there are $\binom{6}{4} = 15$ possible "square" input–output pairs. Perform input–output linearization for some of these sets. Note the vector relative degree in all cases. Verify that the zero dynamics are unstable! *For definiteness you may wish to stick to the outputs being $p \in \mathbb{R}^3$, ψ.*

To set up for approximate linearization, we rewrite the system equations as

$$
\ddot{\bar{p}} = R \begin{bmatrix} -\bar{T}_M \sin a_{1s} \\ \bar{T}_M \sin b_{1s} - \bar{T}_T \\ -\bar{T}_M \cos a_{1s} \cos b_{1s} \end{bmatrix} + \begin{bmatrix} 0 \\ 0 \\ 1 \end{bmatrix}, \tag{10.170}
$$

$$
\dot{R} = R\hat{\omega}, \tag{10.171}
$$

$$
\dot{\omega} = \mathcal{I}^{-1}(\tau^b - \omega^b \times \mathcal{I}\omega^b), \tag{10.172}
$$

where $\bar{p} = p/g, \bar{T}_M = T_M/(mg)$ and $\bar{T}_T = T_T/(mg)$. The reason for nonminimum phase zero dynamics is the coupling between rolling (pitching) moments and lateral (longitudinal) acceleration. These couplings are introduced by the presence of a_{1s}, b_{1s} and T_T. To approximately linearize the system neglect the coupling terms. This can be done by assuming that a_{1s}, b_{1s}, $\bar{T}_T/\bar{T}_M$ are small. Therefore, equation

(10.170) can be rewritten as

$$
\ddot{p}_m = R \begin{bmatrix} 0 \\ 0 \\ -\bar{T}_M \end{bmatrix} + \begin{bmatrix} 0 \\ 0 \\ 1 \end{bmatrix}
\tag{10.173}
$$

while equations (10.171) and (10.172) remain the same. Verify that this system is input–output linearizable for each of the set of outputs discussed above.

11

Geometric Nonlinear Control

In this chapter, we give the reader a brief introduction to controllability and observability of nonlinear systems. We discuss topics of constructive controllability for nonlinear systems, that is, algorithms for steering the system from a specified initial state to a specified final state. This topic has been very much at the forefront of research in nonlinear systems recently because of the interest in non-holonomic systems. We give a reader of the power of differential geometric methods for nonlinear control when we discuss input–output expansions for nonlinear systems and disturbance decoupling problems using appropriately defined invariant distributions.

11.1 Controllability Concepts

In this section we consider analytic affine control systems of the form

$$\dot{x} = f(x) + \sum_{i=1}^{m} g_j(x) u_j. \tag{11.1}$$

In equation (11.1) above, $x = (x_1, x_2, \ldots, x_n)^T$, are local coordinates for a smooth manifold M (the state space manifold), and $f, g_1, \ldots, g_m$ are smooth vector fields on M represented in local coordinates. For the rest of this section unless otherwise stated we will work exclusively in a single coordinate patch and will refer to $x \in \mathbb{R}^n$. Also a number of results in this chapter will need the vector fields f, g_i to be **analytic** rather than just smooth. Thus, *the assumption of analytic vector fields and functions will be in force throughout this chapter*. The vector field f is referred to as the *drift* vector field and the vector fields g_j are referred to as the *input* vector fields. The inputs $(u_1, \ldots, u_m)^T$ are constrained to lie in $U \subset \mathbb{R}^m$. Input functions $u_j(\cdot)$ satisfying this constraint are known as admissible input functions or controls. The **controllability**, or **accessibility**, problem consists of determining to which points one can steer trajectories from x_0 at the initial time: The results in controllability first appeared in the work of Sussmann and Jurdjevic [291], Hermann and Krener [136], and Sussmann [289], [290]. See also Nijmeijer and van der Schaft [232] for a nice exposition. In this chapter our use of the phrases "controllability distribution" and "accessibility distribution" are not standard, but are consistent with Murray, Li, and Sastry [225]. We begin with the following definition:

Definition 11.1 Controllable. *The nonlinear system (11.1) is said to be controllable if for any two points x_0, x_1 there exist a time T and an admissible control defined on $[0, T]$ such that for $x(0) = x_0$ we have $x(T) = x_1$.*

The reader is undoubtedly familiar with the result from linear systems theory that the control system

$$\dot{x} = Ax + \sum_{i=1}^{m} b_j u_j \qquad (11.2)$$

is controllable if the following controllability rank condition is satisfied:

$$\text{span}[b_1, \ldots, b_m, Ab_1, \ldots, Ab_m, A^{n-1}b_1, \ldots, A^{n-1}b_m] = n.$$

Now, if $f(x_0) = 0$ in equation (11.1) and

$$A = \frac{\partial f}{\partial x}(x_0), \quad b_j = g_j(x_0) \qquad (11.3)$$

is the linearization of equation (11.1) about x_0, it is then of obvious interest to relate the controllability of the system of (11.1) to that of its linearization (11.2):

Proposition 11.2 Local Controllability from Controllability of the Linearization. *Consider the system of (11.1) with x_0 such that $f(x_0) = 0$. Let the admissible set of controls U include a neighborhood of the origin. Assume that the linearization (11.2) is controllable. Then for every $T > 0$ the set of points that can be reached from x_0 contains a neighborhood of x_0.*

Proof: Since the linearization is completely controllable, there exist input functions, $v^i(t)$ defined on $[0, T]$ such that they steer the linearized system of (11.2) to the coordinate directions $(1, 0, \ldots, 0)^T, \ldots, (0, 0, \ldots, 1)^T$. Consider the input $u(t)$ defined for $\xi \in \mathbb{R}, i = 1, \ldots, n$, by

$$u(t, \xi_1, \ldots, \xi_n) := \xi_1 v^1(t) + \cdots + \xi_n v^n(t).$$

For ξ_i small enough, it follows that $|u(t, \xi_1, \ldots, \xi_n)| < \epsilon$. Now apply this parameterized input to the system (11.1) and consider the map

$$\xi \mapsto x(T, \xi)$$

defined for ξ near zero. We will show that the matrix

$$Z(T) := \frac{\partial x}{\partial \xi}(T, \xi)|_{\xi=0}$$

is nonsingular. Then, the conclusion of the theorem follows from the implicit function theorem applied to the map $x(T, \xi)$. To do this we first note that $x(t, \xi)$ satisfies equation (11.1), namely,

$$\dot{x}(t, \xi) = f(x(t, \xi)) + \sum_{j=1}^{m} g_j(x(t, \xi))u_j(t, \xi).$$

Differentiating this equation with respect to ξ yields at $\xi = 0$:

$$\dot{Z}(t) = AZ(t) + B[v^1 \cdots v^n], \quad Z(0) = 0,$$

with A, B defined as in the linearization equation (11.3). The fact that $Z(T) = I$ follows from the definition of the controls $v^i(t)$. Since $Z(T)$ is nonsingular, the proof is complete.   □

The preceding proposition covers the situation where the underlying linear system is controllable. There are a large number of systems in which this fails to be the case. In the next subsection we consider the controllability of drift free control systems, i.e., $f(x) \equiv 0$. Then since $A = 0$, the linearization is not controllable unless $m > n$. Before we do this, we introduce a discussion of the use of Lie brackets in studying controllability. Consider the following two input drift-free control system:

$$\dot{x} = g_1(x)u_1 + g_2(x)u_2. \tag{11.4}$$

It is clear that from a point $x_0 \in \mathbb{R}^n$ we can steer in all directions contained in the subspace of the tangent space $T_{x_0}\mathbb{R}^n \simeq \mathbb{R}^n$ given by

$$\text{span}\{g_1(x_0), g_2(x_0)\}$$

simply by using constant inputs. It may also be possible to steer along other directions using piecewise constant input functions as follows: Consider the following example which is also a way of motivating the definition of the Lie bracket which we have seen in Chapter 8.

$$u(t) = \begin{cases} (1, 0), & t \in [0, h[, \\ (0, 1), & t \in [h, 2h[, \\ (-1, 0), & t \in [2h, 3h[, \\ (0, -1), & t \in [3h, 4h[. \end{cases}$$

For h small we evaluate the Taylor series expansion formula for x. Indeed, we have

$$x(h) = x(0) + h\dot{x}(0) + \frac{1}{2}h^2\ddot{x}(0) + \cdots$$

$$= x_0 + hg_1(x_0) + \frac{1}{2}h^2\frac{dg_1}{dx}(x_0)g_1(x_0) + \cdots.$$

Similarly, continuing, we get

$$x(2h) = x(h) + hg_2(x(h)) + \frac{1}{2}h^2\frac{dg_2}{dx}g_2(x(h))$$

$$= x_0 + h(g_1(x_0) + g_2(x_0))$$

$$+ h^2\left(\frac{1}{2}\frac{dg_1}{dx}(x_0)g_1(x_0) + \frac{dg_2}{dx}(x_0)g_1(x_0) + \frac{1}{2}\frac{dg_2}{dx}(x_0)g_2(x_0) + \cdots\right).$$

Here we have used the Taylor series expansion of $g_2(x_0 + hg_1(x_0)) = g_2(x_0) + h\frac{dg_2}{dx}(x_0)g_1(x_0) + \cdots$. Continuing this one step further yields (the details are interesting to verify) that

$$x(3h) = x_0 + hg_2 + h^2\left(\frac{dg_2}{dx}g_1 - \frac{dg_1}{dx}g_2 + \frac{1}{2}\frac{dg_2}{dx}g_2\right) + \cdots.$$

Finally, we have

$$x(4h) = x_0 + h^2\left(\frac{dg_2}{dx}g_1 - \frac{dg_1}{dx}g_2\right) + \cdots$$

Thus, up to terms of order h^2, we get a change in x along the direction of

$$\left(\frac{dg_2}{dx}(x_0)g_1(x_0) - \frac{dg_1}{dx}(x_0)g_2(x_0)\right).$$

This is precisely the direction that we have defined to be the **Lie bracket** between g_1, g_2 namely $[g_1, g_2]$. Thus if $[g_1, g_2]$ is not in the span of g_1, g_2, it follows that we can steer along this direction. It is also not difficult to imagine that by choosing more elaborate switching between the inputs that we could steer the system along higher level brackets, for example $[g_1, [g_1, g_2]]$, $[[g_1, g_2], [g_1, [g_1, g_2]]]$, and so on. Nevertheless, the preceding analysis is somewhat formal. Also the calculations yield only approximate steering control laws. For more explicit steering control schemes we need a little extra machinery, which we will now develop.

11.2   Drift–Free Control Systems

Consider control systems of the form

$$\dot{x} = g_1(x)u_1 + \cdots + g_m(x)u_m. \tag{11.5}$$

We will assume that the $g_i(x)$ are complete in the sense that their flows are defined for all time. In Section 11.5.4 we will study the case where there is non-zero drift $f(x)$. Following the formulation of Hermann and Krener [136], we present the controllability results in this context.

Definition 11.3 Locally Controllable.   *The system of equation (11.5) is said to be* small time locally controllable at x_0 *if given an open subset $V \subset \mathbb{R}^n$ and $x_0, x_1 \in V$, if for all $T > 0$, there exists an admissible control $u : [0, T] \mapsto \mathbb{R}^n$ such that the system can be steered from x_0 to x_1 with $x(t) \in V$ for all $t \in [0, T]$.*

The *reachable set from x_0 in T seconds* is defined by

$$R^V(x_0, T) = \{x \in \mathbb{R}^n : \exists \text{ admissible } u : [0, T] \mapsto \mathbb{R}^m \text{ steering}$$

$$x(0) = x_0 \text{ to } x(T) = x \text{ with } x(t) \in V \; \forall \, t \in [0, T]\}.$$

The set of states that can be reached in up to T seconds is also defined as

$$R^V(x_0, \leq T) = \cup_{0 \leq \tau \leq T} R^V(x_0, \tau).$$

These definitions enable us to characterize local controllability as being equivalent to $R^V(x_0, \leq T)$ containing a neighborhood of x_0 for all neighborhoods V of x_0 and $T > 0$. This, in turn, may be characterized by defining $\Delta = \text{span}\{g_1, \ldots, g_m\}$ and its involutive closure $\bar{\Delta}$, that is, the smallest involutive distribution containing Δ:

$$\bar{\Delta} = \text{span}\{[X_k, [X_{k-1}, [\cdots, [X_2, X_1] \cdots]]], \; k = 2, 3, \ldots\}.$$

Then we have the following characterization of reachable sets:

Theorem 11.4 Characterization of Reachable Sets. *If $\bar{\Delta} = \mathbb{R}^n$ for all x in a neighborhood of x_0, then for any $T > 0$ and any neighborhood V of x_0*

$$\text{int } R^V(x_0, \leq T) \neq \varnothing,$$

that is, the reachable set has nonempty interior.

Proof: The proof follows by recursion. For the first step, choose some vector field $f_1 \in \bar{\Delta}$. For small ϵ_1 define the manifold

$$N_1 = \{\phi_{f_1}(t_1, x) : |t_1| < \epsilon_1\}.$$

By choosing δ_1 sufficiently small we may guarantee that $N_1 \subset V$. For the k-th step, assume that $N_k \subset V$ is a k-dimensional manifold. If $k < n$ then there exists some $x \in N_k$ and $f_{k+1} \in \bar{\Delta}$ such that $f_{k+1} \notin T_x N_k$, since if this were not so, then $\bar{\Delta}_x \subset T_x N_k$ in some neighborhood of x_0. This would contradict the fact that the dimension of $\bar{\Delta}$ is greater than the dimension of N_k. For ϵ_{k+1} small enough define

$$N_{k+1} = \{\phi_{f_{k+1}}(t_{k+1}, x) \circ \cdots \circ \phi_{f_1}(t_1, x) : |t_i| < \delta_i, \; i = 1, \ldots, k+1\} \subset V.$$

From arguments similar to those in the proof of the Frobenius theorem 8.15, it can be verified that N_{k+1} is a $(k+1)$ dimensional manifold, since the f_i are linearly independent. The construction is depicted in Figure (11.1). If $k = n$ then N_k is an n-dimensional manifold, and by construction $N_k \subset R^V(x_0, \delta_1 + \cdots + \delta_k)$. Thus $R^V(x_0, \delta)$ contains an open set in its interior. The conclusion of the theorem follows since δ can be chosen arbitrarily small. $\square$

Theorem 11.5 Controllability Condition. *The interior of $R^V(x_0, \leq T) \neq \varnothing$ for all neighborhoods V of x_0 and $T > 0$ if and only if the system (11.5) is locally controllable at x_0.*

Proof: The sufficiency follows from the definition of local controllability. To prove necessity, we need to show that $R^V(x_0, \leq T)$ contains a neighborhood of

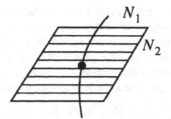

FIGURE 11.1. Illustrating the proof of local controllability

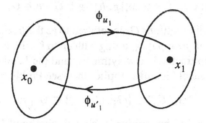

FIGURE 11.2. Two step steering to a neighborhood of a point

the origin. Choose, an input u_1 that steers x_0 to some $x_1 \in R^V(x_0, \leq T/2)$ staying inside V. Denote the flow corresponding to this input ϕ_u. For $T/2 \leq t \leq T$ choose an input $u'_1(t) = -u_1(T-t)$. Since the system has no drift, the input u' reverses the flow, and $\phi_{u'_1}(T, x) = \phi_{u_1}(T/2, x)^{-1}$. By continuity of the flow, $\phi_{u'_1}(T, x)$ maps a neighborhood of x_1 into a neighborhood of x_0. By concatenating inputs that steer a neighborhood of x_0 to a neighborhood of x_1 and the flow of $\phi_{u'_1}$ we see that we can steer the system from x_0 to a neighborhood of x_0 in T seconds. The proof is shown in Figure (11.2). □

Thus, we have seen that for drift free control systems, local controllability is assured by the *controllability rank condition* given by

$$\text{dimension } \bar{\Delta} = \text{span}\{g_1, \ldots, g_m,$$
$$[g_1, g_2], [g_1, g_3], \ldots,$$
$$[g_1, [g_2, g_3]], [g_2, [g_3, g_1]], \ldots\}$$
$$= n.$$

Thus in principle, given two points one can steer between them by covering the line joining them by a finite number of ϵ balls and applying the preceding theorem for steering between points in each of these balls. We say in principle, because this procedure is laborious and not intuitive, though the preceding proofs suggest that piecewise constant controls would do the job of steering the nonlinear system from one location to another. We will return to this method of steering nonlinear systems in the next section.

Here, we will now develop some concepts from the theory of Lie algebras. The level of difficulty in steering controllable systems is proportional to the order of Lie brackets that one has to compute starting from the initial distribution Δ in order to arrive at $\bar{\Delta}$. Thus, we will study the growth of this distribution under different orders of Lie brackets. Given a distribution $\Delta = \mathrm{span}\{g_1, \ldots, g_m\}$ associated with a drift-free control system, define a chain of distributions G_i by $G_1 = \Delta$ and

$$G_i = G_{i-1} + [G_1, G_{i-1}],$$

where

$$[G_1, G_{i-1}] = \mathrm{span}\{[g, h] : g \in G_1, h \in G_{i-1}\}.$$

This chain of nested distributions G_i is referred to as the *filtration* associated with Δ. Each G_i consists of the inputs g_i along with up to $i - 1$ Lie brackets. Of course it is important to keep track of skew symmetry and the Jacobi identity properties of Lie brackets. The Jacobi identity implies that (see Problem 11.3)

$$[G_i, G_j] \subset [G_1, G_{i+j-1}] \subset G_{i+j}. \tag{11.6}$$

The filtration G_i is said to be *regular* if in a neighborhood U of x_0, each of the distributions G_i is regular (has constant rank in U). A control system of the form (11.5) is said to be *regular* if the associated filtration is regular. If a filtration is regular, then at every step of the construction G_i either gains dimension or the construction terminates. If rank $G_{i+1} = $ rank G_i, then G_i is involutive, and hence $G_{i+j} = G_i$ for all j. Since for all i, $G_i \subset T\mathbb{R}^n$ it follows that for a regular filtration, there is an integer $p < n$ such that $G_i = G_{p+1}$ for all $i \geq p + 1$. The smallest such integer p is referred to as the *degree of nonholonomy* of the distribution. The following theorem is a simplified version of a celebrated theorem referred to as Chow's Theorem (which holds for non-regular systems as well).

Theorem 11.6 Chow's Theorem for Regular Systems. *For regular systems of the form of (11.5) there exist admissible controls to steer the system between two given arbitrary points $x_0, x_1 \in U$ iff for some p*

$$G_p(x) = T\mathbb{R}^n \simeq \mathbb{R}^n$$

for all $x \in U$.

Proof: (Necessity) Suppose that $G_p(x) \simeq \mathbb{R}^k$ for some $k < n$. Since the system is regular, the distribution G_p is regular and is involutive. Let N be an integral manifold of G_p through x_0 (note that we are invoking the Frobenius theorem here!). Then if $x_1 \in U$ does not lie on N it is not possible to find an admissible control to steer the system to x_1, since trajectories of (11.5) are confined to N.

(Sufficiency) Join the points x_0, x_1 by a straight line. Cover the line by a finite number of open balls of radius ϵ. Since the system is locally controllable it is possible to steer between arbitrary points (see Theorem 11.5). Now we can steer the system from x_0 to x_1 by steering within the open sets and concatenating the resultant trajectories.   □

A filtration for which $G_p \simeq \mathbb{R}^n$ is said to be *maximally nonholonomic*. The dimensions of the distributions of a regular filtration are recorded as a growth vector $r \in Z^{p+1}$ with

$$r_i = \text{rank } G_i.$$

The *relative growth vector* $\sigma \in Z^{p+1}$ is defined by setting $\sigma_i = r_i - r_{i-1}$ with $r_0 := 0$. The growth vector is bounded above at each step: for example since there are only $m(m - 1)/2$ distinct pairs of the m original g_i, it follows that

$$r_2 \le \frac{m(m - 1)}{2}.$$

For other terms in the filtration one has to be more careful in keeping track of the maximum possible new elements for two reasons:

(i) Skew symmetry

$$[f, g] = -[g, f].$$

(ii) The Jacobi identity

$$[f_1, [f_2, f_3]] + [f_2, [f_3, f_1]] + [f_3, [f_1, f_2]] = 0.$$

A careful calculation (see Problem 11.1) yields that if $\bar{\sigma}_i$ is the maximum growth at the i th stage then

$$\bar{\sigma}_i = \frac{1}{i}\left((\bar{\sigma}_1)^i - \sum_{j|i, j<i} j\bar{\sigma}_j \right), \quad i > 1. \tag{11.7}$$

The notation $j|i, j < i$ means all integers less than i which divide i. If $\sigma_i = \bar{\sigma}_i$ for all i, we say that the filtration has *maximum growth*.

The Philip Hall basis is a systematic way of generating a basis for a Lie algebra generated by some $g_1, \ldots, g_m$, which takes into account skew symmetry and the Jacobi identity. Thus, let $\mathcal{L} (g_1, \ldots, g_m)$ be the Lie algebra generated by a set of vector fields $g_1, \ldots, g_m$. One approach to equipping $\mathcal{L} (g_1, \ldots, g_m)$ with a basis is to list all the generators and all of their Lie products. The problem is that not all Lie products are linearly independent because of skew symmetry and the Jacobi identity.

Given a set of generators $\{g_1, \cdots, g_m\}$, we define the length of a Lie product recursively as

$$l(g_i) = 1, \quad i = 1, \ldots, m,$$

$$l([A, B]) = l(A) + l(B),$$

where A and B are themselves Lie products. Alternatively, $l(A)$ is the number of generators in the expansion for A. A Lie algebra is *nilpotent* if there exists an integer k such that all Lie products of length greater than k are zero. The integer k is called the *order of nilpotency*.

A *Philip Hall basis* is an ordered set of Lie products $H = \{B_i\}$ satisfying:

1. $g_i \in H, i = 1, \ldots, m$.
2. If $l(B_i) < l(B_j)$, then $B_i < B_j$.
3. $[B_i, B_j] \in H$ if and only if

   a. $B_i, B_j \in H$ and $B_i < B_j$,
   b. either $B_j = g_k$ for some k or $B_j = [B_l, B_r]$ with $B_l, B_r \in H$ and $B_l \leq B_i$.

The proof that a Philip Hall basis is indeed a basis for the Lie algebra generated by $\{g_1, \ldots, g_m\}$ is beyond the scope of this book and may be found in [127] and [270]. A Philip Hall basis that is nilpotent of order k (meaning to say that all Lie products of order $k + 1$ and higher are zero) can be constructed from a set of generators using this definition. The simplest approach is to construct all possible Lie products with length less than k and use the definition to eliminate elements which fail to satisfy one of the properties. In practice, the basis can be built in such a way that only condition 3 need be checked.

Example 11.7 Philip Hall Basis of Order 3. *A basis for the nilpotent Lie algebra of order 3 generated by g_1, g_2, g_3 is*

$$
\begin{array}{llll}
g_1, & g_2, & g_3, & \\
[g_1, g_2], & [g_2, g_3]], & [g_3, g_1], & \\
[g_1, [g_1, g_2]], & [g_1, [g_1, g_3]], & [g_2, [g_1, g_2]], & [g_2, [g_1, g_3]], \\
[g_2, [g_2, g_3]], & [g_3, [g_1, g_2]], & [g_3, [g_1, g_3]] & [g_3, [g_2, g_3]].
\end{array}
$$

Note that $[g_1, [g_2, g_3]]$ *does not appear, since*

$$[g_1, [g_2, g_3]] + [g_2, [g_3, g_1]] + [g_3, [g_1, g_2]] = 0$$

by the Jacobi identity and the second two terms in the formula are already present.

11.3   Steering of Drift–Free Nonholonomic Systems

In this section we introduce the reader to some of the general approaches to steering drift-free control systems that are *maximally nonholonomic*, that is to say the filtration generated by the input vector fields saturates with $G_p \simeq \mathbb{R}^n$ for some p. In the sequel we refer to these simply as *nonholonomic*. We first outline three approaches:

Optimal Control

Perhaps the best formulated method for finding trajectories of a general control system is the use of optimal control. By attaching a cost functional to each trajectory, we can limit our search to trajectories which minimize a cost function. Typical cost functions might be the length of the path (in some appropriate metric), the control cost, or the time required to execute the trajectory. If the system has bounds on the magnitudes of the inputs, it makes sense to solve the motion planning problem in *minimum time*. It is well known that for many problems, when the set of allowable inputs is convex, then the time-optimal paths result from saturating the inputs at all times (this is often referred to as *bang-bang* control). The inputs may change between one set of values and another at a possibly infinite number of *switching times*. Choosing an optimal trajectory is then equivalent to choosing the switching times and the values of the inputs between switching times.

Piecewise Constant Inputs

Related to the bang–bang trajectories of optimal control, it is also possible to steer nonholonomic control systems using piecewise constant inputs. Perhaps the most naive way of using constant inputs is to pick a time interval and generate a graph by applying all possible sequences of inputs (discretized to finitely many values). Each node on the graph corresponds to a configuration, and branches indicate the choice of a fixed control over the time interval. The size of the graph grows as m^d, where m is the number of input combinations considered at each step and d is the number of steps. Since we do not know how long to search, the amount of computer memory required by such an algorithm can be very large. Also, we are likely not to hit our goal exactly, so some post-processing must be done if exact maneuvering is needed.

Recently, a very elegant and general motion planning method using piecewise constant inputs has been developed by Lafferriere and Sussmann [173]. A different so-called multi-rate method inspired by techniques from discrete time nonlinear control using controls held constant over different intervals of time is interesting as well [215] Consider the case of a *nilpotent* system, that is one whose controllability distribution Δ is nilpotent of order k for some k (that is, if Δ is spanned by $g_1, \ldots, g_p$, then all the Lie products with more than k terms vanish). The advantage of nilpotent Lie distributions is that certain computations are greatly simplified, as we shall see in Section 11.5. The main tool in their method is the Campbell–Baker–Hausdorff formula. If the system is not nilpotent, it can be shown that if the initial and final points are close, then the algorithm of Lafferriere and Sussmann moves the original system closer to the goal by at least half. By breaking the path up into small pieces, we can move arbitrarily close to the goal with repeated applications of the algorithm.

Canonical Paths

A third approach to steering nonholonomic control systems is by choosing a family of paths that can be used to produce desired motions. For example, we might consider paths for a car that cause a net rotation of any angle or a translation in the direction that the car is facing. We can then move to any configuration by reorienting the car, driving forward, and again reorienting the car. The path used to cause a net rotation might consist of a set of parameterized piecewise constant inputs or a heuristic trajectory. The set of canonical paths used for a given problem is usually specific to that problem. In some cases the paths may be derived from some unifying principle. For example, if we could solve the optimal control problem in closed form, these optimal paths would form a set of canonical paths. In the case of time-optimal control, we might consider paths corresponding to saturated inputs as canonical paths, but since it is not clear how to combine these paths to get a specific motion, we distinguish these lower level paths from canonical paths. Canonical paths have been used by Li and Canny to study the motion of a spherical fingertip on an object [183].

We now give a bit more detail about some of the various methods of nonholonomic motion planning being pursued in the literature. The following subsections are not organized according to the different approaches, but from a pedagogical point of view. In Section 11.4, we study the steering of "model" versions of our nonholonomic system,

$$\dot{q} = g_1(q)u_1 + \cdots + g_m(q)u_m. \tag{11.8}$$

By model systems we mean those that are in some sense canonical. We use some results from optimal control to explicitly generate optimal inputs for a class of systems. The class of systems that we consider, called first-order control systems, have sinusoids as optimal steering inputs. Motivated by this observation, we explore the use of sinusoidal input signals for steering second- and higher-order model control systems. This study takes us forward to a very important model class which we refer to as chained form systems.

In Section 11.5 we begin by applying the use of sinusoidal inputs to steering systems that are not in a model form. We do so by using some elementary Fourier analysis in some cases of systems that resemble chained form systems. The question of when a given system can be converted into a chained form system is touched upon as well. Then we move on to the use of approximations of the optimal input class by the Ritz approximation technique. Finally, we discuss the use of piecewise constant inputs to solve the general motion planning problem.

11.4   Steering Model Control Systems Using Sinusoids

In this section we study techniques for steering certain "model" control systems; that is, systems which are, in a certain sense canonical. The use of sinusoids at integrally related frequencies is motivated by the results of Brockett [46] in the context of optimally steering a class of systems. This section begins with a review of his results for a class of systems whose degree of nonholonomy is two and growth vector is $(m, m(m + 1)/2)$. The technique is then extended to certain other classes of model control systems, with specific attention being paid to a class of systems referred to as chained form systems. Techniques for steering using methods other than sinusoids and systems other than the model systems considered in this section are deferred to Section 11.5.

First-Order Controllable Systems: Brockett's System

By a *first-order controllable* system, we mean a control system of the form

$$\dot{q} = g_1(q)u_1 + \cdots + g_m(q)u_m,$$

where the vector fields $g_i(q), i = 1, \ldots, m$, and their first-order Lie brackets $[g_j, g_k], j < k, k = 1, \ldots, m$ are linearly independent and furthermore, we have that

$$T_q\mathbb{R}^n = \text{span}\{g_i, [g_j, g_k] : i, j, k = 1, \ldots, m\}.$$

In particular, this implies that $n = m + m(m - 1)/2 = m(m + 1)/2$. A very important class of model control systems that satisfy this condition was proposed by Brockett. This system has origins in the classical Heisenberg system. We begin with a discussion of this class for the case that $m = 2$ and $n = m(m + 1)/2 = 3$,

$$\dot{q}_1 = u_1,$$
$$\dot{q}_2 = u_2, \tag{11.9}$$
$$\dot{q}_3 = q_1 u_2 - q_2 u_1.$$

For this system,

$$g_1 = \begin{bmatrix} 1 \\ 0 \\ -q_2 \end{bmatrix}, \quad g_2 = \begin{bmatrix} 0 \\ 1 \\ q_1 \end{bmatrix}, \quad [g_1, g_2] = \begin{bmatrix} 0 \\ 0 \\ 2 \end{bmatrix}.$$

Thus the system is maximally nonholonomic with degree of nonholonomy 2 and growth vector $(2, 3)$. We will consider the problem of steering system (11.9) from $q_0 \in \mathbb{R}^3$ at $t = 0$ to $q_f \in \mathbb{R}^3$ at $t = 1$. In fact, we will do so as to minimize the least squares control cost given by

$$\int_0^1 |u|^2 \, dt.$$

Using the fact that $\dot{q}_i = u_i$ for $i = 1, 2$, an equivalent description of the last equation in (11.9) is as a constraint of the form

$$\dot{q}_3 = q_1 \dot{q}_2 - q_2 \dot{q}_1.$$

Similarly, the Lagrangian to be minimized can be written as $\dot{q}_1^2 + \dot{q}_2^2$. Using a Lagrange multiplier $\lambda(t)$ for the constraint, we augment the Lagrangian to be minimized as follows:

$$L(q, \dot{q}) = (\dot{q}_1^2 + \dot{q}_2^2) + \lambda(\dot{q}_3 - q_1\dot{q}_2 + q_2\dot{q}_1).$$

The method for minimizing this constrained Lagrangian is to use the classical calculus of variations for the Lagrangian $L(q, \dot{q})$ above, with the control system written as a constraint with a Lagrange multiplier (the reader wishing to learn more about optimal control may consult one of several nice books on the subject, such as [334]). There it is shown that stationary solutions satisfy the Euler–Lagrange equations. The Lagrange multiplier $\lambda(t)$ is determined using the form of the constraint. The fact that the equations are precisely the Euler–Lagrange equations of dynamics should come as no surprise when one considers that the dynamical equations may be derived from a least-action principle. The *Euler–Lagrange equations* for minimizing the Lagrangian of our optimal control problem are

$$\frac{d}{dt}\left(\frac{\partial L(q, \dot{q})}{\partial \dot{q}_i}\right) - \frac{\partial L(q, \dot{q})}{\partial q_i} = 0,$$

or equivalently,

$$\ddot{q}_1 + \lambda \dot{q}_2 = 0,$$
$$\ddot{q}_2 - \lambda \dot{q}_1 = 0, \qquad\qquad (11.10)$$
$$\dot{\lambda} = 0.$$

Equation (11.10) establishes that $\lambda(t)$ is constant and, in fact, that the optimal inputs satisfy the equation

$$\begin{bmatrix} \dot{u}_1 \\ \dot{u}_2 \end{bmatrix} = \begin{bmatrix} 0 & -\lambda \\ \lambda & 0 \end{bmatrix}\begin{bmatrix} u_1 \\ u_2 \end{bmatrix} := \Lambda \begin{bmatrix} u_1 \\ u_2 \end{bmatrix}.$$

Note that the matrix Λ is skew symmetric with λ constant, so that the optimal inputs are sinusoids at frequency λ; thus,

$$\begin{bmatrix} u_1(t) \\ u_2(t) \end{bmatrix} = \begin{bmatrix} \cos \lambda t & -\sin \lambda t \\ \sin \lambda t & \cos \lambda t \end{bmatrix}\begin{bmatrix} u_1(0) \\ u_2(0) \end{bmatrix} := e^{\Lambda t} u(0)$$

Having established this functional form of the optimal controls, given values for q_0 and q_f one can solve for the $u(0)$ and λ required to steer the system optimally. However, from the form of the control system (11.9), it is clear that the states q_1 and q_2 may be steered directly. Thus, it is of greatest interest to steer from $q(0) = (0, 0, 0)$ to $q(1) = (0, 0, a)$. By directly integrating for q_1 and q_2 we have that

$$\begin{bmatrix} q_1(t) \\ q_2(t) \end{bmatrix} = (e^{\Lambda t} - I)\Lambda^{-1} u(0).$$

Since $q_1(1) = q_2(1) = 0$, it follows that $e^{\Lambda} = I$. Thus $\lambda = 2n\pi$, $n = 0, \pm 1, \pm 2, \ldots$. Integrating $\dot{q}_3$ yields

$$q_3(1) = \int_0^1 (q_1 u_2 - q_2 u_1)\, dt = -\frac{1}{\lambda}\left(u_1^2(0) + u_2^2(0)\right) = a$$

Further, the total cost is

$$\int_0^1 |u|^2\, dt = |u(0)|^2 = -\lambda a.$$

Since $\lambda = 2n\pi$, it follows that the minimum cost is achieved for $n = -1$ and that $|u(0)|^2 = 2\pi a$. However, apart from its magnitude, the direction of $u(0) \in \mathbb{R}^2$ is arbitrary. Thus, the optimal input steering the system between the points $(0, 0, 0)$ and $(0, 0, a)$ is a sum of sines and cosines at a frequency 2π (more generally $(2\pi)/T$ if the time period of the steering is T).

The generalization of the system (11.9) to an m-input system is the system

$$\begin{aligned} \dot{q}_i &= u_i, & i = 1, \ldots, m, \\ \dot{q}_{ij} &= q_i u_j - q_j u_i, & i < j = 1, \ldots, m. \end{aligned} \tag{11.11}$$

A slightly more pleasing form of this equation is obtained by forming a skew-symmetric matrix $Y \in so(m)$ with the $-q_{ij}$ as the bottom lower half (below the diagonal) to give a control system in $\mathbb{R}^m \times so(m)$:

$$\begin{aligned} \dot{q} &= u, \\ \dot{Y} &= qu^T - uq^T. \end{aligned} \tag{11.12}$$

The Euler–Lagrange equations for this system are an extension of those for the two–input case, namely

$$\ddot{q} - \Lambda \dot{q} = 0,$$

$$\dot{\Lambda} = 0,$$

where $\Lambda \in so(m)$ is the skew-symmetric matrix of Lagrange multipliers associated with Y. As before, Λ is constant, and the optimal input satisfies

$$\dot{u} = \Lambda u,$$

so that

$$u(t) = e^{\Lambda t} u(0).$$

It follows that $e^{\Lambda t} \in SO(m)$. It is of special interest to determine the nature of the input when $q(0) = q(1) = 0$, $Y(0) = 0$, and $Y(1)$ is a given matrix in $so(m)$. In this context, an amazing fact that has been shown by Brockett [46] is that when m is even and Y is nonsingular, the input has $m/2$ sinusoids at frequencies

$$2\pi, 2 \cdot 2\pi, \ldots, m/2 \cdot 2\pi.$$

If m is odd, then Y is of necessity singular, but if it is of rank $m - 1$, then the input has $(m - 1)/2$ sinusoids at frequencies

$$2\pi, 2 \cdot 2\pi, \ldots, \frac{m - 1}{2} \cdot 2\pi.$$

While the proof of this fact is somewhat involved and would take us afield from what we would like to highlight in this section, we may use the fact to propose the following algorithm for steering systems of the form of equation (11.11):

Algorithm 1 Steering First-Order Canonical Systems.

1. Steer the q_i to their desired values using any input and ignoring the evolution of the q_{ij}.
2. Using sinusoids at integrally related frequencies, find u_0 such that the input steers the q_{ij} to their desired values. By the choice of input, the q_i are unchanged.

The algorithm involves steering the states step by step. The states that are directly controlled (zeroth order) are steered first, and then the first order Lie bracket directions are steered.

Second-Order Controllable systems

Consider systems in which the first level of Lie bracketing is not enough to span $T_q \mathbb{R}^n$. We begin by trying to extend the previous canonical form to the next higher level of bracketing:

$$\dot{q}_i = u_i, \qquad i = 1, \ldots, m,$$
$$\dot{q}_{ij} = q_i u_j, \qquad 1 \le i < j \le m, \qquad \qquad (11.13)$$
$$\dot{q}_{ijk} = q_{ij} u_k, \qquad 1 \le i, j, k \le m \qquad \text{(mod Jacobi identity)}.$$

Because the Jacobi identity imposes relationships between Lie brackets of the form

$$[g_i, [g_j, g_k]] + [g_k, [g_i, g_j]] + [g_j, [g_k, g_i]] = 0$$

for all i, j, k, it follows that not all state variables of the form of q_{ijk} are controllable. For this reason, we refer to the last of the preceding equations as "mod Jacobi identity." Indeed, a straightforward but somewhat laborious computation shows that

$$q_{231} - q_{132} = q_1 q_{23} - q_2 q_{13}.$$

From Problem 11.1 it may be verified that the maximum number of controllable q_{ijk} is

$$\frac{(m+1)m(m-1)}{3}.$$

Constructing the Lagrangian with the same integral cost criterion as before and deriving the Euler–Lagrange equations does not, in general, result in a constant set of Lagrange multipliers. For the case of $m = 2$, Brockett and Dai [48] have shown that the optimal inputs are elliptic functions (see also the next section). However, we can extend Algorithm 1 to this case as follows:

Algorithm 2 Steering Second-Order Model Systems.

1. Steer the q_i to their desired values. This causes drift in all the other states.
2. Steer the q_{ij} to their desired values using integrally related sinusoidal inputs. If the i th input has frequency ω_i, then q_{ij} will have frequency components at $\omega_i \pm \omega_j$. By choosing inputs such that we get frequency components at zero, we can generate net motion in the desired states.

3. Use sinusoidal inputs a second time to move all the previously steered states in a closed loop and generate net motion only in the q_{ijk} direction. This requires careful selection of the input frequencies such that $\omega_i \pm \omega_j \neq 0$ but $\omega_i + \omega_j + \omega_k$ has zero-frequency components.

The required calculations for Step 2 above are identical to those in Algorithm 1. A general calculation of the motion in Step 3 is quite cumbersome, although for specific systems of practical interest the calculations are quite straightforward. For example, if $m = 2$, equation (11.13) becomes

$$\dot{q}_1 = u_1,$$

$$\dot{q}_2 = u_2,$$

$$\dot{q}_{12} = q_1 u_2,$$

$$\dot{q}_{121} = q_{12} u_1,$$

$$\dot{q}_{122} = q_{12} u_2.$$

To steer q_1, q_2, and q_{12} to their desired locations, we apply Algorithm 1. To steer q_{121} independently of the other states, choose $u_1 = a \sin 2\pi t$, $u_2 = b \cos 4\pi t$ to obtain

$$q_{121}(1) = q_{121}(0) - \frac{a^2 b}{16\pi^2}.$$

Similarly, choosing $u_1 = b \cos 4\pi t$, $u_2 = a \sin 2\pi t$ gives

$$q_{122}(1) = q_{122}(0) + \frac{a^2 b}{32\pi^2},$$

and all the other states return to their original values.

Both the algorithms presented above require separate steps to steer in each of the q_{ijk} directions. It is also possible to generate net motion in multiple coordinates simultaneously by using a linear combination of sinusoids and by solving a polynomial equation for the necessary coefficients (see Problem 11.5).

Higher-Order Systems: Chained Form Systems

We now study more general examples of nonholonomic systems and investigate the use of sinusoids for steering such systems. As in the previous section, we may try to generate canonical classes of higher-order systems, i.e., systems where more than one level of Lie brackets is needed to span the tangent space to the configuration space. Such a development is given by Grayson and Grossmann [118], and in [226] Murray and Sastry showed in full generality that it is difficult to use sinusoids to steer such systems. This leads us to specialize to a smaller class of higher-order systems, which we refer to as chained systems, that can be steered using sinusoids at integrally related frequencies. These systems are interesting in their own right as well, since they are duals of a classical construction in differential forms referred to as the Goursat normal form (see Chapter 12). Further, we can convert many other nonlinear systems into chained form systems as we discuss preliminarily in the next section and then in greater detail in the next chapter.

Consider a two-input system of the following form:

$$\dot{q}_1 = u_1,$$
$$\dot{q}_2 = u_2,$$
$$\dot{q}_3 = q_2 u_1,$$
$$\dot{q}_4 = q_3 u_1, \tag{11.14}$$
$$\vdots$$
$$\dot{q}_n = q_{n-1} u_1.$$

In vector field form, equation (11.14) becomes

$$\dot{q} = g_1 u_1 + g_2 u_2$$

with

$$g_1 = \begin{bmatrix} 1 \\ 0 \\ q_2 \\ q_3 \\ \vdots \\ q_{n-1} \end{bmatrix}, \qquad g_2 = \begin{bmatrix} 0 \\ 1 \\ 0 \\ 0 \\ \vdots \\ 0 \end{bmatrix}. \tag{11.15}$$

We define the system (11.14) as a *one-chain system*. The first item is to check the controllability of these systems. Recall that the iterated Lie products as $\mathrm{ad}_{g_1}^k g_2$, defined by

$$\mathrm{ad}_{g_1} g_2 = [g_1, g_2], \qquad \mathrm{ad}_{g_1}^k g_2 = [g_1, \mathrm{ad}_{g_1}^{k-1} g_2] = [g_1, [g_1, \ldots, [g_1, g_2] \ldots]].$$

Lemma 11.8 Lie Bracket Calculations. *For the vector fields in equation (11.15), with $k \geq 1$,*

$$\mathrm{ad}_{g_1}^k g_2 = \begin{bmatrix} 0 \\ \vdots \\ (-1)^k \\ \vdots \\ 0 \end{bmatrix}.$$

(Here the only nonzero entry is in the $(k+2)$-th entry.)

Proof: By induction. Since the first level of brackets is irregular, we begin by expanding $[g_1, g_2]$ and $[g_1, [g_1, g_2]]$ to get

$$[g_1, g_2] = \begin{bmatrix} 0 \\ 0 \\ -1 \\ 0 \\ 0 \\ \vdots \\ 0 \end{bmatrix}, \quad [g_1, [g_1, g_2]] = \begin{bmatrix} 0 \\ 0 \\ 0 \\ 1 \\ 0 \\ \vdots \\ 0 \end{bmatrix}.$$

Now assume that the formula is true for k. Then

$$\mathrm{ad}_{g_1}^{k+1} g_2 = [g_1, \mathrm{ad}_{g_1}^k g_2] = \begin{bmatrix} 0 \\ \vdots \\ 0 \\ (-1)^{k+1} \\ 0 \\ \vdots \\ 0 \end{bmatrix}. \qquad \square$$

Proposition 11.9 Controllability of the One-Chain System. *The one-chain system (11.14) is maximally nonholonomic (controllable).*

Proof: There are n coordinates in equation (11.14) and the n Lie products

$$\{g_1, g_2, \mathrm{ad}_{g_1}^i g_2\}, \qquad 1 \leq i \leq n - 2,$$

are independent by Lemma 11.8. $\qquad \square$

To steer this system, we use sinusoids at integrally related frequencies. Roughly speaking, if we use $u_1 = \sin 2\pi t$ and $u_2 = \cos 2\pi k t$, then $\dot{q}_3$ will have components at frequency $2\pi (k - 1)$, $\dot{q}_4$ at frequency $2\pi (k - 2)$, etc. $\dot{q}_{k+2}$ will have a component at frequency zero and when integrated gives motion in q_{k+2} while all previous variables return to their starting values.

Algorithm 3 Steering Chained Form Systems.

1. Steer q_1 and q_2 to their desired values.
2. For each q_{k+2}, $k \geq 1$, steer q_k to its final value using $u_1 = a \sin 2\pi t$, $u_2 = b \cos 2\pi k t$, where a and b satisfy

$$q_{k+2}(1) - q_{k+2}(0) = \left(\frac{a}{4\pi}\right)^k \frac{b}{k!}.$$

Proposition 11.10 Chained form algorithm. *Algorithm 3 can steer system* *(11.14) to an arbitrary configuration.*

Proof: The proof is constructive. We first show that using $u_1 = a \sin 2\pi t$, $u_2 = b \cos 2\pi k t$ produces motion only in q_{k+2} and not in q_j, $j < k + 2$ after one period, by direct integration. If q_{k-1} has terms at frequency $2\pi n_i$, then q_k has corresponding terms at $2\pi(n_i \pm 1)$ (by expanding products of sinusoids as sums of sinusoids). Since the only way to have $q_i(1) \neq q_i(0)$ is to have q_i have a component at frequency zero, it suffices to keep track only of the lowest frequency component in each variable; higher components will integrate to zero. Direct computation starting from the origin yields

$$q_1 = \frac{a}{2\pi}(1 - \cos 2\pi t),$$

$$q_2 = \frac{b}{2\pi k} \sin 2\pi k t,$$

$$q_3 = \int \frac{ab}{2\pi k} \sin 2\pi k t \sin 2\pi t \, dt,$$

$$= \frac{1}{2} \frac{ab}{2\pi k} \left(\frac{\sin 2\pi(k-1)t}{2\pi(k-1)} - \frac{\sin 2\pi(k+1)t}{2\pi(k+1)} \right),$$

$$q_4 = \frac{1}{2} \frac{a^2 b}{2\pi k \cdot 2\pi(k-1)} \int \sin 2\pi(k-1)t \cdot \sin 2\pi t \, dt + \cdots,$$

$$= \frac{1}{2^2} \frac{a^2 b}{2\pi k \cdot 2\pi(k-1) \cdot 2\pi(k-2)} \sin 2\pi(k-2)t + \cdots,$$

$$\vdots$$

$$q_{k+2} = \int \frac{1}{2^{k-1}} \frac{a^k b}{2\pi k \cdot 2\pi(k-1) \cdots 2\pi} \sin^2 2\pi t \, dt + \cdots,$$

$$= \frac{1}{2^{k-1}} \frac{a^k b}{(2\pi)^k k!} \frac{t}{2} + \cdots.$$

It follows that

$$q_{k+2}(1) = q_{k+2}(0) + \left(\frac{a}{4\pi} \right)^k \frac{b}{k!},$$

and all earlier q_i's are periodic and hence $q_i(1) = q_i(0)$, $i < k$. If the system does not start at the origin, the initial conditions generate extra terms of the form $q_{i-1}(0)u_1$ in the i-th derivative, and this integrates to zero, giving no net contribution. □

For the case of systems with more than two inputs, there is a generalization to *multi-chained* form systems; we refer the reader to [226], [53], [302], and [304].

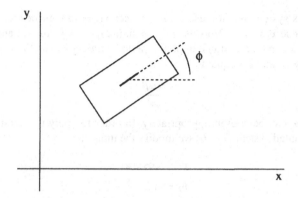

FIGURE 11.3. Steerable unicycle. The unicycle has two independent inputs: the *steering* input controls the angle of the wheel, ϕ; the *driving* input controls the velocity of the cart in the direction of the wheel. The configuration of the cart is its Cartesian location and the wheel angle.

11.5   General Methods for Steering

Model control systems of the kind that we discussed in the previous section will very seldom show up verbatim in applications. In this section, we consider some techniques in motion planning for more general nonholonomic systems.

11.5.1   Fourier Techniques

The methods involving sinusoids at integrally related frequencies can be modified using some elementary Fourier analysis to steer systems which are not in any of the model classes that we have discussed. We will illustrate these notions on two examples.

First, we illustrate the application of this method to the steering of a unicycle as shown in Figure (11.3). If u_1 denotes the driving velocity and u_2 the steering velocity, the functional form of the state equations for this system is

$$\dot{x} = \cos \phi \, u_1,$$

$$\dot{y} = \sin \phi \, u_1, \qquad\qquad (11.16)$$

$$\dot{\phi} = u_2.$$

An approximation to this system is obtained by setting $\cos \phi \approx 1$, $\sin \phi \approx \phi$, and relabeling x as x_1, y as x_3, and ϕ as x_2 to get

$$\dot{x}_1 = u_1,$$

$$\dot{x}_2 = u_2, \qquad\qquad (11.17)$$

$$\dot{x}_3 = x_2 u_1.$$

To steer this system, we first use u_1, u_2 to steer x_1, x_2 to their desired locations; this may cause x_3 to drift. Now use $u_1 = \alpha \sin(\omega t)$, $u_2 = \beta \cos(\omega t)$ and note that after $2\pi/\omega$ seconds, x_1 and x_2 complete a periodic trajectory and the x_3 coordinate advances by an amount equal to

$$\frac{\pi \alpha \beta}{\omega^2}.$$

Now α, β, ω can be chosen appropriately. In order to apply this strategy to the unapproximated system (11.16) we modify the input to

$$v_1 = \cos \phi \, u_1,$$

$$v_2 = u_2.$$

and relabel the states to get

$$\dot{x}_1 = v_1,$$

$$\dot{x}_2 = v_2, \tag{11.18}$$

$$\dot{x}_3 = (\tan x_2)v_1.$$

As before, we steer x_1 and x_2 using v_1, v_2. To steer the third variable, we use $v_1 = \alpha \sin(\omega t)$, $v_2 = \beta \cos(\omega t)$ as before. Then

$$\dot{x}_3 = \tan \left(\frac{\beta}{\omega} \sin \omega t \right) * \alpha \sin(\omega t). \tag{11.19}$$

The value of x_3 after $2\pi/\omega$ seconds is determined by the constant part of the right hand side of (11.19). The constant coefficient is given by

$$\frac{1}{2} \cdot \frac{1}{\pi} \int_{-\pi}^{\pi} \tan \left(\frac{\beta}{\omega} \sin \theta \right) \alpha \sin(\theta) d\theta.$$

Sample trajectories for this scenario are shown in Figure (11.4)

For the second example, we consider a system with degree of nonholonomy 2. An example of such a system is a front wheel drive cart of the form shown in Figure (11.5). As in the case of the previous example u_1 is the driving velocity and u_2 the steering velocity. The equations of this cart are

$$\dot{x} = (\cos \theta \cos \phi) \, u_1,$$

$$\dot{y} = (\sin \theta \cos \phi) \, u_1,$$

$$\dot{\phi} = u_2, \tag{11.20}$$

$$\dot{\theta} = \left(\frac{1}{l} \sin \phi \right) u_1.$$

The form of the equations shows that when $\phi = \pi/2$, the cart cannot be driven forward. As in the previous section an approximation to this system is instructive. Relabeling the variables x, y, ϕ, θ as x_1, x_4, x_2, x_3, setting $l = 1$, and

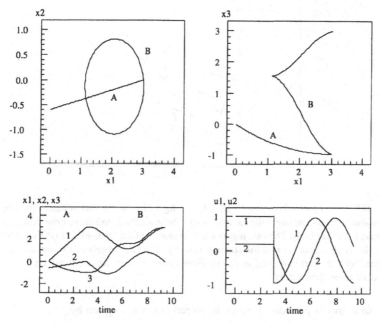

FIGURE 11.4. Sample trajectories for the unicycle. The trajectory shown is a two stage path which moves the unicycle from $(0, -.6, 0)$ to $(3, 0, 3)$. The first portion of the path, labeled A, drives the x_1 and x_2 states to their desired values using a constant input. The second portion, labeled B, uses a periodic input to drive x_3 while bringing the other two states back to their desired values. The top two figures show the states versus x_1; the bottom figures show the states and inputs as functions of time.

approximating sines and cosines as before yields

$$\dot{x}_1 = u_1,$$
$$\dot{x}_2 = u_2,$$
$$\dot{x}_3 = x_2 u_1, \qquad (11.21)$$
$$\dot{x}_4 = x_3 u_1.$$

Note that $\text{span}\{g_1, g_2, [g_1, g_2], [g_1, [g_1, g_2]]\} = \mathbb{R}^n$, so that system is a regular system with degree of nonholonomy 2. It is also easy to verify that this condition also holds for the original system. Steering the states x_1, x_2, x_3 of (11.21) is immediate from the previous section. To steer x_4 note that if

$$u_1 = \alpha \cos \omega t, \qquad u_2 = \beta \cos 2\omega t,$$

then x_1, x_2 and x_3 are all periodic and return to their initial values after $2\pi/\omega$ seconds. Also

$$x_3 = -\frac{\alpha\beta}{4\omega^2} \cos \omega t - \frac{\alpha\beta}{12\omega^2} \cos 3\omega t,$$

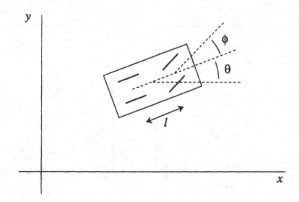

FIGURE 11.5. Front wheel drive cart. The configuration of the cart is determined by its Cartesian location, the angle the car makes with the horizontal, and the steering wheel angle relative to the car body. The two inputs are the velocity of the front wheels (in the direction the wheels are pointing) and the steering velocity. The rear wheels of the cart are always aligned with the cart body and are constrained to move along the line in which they point or rotate about their center.

so that it, too, is periodic. Finally the increment in x_4 is given by

$$-\frac{\pi \alpha^2 \beta}{4\omega^3}.$$

To carry this development through for the unapproximated system define $v_1 = u_1 \cos \theta \cos \phi$ and $v_2 = u_2$. Then with the same relabeling as before, the equations become

$$\dot{x}_1 = v_1,$$
$$\dot{x}_2 = v_2,$$
$$\dot{x}_3 = \frac{\tan x_2}{\cos x_3} v_1, \qquad (11.22)$$
$$\dot{x}_4 = \tan x_3 \, v_1.$$

We refer to such systems as *triangular* but not *strictly triangular*, since $\dot{x}_3$ depends on x_3. By approximating $\cos x_3$ by 1, the equations become strictly triangular; using $v_1 = \alpha \cos \omega t$, $v_2 = \beta \cos 2\omega t$ we can solve for the Fourier series coefficients of x_1, x_2, x_3 and x_4. Note that only the Fourier coefficient corresponding to the zero frequency is needed to get the change in x_4 after one time period. We have a sense now that it is easy to steer drift free systems if they can be put into the "triangular" form discussed earlier using sinuoids. A natural question arises as to whether arbitrary systems of higher order of nonholonomy can be put into triangular form. The answer to this question is not easy, and we refer the reader to the work of Bicchi in this regard (see his survey [25]). Another question is whether systems can be converted into the chained form that we can steer as shown above. The answer to this question is the subject of some discussion in the next section and also the next chapter.

Conversion to Chained Form

In view of the discussion of the previous section, an interesting question to ask is whether it is possible using a change of input and nonlinear transformation of the coordinates to convert a given nonholonomic control system into one of the model forms discussed in the previous section. More precisely, given the system

$$\dot{q} = g_1(q)u_1 + \cdots + g_m(q)u_m,$$

does there exist a matrix $\beta(q) \in \mathbb{R}^{m \times m}$ and a diffeomorphism $\phi : \mathbb{R}^n \to \mathbb{R}^n$ such that with

$$v = \beta(q)u, \qquad z = \phi(q),$$

the system is in chained form in the z coordinates with inputs v? One can give necessary and sufficient conditions to solve this problem (see [223]), but the discussion of these conditions would take us into far too much detail for this section. In [226] conditions for the two-input case, for converting the system to the one-chain form were given. The conditions assume that g_1, g_2 are linearly independent. The system can be converted into one-chain form if the following distributions are all of constant rank and involutive:

$$\Delta_0 = \text{span}\{g_1, g_2, ad_{g_1}g_2, \ldots, ad_{g_1}^{n-2}g_1\},$$
$$\Delta_1 = \text{span}\{g_2, ad_{g_1}g_2, \ldots, ad_{g_1}^{n-2}g_2\},$$
$$\Delta_2 = \text{span}\{g_2, ad_{g_1}g_2, \ldots, ad_{g_1}^{n-3}g_2\},$$

and there exists a function $h_1(q)$ such that

$$dh_1 \cdot \Delta_1 = 0 \quad \text{and} \quad dh_1 \cdot g_1 = 1.$$

If these conditions are met, then a function h_2 independent of h_1 may be chosen so that

$$dh_2 \cdot \Delta_2 = 0.$$

The existence of independent h_1 and h_2 so that $dh_1 \cdot \Delta_1 = 0, dh_2 \cdot \Delta_2 = 0$ is guaranteed by the Frobenius theorem, since $\Delta_2 \subset \Delta_1$ are both involutive distributions. There is, however, an added condition on h_1, namely that $dh_1 \cdot g_1 = 1$. If we can find these functions h_1, h_2, then the map $\phi : q \mapsto z$ and input transformation given by

$$z_1 = h_1,$$
$$v_1 := u_1,$$
$$z_2 = L_{g_1}^{n-2}h_2,$$
$$v_2 := (L_{g_1}^{n-1}h_2)u_1 + (L_{g_2}L_{g_1}^{n-2}h_2)u_2,$$
$$\vdots$$
$$z_{n-1} = L_{g_1}h_2,$$
$$z_n = h_2.$$

yields

$$\dot{z}_1 = v_1,$$
$$\dot{z}_2 = v_2,$$
$$\dot{z}_3 = z_2 v_1,$$
$$\vdots$$
$$\dot{z}_n = z_{n-1} v_1.$$

This procedure can be illustrated on the kinematic model of a car.

Example 11.11 Conversion of the Kinematic Car into Chained Form. *First, we rewrite the kinematic equations by defining a new input $u_1 \leftarrow u_1 \cos\theta \cos\phi$*

$$\dot{x} = u_1,$$
$$\dot{y} = \tan\theta\, u_1,$$
$$\dot{\theta} = \frac{1}{l}\tan\phi\, u_1,$$
$$\dot{\phi} = u_2.$$

Then, with $h_1 = x$ and $h_2 = y$, it is easy to verify that this system satisfies the conditions given above and the change of variables and input are given by

$$z_1 = x,$$
$$u_1 = v_1,$$
$$z_2 = \frac{1}{l}\sec^2\theta \tan\phi,$$
$$u_2 = -\frac{2}{l}\sin^2\phi \tan\theta\, v_1 + l\cos^2\theta\cos^2\phi\, v_2,$$
$$z_3 = \tan\theta,$$
$$z_4 = y.$$

to give a one-chain system.

11.5.2 Optimal Steering of Nonholonomic Systems

In this section, we discuss the least squares optimal control problem for steering a control system of the form

$$\dot{q} = g_1(q)u_1 + \cdots + g_m(q)u_m$$

from q_0 to q_f in 1 second. Thus, we minimize the cost function

$$\frac{1}{2} \int_0^1 |u(t)|^2 \, dt.$$

Our treatment here is, of necessity, somewhat informal; To get all the smoothness hypotheses worked out would be far too large an excursion to make here. We will assume that the steering problem has a solution (by Chow's theorem, this is guaranteed when the controllability Lie algebra generated by $g_1, \ldots, g_m$ is of rank n for all q). We give a heuristic derivation from the calculus of variations of the necessary conditions for optimality. To do so, we incorporate the constraints into the cost function using a Lagrange multiplier function $p(t) \in \mathbb{R}^n$ to get

$$J(q, p, u) = \int_0^1 \left\{ \frac{1}{2} u^T(t)u(t) - p^T \left(\dot{q} - \sum_{i=1}^m g_i(q)u_i \right) \right\} dt. \qquad (11.23)$$

Introduce the *Hamiltonian function*

$$H(q, p, u) = \frac{1}{2} u^T u + p^T \sum_{i=1}^m g_i(q)u_i. \qquad (11.24)$$

Using this definition and integrating the second term of (11.23) by parts yields

$$J(q, p, u) = -p^T(t)q(t) \Big|_0^1 + \int_0^1 \left(H(q, p, u) + \dot{p}^T q \right) dt.$$

Consider the variation in J caused by variations in the control input u,[1]

$$\delta J = -p^T(t)\delta q(t) \Big|_0^1 + \int_0^1 \left(\frac{\partial H}{\partial q} \delta q + \frac{\partial H}{\partial u} \delta u + \dot{p}^T \delta q \right) dt.$$

If the optimal input has been found, a necessary condition for stationarity is that the first variation above be zero for all variations δu and δq:

$$\dot{p} = -\frac{\partial H}{\partial q}, \qquad \frac{\partial H}{\partial u} = 0. \qquad (11.25)$$

From the second of these equations, it follows that the optimal inputs are given by

$$u_i = -p^T g_i(q), \qquad i = 1, \ldots, m. \qquad (11.26)$$

---

[1] In the calculus of variations, one makes a variation in u, namely δu, and calculates the changes in the quantities p and H as δp and δH. See, for example, [334].

Using (11.26) in (11.24) yields the optimal Hamiltonian

$$H^*(q, p) = -\frac{1}{2} \sum_{i=1}^{m} \left(p^T g_i(q) \right)^2 . \tag{11.27}$$

Thus, the optimal control system satisfies *Hamilton's equations*:

$$\dot{q} = \frac{\partial H^*}{\partial p}(q, p),$$

$$\dot{p} = -\frac{\partial H^*}{\partial q}(q, p). \tag{11.28}$$

with boundary conditions $q(0) = q_0$ and $q(1) = q_f$. Using this result, we may derive the following proposition about the structure of the optimal controls:

Proposition 11.12 Constant Norm of Optimal Controls. *For the least squares optimal control problem for the control system*

$$\dot{q} = \sum_{i=1}^{m} g_i(q) u_i$$

which satisfies the controllability rank condition, the norm of the optimal input is constant, that is,

$$|u(t)|^2 = |u(0)|^2 \quad \forall t \in [0, 1].$$

Proof: The formula for the optimal input is given in (11.26). Differentiating it yields

$$\dot{u}_i = -\dot{p}^T g_i(q) - p^T \frac{\partial g_i}{\partial q} \dot{q}.$$

Further, using the Hamiltonian equation for $\dot{p}_k$ given by

$$\dot{p}_k = -\sum_{i=1}^{m} p^T \frac{\partial g_i}{\partial q_k} u_i,$$

it may be verified that the formula for $\dot{u}_i$ is given by

$$\dot{u}_i = \sum_{j=1}^{m} p^T [g_i, g_j] u_j.$$

Collecting these in a matrix gives

$$\dot{u} = \Omega(q, p) u, \tag{11.29}$$

where $\Omega(q, p)$ is a skew-symmetric matrix (i.e., it is in $so(m)$) given by

$$\Omega(q, p) = \begin{bmatrix} 0 & p^T[g_1, g_2] & \cdots & p^T[g_1, g_m] \\ -p^T[g_1, g_2] & 0 & \cdots & p^T[g_2, g_m] \\ \vdots & \vdots & & \vdots \\ -p^T[g_1, g_m] & -p^T[g_2, g_m] & \cdots & 0 \end{bmatrix} .$$

The solution of the linear time-varying equation (11.29) is of the form

$$u(t) = U(t)u(0) \tag{11.30}$$

for some $U(t) \in SO(m)$ (see Problem 11.6). From this fact the statement of the proposition follows.   □

This proposition provides an interesting formula (11.29) for the derivatives of the optimal input and establishes that the norm of the optimal input is constant. This fact can be used to establish that the same optimal input also solves other optimization problems which involve a monotone transformation of the integrand, such as

$$\int_0^1 \sqrt{u^T u} \, dt,$$

as well as some minimum-time problems (see Problem 11.7). This proposition can be used to solve certain optimization problems, such as that for the so-called Engel's system:

Example 11.13 Optimal Inputs for Engel's System.   *This system is of the form*

$$\dot{q}_1 = u_1,$$
$$\dot{q}_2 = u_2,$$
$$\dot{q}_3 = q_1 u_2 - q_2 u_1,$$
$$\dot{q}_4 = q_1^2 u_2.$$

This system has growth vector (2, 3, 4). It may be verified that

$$[g_1, g_2] = \begin{bmatrix} 0 \\ 0 \\ 2 \\ 2q_1 \end{bmatrix},$$

so that the optimal inputs satisfy the differential equation obtained by specializing (11.29), namely

$$\dot{u}_1 = 2(p_3 + q_1 p_4)u_2,$$
$$\dot{u}_2 = -2(p_3 + q_1 p_4)u_1. \tag{11.31}$$

The solution of this equation is of the form

$$u_1(t) = r \sin \alpha(t),$$
$$u_2(t) = r \cos \alpha(t),$$

where $r^2 = u_1^2(0) + u_2^2(0)$. Further, since the optimal Hamiltonian given by

$$H^*(q, p) = -\frac{1}{2}(p_1 - q_2 p_3)^2 - \frac{1}{2}(p_2 + q_1 p_3 + q_1^2 p_4)^2$$

*is independent of q_3 and q_4, it follows that $\dot{p}_3 = \dot{p}_4 = 0$, so that p_3 and p_4 are
constant. Using the functional form of u_1 and u_2 in (11.31) we get*

$$\ddot{\alpha} = 2p_4 r \sin \alpha.$$

*To integrate this equation, multiply both sides by $2\dot{\alpha}$, integrate and define $\delta = 2p_4 r$
to get*

$$(\dot{\alpha})^2 = b - 2\delta \cos \alpha,$$

*where b is a constant of the integration. If $b - 2|\delta| > 0$, this equation can be
written as*

$$\dot{\alpha} = \pm\sqrt{b - 2\delta \cos \alpha} = c\sqrt{1 - k \sin^2 \left(\frac{\alpha}{2}\right)},$$

*for some constants c, k. This last equation may be integrated using elliptic integrals
(see, for example, Lawden [175]) for α using*

$$\int_0^{\alpha/2} \frac{d\sigma}{\sqrt{1 - k^2 \sin^2 \sigma}} = \frac{ct}{2} + d. \tag{11.32}$$

*The left hand side of this equation is an elliptic integral. Hence the optimal inputs
for the Engel's system come from elliptic functions.*

In general, it is difficult to find the solution to the least squares optimal control
problem for steering the system from an initial to a final point. However, it may
be possible to find an approximate solution to the problem by using the so-called
Ritz approximation method. To explain what this means, we begin by choosing an
orthonormal basis for $L^2[0, 1]$, the set of square integrable functions on $[0, 1]$. One
basis is a set of trigonometric functions $\{\psi_0(t), \psi_1(t), \ldots\}$ given by $\psi_0(t) \equiv 1$ and
for $k \geq 1$,

$$\psi_{2k-1}(t) = \sqrt{2} \cos 2k\pi t, \quad \psi_{2k}(t) = \sqrt{2} \sin 2k\pi t, \quad t \in [0, 1].$$

For a choice of integer N, a Ritz approximation to the optimal i-th input is assumed
to be of the form

$$u_i(t) = \sum_{k=0}^{N} \alpha_{ik} \psi_k(t).$$

The plan now is to apply this input to steer the system

$$\dot{q} = g_1(q)u_1 + \cdots + g_m(q)u_m \tag{11.33}$$

from $q(0) = q_0$ to $q(1) = q_f$ and to determine the coefficient vectors, $\alpha_i = (\alpha_{i0}, \alpha_{i1}, \ldots, \alpha_{iN}) \in \mathbb{R}^{N+1}$ for $i = 1, \ldots, m$, so as to minimize the cost function

$$J = \frac{1}{2} \left(\sum_{i=1}^{m} |\alpha_i|^2 + \gamma |q(1) - q_f|^2 \right). \tag{11.34}$$

In equation (11.34) the coefficient $\gamma > 0$ is a penalty term corresponding to
reaching the final state. For large γ the cost function weights the reaching of the

goal heavily. The heart of the Ritz approximation procedure is the hope that for large N and large γ, we will find a u that steers to the final point q_f with low cost.

At this point, we have a finite-dimensional optimization problem, which may be solved by a variety of methods. One method involving a modification of the Newton iteration for minimizing J has been explored in [98] and [99], where a software package called NMPack for this purpose is described. We refer the reader to these papers for details of its application to various examples.

11.5.3 Steering with Piecewise Constant Inputs

In this section, we describe the rudiments of a method for motion planning for general nonholonomic systems due to Lafferriere and Sussmann [173]. The algorithm works for systems whose controllability Lie algebra is *nilpotent of order k*. By way of review, this means that for systems of the form

$$\dot{x} = g_1 u_1 + \cdots + g_m u_m$$

the Lie products between control vector fields of order greater than k are 0; i.e., $[g_{i_1}, \ldots, [g_{i_{p-1}}, g_{i_p}] \ldots]$ is zero when $p > k$. The method proposed in [173] is conceptually straightforward but the details of their method are somewhat involved. The first step is to derive a "canonical system equation" associated with the given control system. The chief new concept involved is that of formal power series in vector fields.

Recall that the flow associated with a vector field g_i was denoted by $\phi_t^{g_i}(q)$, referring to the solution of the differential equation $\dot{q} = g_i(q)$ at time t starting from q at time 0. This flow is referred to as the *formal exponential* of g_i and is denoted

$$e^{t g_i}(q) := \phi_t^{g_i}(q).$$

The usage of the formal exponential is in that we will actually use identities of the form

$$e^{t g_i} = \left(I + t g_i + \frac{t^2}{2!} g_i^2 + \cdots \right),$$

where polynomials like g_i^2 and g_i^3 need to be carefully justified. We will defer this question for the moment and think formally of $e^{t g_i}(q)$ as a diffeomorphism from $\mathbb{R}^n$ to $\mathbb{R}^n$. Now consider a nilpotent Lie algebra of order k generated by the vector fields $g_1, \ldots, g_m$. Recall from Section 11.2 that a Philip Hall basis is a basis for the controllability Lie algebra which has been constructed in such a way as to keep track of skew symmetry and the Jacobi identity associated with the Lie bracket. We define the Philip Hall basis of the controllability Lie algebra generated by $g_1, \ldots, g_m$ to be

$$B_1, B_2, \ldots, B_s.$$

Thus, basis elements are Lie products of order less than or equal to k. In our language of formal power series, we will refer, for instance, to

$$[g_1, g_2] := g_1 g_2 - g_2 g_1 \quad [g_1, [g_2, g_3]] := g_1 g_2 g_3 - g_1 g_3 g_2 - g_2 g_3 g_1 + g_3 g_2 g_1.$$

It is a basic result of nonlinear control, called the *Chen–Fliess series* formula (see [103]), that all flows of the nonlinear control system (11.35):

$$\dot{q} = \sum_{i=1}^{m} g_i(q) u_i, \qquad q(0) = q, \tag{11.35}$$

are of the form

$$S_t(q) = e^{h_s(t) B_s} e^{h_{s-1}(t) B_{s-1}} \cdots e^{h_2(t) B_2} e^{h_1(t) B_1}(q) \tag{11.36}$$

for some suitably chosen functions $h_1, h_2, \ldots, h_s$, known as the Philip Hall coordinates. The meaning of equation (11.36) is as follows: all flows that could possibly be generated by the control system of (11.35) may be obtained by composing flows along the Philip Hall basis elements $B_1, \ldots, B_s$. Furthermore, $S_t(q)$ satisfies a differential equation involving the basis elements, namely,

$$\dot{S}(t) = S(t)(B_1 v_1 + \cdots + B_s v_s), \qquad S(0) = I, \tag{11.37}$$

where $S_t(q)$ has been replaced by $S(t)$ and the inputs $v_1, \ldots, v_s$ are the "fictitious inputs" corresponding to the directions of the Philip Hall basis elements $B_1, \ldots, B_s$. We say "fictitious" since only the first m of the Philip Hall basis elements correspond to $g_1, \ldots, g_m$. The other inputs correspond to Lie bracket elements and will eventually be dropped. Differentiating equation (11.360 yields

$$\dot{S}(t) = \sum_{j=1}^{s} e^{h_s B_s} \cdots e^{h_j B_j} \dot{h}_j B_j e^{h_{j-1} B_{j-1}} \cdots e^{h_1 B_1}$$

$$= \sum_{j=1}^{s} S(t) e^{-h_1 B_1} \cdots e^{-h_{j-1} B_{j-1}} \dot{h}_j B_j e^{h_{j-1} B_{j-1}} \cdots e^{h_1 B_1} \tag{11.38}$$

$$:= \sum_{j=1}^{s} S(t) Ad_{e^{-h_1 B_1} \cdots e^{-h_{j-1} B_{j-1}}} \dot{h}_j B_j.$$

Here we have introduced the notation

$$Ad_{e^{-h_i B_i}} B_j = e^{-h_i B_i} B_j e^{h_i B_i}.$$

Since the controllability Lie algebra is nilpotent of degree k, we can express each one of the elements on the right hand side in terms of the basis elements $B_1, \ldots, B_s$. More specifically, it may be verified that

$$Ad_{e^{-h_1 B_1} \cdots e^{-h_{j-1} B_{j-1}}} = Ad_{e^{-h_1 B_1}} \cdots Ad_{e^{-h_{j-1} B_{j-1}}}. \tag{11.39}$$

Thus each element on the right hand side of (11.38) is a linear combination of $B_1, \ldots, B_s$ and we may express

$$Ad_{e^{-h_1 B_1} \cdots e^{-h_{j-1} B_{j-1}}} \dot{h}_j B_j = \left(\sum_{k=1}^{s} p_{j,k}(h) B_k \right) \dot{h}_j$$

for some polynomials $p_{j,k}(h)$. Using this in the equation (11.37) and equating coefficients of the basis elements B_i yields

$$\sum_{j=1}^{s} p_{j,k}(h)\dot{h}_j = v_k, \qquad k = 1, \ldots, s.$$

These equations are then solved to give the differential equation

$$\dot{h} = Q(h)v, \quad h(0) = 0, \tag{11.40}$$

which is a control system in $\mathbb{R}^s$, called the Chen–Fliess–Sussmann equation, specifying the evolution of the Philip Hall coordinates in response to the "fictitious inputs" $v_1, \ldots, v_s$. It is important to realize that, in general, the dimension s of the Philip Hall coordinates is greater than n, the dimension of the state space of the control system. The initial conditions are $h_i(0) = 0$ corresponding to the identity diffeomorphism at $t = 0$. This equation is the canonical form associated with the nilpotent controllability Lie algebra associated with the given problem.

Example 11.14 Nilpotent System of Degree Three with Two Inputs. *Consider a two-input system which is nilpotent of degree three on $\mathbb{R}^4$. We will assume that g_1, g_2, $[g_1, g_2]$, $[g_1, [g_1, g_2]]$ are linearly independent. As the Philip Hall basis, we have*

$$B_1 = g_1, \quad B_2 = g_2, \quad B_3 = [g_1, g_2],$$
$$B_4 = [g_1, [g_1, g_2]], \quad B_5 = [g_2, [g_1, g_2]].$$

Since $[g_2, [g_1, g_2]] = B_5$ is dependent on B_1, B_2, B_3, B_4 by hypothesis, we have in (11.37) that $v_5 \equiv 0$. An easy calculation shows that the coefficients of the $\dot{h}_j$ on the right hand side of (11.38) are given by

$$\dot{h}_1: \quad B_1,$$
$$\dot{h}_2: \quad B_2 - h_1 B_3 + \frac{1}{2} h_1^2 B_4,$$
$$\dot{h}_3: \quad B_3 - h_2 B_5 - h_1 B_4,$$
$$\dot{h}_4: \quad B_4,$$
$$\dot{h}_5: \quad B_5.$$

For instance, the coefficient of $\dot{h}_2$ is calculated as

$$Ad_{e^{-h_1 B_1}} B_2 = B_2 - h_1[B_1, B_2] + \frac{1}{2} h_1^2 [B_1, [B_1, B_2]]$$
$$= B_2 - h_1 B_3 + \frac{1}{2} h_1^2 B_4.$$

Equating the coefficients of the B_i to v_i with $v_5 = 0$, we get the Chen–Fliess–Sussmann equation

$$\dot{h}_1 = v_1,$$

$$\dot{h}_2 = v_2,$$

$$\dot{h}_3 = h_1 v_2 + v_3, \qquad\qquad (11.41)$$

$$\dot{h}_4 = \frac{1}{2} h_1^2 v_2 + h_1 v_3 + v_4,$$

$$\dot{h}_5 = h_2 v_3 + h_1 h_2 v_2.$$

Note that this system is in $\mathbb{R}^5$, though the state space equations evolve on $\mathbb{R}^4$.

Example 11.15 Two-input Five-state Chained System. *Consider a chained system with two inputs and five states, where the input vector fields are*

$$g_1 = \begin{bmatrix} 1 \\ 0 \\ q_2 \\ q_3 \\ q_4 \end{bmatrix}, \qquad g_2 = \begin{bmatrix} 0 \\ 1 \\ 0 \\ 0 \\ 0 \end{bmatrix}.$$

The system is nilpotent of degree $k = 4$, and the Philip Hall basis vectors are

$$B_1, B_2: \quad g_1 \quad g_2,$$

$$B_3: \quad [g_1, g_2],$$

$$B_4, B_5: \quad [g_1, [g_1, g_2]] \quad [g_2, [g_1, g_2]],$$

$$B_6, B_7: \quad [g_1, [g_1, [g_1, g_2]]], \quad [g_2, [g_1, [g_1, g_2]]],$$

$$B_8: \quad [g_2, [g_2, [g_1, g_2]]].$$

The vector fields g_1, g_2, $g_3 := ad_{g_1} g_2$, $g_4 := ad_{g_1}^2 g_2$, and $g_6 := ad_{g_1}^3 g_2$ span the tangent space of $\mathbb{R}^5$.

Thus, for the Chen–Fliess–Sussmann equations, we have $v_5 = v_7 = v_8 = 0$, and the coefficient of $\dot{h}_j$ is given by

$$Ad_{e^{-h_1 B_1}} \cdots Ad_{e^{-h_{j-1} B_{j-1}}} B_j, \quad j = 1, \ldots .8.$$

We carry out the calculation for $\dot{h}_3$ in detail:

$$Ad_{e^{-h_1 B_1}} Ad_{e^{-h_2 B_2}} B_3$$

$$= Ad_{e^{-h_1 B_1}} \left(B_3 - h_2 B_5 + \frac{1}{2} h_2^2 B_8 \right)$$

$$= B_3 - h_1 B_4 - h_2 B_5 + \frac{1}{2} h_1^2 B_6 + h_1 h_2 B_7 + \frac{1}{2} h_2^2 B_8.$$

The calculations for the remaining terms are carried out in a similar fashion, and the results are given below (we invite the reader to do the calculation herself):

$$\dot{h}_1: \quad B_1,$$

$$\dot{h}_2: \quad B_2 - h_1 B_3 + \frac{1}{2}h_1^2 B_4 - \frac{1}{6}h_1^3 B_6,$$

$$\dot{h}_3: \quad B_3 - h_1 B_4 - h_2 B_5 + \frac{1}{2}h_1^2 B_6 + h_1 h_2 B_7 + \frac{1}{2}h_2^2 B_8,$$

$$\dot{h}_4: \quad B_4 - h_1 B_6 - h_2 B_7,$$

$$\dot{h}_5: \quad B_5 - h_1 B_7 - h_2 B_8,$$

$$\dot{h}_6: \quad B_6,$$

$$\dot{h}_7: \quad B_7,$$

$$\dot{h}_8: \quad B_8.$$

Finally, with $v_5 = v_7 = v_8 = 0$ the differential equation for the $h \in \mathbb{R}^8$ is found to be

$$\dot{h}_1 = v_1,$$

$$\dot{h}_2 = v_2,$$

$$\dot{h}_3 = h_1 v_2 + v_3,$$

$$\dot{h}_4 = \frac{1}{2}h_1^2 v_2 + h_1 v_3 + v_4,$$

$$\dot{h}_5 = h_1 h_2 v_2 + h_2 v_3,$$

$$\dot{h}_6 = \frac{1}{6}h_1^3 v_2 + \frac{1}{2}h_1^2 v_3 + h_1 v_4 + v_6,$$

$$\dot{h}_7 = \frac{1}{2}h_1^2 h_2 v_2 + h_1 h_2 v_3 + h_2 v_4,$$

$$\dot{h}_8 = \frac{1}{2}h_1 h_2^2 v_2 + \frac{1}{2}h_2^2 v_3.$$

with initial condition $h(0) = 0$.

The task of steering the system from q_0 to q_f still remains to be done. This is accomplished by choosing *any* trajectory connecting q_0 to q_f indexed by time t. This is substituted into (11.37) to get the expression for the "fictitious inputs" $v_1, \ldots, v_n$, corresponding to the first n linearly independent vector fields in the Philip Hall basis of the control system. The inputs v_i are said to be fictitious, since they need to be generated by using the "real" inputs $u_i, i = 1, \ldots, m$. To do so involves a generalization of the definition of Lie brackets. For instance, to try to generate an input v_3 corresponding to $[g_1, g_2]$ in the previous example, one follows the definition and uses the concatenation of four segments, each for $\sqrt{\epsilon}$ seconds: $u_1 = 1, u_2 = 0; u_1 = 0, u_2 = 1; u_1 = -1, u_2 = 0; u_1 = 0, u_2 = -1$. The resulting flow is given by

$$e^{\sqrt{\epsilon} g_1} e^{\sqrt{\epsilon} g_2} e^{-\sqrt{\epsilon} g_1} e^{-\sqrt{\epsilon} g_2}. \tag{11.42}$$

The *Campbell–Baker–Hausdorff* formula gives an expression for this in terms of the Philip Hall basis elements. We will make a brief digression to repeat this formula from Chapter 8 (the proof was sketched in the exercises of that chapter and a more exhaustive recursive formulation of the coefficients is given in [312]).

Theorem 11.16 Campbell–Baker–Hausdorff Formula. *Given two smooth vector fields g_1, g_2, the composition of their exponentials is given by*

$$e^{g_1} e^{g_2} = e^{g_1 + g_2 + \frac{1}{2}[g_1, g_2] + \frac{1}{12}([g_1, [g_1, g_2]] - [g_2, [g_1, g_2]]) + \cdots}, \tag{11.43}$$

where the remaining terms may be found by equating terms in the (non-commutative) formal power series on the right- and left-hand sides.

Using the Campbell–Baker–Hausdorff formula for the flow in equation (11.42) gives the Philip Hall coordinates

$$h = (0, 0, \epsilon, h_4(\epsilon)),$$

where $h_4(\epsilon)$ is of order higher than one in ϵ. Thus, the strategy of cycling between u_1 and u_2 not only produces motion along the direction $[g_1, g_2]$, but also along $[g_1, [g_1, g_2]]$. Thus, both v_3 and v_4 are not equal to 0. However, both the h_1 and h_2 variables are unaffected. To generate motion along the v_4 direction, one concatenates an eight-segment input consisting of cycling between the u_1, u_2 for $\sqrt[3]{\epsilon}$ seconds. Since the controllability Lie algebra is nilpotent, this motion does not produce any motion in any of the other bracket directions (in the context of the Campbell–Baker–Hausdorff formula above, the series on the right hand side of (11.43) is finite). This example can be generalized to specify a constructive procedure for generating piecewise constant inputs for steering systems which have nilpotent controllability Lie algebras.

When the controllability Lie algebra is not nilpotent, the foregoing algorithm needs to be modified to steer the system only approximately. The two difficulties in this case are:

1. The Philip Hall basis is not finite.
2. The Campbell–Baker–Hausdorff formula does not terminate after finitely many terms.

One approach that has been suggested in this case is to try to change the input g_i using a transformation of the inputs, that is by choosing a matrix $\beta \in \mathbb{R}^{m \times m}$ so as to make the transformed $\tilde{g}_i$, defined by

$$\begin{bmatrix} \tilde{g}_1 & \tilde{g}_2 & \cdots & \tilde{g}_m \end{bmatrix} = \beta \begin{bmatrix} g_1 & g_2 & \cdots & g_M \end{bmatrix}$$

a nilpotent control system. Other approaches involve "nilpotent approximations" to the given control system. For more details on the algorithm for steering nonholonomic systems using piecewise constant inputs, see the paper of Lafferriere and Sussmann in [184].

An interesting other approach to using piecewise constant controls is motivated by "multi-rate" sampled data control where some controls are held constant for integer multiples of a basic sampling rate (i.e., they are changed at lower frequencies than the basic frequency). This results in more elegant paths than those obtained by the method presented above. To explain this method fully requires a detailed exposition of multi-rate control, we refer the reader to the work of Monaco and Normand Cyrot, and co-authors [110], [216].

11.5.4   Control Systems with Drift

In this section we study control systems with drift, i.e., systems of the form of (11.1)

$$\dot{x} = f(x) + \sum_{i=1}^{m} g_j(x) u_j.$$

It may be thought of as a special case of the drift free systems studied in the previous subsection by defining the new system

$$\dot{x} = \sum_{i=0}^{m} g_j(x) u_j,$$

with the input u_0 attached to the vector field $g_0(x) = f(x)$ frozen at 1. The *controllability Lie Algebra* is defined as follows:

Definition 11.17 Controllability Lie Algebra.   *The* controllability Lie algebra *C of a control system of the form (11.1) is defined to be the span over the ring of smooth functions of elements of the form*

$$[X_k, [X_{k-1}, [\ldots, [X_2, X_1] \ldots]]],$$

where $X_i \in \{f, g_1, \ldots, g_m\}$ and $k = 0, 1, 2, \ldots$.

In the preceding definition recall that a *Lie sub-algebra* of a module is a sub-module that is closed under the *Lie bracket*. The controllability Lie algebra may be used to define the *accessibility distribution* as the distribution

$$C(x) = \text{span}\{X(x) : X \in C\}.$$

From the definition of C it follows that $C(x)$ is an involutive distribution. The controllability Lie algebra is the smallest Lie algebra containing $f, g_1, \ldots, g_m$ which is closed under Lie bracketting which is invariant under $f, g_1, \ldots, g_m$. The controllability distribution contains valuable information about the set of states that are accessible from a given starting point x_0. To develop this, we first develop some machinery:

Definition 11.18 Invariance.   *A distribution Δ is said to be invariant under f if for every $\tau \in \Delta, [f, \tau] \in \Delta$.*

This definition is the nonlinear counterpart of the definition of an invariant subspace invariant under a linear operator. Indeed if $V \subset \mathbb{R}^n$ then we can conceive of it is as a constant distribution and then $AV \subset V$ is equivalent to $[Ax, V] \subset V$ (see Exercises). This definition is useful in representing vector fields as follows:

Theorem 11.19 Representation Theorem.   *Let Δ be a regular involutive distribution of dimension d and let Δ be invariant under f. Then at each point x_0 there exists a neighborhood U and a coordinate transformation $z = \phi(x)$ defined on U in which the vector field has the form*

$$\tilde{f}(z) = \begin{bmatrix} f_1(z_1, \ldots, z_d, z_{d+1}, \ldots, z_n) \\ \vdots \\ f_d(z_1, \ldots, z_d, z_{d+1}, \ldots, z_n) \\ f_{d+1}(z_{d+1}, \ldots, z_n) \\ \vdots \\ f_n(z_{d+1}, \ldots, z_n) \end{bmatrix}. \tag{11.44}$$

Proof: Since the distribution Δ is regular and involutive, it follows that it is integrable and that there exist functions labeled $\phi_{d+1}, \ldots, \phi_n$ such that

$$\text{span}\{d\phi_{d+1}, \ldots, d\phi_n\} = \Delta^\perp. \tag{11.45}$$

Complete the basis to get $\phi_1, \ldots, \phi_d$ and define $z = \phi(x)$. Now if $\tau(x) \in \Delta(x)$ it follows that in the z coordinates

$$\tilde{\tau}(z) = \left[\frac{d\phi}{dx} \tau(x) \right]_{x=\phi^{-1}(z)}$$

has zeros in its last $n - d$ coordinates. This is simply a consequence of the fact that $\tau(x)$ is in the null space of the last $n - d$ rows of $\frac{d\phi}{dx}$. Now choose as basis set for Δ in the z coordinates

$$\tilde{\tau}_i(z) = (0, \ldots, 0, 1, 0, \ldots)^T$$

with the 1 in the i-th entry for $i = 1, \ldots, d$. Now an easy calculation yields that

$$[\tilde{f}, \tilde{\tau}] = -\frac{\partial f}{\partial z_i}.$$

Since $[\tilde{f}, \tilde{\tau}] \in \Delta$ by the definition of invariance it follows that the last $n - d$ rows of $\frac{\partial f}{\partial z_i}$ are identically 0. Thus we have

$$\frac{\partial f_k}{\partial z_i} \equiv 0 \quad \text{for } i = 1, \ldots, d, \quad k = d + 1, \ldots, n.$$

This completes the proof.   □

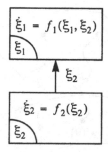

FIGURE 11.6. The triangular decomposition of a given system

Remarks:

1. The preceding representation theorem is the nonlinear extension of a familiar result in linear algebra.
2. From a control theoretic point of view it represents a triangular decomposition of a given system

$$\dot{x} = f(x)$$

into the form

$$\dot{\xi}_1 = f_1(\xi_1, \xi_2),$$
$$\dot{\xi}_2 = f_2(\xi_2),$$

where $\xi_1 = (z_1, \ldots, z_d)^T$, $\xi_2 = (z_{d+1}, \ldots, z_n)^T$. This decomposition is shown in Figure 11.6.
3. Another valuable geometric insight is gained by representing the foliation associated with Δ as the level sets of ξ_2. The preceding theorem states that if two different solution trajectories start with the same value of ξ_2, that is if $x^a(0)$, $x^b(0)$ are such that that $\xi_2(x^a(0)) = \xi_2(x^b(0)) = \xi_2(0)$ then it follows that $\xi_2(x^a(t)) = \xi_2(x^b(t)) = \xi_2(t)$ with $\xi_2(t)$ satisfying

$$\dot{\xi}_2 = f_2(\xi_2)$$

with initial condition $\xi_2(0)$. Thus, the solution of the differential equation carries sheets of the foliation onto other sheets of the foliation. This is illustrated in Figure 11.7.

This machinery is very useful to study local decompositions of control systems of the form of (11.1). Let Δ be a distribution which contains $f, g_1, g_2, \ldots, g_m$ and is invariant under the vector fields $f, g_1, \ldots, g_m$ and which is involutive. Then by an easy extension of the preceding theorem it is possibly to find coordinates such that the control system may be represented in coordinates as

$$\dot{\xi}_1 = f_1(\xi_1, \xi_2) + \sum_{i=1}^{m} g_{1i}(\xi_1, \xi_2) u_i, \qquad (11.46)$$
$$\dot{\xi}_2 = 0.$$

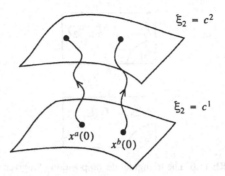

FIGURE 11.7. Flow leaving the foliation intact

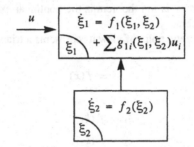

FIGURE 11.8. Controllable and uncontrollable parts of the controllable system

A candidate for the involutive distribution Δ which satisfies the preceding requirements is the controllability distribution $C(x)$ which was defined to be the involutive closure of $f, g_1, \ldots, g_m$ and is hence involutive by definition. The decomposition of equation (11.46) is visualized in Figure (11.8).

The preceding discussion shows that the ξ_2 states are in some sense to be thought of as the *uncontrollable states* in that the dynamics of the ξ_2 variables are unaffected by the choice of the input. The number of ξ_2 states is equal to $n-$ dimension of $C(x)$. By choosing $C(x)$ to be the smallest invariant distribution containing the f, g_i it follows that we have tried to extract as many ξ_2 variables as possible. However, since f is the drift vector field, which is not steerable using controls it is not completely clear in which directions it is possible to steer the control system in the distribution $C(x)$. We now define the set of states accessible from a given state x_0:

Definition 11.20 Reachable Set from x_0.   *Let* $R^U(x_0, T) \subset \mathbb{R}^n$ *to be the subset of all states accessible from state x_0 in time T with the trajectories being confined to a neighborhood U of x_0. This is called the* reachable set *from x_0.*

In order to categorize the reachable set we define the *accesibility Lie algebra*:

Definition 11.21 Accessibility Lie Algebra. *The* accessibility Lie algebra $\mathcal{A}$ *of a control system of the form (11.1) is defined to be the span over the ring of smooth functions of elements of the form*

$$[X_k, [X_{k-1}, [\ldots, [X_2, X_1]\ldots]]],$$

where $X_1 \in \{g_1, \ldots, g_m\}$, *and* $i \geq 2 X_i \in \{f, g_1, \ldots, g_m\}$ *and* $k = 0, 1, 2, \ldots$.[2]

The *accesibility distribution* is defined as the evaluation of the Lie algebra $A(x)$. The distribution is involutive and g_i invariant. In fact, if the distributions $A(x), C(x)$ are related by

$$C(x) = A(x) + \text{span}(f(x))$$

Further, if both $A(x), C(x)$ are regular, it follows that $\dim C(x) \leq \dim A(x) + 1$. Using the coordinates afforded by this distribution it follows that the system of (11.1) has the form

$$\dot{\xi}_{11} = f_1(\xi_1, \xi_2) + \sum_{i=1}^{m} g_{1i}(\xi_1, \xi_2)u_i,$$

$$\ldots$$

$$\dot{\xi}_{1.d-1} = f_{d-1}(\xi_1, \xi_2) + \sum_{i=1}^{m} g_{d-1,i}(\xi_1, \xi_2)u_i, \tag{11.47}$$

$$\dot{\xi}_{1,d} = f_d(\xi_1, d, \xi_2),$$

$$\dot{\xi}_2 = 0.$$

Note the close resemblance between equations (11.46) and (11.47). In fact, the only difference is the presence of the extra equation for the d-th variable in ξ_1. Indeed, if the last $n - d$ variables ξ_2 start at 0 zero at $t = 0$ they will stay at zero, and the d-th variable will satisfy the autonomous differential equation

$$\dot{\xi}_{1d} = f_d(\xi_{1d}, 0).$$

Constructive controllability results, that is, constructions for generating control laws for steering systems with drift from one point in the state space to another are not easy to come by. Further reading in this topic is indicated in the summary section of this chapter.

11.6 Observability Concepts

Before we introduce the formal definitions, let us introduce some new machinery. Recall that a *covector field* is a smooth mapping from $\mathbb{R}^n \mapsto \mathbb{R}^{n*}$ described by a

---

[2]Note that X_1 is chosen from the set $g_1, g_2, \ldots, g_m$ and does not include f.

row vector of the form $\omega(x) \in \mathbb{R}^{n*}$. A *codistribution* is defined analogously to a distribution (cf. Chapter 8) as

$$\Omega(x) = \text{span}\{\omega_1(x), \ldots, \omega_k(x)\}.$$

Also, recall that in coordinates, the *Lie derivative of a covector field* ω along a vector field f is a new covector field $L_f\omega$ defined by

$$L_f\omega(x) = f^T(x)\left[\frac{\partial \omega^T}{\partial x}\right]^T + \omega(x)\frac{\partial f}{\partial x}. \qquad (11.48)$$

The following is a list of some useful properties of the Lie derivative of a covector field. The proofs are left as exercises:

1. If f is a vector field and λ a smooth function, then

$$L_f d\lambda(x) = dL_f\lambda(x).$$

2. If α, β are smooth functions, f a vector field and ω a covector field, then

$$L_{\alpha f}\beta\omega(x) = \alpha\beta(L_f\omega(x)) + \beta(x)\omega f d\alpha(x) + (L_f\beta(x))\alpha(x)\omega(x).$$

3. If f, g are vector fields and ω a covector field it follows that

$$L_f(\omega g)(x) = L_f\omega(x)g(x) + \omega(x)[f, g](x).$$

A codistribution $\Omega(x)$ is said to be f invariant if the codistribution $L_f\Omega$ defined by

$$L_f\Omega = \text{span}\{L_f\omega : \omega \in \Omega\}.$$

is contained in Ω. The following lemma establishes the relationship between the f invariance of a distribution or a co-distribution and its duals:

Proposition 11.22. *If a smooth distribution Δ is invariant under the vector field f, then so is the codistribution $\Omega = \Delta^\perp$. Conversely, if a smooth codistribution Ω is invariant under f then so is the distribution $\Delta = \Omega^\perp$.*

Proof: Let Δ be a smooth f invariant distribution and let τ be a vector field in Δ. Then by f invariance $[f, \tau]$ is in Δ. By definition of $\Omega = \Delta^\perp$ it follows that if $\omega \in \Omega$, then

$$\omega\tau = \omega[f, \tau] \equiv 0.$$

Further, using the last of the properties of the Lie derivative of a covector field listed above, it follows that

$$L_f\omega\tau = L_f(\omega\tau) - \omega[f, \tau],$$

so that we have that

$$L_f\omega\tau = 0.$$

Since $\tau \in \Delta$ is arbitrary, it follows that $L_f\omega \in \Omega$. This establishes the f invariance of Ω. the converse follows similarly. $\square$

We are now ready to study decompositions of control systems of the form

$$\dot{x} = f(x) + \sum_{i=1}^{m} g_i(x)u_i,$$

$$y_1 = h_1(x),$$

$$\vdots \qquad\qquad\qquad\qquad\qquad (11.49)$$

$$y_m = h_m(x).$$

Our development parallels that of Kou, Elliott, and Tarn [166].

Proposition 11.23 Representation. *Let Δ be a nonsingular involutive distribution of dimension d invariant under the vector fields $f, g_1, \ldots, g_m$. Further, assume that the codistribution spanned by $dh_1, \ldots, dh_m$ is contained in $\Delta^\perp$. Then it is possible to find a local change of coordinates $\xi = \phi(x)$ in a neighborhood of x_0 such that the system has the form*

$$\dot{\xi}_1 = f_1(\xi_1, \xi_2) + \sum_{i=1}^{m} g_{1i}(\xi_1, \xi_2)u_i,$$

$$\dot{\xi}_2 = f_2(\xi_2) + \sum_{i=1}^{m} g_{2i}(\xi_2)u_i \qquad\qquad (11.50)$$

$$y_i = h_i(\xi_2),$$

where $\xi_1 \in \mathbb{R}^d$ and $\xi_2 \in \mathbb{R}^{n-d}$.

The decomposition of this proposition is depicted in Figure 11.9.

11.7 Zero Dynamics Algorithm and Generalized Normal Forms

We began a study of the zero dynamics in Chapter 9 of nonlinear systems that have relative degree (either scalar relative degree for SISO systems or vector rel-

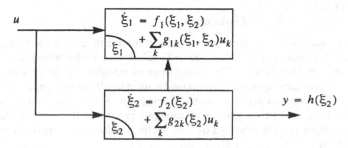

FIGURE 11.9. Decomposition into observable and unobservable states

ative degree for MIMO systems) in the context of input–output linearization. For systems for which vector relative degree accrues after dynamic extension, too, the zero dynamics were well defined since dynamic extension does not alter the zero dynamics of the system. The question that arises then is as to whether the zero dynamics of nonlinear systems may be intrinsically defined. The answer to this question is furnished by the so-called zero dynamics algorithm. This algorithm is a nonlinear counterpart of a corresponding algorithm in the linear case attributed to Wonham [331]. In the following we follow closely the treatment of [149]. The set up is as before; We consider "square" nonlinear control systems of the form

$$\dot{x} = f(x) + g(x)u,$$
$$y = h(x),$$
(11.51)

with m inputs and outputs and $x \in \mathbb{R}^n$. We will also assume that the inputs are all independent, that is to say,

$$\dim(\text{span}\{g_1(x_0), \ldots, g_m(x_0)\}) = m.$$

Further, $x_0 \in \mathbb{R}^n$ is an equilibrium point of the undriven system, i.e., $f(x_0) = 0$, and we will assume that the output is unbiased at this point, namely $h(x_0) = 0$. It is clear that if the state of the system (11.51) is x_0 at $t = 0$ and the input $u(t)$ identically 0 then the output $y(t)$ is identically 0. The question that is to be answered is: *For which other initial conditions besides x_0, does there exist a choice of a suitable input $u(t)$ so as to hold the output $y(t) \equiv 0$?* The answer to this question needs some more definitions:

Definition 11.24 Locally Controlled Invariant. *Given a smooth connected manifold $M \subset \mathbb{R}^n$ with $x_0 \in M$, M is said to be* locally controlled invariant *at x_0 if there exists a choice of smooth feedback law $u : M \mapsto \mathbb{R}^m$ defined in a neighborhood of x_0 such that the vector field*

$$f(x) + g(x)u(x) \in TM_x \ \ \forall x \in nbhd \ of \ x_0 \in M.$$

Definition 11.25 Output Zeroing Manifold. *Given a smooth connected manifold $M \subset \mathbb{R}^n$ with $x_0 \in M$, M is said to be* locally output zeroing *at x_0 if:*

(i) $h(x) \equiv 0$ *for $x \in M$.*
(ii) M *is locally controlled invariant at x_0.*

This second definition is particularly useful in solving the problem at hand, since each output zeroing manifold is a set of initial conditions for which the output can be held identically zero. The zero dynamics manifold of the system is the maximal output zeroing manifold (maximal is used locally in this context to mean that it locally contains all other output zeroing manifolds). The zero dynamics algorithm described below gives a constructive procedure for constructing this maximal output zeroing manifold by constructing a (decreasing) nested chain of manifolds $M_0 \supset M_1 \supset \ldots \supset M_{k^*} = M_{k^*+1} = \cdots$.

Zero Dynamics Algorithm

Step 0: Set $M_0 = \{x : h(x) = 0\}$.

Step k: Regularity Assumption: Assume that for a neighborhood U_{k-1} of x_0, $M_{k-1} \cap U_{k-1}$ is a smooth manifold.

Denote by V_{k-1} the connected component of $M_{k-1} \cap U_{k-1}$ containing x_0. Now define

$$M_k = \{x \in V_{k-1} : \quad f(x) \in \text{span}\{g_1(x), \ldots, g_m(x)\} + T_x V_{k-1}\}.$$

Proposition 11.26 Locally Maximal Output Zeroing Manifold. *Let the regularity assumption of the zero dynamics algorithm be in effect, i.e., assume that there exist neighborhoods U_k of x_0 such that $M_k \cap U_k$ is a smooth manifold. Then there exists $k^* < n$ and some neighborhood U_{k^*} of x_0, such that*

$$M_{k^*+1} = V_{k^*}.$$

Further, assume that span $\{g_1(x), \ldots, g_m(x)\} \cap T_x V_{k^*}$ has constant dimension for all $x \in V_{k^*}$. Then the manifold $M^* = V_{k^*}$ is a locally maximal output zeroing manifold.

Proof: The regularity hypothesis guarantees that the M_k are a chain of submanifolds of decreasing dimension. As a consequence it follows that there exists an integer k^* such that $M_{k^*+1} = V_{k^*}$ for some neighborhood U_{k^*} of x_0. Set $M^* = V_{k^*}$. It follows from the construction of the zero dynamics algorithm that for each $x \in M^*$ there exists $u(x) \in \mathbb{R}^m$ such that

$$f(x) + g(x)u \in T_x M^*,$$

The smoothness of $u(x)$ needs as yet to be established. We express this a little more explicitly by characterizing M^* locally as the zero set of a function $H(x) : \mathbb{R}^n \mapsto \mathbb{R}^{n-\gamma}$, where γ is the dimension of M^*. Then it follows that the preceding equation may be written as

$$dH(x)(f(x) + g(x)u) = 0.$$

or

$$dH(x)f(x) \in \text{span } dH(x)g(x)$$

at each point x of $M^* \cap U^*$ for some U^* a neighborhood of x_0. The hypotheses on the rank of the span of $\{g_1(x), \ldots, g_m(x)\}$ and the span of $\{g_1(x), \ldots, g_m(x)\} \cap T_x M^*$ guarantee that the matrix $dH(x)g(x)$ has constant rank on M^* in a neighborhood of x_0. Thus in a neighborhood of x_0 we can find a smooth mapping $u^* : U^*$ (a neighborhood of x_0) $\mapsto \mathbb{R}^m$ such that

$$f(x) + g(x)u^*(x) \in T_x M^*.$$

This establishes that at the conclusion of the zero dynamics algorithm we have a controlled invariant manifold.

To show the local maximality of M^*, consider another controlled invariant output zeroing manifold Z. It is immediate that $Z \subset M_0 = \{x : h(x) = 0\}$. The rest of the proof proceeds by recursion: we show that it $Z \subset V_{k-1} \Rightarrow Z \subset M_k$. Indeed note that since Z is controlled invariant we must have that

$$\forall x \in Z \Rightarrow f(x) \in \text{span}\{g_1, \ldots, g_m\} + T_x Z$$
$$\Rightarrow f(x) \in \text{span}\{g_1, \ldots, g_m\} + T_x M_{k-1}$$
$$\Rightarrow x \in M_k.$$

Thus $Z \subset M^*$ locally, completing the proof.   □

The outcome of the zero dynamics algorithm is refered to as the zero dynamics manifold. Since there exists a choice of feedback control law $u^*(x)$ to make M^* unique, it follows that the vector field

$$f(x) + g(x)u^*(x)$$

may be restricted to M^*. This restriction of $f + gu^*$ to M^* is an intrinsic definition of the zero dynamics. It is important at this juncture to note that this development has not assumed any minimality hypothesis on the original system. Thus, for example, if the original system of (11.51) was not observable, that is there exists (cf. Proposition 11.23 an f, g-invariant regular distribution Δ a strict subset of $\mathbb{R}^n$ with $dh \subset \Delta^\perp$ then it follows that the integral manifold of Δ is contained in M^*. The interpretation of the zero dynamics algorithm in the terminology of observability of Proposition 11.23 is that it locally produces the *largest* manifold in the zero set of h, whose tangent space annihilates the codistribution $dh^\perp$, which can be made unobservable by choice of u^*. Since states that are unobservable to begin with already belong to this category, it follows that the zero dynamics algorithm will capture these unless explicitly disallowed. In general the choice of the control which makes M^* invariant is not unique. The following proposition gives sufficient conditions for this to occur:

Proposition 11.27 Uniqueness of the Zero Dynamics. *Consider the zero dynamics alogrithm with all the hypotheses of Proposition 11.26 in effect and in addition*

$$\text{span}\{g_1(x_0), \ldots, g_m(x_0)\} \cap T_{x_0} M^* = 0.$$

Then the control law u^* needed to render M^* invariant is unique.

Proof: By the hypothesis of this proposition the matrix $dH(x)g(x)$ has rank m near x_0, since $g(x)$ has rank m near x_0 by the hypothesis of the previous proposition and span $g(x) \cap dH(x)$ is 0 by the hypothesis of the current proposition. As a consequence the equation

$$dH(x)f(x) + dH(x)g(x)u^*(x)$$

has a unique solution $u^*(x)$.   □

The outcome of the previous two propositions allows us to precisely define **the local zero dynamics** of the system to be uniquely the dynamics of $f(x) +$

$g(x)u^*(x)$ restricted to M^*. By abuse of notation we will refer to this as $f^*(x)$. Thus the dynamical system

$$\dot{x} = f^*(x), \quad x \in M^*, \tag{11.52}$$

is referred to as the zero dynamics of the system under the hypotheses of the preceding two propositions. Of course, if any of these hypotheses are not met, there may be several candidates for the zero dynamics (and this is a point that needs to be considered with care!). In the exercises we will work through the linear counterpart of this definition (cf. Problem 11.19). A point of difference from the linear case is that the "zeros" of a nonlinear system are not represented by a list of numbers but by a whole dynamical system. Of course, in the linear case there is a notion of zeros and zero directions (corresponding to the eigendirections of the system of (11.52) in this case).

The zero dynamics algorithm just discussed is a little abstract because the construction involves finding manifolds M_k at each step satisfying certain conditions. Since the construction is local the manifolds can be described locally as zero sets of certain functions. The derivation of these functions is now described. To aid insight into this, we will work through the construction for two steps of the algorithm:

Step 0: $M_0 = \{x : h(x) = 0\}$. To guarantee that M_0 is a manifold we will assume that the differential dh has constant rank s_0 near x_0 on M_0. Then for x near x_0, M_0 is a manifold of dimension $n - s_0$. If $s_0 < m$, we will assume, without loss of generality that the first s_0 rows of dh are linearly independent near x_0. Indeed, this is done by rearrangement of the outputs, if necessary. The selection matrix $S_0 \in \mathbb{R}^{s_0 \times m}$ is of the form

$$S_0 = [I \quad 0].$$

Define the function

$$H_0(x) = S_0 h(x)$$

so that

$$M_0 \cap U_0 = \{x \in U_0 : H_0(x) = 0\}$$

for some neighborhood U_0 of x_0. Define V_0 to be the connected piece of $M_0 \cap U_0$ which contains x_0.

Step 1: To find the set of points in V_0 where $f(x)$ lies in $T_x M_0 + \text{span}(g(x))$ is equivalent to finding the set of $X \in M_0 \cap U_0$ where the equation

$$dH_0(x)(f(x) + g(x)u) = 0$$

may be solved for u (convince yourself of this point). The preceding equation is rewritten as

$$L_f H_0(x) + L_g H_0(x)u = 0. \tag{11.53}$$

If one assumes that the matrix $L_g H_0(x)$ has constant rank, say r_0 on V_0, then it follows that there is a matrix $R_0(x) \in \mathbb{R}^{(s_0 - r_0) \times s_0}$ such that

$$R_0(x)L_g H_0(x) = 0$$

in another (perhaps smaller) neighborhood of x_0. Thus equation (11.53) has a solution for those x for which

$$R_0(x)L_f H_0(x) = 0.$$

Thus in a neighborhood of x_0, M_1 may be expressed as

$$M_1 \cap U_1 = \{x \in U_1 : H_0(x) = 0 \quad \Phi_0(x) := R_0(x)L_f H_0(x) = 0\}.$$

Now if the mapping consisting of (H_0, Φ_0); $\mathbb{R}^n \mapsto \mathbb{R}^{s_0 + (s_0 - r_0)}$ has rank $s_0 + s_1$ then a selection matrix S_1 can be used to choose the s_1 linearly dependent rows of $\Phi_0(x)$ and we define

$$H_1(x) = \begin{pmatrix} H_0(x) \\ S_1 \Phi_0(x) \end{pmatrix}.$$

Step k: At the k-th step, we begin the iteration with $H_k : \mathbb{R}^n \mapsto \mathbb{R}^{s_0 + \cdots + s_k}$,

$$H_k(x) = \begin{pmatrix} H_{k-1}(x) \\ S_k \Phi_{k-1}(x) \end{pmatrix},$$

and with

$$M_k \cap U_k = \{x \in U_k : H_k(x) = 0\}.$$

The mapping H_k is assumed to have constant rank $s_0 + s_1 + \cdots + s_k$ on U_k, guaranteeing that M_k is a manifold of dimension $n - s_0 - s_1 - \cdots - s_k$. To find M_{k+1} one has to look at the equation

$$L_f H_k + L_g H_k u = 0.$$

If the matrix $L_g H_k(x)$ has constant rank r_k for all $x \in V_k$ the connected component of $M_k \cap U_k$ containing x_0 it is possible to find a matrix $R_k(x)$ of smooth functions whose $(s_0 + \cdots + s_k - r_k)$ rows locally span the left null space of $L_g H_k(x)$, namely,

$$R_k(x)L_g H_k(x) = 0 \qquad (11.54)$$

Given the recursive definition of $H_k(x)$ and the definition of $R_{k-1}(x)$ as the left null space of $L_g H_{k-1}$, we may choose $R_k(x)$ to be

$$R_k(x) = \begin{bmatrix} R_{k-1}(x) & 0 \\ P_{k-1}(x) & Q_{k-1}(x) \end{bmatrix},$$

It follows that equation (11.54) has a solution for those x for which we have

$$R_k(x)L_f H_k(x) = 0.$$

This in turn corresponds to

$$\begin{bmatrix} R_{k-1}L_f H_{k-1} \\ P_{k-1}L_f H_{k-1} + Q_{k-1}L_f S_{k-1}\Phi_{k-1} \end{bmatrix} = 0$$

Since the first equation above was satisfied in the previous step of the recursion it follows that

$$\Phi_k := P_{k-1}L_f H_{k-1} + Q_{k-1}L_f S_{k-1}\Phi_{k-1} = 0$$

needs to be satisfied for equation (11.54) to have a solution. The number of new constraints introduced is $s_k - r_k + r_{k-1}$. As in the preceding steps S_{k+1} is defined by defining the rank of

$$\left(\; H_k, \Phi_k \; \right)$$

to be $\sum_{i=0}^{k+1} s_i$. The selection matrix S_{k+1} picks the s_{k+1} linearly independent rows of Φ_{k+1}.

Summary: Two kinds of regularity assumptions were made above:

- To guarantee that the M_k are manifolds, it was assumed that the rank of $dH_k(x)$ was constant on the $M_k \cap U_k$. Further, since the rank of dH_k is $\sum_{i=0}^{k} s_i$ it follows that

$$s_k = \dim(M_{k-1}) - \dim(M_k),$$

- To guarantee that $L_g H_k$ has constant rank r_k for $x \in V_k$ so that a smooth left null space $R_k(x)$ may be found we assume that

$$r_k = \dim(\mathrm{span}\{g_1(x), \ldots, g_m(x)\})$$
$$- \dim(\mathrm{span}\{g_1(x), \ldots, g_m(x)\} \cap T_x M_k).$$

The hypothesis for obtaining a unique zero dynamics, namely

$$\mathrm{span}\{g_1(x_0), \ldots, g_m(x_0)\} \cap T_{x_0} M^* = \{0\}$$

may now be expressed as

$$r_{k^*} = m.$$

Now, following the development in [149], once again we develop a "generalized normal form" for nonlinear systems which satisfy the three assumptions necessary for the implementation of the zero dynamics algorithm of the preceding section:

1. $dH(x)$ has constant rank s_0 in a neighborhood of x_0, and for choice of matrices $R_0, R_1, \ldots, R_{k^*-1}$, with the R_k chosen to be such that $R_k L_g H_{k-1} = 0$, the differentials of the mappings H_k defined by

$$H_k = \left(\begin{array}{c} H_{k-1} \\ \Phi_{k-1} \end{array} \right)$$

have constant rank $\sum_{i=0}^{k} s_i$ for x in a neighborhood of x_0.
2. The matrices $L_g H_k(x)$ have constant rank r_k for all $x \in M_k$ around x_0 for $0 \le k \le k^* - 1$.
3. The matrix $L_g H_{k^*}(x)$ has rank m.

A nonlinear system of the form (11.51) that satisfies these conditions is said to have x_0 as a *regular point of the zero dynamics algorithm*. The nature of the regularity hypothesis varies somewhat in the literature, for example [90] require that the matrices $L_g H_k$ have rank r_k not just on M_k but in an open set around x_0. This is referred to as a *strong regularity condition for the zero dynamics algorithm*.

It is useful to note that as a consequence of the first of the three assumptions spelt out above, there is no necessity for the selection matrices (see also Proposition 3.34). An immediate consequence of this is that the differentials of the $\sum_{i=0}^{k^*} s_i$ entries of

$$H_{k^*} = \mathrm{col}(h(x), \Phi_0(x), \ldots, \Phi_{k^*-1}(x))$$

are linearly independent. These will thus serve as a partial change of coordinates for the generalized normal form. The rest of this section is devoted to a discussion of the appearance of the system dynamics in the coordinates given by this transformation. A three input, three output conceptual example worked through by Isidori is particularly revealing of the structure of the normal form and we give it here:

Example 11.28 Three Input, Three Output Example Due to Isidori [149].
Step 0: $M_0 = \{x : h_1 = h_2 = h_3 = 0\}$ *is assumed to be a* $n - 3$ *dimensional manifold as a consequence of the regularity of 0 for the h_i.*
Step 1:

$$L_g h = \begin{bmatrix} L_g h_1 \\ L_g h_2 \\ 0 \end{bmatrix}.$$

Further, we will assume that $L_g h$ has rank 1 near x_0. Thus we may write

$$L_g h_2(x) = -\gamma(x) L_g h_1(x) + \sigma_2(x)$$

where $\sigma_2(x) = 0$ for $x \in M_0$. Note that strong regularity would guarantee that $\sigma_2 \equiv 0$. The preceding helps determine specific values for $R_0(x)$, $\Phi_0(x)$ as

$$R_0(x) = \begin{bmatrix} \gamma(x) & 1 & 0 \\ 0 & 0 & 1 \end{bmatrix}, \quad \Phi_0(x) = \begin{bmatrix} \gamma L_f h_1 + L_f h_2 \\ L_f h_3 \end{bmatrix} = \begin{bmatrix} \phi_2 \\ \phi_3 \end{bmatrix}.$$

Step 2: $M_1 = \{x \in M_0 : \phi_2 = \phi_3 = 0\}$ *is assumed to be an* $n - 5$ *dimensional manifold. Now consider the matrix*

$$L_g H_1 = \begin{bmatrix} L_g h \\ L_g \Phi_0 \end{bmatrix}$$

and assume it to have rank 2. Since the second row is dependent on the first row for $x \in M_0$ and the third row is zero, we will assume that the first and the fourth rows are independent. Then, the fifth row can be expressed in terms of the first and fourth as

$$L_g \phi_3 = -\delta_1 L_g h_1 - \delta_2 L_g \phi_2 + \sigma_3$$

with $\sigma_3(x) = 0$ for $x \in M_1$ (respectively $\equiv 0$) by the regularity (strong regularity) hypothesis. Then we may choose

$$R_1(x) = \begin{bmatrix} R_0 & \begin{bmatrix} 0 & 0 \end{bmatrix} \\ \begin{bmatrix} \delta_1 & 0 & 0 \end{bmatrix} & \begin{bmatrix} \delta_2 & 1 \end{bmatrix} \end{bmatrix},$$

so that

$$\Phi_1(x) = \delta_1 L_f h_1 + \delta_2 L_f \phi_2 + L_f \phi_3 = \psi_3(x).$$

Step 3: $M_2 = \{x \in M_1 : \psi_3(x) = 0\}$ *is assumed to be a manifold of dimension $n - 6$ by the regularity assumption for the zero dynamics algorithm. The matrix $L_g H_2 \in \mathbb{R}^{6\times 3}$ defined by*

$$L_g H_2 = \begin{bmatrix} L_g H_1 \\ L_g \Phi_1 \end{bmatrix}$$

is assumed to have rank 3 at x_0, so that its first, fourth, and sixth rows are linearly independent. Thus the algorithm terminates at this step, and the zero dynamics manifold

$$M^* = M_2 = \{x : h_1(x) = h_2(x) = h_3(x) = \phi_2(x) = \phi_3(x) = \psi_3(x) = 0\}$$

Further, the control $u^*(x)$ is obtained by solving the equation

$$L_f H_2 + L_g H_2 u^*(x) = 0.$$

There are 6 scalar equations in the above equation, but the construction of M^* automatically fulfills the second, third, and fifth equations. Thus we may obtain u^* from the remaining equations to be

$$u^*(x) = - \begin{bmatrix} L_g h_1 \\ L_g \phi_2 \\ L_g \psi_3 \end{bmatrix}^{-1} \begin{bmatrix} L_f h_1 \\ L_f \phi_2 \\ L_f \psi_3 \end{bmatrix}.$$

The matrix inverse above exists because of the hypothesis on the first, fourth and sixth row of $L_g H_2$. By the regularity hypothesis the differentials of $h_1, h_2, h_3, \phi_2, \phi_3, \psi_3$ are linearly independent at x_0 and thus qualify for a local, partial change of coordinates. Complementary coordinates $\eta \in \mathbb{R}^{n-6}$ may be chosen with the requirement that $\eta(x_0) = 0$. In these coordinates the system *dynamics are given by*

$$\dot{y}_1 = L_f h_1 + L_g h_1 u,$$

$$\dot{y}_2 = L_f h_2 + L_g h_2 u \qquad\qquad = L_f h_2 - \gamma L_g h_1 u + \sigma_2 u$$

$$= \phi_2 - \gamma(L_f h_1 + L_g h_1 u) + \sigma_2 u = \phi_2 - \gamma \dot{y}_1 + \sigma_2 u,$$

$$\dot{y}_3 = L_f h_3 \qquad\qquad\qquad = \phi_3,$$

$$\dot{\phi}_2 = L_f \phi_2 + L_g \phi_2 u,$$

$$\dot{\phi}_3 = L_f \phi_3 + L_g \phi_3 u \qquad\qquad = L_f \phi_3 - (\delta_1 L_g h_1 + \delta_2 L_g \phi_2)u + \sigma_3 u$$

$$= \psi_3 - \delta_1 \dot{y}_1 - \delta_2 \dot{y}_2 + \sigma_3 u,$$

$$\dot{\psi}_3 = L_f \psi_3 + L_g \psi_3 u,$$

with the remaining variables satisfying

$$\dot{\eta} = p(y_1, y_2, y_3, \phi_2, \phi_3, \psi_3, \eta) + q(y_1, y_2, y_3, \phi_2, \phi_3, \psi_3, \eta)u.$$

This is the generalized normal form of the three input, three output system under the foregoing assumptions on its zero dynamics algorithm. Several interesting points may be noted: If strong regularity rather than regularity holds, then the $\sigma_2, \sigma_3 \equiv 0$ and the form of the preceding equations differs from the normal form for a MIMO system with vector relative degree only in-as-much-as the presence of the terms $\gamma \dot{y}_1$ and $\delta_1 \dot{y}_1, \delta_2 \dot{y}_2$ in addition to the chain of integrators (one long for y_1, two long for y_2 and three long for y_3). To derive the zero dynamics we set $u = u^*$ and $x \in M^*$. As a consequence we get $y_1 = y_2 = y_3 = \phi_2 = \phi_3 = \psi_3 \equiv 0$, and the residual dynamics are the dynamics of the η coordinates parameterizing M^* locally

$$\dot{\eta} = p(0, 0, 0, 0, 0, 0, \eta) + q(0, 0, 0, 0, 0, 0, \eta)u^*(0, 0, 0, 0, 0, 0, \eta).$$

To extend the results of the preceding example to a general nonlinear control system, one needs to organize the ordering of the entries of the new functions Φ_k at every stage so as to explicitly keep track of the chains of integrators associated with each output. The key to this ordering is to move those entries of Φ_k for which the rows of $d\Phi_k g$ are linearly dependent on the previous $d\Phi_i g$ rows to the bottom. The details of this are left to Problem 11.21.

11.8   Input–Output Expansions for Nonlinear Systems

In this section we take a more geometric view to deriving input output Volterra type series for nonlinear control systems. This approach is complementary to the more algebraic approach taken in Chapter 4 in the use of Peano–Baker series for Carleman linearizations and bilinearizations of control systems. In the control literature these series are referred to as Chen–Fliess series. Consider the input affine nonlinear control system of (11.51):

$$\dot{x} = f(x) + \sum_{i=1}^{m} g_i(x)u_i, \quad x(0) = x_0,$$

$$y_j = h_j(x), \quad 1 \le j \le p.$$

Associated with this system and the inputs define the following iterated integrals:

$$\xi_0(t) = t,$$

$$\xi_j(t) = \int_0^t u_j(\tau)d\tau,$$

$$\int_0^t d\xi_{i_k} \ldots d\xi_{i_0} = \int_0^t d\xi_{i_k}(\tau) \int_0^\tau d\xi_{i_{k-1}} \cdots d\xi_{i_0}.$$

The last entry in the formula above is an iterated integral defined for multi-indices of the form $i_k i_{k-1} \ldots i_0$. It is easy to see that if the inputs are bounded, $|u_j(\cdot)| \le M$, then

$$\int_0^t d\xi_{i_k} \cdots d\xi_{i_0} \le \frac{(Mt)^{k+1}}{(k+1)!}.$$

We will be interested in series of the form

$$v(t) = \lambda(x_0) + \sum_{k=0}^{m} \sum_{i_0,\ldots,i_k=0}^{m} c(i_k, \ldots, i_0) \int_0^t d\xi_{i_k} \ldots d\xi_{i_0}. \tag{11.55}$$

which, in the light of the previous bound, are convergent when the coefficients satisfy the following growth condition

$$|c(i_k, \ldots, i_0)| \le M_1 M_2^{k+1}. \tag{11.56}$$

For a given set of vector fields $g_0 = f, g_1, \ldots, g_m$, we will be interested in functionals with $c(i_k, \cdots, i_0) = L_{g_{i_0}} \cdots L_{g_{i_k}} \lambda(x_0)$ for some smooth function $\lambda : \mathbb{R}^n \mapsto \mathbb{R}$. The following theorem needs an interesting fact about series of the form (11.55), whose proof we leave to the Exercises (Problem 11.24). This fact establishes that series of this kind, the Chen–Fliess series behave like formal power series!

Proposition 11.29 Formal Power Series Property of Chen–Fliess Series. *Let* $v_1(t), \ldots, v_k(t)$ *be series of the form of (11.55) corresponding to functions* $\lambda_1, \ldots, \lambda_k$ *respectively. Then, for an analytic function* $\mu : \mathbb{R}^k \mapsto \mathbb{R}$, *we have that*

$$\mu(v_1(t), \ldots, v_k(t)) = \mu(\lambda_1, \ldots, \lambda_k)(x_0)$$

$$+ \sum_{k=0}^{m} \sum_{i_0,\ldots,i_k=0}^{m} L_{g_{i_0}} \cdots L_{g_{i_k}} \mu(\lambda_1, \ldots, \lambda_k)(x_0) \int_0^t d\xi_{i_k} \cdots d\xi_{i_0}. \tag{11.57}$$

Theorem 11.30 Chen–Fliess Series for Nonlinear Control Systems. *Consider the nonlinear control system (11.51) with inputs* $u_i(\cdot)$ *bounded by* M *for* $i =$

$1, \ldots, m$. Then, for T sufficiently small, the j-th output $y_j(t)$, $t \in [0, T]$, can be expressed as

$$y_j(t) = h_j(x_0) + \sum_{k=0}^{\infty} \sum_{i_0 \ldots i_k = 0}^{m} L_{g_{i_0}} \cdots L_{g_{i_k}} h_j(x_0) \int_0^t d\xi_{i_k} \cdots d\xi_{i_0}, \qquad (11.58)$$

with the understanding that $f = g_0$.

Proof: We first show that the j-th component of the solution of the differential equation in (11.51) satisfies

$$x_j(t) = x_j(x_0) + \sum_{k=0}^{\infty} \sum_{i_0 \ldots i_k = 0}^{m} L_{g_{i_0}} \cdots L_{g_{i_k}} x_j(x_0) \int_0^t d\xi_{i_k} \cdots d\xi_{i_0}, \qquad (11.59)$$

where the function $x_j(\cdot)$ simply extracts the j-th component of its argument. We do this by invoking the existence and uniqueness theorem of differential equations after checking that it satisfies the correct initial condition (which is obvious) and that its derivative satisfies the differential equation, which we do now. First notice that

$$\frac{d}{dt} \int_0^t d\xi_0 d\xi_{i_{k-1}} \cdots d\xi_{i_0} = \int_0^t d\xi_{i_{k-1}} \cdots d\xi_{i_0},$$
$$\frac{d}{dt} \int_0^t d\xi_i d\xi_{i_{k-1}} \cdots d\xi_{i_0} = u_i(t) \int_0^t d\xi_{i_{k-1}} \cdots d\xi_{i_0}. \qquad (11.60)$$

Now differentiating (11.59) with respect to time yields

$$\dot{x}_j(t) = L_f x_j(x_0) + \sum_{k=1}^{\infty} \sum_{i_0 \ldots i_k = 0}^{m} L_{g_{i_0}} \ldots L_{g_{i_k}} L_f x_j(x_0) \int_0^t d\xi_{i_{k-1}} \ldots d\xi_{i_0}$$

$$= \sum_{i=1}^{m} \left(L_{g_i} x_j(x_0) \right) u_i(t)$$

$$+ \sum_{i=1}^{m} \left(\sum_{k=1}^{\infty} \sum_{i_0 \ldots i_k = 0}^{m} L_{g_{i_0}} \cdots L_{g_{i_k}} L_{g_i} x_j(x_0) \int_0^t d\xi_{i_{k-1}} \cdots d\xi_{i_0} \right) u_i(t).$$

Using the fact that $L_f x_j = f_j(x)$ and the formal power series property of the Chen–Fliess series, the first set of terms may be written as

$$f_j(x_0) + \sum_{k=1}^{\infty} \sum_{i_0 \ldots i_k = 0}^{m} L_{g_{i_0}} \cdots L_{g_{i_k}} f_j(x_0) \int_0^t d\xi_{i_{k-1}} \cdots d\xi_{i_0}$$

$$= f_j(x_1(t), \ldots, x_n(t)).$$

A similar substitution to the second set of terms yields

$$\dot{x}_j(t) = f_j(x_1(t), \ldots, x_n(t)) + \sum_{i=1}^{m} g_{ij}(x_1(t), \ldots, x_n(t)) u_i(t),$$

where $g_{ij}(x)$ stands for the j-th component of the vector field g_i. Completing this calculation yields that x satisfies the differential equation in (11.51). Since it satisfies the correct initial condition, it is *the* solution of the differential equation. The assertion of the theorem, namely (11.58), now follows from the power series property of Chen–Fliess series applied to (11.59). □

Remark: One of the major consequences of this theorem is that we can give necessary and sufficient conditions for an output y_j to be unaffected by an input u_i, namely,

$$L_{g_i} h_j(x) = 0,$$
$$L_{g_i} L_{\tau_1} \cdots L_{\tau_r} h_j(x) = 0,$$
(11.61)

for all choices of vector fields $\tau_1, \ldots, \tau_r$ in the set $\{f, g_1, \ldots, g_m\}$. The condition (11.61) may be given a more geometric interpretation. To do this define the *observability codistribution* of the output y_j associated with the system (11.51) by

$$\Omega_j(x) = \text{span}\{d\lambda : \lambda = L_{g_{i_0}} \cdots L_{g_{i_k}} h_j, 0 \le i_k \le m, 0 \le k \le \infty\}.$$
(11.62)

We write $\Omega_j(x)$ to emphasize the dependence of the codistribution on x. Then the statement (11.61) is equivalent to

$$g_i \in \Omega_j^\perp.$$
(11.63)

From the form of the definition of Ω_j it follows that it is invariant under $f, g_1, \ldots, g_m$. Now, if (11.63) is true, it follows that the smallest distribution $A(x)$ (called the accessibility distribution earlier in this chapter) containing $g_1, \ldots, g_m$ which is invariant under $f, g_1, \ldots, g_m$ is also in $\Omega_j^\perp$:

$$A(x) \subset \Omega_j^\perp(x).$$
(11.64)

In turn, since

$$\Omega_j^\perp \subset (\text{span}\{dh_j\})^\perp,$$

it follows that (11.61) implies that the smallest distribution $A_i(x)$ containing g_i, which is invariant under $f, g_1, \ldots, g_m$, is in $\{\text{span}\{dh_j\}\}^\perp$:

$$A_i(x) \subset (\text{span}\{dh_j\})^\perp.$$
(11.65)

Thus, we have shown that (11.63) implies (11.64) which in turn implies (11.65). We will now show that the condition (11.65) implies condition (11.63), showing that the three conditions are equivalent. Indeed, $[f\tau, g_i](x) \in A_i(x)$ for all $\tau \in \{f, g_1, \ldots, g_m\}$. By condition (11.65) it follows that

$$0 = L_{[\tau, g_i]} h_j = L_\tau L_{g_i} h_j - L_{g_i} L_\tau h_j.$$

Again, since $g_i \in A_i$ it follows that $L_{g_i} h_j = 0$ so that

$$L_{g_i} L_\tau h_j = 0 \quad \Rightarrow g_i \in (\text{span}\{dL_\tau h_j\})^\perp$$

Iterating on this argument we will arrive at

$$g_i \in \Omega_j^\perp.$$

Now, if Δ_i is any distribution containing g_i and invariant under $f, g_1, \ldots, g_m$ and contained in $(\text{span}\{dh_j\})^{\perp}$, it follows that

$$A_i(x) \subset \Delta_i(x) \subset (\text{span}\{dh_j\})^{\perp}. \tag{11.66}$$

We collect this discussion into the following theorem:

Theorem 11.31 Decoupling Output y_j from Input u_i. *The output y_j of the system (11.51) is unaffected by the input u_i if and only if there exists a distribution $\Delta_i \supset g_i$ that is invariant under $f, g_1, \ldots, g_m$ and further more*

$$\Delta_i(x) \subset (\text{span}\{dh_j\})^{\perp}.$$

11.9   Controlled Invariant Distributions and Disturbance Decoupling

In this section we give a definition of controlled invariant distributions and controllability distributions of the control affine nonlinear system of (11.51), with equal numbers of inputs and outputs, namely,

$$\dot{x} = f(x) + g(x)u,$$

$$y = h(x),$$

with $f, g_1, g_2, \ldots, g_m$ all smooth vector fields and $h : \mathbb{R}^n \mapsto \mathbb{R}^m$ a smooth map. We show that controlled invariant distributions can be used to solve problems such as disturbance decoupling and input–output decoupling of nonlinear systems.

Definition 11.32 Controlled Invariant Distributions. *A (locally) controlled invariant distribution of (11.51) is a distribution $\Delta \subset T\mathbb{R}^n$ defined on a neighborhood U of the origin such that there exist $\alpha(x) : \mathbb{R}^n \mapsto \mathbb{R}^m, \beta(x) \in \mathbb{R}^{n \times m}$ such that Δ is invariant under the closed loop state feedback vector fields*

$$\tilde{f}(x) = f(x) + g(x)\alpha(x),$$

$$\tilde{g}_i(x) = \sum_{j=1}^{m} g_j(x)\beta_{ji}(x), \quad i = 1, \ldots m, \tag{11.67}$$

that is to say, that Δ is invariant under

$$[\tilde{f}, \Delta](x) \subset \Delta(x), \quad [\tilde{g}_i, \Delta](x) \subset \Delta(x), \quad i = 1, \ldots, m.$$

Thus, a distribution Δ is said to be controlled invariant if it can be rendered invariant by a state feedback of the form

$$u = \alpha(x) + \beta(x)v.$$

The following proposition gives a characterization of controlled invariant distributions. We define the distribution

$$G = \text{span}\{g_1, \ldots, g_m\}.$$

Proposition 11.33 Characterization of Controlled Invariant Distributions.
Let Δ be a controlled invariant distribution. Suppose that Δ, G, $\Delta + G$ are all regular on a neighborhood U of the origin. Then Δ is controlled invariant iff

$$[f, \Delta] \subset \Delta + G, \quad [g_i, \Delta] \subset \Delta + G. \tag{11.68}$$

Proof: We begin with the necessity of the condition (11.68). Suppose that Δ is controlled invariant, then there exist $\alpha(x)$, $\beta(x)$ such that (11.67) holds. Then, we have that for some vector field $\tau \in \Delta$,

$$[\tilde{f}, \tau] = [f + g\alpha, \tau] = [f, \tau] + \sum_{j=1}^{m} \alpha_j [g_j, \tau] - \sum_{j=1}^{m} (L_\tau \alpha_j) g_j$$

and

$$[\tilde{g}_i, \tau] = \sum_{j=1}^{m} [g_j \beta_{ji}, \tau] = \sum_{j=1}^{m} [g_j, \tau] \beta_{ji} - \sum_{j=1}^{m} (L_\tau \beta_{ji}) g_j.$$

In the second equation above, since $\beta(x)$ is invertible for $x \in U$ it follows that we can solve for $[g_j, \tau]$ and using $[\tilde{g}_i, \tau] \in \Delta$ it follows that

$$[g_j, \tau] \in \Delta + G, \quad j = 1, \ldots, m.$$

We can use this in the first equation, along with $[\tilde{f}, \tau] \in \Delta$ to establish that

$$[f, \tau] \in \Delta + G.$$

The proof of the sufficiency is more involved and we refer the reader to Isidori [149], Chapter 6 for the details. $\qquad \square$

The notion of controlled invariant distributions is of particular interest to render the outputs of the system independent of the effect of certain inputs. Let us revisit the disturbance decoupling problem that we first encountered in Chapter 9. We repeat it here: Consider the system

$$\dot{x} = f(x) + \sum_{i=1}^{m} g_i(x) u_i + p(x) w,$$
$$y = h(x), \tag{11.69}$$

where $w(t) \in \mathbb{R}$ represents a disturbance input that is to be decoupled from the outputs y using a state feedback of the form

$$u = \alpha(x) + \beta(x) v,$$

which yields the closed loop system

$$\dot{x} = f(x) + g(x)\alpha(x) + g(x)\beta(x)v + p(x)w,$$
$$= \tilde{f}(x) + \tilde{g}(x)v + p(x)w, \tag{11.70}$$
$$y = h(x).$$

Using the decoupling theorem 11.31 of the preceding section, we see that the disturbance w is decoupled from all the outputs if there exists a distribution $\Delta(x) \supset p(x)$ invariant under $\tilde{f}, \tilde{g}_1, \tilde{g}_2, \ldots, \tilde{g}_m$ such that

$$\text{span}\{p(x)\} \subset \Delta \subset (\text{span}\{dh_1, \ldots, dh_p\})^{\perp}. \tag{11.71}$$

From the preceding proposition 11.33 it follows that Δ is a controlled invariant distribution. Thus, the disturbance decoupling problem is solvable if there exists a controlled invariant distribution Δ in

$$\ker(dh) = \cap_{i=1}^{p} \ker(dh_j) = (\text{span}\{dh_1, \ldots, dh_p\})^{\perp}$$

containing the disturbance vector field p. In fact, there is a conceptual algorithm, referred to as the controlled invariant distribution algorithm (strongly motivated by the controlled invariant subspace algorithm), to obtain the *largest*, or *maximal* controlled invariant distribution algorithm Δ^* in the kernel of dh as defined above. In this case the necessary and sufficient condition for disturbance decoupling is a small modification of (11.71):

$$\text{span}\{p(x)\} \subset \Delta^*. \tag{11.72}$$

We do not present the proof of the existence of this maximal distribution or show the connection between this distribution Δ^* and the necessary and sufficient conditions for disturbance decoupling which we derive in Chapter 9 (9.107) for systems that have vector relative degree, but refer the reader to [149] or [232] for more details on the nonlinear counterpart of the geometric control theory begun by Wonham [331]. This theory is very rich and powerful and may also be used to give necessary and sufficient conditions for decoupling of a MIMO system (generalizing the results of Chapter 9) using the concept of *controllability distributions*. However, from a computational point of view the design procedures most amenable to computer aided designs are those discussed in Chapter 9.

11.10 Summary

In this chapter we have given the reader a very brief snapshot of some of the exciting new directions in nonlinear control with the studies of constructive controllability and nonholonomy. There is a lot more on this topic in addition to the dual "differential form" point of view, which we study in the next chapter. An excellent recent survey of new methods and directions is in the paper by Kolmanovsky and McClamroch [164]. Another body of interesting work in this area is the control of underactuated systems, see for example [272; 284]. Constructive controllability methods for systems with drift are not easy to come by. Some papers to study here include Bloch, Reyhanoglu, and McClamroch [31], Kapitnavosky, Goldberg, and Mills [160]. An interesting approach to using sampled data techniques to systems whose controllability distributions are nilpotent or nilpotentizable (see [137] and [161]) was proposed by Di Giamberardino, Monaco, and Normand Cyrot in [111]. A survey of these methods as well as some new methods, along with applications to some interesting systems including the "diver" is in a paper by Godhvan, Balluchi, Crawford, and Sastry [115]. See also [72]. The control of Hamiltonian systems with symmetries is an interesting direction of work begun by van der Schaft (see [232]) and continuing with [309; 74]. There has been recent activity on the use of geometric methods to study controllability of dynamic models of nonholonomic systems with symmetries, see for example the work of Montgomery on falling cats, Yang Mills fields, and other exotic applications [217], Bloch and Crouch [27], Sarychev and Nijmeijer [258], and Bloch et al [29]. See also the forthcoming monograph of Montgomery on nonholonomy in mechanics and optimal control. For application to problems of gait control in locomotion see for example Chirikjian, Ostrowski and Burdick [63; 237].

We gave a reader a brief flavor of nonlinear differential geometric approaches to the solution of problems that were proposed in a linear context by Wonham [331]. In a historical sense much of the recent interest in these methods may be traced to a paper by Isidori, Krener, et al. [150], which paralleled the linear development for the nonlinear case. Constructive algorithms for mechanizing these calculations remain an open area, and the zero dynamics algorithm is indispensable in this regard. The reader is invited to study [149], [232] for more details on this topic.

11.11 Exercises

Problem 11.1. Prove the formula for the maximum growth of the distributions of a regular filtration, namely (11.7).

Problem 11.2 Car with N trailers [174]. The figure below shows a car with N trailers attached. We attach the hitch of each trailer to the center of the rear axle of the previous trailer. The wheels of the individual trailers are aligned with the body of the trailer. The constraints are again based on allowing the wheels only to roll and spin, but not slip. The dimension of the state space is $N + 4$ with two controls.

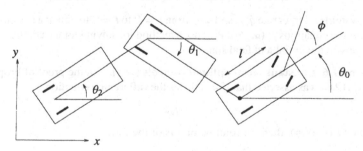

Parameterize the configuration by the states of the automobile plus the angle of each of the trailers with respect to the horizontal. Show that the control equation for the system has the form

$$\dot{x} = \cos\theta_0\, u_1,$$

$$\dot{y} = \sin\theta_0\, u_1,$$

$$\dot{\phi} = u_2,$$

$$\dot{\theta}_0 = \frac{1}{l}\tan\phi\, u_1,$$

$$\dot{\theta}_i = \frac{1}{d_i}\left(\prod_{j=1}^{i-1}\cos(\theta_{j-1}-\theta_j)\right)\sin(\theta_{i-1}-\theta_i)u_1.$$

Problem 11.3. Use induction and Jacobi's identity to prove that

$$[\Delta_i, \Delta_j] \subset [\Delta_1, \Delta_{i+j-1}] \subset \Delta_{i+j},$$

where $\Delta = \Delta_1 \subset \Delta_2 \subset \cdots$ is a filtration associated with a distribution Δ.

Problem 11.4. Let $\Delta_i, i = 1, \ldots, \kappa$, be a regular filtration associated with a distribution. Show that if rank$(\Delta_{i+1}) =$ rank(Δ_i) then Δ_i is involutive. (Hint: use Exercise 11.3).

Problem 11.5 Steering by sinusoids [225].   Consider the following system

$$\dot{q}_1 = u_1,$$

$$\dot{q}_2 = u_2,$$

$$\dot{q}_{12} = q_1 u_2,$$

$$\dot{q}_{121} = q_{12} u_1,$$

$$\dot{q}_{122} = q_{12} u_2.$$

Apply the inputs

$$u_1 = a_1 \sin 2\pi t + a_2 \cos 2\pi t + a_3 \sin 4\pi t + a_4 \cos 4\pi t,$$

$$u_2 = b_1 \sin 2\pi t + b_2 \cos 2\pi t + b_3 \sin 4\pi t + b_4 \cos 4\pi t,$$

to this system and integrate $\dot{q}_{121}$ and $\dot{q}_{122}$ from $t = 0$ to $t = 1$ to obtain a system of polynominal equations in (a_i, b_i). Propose a method for solving for the coefficients (a_i, b_i) given the initial and final states.

Problem 11.6 Constant-norm optimal controls [264].   In the proof of Proposition 11.12, it was asserted that if u satisfies the differential equation

$$\dot{u} = \Omega(t)u$$

for some $\Omega \in so(m)$, then the solution of u is of the form

$$u(t) = U(t)u(0)$$

for some $U(t) \in SO(m)$. Prove this assertion.

Problem 11.7 Minimum-time steering problem.   Use the results of Proposition 11.12 to show that the optimal input of that proposition normalized by $|u(0)|$, that is,

$$\frac{u(t)}{|u(0)|},$$

solves the minimum time steering problem to steer the system from $q(0) = q_0$ to $q(T) = q_f$ subject to the constraint that $|u(t)|^2 \leq 1$ for all t.

Problem 11.8 Heisenberg Control System [225].   Consider the control system

$$\dot{q} = u,$$
$$\dot{Y} = qu^T - uq^T,$$

where $u \in \mathbb{R}^m$ and $Y \in so(m)$.

1. Derive the Euler-Lagrange equations for the system, by minimizing the following integral

$$\frac{1}{2} \int_0^1 u^T u \, dt.$$

2. For the boundary conditions $q(0) = q(1) = 0$, $Y(0) = 0$, and $Y(1) = \hat{y}$ for some $y \in \mathbb{R}^3$, solve the Euler–Lagrange equations to obtain the optimal inputs u.
3. Find the input u to steer the system from $(0, 0)$ to $(0, \tilde{Y}) \in \mathbb{R}^m \times so(m)$.

Problem 11.9. Extend the method used to find the optimal inputs for Engel's system to find optimal inputs for the system of Problem 11.8.

Problem 11.10 Optimal controls for systems with drift [264].   Consider the least squares optimal input steering problem for a system with drift:

$$\dot{q} = f(q) + \sum_{i=1}^{m} g_i(q)u_i.$$

1. Find the expression for the optimal Hamiltonian and prove that the optimal inputs satisfy the differential equation

$$\dot{u} = \Omega(q, p)u + \begin{bmatrix} p^T[f, g_1] \\ \vdots \\ p^T[f, g_m] \end{bmatrix} \tag{11.73}$$

with $\Omega(q, p)$ defined as in (11.29).
2. Find the second derivatives of the optimal inputs given in equation (11.73).

Problem 11.11 Conversion to chained form [225], [205]. Show that the following system

$$\dot{q}_1 = u_1,$$

$$\dot{q}_2 = u_2,$$

$$\dot{q}_3 = q_1 u_2,$$

$$\dot{q}_4 = \frac{1}{2} q_1^2 u_2,$$

$$\dot{q}_5 = q_1 q_2 u_2,$$

$$\dot{q}_6 = \frac{1}{6} q_1^3 u_1,$$

$$\dot{q}_7 = \frac{1}{2} q_1^2 q_2 u_2,$$

$$\dot{q}_8 = \frac{1}{2} q_1 q_2^2 u_2,$$

is controllable and nilpotent of degree four. Can you find a nonlinear change of coordinates to transform this system into a one-chained form?

Problem 11.12. Show that the following system is controllable and nilpotent:

$$\dot{q}_1 = u_1,$$

$$\dot{q}_2 = u_2,$$

$$\dot{q}_3 = q_1 u_2 - q_2 u_1,$$

$$\dot{q}_4 = q_1^2 u_2,$$

$$\dot{q}_5 = q_2^2 u_1.$$

Problem 11.13 Chen–Fliess expansion for chained form systems [225]. Consider the one-chain system in equation (11.14). For the Philip Hall basis g_1, g_2, $ad_{g_1}^k g_2, k = 1, \ldots, n - 1$, derive the Chen–Fliess–Sussmann equation.

Problem 11.14. Consider the following system

$$\dot{q}_1 = u_1,$$

$$\dot{q}_2 = u_2,$$

$$\dot{q}_3 = q_1 u_2 - q_2 u_1,$$

$$\dot{q}_4 = q_1^2 u_2.$$

1. Derive the Chen–Fliess–Sussmann equation.
2. Assuming that the inputs are of the form

$$u_1 = a_0 + a_1 \cos 2\pi t + a_2 \sin 2\pi t,$$

$$u_2 = b_0 + b_1 \cos 2\pi t + b_2 \sin 2\pi t,$$

compute the polynomial equation for the amplitude parameters in terms of the initial and final states.

Problem 11.15. Prove that if $V \subset \mathbb{R}^n$ then we can conceive of V as a constant distribution and then $AV \subset V$ is equivalent to $[Ax, V] \subset V$.

Problem 11.16. If V is an A-invariant subspace, prove that it can be represented in suitable coordinates as

$$\begin{bmatrix} A_{11} & A_{12} \\ 0 & A_{22} \end{bmatrix}.$$

Problem 11.17. Show that if a distribution Δ is invariant under vector fields f_1, f_2 it is also invariant under $[f_1, f_2]$. Hint: Use the Jacobi identity.

Problem 11.18. Prove the following facts about the properties of the Lie derivative of a covector field:

1. If f is a vector field and λ a smooth function, then

$$L_f d\lambda(x) = dL_f \lambda(x).$$

2. If α, β are smooth functions, f a vector field and ω a covector field then

$$L_{\alpha f} \beta \omega(x) = \alpha \beta (L_f \omega(x)) + \beta(x)\omega f d\alpha(x) + (L_f \beta(x))\alpha(x)\omega(x).$$

3. If f, g are vector fields and ω a covector field, it follows that

$$L_f \omega g(x) = L_f \omega(x)g(x) + \omega(x)[f, g](x).$$

Problem 11.19. For the minimal linear system

$$\dot{x} = Ax + Bu,$$
$$y = Cx,$$

specialize the definition of a controlled invariant manifold (subspace) and an output zeroing manifold. Assuming that $B \in \mathbb{R}^{n \times m}$ is of full rank characterize the zero dynamics. Use Problem (11.16) to represent A, B, C in coordinates adapted to the subspace $TM^* = M^*$. Convince yourself that the eigenvalues of the (linear) zero dynamics are indeed what you mean by the *transmission zeros* of the linear system.

Problem 11.20. Consider a MIMO system with vector relative degree $(\gamma_1, \ldots, \gamma_m)$. Assume that $\gamma_1 \leq \gamma_2 \leq \cdots \leq \gamma_m$. We now have two ways of deriving the zero dynamics of this system: the procedure given in Chapter 9 and the zero dynamics algorithm of this chapter. Show that they give the same answer. You may impose the regularity conditions and other assumptions as needed.

Problem 11.21 Generalized normal form [149]. Derive the generalized normal form following the steps of the zero dynamics algorithm and Example 11.28 for an arbitrary nonlinear control system (provided that the regularity conditions for the zero dynamics algorithm are met).

Problem 11.22 Exact model matching [89; 88]. An extremely important control problem is to match the plant output to the output of a reference model. This problem examines a method proposed by Di Benedetto to cast this problem into the framework of the zero dynamics algorithm. To this end consider a reference model with m inputs, m outputs and n_M states described by

$$\dot{z} = f_M(z) + g_M(z)v,$$

$$y_M = h_M(z).$$

Further, define the extended system consisting of the plant and model Σ^E given by

$$\dot{x}^E = f^E(x^E) + \hat{g}(x^E)u + \hat{p}(x^E)v,$$
$$y^E = h^E(x^E),$$
(11.74)

with state $(x^E)^T := (x^T, z^T) \in \mathbb{R}^n \times \mathbb{R}^{n_M}$, inputs u, v and

$$f^E(x^E) = \begin{bmatrix} f(x) \\ f_M(z) \end{bmatrix}, \hat{g}(x^E) = \begin{bmatrix} g(x) \\ 0 \end{bmatrix}, \hat{p}(x^E) = \begin{bmatrix} 0 \\ g_M(z) \end{bmatrix},$$

$$h^E(x^E) = h(x) - h_M(z).$$

Further define

$$g^E(x^E) = \begin{bmatrix} \hat{g}(x^E) & \hat{p}(x^E) \end{bmatrix}.$$

Also, define the dynamical system with state x^E, input u, and output y^E described by the triple $(f^E, \hat{g}, h^E)$ to be $\hat{\Sigma}$. Now consider a point $x_0^E = (x_0, z_0)$ which is an equilibrium point of f^E and also produces zero output for the system $\hat{\Sigma}$, i.e.,

$$f^E(x_0^E) = 0, \quad h^E(x_0^E) = 0.$$

Assume that x_0^E is a regular point for the zero dynamics algorithm applied to $\hat{\Sigma}$. Let $\hat{M}_k$ denote the manifold defined at step k of the zero dynamics algorithm and $\hat{M}^*$ the zero dynamics manifold obtained at the conclusion of the algorithm. Show that there exists a choice of control law $u(x^E, v)$ and initial conditions x^E for which the output of the extended system is identically zero if:

$$\text{span}\{\hat{p}(x^E)\} \subset T_{x^E}\hat{M}_k + \text{span}\{\hat{g}(x^E)\}$$

in a neighborhood of x_0^E in $\hat{M}_k$ for all k.

Problem 11.23 Adaptive nonlinear control [259; 266]. Consider the SISO nonlinear system

$$\dot{x} = f_0(x) + g_0(x)u + \sum_{i=1}^{p} \theta_i^* f_i(x),$$
(11.75)

$$y = h(x).$$

The functions $f_0, g_0, f_i(x)$ are considered known, but the parameters $\theta^* \in \mathbb{R}^p$ are fixed and unknown. These are referred to as the *true* parameters. We will explore a control law for making y track a given *bounded* trajectory $y_d(t)$. Assume that $L_{g_0}h(x)$ is bounded away from zero and the control law is the following modification of the standard relative degree one asymptotic tracking control law

$$u = \frac{1}{L_{g_0}h}\left(-L_{f_0}h + \dot{y}_d + \alpha(y_d - y) - \sum_{i=1}^{p}\hat{\theta}_i(t)L_{f_i}h\right). \tag{11.76}$$

Here $\hat{\theta}_i(t)$ is our estimate of θ_i^* at time t. Prove (using the simplest quadratic Lyapunov function of $y - y_d$, $\hat{\theta} - \theta^*$ that you can think of) that under the parameter update law,

$$\dot{\hat{\theta}} = -\begin{bmatrix} L_{f_1}h \\ \vdots \\ L_{f_p}h \end{bmatrix}(y - y_d) \tag{11.77}$$

for all initial conditions $\hat{\theta}(0)$ and $x(0)$, $y(t) - y_d(t) \to 0$ as $t \to \infty$. Note that you are *not* required to prove anything about $\hat{\theta}(t)$!

Sketch the proof of the fact that if the true system of (11.75) is exponentially minimum phase, then $x(\cdot)$ is bounded as well. Can you modify the control law (11.76) and the parameter update law (11.77) for the case that the right-hand side of the system (11.75) has terms of the form

$$\sum_{i=1}^{p}(f_i(x) + g_i(x)u)\theta_i^*$$

with f_i, g_i known and $\theta^* \in \mathbb{R}^p$ unknown? You can assume here that

$$L_{g_0}h + \sum_{i=1}^{p}\hat{\theta}_i(t)L_{g_i}h$$

is invertible.

Problem 11.24 Formal power series property of Chen–Fliess series. Prove Proposition 11.29. You may wish to start not with arbitrary analytic functions $\mu(v_1, \ldots, v_k)$, but instead start with functions of the form $v_1v_2, v_1v_3, \ldots$ etc. to arrive at the proof.

12
Exterior Differential Systems in Control

This chapter has been extensively based on the research work of (and papers written with) Richard Murray, Dawn Tilbury, and Linda Bushnell. The text of the chapter follows notes written with George Pappas, John Lygeros, and Dawn Tilbury [241].

12.1   Introduction

The vast majority of the mathematically oriented literature in the areas of robotics and control has been heavily influenced by a differential geometric "vector field" point of view, which we have seen in the earlier chapters. In recent years, however, a small but influential trend has begun in the literature on the use of other methods, such as exterior differential systems [119; 108] for the analysis of nonlinear control systems and nonlinear implicit systems. In this chapter we present some key results from the theory of exterior differential systems and their application to current and challenging problems in robotics and control. The area of exterior differential systems has a long history. The early theory in this area sprung from the work of Darboux, Lie, Engel, Pfaff, Carnot, and Caratheodory on the structure of systems with non-integrable linear constraints on the velocities of their configuration variables, the so-called nonholonomic control systems (for a good development of this see [49]). This was followed by the work of Goursat and Cartan, which is considered to contain some of the finest achievements of the middle part of this century on exterior differential systems. In parallel has been an effort to develop connections between exterior differential systems and the calculus of variations (see [119]).

Our attention was first attracted to exterior differential systems through their applications in path planning for nonholonomic control systems, where initial results were for the problem of steering a car with trailers [302], [223], the so-called "parallel parking a car with N trailers problem." This involved the transformation of the system of nonholonomic rolling without slipping constraints on each pair of wheels into a canonical form, the so-called Goursat normal form. This program continued with another example, the parallel parking of a fire truck [53], which in turn was generalized to a multi-steering N trailer system. In [304] we showed how the multi-steering N trailer system could be converted into a generalized Goursat normal form, which was easy to steer. The full analysis of the system from the exterior differential systems point of view was made in [303]. The vector field point of view to these problems was studied in Chapter 11.

In parallel with this activity in nonholonomic motion planning, there has been considerable activity in the nonlinear control community on the problem of exactly linearizing a nonlinear control system using (possibly dynamic) state feedback and change of coordinates. The first results in this direction were necessary and sufficient conditions for exact linearization of a nonlinear control system using static state feedback, which was presented in Chapter 9. It was shown there that a system that satisfies these conditions can be transformed into a special canonical form, the so called Brunovsky normal form. As pointed out by Gardner and Shadwick in [108], this normal form is very close to the Goursat normal form for exterior differential systems. The problem of dynamic state feedback linearization, on the other hand, remained largely open, despite some early results by [59; 276].

This chapter is divided into three parts. Section 12.2 contains the necessary mathematical background on exterior differential systems. Section 12.3 describes some of the important normal forms for exterior differential systems: the Engel, Pfaff, Caratheodory, Goursat, and extended Goursat normal forms. It is shown how certain important robotic systems can be converted to these normal forms. Section 12.4 discusses some of the connections between the exterior differential systems formalism, specialized to the case of control systems, and the vector field approach currently popular in nonlinear control.

12.2   Introduction to Exterior Differential Systems

In this section we will introduce the concept of an exterior differential system. To this end, we first introduce multilinear algebra, including the tensor and wedge products, and exterior algebra. Once we have defined the exterior derivative, we will study many of its important properties. We then review the Frobenius theorem, for both vector fields and forms, and finally define an exterior differential system. In order to keep the development compact, we have omitted a number of proofs, and we refer the reader to the excellent books of Munkres [221, Chapter 6], Abraham,

Marsden, and Ratiu [2], and Spivak [282] on whose presentations the present development is based.

12.2.1 Multilinear Algebra

The Dual Space of a Vector Space

Many of the ideas underlying the theory of multilinear algebra involve duality and the notion of the dual space to a vector space.

Definition 12.1 Dual Space, Covectors. *Let $(V, \mathbb{R})$ denote a finite dimensional vector space over $\mathbb{R}$. The* dual space *associated with $(V, \mathbb{R})$ is defined as the space of all linear mappings $f : V \to \mathbb{R}$. The dual space of V is denoted as V^* and the elements of V^* are called* covectors. *V^* is a vector space over $\mathbb{R}$ with $dim(V^*) = dim(V)$ for the operations of addition and scalar multiplication defined by:*

$$(\alpha + \beta)(v) = \alpha(v) + \beta(v),$$

$$(c\alpha)(v) = c \cdot \alpha(v).$$

Furthermore, if $\{v_1, \ldots, v_n\}$ is a set of basis vectors for V, then the set of linear functions $\phi^i : V \to \mathbb{R}$, $1 \le i \le n$, defined by:

$$\phi^i(v_j) = \begin{cases} 0 & if\ i \neq j, \\ 1 & if\ i = j, \end{cases}$$

form a basis of V^* called the dual basis.

Let $V = \mathbb{R}^n$ with the standard basis $e_1, \ldots, e_n$ and let $\phi^1, \ldots, \phi^n$ be the dual basis. If

$$x \in \mathbb{R}^n = \sum_{j=1}^{n} x_j e_j,$$

then evaluating each function in the dual basis at x gives

$$\phi^i(x) = \phi^i\left(\sum_{j=1}^{n} x_j e_j\right) = \sum_{j=1}^{n} x_j \phi^i(e_j) = x_i.$$

Since the functions $\phi^1, \ldots, \phi^n$ form a basis for V^*, a general covector in $(\mathbb{R}^n)^*$ is of the form $f = \alpha_1 \phi^1 + \cdots + \alpha_n \phi^n$. Evaluating this covector at the point x gives $f(x) = \alpha_1 x_1 + \cdots + \alpha_n x_n$. If we think of a vector as a column matrix and a covector as a row matrix, then

$$f(x) = [\alpha_1 \ldots \alpha_n] \begin{bmatrix} x_1 \\ \vdots \\ x_n \end{bmatrix}.$$

Definition 12.2 Annihilator.   *Given a subspace $W \subset V$ its annihilator is the subspace $W^\perp \subset V^*$ defined by*

$$W^\perp := \{\alpha \in V^* \mid \alpha(v) = 0 \; \forall \, v \in W\}$$

Given a subspace $X \subset V^*$, its annihilator is the subspace $X^\perp \subset V$ defined by

$$X^\perp := \{v \in V \mid \alpha(v) = 0 \; \forall \, \alpha \in X\}$$

A linear mapping $F : V_1 \mapsto V_2$ between any two vector spaces induces a linear mapping between their dual spaces.

Definition 12.3 Dual Map.   *Given a linear mapping $F : V_1 \to V_2$, its dual map is the linear mapping $F^* : V_2^* \to V_1^*$ defined by*

$$(F^*(\alpha))(v) = \alpha(F(v)), \; \forall \, \alpha \in V_2^*, \; v \in V_1$$

Tensors

Let $V_1, \ldots, V_k$ be a collection of real vector spaces. A function $f : V_1 \times \ldots \times V_k \mapsto \mathbb{R}$ is said to be linear in the ith variable if the function $T : V_i \mapsto \mathbb{R}$ defined for fixed v_j, $j \neq i$ as $T(v_i) = f(v_1, \ldots, v_{i-1}, v_i, v_{i+1}, \ldots, v_k)$ is linear. The function f is called multilinear if it is linear in each variable, with the other variables held fixed. A multilinear function $T : V^k \to \mathbb{R}$ is called a *covariant tensor of order k* or simply a *k-tensor*. The set of all k-tensors on V is denoted by $\mathcal{L}^k(V)$. Note that $\mathcal{L}^1(V) = V^*$, the dual space of V. Therefore, we can think of covariant tensors as generalized covectors. The inner product of two vectors is an example of a 2-tensor. Another important example of a multilinear function is the determinant. If $x_1, x_2, \ldots, x_n$ are n column vectors in $\mathbb{R}^n$, then

$$f(x_1, x_2, \ldots, x_n) = \det[x_1 \; x_2 \; \ldots \; x_n]$$

is multilinear by the properties of the determinant.

As in the case of V^*, each $\mathcal{L}^k(V)$ can be made into a vector space. If for $S, T \in \mathcal{L}^k(V)$ and $c \in \mathbb{R}$ we define addition and scalar multiplication by:

$$(S + T)(v_1, \ldots, v_k) = S(v_1, \ldots, v_k) + T(v_1, \ldots, v_k),$$

$$(cT)(v_1, \ldots, v_k) = c \cdot T(v_1, \ldots, v_k),$$

then, the set of all k-tensors on V, $\mathcal{L}^k(V)$, is a real vector space. Because of their multilinear structure, two tensors are equal if they agree on any set of basis elements. Indeed, let $a_1, \ldots, a_n$ be a basis for V. Let $f, g : V^k \mapsto \mathbb{R}$ be k-tensors on V. If $f(a_{i_1}, \ldots, a_{i_k}) = g(a_{i_1}, \ldots, a_{i_k})$ for every k-tuple $I = (i_1, \ldots, i_k) \in \{1, 2, \ldots, n\}^k$, then $f = g$. This allows us to construct a basis for the space $\mathcal{L}^k(V)$ as follows:

Theorem 12.4.   *Let $a_1, \ldots, a_n$ be a basis for V. Let $I = (i_1, \ldots, i_k) \in \{1, 2, \ldots, n\}^k$. Then there is a unique tensor ϕ^I on V such that for every k-tuple*

$J = (j_1, \ldots, j_k) \in \{1, 2, \ldots, n\}^k$,

$$\phi^I(a_{j_1}, \ldots, a_{j_k}) = \begin{cases} 0 & \text{if } I \neq J \\ 1 & \text{if } I = J. \end{cases}$$

The collection of all such ϕ^I forms a basis for $\mathcal{L}^k(V)$.

Proof: Uniqueness follows from the basis construction which we have just described. To construct the functions ϕ^I, we start with a basis for V^*, $\phi^i : V \mapsto \mathbb{R}$, defined by $\phi^i(a_j) = \delta_{ij}$. We then define each ϕ^I by

$$\phi^I = \phi^{i_1}(v_1) \cdot \phi^{i_2}(v_2) \cdot \ldots \cdot \phi^{i_k}(v_k) \tag{12.1}$$

and claim that these ϕ^I form a basis for $\mathcal{L}^k(V)$. To show this, we select an arbitrary k-tensor $f \in \mathcal{L}^k(V)$ and define the scalars $\alpha_I := f(a_{i_1}, \ldots, a_{i_k})$. Next, we define a k-tensor

$$g = \sum_J \alpha_J \phi^J, \tag{12.2}$$

where $J \in \{1, \ldots, n\}^k$. Then by the uniqueness of the representation of a vector in terms of a basis, $f \equiv g$. □

Since there are n^k distinct k-tuples from the set $\{1, \ldots, n\}$ the space $\mathcal{L}^k(V)$ has dimension n^k.

Tensor Products

Definition 12.5 Tensor Products. *Let $f \in \mathcal{L}^k(V)$ and $g \in \mathcal{L}^\ell(V)$. The tensor product $f \otimes g$ of f and g is a tensor in $\mathcal{L}^{k+\ell}(V)$ defined by*

$$(f \otimes g)(v_1, \ldots, v_{k+\ell}) := f(v_1, \ldots, v_k) \cdot g(v_{k+1}, \ldots, v_{k+\ell}).$$

The following theorem is elementary:

Theorem 12.6 Properties of Tensor Products. *Let f, g, h be tensors on V and $c \in \mathbb{R}$. The following properties hold:*

1. *Associativity $f \otimes (g \otimes h) = (f \otimes g) \otimes h$.*
2. *Homogeneity $cf \otimes g = c(f \otimes g) = f \otimes cg$.*
3. *Distributivity $(f + g) \otimes h = f \otimes h + g \otimes h$.*
4. *Given a basis $a_1, \ldots, a_n$ for V, the corresponding basis tensors satisfy $\phi^I = \phi^{i_1} \otimes \phi^{i_2} \otimes \cdots \otimes \phi^{i_k}$.*

We can also define the tensor product of two subspaces $U, W \subset V^*$ by

$$U \otimes W := \text{span}\{x \in \mathcal{L}^2(V) \mid x = u \otimes w, \, u \in U, \, w \in W\}.$$

From Theorem 12.4 we can conclude that $V^* \otimes V^* = \mathcal{L}^2(V)$. More generally:

$$\underbrace{V^* \otimes \cdots \otimes V^*}_{k\text{-times}} = \otimes^k V^* = \mathcal{L}^k(V).$$

Alternating Tensors

Before introducing alternating tensors, we present some facts about permutations.

Definition 12.7 Elementary Permutations.   *A permutation of the set of integers* $\{1, 2, \ldots, k\}$ *is an injective function* σ *mapping this set into itself. The set of all permutations* σ *is a group under function composition called the* symmetric group *on* $\{1, \ldots, k\}$ *and is denoted by* S_k. *Given* $1 \leq i < k$, *a permutation* e_i *is called* elementary *if given some* $i \in \{1, 2, \ldots, k\}$ *we have*

$$e_i(j) = j \quad \text{for} \quad j \neq i, i+1,$$
$$e_i(i) = i + 1,$$
$$e_i(i + 1) = i.$$

An elementary permutation leaves the set intact except for one pair of consecutive elements i and $i + 1$ which are switched. The space S_k is of cardinality $k!$; its elements can be written as the composition of elementary permutations.

Definition 12.8 Even, Odd Permutations.   *Let* $\sigma \in S_k$. *Consider the set of all pairs of integers* i, j *from the set* $\{1, \ldots, k\}$ *for which* $i < j$ *and* $\sigma(i) > \sigma(j)$. *Each such pair is called an* inversion in σ. *The* sign *of* σ *is defined to be the number* -1 *if the number of inversions is odd and* $+1$ *if it is even. We call* σ *an* odd *or even* permutation respectively. *The sign of* σ *is denoted by* $\text{sgn}(\sigma)$.

Here is how we calculate the sign of arbitrary permutations: Let $\sigma, \tau \in S_k$. Then

1. If σ is the composition of m elementary permutations then $\text{sgn}(\sigma) = (-1)^m$
2. $\text{sgn}(\sigma \circ \tau) = \text{sgn}(\sigma) \cdot \text{sgn}(\tau)$
3. $\text{sgn}(\sigma^{-1}) = \text{sgn}(\sigma)$
4. If $p \neq q$, and if τ is the permutation that exchanges p and q and leaves all other integers fixed, then $\text{sgn}(\tau) = -1$

We are now ready to define alternating tensors.

Definition 12.9 . Alternating Tensors.   *Let* f *be an arbitrary k-tensor on* V. *If* σ *is a permutation of* $\{1, \ldots, k\}$, *we define* f^σ *by the equation*

$$f^\sigma(v_1, \ldots, v_k) = f(v_{\sigma(1)}, \ldots, v_{\sigma(k)}).$$

Since f *is linear in each of its variables, so is* f^σ. *The tensor* f *is said to be* symmetric *if* $f = f^e$ *for each elementary permutation* e, *and it is said to be* alternating *if* $f = -f^e$ *for every elementary permutation* e.

We will denote the set of all alternating k-tensors on V by $\Lambda^k(V^*)$. The reason for this notation will be apparent when we introduce the wedge product in the next section. One can verify that the sum of two alternating tensors is alternating and a scalar multiple of an alternating tensor is alternating. Therefore, $\Lambda^k(V^*)$ is a linear subspace of the space $\mathcal{L}^k(V)$ of all k-tensors on V. In the special case of $\mathcal{L}^1(V)$, elementary permutations cannot be performed and therefore every 1-

tensor is vacuously alternating, i.e., $\Lambda^1(V^*) = \mathcal{L}^1(V) = V^*$. For completeness, we define $\Lambda^0(V^*) = \mathbb{R}$.

Elementary tensors are not alternating but the linear combination

$$f = \phi^i \otimes \phi^j - \phi^j \otimes \phi^i$$

is alternating. To see this, let $V = \mathbb{R}^n$ and let ϕ^i be the usual dual basis. Then

$$f(x, y) = x_i y_j - x_j y_i = \det \begin{bmatrix} x_i & y_i \\ x_j & y_j \end{bmatrix}$$

and it is easily seen that $f(x, y) = -f(y, x)$. Similarly, the function

$$g(x, y, z) = \det \begin{bmatrix} x_i & y_i & z_i \\ x_j & y_j & z_j \\ x_k & y_k & z_k \end{bmatrix}$$

is an alternating 3-tensor.

In order to obtain a basis for the linear space $\Lambda^k(V^*)$, we start with the following lemma:

Lemma 12.10 Properties of Alternating Tensors. *Let f be a k-tensor on V and $\sigma, \tau \in S_k$ be permutations. Then*

1. *The transformation $f \longrightarrow f^\sigma$ is a linear transformation from $\mathcal{L}^k(V^*)$ to $\mathcal{L}^k(V^*)$. It has the property that for all $\sigma, \tau \in S_k$*

$$(f^\sigma)^\tau = f^{\tau \circ \sigma}$$

,
2. *The tensor f is alternating if and only if $f^\sigma = \mathrm{sgn}(\sigma) \cdot f$ for all $\sigma \in S_k$.*
3. *If f is alternating and if $v_p = v_q$ with $p \neq q$ then $f(v_1, \ldots, v_k) = 0$.*

Proof: The linearity property is obvious since $(af + bg)^\sigma = af^\sigma + bg^\sigma$. Now,

$$\begin{aligned} (f^\sigma)^\tau(v_1, \ldots, v_k) &= f^\sigma(v_{\tau(1)}, \ldots, v_{\tau(k)}) \\ &= f(v_{\tau(\sigma(1))}, \ldots, v_{\tau(\sigma(k))}) \\ &= f^{\tau \circ \sigma}(v_1, \ldots, v_k), \end{aligned}$$

Let σ be an arbitrary permutation, $\sigma = \sigma_1 \circ \sigma_2 \circ \cdots \circ \sigma_m$, where each σ_i is an elementary permutation. Then, we have that

$$\begin{aligned} f^\sigma &= f^{\sigma_1 \circ \sigma_2 \circ \ldots \circ \sigma_m} \\ &= ((\ldots (f^{\sigma_m}) \ldots)^{\sigma_2})^{\sigma_1} \\ &= (-1)^m \cdot f \\ &= \mathrm{sgn}(\sigma) \cdot f. \end{aligned}$$

Now suppose $v_p = v_q$ and $p \neq q$. Let τ be a permutation that exchanges p and q. Since $v_p = v_q$, $f^\tau(v_1, \ldots, v_k) = f(v_1, \ldots, v_k)$. Since f is an alternating tensor and $\text{sgn}(\tau) = -1$, $f^\tau(v_1, \ldots, v_k) = -f(v_1, \ldots, v_k)$. Therefore, $f(v_1, \ldots, v_k) = 0$.   □

Lemma 12.10 implies that if $k > n$, the space $\Lambda^k(V^*)$ is trivial since one of the basis elements must appear in the k-tuple more than once. Hence for $k > n$, $\Lambda^k(V^*) = 0$. We have also seen that for $k = 1$ we have $\Lambda^1(V^*) = \mathcal{L}^1(V) = V^*$, and therefore one can use the dual basis as a basis for $\Lambda^1(V^*)$. In order to specify an alternating tensor for $1 < k \leq n$ we simply need to define it on an ascending k-tuple of basis elements since, from Lemma 12.10, every other combination can be obtained by permuting the k-tuple. Thus, we have that if $a_1, a_2, \ldots, a_n$ is a basis for V and f, g are alternating k-tensors on V with

$$f(a_{i_1}, a_{i_2}, \ldots, a_{i_k}) = g(a_{i_1}, a_{i_2}, \ldots, a_{i_k})$$

for every ascending k-tuple of integers $(i_1, \ldots, i_k) \in \{1, 2, \ldots, n\}^k$, then $f = g$.

Theorem 12.11. *Let $a_1, \ldots, a_n$ be a basis for V. Let $I = (i_1, \ldots, i_k) \in \{1, 2, \ldots, n\}^k$ be an ascending k-tuple. There is a unique alternating k-tensor ψ^I on V such that for every ascending k-tuple $J = (j_1, \ldots, j_k) \in \{1, 2, \ldots, n\}^k$,*

$$\psi^I(a_{j_1}, \ldots, a_{j_k}) = \begin{cases} 0 & \text{if } J \neq I, \\ 1 & \text{if } J = I. \end{cases}$$

The tensors ψ^I form a basis for $\Lambda^k(V^*)$ and satisfy the formula

$$\psi^I = \sum_{\sigma \in S_k} \text{sgn}(\sigma)(\phi^I)^\sigma.$$

The tensors ψ^I are called elementary alternating k-tensors on V corresponding to the basis $a_1, \ldots, a_n$ of V. Every alternating k-tensor f may be uniquely expressed as $f = \sum_J d_J \psi^J$, where J indicates that summation extends over all ascending k-tuples. The dimension of $\Lambda^1(V^*)$ is simply n; its basis is the standard basis for V^*. If $k > 1$, then we need to find the number of possible ascending k-tuples from the set $\{1, 2, \ldots, n\}$. Since if we choose k elements from a set of n elements there is only one way to put them in ascending order, the number of ascending k-tuples, and therefore the dimension of $\Lambda^k(V^*)$, is:

$$\dim(\Lambda^k(V^*)) = \binom{n}{k} = \frac{n!}{k!(n-k)!}.$$

The Wedge Product

Just as we defined the tensor product operation in the set of all tensors on a vector space V, we can define an analogous product operation, the wedge product, in the space of alternating tensors. The tensor product alone will not suffice, since even if $f \in \Lambda^k(V^*)$ and $g \in \Lambda^l(V^*)$ are alternating, their tensor product $f \otimes g \in \mathcal{L}^{k+l}(V)$ need not be alternating. We therefore construct an alternating operator taking k-tensors to alternating k-tensors.

Theorem 12.12. *For any tensor* $f \in \mathcal{L}^k(V)$, *define* $Alt : \mathcal{L}^k(V) \to \Lambda^k(V^*)$ *by:*

$$Alt(f) = \frac{1}{k!} \sum_{\sigma \in S_k} \mathrm{sgn}(\sigma) f^{\sigma} \qquad (12.3)$$

Then $Alt(f) \in \Lambda^k(V^*)$, *and if* $f \in \Lambda^k(V^*)$ *then* $Alt(f) = f$.

Proof: The fact that $Alt(f) \in \Lambda^k(V^*)$ is a consequence of Lemma 12.10, parts (1) and (2). Simply expanding the summation for $f \in \Lambda^k(V^*)$ yields that $Alt(f) = f$. $\qquad\qquad\Box$

Example 12.13. *Let* $f(x, y)$ *be any 2-tensor. By using the alternating operator we obtain,*

$$Alt(f) = \frac{1}{2}(f(x, y) - f(y, x)),$$

which is clearly alternating. Similarly, for any 3-tensor $g(x, y, z)$ *we have:*

$$Alt(g) = \frac{1}{6}(g(x, y, z) + g(y, z, x) + g(z, x, y) - g(y, x, z) - g(z, y, x) - g(x, z, y)),$$

which can be easily checked to be alternating.

Definition 12.14 Wedge Product. *Given* $f \in \Lambda^k(V^*)$ *and* $g \in \Lambda^l(V^*)$, *we define the* wedge product, *or exterior product,* $f \wedge g \in \Lambda^{k+l}(V^*)$ *by*

$$f \wedge g = \frac{(k+l)!}{k!\, l!} Alt(f \otimes g).$$

The somewhat complicated normalization constant is required, since we would like the wedge product to be associative and $Alt(f) = f$ if f is already alternating. Since alternating tensors of order zero are elements of $\mathbb{R}$, we define the wedge product of an alternating 0-tensor and any alternating k-tensor by the usual scalar multiplication. The following theorem lists some important properties of the wedge product.

Theorem 12.15 Properties of Wedge Product. *Let* $f \in \Lambda^k(V^*), g \in \Lambda^l(V^*)$, *and* $h \in \Lambda^m(V^*)$. *The wedge product satsifies the following properties:*

1. *Associativity:* $f \wedge (g \wedge h) = (f \wedge g) \wedge h$.
2. *Homogeneity:* $cf \wedge g = c(f \wedge g) = f \wedge cg$,
3. *Distributivity:* $(f + g) \wedge h = f \wedge h + g \wedge h$, $h \wedge (f + g) = h \wedge f + h \wedge g$,
4. *Skew-commutativity:*[1], $g \wedge f = (-1)^{kl} f \wedge g$.

Proof: Properties (2), (3), and (4) follow directly from the definitions of the alternating operator and the tensor product. Associativity, property (1), requires a few more manipulations (see Spivak [281]). $\qquad\qquad\Box$

---

[1] also called anti-commutativity

An elegant basis for $\Lambda^k(V^*)$ can be formed using that for the dual basis for V. Thus, given a basis $a_1, \ldots, a_n$ for a vector space V, let $\phi^1, \ldots, \phi^n$ denote its dual basis and ψ^I the corresponding elementary alternating tensors. If $I = (i_1, \ldots, i_k)$ is any ascending k-tuple of integers, then

$$\psi^I = \phi^{i_1} \wedge \phi^{i_2} \wedge \cdots \wedge \phi^{i_k}. \tag{12.4}$$

Using the basis in (12.4), any alternating k-tensor $f \in \Lambda^k(V^*)$ may be expressed in terms of the dual basis $\phi^1, \ldots, \phi^n$ as

$$f = \sum_J d_{j_1, \ldots, j_k} \phi^{j_1} \wedge \phi^{j_2} \wedge \cdots \wedge \phi^{j_k}$$

for all ascending k-tuples $J = (j_1, \ldots, j_k)$ and some scalars $d_{j_1, \ldots, j_k}$. If we require the coefficients to be skew-symmetric, i.e.,

$$d_{i_1, \ldots, i_l, i_{l+1}, \ldots, i_k} = -d_{i_1, \ldots, i_{l+1}, i_l, \ldots, i_k}$$

for all $l \in \{1, \ldots, k - 1\}$, we can extend this summation over all k-tuples:

$$f = \frac{1}{k!} \sum_{i_1, \ldots, i_k = 1}^{n} d_{i_1, \ldots, i_k} \phi^{i_1} \wedge \phi^{i_2} \wedge \ldots \wedge \phi^{i_k}. \tag{12.5}$$

The wedge product provides a convenient way to check whether a set of 1-tensors is linearly independent.

Theorem 12.16 Independence of Alternating 1-tensors. *If $\omega^1, \ldots, \omega^k$ are 1-tensors over V, then*

$$\omega^1 \wedge \omega^2 \wedge \cdots \wedge \omega^k = 0$$

if and only if $\omega^1, \ldots, \omega^k$ are linearly dependent.

Proof: Suppose that $\omega^1, \ldots, \omega^k$ are linearly independent, and pick $\alpha^{k+1}, \ldots, \alpha^n$ to complete a basis for V^*. From the construction of the basis of equation (12.4), we know that $\omega^1 \wedge \omega^2 \wedge \cdots \wedge \omega^k$ is a basis element for $\Lambda^k(V^*)$. Therefore, it must be nonzero. If $\omega^1, \ldots, \omega^k$ are linearly dependent, then at least one of the them can be written as a linear combination of the others. Without loss of generality, assume that

$$\omega^k = \sum_{i=1}^{k-1} c_i \omega^i.$$

From this we get that

$$\omega^1 \wedge \omega^2 \wedge \cdots \wedge \omega^k = \omega^1 \wedge \omega^2 \wedge \cdots \wedge \omega^{k-1} \wedge \left(\sum_{i=1}^{k-1} c_i \omega^i \right) = 0$$

by the skew-commutativity of the wedge product. $\square$

Theorem 12.16 allows us to give a geometric interpretation to a nonzero k-tensor

$$\omega^1 \wedge \omega^2 \wedge \cdots \wedge \omega^k \neq 0$$

by associating it with the subspace

$$W := \operatorname{span}\{\omega^1, \dots, \omega^k\} \subset V^*.$$

An obvious question that arises is: What happens if we select a different basis for W?

Theorem 12.17. *Given a subspace $W \subset V^*$ and two sets of linearly independent 1-tensors that span W, there exists a nonzero scalar $c \in \mathbb{R}$ such that*

$$c \cdot \omega^1 \wedge \omega^2 \wedge \cdots \wedge \omega^k = \alpha^1 \wedge \alpha^2 \wedge \cdots \wedge \alpha^k \neq 0.$$

Proof: Each α^i can be written as a linear combination of the ω^i

$$\alpha^i = \sum_{j=1}^{k} a_{ij} \omega^j.$$

Therefore, the product of the α^i can be written as

$$\alpha^1 \wedge \alpha^2 \wedge \cdots \wedge \alpha^k = \left(\sum_{j=1}^{k} a_{1j} \omega^j \right) \wedge \cdots \wedge \left(\sum_{j=1}^{k} a_{kj} \omega^j \right).$$

Multiplying this out gives

$$\alpha^1 \wedge \alpha^2 \wedge \cdots \wedge \alpha^k = \sum_{i_1, \dots, i_k = 1}^{n} b_{i_1, \dots, i_k} \omega^{i_1} \wedge \omega^{i_2} \wedge \cdots \wedge \omega^{i_k}.$$

The claim follows by Theorem 12.16 and the skew commutativity of the wedge product.   □

Definition 12.18 Decomposable Tensors. *A k-tensor $\xi \in \Lambda^k(V^*)$ is decomposable if there exist $x^1, x^2, \dots, x^k \in \Lambda^1(V^*)$ such that $\xi = x^1 \wedge x^2 \wedge \cdots \wedge x^k$.*

Note that if ξ is decomposable, then we must have $\xi \wedge \xi = 0$. The reason is that we should be able to express ξ as $\xi = \alpha^1 \wedge \alpha^2 \wedge \cdots \wedge \alpha^k$ for some basis vectors $\{\alpha^1, \alpha^2, \dots, \alpha^n\}$, and therefore $\xi \wedge \xi = \alpha^1 \wedge \alpha^2 \wedge \cdots \wedge \alpha^k \wedge \alpha^1 \wedge \alpha^2 \wedge \cdots \wedge \alpha^k = 0$. Not all $\xi \in \Lambda^k(V^*)$ are decomposable, as demonstrated in the following example: Let $\xi = \phi^1 \wedge \phi^2 + \phi^3 \wedge \phi^4 \in \Lambda^2((\mathbb{R}^4)^*)$. Then $\xi \wedge \xi = 2\phi^1 \wedge \phi^2 \wedge \phi^3 \wedge \phi^4 \neq 0$. Therefore, ξ is not decomposable. Note that $\xi \wedge \xi = 0$ is a necessary but not a sufficient condition for ξ to be decomposable. For example, if ξ is an odd alternating tensors (say of dimension $2k + 1$), then

$$\xi \wedge \xi = (-1)^{(2k+1)^2} \xi \wedge \xi = 0.$$

If an alternating k-tensor ξ is not decomposable, it may still be possible to factor out a 1-tensor from every term in the summation which defines it. For example, let $\xi = \phi^1 \wedge \phi^2 \wedge \phi^5 + \phi^3 \wedge \phi^4 \wedge \phi^5 \in \Lambda^3((\mathbb{R}^5)^*)$. From the previous example, we know that this tensor is not decomposable, but the 1-tensor ϕ^5 can clearly be factored from every term

$$\xi = (\phi^1 \wedge \phi^2 + \phi^3 \wedge \phi^4) \wedge \phi^5 = \hat{\xi} \wedge \phi^5.$$

Definition 12.19 Divisor Space. *Let $\xi \in \Lambda^k(V^*)$. The subspace $L_\xi \subset V^*$ defined by*

$$L_\xi := \{\omega \in V^* \mid \xi = \hat{\xi} \wedge \omega \text{ for some } \hat{\xi} \in \Lambda^{k-1}(V^*)\}$$

is called the divisor space *of ξ. Any $\omega \in L_\xi$ is called a* divisor *of ξ.*

Theorem 12.20. *A 1-tensor $\omega \in V^*$ is a divisor of $\xi \in \Lambda^k(V^*)$ if and only if $\omega \wedge \xi \equiv 0$.*

Proof: Pick a basis $\phi^1, \phi^2, \ldots, \phi^n$ for V^* such that $\omega = \phi^1$. With respect to this basis, ξ can be written as

$$\xi = \sum_J d_{j_1, \ldots, j_k} \phi^{j_1} \wedge \phi^{j_2} \wedge \cdots \wedge \phi^{j_k}, \tag{12.6}$$

for all ascending k-tuples $J = (j_1, \ldots, j_k)$ and some scalars, $d_{j_1, \ldots, j_k}$. If ω is a divisor of ξ, then it must be contained in each nonzero term of this summation. Therefore $\omega \wedge \xi \equiv 0$. On the other hand, if $\omega \wedge \xi \equiv 0$, then every nonzero term of ξ must contain ω. Otherwise, we would have $\omega \wedge \phi^{j_1} \wedge \cdots \wedge \phi^{j_k} = \phi^1 \wedge \phi^{j_1} \wedge \cdots \wedge \phi^{j_k}$ for $j_1, \ldots, j_k \neq 1$ which is a basis element of $\Lambda^{k+1}(V^*)$ and therefore nonzero. $\square$

If we select a basis $\phi^1, \phi^2, \ldots, \phi^n$ for V^* such that span $\{\phi^1, \phi^2, \ldots, \phi^l\} = L_\xi$, then, ξ can be written as $\xi = \hat{\xi} \wedge \phi^1 \wedge \cdots \wedge \phi^l$, where $\hat{\xi} \in \Lambda^{k-l}(V^*)$ is not decomposable and involves only $\phi^{l+1}, \ldots, \phi^n$.

The Interior Product

Definition 12.21 Interior Product. *The* interior product *is a linear mapping $\lrcorner : V \times \mathcal{L}^k(V) \to \mathcal{L}^{k-1}(V)$ which operates on a a vector $v \in V$ and a tensor $T \in \mathcal{L}^k(V)$ and produces a tensor $(v \lrcorner T) \in \mathcal{L}^{k-1}(V)$ defined by*

$$(v \lrcorner T)(v_1, \ldots, v_{k-1}) := T(v, v_1, \ldots, v_{k-1})$$

It may be verified (see for example Abraham et al. [2, page 429]) that if $a, b, c, d \in \mathbb{R}$, $v, w \in V$, $g, h \in \mathcal{L}^k(V)$ are k-tensors, and $r \in \Lambda^s(V^*)$, $f \in \Lambda^m(V^*)$ are alternating tensors, then we have the following identities:

1. Bilinearity:

$$(av + bw) \lrcorner g = a(v \lrcorner g) + b(w \lrcorner g)$$

$$v \lrcorner (cg + dh) = c(v \lrcorner g) + d(v \lrcorner h)$$

2. $v \lrcorner (f \wedge r) = (v \lrcorner f) + (-1)^m f \wedge (v \lrcorner g)$

Theorem 12.22 Test for Independence of Alternating k-tensors. *Let $a_1, \ldots, a_n$ be a basis for V. Then the value of an alternating k-tensor $\omega \in \Lambda^k(V^*)$ is independent of a basis element a_i if and only if $a_i \lrcorner \omega \equiv 0$.*

Proof: Let $\phi^1, \ldots, \phi^n$ be the dual basis to $a_1, \ldots, a_n$. Then ω can be written with respect to the dual basis as

$$\omega = \sum_J d_J \phi^{j_1} \wedge \phi^{j_2} \wedge \cdots \wedge \phi^{j_k} = \sum_J d_J \psi^J,$$

where the sum is taken oven all ascending k-tuples J. If a basis element ψ^J does not contain ϕ_i, then clearly, $a_i \lrcorner \psi^J \equiv 0$. If a basis element contains ϕ_i, then $a_i \lrcorner \phi^{j_1} \wedge \phi^{j_2} \wedge \cdots \wedge \phi^{j_k} \neq 0$ because a_i can always be matched with ϕ_i through a permutation that affects only the sign. Consequently, $(a_l \lrcorner \omega) \equiv 0$ if and only if the coefficients d_J of all the terms containing ϕ^i are zero.   □

Definition 12.23 Associated Space. *Let $\omega \in \Lambda^k(V^*)$ be an alternating k-tensor. The space consisting of all vectors of which the value of ω is independent is called the* associated space *of ω:*

$$A_\omega := \{v \in V \mid v \lrcorner \omega \equiv 0\}$$

The dual associated space *of ω is defined as $A_\omega^\perp \subset V^*$.*

Recall that the divisor space L_ω of an alternating k-tensor ω contains all the 1-tensors which can be factored from every term of ω. The dual associated space $A_\omega^\perp$ contains all the 1-tensors which are contained in at least one term of ω. Therefore, $L_\omega \subset A_\omega^\perp$.

Theorem 12.24 Characterization of Associated Space. *The following statements are equivalent:*

1. *An alternating k-tensor $\omega \in \Lambda^k(V^*)$ is decomposable.*
2. *The divisor space L_ω has dimension k.*
3. *The dual associated space $A_\omega^\perp$ has dimension k.*
4. *$L_\omega = A_\omega^\perp$.*

Proof: (1) $\Leftrightarrow$ (2) If ω is decomposable, then there exists a set of basis vectors $\phi^1, \phi^2, \ldots, \phi^n$ for V^* such that $\omega = \phi^1 \wedge \cdots \wedge \phi^k$. Therefore $L_\omega = \text{span}\{\phi^1, \phi^2, \ldots, \phi^k\}$ which has dimension k. Conversely, if L_ω has dimension k, then k terms can be factored from ω. Since ω is a k-tensor, it must be decomposable.

(1) $\Leftrightarrow$ (3) Let $a_1, \ldots, a_n$ be the basis of V which is dual to $\phi^1, \phi^2, \ldots, \phi^n$. Since $\omega = \phi^1 \wedge \cdots \wedge \phi^k$, ω is not a function of $a_{k+1}, \ldots, a_n$. Therefore,

$$A_\omega = \text{span}\{a_{k+1}, \ldots, a_n\}.$$

This implies that $A_\omega^\perp$ has dimension k. Conversely, if $A_\omega^\perp$ has dimension k, then A_ω has dimension $n - k$, which means that ω is an alternating k-tensor that is a function of k variables. Therefore, it must have the form $\omega = \phi^1 \wedge \cdots \wedge \phi^k$, for some linearly independent $\phi^1, \phi^2, \ldots, \phi^k$ in V^*.

(2)&(3) $\Leftrightarrow$ (4) It is always true that $L_\omega \subset A_\omega^\perp$. Therefore if $\dim(L_\omega) = \dim(A_\omega^\perp)$ then $L_\omega = A_\omega^\perp$. It is also always true that $0 \leq \dim(L_\omega) \leq k$ and $k \leq \dim(A_\omega^\perp) \leq n$. Therefore, $L_\omega = A_\omega^\perp$ implies that $\dim(L_\omega) = \dim(A_\omega^\perp) = k$.

□

The Pullback of a Linear Transformation

Let T be a linear map from a vector space V to a vector space W. Let f be a multilinear function on W. We may now define a multilinear function on V as follows:

Definition 12.25 Pullback of a Linear Transformation. *Let* $T : V \to W$ *be a linear transformation. The* dual *or* pull back transformation

$$T^* : \mathcal{L}^k(W) \to \mathcal{L}^k(V)$$

is defined for all $f \in \mathcal{L}^k(W)$ *by*

$$(T^* f)(v_1, \ldots, v_k) := f(T(v_1), \ldots, T(v_k)).$$

Note that $T^* f$ is multilinear since T is a linear transformation. It is easy to check that $T * f$ satisfies the following properties:

1. T^* is linear.
2. $T^*(f \otimes g) = T^* f \otimes T^* g$.
3. If $S : W \to X$ is linear, then $(S \circ T)^* f = T^*(S^* f)$.

Theorem 12.26. *Let* $T : V \longrightarrow W$ *be a linear transformation. If* f *is an alternating tensor on* W *then* $T^* f$ *is an alternating tensor on* V, *and*

$$T^*(f \wedge g) = T^* f \wedge T^* g$$

Proof: See Exercises.

Algebras and Ideals

In Sections 12.2.1 and 12.2.1 we introduced the wedge product and interior product and demonstrated some of their properties. We now look more closely at the algebraic structure these operations impart to the space of alternating tensors. We begin by introducing some algebraic structures which will be used in the development of the exterior algebra. The definition of an algebra was also given in Chapter 3.

Definition 12.27 Algebra. *An* algebra, $(V, \odot)$ *is a vector space* V *over a field (here we will use* $\mathbb{R}$ *as the underlying field), together with a multiplication operation* $\odot : V \times V \to V$ *that for every scalar* $\alpha \in \mathbb{R}$ *and* $a, b \in V$ *satisfies* $\alpha(a \odot b) = (\alpha a) \odot b = a \odot (\alpha b)$. *If there exists an element* $e \in V$ *such that* $x \odot e = e \odot x = x$ *for all* $x \in V$, *then* e *is unique and is called the* identity *element.*

Definition 12.28 Algebraic Ideal. *Given an algebra* $(V, \odot)$, *a subspace* $W \subset V$ *is called an* algebraic ideal[2] *if* $x \in W$, $y \in V$ *implies that* $x \odot y, y \odot x \in W$.

---

[2]The algebraic ideal is the ideal of the algebra considered as a ring. Furthermore, since this ring has an identity, any ideal must be a subspace of the algebra considered as a vector space.

Recall that if W is an ideal and $x, y \in W$, then $x + y \in W$, since W is a subspace. It can be easily verified that the intersection of ideals is also an ideal. Using this fact we have the following definition.

Definition 12.29 Minimal Ideal. *Let $(V, \odot)$ be an algebra. Let the set $A :=$ $\{a_i \in V, 1 \le i \le K\}$ be any finite collection of linearly independent elements in V. Let S be the set of all ideals containing A, i.e.,*

$$S := \{I \subset V \mid I \text{ is an ideal and } A \subset I\}.$$

The ideal I_A generated by A is defined as

$$I_A = \bigcap_{I \in S} I$$

and is the minimal ideal in S containing A.

If $(V, \odot)$ has an identity element, then the ideal generated by a finite set of elements can be represented in a simple form.

Theorem 12.30. *Let $(V, \odot)$ be an algebra with an identity element, $A := \{a_i \in V, 1 \le i \le K\}$ a finite collection of elements in V, and I_A the ideal generated by A. Then for each $x \in I_A$, there exist vectors $v_1, \ldots, v_K$ such that*

$$x = v_1 \odot a_1 + v_2 \odot a_2 + \cdots + v_K \odot a_K.$$

Proof: See Hungerford [146, pages 123–124].

Definition 12.31 Equivalence mod I. *Let $(V, \odot)$ be an algebra, and $I \subset V$ an ideal. Two vectors $x, y \in V$ are said to be* equivalent mod I *if and only if $x - y \in I$. This equivalence is denoted*

$$x \equiv y \bmod I.$$

If $(V, \odot)$ has an identity element the above definition implies that $x \equiv y \bmod I$ if and only if:

$$x - y = \sum_{i=1}^{K} \theta_i \odot \alpha_i$$

for some $\theta_K \in V$. It is customary to denote this by $x \equiv y \bmod \alpha_1, \ldots, \alpha_K$, where the mod operation is performed over the ideal generated by $\alpha_1, \ldots, \alpha_K$.

The Exterior Algebra of a Vector Space

Although the space $\Lambda^k(V^*)$ is a vector space with a multiplication operation, the wedge product of two alternating k-tensors is not a k-tensor. Therefore, $\Lambda^k(V^*)$ is not an algebra under the wedge product. If we consider, however, the direct sum of the space of all alternating tensors we obtain,

$$\Lambda(V^*) = \Lambda^0(V^*) \oplus \Lambda^1(V^*) \oplus \Lambda^2(V^*) \oplus \cdots \oplus \Lambda^n(V^*).$$

Any $\xi \in \Lambda(V^*)$ may be written as $\xi = \xi_0 + \xi_1 + \cdots + \xi_n$, where each $\xi_p \in \Lambda^p(V^*)$. $\Lambda(V^*)$ clearly is a vector space, and is also closed under exterior multiplication. It is therefore an algebra.

Definition 12.32 Exterior Algebra. $(\Lambda(V^*), \wedge)$ *is an algebra, called the exterior algebra over V^*.*

Since $(\Lambda(V^*), \wedge)$ has the identity element $1 \in \Lambda^0(V^*)$, Theorem 12.30 implies that the ideal generated by a finite set $\Sigma := \{\alpha^i \in \Lambda(V^*), 1 \leq i \leq K\}$ can be written as

$$I_\Sigma = \left\{ \pi \in \Lambda(V^*) \mid \pi = \sum_{i=1}^{K} \theta^i \wedge \alpha^i, \theta^i \in \Lambda(V^*) \right\}.$$

Given an arbitrary set Σ of linearly independent generators, it may also be possible to generate I_Σ with a smaller set of generators Σ'.

Systems of Exterior Equations

In the preceding sections we have developed an algebra of alternating multilinear functions over a vector space. We will now apply these ideas to solve a system of equations in the form

$$\alpha^1 = 0, \ldots, \alpha^K = 0,$$

where each $\alpha^i \in \Lambda(V^*)$ is an alternating tensor. First we need to clarify what constitutes a "solution" to these equations.

Definition 12.33 Systems of Exterior Equations. *A system of exterior equations on V is a finite set of linearly independent equations*

$$\alpha^1 = 0, \ldots, \alpha^K = 0,$$

where each α^i is in $\Lambda^k(V^*)$ for some $1 \leq k \leq n$. A solution to a system of exterior equations is any subspace $W \subset V$ such that

$$\alpha^1|_W \equiv 0, \ldots, \alpha^K|_W \equiv 0,$$

where $\alpha|_W$ stands for $\alpha(v_1, \ldots, v_k)$ for all $v_1, \ldots, v_k \in W$.

A system of exterior equations generally does not have a unique solution, since any subspace $W_1 \subset W$ will satisfy $\alpha|_{W_1} \equiv 0$ if $\alpha|_W \equiv 0$. A central fact concerning systems of exterior equations is given by the following theorem:

Theorem 12.34 Solution of an Exterior System. *Given a system of exterior equations $\alpha^1 = 0, \ldots, \alpha^K = 0$ and the corresponding ideal I_Σ generated by the collection of alternating tensors $\Sigma := \{\alpha^1, \ldots, \alpha^K\}$, a subspace W solves the system of exterior equations if and only if it also satisfies $\pi|_W \equiv 0$ for every $\pi \in I_\Sigma$.*

Proof: Clearly, $\pi|_W \equiv 0$ for every $\pi \in I_\Sigma$ implies $\alpha^i|_W = 0$ as $\alpha^i \in I_\Sigma$. Conversely, if $\pi \in I_\Sigma$, then $\pi = \sum_{i=1}^{K} \theta^i \wedge \alpha^i$ for some $\theta^i \in \Lambda(V^*)$. Therefore, $\alpha^i|_W = 0$ implies that $\pi|_W \equiv 0$. $\qquad \square$

This result allows us to treat the system of exterior equations, the set of generators for the ideal, and the algebraic ideal as essentially equivalent objects. We may sometimes abuse notation and confuse a system of equations with its corresponding generator set and a generator set with its corresponding ideal. When it is important to distinguish them, we will explicitly write out the system of exterior equations, and denote the set of generators by Σ and the ideal which they generate by I_Σ. Recall that an algebraic ideal was defined in a coordinate-free way as a subspace of the algebra satisfying certain closure properties. Thus the ideal has an intrinsic geometric meaning, and we can think of two sets of generators as representing the same system of exterior equations if they generate the same algebraic ideal.

Definition 12.35 Equivalent Systems.   *Two sets of generators, Σ_1 and Σ_2 which generate the same ideal, i.e. $I_{\Sigma_1} = I_{\Sigma_2}$, are said to be* algebraically equivalent.

We will exploit this notion of equivalence to represent a system of exterior equations in a simplified form. In order to do this, we need a few preliminary definitions.

Definition 12.36 Associated Space of an Ideal.   *Let Σ be a system of exterior equations and I_Σ the ideal which it generates. The* associated space of the ideal I_Σ *is defined by:*

$$A(I_\Sigma) := \{v \in V \mid v \lrcorner \alpha \in I_\Sigma \; \forall \, \alpha \in I_\Sigma\},$$

that is, for all v in the associated space and α in the ideal, $v \lrcorner \alpha \equiv 0$ mod I_Σ. The dual associated space, *or* retracting space *of the ideal is defined by $A(I_\Sigma)^\perp$ and denoted by $C(I_\Sigma) \subset V^*$.*

Once we have determined the retracting space $C(I_\Sigma)$, we can find an algebraically equivalent system Σ' that is a subset of $\Lambda(C(I_\Sigma))$, the exterior algebra over the retracting space.

Theorem 12.37 Characterization of Retracting Space.   *Let Σ be a system of exterior equations and I_Σ its corresponding algebraic ideal. Then there exists an algebraically equivalent system Σ' such that $\Sigma' \subset \Lambda(C(I_\Sigma))$.*

Proof: Let $v_1, \ldots, v_n$ be a basis for V, and $\phi^1, \ldots, \phi^n$ be the dual basis, selected such that $v_{r+1}, \ldots, v_n$ span $A(I_\Sigma)$. Consequently, $\phi^1, \ldots, \phi^r$ must span $C(I_\Sigma)$. The proof is by induction. First, let α be any one-tensor in I_Σ. With respect to the chosen basis, α can be written as

$$\alpha = \sum_{i=1}^{n} a_i \phi^i.$$

Since $v \lrcorner \alpha \equiv 0$ mod I_Σ for all $v \in A(I_\Sigma)$ by the definition of the associated space, we must have $a_i = 0$ for $i = r + 1, \ldots, n$. Therefore,

$$\alpha = \sum_{i=1}^{r} a_i \phi^i.$$

So all the 1-tensors in Σ are contained in $\Lambda^1(C(I_\Sigma))$.

Now suppose that all tensors of degree $\leq k$ in I_Σ are contained in $\Lambda(C(I_\Sigma))$. Let α be any $(k+1)$ tensor in I_Σ. Consider the tensor

$$\alpha' = \alpha - \phi^{r+1} \wedge (v_{r+1} \lrcorner \alpha).$$

The term $v_{r+1} \lrcorner \alpha$ is a k-tensor in I_Σ by the definition of associated space, and thus, by the induction hypothesis, it must be in $\Lambda(C(I_\Sigma))$. The wedge product of this term with ϕ^{r+1} is also clearly in $\Lambda(C(I_\Sigma))$. Furthermore,

$$v_{r+1} \lrcorner \alpha' = v_{r+1} \lrcorner \alpha - (v_{r+1} \lrcorner \phi^{r+1}) \wedge (v_{r+1} \lrcorner \alpha) + \phi^{r+1} \wedge (v_{r+1} \lrcorner (v_{r+1} \lrcorner \alpha)) \equiv 0$$

By Theorem 12.22, α' has no terms involving ϕ^{r+1}.

If we now replace α with α' the ideal generated will be unchanged since

$$\theta \wedge \alpha = \theta \wedge \alpha' + \theta \wedge \phi^{r+1} \wedge (v_{r+1} \lrcorner \alpha).$$

and $v_{r+1} \lrcorner \alpha \in I_\Sigma$.

We can continue this process for $v_{r+2}, \ldots, v_n$ to produce an $\hat\alpha$ that is a generator of I_Σ and is an element of $\Lambda(C(I_\Sigma))$. $\square$

Example 12.38. *Let* $v_1, \ldots, v_6$ *be a basis for* $\mathbb{R}^6$ *and let* $\theta^1, \ldots, \theta^6$ *be the dual basis. Consider the system of exterior equations*

$$\alpha^1 = \theta^1 \wedge \theta^3 = 0,$$

$$\alpha^2 = \theta^1 \wedge \theta^4 = 0,$$

$$\alpha^3 = \theta^1 \wedge \theta^2 - \theta^3 \wedge \theta^4 = 0,$$

$$\alpha^4 = \theta^1 \wedge \theta^2 \wedge \theta^5 - \theta^3 \wedge \theta^4 \wedge \theta^6 = 0.$$

Let I_Σ *be the ideal generated by* $\Sigma = \{\alpha^1, \alpha^2, \alpha^3, \alpha^4\}$ *and* $A(I_\Sigma)$ *the associated space of* I_Σ. *Because* I_Σ *contains no 1-tensors, we can infer that for all* $v \in A(I_\Sigma)$, *we have* $v \lrcorner \alpha^1 = 0$, $v \lrcorner \alpha^2 = 0$, *and* $v \lrcorner \alpha^3 = 0$. *Expanding the first equation, we get*

$$
\begin{aligned}
v \lrcorner \alpha^1 &= v \lrcorner (\theta^1 \wedge \theta^3) \\
&= (v \lrcorner \theta^1) \wedge \theta^3 + (-1)^1 \theta^1 \wedge (v \lrcorner \theta^3) \\
&= \theta^1(v)\theta^3 - \theta^3(v)\theta^1 = 0.
\end{aligned}
$$

which implies that $\theta^1(v) = 0$ *and* $\theta^3(v) = 0$. *Similarly,*

$$v \lrcorner \alpha^2 = \theta^1(v)\theta^4 - \theta^4(v)\theta^1 = 0,$$

$$v \lrcorner \alpha^3 = \theta^1(v)\theta^2 - \theta^2(v)\theta^1 - \theta^3(v)\theta^4 + \theta^4(v)\theta^3 = 0.$$

implying that $\theta^2(v) = 0$ *and* $\theta^4(v) = 0$. *Therefore, we can conclude that*

$$A(I_\Sigma) \subset \text{span}\{v_5, v_6\}.$$

Evaluating the expression $v \lrcorner \alpha^4 \subset I_\Sigma$ gives

$$v \lrcorner \alpha^4 = (v \lrcorner (\theta^1 \wedge \theta^2)) \wedge \theta^5 + (-1)^2 (\theta^1 \wedge \theta^2) \wedge (v \lrcorner \theta^5)$$
$$- (v \lrcorner (\theta^3 \wedge \theta^4)) \wedge \theta^6 - (-1)^2 (\theta^3 \wedge \theta^4) \wedge (v \lrcorner \theta^6)$$
$$= \theta^5(v)\theta^1 \wedge \theta^2 - \theta^6(v)\theta^3 \wedge \theta^4$$
$$= a(\theta^1 \wedge \theta^3) + b(\theta^1 \wedge \theta^4) + c(\theta^1 \wedge \theta^2 - \theta^3 \wedge \theta^4).$$

Equating coefficients, we find that

$$\theta^5(v) = \theta^6(v) = c, \ \forall \, v \in A(I_\Sigma).$$

Now, v must be of the form $v = xv_5 + yv_6$, so we get

$$\theta^5(xv_5 + yv_6) = x = c,$$
$$\theta^6(xv_5 + yv_6) = y = c.$$

s Therefore, $A(I_\Sigma) = \mathrm{span}\{(v_5 + v_6)\}$. If we select as a new basis for $\mathbb{R}^6$ the vectors

$$w_i = v_i, \quad i = 1, \ldots, 4, \quad w_5 = v_5 - v_6, \quad w_6 = v_5 + v_6,$$

then the new dual basis becomes

$$\gamma^i = \theta^i, \quad i = 1, \ldots, 4, \quad \gamma^5 = \frac{\theta^5 - \theta^6}{2}, \quad \gamma^6 = \frac{\theta^5 + \theta^6}{2}.$$

With respect to this new basis, the retracting space $C(I_\Sigma)$ is given by

$$C(I_\Sigma) = \mathrm{span}\{\gamma^1, \ldots, \gamma^5\}$$

In these coordinates, the generator set becomes

$$\Sigma' = \{\gamma^1 \wedge \gamma^3, \ \gamma^1 \wedge \gamma^4, \ \gamma^1 \wedge \gamma^2 - \gamma^3 \wedge \gamma^4, \ \gamma^1 \wedge \gamma^2 \wedge \gamma^5\} \subset \Lambda(C(I_\Sigma)).$$

We conclude this section on exterior algebra with a theorem that allows us to find the dimension of the retracting space in the special case where the generators of the ideal are a collection of 1-tensors together with a single alternating 2-tensor.

Theorem 12.39. *Let I_Σ be an ideal generated by the set*

$$\Sigma = \{\omega^1, \ldots, \omega^s, \Omega\}$$

where $\omega^i \in V^*$ and $\Omega \in \Lambda^2(V^*)$. Let r be the smallest integer such that

$$(\Omega)^{r+1} \wedge \omega^1 \wedge \cdots \wedge \omega^s = 0$$

Then the retracting space $C(I_\Sigma)$ is of dimension $2r + s$.

Proof: See Bryant et al. [49, pages 11–12].

12.2.2   Forms

Since the tangent space to a differentiable manifold at each point is a vector space, we can apply to it the multilinear algebra presented in the previous section. Recall the definition of a differentiable manifold developed in Chapters 3 and 8. Let M be a manifold of dimension n, and (U, x) be a chart containing the point p. Recall that $x : M \mapsto \mathbb{R}^n$ is defined in a neighborhood of $p \in M$. In this chart we can associate the tangent vectors

$$\frac{\partial}{\partial x^1}, \ldots, \frac{\partial}{\partial x^n} \tag{12.7}$$

defined by

$$\frac{\partial}{\partial x^i}(f) = \frac{\partial(f \circ x^{-1})}{\partial r^i} \tag{12.8}$$

for any smooth function $f \in C^\infty(p)$. The space $T_p M$ is an n-dimensional vector space and if (U, x) is a local chart around p, then the tangent vectors

$$\frac{\partial}{\partial x^1}, \ldots, \frac{\partial}{\partial x^n} \tag{12.9}$$

form a basis for $T_p M$. Thus, if X_p is a tangent vector at p, then

$$X_p = \sum_{i=1}^{n} a_i \frac{\partial}{\partial x^i}, \tag{12.10}$$

where $a_1, \ldots, a_n$ are real numbers. Recall also that a *vector field* on M is a continuous function F that associates a tangent vector from $T_p M$ to each point p of M. If F is of class C^∞, it is called a smooth section of TM or a smooth vector field. An *integral curve* of a vector field F is a curve $c : (-\varepsilon, \varepsilon) \mapsto M$ such that

$$\dot{c}(t) = F(c(t)) \in T_{c(t)} M$$

for all $t \in (-\varepsilon, \varepsilon)$. Thus, a local expression for a vector field F in the chart (U, x) is

$$F(p) = \sum_{i=1}^{n} a_i(p) \frac{\partial}{\partial x^i} \tag{12.11}$$

The vector field F is C^∞ if and only if the scalar functions $a_i : M \mapsto \mathbb{R}$ are C^∞.

Tensor Fields

Since the tangent space to a manifold at a point is a vector space, we can apply all the multilinear algebra that we presented in the previous section to it. The dual space of $T_p M$ at each $p \in M$ is called the *cotangent space* to the manifold M at p and is denoted by $T_p^* M$. The collection of all cotangent spaces, called the cotangent bundle, is

$$T^* M := \bigcup_{p \in M} T_p^* M, \tag{12.12}$$

is called the *cotangent bundle*. Similarly, we can form the bundles

$$\mathcal{L}^k(M) := \bigcup_{p \in M} \mathcal{L}^k(T_p M). \tag{12.13}$$

$$\Lambda^k(M) := \bigcup_{p \in M} \Lambda^k(T_p^* M). \tag{12.14}$$

Tensor fields are constructed on a manifold M by assigning to each point p of the manifold a tensor. A *k-tensor field* on M is a section of $\mathcal{L}^k(M)$, i.e., a function ω assigning to every $p \in M$ a k-tensor $\omega(p) \in \mathcal{L}^k(T_p M)$. At some point $p \in M$, $\omega(p)$ is a function mapping k-tuples of tangent vectors of $T_p M$ to $\mathbb{R}$, that is $\omega(p)(X_1, X_2, \ldots, X_k) \in \mathbb{R}$ is a multilinear function of tangent vectors $X_1, \ldots, X_k \in T_p M$. In particular, if ω is a section of $\Lambda^k(M)$ then ω is called a differential form of order k or a k-form on M. In this case, $\omega(p)$ is an alternating k-tensor at each point $p \in M$. The space of all k-forms on a manifold M will be denoted by $\Omega^k(M)$, and the space of all forms on M is simply

$$\Omega(M) := \Omega^0(M) \oplus \cdots \oplus \Omega^n(M). \tag{12.15}$$

At each point $p \in M$, let

$$\frac{\partial}{\partial x^1}, \ldots, \frac{\partial}{\partial x^n} \tag{12.16}$$

be the basis for $T_p M$. Let the 1-forms ϕ^i be the dual basis to these basis tangent vectors, i.e.,

$$\phi^i(p)\left(\frac{\partial}{\partial x^j}\right) = \delta_{ij}. \tag{12.17}$$

Recall that the forms $\phi^I = \phi^{i_1} \otimes \phi^{i_2} \otimes \cdots \otimes \phi^{i_k}$ for multi-index $I = (i_1, \ldots, i_k)$ form a basis for $\mathcal{L}^k(T_p M)$. Similarly, given an ascending multi-index $I = (i_1, \ldots, i_k)$, the k-forms $\psi^I = \phi^{i_1} \wedge \phi^{i_2} \wedge \cdots \wedge \phi^{i_k}$ form a basis for $\Lambda^k(T_p M)$. If ω is a k-tensor on M, it can be uniquely written as

$$\omega(p) = \sum_I b_I(p)\phi^I(p) \tag{12.18}$$

for multi-indices I and scalar functions $b_I(p)$. The k-form α can be written uniquely as

$$\alpha(p) = \sum_I c_I(p)\psi^I(p) \tag{12.19}$$

for ascending multi-indices I and scalar functions c_I. The k-tensor ω and k-form α are of class C^∞ if and only if the functions b_I and c_I are of class C^∞ respectively. Given two forms $\omega \in \Omega^k(M)$, $\theta \in \Omega^l(M)$, we have

$$\omega = \sum_I b_I \psi^I,$$

$$\theta = \sum_J c_J \psi^J,$$

$$\omega \wedge \theta = \sum_I \sum_J b_I c_J \psi^I \wedge \psi^J.$$

Recall that we have defined $\Lambda^0(T_pM) = \mathbb{R}$. As a result, the space of differential forms of order 0 on M is simply the space of all functions $f : M \longrightarrow \mathbb{R}$ and the wedge product of $f \in \Omega^0(M)$ and $\omega \in \Omega^k(M)$ is defined as

$$(w \wedge f)(p) = (f \wedge w)(p) = f(p) \cdot w(p). \tag{12.20}$$

The Exterior Derivative

Recall that a 0-form on a manifold M is a function $f : M \to \mathbb{R}$. The differential df of a 0-form f is defined pointwise as the 1-form

$$df(p)(X_p) = X_p(f). \tag{12.21}$$

It acts on a vector field X_p to give the directional derivative of f in the direction of X_p at p. As X_p is a linear operator, the operator d is also linear, that is, if a, b are real numbers,

$$d(af + bg) = a \cdot df + b \cdot dg. \tag{12.22}$$

The operator d provides a new way of expressing the elementary 1-forms $\phi^i(p)$ on T_pM. Let $x : M \longrightarrow \mathbb{R}^n$ be the coordinate function in a neighborhood of p. Consider the differentials of the coordinate functions

$$dx^i(p)(X_p) = X_p(x^i). \tag{12.23}$$

If we evaluate the differentials dx^i at the basis tangent vectors of T_pM we obtain,

$$dx^i(p)\left(\frac{\partial}{\partial x^j}\right) = \delta_{ij}, \tag{12.24}$$

and therefore the $dx^i(p)$ are the dual basis of T_pM. Since the $\phi^i(p)$ are also the dual basis, $dx^i(p) = \phi^i(p)$. Thus the differentials $dx^i(p)$ span $\mathcal{L}^1(T_pM)$ and from our previous results, any k-tensor ω can be uniquely written as

$$\omega(p) = \sum_I b_I(p)dx^I(p) = \sum_I b_I(p)dx^{i_1} \otimes \cdots \otimes dx^{i_k} \tag{12.25}$$

for multi-indices $I = \{i_1, i_2, \ldots, i_k\}$. Similarly, any k-form can be uniquely written as

$$\omega(p) = \sum_I c_I(p)dx^I(p) = \sum_I c_I(p)dx^{i_1} \wedge \cdots \wedge dx^{i_k} \tag{12.26}$$

for ascending multi-indices $I = \{i_1, i_2, \ldots, i_k\}$. Using this basis we have that for a 0-form,

$$df = \sum_{i=1}^n \frac{\partial f}{\partial x^i}dx^i. \tag{12.27}$$

More generally, we can define an operator $d : \Omega^k(M) \mapsto \Omega^{k+1}(M)$ that takes k-forms to $(k+1)$-forms.

Definition 12.40 Exterior Derivative. *Let ω be a k-form on a manifold M whose representation in a chart (U, x) is given by*

$$\omega = \sum_I \omega_I \, dx^I \tag{12.28}$$

for ascending multi-indices I. The exterior derivative *or* differential operator, d, *is a linear map taking the k-form ω to the $(k+1)$-form $d\omega$ by*

$$d\omega = \sum_I d\omega_I \wedge dx^I. \tag{12.29}$$

Notice that the ω_I are smooth functions (0-forms) whose differential $d\omega_I$ has already been defined as

$$d\omega_I = \sum_{j=1}^n \frac{\partial \omega_I}{\partial x^j} dx^j. \tag{12.30}$$

Therefore, for any k-form ω,

$$d\omega = \sum_I \sum_{j=1}^n \frac{\partial \omega_I}{\partial x^j} dx^j \wedge dx^I. \tag{12.31}$$

From the definition, this operator is certainly linear. We now prove that this differential operator is a true generalization of the operator taking 0-forms to 1-forms, satisfies some important properties, and is the unique operator with those properties.

Theorem 12.41 Properties of Exterior Derivative. *Let M be a manifold and let $p \in M$. Then the exterior derivative is the unique linear operator*

$$d : \Omega^k(M) \to \Omega^{k+1}(M) \tag{12.32}$$

for $k \geq 0$ that satisfies:

1. *If f is a 0-form, then df is the 1-form*

$$df(p)(X_p) = X_p(f). \tag{12.33}$$

2. *If $\omega^1 \in \Omega^k(M), \omega^2 \in \Omega^\ell(M)$ then*

$$d(\omega^1 \wedge \omega^2) = d\omega^1 \wedge \omega^2 + (-1)^k \omega^1 \wedge d\omega^2. \tag{12.34}$$

3. *For every form ω, $d(d\omega) = 0$.*

Proof: Property (1) can be easily checked from the definition of the exterior derivative. For property (2), it suffices to consider the case $\omega^1 = f dx^I$ and $\omega^2 = g dx^J$ in some chart (U, x), because of linearity of the exterior derivative.

$$
\begin{aligned}
d(\omega^1 \wedge \omega^2) &= d(fg) \wedge dx^I \wedge dx^J \\
&= g\,df \wedge dx^I \wedge dx^J + f\,dg \wedge dx^I \wedge dx^J \\
&= d\omega^1 \wedge \omega^2 + (-1)^k f dx^I \wedge dg \wedge dx^J \\
&= d\omega^1 \wedge \omega^2 + (-1)^k \omega^1 \wedge d\omega^2.
\end{aligned}
$$

For property (3), it again suffices to consider the case $\omega = f dx^I$ because of linearity. Since f is a 0-form,

$$d(df) = d\left(\sum_{j=1}^{n} \frac{\partial f}{\partial x^j}\right) dx^j = \sum_{i=1}^{n}\sum_{j=1}^{n} \frac{\partial}{\partial x^i}\frac{\partial f}{\partial x^j} dx^i \wedge dx^j$$
$$= \sum_{i<j}\left(\frac{\partial}{\partial x^i}\frac{\partial f}{\partial x^j} - \frac{\partial}{\partial x^j}\frac{\partial f}{\partial x^i}\right) dx^i \wedge dx^j = 0.$$

We therefore have $d(df) = 0$ by the equality of mixed partial derivatives and the fact that $dx^i \wedge dx^i = 0$. If $\omega = f dx^I$ is a k-form, then $d\omega = df \wedge dx^I + f \wedge d(dx^I)$ by property (2), and since

$$d(dx^I) = d(1 \wedge dx^I) = d(1) \wedge dx^I = 0, \tag{12.35}$$

we get

$$d(d\omega) = d(df) \wedge dx^I - df \wedge d(dx^I) = 0. \tag{12.36}$$

To show that d is the unique such operator, we assume that d' is another linear operator with the same properties and then show that $d = d'$. Consider again a k-form $\omega = f dx^I$. Since d' satisfies property (2), we have

$$d'(f dx^I) = d'f \wedge dx^I + f \wedge d'(dx^I). \tag{12.37}$$

From the above formula we see that if we can show that $d'(dx^I) = 0$, then we will get

$$d'(f dx^I) = d'f \wedge dx^I = df \wedge dx^I = d(f dx^I), \tag{12.38}$$

because $d'f = df$ by property (1), and that will complete the proof. We therefore want to show that

$$d'(dx^{i_1} \wedge \cdots \wedge dx^{i_k}) = 0, \tag{12.39}$$

But since both d and d' satisfy property (1) we have

$$dx^I = dx^{i_1} \wedge \cdots \wedge dx^{i_k} = d'x^{i_1} \wedge \cdots \wedge d'x^{i_k} = d'x^I, \tag{12.40}$$

since the coordinate functions x^i are 0-forms. Then

$$d'(dx^{i_1} \wedge \cdots \wedge dx^{i_k}) = d'(d'x^{i_1} \wedge \cdots \wedge d'x^{i_k}) = 0. \tag{12.41}$$

since d' satisfies property (3). □

Now let $f : M \mapsto N$ be a smooth map between two manifolds. We have seen that the push-forward map, f_*, is a linear transformation from $T_p M$ to $T_{f(p)} N$. Therefore given tensors or forms on $T_{f(p)} N$ we can use the pullback transformation[3], f^*, in order to define tensors or forms on $T_p M$. The next theorem shows that the exterior derivative and the pull back transformation commute.

---

[3] to be consistent with our previous notation, we should write $(f_*)^*$ to denote the pullback of f_*. However we simply denote it by f^*.

Theorem 12.42. *Let* $f : M \mapsto N$ *be a smooth map between manifolds. If* ω *is a* k-*form on* N *then*

$$f^*(d\omega) = d(f^*\omega). \qquad (12.42)$$

Proof: See Exercises.

From a historical standpoint, the language of forms arose from the study of integration on manifolds. In addition to generalizing certain notions of vector calculus in $\mathbb{R}^3$ to higher dimensions and arbitrary manifolds, forms provided an elegant reformulation of many of the original theorems regarding vector and scalar fields in $\mathbb{R}^3$. As a result, considerable physical insight can be gained in this context by studying the relationships between vector fields and scalar fields in $\mathbb{R}^3$ and differential forms. This is explored in the Exercises.

Closed and Exact Forms

The d-operator may be used to define two classes of forms of particular interest.

Definition 12.43 Exact Forms. *A* k-*form* $\omega \in \Omega^k(M)$ *is said to be* closed *if* $d\omega \equiv 0$.

Definition 12.44 Exact Forms. *A* k-*form* $\omega \in \Omega^k(M)$ *with* $k > 0$ *is* exact *if there exists a* $(k-1)$-*form* θ *such that* $\omega = d\theta$. *A* 0-*form is exact on any open set if it is constant on that set.*

Clearly, since for any form θ, $d(d\theta) = 0$, every exact form is closed. However, not all closed forms are exact. We will have a great deal to say about closed and exact forms in what follows.

The Interior Product

We can define the interior product of a tensor field and a vector field pointwise as the interior product of a tensor and a tangent vector.

Definition 12.45 Interior Product. *Given a* k-*form* $\omega \in \Omega^k(M)$ *and a vector field* X *the* interior product *or* anti-derivation *of* ω *with* X *is a* $(k-1)$-*form defined pointwise by*

$$(X(p) \lrcorner \omega(p))(v_1, \ldots, v_{k-1}) = \omega(p)(X(p), v_1, \ldots, v_{k-1}). \qquad (12.43)$$

Recall from Chapter 8 that given a function $h : M \mapsto \mathbb{R}$, the *Lie derivative* of h along the vector field X is denoted as $L_X h$ and is defined by

$$L_X h = X(h) = X \lrcorner dh. \qquad (12.44)$$

Also, recall that given two vector fields X and Y, their *Lie bracket* is defined to be the vector field such that for each $h \in C^\infty(p)$ we have

$$[X, Y](h) = X(Y(h)) - Y(X(h)) = X \lrcorner d(Y \lrcorner dh) - Y \lrcorner d(X \lrcorner dh). \qquad (12.45)$$

Thus, if we choose the coordinate functions x^i, we get

$$[X, Y](x) = \frac{\partial Y}{\partial x} X(x) - \frac{\partial X}{\partial x} Y(x). \qquad (12.46)$$

The following lemma establishes a relation between the exterior derivative and Lie brackets.

Lemma 12.46 Cartan's Magic Formula. *Let* $\omega \in \Omega^1(M)$ *and* X, Y *smooth vector fields. Then*

$$d\omega(X, Y) = X(\omega(Y)) - Y(\omega(x)) - \omega([X, Y]). \qquad (12.47)$$

Proof: Because of linearity, it is adequate to consider $\omega = f\,dg$ where f, g are functions. The left hand side of the above formula is

$$d\omega(X, Y) = df \wedge dg(X, Y)$$
$$= df(X) \cdot dg(Y) - df(Y) \cdot dg(X)$$
$$= X(f) \cdot Y(g) - Y(f) \cdot X(g),$$

while the right hand side is

$$X(\omega(Y)) - Y(\omega(X)) - \omega([X, Y])$$
$$= X(fY(g)) - Y(fX(g)) - f(XY(g) - YX(g))$$
$$= X(f) \cdot Y(g) - Y(f) \cdot X(g)$$

which completes the proof. $\qquad\qquad\qquad\qquad\qquad\qquad\qquad\qquad\qquad$ □

Codistributions

Recall from Chapter 8 that a *smooth distribution* smoothly associates a subspace of the tangent space at each point $p \in M$. It is represented as the span of d smooth vector fields, as follows

$$\Delta = \text{span}\{f_1, \ldots, f_d\}. \qquad (12.48)$$

The *dimension* of the distribution at a point is defined to be the dimension of the subspace $\Delta(p)$. A distribution is *regular* if its dimension does not vary with p. Similarly, recall from Chapter 8 that one may also assign to each point of the manifold a set of 1-forms. The span of these 1-forms at each point will be a subspace of the cotangent space $T_p^* M$. This assignment is called a *codistribution* and is denoted by $\Theta(p) = \text{span}\{\omega_1(p), \ldots, \omega_d(p)\}$ or, dropping the dependence on the point p,

$$\Theta = \text{span}\{\omega_1, \ldots, \omega_d\}, \qquad (12.49)$$

where $\omega_1, \ldots, \omega_d$ are the 1-forms which generate this codistribution. As discussed in Chapter 8, there is a notion of duality between distributions and codistributions which allows us to construct codistributions from distributions and vice versa. Given a distribution Δ, for each p in a neighborhood U, consider all the 1-forms which pointwise annihilate all vectors in $\Delta(p)$,

$$\Delta^{\perp}(p) = \text{span}\{\omega(p) \in T_p^* M : \omega(p)(f) = 0 \;\forall f \in \Delta(p)\}. \qquad (12.50)$$

Clearly, $\Delta^{\perp}(p)$ is a subspace of $T_p^* M$ and is therefore a codistribution. We call $\Delta^{\perp}$ the annihilator or dual of Δ. Conversely, given a codistribution Θ, we construct the *annihilating* or *dual* distribution pointwise as

$$\Theta^{\perp}(p) = \text{span}\{v \in T_p M : \omega(p)(v) = 0 \,\forall \omega(p) \in \Omega(p)\}. \tag{12.51}$$

If N is an integral manifold of a distribution Δ and v is a vector in the distribution Δ at a point p (and consequently in $T_p N$), then for any $\alpha \in \Delta^{\perp}, \alpha(p)(v) = 0$. Notice that this must also be true for any integral curve of the distribution. Therefore given a codistribution $\Theta = \text{span}\{\omega_1, \ldots, \omega_s\}$, an integral curve of the codistribution is a curve $c(t)$ whose tangent $c'(t)$ at each point satisfies, for $i = 1, \ldots, s$,

$$\omega_i(c(t))(c'(t)) = 0. \tag{12.52}$$

Example 12.47 Unicycle.   *Consider the following kinematic model of a unicycle*

$$\dot{x} = u_1 \cos \theta \tag{12.53}$$
$$\dot{y} = u_1 \sin \theta \tag{12.54}$$
$$\dot{\theta} = u_2 \tag{12.55}$$

which can be written as

$$\begin{bmatrix} \dot{x} \\ \dot{y} \\ \dot{\theta} \end{bmatrix} = \begin{bmatrix} \cos \theta \\ \sin \theta \\ 0 \end{bmatrix} u_1 + \begin{bmatrix} 0 \\ 0 \\ 1 \end{bmatrix} u_2 = f_1 u_1 + f_2 u_2. \tag{12.56}$$

The corresponding control distribution is

$$\Delta(x) = \text{span} \left\{ \begin{bmatrix} \cos \theta \\ \sin \theta \\ 0 \end{bmatrix}, \begin{bmatrix} 0 \\ 0 \\ 1 \end{bmatrix} \right\} \tag{12.57}$$

while the dual codistribution is

$$\Delta^{\perp} = \text{span}\{\omega\} \tag{12.58}$$

where $\omega = \sin \theta \, dx - \cos \theta \, dy + 0 \, d\theta$, *the nonholonomic constraint of rolling without slipping.*

12.2.3   Exterior Differential Systems

The Exterior Algebra on a Manifold

The space of all forms on a manifold M,

$$\Omega(M) = \Omega^0(M) \oplus \cdots \oplus \Omega^n(M),$$

together with the wedge product is called the *exterior algebra on M*. An algebraic ideal of this algebra is defined as in Section 12.2.1 as a subspace I such that if $\alpha \in I$ then $\alpha \wedge \beta \in I$ for any $\beta \in \Omega(M)$.

Definition 12.48 Closed Ideals. *An ideal* $I \subset \Omega(M)$ *is said to be* closed *with respect to exterior differentiation if and only if*

$$\alpha \in I \mapsto d\alpha \in I,$$

or more compactly $dI \subset I$. *An algebraic ideal which is closed with respect to exterior differentiation is called a* differential ideal.

A finite collection of forms, $\Sigma := \{\alpha^1, \ldots, \alpha^K\}$ generates an algebraic ideal

$$I_\Sigma := \left\{ \omega \in \Omega(M) \mid \omega = \sum_{i=1}^{K} \theta^i \wedge \alpha^i \text{ for some } \theta^i \in \Omega(M) \right\}.$$

We can also talk about the differential ideal generated by Σ. Thus, if S_d denote the collection of all differential ideals containing Σ. Then the *differential ideal generated by* Σ is defined to be the smallest differential ideal containing Σ

$$\mathcal{I}_\Sigma := \bigcap_{I \in S_d} I.$$

Theorem 12.49. *Let* Σ *be a finite collection of forms, and let* $\mathcal{I}_\Sigma$ *denote the differential ideal generated by* Σ. *Define the collection*

$$\Sigma' = \Sigma \cup d\Sigma$$

and denote the algebraic ideal which it generates by $I_{\Sigma'}$. *Then*

$$\mathcal{I}_\Sigma = I_{\Sigma'}.$$

Proof: By definition, $\mathcal{I}_\Sigma$ is closed with respect to exterior differentiation, so $\Sigma' \subset \mathcal{I}_\Sigma$. Consequently, $I_{\Sigma'} \subset \mathcal{I}_\Sigma$. The ideal $I_{\Sigma'}$ is a closed with respect to exterior differentiation and contains Σ by construction. Therefore, from the definition of $\mathcal{I}_\Sigma$ we have that $\mathcal{I}_\Sigma \subset I_{\Sigma'}$.   □

The *associated space* and *retracting space* of an ideal in $\Omega(M)$ are defined pointwise as in Section 12.2.1. The associated space of $\mathcal{I}_\Sigma$ is called the *Cauchy characteristic distribution* and is denoted by $A(\mathcal{I}_\Sigma)$.

Exterior Differential Systems

In Section 12.2.1 we introduced systems of exterior equations on a vector space V and characterized their solutions as subspaces of V. We are now ready to define a similar notion for a collection of differential forms defined on a manifold M. The basic problem will be to study the integral submanifolds of M which satisfy the constraints represented by the exterior differential system.

Definition 12.50 Solution of an Exterior Differential System. *An exterior differential system is a finite collection of equations*

$$\alpha^1 = 0, \ldots, \alpha^r = 0,$$

where each $\alpha^i \in \Omega^k(M)$ *is a smooth k-form. A solution to an exterior differential system is any submanifold N of M which satisfies* $\alpha^i(x)|_{T_x N} \equiv 0$ *for all* $x \in N$ *and all* $i \in \{1, \ldots, r\}$.

An exterior differential system can be viewed pointwise as a system of exterior equations on $T_p M$. In view of this, one might expect that a solution would be defined as a distribution on the manifold. The trouble with this approach is that most distributions are not integrable, and we want our solution set to be a collection of integral submanifolds. Therefore, we will restrict our solution set to integrable distributions.

Theorem 12.51 Solvability of an Exterior Differential System. *Given an exterior differential system*

$$\alpha^1 = 0, \ldots, \alpha^K = 0 \tag{12.59}$$

and the corresponding differential ideal $\mathcal{I}_\Sigma$ generated by the collection of forms

$$\Sigma := \{\alpha^1, \ldots, \alpha^K\}, \tag{12.60}$$

an integral submanifold N of M solves the system of exterior equations if and only if it also solves the equation $\pi = 0$ for every $\pi \in \mathcal{I}_\Sigma$.

Proof: If an integral submanifold N of M is a solution to Σ, then for all $x \in N$ and all $i \in 1, \ldots, K$,

$$\alpha^i(x)|_{T_x N} \equiv 0.$$

Taking the exterior derivative gives

$$d\alpha^i(x)|_{T_x N} \equiv 0.$$

Therefore, the submanifold also satisfies the exterior differential system

$$\alpha^1 = 0, \ldots, \alpha^K = 0, \ d\alpha^1 = 0, \ldots, d\alpha^K = 0.$$

From Theorem 12.49 we know that the differential ideal generated by Σ is equal to the algebraic ideal generated by the above system. Therefore, Theorem 12.34 tells us that every solution N to Σ is also a solution for every element of $\mathcal{I}_\Sigma$. Conversely, if N solves the equation $\pi = 0$ for every $\pi \in \mathcal{I}_\Sigma$ then in particular it must solve Σ. □

This theorem allows us to work either with the generators of an ideal or with the ideal itself. In fact some authors define exterior differential systems as differential ideals of $\Omega(M)$. Because a set of generators Σ generates both a differential ideal $\mathcal{I}_\Sigma$ and a algebraic ideal I_Σ, we can define two different notions of equivalence for exterior differential systems. Two exterior differential systems Σ_1 and Σ_2 are said to be *algebraically equivalent* if they generate the same algebraic ideal. i.e., $I_{\Sigma_1} = I_{\Sigma_2}$. Two exterior differential systems Σ_1 and Σ_2 are said to be *equivalent* if they generate the same differential ideal. i.e., $\mathcal{I}_{\Sigma_1} = \mathcal{I}_{\Sigma_2}$. Intuitively, we want to think of two exterior differential systems as equivalent if they have the same solution set. Therefore, we will usually discuss equivalence in the latter sense.

Pfaffian Systems

Pfaffian systems are of particular interest because they can be used to represent a set of first-order ordinary differential equations.

Definition 12.52 Pfaffian System. *An exterior differential system of the form*

$$\alpha^1 = \alpha^2 = \cdots = \alpha^s = 0,$$

where the α^i are independent 1-forms on an n-dimensional manifold, is called a Pfaffian system of codimension $n - s$. If $\{\alpha^1, \ldots, \alpha^n\}$ is a basis for $\Omega^1(M)$, then the set $\{\alpha^{s+1}, \ldots, \alpha^n\}$ is called a complement *to the Pfaffian system.*

An *independence condition* is a one-form τ that is required to be nonzero along integral curves of the Pfaffian system. That is, if $\alpha^i(c(t))(c'(t)) = 0$, then $\tau(c(t))(c'(t)) \neq 0$. The 1-forms $\alpha^1, \ldots, \alpha^s$, generate the algebraic ideal

$$I = \{\sigma \in \Omega(M) : \sigma \wedge \alpha^1 \wedge \ldots \wedge \alpha^s = 0\}.$$

The algebraic ideal generated by the 1-forms α^i is also a differential ideal if the following conditions are satisfied.

Definition 12.53 Frobenius Condition. *A set of linearly independent 1-forms $\alpha^1, \ldots, \alpha^s$ in a neighborhood of a point is said to satisfy the* Frobenius condition *if one of the following equivalent conditions holds:*

1. $d\alpha^i$ *is a linear combination of* $\alpha^1, \ldots, \alpha^s$.
2. $d\alpha^i \wedge \alpha^1 \wedge \cdots \wedge \alpha^s = 0$ *for* $1 \le i \le s$.
3. $d\alpha^i = \sum_{j=1}^{s} \theta^j \wedge \alpha^j$.

When $d\alpha^i$ is a linear combination of $\alpha^1, \ldots, \alpha^s$, the following expression is frequently used

$$d\alpha^i \equiv 0 \bmod \alpha^1, \ldots, \alpha^s \; 1 \le i \le s$$

where the mod operation is implicitly performed over the algebraic ideal generated by the α^i.

Example 12.54 Unicycle Continued. *We will illustrate the above concepts for the unicycle. Recall that the unicycle can be described by the codistribution*

$$\omega = \sin\theta \, dx - \cos\theta \, dy + 0 \, d\theta.$$

The exterior derivative of ω is

$$d\omega = \cos\theta \, d\theta \wedge dx + \sin\theta \, d\theta \wedge dy,$$

and therefore

$$d\omega \wedge \omega = -\cos^2 d\theta \wedge dx \wedge dy + \sin^2 d\theta \wedge dy \wedge dx = -dx \wedge dy \wedge d\theta \neq 0.$$

Since the second condition of Definition 12.53 is not satisfied, I is not a differential ideal.

Theorem 12.55 Frobenius Theorem for Codistributions. *Let I be an algebraic ideal generated by the independent 1-forms $\alpha^1, \ldots, \alpha^s$ which satisfies the Frobenius condition. Then in a neighborhood of x there exist functions h^i with $1 \le i \le s$ such that*

$$I = \{\alpha^1, \ldots, \alpha^s\} = \{dh^1, \ldots, dh^s\}.$$

Proof: See Bryant et al. [49, pages 27–29].

For more general exterior differential systems we have the following integrability results.

Theorem 12.56. *If the Cauchy characteristic distribution $A(\mathcal{I}_\Sigma)$ of $\mathcal{I}_\Sigma$ has constant dimension r in a neighborhood of x, then the distribution $A(\mathcal{I}_\Sigma)$ is integrable.*

Proof: See Bryant et al. [49, page 31].

Theorem 12.57. *Let $\mathcal{I}$ be a differential ideal whose retracting space $C(\mathcal{I})$ has a constant dimension $n - r$. There is a neighborhood in which there are coordinates $y^1, \ldots, y^n$ such that $\mathcal{I}$ has a set of generators which are forms in $y^1, \ldots, y^{n-r}$.*

Proof: See Bryant et al. [49, pages 31–33].

Derived Flags

If the algebraic ideal generated by a Pfaffian system does not satisfy the Frobenius condition, then it is not a differential ideal. However, there may exist a differential ideal which is a subset of the algebraic ideal. This subideal can be found by taking the derived flag of the Pfaffian system. Let $I^{(0)} = \{\omega^1, \ldots, \omega^s\}$ be the algebraic ideal generated by independent 1-forms $\omega^1, \ldots, \omega^s$. We define $I^{(1)}$ as

$$I^{(1)} = \{\lambda \in I^{(0)} : d\lambda \equiv 0 \bmod I^{(0)}\} \subset I^{(0)}.$$

The ideal $I^{(1)}$ is called the *first derived system*. The analogue of the first derived system from the distribution point of view is given by the following theorem.

Theorem 12.58. *If $I^{(0)} = \Delta^\perp$, then $I^{(1)} = (\Delta + [\Delta, \Delta])^\perp$.*

Proof: Let $I^{(0)}$ be spanned by 1-forms $\omega^1, \ldots, \omega^s$ and let Δ be its annihilating distribution. By definition we have that

$$I^{(1)} = \{\omega \in I^{(0)} : d\omega \equiv 0 \bmod I^{(0)}\}.$$

Let $\eta \in I^{(1)}$. Therefore, $d\eta \equiv 0 \bmod I^{(0)}$ which means that

$$d\eta = \sum_{j=1}^{s} \theta^j \wedge \omega^j$$

for some forms θ^j. Now let X, Y be vector fields in Δ. Since Δ is the annihilating distribution of $I^{(0)}$, $\omega^j(X) = \omega^j(Y) \doteq 0$. Also, $\eta \in I^{(1)} \subset I^{(0)}$, and therefore $\eta(X) = \eta(Y) = 0$. Now, using the expression for $d\eta$,

$$d\eta(X, Y) = \sum_{j=1}^{s} \theta^j \wedge \omega^j(X, Y)$$

$$= \sum_{j=1}^{s} \theta^j(X)\omega^j(Y) - \theta^j(Y)\omega^j(X)$$

$$= 0,$$

Cartan's magic formula gives

$$d\eta(X, Y) = X\eta(Y) - Y\eta(X) - \eta([X, Y]) = 0,$$

and therefore

$$\eta([X, Y]) = 0,$$

which means that η annihilates any vector fields belonging in $[\Delta, \Delta]$ in addition to any vector fields in Δ. Therefore, $\eta \in (\Delta + [\Delta, \Delta])^\perp$, and thus

$$I^{(1)} \subset (\Delta + [\Delta, \Delta])^\perp.$$

To show the other inclusion, let $\eta \in (\Delta + [\Delta, \Delta])^\perp$ and let X, Y be vector fields in Δ. Cartan's magic formula gives

$$d\eta(X, Y) = X\eta(Y) - Y\eta(X) - \eta([X, Y]) = 0$$

and therefore $d\eta = 0 \mod I^{(0)}$ which means that $\eta \in I^{(1)}$. Thus $(\Delta + [\Delta, \Delta])^\perp \subset I^{(1)}$, and therefore $(\Delta + [\Delta, \Delta])^\perp = I^{(1)}$. □

One may inductively continue this procedure of obtaining derived systems and define

$$I^{(2)} = \{\lambda \in I^{(1)} : d\lambda \equiv 0 \mod I^{(1)}\} \subset I^{(1)}$$

or, in general,

$$I^{(k+1)} = \{\lambda \in I^{(k)} : d\lambda \equiv 0 \mod I^{(k)}\} \subset I^{(k)}.$$

This procedure results in a nested sequence of codistributions

$$I^{(k)} \subset I^{(k-1)} \subset \cdots \subset I^{(1)} \subset I^{(0)}. \tag{12.61}$$

We can also generalize Theorem 12.58. If we define $\Delta_0 = (I^{(0)})^\perp$, $\Delta_1 = (I^{(1)})^\perp$, and in general $\Delta_k = (I^{(k)})^\perp$, then it is not hard to show that if $I^{(k)} = \Delta_k^\perp$ then $I^{(k+1)} = (\Delta_k + [\Delta_k, \Delta_k])^\perp$. The proof of this fact is similar to the proof of Theorem 12.58 but uses a more general form of Cartan's magic formula. The sequence of decreasing codistributions (12.61), called the *derived flag* of $I^{(0)}$, is associated with a sequence of increasing distributions, called the *filtration* of Δ_0,

$$\Delta_k \supset \Delta_{k-1} \supset \cdots \supset \Delta_1 \supset \Delta_0.$$

If the dimension of each codistribution is constant then there will be an integer N such that $I^{(N)} = I^{(N+1)}$. This integer N is called the derived length of I. A basis for a codistribution I is simply a set of generators for I. A basis of 1-forms α^j for I is said to be *adapted to the derived flag* if a basis for each derived system $I^{(j)}$ can be chosen to be some subset of the α^j's. The codistribution $I^{(N)}$ is always integrable by definition since

$$dI^{(N)} \equiv 0 \mod I^{(N)}.$$

The codistribution $I^{(N)}$ is the *largest integrable subsystem* in I. Therefore, if $I^{(N)} \neq \{0\}$ then there exist functions $h^1, \ldots, h^r$ such that $\{dh^1, \ldots, dh^r\} \subset I$. As a result, if a Pfaffian system contains an integrable subsystem $I^{(N)} \neq 0$, which is spanned by the 1-forms $dh^1, \ldots, dh^r$, then the integral curves of the system are constrained to satisfy the following equations for some constants k_i.

$$dh^i = 0 \Longrightarrow h^i = k_i, \quad \text{for} \quad 1 \leq i \leq r,$$

or equivalently, trajectories of the system must lie on the manifold,

$$M = \{x : h^i(x) = k_i \quad \text{for} \quad 1 \leq i \leq r\}.$$

In particular, this implies that if $I^{(N)} \neq 0$, it is not possible to find an integral curve of the Pfaffian system which connects a configuration $x(0) = x_0$ to another configuration $x(1) = x_1$ unless the initial and final configurations satisfy

$$h^i(x_0) = h^i(x_1) \quad \text{for} \quad 1 \leq i \leq r.$$

12.3   Normal Forms

Now that we have defined an exterior differential system and introduced some tools for analyzing them, we are ready to study some important normal forms for exterior differential systems. We will restrict ourselves to Pfaffian systems. The first normal form which we introduce, the Pfaffian form, is restricted to systems of only one equation. The Engel form applies to two equations on a four-dimensional space, and the Goursat form is for $n - 2$ equations on an n-dimensional space. The extended Goursat normal form is defined for systems with codimension greater than two. The Goursat normal forms can be thought of as the generalization of linear systems. Their study will lead us to the study of linearization of control systems in Section 12.4.

12.3.1   The Goursat Normal Form

Systems of One Equation

We will first study Pfaffian systems of codimension $n - 1$, or systems consisting of a single equation

$$\alpha = 0$$

where α is a 1-form on a manifold M. In some chart (U, x) of a point $p \in M$ the equation can be expressed as

$$a_1(x)dx^1 + a_2(x)dx^2 + \cdots + a_n(x)dx^n = 0.$$

In order to understand the integral manifolds of this equation we will attempt to express α in a normal form by performing a coordinate transformation.

Definition 12.59   Rank of a Form. *Let $\alpha \in \Omega^1(M)$. The integer r defined by*

$$(d\alpha)^r \wedge \alpha \neq 0$$
$$(d\alpha)^{r+1} \wedge \alpha = 0$$

is called the rank *of α.*

The following theorem allows us, under a rank condition, to write α in a normal form.

Theorem 12.60 Pfaff Theorem. *Let* $\alpha \in \Omega^1(M)$ *have constant rank* r *in a neighborhood of* p. *Then there exists a coordinate chart* (U, z) *such that in these coordinates,* $\alpha = dz^1 + z^2 dz^3 + \cdots + z^{2r} dz^{2r+1}$.

Proof: Let $\mathcal{I}$ be the differential ideal generated by α. From Theorem 12.39 the retracting space of $\mathcal{I}$ is of dimension $2r + 1$. By Theorem 12.57 there exist local coordinates $y^1, \ldots, y^n$ such that $\mathcal{I}$ has a set of generators in $y^1, \ldots, y^{2r+1}$. Then, by dimension count, any function f_1 of those $2r + 1$ coordinates results in

$$(d\alpha)^r \wedge \alpha \wedge df_1 = 0.$$

Now let $\mathcal{I}_1$ be the ideal generated by $\{df_1, \alpha, d\alpha\}$. If $r = 0$, then the result follows from the Frobenius theorem, Theorem 12.55. If $r > 0$, the forms df_1 and α must be linearly independent, since α is not integrable. Applying Theorem 12.39 to $\mathcal{I}_1$, let r_1 be the smallest integer such that

$$(d\alpha)^{r_1+1} \wedge \alpha \wedge df_1 = 0.$$

Clearly, $r_1 + 1 \leq r$. Furthermore, the equality sign must hold because $(d\alpha)^r \wedge \alpha \neq 0$. Applying Theorem 12.57 to $\mathcal{I}_1$ there exists a function f_2 such that

$$(d\alpha)^{r-1} \wedge \alpha \wedge df_1 \wedge df_2 = 0.$$

Repeating this process, we find r functions $f_1, f_2, \ldots, f_r$ satisfying

$$d\alpha \wedge \alpha \wedge df_1 \wedge df_2 \wedge \cdots \wedge df_r = 0,$$
$$\alpha \wedge df_1 \wedge df_2 \wedge \cdots \wedge df_r \neq 0.$$

Finally, let $\mathcal{I}_r$ be the ideal $\{df_1, \ldots, df_r, \alpha, d\alpha\}$. Its retraction space $C(\mathcal{I}_r)$ is of dimension $r + 1$. There is a function f_{r+1} such that

$$\alpha \wedge df_1 \wedge df_2 \cdots \wedge df_{r+1} = 0,$$
$$df_1 \wedge df_2 \wedge \cdots \wedge df_{r+1} \neq 0.$$

By modifying α by a factor, we can write

$$\alpha = df_{r+1} + g_1 df_1 + \cdots + g_r df_r.$$

Because $(d\alpha)^r \wedge \alpha \neq 0$, the functions $f_1, \ldots, f_{r+1}, g_1, \ldots, g_r$ are independent. The result then follows by setting

$$z^1 = f_{r+1} \qquad z^{2i} = g_i \qquad z^{2i+1} = f_i k$$

for $1 \leq i \leq r$. $\square$

Example 12.61 Unicycle in Pfaff Form. *Consider the unicycle example described by the codistribution* $I = \{\alpha\}$, *where* $\alpha = \sin\theta dx - \cos\theta dy$. *We can immediately see that*

$$d\alpha = \cos\theta \, d\theta \wedge dx + \sin\theta \, d\theta \wedge dy$$

and that

$$d\alpha \wedge \alpha = d\theta \wedge dy \wedge dx \neq 0,$$

$$(d\alpha)^2 \wedge \alpha = 0.$$

Therefore α has rank 1 and by Pfaff's Theorem there exist coordinates z^1, z^2, z^3 such that

$$\alpha = dz^1 + z^2 dz^3.$$

In this example we trivially obtain

$$\alpha = dy + (-\tan\theta)dx.$$

The following theorem is similar to Pfaff's theorem and simply expresses the result in a more symmetric form.

Theorem 12.62 Symmetric Version of Pfaff Theorem. *Given any $\alpha \in \Omega^1(M)$ with constant rank r in a neighborhood U of p, there exist coordinates $z, y^1, \ldots, y^r, x^1, \ldots, x^r$ such that*

$$\alpha = dz + \frac{1}{2}\sum_{i=1}^{r}(y^i dx^i - x^i dy^i).$$

Proof: The coordinate transformation

$$z^1 = z - \frac{1}{2}\sum_{i=1}^{r}x^i y^i,$$

$$z^{2i} = y^i \qquad 1 \leq i \leq r,$$

$$z^{2i+1} = x^i \qquad 1 \leq i \leq r.$$

reduces the above theorem to the Pfaff theorem.   □

The Pfaffian system $\alpha = 0$ on a manifold M is said to have the *local accessibility property* if every point $x \in M$ has a neighborhood U such that every point in U can be joined to x by an integral curve. The following theorem answers the question of when this Pfaffian system has the local accessibility property. The reader should reconcile this definition with the definition of controllability and accessibility in Section 11.1.

Theorem 12.63 Caratheodory Theorem. *The Pfaffian system*

$$\alpha = 0,$$

where α has constant rank, has the local accessibility property if and only if

$$\alpha \wedge d\alpha \neq 0.$$

Proof: The above condition simply says that the rank of α must be greater than or equal to 1. If α has zero rank then $d\alpha \wedge \alpha = 0$, and therefore by the Frobenius theorem (Theorem 12.55), we can write

$$\alpha = dh = 0$$

for some function h. The integral curves are of the form $h = c$ for any arbitrary constant c. Since we can only join points $p, q \in M$ for which $h(p) = h(q)$, we do not have the local accessibility property.

Conversely, let α have rank $r \geq 1$. From Theorem 12.62, we can find coordinates $z, x^1, \ldots, x^r, y^1, \ldots, y^r, u^1, \ldots, u^s$ in some neighborhood U with $2r + s + 1 = \dim M$ such that

$$\alpha = dz + \frac{1}{2} \sum_{i=1}^{r} (y^i dx^i - x^i dy^i) = 0,$$

and therefore

$$dz = \frac{1}{2} \sum_{i=1}^{r} (x^i dy^i - y^i dx^i).$$

Given any two points $p, q \in U$ we must find integral curves $c : [0, 1] \longrightarrow U$ with $c(0) = p$ and $c(1) = q$. Since we are working locally, we can assume that the initial point p is the origin: $z(p) = x^i(p) = y^i(p) = u^i(p) = 0$. Let the final point q be defined by $z(q) = z_1, x^i(q) = x^i_1, y^i(q) = y^i_1, u^i(q) = u^i_1$. Because the expression of the one-form α does not depend on the u^i coordinates, we can choose the curve $t u^i_1$ to connect the u^i coordinates of p and q.

In the (x^i, y^i) plane there are many curves $(x^i(t), y^i(t))$ that join the origin with the desired point (x^i_1, y^i_1). We need to find one which steers the z coordinate to z_1. In order to satisfy the equation $\alpha = 0$, we must have that

$$dz = \frac{1}{2} \sum_{i=1}^{r} (x^i dy^i - y^i dx^i).$$

Integrating this equation gives

$$z(t) = \frac{1}{2} \int_0^t \sum_{i=1}^{r} \left(x^i \frac{dy^i}{dt} - y^i \frac{dx^i}{dt} \right) dt = \frac{1}{2} \sum_{i=1}^{r} A_i,$$

where A_i is the area enclosed by the curve $(x^i(t), y^i(t))$ and the chord joining the origin with (x^i_1, y^i_1). To reach the point q, the curve $(x^i(t), y^i(t))$ must satisfy $z(1) = z_1$. Geometrically, it is clear that a curve $(x^i(t), y^i(t))$ linking the points p and q while enclosing the area prescribed by z_1 will always exist. Thus, the integral curve $c(t)$ given by

$$(z(t), x^1(t), \ldots, x^r(t), y^1(t), \ldots, y^r(t), t u^1(t), \ldots, t u^s(t))$$

has $c(0) = p$ and $c(1) = q$ and satisfies the equation $\alpha = 0$, and the system therefore has the local accessibility property.   □

Codimension Two Systems

We now consider Pfaffian systems of codimension two. We are again interested in performing coordinate changes so that the generators of these Pfaffian systems are in some normal form.

Theorem 12.64 Engel's Theorem.   *Let I be a dimension two codistribution, spanned by*

$$I = \text{span}\{\alpha^1, \alpha^2\}$$

of four variables. If the derived flag satisfies

$$\dim I^{(1)} = 1,$$

$$\dim I^{(2)} = 0,$$

then there exist coordinates z^1, z^2, z^3, z^4 such that

$$I = \{dz^4 - z^3 dz^1, dz^3 - z^2 dz^1\}.$$

Proof: Choose a basis for I that is adapted to the derived flag; that is $I^{(0)} = I = \{\alpha^1, \alpha^2\}$, $I^{(1)} = \{\alpha^1\}$, and $I^{(2)} = \{0\}$. Choose α^3 and α^4 to complete the basis. Since $I^{(2)} = \{0\}$ we have

$$d\alpha^1 \wedge \alpha^1 \neq 0,$$

while

$$(d\alpha^1)^2 \wedge \alpha^1 = 0,$$

since it is a 5-form on a 4-dimensional space. Therefore, α^1 has rank 1. By Pfaff's Theorem, we know that there exists a coordinate change such that

$$\alpha^1 = dz^4 - z^3 dz^1.$$

Taking the exterior derivative, we have that

$$d\alpha^1 = -dz^3 \wedge dz^1 = dz^1 \wedge dz^3.$$

Now, since $\alpha^1 \in I^{(1)}$, the definition of the first derived system will imply that

$$d\alpha^1 \wedge \alpha^1 \wedge \alpha^2 = 0,$$

and thus

$$dz^1 \wedge dz^3 \wedge \alpha^1 \wedge \alpha^2 = 0.$$

Therefore, α^2 must be a linear combination of dz^1, dz^3 and α^1:

$$\alpha^2 \equiv a(x)dz^3 + b(x)dz^1 \bmod \alpha^1.$$

By definition, this means that

$$\alpha^2 + \lambda(x)\alpha^1 = a(x)dz^3 + b(x)dz^1.$$

Now if either $a(x) = 0$ or $b(x) = 0$, then $d\alpha^2 \wedge \alpha^1 \wedge \alpha^2 = 0$ and thus the flag assumptions are violated. Thus $a(x) \neq 0$. and therefore

$$\frac{1}{a(x)}\alpha^2 + \frac{\lambda(x)}{a(x)}\alpha^1 = dz^3 + \frac{b(x)}{a(x)}dz^1,$$

and if we set $z^2 = -\frac{b(x)}{a(x)}$ then

$$\frac{1}{a(x)}\alpha^2 + \frac{\lambda(x)}{a(x)}\alpha^1 = dz^3 - z^2 dz^1,$$

and thus

$$I = \{\alpha^1, \alpha^2\} = \left\{\alpha^1, \frac{1}{a(x)}\alpha^2 + \frac{\lambda(x)}{a(x)}\alpha^1\right\} = \{dz^4 - z^3 dz^1, dz^3 - z^2 dz^1\}. \quad \square$$

It should be noted that the only place the dimension assumption is used in the proof is to guarantee that $(d\alpha^1)^2 \wedge \alpha^1 = 0$. If α^1 has rank 1, this equality holds by definition.

Corollary 12.65. *Let $I = \{\alpha^1, \alpha^2\}$ be a two-dimensional codistribution. If the derived flag satisfies* $\dim I^{(1)} = 1$ *and* $\dim I^{(2)} = 0$ *and* $\alpha^1 \in I^{(1)}$ *has rank 1, then there exist coordinates* z^1, z^2, z^3, z^4 *such that*

$$I = \{dz^4 - z^3 dz^1, dz^3 - z^2 dz^1\}.$$

Proof: The corollary follows by the proof of Engel's Theorem.

Engel's theorem can be generalized to a system with n configuration variables and $n - 2$ constraints.

Theorem 12.66 Goursat Normal Form. *Let I be a Pfaffian system spanned by s 1-forms,*

$$I = \{\alpha^1, \ldots, \alpha^s\},$$

on a space of dimension $n = s + 2$. Suppose that there exists an integrable form π with $\pi \neq 0 \bmod I$ satisfying the Goursat congruences,

$$d\alpha^i \equiv -\alpha^{i+1} \wedge \pi \bmod \alpha^1, \ldots, \alpha^i, \quad 1 \leq i \leq s - 1,$$

$$d\alpha^s \not\equiv 0 \bmod I. \tag{12.62}$$

Then there exists a coordinate system $z^1, z^2, \ldots, z^n$ in which the Pfaffian system is in Goursat normal form:

$$I = \{dz^3 - z^2 dz^1, dz^4 - z^3 dz^1, \ldots, dz^n - z^{n-1} dz^1\}.$$

Proof: The Goursat congruences can be expressed as

$$d\alpha^1 \equiv -\alpha^2 \wedge \pi \bmod \alpha^1,$$

$$d\alpha^2 \equiv -\alpha^3 \wedge \pi \bmod \alpha^1, \alpha^2,$$

$$\vdots$$

$$d\alpha^{s-1} \equiv -\alpha^s \wedge \pi \bmod \alpha^1, \alpha^2, \ldots, \alpha^s,$$

$$d\alpha^s \equiv -\alpha^{s+1} \wedge \pi \bmod \alpha^1, \alpha^2, \ldots, \alpha^s,$$

where $\alpha^{s+1} \notin I$. It can be shown that $\{\alpha^{s+1}, \pi\}$ must form a complement to I. This basis satisfies the Goursat congruences and is adapted to the derived flag of

I:

$$I^{(0)} = \{\alpha^1, \alpha^2, \ldots, \alpha^s\},,$$

$$I^{(1)} = \{\alpha^1, \ldots, \alpha^{s-1}\}$$

$$\vdots$$

$$I^{(s-1)} = \{\alpha^1\},$$

$$I^{(s)} = \{0\}.$$

From the Goursat congruences,

$$d\alpha^1 \equiv -\alpha^2 \wedge \pi \bmod \alpha^1,$$

which means that

$$d\alpha^1 = -\alpha^2 \wedge \pi + \alpha^1 \wedge \eta$$

for some one-form η. But then we have that

$$d\alpha^1 \wedge \alpha^1 = -\alpha^2 \wedge \pi \wedge \alpha^1 \neq 0,$$

$$(d\alpha^1)^2 \wedge \alpha^1 = 0,$$

which means that α^1 has rank 1. We can therefore apply Pfaff's Theorem and express α^1 as

$$\alpha^1 = dz^n - z^{n-1}dz^1$$

for some choice of z^1, z^{n-1}, z^n. Furthermore, by Corollary 12.65 we can express α^2 as

$$\alpha^2 = dz^{n-1} - z^{n-2}dz^1. \tag{12.63}$$

In these new coordinates we have

$$d\alpha^1 \wedge \alpha^1 = -dz^{n-1} \wedge dz^1 \wedge dz^n.$$

Now we have that

$$d\alpha^1 \wedge \alpha^1 \wedge \pi = \pi \wedge (-dz^{n-1} \wedge dz^1 \wedge dz^n) = \pi \wedge (-\alpha^2 \wedge \pi \wedge \alpha^1) = 0,$$

and therefore π is a linear combination of dz^1, dz^{n-1}, dz^n. Noting that $dz^{n-1} \equiv z^{n-2}dz^1 \bmod \alpha^1, \alpha^2$,

$$\pi = adz^1 + bdz^{n-1} + cdz^n,$$

$$= adz^1 + bz^{n-2}dz^1 + cz^{n-1}dz^1 \bmod \alpha^1, \alpha^2$$

where $\psi = a + bz^{n-2} + cz^{n-1}$ is nonzero, since we have assumed that $\pi \neq 0 \bmod I$. From the Goursat congruences we have that

$$d\alpha^2 = -\alpha^3 \wedge \pi \bmod \alpha^1, \alpha^2,$$

while from equation (12.63) we have

$$d\alpha^2 = -dz^{n-2} \wedge dz^1,$$

and thus

$$-dz^{n-2} \wedge dz^1 = -\alpha^3 \wedge \pi \bmod \alpha^1, \alpha^2,$$

which means that

$$\alpha^3 = \lambda(x)dz^{n-2} \bmod dz^1, \alpha^1, \alpha^2,$$

for a nonzero function $\lambda(x)$. Therefore we can rewrite this as

$$\alpha^3 = dz^{n-2} - \frac{1}{\lambda(x)}dz^1 \bmod \alpha^1, \alpha^2,$$

and if we set $z^{n-3} = 1/\lambda(x)$ we have

$$\alpha^3 = dz^{n-2} - z^{n-3}dz^1 \bmod \alpha^1, \alpha^2,$$

and we can therefore let

$$\alpha^3 = dz^{n-2} - z^{n-3}dz^1.$$

If we inductively continue this procedure using the Goursat congruences we obtain

$$\alpha^4 = dz^{n-3} - z^{n-4}dz^1,$$

$$\vdots$$

$$\alpha^s = dz^3 - z^2dz^1.$$

Now, from the Goursat congruences we have that

$$d\alpha^s \neq 0 \bmod I.$$

and therefore

$$\alpha^1 \wedge \alpha^2 \wedge \cdots \wedge \alpha^s \wedge d\alpha^s \neq 0.$$

If we substitute the α^i in the above expression we obtain

$$dz^1 \wedge dz^2 \wedge \cdots \wedge dz^n \neq 0,$$

and therefore the functions $z^1, \ldots, z^n$ can serve as a local coordinate system. □

The following example illustrates the power of the Goursat's theorem by applying it in order to linearize a nonlinear system. A more systematic approach to the feedback linearization problem can be found in the paper by Gardner and Shadwick [108]. Note that the integral curves of a system in Goursat normal form are completely determined by two arbitrary functions in one variable and their derivatives. For example, once $z^1(\tau)$ and $z^s(\tau)$ are known, all of the other coordinates are determined from

$$z^i = \frac{\dot{z}^{i+1}(\tau)}{\dot{z}^1(\tau)},$$

where the dot indicate the standard derivative with respect to the independent variable τ. Because of this property, these two coordinates are sometimes referred to as *linearizing outputs* for the Pfaffian system.

Example 12.67 Feedback Linearization by Goursat Normal Form. *Consider the following nonlinear system with s configuration variables and a single input:*

$$\dot{x}_1 = f_1(x_1, \ldots, x_s, u),$$
$$\dot{x}_2 = f_2(x_1, \ldots, x_s, u),$$
$$\vdots$$
$$\dot{x}_s = f_s(x_1, \ldots, x_s, u).$$

Equivalently, we can look at following Pfaffian system,

$$I = \{dx^i - f_i(x^1, \ldots, x^s, u)dt\}, \quad 1 \le i \le s.$$

The system is of codimension 2 since we have s constraints and s + 2 variables, namely $x^1, \ldots, x^s, u, t$. Assume that the form $\pi = dt$ satisfies the Goursat congruences. Then by Goursat's theorem there exists a coordinate transformation $z = \Phi(x, u, t)$ such that I is generated by

$$I = \{dz^3 - z^2dz^1, dz^4 - z^3dz^1, \ldots, dz^{s+2} - z^{s+1}dz^1\}.$$

The annihilating distribution of the above codistribution is

$$\dot{z}^1 = v_1,$$
$$\dot{z}^2 = v_2,$$
$$\dot{z}^3 = z^2 v_1,$$
$$\vdots$$
$$\dot{z}^{s+2} = z^{s+1} v_1,$$

which, if we set $v_1 = 1$, is clearly a linear system. If it turns out that the z^1 coordinate corresponds to time in the original coordinates, that is, $z^1 = t$, then the connection becomes even more clear. Goursat's theorem can thus be used to linearize single-input nonlinear systems that satisfy the Goursat congruences. These and other issues related to control systems will be explored more fully in Section 12.4.

12.3.2   The n-trailer Pfaffian System

In this section we will show how the system of a mobile robot towing n trailers can be represented as a Pfaffian system. As we saw in the unicycle example, the constraint that a wheel rolls without slipping can be represented as a one-form on the configuration manifold. The velocity of the n-trailer system is constrained in n directions corresponding to the n axles of wheels. A basis for this constraint codistribution (or equivalently, the Pfaffian system) is found by writing down the rolling-without-slipping conditions for all n axles.

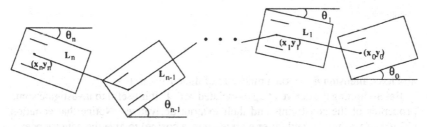

FIGURE 12.1. A mobile robot with n trailers

The System of Rolling Constraints and Its Derived Flag

Consider a single-axle mobile robot[4] with n trailers attached, as sketched in Figure 12.1. Each trailer is attached to the body in front of it by a rigid bar, and the rear set of wheels of each body is constrained to roll without slipping. The trailers are assumed to be identical, with possibly different link lengths L_i. The x, y coordinates of the midpoint between the two wheels on the i-th axle are referred to as (x^i, y^i) and the hitch angles (all measured with respect to the horizontal) are given by θ^i. The connections between the bodies give rise to the following relations:

$$x^{i-1} = x^i + L_i \cos \theta^i,$$
$$y^{i-1} = y^i + L_i \sin \theta^i, \quad i = 1, 2, \ldots, n. \tag{12.64}$$

Thus, it follows that the space parameterized by the coordinates

$$(x^0, y^0, \theta^0, \ldots, x^n, y^n, \theta^n) \in \mathbb{R}^{2n+2} \times (S^1)^{n+1}$$

is not reachable. Taking into account the connection relations (12.64) any one of the Cartesian positions x^i, y^i together with all the hitch angles $\theta^0, \ldots, \theta^n$ will completely represent the configuration of the system. The configuration space is thus $M = \mathbb{R}^2 \times (S^1)^{n+1}$ and has dimension $n + 3$. In any neighborhood, the configuration space can be parameterized by $\mathbb{R}^{n+3}$. The velocity constraints on the system arise from constraining the wheels of the robot and trailers to roll without slipping; the velocity of each body in the direction perpendicular to its wheels must be zero. Each pair of wheels is modeled as a single wheel at the midpoint of the axle. Each velocity constraint can be written as a one-form,

$$\alpha^i = \sin \theta^i dx^i - \cos \theta^i dy^i, \quad i = 0, \ldots, n. \tag{12.65}$$

The one-forms $\alpha^0, \alpha^1, \ldots, \alpha^n$ represent the constraints that the wheels of the zeroth trailer (i.e. the cab), the first trailer, $\ldots$, the n-th trailer, respectively, roll without slipping. The Pfaffian system corresponding to this mobile robot system is generated by the codistribution spanned by all of the rolling without slipping

---

[4]For example, see the Hilare family of mobile robots at the laboratory LAAS in Toulouse [60; 113].

constraints,

$$I = \{\alpha^0, \ldots, \alpha^n\}, \tag{12.66}$$

and has dimension $n + 1$ on a manifold of dimension $n + 3$.

Before finding the derived flag associated with I, it is useful to investigate some properties of the constraints and their exterior derivatives. Notice that equation (12.65) can be rearranged (after a division by a cosine) to give the congruence:

$$dy^i \equiv \tan \theta^i dx^i \bmod \alpha^i. \tag{12.67}$$

This division by a cosine introduces a singularity; the resulting coordinate transformation will not be valid at points where $\theta^i = \pm\pi/2$. See also Subsection 12.3.2 at the end of this section for a brief discussion of singularities.

All of the (x^i, y^i) are related by the hitch relationships. The exterior derivatives of these relationships can be taken,

$$\begin{aligned} x^{i-1} &= x^i + L_i \cos \theta^i, & dx^{i-1} &= dx^i - L_i \sin \theta^i d\theta^i, \\ y^{i-1} &= y^i + L_i \sin \theta^i, & \implies & dy^{i-1} &= dy^i + L_i \cos \theta^i d\theta^i, \end{aligned}$$

and these expressions can then be substituted into the formula for α^{i-1} from 12.65, allowing the constraint for the $(i-1)$-st axle to be rewritten as

$$\begin{aligned} \alpha^{i-1} &= \sin \theta^{i-1} dx^{i-1} - \cos \theta^{i-1} dy^{i-1} \\ &= \sin \theta^{i-1} dx^i - \cos \theta^{i-1} dy^i - L_i \cos(\theta^{i-1} - \theta^i) d\theta^i \\ &\equiv (\sin \theta^{i-1} - \tan \theta^i \cos \theta^{i-1}) dx^i - L_i \cos(\theta^{i-1} - \theta^i) d\theta^i \bmod \alpha^i \\ &\equiv \sec \theta^i \sin(\theta^{i-1} - \theta^i) dx^i - L_i \cos(\theta^{i-1} - \theta^i) d\theta^i \bmod \alpha^i \tag{12.68} \end{aligned}$$

after an application of the congruence (12.67). A rearrangement of terms and a division by the cosine in equation (12.68) will give the congruence

$$\begin{aligned} d\theta^i &\equiv \frac{1}{L_i} \sec \theta^i \tan(\theta^{i-1} - \theta^i) dx^i \bmod \alpha^i, \alpha^{i-1} \\ &\equiv f_{\theta^i} dx^i \bmod \alpha^i, \alpha^{i-1}. \end{aligned} \tag{12.69}$$

The exact form of the function f_{θ^i} is unimportant; what will be needed is the relationship between $d\theta^i$ and dx^i. The first lemma relates the exterior derivatives of the x coordinates,

Lemma 12.68. *The exterior derivatives of any of the x variables are congruent modulo the Pfaffian system, that is: $dx^i \equiv f_{x^{i,j}} dx^j \bmod I$.*

Proof: For two adjacent axles, the relationship between the x coordinates is given by the hitching,

$$x^{i-1} = x^i + L_i \cos \theta^i,$$

$$dx^{i-1} = dx^i - L_i \sin \theta^i d\theta^i,$$

$$\equiv (1 - L_i \sin \theta^i f_{\theta^i}) dx^i \bmod \alpha^{i-1}, \alpha^i,$$

$$\equiv f_{x^{i-1,i}} dx^i \bmod \alpha^{i-1}, \alpha^i.$$

The congruence (12.69) was applied in the calculation above.   □

A complement to the Pfaffian system $I = \{\alpha^0, \dots, \alpha^n\}$ is given by

$$\{d\theta^0, dx^i\} \tag{12.70}$$

for any x^i, since by Lemma 12.68 their exterior derivatives are congruent modulo the system, and the complement is defined only modulo the system. These two one-forms, together with the codistribution I, form a basis for the space of all one-forms on the configuration manifold, or $\Omega^1(M)$.

Now consider the exterior derivative of the constraint corresponding to the i-th axle,

$$\alpha^i = \sin \theta^i dx^i - \cos \theta^i dy^i,$$

$$d\alpha^i = d\theta^i \wedge (\cos \theta^i dx^i + \sin \theta^i dy^i)$$

$$\equiv d\theta^i \wedge dx^i (\cos \theta^i + \sin \theta^i \tan \theta^i) \bmod \alpha^i \tag{12.71}$$

$$\equiv d\theta^i \wedge dx^i (\sec \theta^i) \bmod \alpha^i$$

$$\equiv 0 \bmod \alpha^i, \alpha^{i-1},$$

using (12.69). Thus, the exterior derivative of the constraint corresponding to the i-th axle is congruent to zero modulo itself and the constraint corresponding to the axle directly in front of it. The congruences (12.67) and (12.69) were useful in deriving this result.

Theorem 12.69 Derived Flag for the n-trailer Pfaffian System. *Consider the Pfaffian system of the n-trailer system given by (12.66) with the one forms α^i defined by equations (12.65). The one-forms α^i are adapted to the derived flag in the following sense:*

$$I^{(0)} = \{\alpha^0, \alpha^1, \dots, \alpha^n\},$$

$$I^{(1)} = \{\alpha^1, \dots, \alpha^n\},$$

$$\vdots \tag{12.72}$$

$$I^{(n)} = \{\alpha^n\},$$

$$I^{(n+1)} = \{0\}.$$

Proof: The proof is merely a repeated application of equation (12.71). Noting that the exterior derivative of the i-th constraint is equal to zero modulo itself and the constraint corresponding to the axle directly in front of it, it is simple to check that the derived flag has the form given in equation (12.72).   □

Note that $I^{(n+1)} = \{0\}$ implies that there is no integrable subsystem contained in the constraints which define n-trailer Pfaffian system.

Conversion to Goursat Normal Form

In the preceding subsection it was shown that the basis $\{\alpha^0, \ldots, \alpha^n\}$ defined in equation (12.65) is adapted to its derived flag in the sense of (12.72). It remains to be checked whether the α^i satisfy the Goursat congruences and if they do, to find a transformation that puts them into the Goursat canonical form. The following theorem guarantees the existence of such a transformation.

Theorem 12.70 Goursat Congruences for the n-Trailer System. *Consider the Pfaffian system $I = \{\alpha^0, \ldots, \alpha^n\}$ associated with the n-trailer system (12.66) with the one-forms α^i defined by equation (12.65). There exists a change of basis of the one-forms α^i to $\tilde{\alpha}^i$ which preserves the adapted structure, and a one-form π which satisfies the Goursat congruences for this new basis*

$$d\tilde{\alpha}^i \equiv -\tilde{\alpha}^{i-1} \wedge \pi \bmod \tilde{\alpha}^i, \ldots, \tilde{\alpha}^n, \qquad i = 1, \ldots, n,$$

$$d\tilde{\alpha}^0 \not\equiv 0 \bmod I.$$

A one-form which satisfies these congruences is given by

$$\pi = dx^n. \tag{12.73}$$

Proof: First of all, consider the original basis of constraints. The expression for α^i can be written in the configuration space coordinates from equation (12.65) together with the connection relations (12.64) and some bookkeeping as:

$$\alpha^i = \sin\theta^i dx^n - \cos\theta^i dy^n - \sum_{k=i+1}^{n} L_k \cos(\theta^i - \theta^k) d\theta^k. \tag{12.74}$$

Before beginning the main part of the proof, it will be helpful to define a new basis of constraints $\tilde{\alpha}^i$, which is also adapted to the derived flag but is somewhat simpler to work with. Each $\tilde{\alpha}^i$ will have only two terms. Although the last constraint already has only two terms, it will be scaled by a factor,

$$\tilde{\alpha}^n = \sec\theta^n \alpha^n = \tan\theta^n dx^n - dy^n. \tag{12.75}$$

Note that a rearrangement of terms will give the congruence

$$dy^n \equiv \tan\theta^n dx^n \bmod \tilde{\alpha}^n. \tag{12.76}$$

Now consider the next-to-last constraint, α^{n-1}, and apply the preceding congruence:

$$\alpha^{n-1} = \sin\theta^{n-1} dx^n - \cos\theta^{n-1} dy^n - L_n \cos(\theta^n - \theta^{n-1}) d\theta^n,$$

$$\equiv \sec\theta^n \sin(\theta^{n-1} - \theta^n) dx^n - L_n \cos(\theta^{n-1} - \theta^n) d\theta^n \bmod \tilde{\alpha}^n.$$

$$\tag{12.77}$$

Dividing once again by a cosine, the new basis element $\tilde{\alpha}^{n-1}$ is defined as

$$\tilde{\alpha}^{n-1} = \sec\theta^n \tan(\theta^{n-1} - \theta^n)dx^n - L_n d\theta^n. \qquad (12.78)$$

Thus, $\tilde{\alpha}^{n-1} \equiv f_{\alpha^{n-1}}\alpha^{n-1}$ mod α^n. Also, the exterior derivative $d\theta^n$ is related to dx^n by the congruence

$$d\theta^n \equiv \frac{1}{L_n} \sec\theta^n \tan(\theta^{n-1} - \theta^n)dx^n \text{ mod } \tilde{\alpha}^{n-1}. \qquad (12.79)$$

This procedure of eliminating the terms $dy^n, d\theta^n, \ldots, d\theta^i$ from α^{i+1} can be continued.

Lemma 12.71. *A new basis of constraints $\tilde{\alpha}^i$ of the form*

$$\tilde{\alpha}^n = \tan\theta^n dx^n - dy^n,$$

$$\tilde{\alpha}^i = \sec\theta^n \sec(\theta^{n-1} - \theta^n)\ldots\sec(\theta^{i+1} - \theta^{i+2})$$

$$\times \tan(\theta^i - \theta^{i+1})dx^n - L_{i+1}d\theta^{i+1},$$

$$i = 0, \ldots, n-1, \qquad (12.80)$$

is related to the original basis of constraints α^i through the following congruences:

$$\tilde{\alpha}^i \equiv f_{\alpha^i}\alpha^i \text{ mod } \alpha^{i+1}, \ldots, \alpha^n, \qquad (12.81)$$

and thus the basis $\tilde{\alpha}^i$ is also adapted to the derived flag.

Proof of Lemma: Note that by the definition of $\tilde{\alpha}^i$, the exterior derivative $d\theta^{i+1}$ is related to dx^n by the congruence

$$d\theta^{i+1} \equiv \frac{1}{L_{i+1}} \sec\theta^n \sec(\theta^{n-1} - \theta^n)\cdots\sec(\theta^{i+1} - \theta^{i+2})$$

$$\times \tan(\theta^i - \theta^{i+1})dx^n \text{ mod } \tilde{\alpha}^i. \qquad (12.82)$$

The lemma is proved by induction. It has already been shown that $\tilde{\alpha}^n = f_{\alpha^n}\alpha^n$ and $\tilde{\alpha}^{n-1} \equiv f_{\alpha^{n-1}}\alpha^{n-1}$ mod $\tilde{\alpha}^n$. Assume that $\tilde{\alpha}^i \equiv f_{\alpha^i}\alpha^i$ mod $\alpha^{i+1}, \ldots, \alpha^n$ for $i = j+1, \ldots, n$. Consider $\tilde{\alpha}^j$ as defined by equation 12.80,

$$\tilde{\alpha}^j = \sec\theta^n \sec(\theta^{n-1} - \theta^n)\ldots\sec(\theta^{j+1} - \theta^{j+2})$$

$$\times \tan(\theta^j - \theta^{j+1})dx^n - L_{j+1}d\theta^{j+1}. \qquad (12.83)$$

Recall from equation (12.74) that α^j has the form

$$\alpha^j = \sin\theta^j dx^n - \cos\theta^j dy^n - \sum_{k=j+1}^n L_k \cos(\theta^j - \theta^k)d\theta^k. \qquad (12.84)$$

Now, applying the congruences

$$dy^n \equiv \tan\theta^n dx^n \text{ mod } \tilde{\alpha}^n,$$

$$d\theta^i \equiv \frac{1}{L_i} \sec\theta^n \sec(\theta^{n-1} - \theta^n)\ldots\sec(\theta^i - \theta^{i+1})\tan(\theta^{i-1} - \theta^i)dx^n \text{ mod } \tilde{\alpha}^{i-1},$$

to the expression for α^j, and expanding the summation, yields

$$
\begin{aligned}
\alpha^j \equiv{} & \sin\theta^j dx^n - \cos\theta^j \tan\theta^n dx^n - L_{j+1}\cos(\theta^j - \theta^{j+1})d\theta^{j+1} \\
& - \cos(\theta^j - \theta^{j+2})\sec\theta^n \sec(\theta^{n-1} - \theta^n)\ldots \\
& \times \sec(\theta^{j+2} - \theta^{j+3})\tan(\theta^{j+1} - \theta^{j+2})dx^n \\
& - \cdots \\
& - \cos(\theta^j - \theta^{n-1})\sec\theta^n \sec(\theta^{n-1} - \theta^n)\tan(\theta^{n-2} - \theta^{n-1})dx^n \\
& - \cos(\theta^j - \theta^n)\sec\theta^n \tan(\theta^{n-1} - \theta^n)dx^n \\
& \qquad \operatorname{mod} \tilde{\alpha}^{j+1}, \ldots, \tilde{\alpha}^{n-2}, \tilde{\alpha}^{n-1}, \tilde{\alpha}^n.
\end{aligned}
\tag{12.85}
$$

After some trigonometric simplification, it can be seen that this equation reads as

$$
\begin{aligned}
\alpha^j \equiv{} & \sin(\theta^j - \theta^{j+1})\sec\theta^n \sec(\theta^{n-1} - \theta^n)\ldots\sec(\theta^{j+1} - \theta^{j+2})dx^n \\
& - L_{j+1}\cos(\theta^j - \theta^{j+1})d\theta^{j+1} \operatorname{mod} \tilde{\alpha}^{j+1}, \ldots, \tilde{\alpha}^{n-2}, \tilde{\alpha}^{n-1}, \tilde{\alpha}^n \\
\equiv{} & \cos(\theta^j - \theta^{j+1})\tilde{\alpha}^j \operatorname{mod} \tilde{\alpha}^{j+1}, \ldots, \tilde{\alpha}^{n-2}, \tilde{\alpha}^{n-1}, \tilde{\alpha}^n,
\end{aligned}
\tag{12.86}
$$

and the lemma is proved.   □

The basis $\tilde{\alpha}^i$ will now be scaled to find the basis $\bar{\alpha}^i$ which will satisfy the congruences (12.62). Once again, the procedure will start with the last congruence, $\bar{\alpha}^n$. The exterior derivative of $\tilde{\alpha}^n$ is given by

$$
d\tilde{\alpha}^n = \sec^2\theta^n d\theta^n \wedge dx^n.
\tag{12.87}
$$

Looking at the expression for $\tilde{\alpha}^{n-1}$ given in equation (12.78), it can be seen that π should be chosen to be some multiple of dx^n or $d\theta^n$. In fact, either $\pi = dx^n$ or $\pi = d\theta^n$ will work, although the computations are different for each case. The calculations here are for choosing $\pi = dx^n$. Choosing the new basis element $\bar{\alpha}^{n-1}$ as

$$
\bar{\alpha}^{n-1} = \frac{1}{L_n}\sec^2\theta^n \tilde{\alpha}^{n-1},
\tag{12.88}
$$

will result in the desired congruence,

$$
d\bar{\alpha}^n \equiv -\bar{\alpha}^{n-1} \wedge \pi \operatorname{mod} \alpha^n.
\tag{12.89}
$$

Now consider the exterior derivative of $\bar{\alpha}^{n-1}$,

$$
\begin{aligned}
d\bar{\alpha}^{n-1} &= d\left(\frac{1}{L_n}\sec^3\theta^n \tan(\theta^{n-1} - \theta^n)dx^n - L_n \sec^2\theta^n d\theta^n\right) \\
&\equiv \frac{1}{L_n}\sec^3\theta^n \sec^2(\theta^{n-1} - \theta^n)d\theta^{n-1} \wedge dx^n \operatorname{mod} \tilde{\alpha}^{n-1},
\end{aligned}
\tag{12.90}
$$

since any terms $d\theta^n \wedge dx^n$ are congruent to $0 \operatorname{mod} \tilde{\alpha}^{n-1}$. Thus, in order to achieve the next Goursat congruence $d\bar{\alpha}^{n-1} \equiv \bar{\alpha}^{n-2} \wedge \pi$, the new basis element $\bar{\alpha}^{n-2}$

should be chosen as

$$\tilde{\alpha}^{n-2} = \frac{1}{L_n L_{n-1}} \sec^3 \theta^n \sec^2(\theta^{n-1} - \theta^n) \tilde{\alpha}^{n-2}. \tag{12.91}$$

In general, the new basis is defined by

$$\tilde{\alpha}^i = \frac{1}{L_n L_{n-1} \cdots L_{i+1}} \sec^{n-i+1} \theta^n \sec^{n-i}(\theta^{n-1} - \theta^n) \cdots$$
$$\times \sec^3(\theta^{i+2} - \theta^{i+3}) \sec^2(\theta^{i+1} - \theta^{i+1}) \tilde{\alpha}^i.$$

It has already been shown that the congruences hold for $i = n$ and $i = n - 1$. Assume that the congruences

$$d\tilde{\alpha}^i = -\tilde{\alpha}^{i-1} \wedge \pi \bmod \alpha^i, \dots, \alpha^n. \tag{12.92}$$

hold for $i = j + 1, \dots, n$. Consider the exterior derivative of $\tilde{\alpha}^j$,

$$d\tilde{\alpha}^j = d\left[\left(\frac{1}{L_n L_{n-1} \dots L_{j+1}} \sec^{n-j+1} \theta^n \sec^{n-j}(\theta^{n-1} - \theta^n) \cdots \right. \right.$$
$$\left. \left. \times \sec^3(\theta^{j+2} - \theta^{j+3}) \sec^2(\theta^{j+1} - \theta^{j+2}) \right) \tilde{\alpha}^i \right]. \tag{12.93}$$

Before calculating all of the terms, recall that the following congruences hold:

$$d\theta^i \wedge dx^n \equiv 0 \bmod \tilde{\alpha}^{i-1},$$
$$d\theta^i \wedge d\theta^k \equiv 0 \bmod \tilde{\alpha}^{i-1}, \tilde{\alpha}^{k-1}, \tag{12.94}$$

and thus the only term in $d\tilde{\alpha}^j \bmod \tilde{\alpha}^j, \dots, \tilde{\alpha}^n$ will be a multiple of $d\theta^j \wedge dx^n$,

$$d\tilde{\alpha}^j \equiv \frac{1}{L_n L_{n-1} \cdots L_{j+1}} \sec^{n-j+2} \theta^n \sec^{n-j+1}(\theta^{n-1} - \theta^n) \cdots$$
$$\times \sec^4(\theta^{j+2} - \theta^{j+3}) \sec^3(\theta^{j+1} - \theta^{j+2}) \sec^2(\theta^j - \theta^{j+1}) \tag{12.95}$$
$$\bmod \tilde{\alpha}^j, \dots, \tilde{\alpha}^n$$
$$\equiv \tilde{\alpha}^{j-1} \wedge \pi \bmod \tilde{\alpha}^j, \dots, \tilde{\alpha}^n.$$

This completes the proof that the Goursat congruences are satisfied.   □

Since the one-forms $\tilde{\alpha}^i$ do satisfy the Goursat congruences, a coordinate transformation into Goursat normal form can be found. As seen in the proof of Goursat's theorem, the one-form α^n in the last nonzero derived system has rank 1. We can therefore use Pfaff's theorem to find functions f_1 and f_2 which satisfy the following equations

$$d\alpha^n \wedge \alpha^n \wedge df_1 = 0, \qquad \alpha^n \wedge df_1 \neq 0,$$
$$\text{and}$$
$$\alpha^n \wedge df_1 \wedge df_2 = 0, \qquad df_1 \wedge df_2 \neq 0. \tag{12.96}$$

The constraint corresponding to the last axle is once again given by[5]

$$\alpha^n = \sin\theta^n dx^n - \cos\theta^n dy^n, \tag{12.98}$$

and its exterior derivative has the form

$$d\alpha^n = -\cos\theta^n dx^n \wedge d\theta^n - \sin\theta^n dy^n \wedge d\theta^n, \tag{12.99}$$

It follows that the exterior product of these two quantities is given by

$$d\alpha^n \wedge \alpha^n = -dx^n \wedge dy^n \wedge d\theta^n. \tag{12.100}$$

By the first equation of (12.96), f_1 may be chosen to be *any* function of x^n, y^n, θ^n *exclusively*.

One solution of the equations (12.96) is explained here, and another is relegated to the exercises (Problem 12.7).

Coordinates to Convert the n-trailer System into Goursat Normal Form

Motivated by Sørdalen [280], f_1 can be chosen to be x^n. The second equation of (12.96) then becomes

$$\sin\theta^n dx^n \wedge dy^n \wedge df_2 = 0, \tag{12.101}$$

with the proviso that $df_1 \wedge df_2 \neq 0$. A *non-unique* choice of f_2 is

$$f_2 = y^n. \tag{12.102}$$

The change of coordinates is defined by:

$$z_1 = f_1(x) = x^n,$$
$$z_{n+3} = f_2(x) = y^n.$$

The one form α^n may be written by dividing through by $\sin\theta^n$ as

$$\alpha^n = dy^n - \tan\theta^n dx^n = dz_{n+3} - z_{n+2}dz_1,$$

giving $z_{n+1} = \tan\theta^n$. The remaining coordinates are found by solving the equations

$$\alpha^i \equiv dz_{i+3} - z_{i+2}dz_1 \bmod \alpha^{i+1}, \ldots, \alpha^n \tag{12.103}$$

for $i = n-1, \ldots, 1$. In fact, because $dz_1 = \pi$ as chosen in the proof of Theorem 12.70, the one-forms $\bar{\alpha}^i$ already satisfy these equations, so that the coordinates z_i are given by the coefficients of dx^n in the expression for the $\bar{\alpha}^i$.

---

[5]The basis that satisfies the Goursat congruences was a scaled version of the original basis, $\bar{\alpha}^n = f_{\alpha^n}\alpha^n$. However, it can be checked that

$$d\bar{\alpha}^n \wedge \bar{\alpha}^n = (df_{\alpha^n} \wedge \alpha^n + f_{\alpha^n}d\alpha^n) \wedge f_{\alpha^n}\alpha^n$$
$$= (f_{\alpha^n})^2 d\alpha^n \wedge \alpha^n, \tag{12.97}$$

and thus a function f_1 will satisfy $d\alpha^n \wedge \alpha^n \wedge df_1 = 0$ if and only if $d\bar{\alpha}^n \wedge \bar{\alpha}^n \wedge df_1 = 0$.

Singularities

There are two types of singularities associated with the transformation into Goursat form. At $\theta^n = \pi/2$, for example, the transformation will be singular, but this singularity can be avoided by choosing another coordinate chart at the singular point (such as by interchanging x and y, using the $SE(2)$ symmetry of the system). A singularity also occurs when the angle between two adjacent axles is equal to $\pi/2$; at this point, some of the codistributions in the derived flag will lose rank. The derived flag is not defined at these points; nor is the transformation. There are no singularities of the second type for the unicycle ($n = 0$) or for the front-wheel drive car ($n = 1$).

Once the constraints are in the Goursat normal form, paths can be found which connect any two desired configurations. See Tilbury, Murray, and Sastry [302] and Problem 12.8 for details.

12.3.3   The Extended Goursat Normal Form

While the Goursat normal form is powerful, it is restricted to Pfaffian systems of codimension two. In order to study Pfaffian systems of higher codimension, we present the extended Goursat normal form. Whereas the Goursat normal form can be thought of as a single chain of integrators, the extended Goursat form consists of many chains of integrators. Consider the following definition:

Definition 12.72 Extended Goursat Normal Form. *A Pfaffian system I on* $\mathbb{R}^{n+m+1}$ *of codimension* $m + 1$ *is in* extended Goursat normal form *if it is generated by n constraints of the form*

$$I = \{dz_i^j - z_{i+1}^j dz^0 : i = 1, \ldots, s_j; \, j = 1, \ldots, m\}. \tag{12.104}$$

This is a direct extension of the Goursat normal form, and all integral curves of (12.104) are determined by the $m + 1$ functions $z^0(t), z_1^1(t), \ldots, z_1^m(t)$ and their derivatives with respect to the parameter t. The notation has been changed slightly; the canonical constraints are now $dz_i^j - z_{i+1}^j dz^0$, whereas before they were $dz^i - z^{i-1} dz_1$. For the Goursat form, the constraint in the last nontrivial derived system was $dz^n - z^{n-1} dz^1$; in the extended Goursat normal form, it will be $dz_1^j - z_2^j dz^0$. We refer to the set of constraints with superscript j as the j-th *tower* (the reason for this name will become clear after we compute the derived flag).

Conditions for converting a Pfaffian system to extended Goursat normal form are given by the following theorem:

Theorem 12.73 Extended Goursat Normal Form. *Let I be a Pfaffian system of codimension $m + 1$. If (and only if) there exists a set of generators $\{\alpha_i^j : i = 1, \ldots, s_j; \, j = 1, \ldots, m\}$ for I and an integrable one-form π such that for all j,*

$$
\begin{aligned}
d\alpha_i^j &\equiv -\alpha_{i+1}^j \wedge \pi, \quad \mod I^{(s_j - i)} \quad i = 1, \ldots, s_j - 1, \\
d\alpha_{s_j}^j &\not\equiv 0 \quad \mod I,
\end{aligned} \tag{12.105}
$$

then there exists a set of coordinates z such that I is in extended Goursat normal form,

$$I = \{dz_i^j - z_{i+1}^j dz^0 : i = 1, \ldots, s_j; \; j = 1, \ldots, m\}.$$

Proof: If the Pfaffian system is already in extended Goursat normal form, the congruences are satisfied with $\pi = dz^0$ (which is integrable) and the basis of constraints $\alpha_i^j = dz_i^j - z_{i+1}^j dz^0$.

Now assume that a basis of constraints for I has been found which satisfies the congruences (12.105). It is easily checked that this basis is adapted to the derived flag, that is,

$$I^{(k)} = \{\alpha_i^j : i = 1, \ldots, s_j - k; \; j = 1, \ldots, m\}.$$

The coordinates z which comprise the Goursat normal form can now be constructed.

Since π is integrable, any first integral of π can be used for the coordinate z^0. If necessary, the constraints α_i^j can be scaled so that the congruences (12.105) are satisfied with dz^0:

$$d\alpha_i^j \equiv -\alpha_{i+1}^j \wedge dz^0, \quad \mod I^{(s_j - i)} \quad i = 1, \ldots, s_j - 1,$$

and the constraints can be renumbered so that $s_1 \geq s_2 \geq \cdots \geq s_m$.

Consider the last nontrivial derived system, $I^{(s_1-1)}$. The one-forms $\alpha_1^1, \ldots, \alpha_1^{r_1}$ form a basis for this codistribution, where $s_1 = s_2 = \cdots = s_{r_1}$. From the fact that

$$d\alpha_1^j \equiv -\alpha_2^j \wedge dz^0 \mod I^{(s_1-1)},$$

it follows that the one-forms $\alpha_1^1, \ldots, \alpha_1^{r_1}$ satisfy the Frobenius condition

$$d\alpha_1^j \wedge \alpha_1^1 \wedge \cdots \wedge \alpha_1^{r_1} \wedge dz^0 = 0,$$

and thus, by the Frobenius theorem, coordinates $z_1^1, \ldots, z_1^{r_1}$ can be found such that

$$\begin{bmatrix} \alpha_1^1 \\ \vdots \\ \alpha_1^{r_1} \end{bmatrix} = A \begin{bmatrix} dz_1^1 \\ \vdots \\ dz_1^{r_1} \end{bmatrix} + B dz^0.$$

The matrix A must be nonsingular, since the α_1^j's are a basis for $I^{(s_1-1)}$ and they are independent of dz^0. Therefore, a new basis $\bar{\alpha}_1^j$ can be defined as

$$\begin{bmatrix} \bar{\alpha}_1^1 \\ \vdots \\ \bar{\alpha}_1^{r_1} \end{bmatrix} := A^{-1} \begin{bmatrix} \alpha_1^1 \\ \vdots \\ \alpha_1^{r_1} \end{bmatrix} = \begin{bmatrix} dz_1^1 \\ \vdots \\ dz_1^{r_1} \end{bmatrix} + (A^{-1}B) dz^0,$$

and the coordinates $z_2^j := -(A^{-1}B)_j$ are defined such that the one-forms $\bar{\alpha}_1^j$ have the form

$$\bar{\alpha}_1^j = dz_1^j - z_2^j dz^0$$

for $j = 1, \ldots, r_1$. In these coordinates, the exterior derivative of $\bar{\alpha}_1^j$ is equal to

$$d\bar{\alpha}_1^j = -dz_2^j \wedge dz^0.$$

If there were some coordinate z_2^k which could be expressed as a function of the other z_2^j's and z_1^j's, then there would be some linear combination of the $\bar{\alpha}_1^j$'s whose exterior derivative would be zero modulo $I^{(s_1-1)}$, which is a contradiction. Thus, this is a valid choice of coordinates. By the proof of the standard Goursat theorem, all of the coordinates in the j-th tower can be found from z_1^j and z^0. By the above procedure, all the coordinates in the first r_1 towers can be found.

To find the coordinates for the other towers, the lowest derived systems in which they appear must be considered. The coordinates for the longest towers were found first, next those for the next-longest tower(s) will be found. Consider the smallest integer k such that $\dim I^{(s_1-k)} > kr_1$; more towers will appear at this level. A basis for $I^{(s_1-k)}$ is

$$\{\bar{\alpha}_1^1, \ldots, \bar{\alpha}_k^1, \ldots, \bar{\alpha}_1^{r_1}, \ldots, \bar{\alpha}_k^{r_1}, \alpha_1^{r_1+1}, \ldots, \alpha_1^{r_1+r_2}\},$$

where $\bar{\alpha}_i^j = dz_i^j - z_{i+1}^j dz^0$ for $j = 1, \ldots, r_1$, as found in the first step, and α_1^j for $j = r_1 + 1, \ldots, r_2$ are the one-forms which satisfy the congruences (12.105) and are adapted to the derived flag. The lengths of these towers are $s_{r_1+1} = \cdots = s_{r_1+r_2} = s_1 - k + 1$. For notational convenience, define $z_{(k)}^j := (z_1^j, \ldots, z_k^j)$ for $j = 1, \ldots, r_1$.

By the Goursat congruences, $d\alpha_1^j \equiv -\alpha_2^j \wedge dz^0 \mod I^{(s_1-k)}$ for $j = r_1 + 1, \ldots, r_1 + r_2$. Thus the Frobenius condition

$$d\alpha_1^j \wedge \alpha_1^{r_1+1} \wedge \cdots \wedge \alpha_1^{r_1+r_2} \wedge dz_1^1 \wedge \cdots \wedge dz_k^1 \wedge \cdots \wedge dz_1^{r_1} \wedge \cdots \wedge dz_k^{r_1} \wedge dz^0 = 0$$

is satisfied for $j = r_1 + 1, \ldots, r_1 + r_2$. Using the Frobenius Theorem, new coordinates $z_1^{r_1+1}, \ldots, z_1^{r_1+r_2}$ can be found such that

$$\begin{bmatrix} \alpha_1^{r_1+1} \\ \vdots \\ \alpha_1^{r_1+r_2} \end{bmatrix} = A \begin{bmatrix} dz_1^{r_1+1} \\ \vdots \\ dz_1^{r_1+r_2} \end{bmatrix} + B\,dz^0 + C \begin{bmatrix} dz_{(k)}^1 \\ \vdots \\ dz_{(k)}^{r_1} \end{bmatrix}.$$

Since the congruences are defined only up to mod $I^{(s_1-k)}$, the last group of terms (those multiplied by the matrix C) can be eliminated by adding in the appropriate multiples of $\bar{\alpha}_i^j = dz_i^j - z_{i+1}^j dz^0$ for $j = 1, \ldots, r_1$ and $i = 1, \ldots, k$. This will change the B matrix, leaving the equation

$$\begin{bmatrix} \tilde{\alpha}_1^{r_1+1} \\ \vdots \\ \tilde{\alpha}_1^{r_1+r_2} \end{bmatrix} = A \begin{bmatrix} dz_1^{r_1+1} \\ \vdots \\ dz_1^{r_1+r_2} \end{bmatrix} + \tilde{B}\,dz^0.$$

Again, note that A must be nonsingular, because the α_1^j's are linearly independent mod $I^{(s_1-k)}$ and also independent of dz^0. Define

$$
\begin{bmatrix} \bar{\alpha}_1^{r_1+1} \\ \vdots \\ \bar{\alpha}_1^{r_1+r_2} \end{bmatrix} := A^{-1} \begin{bmatrix} \tilde{\alpha}_1^{r_1+1} \\ \vdots \\ \tilde{\alpha}_1^{r_1+r_2} \end{bmatrix} = \begin{bmatrix} dz_1^{r_1+1} \\ \vdots \\ dz_1^{r_1+r_2} \end{bmatrix} + (A^{-1}\tilde{B})dz^0
$$

and then define the coordinates $z_2^j := -(A^{-1}\tilde{B})_j$ for $j = r_1+1,\ldots,r_1+r_2$ such that $\bar{\alpha}_1^j = dz_1^j - z_2^j dz^0$. Again, by the standard Goursat Theorem, all of the coordinates in the towers $r_1+1,\ldots,r_1+r_2$ are now defined. The coordinates for the rest of the towers are defined in a manner exactly analogous to that of the second-longest tower.   □

If the one-form π which satisfies the congruences (12.105) is not integrable, then the Frobenius Theorem cannot be used to find the coordinates. In the special case where $s_1 > s_2$, that is, there is one tower which is strictly longer than the others, it can be shown that if there exists *any* π which satisfies the congruences, then there also exists an *integrable* π' which also satisfies the congruences (with a rescaling of the basis forms); see [52; 222]. However, if $s_1 = s_2$, or there are at least two towers which are longest, this is no longer true. Thus, the assumption that π is integrable is necessary for the general case.

If I can be converted to extended Goursat normal form, then the derived flag of I has the structure

$$
\begin{aligned}
I &= \{\alpha_1^1, \quad \cdots \quad \cdots \quad \alpha_{s_1-1}^1, \ \alpha_{s_1}^1, \ \cdots \ \alpha_1^m, \ \cdots \ \alpha_{s_m}^m\}, \\
I^{(1)} &= \{\alpha_1^1, \quad \cdots \quad \cdots \quad \alpha_{s_1-1}^1, \quad \cdots \ \alpha_1^m, \ \cdots\}, \\
&\ \ \vdots \qquad\qquad \ddots \qquad\qquad\quad \vdots \qquad \ddots \\
I^{(s_m-1)} &= \{\alpha_1^1, \quad \cdots \ \alpha_{s_1-s_m+1}^1, \qquad\qquad \cdots \ \alpha_1^m\}, \\
&\ \ \vdots \qquad \ddots \\
I^{(s_1-2)} &= \{\alpha_1^1, \ \alpha_2^1\}, \\
I^{(s_1-1)} &= \{\alpha_1^1\}, \\
I^{(s_1)} &= \{0 \ \},
\end{aligned}
$$

where the forms in each level have been arranged to show the different towers. The superscripts j indicate the tower to which each form belongs, and the subscripts i index the position of the form within the j-th tower. There are s_j forms in the j-th tower. An algorithm for converting systems to extended Goursat normal form is given in [52].

Another version of the extended Goursat normal form theorem is given here, which is easier to check, since it does not require finding a basis which satisfies the congruences but only one which is adapted to the derived flag. One special case of this theorem is proven in [273].

Theorem 12.74 Extended Goursat Normal Form. *A Pfaffian system I of codimension $m + 1$ on $\mathbb{R}^{n+m+1}$ can be converted to extended Goursat normal form if and only if $I^{(N)} = \{0\}$ for some N and there exists a one-form π such that $\{I^{(k)}, \pi\}$ is integrable for $k = 0, \dots, N - 1$.*

Proof: The "only-if" part is easily shown by taking $\pi = dz^0$ and noting that

$$I^{(k)} = \{dz_i^j - z_{i+1}^j dz^0 : i = 1, \dots, s_j - k; \, j = 1, \dots, m\},$$

$$\{I^{(k)}, \pi\} = \{dz_i^j, dz^0 : i = k + 1, \dots, s_j; \, j = 1, \dots, m\},$$

which is integrable for each k.

Now assume that such a π exists. After the derived flag of the system, $I =:$ $I^{(0)} \supset I^{(1)} \supset \cdots \supset I^{(s_1)} = \{0\}$, has been found, a basis which is adapted to the derived flag and which satisfies the Goursat congruences (12.105) can be iteratively constructed.

The lengths of each tower are determined from the dimensions of the derived flag. Indeed, the longest tower of forms has length s_1. If the dimension of $I^{(s_1 - 1)}$ is r_1, then there are r_1 towers each of which has length s_1; and we have $s_1 = s_2 = \cdots = s_{r_1}$. Now, if the dimension of $I^{(s_1 - 2)}$ is $2r_1 + r_2$, then there are r_2 towers with length $s_1 - 1$, and $s_{r_1 + 1} = \cdots = s_{r_1 + r_2} = s_1 - 1$. Each s_j is found similarly.

A π which satisfies the conditions must be in the complement of I, for if π were in I, then $\{I, \pi\}$ integrable means that I is integrable, and this contradicts the assumption that $I^{(N)} = \{0\}$ for some N.

Consider the last nontrivial derived system, $I^{(s_1 - 1)}$. Let $\{\alpha_1^1, \dots, \alpha_1^{r_1}\}$ be a basis for $I^{(s_1 - 1)}$. The definition of the derived flag, specifically $I^{(s_1)} = \{0\}$, implies that

$$d\alpha_1^j \not\equiv 0 \mod I^{(s_1 - 1)}, \quad j = 1, \dots, r_1. \tag{12.106}$$

Also, the assumption that $\{I^{(k)}, \pi\}$ is integrable gives the congruence

$$d\alpha_1^j \equiv 0 \mod \{I^{(s_1 - 1)}, \pi\}, \quad j = 1, \dots, r_1. \tag{12.107}$$

Combining equations (12.106) and (12.107), the congruence

$$d\alpha_1^j \equiv \pi \wedge \beta^j \mod I^{(s_1 - 1)}, \quad j = 1, \dots, r_1, \tag{12.108}$$

must be satisfied for some $\beta^j \not\equiv 0 \mod I^{(s_1 - 1)}$.

Now, from the definition of the derived flag,

$$d\alpha_1^j \equiv 0, \mod I^{(s_1 - 2)} \quad j = 1, \dots, r_1,$$

which when combined with (12.108) implies that β^j is in $I^{(s_1 - 2)}$.

Claim $\beta^1, \dots, \beta^{r_1}$ are linearly independent mod $I^{(s_1 - 1)}$.

Proof of Claim: The proof is by contradiction. Suppose there exists some combination of the β^j's, say

$$\beta = b_1 \beta^1 + \cdots + b_{r_1} \beta^{r_1} \equiv 0 \mod I^{(s_1 - 1)},$$

with not all of the b_j's equal to zero. Consider $\alpha = b_1 \alpha_1^1 + \cdots + b_{r_1} \alpha_1^{r_1}$. This one-form α is non-zero because the α_1^j are a basis for $I^{(s_1 - 1)}$. The exterior derivative of α can be found by the product rule,

$$d\alpha = \sum_{j=1}^{r_1} b_j d\alpha_1^j + \sum_{j=1}^{r_1} db_j \wedge \alpha_1^j$$

$$\equiv \sum_{j=1}^{r_1} b_j (\pi \wedge \beta^j) \bmod I^{(s_1-1)}$$

$$\equiv \pi \wedge \left(\sum_{j=1}^{r_1} b_j \beta^j \right) \bmod I^{(s_1-1)}$$

$$\equiv 0 \bmod I^{(s_1-1)},$$

which implies that α is in $I^{(s_1)}$. However, this contradicts the assumption that $I^{(s_1)} = \{0\}$. Thus the β^j's must be linearly independent mod $I^{(s_1-1)}$.

Define $\alpha_2^j := \beta^j$ for $j = 1, \ldots, r_1$. Note that these basis elements satisfy the first level of Goursat congruences, that is:

$$d\alpha_1^j \equiv -\alpha_2^j \wedge \pi, \bmod I^{(s_1-1)} \quad j = 1, \ldots, r_1.$$

If the dimension of $I^{(s_1-2)}$ is greater than $2r_1$, then one-forms $\alpha_1^{r_1+1}, \ldots, \alpha_1^{r_1+r_2}$ are chosen such that

$$\{\alpha_1^1, \ldots, \alpha_1^{r_1}, \alpha_2^1, \ldots, \alpha_2^{r_1}, \alpha_1^{r_1+1}, \ldots, \alpha_1^{r_1+r_2}\}$$

is a basis for $I^{(s_1-2)}$.

For the induction step, assume that a basis for $I^{(i)}$ has been found,

$$\{\alpha_1^1, \ldots, \alpha_{k_1}^1, \alpha_1^2, \ldots, \alpha_{k_2}^2, \ldots, \alpha_1^c, \ldots, \alpha_{k_c}^c\}$$

that satisfies the Goursat congruences up to this level:

$$d\alpha_k^j = -\alpha_{k+1}^j \wedge \pi \bmod I^{(s_j-k)}, \quad k = 1, \ldots, k_j - 1, \quad j = 1, \ldots, c.$$

Note that c towers of forms have appeared in $I^{(i)}$. Consider only the last form in each tower that appears in $I^{(i)}$, that is $\alpha_{k_j}^j$, $j = 1, \ldots, c$. By the construction of this basis (or from the Goursat congruences), $\alpha_{k_j}^j$ is in $I^{(i)}$ but is not in $I^{(i+1)}$, thus

$$d\alpha_{k_j}^j \not\equiv 0 \bmod I^{(i)} \quad j = 1, \ldots, c.$$

The assumption that $\{I^{(i)}, \pi\}$ is integrable ensures

$$d\alpha_{k_j}^j \equiv 0 \bmod \{I^{(i)}, \pi\}, \quad j = 1, \ldots, c.$$

Thus $d\alpha_{k_j}^j$ must be a multiple of π mod $I^{(i)}$,

$$d\alpha_{k_j}^j \equiv \pi \wedge \beta^j \bmod I^{(i)}, \quad j = 1, \ldots, c,$$

for some $\beta^j \not\equiv 0 \bmod I^{(i)}$. From the fact that $\alpha_{k_j}^k$ is in $I^{(i)}$ and the definition of the derived flag,

$$d\alpha_{k_j}^j \equiv 0 \bmod I^{(i-1)}, \quad j = 1, \ldots, c,$$

which implies that $\beta^j \in I^{(i-1)}$. By a similar argument to the claim above, it can be shown that the β^j's are independent mod $I^{(i)}$. Define $\alpha^j_{k_j+1} := \beta^j$, and thus

$$\{\alpha^1_1, \ldots, \alpha^1_{k_1+1}, \alpha^2_1, \ldots, \alpha^2_{k_2+1}, \ldots, \alpha^c_1, \ldots, \alpha^c_{k_c+1}\}$$

forms part of a basis of $I^{(i-1)}$. If the dimension of $I^{(i-1)}$ is greater than $k_1 + k_2 + \cdots + k_c + c$, then complete the basis of $I^{(i-1)}$ with any linearly independent one-forms $\alpha^{c+1}_1, \ldots, \alpha^{c+r_c}_1$ such that

$$\{\alpha^1_1, \ldots, \alpha^1_{k_1+1}, \alpha^2_1, \ldots, \alpha^2_{k_2+1}, \ldots, \alpha^c_1, \ldots, \alpha^c_{k_c+1}, \alpha^{c+1}_1, \ldots, \alpha^{c+r_c}_1\}$$

is a basis for $I^{(i-1)}$.

Repeated application of this procedure will construct a basis for I that is not only adapted to the derived flag, but also satisfies the Goursat congruences.

By assumption, π is integrable mod the last nontrivial derived system, $I^{(s_1-1)}$. Looking at the congruences (12.105), we see that any integrable one-form π' that is congruent to π up to a scaling factor f,

$$\pi' = dz^0 \equiv f\pi \mod I^{(s_1-1)},$$

will satisfy the same set of congruences up to a rescaling of the constraint basis by multiples of this factor f.   □

12.4 Control Systems

The examples considered in Section 12.3 mobile robots towing trailers required purely kinematic models, which had no drift terms, and the variable representing time needed no special consideration. Because of this, and the fact the velocity constraints could be represented as one-forms, exterior differential systems are particularly appropriate for their analysis. Nonlinear control systems have traditionally been defined in terms of distributions of vector fields on manifolds. Because of the duality between vector fields and one-forms, as seen in Section 12.2.1, a control system can also be defined as a Pfaffian system on a manifold and analyzed using techniques from exterior differential systems. In this section we will present some results on linearization for nonlinear control systems and also examine the connections between the two different formalisms of vector fields and one-forms.

We will consider the nonlinear dynamical system:

$$\dot{x} = f(x, u) \tag{12.109}$$

where $x \in \mathbb{R}^n$, $u \in \mathbb{R}^m$, and f is a smooth map

$$f : \mathbb{R}^n \times \mathbb{R}^m \mapsto T\mathbb{R}^n,$$
$$(x, u) \mapsto f(x, u) \in T_x\mathbb{R}^n.$$

A very important special case of system (12.109) is the one where the input enters affinely in the dynamics:

$$\dot{x} = f(x) + g(x)u, \tag{12.110}$$

where $g(x) = [g_1(x) \ldots g_m(x)]$ and $g_i(x)$ are smooth vector fields. Most of the results presented here will be concerned with systems belonging to this class, even though some can be extended to the the more general case (12.109).

In Chapter 9 we have established conditions under which the dynamics of (12.109) and (12.110) are adequately described by those of a linear system

$$\dot{\hat{x}} = A\hat{x} + B\hat{u}, \tag{12.111}$$

where $\hat{x} \in \mathbb{R}^{\hat{n}}$, $\hat{u} \in \mathbb{R}^m$, $A \in \mathbb{R}^{\hat{n} \times \hat{n}}$, and $B \in \mathbb{R}^{\hat{n} \times m}$ with $\hat{n} \geq n$.

The problem of linearization can also be approached from the point of view of exterior differential systems. Note that any control system of the form (12.109) can also be thought of as a Pfaffian system of codimension $m + 1$ in $\mathbb{R}^{n+m+1}$. The corresponding ideal is generated by the codistribution

$$I = \{dx_i - f_i(x, u)dt : i = 1, \ldots, n\}. \tag{12.112}$$

The $n + m + 1$ variables for the Pfaffian system correspond to the n states, m inputs and time t. For the special case of the affine system (12.110) the co-distribution becomes:

$$I = \left\{ dx_i - \left(f_i(x) + \sum_{j=1}^{m} g_{ij}(x)u_j \right) dt : i = 1, \ldots, n \right\}. \tag{12.113}$$

In this light, the extended Goursat normal form looks remarkably similar to the MIMO normal form (also known as Brunovksy normal form) with (Kronecker) indices s_j, $j = 1, \ldots, m$ (see Problem 9.14). Indeed, if we identify coordinates z^0, $z_{s_j+1}^j$, $j = 1, \ldots, m$ in the Goursat normal form with t, u_j, $j = 1, \ldots, m$, the Pfaffian system becomes equivalent (in vector field notation) to a collection of m chains of integrators, each one of length s_j and terminating with an input in the right-hand side. With this in mind, Theorems 12.73 and 12.74, which provide conditions under which a Pfaffian system can be transformed to extended Goursat normal form, can be viewed as linearization theorems with the additional restriction that $\pi = dt$. An equivalent formulation of the conditions of Theorem 12.66 involving the annihilating distributions is given by Murray [223]. The result is restricted to Pfaffian systems of codimension two.

Theorem 12.75. *Given a 2-dimensional distribution Δ, construct two filtrations:*

$$E_0 = \Delta, \qquad\qquad F_0 = \Delta,$$
$$E_{i+1} = E_i + [E_i, E_i], \qquad F_{i+1} = F_i + [F_i, F_0].$$

If all the distributions are of constant rank and:

$$\dim E_i = \dim F_i = i + 2, \quad i = 0, \ldots, n - 2,$$

Then there exists a local basis $\{\alpha^1, \ldots, \alpha^s\}$ and a one-form π such that the Goursat congruences are satisfied for the differential system $I = \Delta^\perp$.

Proof: In [223].

In [223] this theorem is shown to be equivalent to Theorem 12.66. However, there is no known analog of Theorem 12.75 to the extended Goursat case covered by Theorems 12.73 and 12.74. We will now explicitly work through the connection between the classical static feedback linearization theorem (Theorem 9.16) and the extended Goursat normal form theorem (Theorem 12.74).

Proposition 12.76 Connections Between Vector Field and Pfaffian System Approaches. *The control system (12.110) satisfies the conditions of Theorem 9.16 if and only if the corresponding Pfaffian system (12.113) satisfies the conditions of Theorem 12.74 for $\pi = dt$.*

Proof: Consider control system (12.110) and the equivalent Pfaffian system (12.113). For simplicity, we will consider the case $m = 2$. The Pfaffian system $I^{(0)}$ and its annihilating distribution Δ_0 are given by

$$I^{(0)} = \{dx_i - (f_i(x) + g_{i1}(x)u_1 + g_{i2}(x)u_2)dt : i = 1, \ldots, n\},$$

$$\Delta_0 = (I^{(0)})^\perp = \left\{ \begin{bmatrix} 0 \\ 1 \\ 0 \\ 0 \end{bmatrix}, \begin{bmatrix} 0 \\ 0 \\ 1 \\ 0 \end{bmatrix}, \begin{bmatrix} 1 \\ 0 \\ 0 \\ f + g_1 u_1 + g_2 u_2 \end{bmatrix} \right\}.$$

As the notation suggests, the top three entries in each vector field in the distribution Δ_0 are scalars (corresponding to the coordinates t, u_1 and u_2) while the bottom entry is a column vector of dimension n. We will construct the derived flag $I^{(0)} \supset I^{(1)} \supset \cdots \supset I^{(N)}$ and the corresponding orthogonal filtration $\Delta_0 \subset \Delta_1 \subset \cdots \subset \Delta_N$. We will write $\hat{I}^{(i)} = \{I^{(i)}, dt\}$ and $\hat{\Delta}_i = (\hat{I}^{(i)})^\perp$. We will go through the conditions of Theorem 12.74 step by step, assuming $\pi = dt$:

Step 0: As above:

$$I^{(0)} = \{dx_i - (f_i(x) + g_{i1}(x)u_1 + g_{i2}(x)u_2)dt : i = 1, \ldots, n\},$$

$$\Delta_0 = (I^{(0)})^\perp = \left\{ \begin{bmatrix} 0 \\ 1 \\ 0 \\ 0 \end{bmatrix}, \begin{bmatrix} 0 \\ 0 \\ 1 \\ 0 \end{bmatrix}, \begin{bmatrix} 1 \\ 0 \\ 0 \\ f + g_1 u_1 + g_2 u_2 \end{bmatrix} \right\}$$

$$= \{v_1, v_2, v_3\}.$$

The condition of Theorem 12.74 requires that $\hat{I}^{(0)} = \{I^{(0)}, dt\}$ be integrable. Its annihilator is $\hat{\Delta}_0 = \{v_1, v_2\}$ which is indeed involutive, since $[v_1, v_2] = 0$ are constant vector fields.

Step 1: It is easy to show that

$$[v_1, v_2] = 0, \quad [v_1, v_3] = \begin{bmatrix} 0 \\ 0 \\ 0 \\ g_1 \end{bmatrix}, \quad [v_2, v_3] = \begin{bmatrix} 0 \\ 0 \\ 0 \\ g_2 \end{bmatrix}.$$

Therefore,

$$I^{(1)} = \{\alpha \in I^{(0)} : d\alpha \equiv 0 \bmod I^{(0)}\},$$

$$\Delta_1 = (I^{(1)})^\perp = \left\{ \begin{bmatrix} 0 \\ 1 \\ 0 \\ 0 \end{bmatrix}, \begin{bmatrix} 0 \\ 0 \\ 1 \\ 0 \end{bmatrix}, \begin{bmatrix} 1 \\ 0 \\ 0 \\ f + g_1 u_1 + g_2 u_2 \end{bmatrix}, \begin{bmatrix} 0 \\ 0 \\ 0 \\ g_1 \end{bmatrix}, \begin{bmatrix} 0 \\ 0 \\ 0 \\ g_2 \end{bmatrix} \right\}$$

$$= \{v_1, v_2, v_3, v_4, v_5\}.$$

The condition of Theorem 12.74 requires that $\hat{I}^{(1)} = \{I^{(1)}, dt\}$ be integrable. To check this, consider its annihilator $\hat{\Delta}_1 = \{v_1, v_2, v_4, v_5\}$. Then, note that $[v_1, v_2] = [v_1, v_4] = [v_1, v_5] = [v_2, v_4] = [v_2, v_5] = 0$ and

$$[v_4, v_5] = \begin{bmatrix} 0 \\ 0 \\ 0 \\ [g_1, g_2] \end{bmatrix}.$$

Therefore $\hat{\Delta}_1$ is involutive if and only if $[g_1, g_2]$ is in the span of $\{g_1, g_2\}$. The condition of Theorem 12.74 holds for the first iteration of the derived flag if and only if distribution G_0 of Theorem 9.16 is involutive.

Step 2: We compute the bracket of the vector fields v_3 and v_4.

$$[v_3, v_4] = \begin{bmatrix} 0 \\ 0 \\ 0 \\ \operatorname{ad}_f g_1 - [g_1, g_2] u_2 \end{bmatrix}.$$

The computation of $[v_3, v_5]$ is similar. Therefore, assuming that the conditions of Step 1 hold and in particular that $[g_1, g_2] \in \operatorname{span}\{g_1, g_2\}$:

$$I^{(2)} = \{\alpha \in I^{(1)} : d\alpha \equiv 0 \bmod I^{(1)}\},$$

$$\Delta_2 = (I^{(2)})^\perp = \Delta_1 + \left\{ \begin{bmatrix} 1 \\ 0 \\ \vdots \\ 0 \\ ad_f g_1 \end{bmatrix}, \begin{bmatrix} 1 \\ 0 \\ \vdots \\ 0 \\ ad_f g_2 \end{bmatrix} \right\}$$

$$= \{v_i : i = 1, \ldots, 7\}.$$

The condition of Theorem 12.74 requires that $\hat{I}^{(2)}$ be integrable. This is equivalent to $\hat{\Delta}_2 = \{v_1, v_2, v_4, v_5, v_6, v_7\}$ being involutive. As before the only pairs whose involutivity needs to be verified are the ones not involving v_1 and v_2, i.e., the condition is equivalent to $\{g_1, g_2, ad_f g_1, ad_f g_2\}$ being involutive. Overall the condition of Theorem 12.74 holds for the the second iteration of the derived flag if and only if distribution G_1 of Theorem 9.16 is involutive.

Step i: Assume that

$$\Delta_{i-1} = \left\{ \begin{bmatrix} 0 \\ 1 \\ 0 \\ 0 \end{bmatrix}, \begin{bmatrix} 0 \\ 0 \\ 1 \\ 0 \end{bmatrix}, \begin{bmatrix} 1 \\ 0 \\ 0 \\ f + g_1 u_1 + g_2 u_2 \end{bmatrix}, \begin{bmatrix} 1 \\ 0 \\ 0 \\ ad_f^k g_1 \end{bmatrix}, \begin{bmatrix} 1 \\ 0 \\ 0 \\ ad_f^k g_2 \end{bmatrix} \right\}$$

for $0 \le k \le i - 2$. Also assume that $\hat{I}^{(k)}$, $0 \le k \le i - 1$, are integrable, or, equivalently, that $\hat{\Delta}_k$ for $0 \le k \le i - 1$ (which is the same as Δ_k without the third vector field) are involutive, or, equivalently, that $G_k = \{ ad_f^l g_j : 0 \le l \le k, j = 1, 2\}$ for $0 \le k \le i - 2$ are involutive. Construct $\Delta_i = \Delta_{i-1} + [\Delta_{i-1}, \Delta_{i-1}]$. By involutivity of $\hat{\Delta}_{i-1}$ and the construction of the filtration the only terms not already in Δ_{i-1} are ones of the form

$$\left[\begin{bmatrix} 1 \\ 0 \\ 0 \\ f + g_1 u_1 + g_2 u_2 \end{bmatrix}, \begin{bmatrix} 0 \\ 0 \\ 0 \\ ad_f^{i-2} g_1 \end{bmatrix} \right]$$

$$= \begin{bmatrix} 0 \\ 0 \\ 0 \\ ad_f^{i-1} g_1 + [g_1, ad_f^{i-2} g_1] u_1 + [g_2, ad_f^{i-2} g_1] u_2 \end{bmatrix}$$

and similarly for $\operatorname{ad}_f^{i-1} g_2$. By the assumed involutivity of $\hat{\Delta}_{i-1}$ the last two terms are already in Δ_{i-1}. Therefore, we can write

$$
\Delta_i = \Delta_{i-1} + \left\{ \begin{bmatrix} 0 \\ 0 \\ 0 \\ \operatorname{ad}_f^{i-1} g_1 \end{bmatrix}, \begin{bmatrix} 0 \\ 0 \\ 0 \\ \operatorname{ad}_f^{i-1} g_2 \end{bmatrix} \right\}.
$$

The condition of Theorem 12.74 requires that $\hat{I}^{(i)}$ be integrable, or equivalently that $\hat{\Delta}_i$ be involutive. As before, the only pairs that can cause trouble are the ones not involving v_1 and v_2. Hence the condition is equivalent to $G_{i-1} = \{ \operatorname{ad}_f^k g_j : 0 \le k \le i - 1, j = 1, 2 \}$ being involutive.

By induction, the condition of Theorem 12.74 holds for the i-th iteration of the derived flag if and only if distribution G_{i-1} is involutive, i.e., if and only if condition (3) of Theorem 9.16 holds. In addition, note that the dimension of G_i keeps increasing by *at least one* until, for some value (denoted κ in Theorem 9.16) $G_{\kappa+1} = G_\kappa$. The involutivity assumption on G_κ prevents any further increase in dimension after this stage is reached. Since the dimension of G_i is necessarily bounded above by n, the number of steps until saturation, is bounded by the maximum final dimension, $\kappa \le n$. By construction, the dimension of Δ_i is three greater than the dimension of G_{i-1}. Moreover $\Delta_\kappa = \Delta_{\kappa+1}$ and therefore $I^{(\kappa)} = I^{(\kappa+1)}$, i.e., the derived flag stops shrinking after κ steps. The remaining condition of Theorem 12.74, namely that there exists N such that $I^{(N)} = \{0\}$, is equivalent to the existence of κ such that $I^{(\kappa)} = \{0\}$, or that Δ_κ has dimension $n + 3$. As noted above, this is equivalent $G_{\kappa-1}$ having dimension n. Since $\kappa \le n$, this can also be stated as G_{n-1} having dimension n, i.e., condition (2) of Theorem 9.16. The remaining condition of Theorem 9.16, namely that the dimension of G_i is constant for all $0 \le i \le n - 1$, is taken care of by the implicit assumption that all co-distributions in the derived flag have constant dimension.   □

Note that a coordinate transformation in the exterior differential systems context corresponds to a coordinate transformation together with a state feedback in the vector field notation. Because the state space $\mathbb{R}^{n+m+1}$ in the forms context does not discriminate between states, inputs, and time, a coordinate transformation on this larger space can make the inputs in the original coordinates functions of the state in the original coordinates and possibly time. It can be shown (see [108]) that time need not enter into the transformation at all; that is, if the conditions of Theorem 12.73 are satisfied, a time-invariant state feedback and coordinate change can always be found. In addition the coordinate transformation can also be chosen to be independent of both time and input.

Theorems 12.73 and 12.74 in their general form are *not* equivalent to the necessary and sufficient conditions of feedback linearization given in Chapter 9. The extended Goursat theorems allow π to be any integrable one-form and not just dt. Therefore, we expect more systems to match the conditions of Theorems 12.73 and 12.74 than just systems which are feedback linearizable. However, a choice of

π other than dt implies a rescaling of time as a function of the state. Even though this effect is very useful for the case of driftless systems (where the role of time is effectively played by an input), solutions for $\pi \neq dt$ are probably not very useful for linearizing control systems with drift. Because of their generality, Theorems 12.73 and 12.74 are capable of dealing with the more general case of control systems of the form (12.112) (or equivalently (12.109), as well as drift-free systems, which were investigated in Section 12.3. Equivalent conditions for the vector field case have not been thoroughly investigated. These results will be useful in deriving conditions for dynamic full state feedback linearization. Finally, Theorem 12.75 is a very interesting alternative to Theorems 12.73 and 12.74, since it provides a way of determining whether a Pfaffian system can be converted to Goursat normal form just by looking at the annihilating distributions, without having to determine a one-form π or an appropriate basis. Unfortunately a generalization to multi-input systems (or more precisely to the extended Goursat normal form) is not easy to formulate. It should be noted that the conditions on the filtrations are very much like involutivity conditions. It is interesting to try to relate these conditions to the conditions of Theorem 12.74 (the connection to the conditions of Theorem 12.73 is provided in [223]) and see whether a formulation for the extended problem can be constructed in this way.

12.5 Summary

The work on exterior differential systems continues in several different directions. Some important approaches to solving the problem of full state linearizing nonlinear systems using dynamic state feedback were given by Tilbury and Sastry [303]. There has also been a great deal of recent work in making connections between a class of nonlinear systems called differentially flat (and used to characterized systems which can be full state linearized using possibly dynamic state feedback) and exterior differential systems. Differential flat systems are well reviewed in the paper by Fliess et al [104], and the connections between flatness and exterior differential systems is discussed by van Nieuwstadt et al [311], Aranda-Bricaire et al [8], and Pomet [242]. While this literature and its connections to the exterior systems of this chapter is very interesting, we make only some summary comments here: Roughly speaking, a system is said to be differentially flat when the input and state variables can be expressed as meromorphic functions of certain outputs (the flat outputs) and their derivatives. Once these flat outputs have been identified then path planning or tracking can be done in the coordinates given by the flat outputs. While it is easy to see what flat outputs for the n-trailer system, or systems in Goursat or extended Goursat normal form, and several other systems are; there are no general methods, at the current time, for constructively obtaining flat outputs. When systems fail to be flat, they can be sometimes be shown to be approximately flat; see, for example, Tomlin et al [305], and Koo and Sastry [165]. The literature on navigation of n-trailer systems in the presence of obstacles is very large and growing. Some entry points to the literature are in the work of Wen and co-workers [92]. Control and stabilization of chained form systems is also a growing literature: see, for example, Samson [254], Walsh and Bushnell [323], and M'Closkey and Murray [205].

12.6   Exercises

Problem 12.1. Prove Theorem 12.26.

Problem 12.2. Prove Theorem 12.42.

Problem 12.3 Scalar and vector fields.   Given a scalar field (function) f on $\mathbb{R}^3$, there is an obvious transformation to a 0-form $f \in \Omega^0(\mathbb{R}^3)$. Similarly, any vector field in $\mathbb{R}^3$ can be expressed as

$$G(x) = (x; g_1(x)e_1 + g_2(x)e_2 + g_3(x)e_3),$$

where e_i are the standard coordinate vectors for $\mathbb{R}^3$. A corresponding 1-form can be constructed in a very straightforward manner:

$$\omega = g_1(x)dx^1 + g_2(x)dx^2 + g_3(x)dx^3$$

Perhaps less obviously, scalar fields may also be identified with 3-forms and vector fields with 2-forms. Consider the following transformations from scalar and vector fields to forms on $\mathbb{R}^3$ (T_0, T_3 have scalars as their domain and T_1, T_2 vectors in their domain):

$$T_0 : f \mapsto f,$$

$$T_1 : \left(x; \sum g_i e_i\right) \mapsto \sum g_i dx^i,$$

$$T_2 : \left(x; \sum g_i e_i\right) \mapsto g_1 dx^2 \wedge dx^3 + g_2 dx^3 \wedge dx^1 + g_3 dx^1 \wedge dx^2,$$

$$T_3 : f \mapsto f dx^1 \wedge dx^2 \wedge dx^3,$$

Prove that these maps are vector space isomorphisms. This may be generalized to $\mathbb{R}^n$, with vector fields isomorphic to 1-forms and $(n - 1)$ forms, and scalar fields isomorphic to 0-forms and n-forms. Do this generalization.

Problem 12.4 Div, grad, and curl.   Drawing from the familiar results of vector calculus, the three primary operations on vector fields and scalar fields in $\mathbb{R}^3$, the gradient, divergence and the curl, may be defined using the exterior derivative. First we give their standard definitions. Let $f(x)$ be a scalar field on $\mathbb{R}^3$ and $G(x) = (x; \sum g_i(x)e_i)$ a vector field.
The *gradient* of f, ∇f or grad f, is the vector field given by

$$\nabla f(x) = \left(x; \sum_{i=1}^{3} \frac{\partial f}{\partial x^i}(x)e_i\right).$$

The *divergence* of G, $\nabla \cdot G$ or $divG$ is the scalar field given by

$$\nabla \cdot G(x) = \sum_{i=1}^{3} \frac{\partial g_i}{\partial x^i}(x).$$

The *curl* of G, $\nabla \times G$ or $curlG$, is the vector field given by

$$\nabla \times G(x) = \left(x; \left(\frac{\partial g_3}{\partial x^2} - \frac{\partial g_2}{\partial x^3}\right) e_1 + \left(\frac{\partial g_1}{\partial x^3} - \frac{\partial g_3}{\partial x^1}\right) e_2 + \left(\frac{\partial g_2}{\partial x^1} - \frac{\partial g_1}{\partial x^2}\right) e_3\right).$$

The gradient and divergence may be formed analogously in $\mathbb{R}^n$ by extending the summation over all n indices; the curl has no real generalization outside of $\mathbb{R}^3$. Prove using the transformations T_i of the previous problem that the divergence, gradient and curl operators can be expressed in the language of exterior forms:

$$
\begin{array}{ccc}
\text{Scalar Fields} & T_0 \mapsto & \Omega^0(\mathbb{R}^n), \\
\nabla \downarrow & & \downarrow d \\
\text{Vector Fields} & T_1 \mapsto & \Omega^1(\mathbb{R}^n), \\
(\nabla \times) \downarrow & & \downarrow d \qquad\qquad (12.114) \\
\text{Vector Fields} & T_2 \mapsto & \Omega^2(\mathbb{R}^n), \\
(\nabla \cdot) \downarrow & & \downarrow d \\
\text{Scalar Fields} & T_3 \mapsto & \Omega^3(\mathbb{R}^n).
\end{array}
$$

Equivalently,

$$df = T_1(\nabla f),$$
$$d(T_1 G) = T_2(\nabla \times G),$$
$$d(T_2 G) = T_3(\nabla \cdot G).$$

Problem 12.5 Rolling penny: derived flag. Consider the rolling penny system. In addition to the three configuration variables of the unicycle, we also have an angle ϕ describing the orientation of Lincoln's head. The model in this case, assuming for simplicity that the penny has unit radius, is given by

$$\dot{x} = u_1 \cos\theta,$$
$$\dot{y} = u_1 \sin\theta,$$
$$\dot{\theta} = u_2,$$
$$\dot{\phi} = -u_1,$$

which can be written in vector-field notation as

$$
\begin{bmatrix} \dot{x} \\ \dot{y} \\ \dot{\theta} \\ \dot{\phi} \end{bmatrix} = \begin{bmatrix} \cos\theta \\ \sin\theta \\ 0 \\ -1 \end{bmatrix} u_1 + \begin{bmatrix} 0 \\ 0 \\ 1 \\ 0 \end{bmatrix} u_2 = f_1 u_1 + f_2 u_2
$$

Prove that the annihilating codistribution to the distribution $\Delta_0 = \{f_1, f_2\}$ is

$$
I = \Delta^{\perp} = \{\alpha^1, \alpha^2\},
$$

where

$$
\alpha^1 = \cos\theta \, dx + \sin\theta \, dy + 0 \, d\theta + 1 \, d\phi,
$$
$$
\alpha^2 = \sin\theta \, dx - \cos\theta \, dy + 0 \, d\theta + 0 \, d\phi.
$$

Compute the derived flag of this system by taking the exterior derivatives of the constraints. Show that

$$
I^{(1)} = \{\alpha^1\} \quad I^{(2)} = \{0\}.
$$

Thus show that the basis is adapted to the derived flag. Since $I^{(2)} = \{0\}$, an integrable subsystem does not exist, and the system is not constrained to move on some submanifold of $\mathbb{R}^4$.

Problem 12.6 Rolling penny in Engel form.   In Problem 12.5, we saw that the derived flag for the rolling penny system satisfies the conditions of Engel's theorem. After some calculations we obtain

$$
d\alpha^1 \wedge \alpha^1 = -dx \wedge dy \wedge d\theta + \sin\theta \, d\theta \wedge dx \wedge d\phi + \cos\theta \, d\theta \wedge dy \wedge d\phi.
$$

Since $(d\alpha^1)^2 \wedge \alpha^1 = 0$, the rank of α^1 is 1. From Pfaff's Theorem we know that there exists a function f_1 such that

$$
d\alpha^1 \wedge \alpha^1 \wedge df_1 = 0.
$$

We can easily see that the function $f_1 = \theta$ is a solution to this equation. Since the rank of α^1 is 1, we must now search for a function f_2 such that

$$
\alpha^1 \wedge df_1 \wedge df_2 = 0.
$$

Let $f_2 = f_2(x, y, \theta, \phi)$. Verify that a solution to this system of equations is

$$
f_2(x, y, \theta, \phi) = x \cos\theta + y \sin\theta + \phi.
$$

Therefore, following once again the proof of Pfaff's Theorem, we may now choose $z^1 = f_1$ and $z^4 = f_2$ such that

$$
\alpha^1 = dz^4 - z^3 dz^1,
$$

Prove that

$$
z^3 = -x \sin\theta + y \cos\theta.
$$

To transform α^2 into the normal form note that

$$\alpha^2 \equiv [a(x, y, \theta, \phi)dz^3 + b(x, y, \theta, \phi)dz^1]\bmod \alpha^1.$$

Prove that

$$a(x, y, \theta, \phi) = -1,$$

$$b(x, y, \theta, \phi) = -x \cos \theta - y \sin \theta,$$

will satisfy the equation. Complete the transformation of the rolling penny system into Engel normal form, by defining

$$z^2 = -x \cos \theta - y \sin \theta = -\frac{b(x, y, \theta, \phi)}{a(x, y, \theta, \phi)}.$$

Problem 12.7 Coordinates for the n-trailer system consisting of the origin seen from the last trailer. Prove that another choice for f_1 in solving equations (12.96) is to write the coordinates of the origin as seen from the last trailer. This is reminiscent of a transformation used by Samson [253] in a different context, and is given by

$$z_1 := f_1(x) = x^n \cos \theta^n + y^n \sin \theta^n. \tag{12.115}$$

This has the physical interpretation of being the origin of the reference frame when viewed from a coordinate frame attached to the n-th trailer. Derive the formulas for the remaining coordinates $z_2, \ldots, z_{n+2}$ corresponding to this transformation by solving the equations

$$\alpha^i \equiv dz_{i+3} - z_{i+2}dz_1 \bmod \alpha^{i+1}, \ldots, \alpha^n \tag{12.116}$$

for $i = n - 1, \ldots, 1$.

Problem 12.8 Steering the n-trailer system [302]. Once the n-trailer system has been converted into Goursat normal form, show that the control system whose control distribution annihilates I is given (in the z coordinates) by

$$\dot{z}_1 = u_1,$$

$$\dot{z}_2 = u_2,$$

$$\dot{z}_3 = z_2 u_1, \tag{12.117}$$

$$\vdots$$

$$\dot{z}_{n+3} = z_{n+2}u_1$$

Use all the steering methods of Chapter 11, namely, sinusoids, piecewise constant inputs, etc. to steer the n-trailer system for one initial configuration to another. You may wish to see how the trajectories look in the original x_i, y_i, θ_i coordinates.

Problem 12.9 Planar space robot [321]. Consider a simplified model of a planar robot consisting of two arms connected to a central body via revolute joints. If the robot is free-floating, then the law of conservation of angular momentum implies that moving the arms causes the central body to rotate. In the case that the angular momentum is zero, this conservation law can be viewed as a Pfaffian constraint on the system. Let M and I represent the mass and inertia of the central body and let m represent the mass of the arms, which we take to be concentrated at the tips. The revolute joints are located a distance r from the middle of the central body and the links attached to these joints have length l. We let (x_1, y_1) and (x_2, y_2) represent the position of the ends of each of the arms (in terms of θ, ψ_1, and ψ_2). Assuming that the body is free floating in space and that friction is negligible, we can derive the constraints arising from conservation of angular momentum. If the initial angular momentum is zero, then *conservation of angular momentum* ensures that the angular momentum stays zero, giving the constraint equation

$$a_{13}(\psi)\dot{\psi}_1 + a_{23}(\psi)\dot{\psi}_2 + a_{33}(\psi)\dot{\theta} = 0. \qquad (12.118)$$

where a_{ij} can be calculated as

$$a_{11} = a_{22} = ml^2,$$

$$a_{12} = 0,$$

$$a_{13} = ml^2 + mr\cos\psi_1,$$

$$a_{23} = ml^2 + mr\cos\psi_2,$$

$$a_{33} = I + 2ml^2 + 2mr^2 + 2mrl\cos\psi_1 + 2mrl\cos\psi_2.$$

Analyze this Pfaffian constraint and determine whether it can be converted into Goursat normal form.

Problem 12.10 Spherical fingertip rolling on a plane.   When one surface rolls on top of another, the contact is parameterized by 5 variables $q_1, q_2, \ldots, q_5 \in \mathbb{R}$ with four variables representing the two sets of coordinates of the points of contact on the two surfaces. In [225] these equations are derived in general. For the case of a spherical fingertip rolling on a plane, the rolling without slipping constraint equations read

$$\begin{bmatrix} 1 & 0 & -\cos q_5 & \sin q_5 & 0 \\ 0 & \cos q_1 & \sin q_5 & \cos q_5 & 0 \\ 0 & \sin q_1 & 0 & 0 & 1 \end{bmatrix} \dot{q} = 0. \qquad (12.119)$$

Analyze this set of constraints to determine whether they can be converted into generalized Goursat form. Compute the derived flag associated with this system as a start. If it cannot be converted into Goursat form can you think of how you might determine the feasible trajectories of this system? For more details on the dual point of view (vector field) to this system see Bicchi, Marigo, and Pratichizzo [25].

13

New Vistas: Multi-Agent Hybrid Systems

Nonlinear control is very much a growing endeavor, with many new results and techniques being introduced. In the summary sections of each of the preceding chapters we have given the reader a sense of the excitement surrounding each of the new directions. In this chapter, we talk about another area of tremendous recent excitement: hybrid systems. The growth of this area is almost directly attributable to advances in computation, communication, and new methods of distributed sensing and actuation, which makes it critical to have methods for designing and analyzing systems which involve interaction between software and electro-mechanical systems. In this chapter, we give a sense of the research agenda in two areas where this activity is especially current: embedded control and multi-agent distributed control systems.

13.1 Embedded Control and Hybrid Systems

While rapid progress in embedded hardware and software makes plausible ever more ambitious multi-layer, multi-objective, adaptive, nonlinear control systems, adequate design methodologies and design support lag far behind. Consequently, today most of the cost in control system development is spent on ad-hoc, prohibitively expensive systems integration, and validation techniques that rely almost exclusively on exhaustively testing more or less complete versions of complex nonlinear control systems. The newest research direction in control addresses this bottleneck by focusing on *predictive* and *systematic hierarchical design* methodologies for building an analytical foundation based on *hybrid systems* and a practical set of *software design tools* which support the construction, integration, safety and performance analysis, on-line adaptation and off-line functional evolution of multi-agent hierarchical control systems. *Hybrid* systems refer to the distinguishing fundamental characteristics of software-based control systems, namely, the tight coupling and interaction of discrete with continuous phenomena. Hybridness is characteristic of all embedded control systems because it arises from several sources. First, the high-level, abstract protocol layers of *hierarchical* control designs are discrete as to make it easier to manage system complexity and to accommodate linguistic and qualitative information; the low-level, concrete control laws are naturally continuous. Second, while individual feedback control scenarios are naturally modeled as interconnections of modules characterized by their continuous input/output behavior, *multi-modal* control naturally suggests a state-based view, with states representing discrete control modes; software-based control systems typically encompass an integrated mixture of both types. Third, every digital *hardware/software implementation* of a control design is ultimately a discrete approximation that interacts through sensors and actuators with a continuous physical environment. The mathematical treatment of hybrid systems is interesting in that it builds on the preceding framework of nonlinear control, but its mathematics is qualitatively distinct from the mathematics of purely discrete or purely continuous phenomena. Over the past several years, we have begun to build basic formal models (*hybrid automata*) for hybrid systems and to develop methods for hybrid control law design, simulation, and verification. Hybrid automata, in particular, integrate diverse models such as differential equations and state machines in a single formalism with a uniform mathematical semantics and novel algorithms for multi-modal control synthesis and for safety and real-time performance analysis. The control of hybrid systems has also been developed by a number of groups: for an introduction to some of the intellectual ferment in the field, we refer the reader to some recent collections of papers on hybrid systems [6; 4; 7; 135], the April 1998 Special Issue on "Hybrid Systems" of the IEEE Transactions on Automatic Control. See also Branicky, Borkar and Mitter [40], and Lygeros, Tomlin, and Sastry [195].

13.2   Multi-Agent Systems and Hybrid Systems

To a large extent the control theory described in this book investigates the paradigm of "Centralized Control". In this paradigm, sensory information is collected from sensors observing a material process that may be distributed over space. This information is transmitted over a communication network to one center, where the commands that guide the process are calculated and transmitted back to the process actuators that implement those commands. In engineering practice, of course, as soon as the process becomes even moderately large, the central control paradigm breaks down. What we find instead is distributed control: A set of control stations, each of whom receives some data and calculates some of the actions.

Important examples of distributed control are the Air Traffic Management System, the control system of an interconnected power grid, the telephone network, a chemical process control system, and automated highway transportation systems. Although a centralized control paradigm no longer applies here, control engineers have with great success used its theories and its design and analysis tools to build and operate these distributed control systems. There are two reasons why the paradigm succeeded in practice, even when it failed in principle. First, in each case the complexity and scale of the material process grew incrementally and relatively slowly. Each new increment to the process was controlled using the paradigm, and adjustments were slowly made after extensive (but by no means exhaustive) testing to ensure that the new controller worked in relative harmony with the existing controllers. Second, the processes were operated with a considerable degree of "slack." That is, the process was operated well within its performance limits to permit errors in the extrapolation of test results to untested situations and to tolerate a small degree of disharmony among the controllers. However, in each system mentioned above, there were occasions when the material process was stressed to its limits and the disharmony became intolerable, leading to a spectacular loss of efficiency. For example, most air travelers have experienced delays as congestion in one part of the country is transmitted by the control system to other parts. The distributed control system of the interconnected power grid has sometimes failed to respond correctly and caused a small fault in one part of a grid to escalate into a system-wide blackout.

We are now attempting to build control systems for processes that are vastly more complex or that are to be operated much closer to their performance limits in order to achieve much greater efficiency of resource use. The attempt to use the central control paradigm cannot meet this challenge: the material process is already given and it is not practicable to approach its complexity in an incremental fashion as before. Moreover, the communication and computation costs in the central control paradigm would be prohibitive, especially if we insist that the control algorithms be fault-tolerant. What is needed to meet the challenge of control design for a complex, high performance material process, is a new paradigm for distributed control. It must distribute the control functions in a way that avoids the high communication and computation costs of central control, at the same time that it limits complexity. The distributed control must, nevertheless, permit centralized authority over those aspects of the material process that are necessary to achieve the high performance goals. Such a challenge can be met by organizing the distributed control functions in a hierarchical architecture that makes those functions relatively autonomous (which permits using all the tools of central control), while introducing enough coordination and supervision to ensure the harmony of the distributed controllers necessary for high performance. Consistent with this hierarchical organization are sensing hierarchies with fan-in of information from lower to higher levels of the hierarchy and a fan-out of control commands from the higher to the lower levels. Commands and information at the higher levels are usually represented symbolically, calling for discrete event control, while at the lower levels both information and commands are continuous, calling for continuous control laws.

Interactions between these levels involves hybrid control. In addition protocols for coordination between individual agents are frequently symbolic, again making for hybrid control laws. The hybrid control systems approach has been successful in the control of some extremely important multi-agent systems such as automated highway systems [313; 194; 145], air traffic control [306], groups of unmanned aerial vehicles, underwater autonomous vehicles [78], mobile offshore platforms, to give a few examples. We invite the reader to use the tools that she has developed in this book to use in these very exciting new directions.

References

[1] R. Abraham and J. E. Marsden. *Foundations of Mechanics*. Benjamin Cummings, Reading, MA, 1978. Second Edition.

[2] R. Abraham, J. E. Marsden, and T. Ratiu. *Manifolds, Tensor Analysis, and Applications*. Springer-Verlag, second edition, 1983.

[3] R. Abraham and C. Shaw. *Dynamics: The Geometry of Behavior, Part One: Periodic Behavior*. Aerial Press, Santa Cruz, CA, 1984.

[4] R. Alur and T. Henzinger (editors). *Hybrid Systems III: Verification and Control*. Springer Verlag Lecture Notes in Computer Science, Vol. 1066, 1996.

[5] B. D. O. Anderson and S. Vongpanitlerd. *Network Analysis and Synthesis*. Prentice Hall, Englewood Cliffs, NJ, 1973.

[6] P. Antsaklis, W. Kohn, A. Nerode, and S. Sastry (editors). *Hybrid Systems II*. Springer Verlag Lecture Notes in Computer Science, Vol. 999, 1995.

[7] P. Antsaklis, W. Kohn, A. Nerode, and S. Sastry (editors). *Hybrid Systems IV*. Springer Verlag Lecture Notes in Computer Science, Vol. 1273, 1997.

[8] E. Aranda-Bricaire, C. H. Moog, and J-B. Pomet. A linear algebraic framework for dynamic feedback linearization. *IEEE Transactions on Automatic Control*, 40(1):127–132, 1995.

[9] A. Arapostathis, S. Sastry, and P. Varaiya. Global analysis of the swing dynamics of power systems. *IEEE Transactions on Circuits and Systems*, CAS-29:673–679, 1982.

[10] B. Armstrong-Helouvry, P. Dupont, and C. Canudas deWit. A survey of models, analysis tools, and compensation methods for the control of machines with friction. *Automatica*, 30:1083–1088, 1994.

[11] V. I. Arnold. *Ordinary Differential Equations*. MIT Press, Cambridge, MA, 1973.

[12] V. I. Arnold. *Geometrical Methods in the Theory of Ordinary Differential Equations*. Springer Verlag, 1983. (Chapter 5 for Poincaré linearization).

[13] V. I. Arnold. *Catastrophe Theory*. Springer Verlag, 1984.

[14] K. J. Aström and B. Wittenmark. *Adaptive Control*. Addison Wesley, Reading, MA, 1989.

[15] E. W. Bai, L. C. Fu, and S. S. Sastry. Averaging for discrete time and sampled data systems. *IEEE Transactions on Circuits and Systems*, CAS-35:137–148, 1988.

[16] J. Baillieul. Geometric methods for nonlinear optimal control problems. *Journal of Optimiaztion Theory and Applications*, 25(4):519–548, 1978.

[17] A. Banaszuk and J. Hauser. Approximate feedback linearization: a homotopy operator approach. *SIAM Journal of Control and Optimization*, 34:1533–1554, 1996.

[18] J. P. Barbot, N. Pantalos, S. Monaco, and D. Normand-Cyrot. On the control of regular epsilon perturbed nonlinear systems. *International Journal of Control*, 59:1255–1279, 1994.

[19] G. Becker and A. Packard. Robust performance of linear parametrically varying systems using parametrically dependent linear feedback. *Systems and Control Letters*, 23(4):205–215, 1994.

[20] N. Bedrossian and M. Spong. Feedback linearization of robot manipulators and Riemannian curvature. *Journal of Robotic Systems*, 12:541–552, 1995.

[21] S. Behtash and S. S. Sastry. Stabilization of nonlinear systems with uncontrollable linearization. *IEEE Transactions on Automatic Control*, AC-33:585–590, 1988.

[22] A. R. Bergen and R. L. Franks. Justification of the describing function method. *SIAM Journal on Control*, 9:568–589, 1973.

[23] A. R. Bergen and D. J. Hill. A structure preserving model for power systems stability analysis. *IEEE Transactions on Power Apparatus and Systems*, PAS-100:25–35, 1981.

[24] S. Bharadwaj, A. V. Rao, and K. D. Mease. Entry trajectory law via feedback linearization. *Journal of Guidance, Control, and Dynamics*, 21(5):726–732, 1998.

[25] A. Bicchi, A. Marigo, and D. Prattichizzo. Robotic dexterity via nonholonomy. In B. Siciliano and K. Valavanis, editors, *Control Problems in Robotics and Automation*, pages 35–49. Springer-Verlag, 1998.

[26] A. Bloch, R. Brockett, and P. Crouch. Double bracket equations and geodesic flows on asymmetric spaces. *Communications in Mathematical Physics*, 187:357–373, 1997.

[27] A. Bloch and P. Crouch. Nonholonomic control systems on Riemannian manifolds. *SIAM Journal of Control and Optimization*, 33:126–148, 1995.

[28] A. Bloch, P. S. Krishnaprasad, J. Marsden, and G. Sanchez de Alvarez. Stabilization of rigid body dynamics by internal and external torques. *Automatica*, 28:745–756, 1993.

[29] A. Bloch, P. S. Krishnaprasad, J. Marsden, and R. Murray. Nonholonomic mechanical systems with symmetry. *Archive for Rational Mechanics and Analysis*, 136(1):21–99, 1996.

[30] A. Bloch, N. Leonard, and J. Marsden. Stabilization of mechanical systems using controlled Lagrangians. In *Proceedings of the 36th Conference on Decisions and Control, San Diego*, December 1997.

[31] A. Bloch, M. Reyhanoglu, and N. H. McClamroch. Control and stabilization of nonholonomic dynamical systems. *IEEE Transactions on Automatic Control*, AC-38:1746–1753, 1993.

[32] M. Bodson and J. Chiasson. Differential geometric methods for control of electric motors. *International Journal of Robust and Nonlinear Control*, 8:923–954, 1998.

[33] N. N. Bogoliuboff and Y. A. Mitropolskii. *Asymptotic Methods in the theory of nonlinear oscillators*. Gordon Breach, New York, 1961.

[34] W. Boothby. *An introduction to differentiable manifolds and Riemannian geometry*. Academic Press, Orlando, 1986,.

[35] S. Bortoff. Approximate state feedback linearization using spline functions. *Automatica*, 33:1449–1458, 1997.

[36] S. Boyd, R. Balakrishnan, L. El Ghaoui, and E. Feron. *Linear Matrix Inequalities in Systems and Control Theory*. SIAM studies in applied mathematics: Vol. 15. SIAM, Philadelphia, 1994.

[37] S. Boyd, L. Chua, and C. Desoer. Analytical foundations of Volterra series. *IMA Journal of Mathematical Control and Information*, 1:243–282, 1984.

[38] S. Boyd and S. Sastry. Necessary and sufficient conditions for parameter convergence in adaptive control. *Automatica*, 22:629–639, 1986.

[39] S. P. Boyd. *Volterra Series: Engineering Fundamentals*. PhD thesis, Department of Electrical Engineering and Computer Science, University of California, Berkeley, 1980.

[40] M. Branicky, V. Borkar, and S. K. Mitter. A unified framework for hybrid control: model and optimal control theory. *IEEE Transactions on Automatic Control*, AC-43:31–45, 1998.

[41] R. K. Brayton and R. Tong. Stability of dynamical systems: A constructive approach. *IEEE Transactions on Circuits and Systems*, CAS-26:226–234, 1979.

[42] H. Brezis. *Analyse fonctionnelle: theorie et applications*. Masson, New York, 1983.

[43] R. W. Brockett. System theory on group manifolds and coset spaces. *SIAM Journal on Control*, 10:265–284, 1972.

[44] R. W. Brockett. Lie algebras and Lie groups in control theory. In D. Q. Mayne and R. W. Brockett, editors, *Geometric Methods in System Theory*, pages 43–82, Dordrecht, 1973. D. Reidel.

[45] R. W. Brockett. Feedback invariant of nonlinear systems. In *Proceedings of the VIIth IFAC World Congress*, pages 1115–1120, Helsinki, 1978.

[46] R. W. Brockett. Control theory and singular Riemannian geometry. In *New Directions in Applied Mathematics*, pages 11–27. Springer-Verlag, New York, 1981.

[47] R. W. Brockett. Asymptotic stability and feedback stabilization. In R. W. Brockett, R. S. Millman, and H. J. Sussman, editors, *Differential Geometric Control Theory*, pages 181–191. Birkhäuser, 1983.

[48] R. W. Brockett and L. Dai. Non-holonomic kinematics and the role of elliptic functions in constructive controllability. In Z. Li and J. F. Canny, editors, *Nonholonomic Motion Planning*, pages 1–22. Kluwer, 1993.

[49] R. L. Bryant, S. S. Chern, R. B. Gardner, H. L. Goldschmidt, and P. A. Griffiths. *Exterior Differential Systems*. Springer-Verlag, 1991.

[50] A. Bryson and Y. C. Ho. *Applied Optimal Control*. Blaisdell, Waltham, MA, 1969.

[51] F. Bullo, R. Murray, and A. Sarti. Control on the sphere and reduced attitude stabilization. In *Proceedings IFAC, NOLCOS, Tahoe City*, pages 495–501, 1995.

[52] L. Bushnell, D. Tilbury, and S. S. Sastry. Extended Goursat normal forms with applications to nonholonomic motion planning. In *Proceedings of the IEEE Conference on Decision and Control*, pages 3447–3452, San Antonio, Texas, 1993.

[53] L. Bushnell, D. Tilbury, and S. S. Sastry. Steering three-input chained form nonholonomic systems using sinusoids: The fire truck example. *International Journal of Robotics Research*, 14(4):366–381, 1995.

[54] C. I. Byrnes and A. Isidori. Asymptotic stabilization of minimum phase nonlinear systems. *IEEE Transactions on Automatic Control*, 36:1122–1137, 1991.

[55] C. I. Byrnes, A. Isidori, and J. Willems. Passivity, feedback equivalence, and the global stabilization of minimum phase nonlinear systems. *IEEE Transactions on Automatic Control*, 36:1228–1240, 1991.

[56] C. I. Byrnes and Alberto Isidori. A frequency domain philosophy for nonlinear systems, with applications to stabilization and to adaptive control. In *Proceedings of the 23rd Conference on Decision and Control*, pages 1569–1573, Las Vegas, Nevada, 1984.

[57] C. I. Byrnes, F. Delli Priscoli, and A. Isidori. *Output Regulation of Uncertain Nonlinear Systems*. Birkhäuser, Boston, 1997.

[58] M. Camarinha, F. Silva-Leite, and P. Crouch. Splines of class C_k in non-Euclidean spaces. *IMA Journal of Mathematical Control and Information*, 12:399–410, 1995.

[59] B. Charlet, J. Lévine, and R. Marino. Sufficient conditions for dynamic state feedback linearization. *SIAM Journal of Control and Optimization*, 29(1):38–57, 1991.

[60] R. Chatila. Mobile robot navigation: Space modeling and decisional processes. In O. Faugeras and G. Giralt, editors, *Robotics Research: The Third International Symposium*, pages 373–378. MIT Press, 1986.

[61] V. H. L. Cheng. A direct way to stabilize continuous time and discrete time linear time varying systems. *IEEE Transactions on Automatic Control*, 24:641–643, 1979.

[62] H-D. Chiang, F. F. Wu, and P. P. Varaiya. Foundations of the potential energy boundary surface method for power system transient stability analysis. *IEEE Transactions on Circuits and Systems*, CAS-35:712–728, 1988.

[63] G. Chirikjian and J. Burdick. The kinematics of hyperedundant robot locomotion. *IEEE Transactions on Robotics and Automation*, 11(6):781–793, 1995.

[64] D. Cho and J. K. Hedrick. Automotive powertrain modeling for control. *ASME Transactions on Dynamics, Measurement and Control*, 111, 1989.

[65] L. O. Chua and P. M. Lin. *Computer Aided Analysis of Electronic Circuits: Algorithms and Computational Techniques*. Prentice Hall, Englewood Cliffs, NJ, 1975.

[66] L. O. Chua, T. Matsumoto, and S. Ichiraku. Geometric properties of resistive nonlinear n ports: transversality, structural stability, reciprocity and antireciprocity. *IEEE Transactions on Circuits and Systems*, 27:577–603, 1980.

[67] D. Claude, M. Fliess, and A. Isidori. Immersion, directe et par bouclage, d'un système nonlinéaire. *C. R. Academie Science Paris*, 296:237–240, 1983.

[68] E. A. Coddington and N. Levinson. *Theory of Ordinary Differential Equations*. McGraw Hill, New York, 1955.

[69] P. A. Cook. *Nonlinear Dynamical Systems*. Prentice Hall, 1989.

[70] M. Corless and G. Leitmann. Continuous state feedback guarantees uniform ultimate boundedness for uncertain dynamical systems. *IEEE Transactions on Automatic Control*, AC-26:1139–1144, 1981.

[71] J.-M. Coron. Linearized control systems and applications to smooth stabilization. *SIAM Journal of Control and Optimization*, 32:358–386, 1994.

[72] L. Crawford and S. Sastry. Biological motor control approaches for a planar diver. In *Proceedings of the IEEE Conference on Decision and Control*, pages 3881–3886, 1995.

[73] P. Crouch and F. Silva-Leite. The dynamic interpolation problem on Riemannian manifolds, Lie groups, and symmetric spaces. *Journal of Dynamical and Control Systems*, 1:177–202, 1995.

[74] P. E. Crouch, F. Lamnabhi, and A. van der Schaft. Adjoint and Hamiltonian formulations input output differential equations. *IEEE Transactions on Automatic Control*, AC-40:603–615, 1995.

[75] M.L. Curtis. *Matrix Groups*. Springer-Verlag, New York, 1979.

[76] M. Dahleh, A. Pierce, H. Rabitz, and V. Ramkarishnan. Control of molecular motion. *Proceedings of the IEEE*, 84:7–15, 1996.

[77] B. D'Andrea-Novel and J.-M. Coron. Stabilization of a rotating body beam without damping. *IEEE Transactions on Automatic Control*, AC-43:608–618, 1998.

[78] J. Borges de Sousa and A. Gollü. A simulation environment for the coordinated operation of autonomous underwater vehicles. In *Proceedings of the Winter Simulation Conference*, pages 1169–1175, 1997.

[79] C. Canudas de Wit, B. Siciliano, and G. Bastin (editors). *Theory of Robot Control*. Springer Verlag, New York, 1998.

[80] G. Debreu. *Theory of Value: An Axiomatic Analysis of Economic Equilibrium*. Yale University Press, New Haven, 1959.

[81] J. Descusse and C. H. Moog. Decoupling with dynamic compensation for strong invertible affine non-linear systems. *International Journal of Control*, 42:1387–1398, 1985.

[82] C. A. Desoer and H. Haneda. The measure of a matrix as a tool to analyze algorithms for circuit analysis. *IEEE Transactions on Circuit Theory*, CT-13:480–486, 1972.

[83] C. A. Desoer and M. Vidyasagar. *Feedback Systems: Input Output Properties*. Academic Press, 1975.

[84] R. L. Devaney. *An Introduction to Chaotic Dynamical Systems*. Benjamin/Cummings, Menlo Park, CA, 1986.

[85] S. Devasia, D. Chen, and B. Paden. Nonlinear inversion based output tracking. *IEEE Transactions on Automatic Control*, 41:930–942, 1996.

[86] S. Devasia and B. Paden. Exact output tracking for nonlinear time-varying systems. In *Proceedings of the IEEE Conference on Decision and Control*, pages 2346–2355, 1994.

[87] M. D. Di Benedetto and J. W. Grizzle. Intrinsic notions of regularity for local inversion, output nulling and dynamic extension of nonsquare systems. *Control Theory and Advanced Technology*, 6:357–381, 1990.

[88] M. D. Di Benedetto and S. Sastry. Adaptive linearization and model reference control of a class of mimo nonlinear systems. *Journal of Mathematical Systems, Estimation and Control*, 3(1):73–105, 1993.

[89] M.D. Di Benedetto. Nonlinear strong model matching. *IEEE Transactions on Automatic Control*, 35:1351–1355, 1990.

[90] M.D. Di Benedetto, J. W. Grizzle, and C. H. Moog. Rank invariants for nonlinear systems. *SIAM Journal of Control and Optimization*, 27:658–672, 1989.

[91] J. Dieudonné. *Foundations of Modern Analysis, Volume I*. Academic Press, New York, 1969.

[92] A. W. Divelbiss and J.-T. Wen. Nonholonomic motion planning in the presence of obstacles. *IEEE Transactions on Robotics and Automation*, 13(3):443–451, 1997.

[93] O. Egeland and J.-M. Godhavn. Passivity based attitude control of a rigid spacecraft. *IEEE Transactions on Automatic Control*, 39:842–846, 1994.

[94] O. Egeland and K. Pettersen. Free-floating robotic systems. In B. Siciliano and K. Valavanis, editors, *Control Problems in Robotics and Automation*, pages 119–134. Springer-Verlag, 1998.

[95] M. Evans. Bilinear systems with homogeneous input–output maps. *IEEE Transactions on Automatic Control*, AC-28:113–115, 1983.

[96] M. J. Feigenbaum. Qualitative universality for a class of nonlinear transformations. *Journal of Statistical Physics*, 19:25–52, 1978.

[97] N. Fenichel. Geometric singular perturbation theory for ordinary differential equations. *Journal of Differential Equations*, 31:53–98, 1979.

[98] C. Fernandes, L. Gurvits, and Z. Li. Optimal nonholonomic motion planning for a falling cat. In Z. Li and J. Canny, editors, *Nonholonomic Motion Planning*, pages 379–421. Kluwer Academic Publishers, 1993.

[99] C. Fernandes, L. Gurvits, and Z. Li. Near optimal nonholonomic motion planning for a system of coupled rigid bodies. *IEEE Transactions on Automatic Control*, March 1994.

[100] B. Fernandez and K. Hedrick. Control of multivariable non-linear systems by the sliding mode method. *International Journal of Control*, 46:1019–1040, 1987.

[101] A. Feuer and A. S. Morse. Adaptive control for single input single output linear systems. *IEEE Transactions on Automatic Control*, AC-23:557–570, 1978.

[102] A. F. Filippov. Differential equations with discontinuous right hand sides. *Translations of the American Mathematical Society*, 62, 1960.

[103] M. Fliess. Series de Volterra et series formelles non commutatives. *C. R. Academie Science, Paris, Serie A*, 280:965–967, 1975.

[104] M. Fliess, J. Levine, P. Martin, and P. Rouchon. Flatness and defect of nonlinear systems. *International Journal of Control*, 61:1327–1361, 1995.

[105] B. A. Francis. The linear multivariable regulator problem. *SIAM Journal on Control and Optimization*, 14:486–505, 1977.

[106] R. Freeman and P. Kokotović. *Robust Control of Nonlinear Systems*. Birkhäuser, Boston, 1996.

[107] R. Frezza, S. Soatto, and P. Perona. Motion estimation via dynamic vision. *IEEE Transactions on Automatic Control*, AC-41:393–413, 1996.

[108] R. B. Gardner and W. F. Shadwick. The GS algorithm for exact linearization to Brunovsky normal form. *IEEE Transactions on Automatic Control*, 37(2):224–230, 1992.

[109] A. Gelb and W. van der Velde. *Multiple Input Describing Functions and Nonlinear Systems Design*. Mc Graw Hill, New York, 1968.

[110] P. Di Giamberardino, F. Grassini, S. Monaco, and D. Normand-Cyrot. Piecewise continuous control for a car-like robot: implementation and experimental results. In *Proceedings of the IEEE Conference on Decision and Control*, pages 3564–3569, 1996.

[111] P. Di Giamberardino, S. Monaco, and D. Normand-Cyrot. Digital control through finite feedback discretizability. In *IEEE Conference on Robotics and Automation*, pages 3141–3146, 1996.

[112] E. C. Gilbert. Functional expansions for the response of nonlinear differential systems. *IEEE Transactions on Automatic Control*, AC-22:909–921, 1977.

[113] G. Giralt, R. Chatila, and M. Vaisset. An integrated navigation and motion control system for autonomous multisensory mobile robots. In M. Brady and R. Paul, editors, *Robotics Research: The First International Symposium*, pages 191–214. MIT Press, Cambridge, Massachusetts, 1984.

[114] D. N. Godbole and S. S. Sastry. Approximate decoupling and asymptotic tracking for MIMO systems. *IEEE Trans. Automatic Control*, AC-40:441–450, 1995.

[115] J-M. Godhavn, A. Balluchi, L. Crawford, and S. Sastry. Control of nonholonomic systems with drift terms. *Automatica*, 35(5):(to appear), 1999.

[116] G. C. Goodwin and D. Q. Mayne. A parameter estimation perspective to continuous time model reference adaptive control. *Automatica*, 23:57–70, 1987.

[117] G. C. Goodwin and K. S. Sin. *Adaptive Filtering, Prediction and Control*. Prentice Hall, Englewood Cliffs, NJ, 1984.

[118] M. Grayson and R. Grossman. Models for free nilpotent Lie algebras. *Journal of Algebra*, 135(1):177–191, 1990.

[119] P. Griffiths. *Exterior differential systems and the calculus of variations*. Birkhäuser, 1982.

[120] J. Grizzle, M. Di Benedetto, and F. Lamnabhi Lagarrique. Necessary conditions for asymptotic tracking in nonlinear systems. *IEEE Transactions on Automatic Control*, 39:1782–1794, 1994.

[121] J. W. Grizzle. A linear algebraic framework for the analysis of discrete time nonlinear systems. *SIAM Journal of Control and Optimization*, 31:1026–1044, 1993.

[122] J. Guckenheimer and P. J. Holmes. *Nonlinear Oscillations, Dynamical Systems and Bifurcations of Vector Fields*. Springer Verlag, Berlin, 1983.

[123] J. Guckenheimer and P. Worfolk. Instant chaos. *Nonlinearity*, 5:1211–1221, 1992.

[124] W. Hahn. *Stability of Motion*. Springer-Verlag, Berlin, 1967.

[125] J. K. Hale. *Ordinary Differential Equations*. Krieger, Huntington, New York, 1980.

[126] J. K. Hale and H. Koçak. *Dynamics and Bifurcations*. Springer Verlag, 1991.

[127] M. Hall. *The Theory of Groups*. Macmillan, 1959.

[128] P. Hartmann. *Ordinary Differential Equations*. Hartmann, Baltimore, 1973.

[129] J. Hauser and R. Hindman. Aggressive flight maneuvers. In *Proceedings of the IEEE Conference on Decision and Control*, pages 4186–4191, 1997.

[130] J. Hauser, S. Sastry, and P. Kokotović. Nonlinear control via approximate input-output linearization: the ball and beam example. *IEEE Transactions on Automatic Control*, 37:392–398, 1992.

[131] J. Hauser, S. Sastry, and G. Meyer. Nonlinear control design for slightly non-minimum phase systems: Application to V/STOL aircraft. *Automatica,*, 28:665–679, 1992.

[132] U. Helmcke and J. Moore. *Optimization and Dynamical Systems*. Springer Verlag, New York, 1994.

[133] M. Henson and D. Seborg. Adaptive nonlinear control of a pH process. *IEEE Transactions on Control Systems Technology*, 2:169–182, 1994.

[134] M. Henson and D. Seborg. *Nonlinear Process Control*. Prentice Hall, Upper Saddle River, NJ, 1997.

[135] T. Henzinger and S. Sastry (editors). *Hybrid Systems: Computation and Control*. Springer Verlag Lecture Notes in Computer Science, Vol. 1386, 1998.

[136] R. Hermann and A. J. Krener. Nonlinear controllability and observability. *IEEE Transactions on Automatic Control*, AC-22:728–740, 1977.

[137] H. Hermes, A. Lundell, and D. Sullivan. Nilpotent bases for distributions and control systems. *Journal of Differential Equations*, 55:385–400, 1984.

[138] J. Hespanha and A. S. Morse. Certainty equivalence implies detectability. *Systems and Control Letters*, 36:1–13, 1999.

[139] D. J. Hill and P. J. Moylan. Connections between finite gain and asymptotic stability. *IEEE Transactions on Automatic Control*, AC-25:931–936, 1980.

[140] M. Hirsch, C. Pugh, and H. Shub. *Invariant Manifolds*. Springer Lecture Notes in Mathematics, Vol. 583, New York, 1977.

[141] M. W. Hirsch and S. Smale. *Differential equations, Dynamical Systems and Linear Algebra*. Academic Press, New York, 1974.

[142] R. Hirschorn and E. Aranda Bricaire. Global approximate output tracking for nonlinear systems. *IEEE Transactions on Automatic Control*, 43:1389–1398, 1998.

[143] J. J. Hopfield. Neurons, dynamics and computation. *Physics Today*, 47:40–46, 1994.

[144] F. C. Hoppensteadt. Singular perturbations on the infinite time interval. *Transactions of the American Mathematical Society*, 123:521–535, 1966.

[145] H. J. C. Huibert and J. H. van Schuppen. Routing control of a motorway network–a summary. In *Proceedings of the IFAC Nonlinear Control Systems Design Symposium (NOLCOS)*, pages 781–784, Rome, Italy, 1995.

[146] T. W. Hungerford. *Algebra*. Springer-Verlag, 1974.

[147] L. R. Hunt, V. Ramakrishna, and G. Meyer. Stable inversion and parameter variations. *Systems and Control Letters*, 34:203–209, 1998.

[148] L. R. Hunt, R. Su, and G. Meyer. Global transformations of nonlinear systems. *IEEE Transactions on Automatic Control*, AC–28:24–31, 1983.

[149] A. Isidori. *Nonlinear Control Systems*. Springer-Verlag, Berlin, 1989.

[150] A. Isidori, A. J. Krener, C. Gori-Giorgi, and S. Monaco. Nonlinear decoupling via feedback: A differential geometric approach. *IEEE Transactions on Automatic Control*, AC–26:331–345, 1981.

[151] A. Isidori and A. Ruberti. Realization theory of bilinear systems. In D. Q. Mayne and R. W. Brockett, editors, *Geometric Methods in System Theory*, pages 81–130. D. Reidel, Dordrecht, Holland, 1973.

[152] A. Isidori, S. Sastry, P. Kokotović, and C. Byrnes. Singularly perturbed zero dynamics of nonlinear systems. *IEEE Transactions on Automatic Control*, pages 1625–1631, 1992.

[153] N. Jacobsen. *Basic Algebra I*. Freeman, San Francisco, 1974.

[154] B. Jakubczyk and W. Respondek. On linearization of control systems. *Bulletin de L'Academie Polonaise des Sciences, Série des sciences mathématiques*, XXVIII:517–522, 1980.

[155] J.Carr. *Applications of Center Manifold Theory*. Springer Verlag, New York, 1981.

[156] M. Johansson and A. Rantzer. Computation of piecewise quadratic lyapunov functions for hybrid systems. *IEEE Transactions on Automatic Control*, AC-43:555–559, 1998.

[157] V. Jurdjević. *Geometric Control Theory*. Cambridge University Press, Cambridge, UK, 1997.

[158] R. Kadiyala, A. Teel, P. Kokotović, and S. Sastry. Indirect techniques for adaptive input output linearization of nonlinear systems. *International Journal of Control*, 53:193–222, 1991.

[159] W. S. Kang and J. E. Marsden. The Lagrangian and Hamiltonian approach to the dynamics of nonholonomc systems. *Reports on Mathematical Physics*, 40:21–62, 1997.

[160] A. Kapitanovsky, A. Goldenberg, and J. K. Mills. Dynamic control and motion planning for a class of nonlinear systems with drift. *Systems and Control Letters*, 21:363–369, 1993.

[161] M. Kawski. Nilpotent Lie algebras of vector fields. *Journal für die reine und angewandte Mathematik*, 388:1–17, 1988.

[162] H. K. Khalil. *Nonlinear Systems*. Macmillan, New York, 1992.

[163] P. Kokotović, H. Khalil, and J. O'Reilly. *Singular perturbation methods in control: analysis and design*. Academic Press, Orlando, Florida, 1986.

[164] I. Kolmanovsky and N. H. McClamroch. Developments in nonholonomic control problems. *IEEE Control Systems Magazine*, 15:20–36, 1996.

[165] T. J. Koo and S. S. Sastry. Output tracking control design of a helicopter model based on approximate linearization. In *Proceedings of the 37th IEEE Conference on Decision and Control, Tampa, Florida*, December 1998.

[166] S. R. Kou, D. L. Elliott, and T. J. Tarn. Observability of nonlinear systems. *Information and Control*, 22:89–99, 1973.

[167] J.-P. Kramer and K. D. Mease. Near optimal control of altitude and path angle during aerospace plane ascent. *Journal of Guidance, Control, and Dynamics*, 21(5):726–732, 1998.

[168] G. Kreisselmeier. Adaptive observers with an exponential rate of convergence. *IEEE Transactions on Automatic Control*, 22:2–8, 1977.

[169] A. J. Krener. Approximate linearization by state feedback and coordinate change. *Systems and Control Letters*, 5:181–185, 1984.

[170] A. J. Krener. Feedback linearization. In J. Baillieul and J. Willems, editors, *Mathematical Control Theory*. Springer-Verlag, New York, 1998.

[171] M. Krstić, I. Kanellakopoulos, and P. Kokotović. *Nonlinear and Adaptive Control Systems Design*. John Wiley, New York, 1995.

[172] P. Kudva and K. S. Narendra. Synthesis of an adaptive observer using Lyapunov's direct method. *International Journal of Control*, 18:1201–1210, 1973.

[173] G. Lafferriere and H. J. Sussmann. A differential geometric approach to motion planning. In Z. Li and J. F. Canny, editors, *Nonholonomic Motion Planning*, pages 235–270. Kluwer, 1993.

[174] J-P. Laumond. Controllability of a multibody mobile robot. *IEEE Transactions on Robotics and Automation*, 9(6):755–763, 1993.

[175] D. F. Lawden. *Elliptic Functions and Applications*. Springer-Verlag, 1980.

[176] H. G. Lee, A. Arapostathis, and S. Marcus. Linearization of discrete time nonlinear systems. *International Journal of Control*, 45:1803–1822, 1986.

[177] E. Lelarasmee, A. E. Ruehli, and A. L. Sangiovanni-Vincentelli. The waveform relaxation approach to time domain analysis of large scale integrated circuits. *IEEE Transactions on Computer Aided Design*, CAD-1(3):131–145, 1982.

[178] N. Leonard and P. S. Krishnaprasad. Motion control of drift-free, left invariant systems on Lie groups. *IEEE Transactions on Automatic Control*, AC-40:1539–1554, 1995.

[179] J. J. Levin and N. Levinson. Singular perturbations of nonlinear systems of ordinary differential equations and an associated boundary layer equation. *Journal of Rational Mechanics and Analysis*, 3:267–270, 1959.

[180] P. Y. Li and R. Horowitz. Control of smart exercise machines. I. problem formulation and nonadaptive control. *IEEE/ASME Transactions on Mechatronics*, 2:237–247, 1997.

[181] P. Y. Li and R. Horowitz. Control of smart exercise machines. II. self optimizing control. *IEEE/ASME Transactions on Mechatronics*, 2:248–258, 1997.

[182] T. Li and J. Yorke. Period three implies chaos. *American Mathematics Monthly*, pages 985–992, 1975.

[183] Z. Li and J. Canny. Motion of two rigid bodies with rolling constraint. *IEEE Transactions on Robotics and Automation*, 6(1):62–71, 1990.

[184] Z. Li and J. Canny, editors. *Nonholonomic Motion Planning*. Kluwer Academic Publishers, 1993.

[185] D.-C. Liaw and E.-H. Abed. Active control of compressor stall inception: A bifurcation theoretic approach. *Automatica*,, 32:109–115, 1996.

[186] L. Ljung and S. T. Glad. On global identifiability for arbitrary model parameterizations. *Automatica*, 30:265–276, 1994.

[187] E. N. Lorenz. The local structure of a chaotic attractor in four dimensions. *Physica D*, 13:90–104, 1984.

[188] A. Loria, E. Panteley, and H. Nijmeijer. Control of the chaotic Duffing equation with uncertainty in all parameters. *IEEE Transactions on Circuits and Systems*, CAS-45:1252–1255, 1998.

[189] R. Lozano, B. Brogliato, and I. Landau. Passivity and global stabilization of cascaded nonlinear systems. *IEEE Transactions on Automatic Control*, 37:1386–88, 1992.

[190] A. De Luca. Trajectory control of flexible manipulators. In B. Siciliano and K. Valavanis, editors, *Control Problems in Robotics and Automation*, pages 83–104. Springer-Verlag, 1998.

[191] A. De Luca, R. Mattone, and G. Oriolo. Control of redundant robots under end-effector commands: a case study in underactuated systems. *Applied Mathematics and Computer Sciencen*, 7:225–251, 1997.

[192] A. De Luca and G. Ulivi. Full linearization of induction motors via nonlinear state feedback. In *Proceedings of the 26th Conference on Decisions and Control*, pages 1765–1770, December 1987.

[193] A. M. Lyapunov. The general problem of the stability of motion. *International Journal of Control*, 55:531–773, 1992. (this is the English translation of the Russian original published in 1892).

[194] J. Lygeros, D. Godbole, and S. Sastry. A verified hybrid controller for automated vehicles. *IEEE Transactions on Automatic Control*, AC-43:522–539, 1998.

[195] J. Lygeros, C. Tomlin, and S. Sastry. Controllers for reachability specifications for hybrid systems. *Automatica*, 35:349–370, 1999.

[196] Y. Ma, J. Košecká, and S. Sastry. A mathematical theory of camera calibration. Electronics Research Laboratory, Memo M 98/81, University of California, Berkeley, 1998.

[197] Y. Ma, J. Košecká, and S. Sastry. Motion recovery from image sequences: Discrete/differential view point. Electronics Research Laboratory, Memo M 98/11, University of California, Berkeley, 1998.

[198] J. Malmborg and B. Bernhardsson. Control and simulation of hybrid systems. In *Proceedings of the Second World Congress of Nonlinear Analysts, July 1996*, pages 57–64, 1997.

[199] R. Marino and P. Tomei. *Nonlinear Control Design*. Prentice Hall International, UK, 1995.

[200] J. E. Marsden. *Elementary Classical Analysis*. W. H. Freeman and Co., San Francisco, 1974.

[201] J. E. Marsden and M. McCracken. *The Hopf Bifurcation and its Applications*. Springer Verlag, New York, 1976.

[202] C. F. Martin and W. P. Dayawansa. On the existence of a Lyapunov function for a family of switching systems. In *Proceedings of the 35th Conference on Decisions and Control, Kobe, Japan*, pages 1820–1823, December 1996.

[203] R. May. *Stability and Complexity in Model Ecosystems*. Princeton University Press, Princeton, NJ, 1973,.

[204] McDonnell Douglas Corporation. *YAV-8B Simulation and Modeling*, December 1982.

[205] R. M'Closkey and R. M. Murray. Exponential stabilization of driftless nonlinear systems using homogeneous feedback. *IEEE Transactions on Automatic Control*, 42:614–628, 1997.

[206] A. I. Mees. *Dynamics of Feedback Systems*. J. Wiley & Sons, 1981.

[207] G. Meyer and L. Cicolani. Applications of nonlinear systems inverses to automatic flight control design—system concept and flight evaluations. In P. Kent, editor, *AGARDograph 251 on Theory and Applications of Optimal Control in Aerospace Systems*. NATO, 1980.

[208] A. N. Michel. Stability: the common thread in the evolution of feedback control. *IEEE Control Systems Magazine*, 16:50–60, 1996.

[209] A. N. Michel and R. K. Miller. *Qualitative Analysis of Large Scale Dynamical Systems*. Academic Press, New York, 1977.

[210] A. N. Michel and K. Wang. *Dynamical Systems with Saturation Nonlinearities: analysis and design*. Springer Verlag, New York, 1994.

[211] J. Milnor. *Topology from the Differentiable Viewpoint*. University of Virginia Press, Charlottesville, 1976,.

[212] E. F. Miscenko. Asymptotic calculation of periodic solutions of systems containing small parameters in the derivatives. *Translations of the American Mathematical Society, Series 2*, 18:199–230, 1961.

[213] S. Monaco and D. Normand-Cyrot. Invariant distributions for discrete time nonlinear systems. *Systems and Control Letters*, 5:191–196, 1985.

[214] S. Monaco and D. Normand-Cyrot. Functional expansions for nonlinear discrete time systems. *Journal of Mathematical Systems Theory*, 21:235–254, 1989.

[215] S. Monaco and D. Normand-Cyrot. An introduction to motion planning under multirate digital control. In *Proceedings of the IEEE Conference on Decision and Control*, pages 1780–1785, Tucson, Arizona, 1992.

[216] S. Monaco and D. Normand-Cyrot. On nonlinear digital control. In A. J. Fossard and D. Normand-Cyrot, editors, *Nonlinear Systems, Volume 3: Control*, pages 127–155. Chapman and Hall, 1997.

[217] R. Montgomery. Isoholonomic problems and some applications. *Communications in Mathematical Physics*, 128:565–592, 1990.

[218] R. Montgomery. Singular extremals on Lie groups. *Mathematics of Control, Signals, and Systems*, 7:217–234, 1994.

[219] A. S. Morse. Towards a unified theory of parameter adaptive control: Part II: Certainty equivalence and implicit tuning. *IEEE Transactions on Automatic Control*, AC-37:15–29, 1992.

[220] P. J. Moylan and D. J. Hill. Stability criteria for large scale systems. *IEEE Transactions on Automatic Control*, 23:143–149, 1978.

[221] J. R. Munkres. *Analysis on Manifolds*. Addison-Wesley, 1991.

[222] R. M. Murray. Applications and extensions of Goursat normal form to control of nonlinear systems. In *Proceedings of the IEEE Conference on Decision and Control*, pages 3425–3430, 1993.

[223] R. M. Murray. Nilpotent bases for a class of non-integrable distributions with applications to trajectory generation for nonholonomic systems. *Mathematics of Control, Signals, and Systems: MCSS*, 7(1):58–75, 1995.

[224] R. M. Murray. Nonlinear control of mechanical systems: a Lagrangian perspective. *Annual Reviews in Control*, 21:31–42, 1997.

[225] R. M. Murray, Z. Li, and S. S. Sastry. *A Mathematical Introduction to Robotic Manipulation*. CRC Press, 1994.

[226] R. M. Murray and S. S. Sastry. Nonholonomic motion planning: Steering using sinusoids. *IEEE Transactions on Automatic Control*, 38(5):700–716, 1993.

[227] P. Habets N. Rouche and M. Laloy. *Stability Theory by Lyapunov's direct method*. Springer Verlag, New York, 1977.

[228] K. S. Narendra and A. Annaswamy. *Stable Adaptive Systems*. Prentice Hall, Englewood Cliffs, NJ, 1989.

[229] V. V. Nemytskii and V. V. Stephanov. *Qualitative Theory of Differential Equations*. Dover, New York, 1989.

[230] H. Nijmeijer. Control of chaos and synchronization. *Systems and Control Letters*, 31:259–262, 1997.

[231] H. Nijmeijer and H. Berghuis. Lyapunov control in robotic systems: tracking and chaotic dynamics. *Applied Mathematics and Computer Science*, 5:373–389, 1995.

[232] H. Nijmeijer and A. van der Schaft. *Nonlinear Dynamical Control Systems*. Springer Verlag, New York, 1990.

[233] B. Noble and J. Daniel. *Applied Linear Algebra*. Prentice Hall, Englewood Cliffs, 1977.

[234] S. M. Oliva and C. N. Nett. A general nonlinear analysis of a second-order one-dimensional, theoretical compression system model. In *Proceedings of the American Control Conference, Atlanta, GA*, pages 3158–3165, June 1991.

[235] R. Ortega. Passivity properties for stabilization of cascaded nonlinear systems. *Automatica,*, 27:423–424, 1989.

[236] R. Ortega, A. Loria, R. Kelly, and L. Praly. On passivity-based output feedback global stabilization of Euler-Lagrange systems. *International Journal of Robust and Nonlinear Control*, 5:313–323, 1995.

[237] J. Ostrowski and J. Burdick. The geometric mechanics of undulatory robotic locomotion. *International Journal of Robotics Research*, 17(7):683–701, 1998.

[238] A. K. Packard. Gain scheduling via linear fractional transformations. *Systems and Control Letters*, 22(2):79–92, 1994.

[239] M. A. Pai. *Energy Function Analysis for Power System Stability*. Kluwer Academic Press, Boston, 1989.

[240] R. S. Palais. Natural operations on differential forms. *Transactions of the American Mathematical Society*, 92:125–141, 1959.

[241] G. Pappas, J. Lygeros, D. Tilbury, and S. Sastry. Exterior differential systems in robotics and control. In J. Baillieul, S. Sastry, and H. Sussmann, editors, *Essays in Mathematical Robotics, IMA Volumes, No. 104*. Springer-Verlag, New York, 1998.

[242] J-B. Pomet. A differential geometric setting for dynamic equivalence and dynamic linearization. Technical Report RR-2312, INRIA, Sophia Antipolis, July 1994.

[243] M. Postnikov. *Lectures in Geometry: Lie Groups and Lie Algebras*. MIR Publishers, 1986.

[244] T. Poston and I. Stewart. *Catastrophe Theory and Its Applications*. Pitman, London, 1978.

[245] Baron Rayleigh. *The Theory of Sound: Vols. I and II*. Dover (1945 edition), 1887.

[246] W. Rudin. *Real and Complex Analysis*. McGraw Hill, 1974.

[247] D. Ruelle. *Elements of Differentiable Dynamics and Bifurcation Theory*. Academic Press, 1989.

[248] W. J. Rugh. *Nonlinear System Theory: the Volterra / Wiener Approach*. Johns Hopkins University Press, Baltimore, 1981.

[249] M. G. Safonov. *Stability and Robustness of Multivariable Feedback Systems*. MIT Press, Cambridge, MA, 1980.

[250] F. M. Salam and S. Sastry. Dynamics of the forced Josephson junction circuit: Regions of chaos. *IEEE Transactions on Circuits and Systems*, CAS-32:784–796, 1985.

[251] F. M. A. Salam, J. E. Marsden, and P. P. Varaiya. Chaos and Arnold diffusions in dynamical systems. *IEEE Transactions on Circuits and Systems*, CAS-30:697–708, 1983.

[252] F. M. A. Salam, J. E. Marsden, and P. P. Varaiya. Arnold diffusion in the swing equation. *IEEE Transactions on Circuits and Systems*, CAS-31:673–688, 1984.

[253] C. Samson. Velocity and torque feedback control of a nonholonomic cart. In *International Workshop in Adaptive and Nonlinear Control: Issues in Robotics*, pages 125–151, 1990.

[254] C. Samson. Control of chained form systems: applications to path following and time-varying point-stabilization of mobile robots. *IEEE Transactions on Automatic Control*, 40(1):64–77, 1995.

[255] I. W. Sandberg. A frequency domain condition for stability of feedback systems containing a single time varying nonlinear element. *Bell Systems Technical Journal*, 43:1601–1608, 1964.

[256] I. W. Sandberg. On the l_2 boundedness of solutions of nonlinear functional equations. *Bell Systems Technical Journal*, 43:1581–1599, 1964.

[257] A. Sarti, G. Walsh, and S. S. Sastry. Steering left-invariant control systems on matrix Lie groups. In *Proceedings of the 1993 IEEE Conference on Decision and Control*, December 1993.

[258] A. V. Sarychev and H. Nijmeijer. Extremal controls for chained systesm. *Journal of Dynamical and Control Systems*, 2:503–527, 1996.

[259] S. Sastry and M. Bodson. *Adaptive Systems: Stability, Convergence and Robustness*. Prentice Hall, Englewood Cliffs, New Jersey, 1989.

[260] S. Sastry and C. A. Desoer. The robustness of controllability and observability of linear time varying systems. *IEEE Transactions on Automatic Control*, AC-27:933–939, 1982.

[261] S. Sastry, J. Hauser, and P. Kokotović. Zero dynamics of regularly perturbed systems may be singularly perturbed. *Systems and Control Letters*, 14:299–314, 1989.

[262] S. S. Sastry. Effects of small noise on implicitly defined nonlinear systems. *IEEE Transactions on Circuits and Systems*, 30:651–663, 1983.

[263] S. S. Sastry and C. A. Desoer. Jump behavior of circuits and systems. *IEEE Transactions on Circuits and Systems*, 28:1109–1124, 1981.

[264] S. S. Sastry and R. Montgomery. The structure of optimal controls for a steering problem. In *IFAC Workshop on Nonlinear Control*, pages 385–390, 1992.

[265] S. S. Sastry and P. P. Varaiya. Hierarchical stability and alert state steering control of interconnected power systems. *IEEE Transactions on Circuits and Systems*, CAS-27:1102–1112, 1980.

[266] S.S. Sastry and A. Isidori. Adaptive control of linearizable systems. *IEEE Transactions on Automatic Control*, 34:1123 – 1131, 1989.

[267] M. Schetzen. *The Volterra and Wiener Theories of Nonlinear Systems*. John Wiley, New York, 1980.

[268] L. Sciavicco and B. Siciliano. *Modeling and Control of Robot Manipulators*. Mc Graw Hill, New York, 1998.

[269] R. Sepulchre, M. Janković, and P. Kokotović. *Constructive Nonlinear Control*. Springer Verlag, London, 1997.

[270] J-P. Serre. *Lie Algebras and Lie groups*. W. A. Benjamin, New York, 1965.

[271] P. R. Sethna. Method of averaging for systems bounded for positive time. *Journal of Mathematical Analysis and Applications*, 41:621–631, 1973.

[272] D. Seto and J. Baillieul. Control problems in superarticulated mechanical systems. *IEEE Transactions on Automatic Control*, 39:2442–2453, 1994.

[273] W. F. Shadwick and W. M. Sluis. Dynamic feedback for classical geometries. Technical Report FI93-CT23, The Fields Institute, Ontario, Canada, 1993.

[274] M. A. Shubov, C. F. Martin, J. P. Dauer, and B. P. Belinsky. Exact controllability of the damped wave equation. *SIAM Journal on Control and Optimization*, 35:1773–1789, 1997.

[275] L. E. Sigler. *Algebra*. Springer Verlag, New York, 1976.

[276] W. M. Sluis. *Absolute Equivalence and its Applications to Control Theory*. PhD thesis, University of Waterloo, 1992.

[277] S. Smale. Differential dynamical systems. *Bulletin of the American Mathematical Society*, 73:747–817, 1967.

[278] D. R. Smart. *Fixed Point Theorems*. Cambridge University Press, 1973.

[279] E. Sontag. *Mathematical Control theory: Deterministic Finite Dimensional Systems*. Springer Verlag, New York, 1990.

[280] O. J. Sørdalen. Conversion of the kinematics of a car with N trailers into a chained form. In *Proceedings of the IEEE International Conference on Robotics and Automation*, pages 382–387, Atlanta, Georgia, 1993.

[281] M. Spivak. *Calculus on Manifolds*. Addison-Wesley, 1965.

[282] M. Spivak. *A Comprehensive Introduction to Differential Geometry*, volume One. Publish or Perish, Inc., Houston, second edition, 1979.

[283] M. Spong. The swingup control problem for the Acrobot. *IEEE Control Systems Society Magazine*, 15:49–55, 1995.

[284] M. Spong. Underactuated mechanical systems. In B. Siciliano and K. Valavanis, editors, *Control Problems in Robotics and Automation*, pages 135–149. Springer-Verlag, 1998.

[285] M. Spong and M. Vidyasagar. *Robot Dynamics and Control*. John Wiley and Sons, New York, 1989.

[286] J. J. Stoker. *Nonlinear Vibrations*. Wiley Interscience, 1950.

[287] G. Strang. *Linear Algebra and Its Applications*. Harcourt Brace Jovanovich, San Diego, 1988.

[288] R. S. Strichartz. The Campbell–Baker–Hausdorff–Dynkin formula and solutions of differential equations. *Journal of Functional Analysis*, 72(2):320–345, 1987.

[289] H. J. Sussmann. Lie brackets, real analyticity and geometric control. In Roger W. Brocket, Richard S. Millman, and Héctor J. Sussmann, editors, *Differential Geometric Control Theory*, pages 1–116. Birkhäuser, Boston, 1983.

[290] H. J. Sussmann. On a general theorem on local controllability. *SIAM Journal of Control and Optimization*, 25:158–194, 1987.

[291] H. J. Sussmann and V. Jurdjevic. Controllability of nonlinear systems. *Journal of Differential Equations*, 12:95–116, 1972.

[292] D. Swaroop, J. Gerdes, P. Yip, and K. Hedrick. Dynamic surface control of nonlinear systems. In *Proceedings of the American Control Conference, Albuquerque, NM*, pages 3028–3034, June 1997.

[293] F. Takens. Constrained equations, a study of implicit differential equations and their discontinuous solutions. In A. Manning, editor, *Dynamical Systems, Warwick 1974*, pages 143–233. Springer Lecture Notes in Mathematics, Volume 468, 1974.

[294] T. J. Tarn, J. W. Clark, and G. M. Huang. Controllability of quantum mechanical systems. In A. Blaquiere, editor, *Modeling and Control of Systems*, pages 840–855. Springer Verlag, 1995.

[295] T. J. Tarn, G. Huang, and J. W. Clark. Modeling of quantum mechanical control systems. *Mathematical Modeling*, 1:109–121, 1980.

[296] O. Taussky. A remark on a theorem of Lyapunov. *Journal of Mathematical Analysis and Applications*, 2:105–107, 1961.

[297] A. Teel and L. Praly. Tools for semiglobal stabilization by partial state and output feedback. *SIAM Journal of Control and Optimization*, 33:1443–1488, 1995.

[298] A. R. Teel. Towards larger domains of attraction for local nonlinear control schemes. In *Proceedings of the 1st European Control Conference*, pages 638–642, July 1991.

[299] A. R. Teel. Semi-global stabilization of the ball and beam systems using output feedback. In *Proceedings of the American Control Conference, San Francisco, CA*, pages 2577–2581, June 1993.

[300] A. Tesi, E-H. Abed, R. Genesio, and H. O. Wang. Harmonic balance analysis of period doubling bifurcations and its implications for the control of nonlinear systems. *Automatica,*, 32:1255–1271, 1996.

[301] R. Thom. *Structural Stability and Morphogenesis*. W. A. Benjamin, New York, 1973. see especially Chapter 5, pp. 60–90.

[302] D. Tilbury, R. Murray, and S. Sastry. Trajectory generation for the N-trailer problem using Goursat normal form. *IEEE Transactions on Automatic Control*, 40(5):802–819, 1995.

[303] D. Tilbury and S. Sastry. The multi-steering n-trailer system: A case study of Goursat normal forms and prolongations. *International Jounal of Robust and Nonlinear Control*, 5(4):343–364, 1995.

[304] D. Tilbury, O. Sørdalen, L. Bushnell, and S. Sastry. A multi-steering trailer system: Conversion into chained form using dynamic feedback. *IEEE Transactions on Robotics and Automation*, 11(6):807–818, 1995.

[305] C. Tomlin, J. Lygeros, L. Benvenuti, and S. Sastry. Output tracking for a nonminimum phase CTOL aircraft model. In *Proceedings of the IEEE Conference on Decision and Control*, pages 1867–1872, 1995.

[306] C. J. Tomlin, G. Pappas, and S. Sastry. Conflict resolution for air traffic management: A case study in multi-agent hybrid systems. *IEEE Transactions on Automatic Control*, AC-43:509–521, 1998.

[307] C. J. Tomlin and S. S. Sastry. Bounded tracking for non-minimum phase systems with fast zero dynamics. *International Journal of Control*, 68:819–847, 1998.

[308] C. J. Tomlin and S. S. Sastry. Switching through singularities. *Systems and Control Letters*, 35:145–154, 1998.

[309] A. van der Schaft and B. M. Maschke. On the Hamiltonian formulation of nonholonomic mechanical systems. *Reports on Mathematical Physics*, 34:225–233, 1994.

[310] T. van Duzer and C. W. Turner. *Principles of Superconductive Devices and Circuits*. Elsevier, New York, 1983.

[311] M. van Nieuwstadt, M. Rathinam, and R. M. Murray. Differential flatness and absolute equivalence of nonlinear control systems. *SIAM Journal of Control*, 36(4):1225–1239, 1998.

[312] V. S. Varadarajan. *Lie Groups, Lie Algebras, and Their Representations*. Springer-Verlag, 1984.

[313] P. P. Varaiya. Smart cars on smart roads: problems of control. *IEEE Transactions on Automatic Control*, AC-38:195–207, 1993.

[314] P. P. Varaiya, F. F. Wu, and R. L. Chen. Direct methods for transient stability analysis of power systems: Recent results. *Proceedings of the IEEE*, 73:1703–1715, 1985.

[315] H. L. Varian. *Microeconomic Analysis*. Norton, 1978.

[316] M. Vidyasagar. *Input-Output Analysis of Large Scale Interconnected Systems*. Springer Verlag, New York, 1981.

[317] M. Vidyasagar. *Nonlinear Systems Analysis*. Second Edition, Prentice-Hall Electrical Engineering Series. Prentice-Hall, Inc., Englewood Cliffs, N.J., 1993.

[318] M. Vidyasagar and A. Vannelli. New relationships between input-output and Lyapunov stability. *IEEE Transactions on Automatic Control*, AC-27:481–483, 1982.

[319] V. Volterra. *Theory of Functionals and of Integral and Integro-Differential Equations*. Dover, New York, 1958. (this is the English translation of a Spanish original published in 1927).

[320] G. Walsh, A. Sarti, and S. S. Sastry. Algorithms for steering on the group of rotations. In *Proceedings of the 1993 IEEE Automatic Control Conference*, March 1993.

[321] G. Walsh and S. S. Sastry. On reorienting rigid linked bodies using internal motions. *IEEE Transactions on Robotics and Automation*, 11:139–146, 1995.

[322] G. Walsh, D. Tilbury, S. Sastry, R. Murray, and J.-P. Laumond. Stabilization of trajectories for systems with nonholonomic constraints. *IEEE Transactions on Automatic Control*, 39(1):216–222, 1994.

[323] G. C. Walsh and L. G. Bushnell. Stabilization of multiple chained form control systems. *Systems and Control Letters*, 25:227–234, 1995.

[324] H. O. Wang and E.-H. Abed. Bifurcation control of a chaotic system. *Automatica,*, 31:1213–1226, 1995.

[325] F. Warner. *Foundations of Differential Manifolds and Lie groups*. Scott, Foresman, Hill, 1979.

[326] J. Wen and K. Kreutz-Delgado. The attitude control problem. *IEEE Transactions on Automatic Control*, 36:1148–1162, 1991.

[327] N. Wiener. Generalized harmonic analysis. *Acta Mathematica*, 55:117–258, 1930.

[328] S. Wiggins. *Global Bifurcations and Chaos*. Springer Verlag, Berlin, 1988.

[329] S. Wiggins. *Introduction to Applied Nonlinear Dynamical Systems and Chaos*. Springer Verlag, Berlin, 1990.

[330] J. C. Willems. *The Analysis of Feedback Systems*. MIT Press, Cambridge, MA, 1972.

[331] W. M. Wonham. *Linear Multivariable Control: A Geometric Approach*. Springer Verlag, 1979.

[332] C. W. Wu and L. O. Chua. On the generality of the unfolded Chua circuit. *International Journal of Bifurcation Theory and Chaos*, 6:801–832, 1996.

[333] F. F. Wu and C. A. Desoer. Global inverse function theorem. *IEEE Transactions on Circuit Theory*, CT-72:199–201, 1972.

[334] L. C. Young. *Lectures on the Calculus of Variations and Optimal Control Theory*. Chelsea, New York, second edition, 1980.

[335] G. Zames. On the input output stability of time varying nonlinear feedback systems, part I. *IEEE Transactions on Automatic Control*, 11:228–238, 1966.

[336] G. Zames. On the input output stability of time varying nonlinear feedback systems, part II. *IEEE Transactions on Automatic Control*, 11:465–476, 1966.

[337] E. C. Zeeman. *Catastrophe Theory*. Addison Wesley, Reading, MA, 1977.

Index

Interdisciplinary Applied Mathematics